I0075131

Team Steps Guide to Effective Problem Solving

A Step by Step Process to Turn Problems into Solutions

Lisa Turner

Team Steps Guide to Effective Problem Solving

A Step by Step Process to Turn Problems into Solutions

Turner Creek Publishing
515 Barlow Fields Dr
Hayesville NC 28904
(828) 414 4550

Cover photo and design by Lisa Turner

Printed in the United States of America

ISBN 978-0-9970723-1-0

"If it can be thought, it can be done, a problem can be overcome."

~ E.A. Bucchianeri

INTRODUCTION

Companies have used the Team Steps Guide since 1982 to generate cost savings and profit driving innovation in manufacturing and services. Built on the quality concepts presented by W. Edward Deming, Joseph M. Juran and Philip B. Crosby, the essence of the success boils down to the concept that the people doing the work know best how to improve the work.

Since the early 1980s thousands of employee teams have used this guide to generate both incremental and breakthrough improvements in their work areas totaling millions of dollars. Just as important, these teams have enjoyed the recognition and workplace improvements that these projects generated.

When teams fail or when teams produce minimal results, it is usually because they are overwhelmed by too much information about what they are supposed to do. If two people assemble a bicycle and have never done it before, the person who refers to the set of instructions will have a higher quality product in a faster amount of time than the person who sets the parts about in the workshop and experiments with assembly. This guide works because it simplifies the problem solving process by providing a roadmap for action.

A team facilitator will maximize accomplishment by suggesting and explaining problem identification and analysis tools and ensuring that the team is utilizing all the advantages of appropriate methods, from inviting in guest experts to applying the necessary rigor to the process.

You can use this process with any team type. For example, if you are using Lean Manufacturing teams, use the tools you have available (eliminating waste, reducing set-up time, 5S, etc.) to address the root causes you identify and implement solutions.

In today's environment of tight profit margins, companies want maximum improvement in the minimum amount of time. Using this guide will help deliver the fastest and most effective results.

Author's Note: More than thirty years after writing this guide, the same process applies but the tools available have multiplied. Increasing complexity can discourage the team. My advice is to follow the process using the simplest tools and employ a facilitator, sensei, or coach to show and guide the team through the tools and techniques that best fit the problem.

"None of us are as smart as all of us."

~ Japanese Proverb

Table of Contents

FACILITATORS: HOW TO USE THIS GUIDE

Introduce team leaders to the guide in a short training session. New team leaders should be taken through a sample problem from start to finish using the guide with a demonstration of each tool.

Then introduce the team members to the guide. Each member should receive a copy and follow along as the checklist and the tools are used.

After the team has completed an entire project with tools demonstration from the leader, facilitator, or sensei, the team will be ready to use the guide on their own.

START
Brainstorm/ Select Project
Quick Improvements
Collect & Analyze Data
Experiments Root Causes 5 Whys
Implement Benefits Link to KPIs
Document Follow-up Verify
Reflect Learn Improve
TEAM STEPS GUIDE

Action Steps

Start here!

1. Find a project within your own area of work using the brainstorming technique.

    a.□ Review the rules of Brainstorming on page 19.
    b.□ Brainstorm.
    c.□ Quick Hits – Pull off the items that you, your supervisor, or your engineers can action immediately. These are items where the solution appears self-evident (ex., "Need additional #6 screws.")
    d.□ Round One Consensus – multi-voting – discussion ok but not encouraged due to time factor.
    e.□ Decide how many votes earn a circle around the item.
    f. □ Discuss the circled items. Use Criteria for Problem Selection on page 21.
    g.□ Proceed with Round Two voting (each team member has one vote) OR allow to "incubate" and return to the list later with more ideas. Remember that you will be able to return to this list for more project ideas in the future.

2. Quick Improvements – similar to Quick Hits, these are items on your Brainstorming list that do not require data collection or experiments. These range from "go do" to finding someone from another department to help you get something done.

3. Assigned Projects – Your manager or another area's kaizen output might be perfect for your

group's expertise. In this case, simply begin your process on Step 4.

4. Data Collection and Analysis
    a. □ Make a checklist of problems or issues that are associated with the problem you or others identified.
    b. □ Discussion – clarification.
    c. □ Break checklist of issues into categories if possible (i.e., supplier issues, hardware, work instructions, etc.).
    d. □ Assign members to collect data on each checklist item or category.
    e. □ Determine the time period for this data collection (the "before" metrics). Both historical data from available databases and actual in process measurements will yield the best analysis. The Check Sheet and Milestone Charts (see Tools section) are useful for this.
    f. □ Invite subject matter experts in to the next meeting if you need more information or a different perspective.

5. Data Formatting and Charting
    a. □ Arrange the data from your checklist, check sheet, or printouts into a graph (See Tools section).
    b. □ Ask your facilitator or coach to help you depict your data. You can use a check sheet, a Pareto, a line graph, histogram, etc.
    c. □ You should, at a minimum, show the number of problem items (quantity) and the cost of these items (defects). Other items to track which may be relevant are time, yield

percentages, customer returns, etc. Ask your facilitator, supervisor, or subject matter expert to help you with the cost data.

6. Goal Statement and Action Plan
   a. □ The team should agree on the problem to analyze (you may want to take another look at "Criteria for Problem Selection" on page 21).
   b. □ The team should decide what their goal(s) will be in terms of solving the problem. Example: "The XYZ units take 30 minutes each to get out of the potting fixtures because they are sticking." A goal would be: "Reduce the time in potting fixtures by 75% over the next 30 days (by solving this problem)."
   c. □ Set up a timeline for the solution and a tentative date for your management presentation or report out to your manager.

7. Basic Cause and Effect Analysis
   a. □ State the problem as specifically as possible.
   b. □ Use the 4 M's (manpower, materials, methods, and machines) <u>or any other classifications</u> that work well for your problem. Plant, People, Policies, and Procedures is sometimes used in office areas. See other ideas in the sample charts on page 25 and 27.
   c. □ Brainstorm causes. Tell the scribe where to put the ideas as they come up. Branch off from any category where desired. <u>Remember the five Ws</u>

and H: Why, Where, When, What, Who, and How.

d. □ Discuss the causes and the importance of each in contributing to the problem. Speak up! Leader directed questioning may be useful.

e. □ Use the Five Whys (also written as "5Y") on page 31 to explore root causes as you discuss the causes.

f. □ Consider asking whether data is available or not for the causes. At the least it will spark discussion. At the most it will lead you to the root causes much faster. See example format on page 25.

g. □ Multi-vote for the top causes. The team should decide on how many votes each member gets. Circle the causes that get the most votes.

h. □ Second round voting - each member has one vote. This round will give you your "top" cause; you may want to keep the top five on the list to return to for more analysis. Problems usually have more than one contributing cause, so you'll want to keep these handy for the analysis.

8. Process Cause and Effect Analysis - use this variation (see Tools section) if the problem is heavily process based. Use the same technique as above, but use boxes to the left of the problem statement with process steps. Then brainstorm causes for one or more of these boxes. You do not need to use every box.

9. Identifying Root Cause - the Five Whys technique is very useful in getting to the real

cause of your problem. Ask "Why?" multiple times and get thorough answers. See Tools section.

10. Cause Verification - Team discussion: how are you going to determine if the top cause or causes are really the cause of the problem? Consider inviting subject matter experts to your discussion. If you determine that you are on the wrong track after some analysis, you can return to your cause and effect diagram and get another start.

11. Recommended Solution
    a. □ Brainstorm possible solutions to solve the cause of the problem. While brainstorming, do not discuss or comment except to clarify the solution so that it can be written down.
    b. □ When all idea output has been exhausted, the team should conduct a quick first round vote; each member can vote multiple times. Circle the top solutions in the voting.
    c. □ Discuss the top solutions. Come to consensus on the top 3 solutions, and choose the one to work on first.
    d. □ Brainstorm how the solution will be implemented. Use the 5 W's and H and invite a subject matter expert (engineer, accountant, manager) to your meeting to obtain the most creative ideas. This step is important; you may want to let the ideas incubate and go to it again a little later before finalizing.
    e. □ The scribe should make note of the team's recommended solution(s) for the presentation or report-out.

Describe the solution in enough detail so that it can be easily understood.

12. Solution Implementation and Verification
    a. □ Ask what the best way is to test and or implement your solution(s).
    b. □ Determine what data you will need to collect – before and after cycle time, scrap rates, yield rates, etc. and assemble a timeline for measurement
    c. □ Capture the anticipated benefits as well as the cost of the solution.
    d. □ Make sure your solution meets or exceeds the requirements of your goal statement.

13. Project Write-up and/or Report-out.
    a. □ Get together and decide what information should be presented. Using the charts you already made are the easiest and simplest.
    b. □ If desired, a "4-Up" summary can be used as a handout. Your facilitator or leader can help you. See the following example.

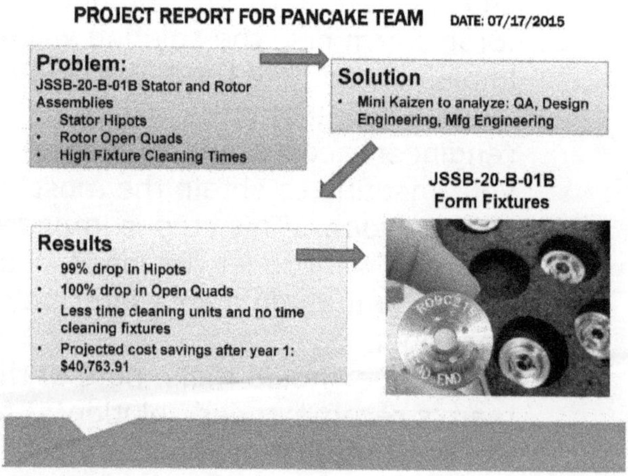
PROJECT REPORT FOR PANCAKE TEAM
DATE: 07/17/2015
Problem:
JSSB-20-B-01B Stator and Rotor Assemblies
• Stator Hipots
• Rotor Open Quads
• High Fixture Cleaning Times
Solution
• Mini Kaizen to analyze: QA, Design Engineering, Mfg Engineering
JSSB-20-B-01B
Form Fixtures
Results
• 99% drop in Hipots
• 100% drop in Open Quads
• Less time cleaning units and no time cleaning fixtures
• Projected cost savings after year 1: $40,763.91

c.□ While the 4-Up is simple, your team methods may require more documentation. In this case, develop the materials as you go (charter, A3, visual controls, etc.).

14. Post Presentation Follow-up
    a.□ Review comments. Could the presentation have been better?
    b.□ Keep a record of your presentation and metrics.
    c.□ Collect data for the timeline you established. At the end of this timeline, ask if you got the results you wanted. If results were substantially different from what you expected, you should report this as a follow-up.
    d.□ Keep in mind that once you have completed a project you will have a tendency not to return to it. You should follow up on the success of your solution several times before putting it away in a file cabinet! See A Note About Chronic Problems on page 41.

NOTES:

Brainstorming Guide

Brainstorming is an excellent creativity tool to discover and prioritize a set of ideas. While most people believe they know how to brainstorm, the truth is that most of the time the group is judging the ideas as they are being written down. The process bogs down and this limits creativity and the flow of ideas. If you want full benefit from the technique, follow these tips:

- ☐ State the topic to be brainstormed as specifically as possible. Use a large sheet of paper that everyone can see.
- ☐ Each group member offers one idea per turn in rotation around the room.
- ☐ If a team member does not have an idea, they simply say "pass."
- ☐ <u>Do not evaluate the ideas now</u>; discussion should be limited to clarifying the idea so that it can be written down.
- ☐ Strive for quantity of ideas; do not be concerned about whether they make sense or not, or even relate to the topic.
- ☐ After everyone has said "pass," this stage is complete.
- ☐ The leader should take a quick multi-vote, asking the team how many votes per member should be used. Three to five votes at this stage works well. <u>Limit discussion</u> during this voting to save time, as the next step will include discussion.

- ☐ The group may want to let the ideas "incubate" and come back to the list later, or add to it in between meetings.
- ☐ Circle the ideas that get the most votes. The group decides how many should be included in this identification process.
- ☐ Now is the time to discuss the circled ideas! Spend as much time as you need. Allow a free range of discussion and flow of ideas on the items. Some of the information may be worth saving on a separate "parking lot" sheet of paper for a return visit or use elsewhere.
- ☐ Now vote again, with one vote per member if picking one idea. It's a great idea to return to this list for the next project.

NOTES:

Criteria for Problem Selection

Before making your final project selection, consider the following tips:

- ☐ Is this project or issue under your own control and in your own area?
- ☐ If you are new to this process, consider whether this is a good size project to start on. If it is too large, you can always come back to it when you have more experience.
- ☐ Can you measure, or collect data on the issue? This means asking questions such as, "How many are good/bad, over a certain amount of time," and defining problem defects or characteristics.
- ☐ Ask yourselves: will solving this problem:
    - ☐ Improve quality?
    - ☐ Save time?
    - ☐ Save money?
    - ☐ Improve safety?
    - ☐ Reduce inventory?
    - ☐ Make the task easier to do?
    - ☐ Reduce waste?
    - ☐ Improve flow?

☐ Use a Problem Selection Table, or Decision Matrix, if you have lots of choices or multiple criteria. Using this table will improve the logic of your choice, and it's fun to do. Use the following steps.

- ✓ To make your matrix, first decide on what criteria you will use. If you are selecting an improvement opportunity, for example, you could use Within our Control, Financial Payback, Delight the Customer, Low Cost to Implement.

✓ Place the criteria and weights across the top. The weights are arbitrary – use what makes sense. 1-5 is simple. Then enter the top 5 causes circled from your brainstorming list on each row.

Problem Selection or Decision Matrix

PROBLEM: Part Exceeds Testing Time Allotted by 60% slowing production

Criteria →

↓ **Issues**

Issues	Resource Availability =3	Easy to Fix =1	Low Cost to Implement = 2	Addresses Root Cause =4	Speed of Fix =3	**SCORE**
Procedures Not Clear	2	1	3	3	2	34
Lack of Operator Training	2	2	3	1	1	27
Equipment Not Working	3	3	3	3	1	60 ☆
Not First In First Out	1	1	3	1	1	20
Wrong Tester Used	3	2	2	3	3	42 ☆
Internal Customer Waits	3	1	2	1	2	27

In this case, "Equipment Not Working" and "Wrong Tester Used" would be explored first

Issue Ratings: Important = 3 Average Importance = 2 Not Very Important = 1

[To get score, multiply criteria rating times issues rating]

List your options or ideas as rows, and the factors you need consider as columns. Score each option/factor combination, weight this score by the relative importance of the criteria item, and add these scores up to give an overall score for each option

✓ While looking at each criteria, use an issue ranking from the team to fill in each intersection box. In the example above,

1-3 is used but you can change that if you wish.

- ✓ You can also decide what criteria is appropriate depending upon what you are trying to accomplish. If you are trying to select a solution, try criteria such as, Root Cause Fully Addressed, High Customer Value, Low Cost to Implement, Amount of Time to Implement, etc.

- ✓ The decision matrix works well for prioritizing any set of actions and helps the team think through the ideas and their value.

- ✓ The important thing to remember is to design the table as simply or as complex as you wish, depending entirely on what you are trying to do. Weightings and ratings are up to the team, so there is maximum flexibility.

- ✓ This technique, by the way, can be used for decisions that you make personally, both in and out of the workplace. When you have a tough decision to make, consider finding a pocket of quiet time with a notebook and go through the exercise. If nothing else, it will provide insight and reflection on your decision choices.

Tools Review

Basic Cause & Effect Diagram (Fishbone)

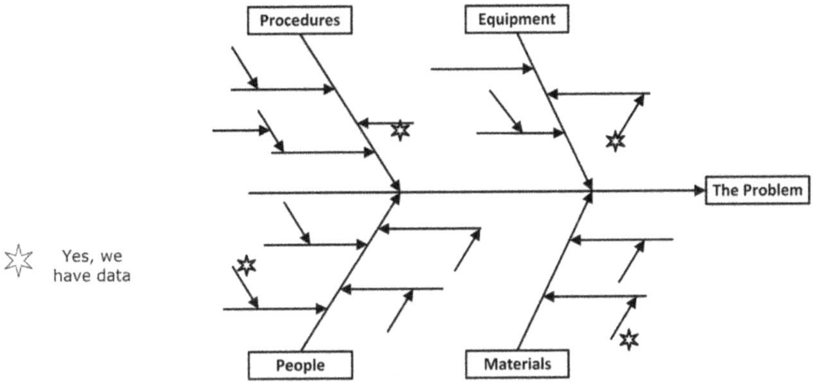

Use whatever categories best relate to the problem. Branch off as many times as you need to in order to capture everything.

The Cause and Effect Diagram, also called a "fishbone" diagram, is one of the most useful tools in your problem solving kit. Use this method not only for sorting and categorizing causes, but also for brainstorming solutions to your problem.

As you write the causes or solutions, ask "WHY?" repeatedly and continue on to sub categories. Keep asking why until you've exhausted all ideas. Let it all incubate and return to it later with another round of brainstorming.

Tools Review – continued

When you've exhausted ideas and begin your voting for top causes (or solutions), one great trick is to ask whether you have data for the idea. If yes, place a star next to the idea or highlight it. This will be very helpful during research and evaluation.

Process C&E

Process Cause and Effect Diagram Example

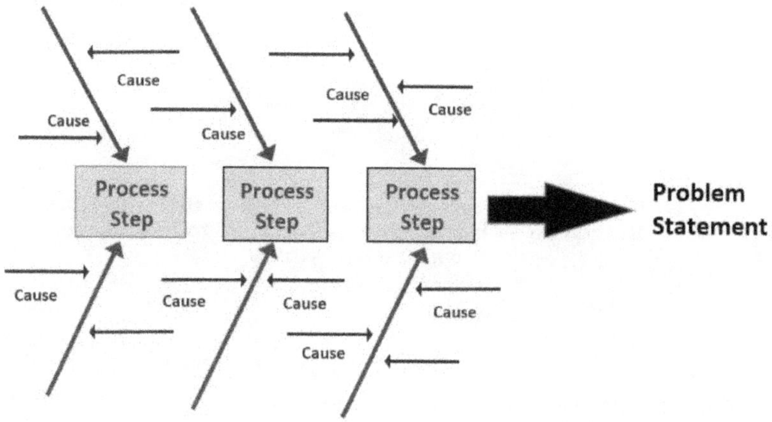

Use the Process Cause and Effect diagram above for problems that are more transactional. This method works especially well for administrative areas.

NOTES:

Tools Review - continued

EXAMPLE: Completed C&E

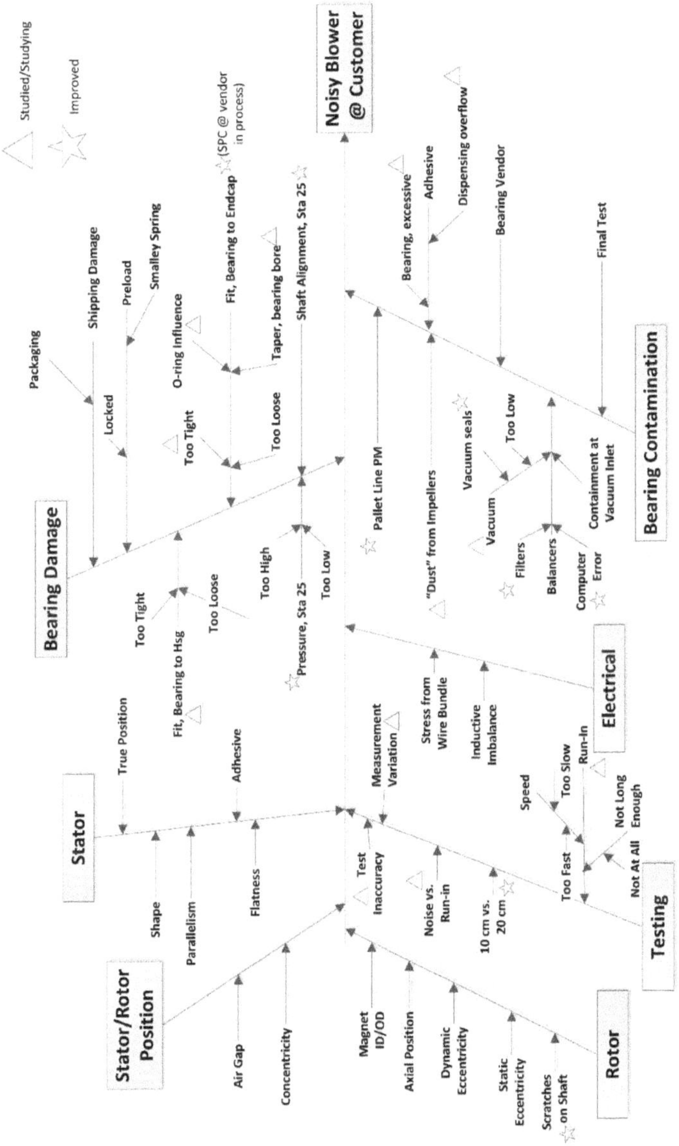

Courtesy of Jerry D. Kilpatrick

Tools Review – continued

CHECK SHEET EXAMPLE

PROBLEMS IN OUR AREA **TEAM: Problem Solvers** **DATE:**

DEFECTS/DAY	1	2	3	4	5	6	7	TOTAL
Loose Hardware	7	4	2	5	8	4	3	33
Rev. Level Wrong	2	3	2	5	2	1	2	17
Missing part	2	1	3	1	2	1	1	11
Wrong Part	3	5	2	1	1	3	1	16
Part # Missing	1	0	1	3	3	1	1	10
TOTAL DEFECTS	15	13	10	15	16	10	8	87

NOTES:

Tools Review – continued

The "Pareto Principle," also known as the "80/20" Rule, states that roughly 80% of the effects come from 20% of the causes. This is a very handy principle to remember as you review your data.

The Pareto Chart is used to graphically summarize and display the relative importance of the differences between groups of data. It will help you prioritize action by determining what to look at first.

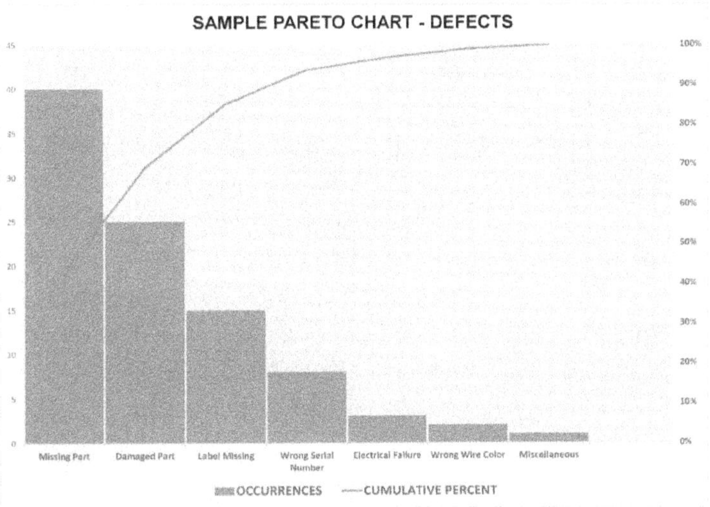

DEFECT DATA

PROBLEM AREA	OCCURRENCES	PERCENT OF TOTAL	CUMULATIVE PERCE
Missing Part	40	42.55%	42.55%
Damaged Part	25	26.60%	69.15%
Label Missing	15	15.96%	85.11%
Wrong Serial Number	8	8.51%	93.62%
Electrical Failure	3	3.19%	96.81%
Wrong Wire Color	2	2.13%	98.94%
Miscellaneous	1	1.06%	100.00%

NOTES:

Tools Review - continued

Example of a Milestone or Gantt Chart

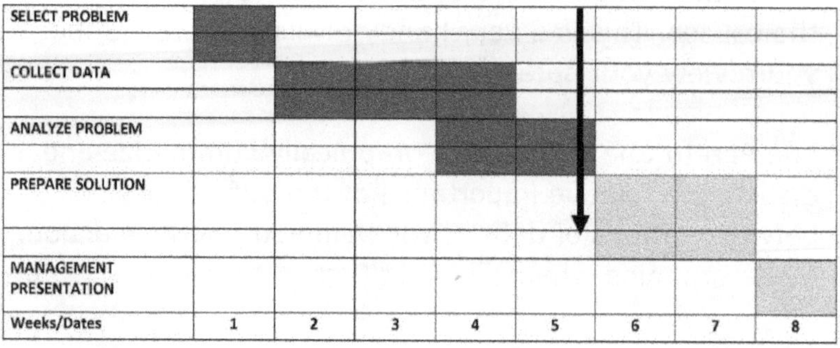

Use a Milestone or Gantt chart to depict where you are on your project. This can be done by hand on a chart if the team does not have access to a PC. The chart above was done in Microsoft Word. You don't need anything fancy.

The horizontal bars are filled in to indicate how much of the project has been completed to date. The vertical line represents where you are right now. This way it is easy to see whether you are on schedule or not. In this example, the team is slightly ahead of schedule.

NOTES:

Tools Review – continued

ASKING THE FIVE WHYS

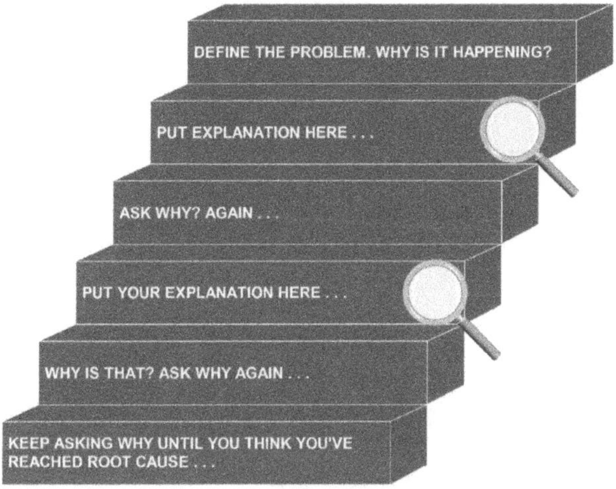
DEFINE THE PROBLEM. WHY IS IT HAPPENING?
PUT EXPLANATION HERE . . .
ASK WHY? AGAIN . . .
PUT YOUR EXPLANATION HERE . . .
WHY IS THAT? ASK WHY AGAIN . . .
KEEP ASKING WHY UNTIL YOU THINK YOU'VE REACHED ROOT CAUSE . . .

ASKING THE FIVE WHYS IS AN EFFECTIVE WAY TO IDENTIFY THE ROOT CAUSES FOR A PARTICULAR PROBLEM. USE THE SAMPLE PROCESS TO THE LEFT WITH SHEETS OF PAPER TO BEGIN THE PROCESS.

TAKE YOUR TIME AND IF YOU FIND YOURSELF WITH AN ANSWER OR RESULT THAT YOU DON'T HAVE ANY CONTROL OVER, BACK UP TO THE PREVIOUS QUESTION AND BEGIN AGAIN. PERSEVERE!

And keep going!

NOTES:

Creativity Tips and Shortcuts

Sometimes our single-minded focus on our day to day jobs blind us from seeing solutions that are right in front of us. When you get ready to consider solutions to a problem, review the following tips to disengage your brain from the day to day and get it hooked up with big landscape thinking.

- ☐ **Use the tools**: The tools draw us into effective and objective problem solving. The tools provide time tested shortcuts to efficient "solutioning" of even the most chronic and troublesome of problems. Using the tools help us to stop jumping to the solutions we think are the right ones when we do not have the data. Don't assume you know the answer.

- ☐ **Don't shortcut the ideas**: Use brainstorming properly and let ALL the ideas out on paper. Don't stop to judge each idea as you brainstorm. This is the most common error in brainstorming and leads to "conclusion jumping" - a data limiting exercise.

- ☐ **Reverse Engineer**: Make new connections using "what if" questions, wishful thinking, cross connecting ideas, daydreaming, working backwards, using negative questions, and using the 5 whys and H (Why, Where, When, What, Who, and How) and the Five Whys.

- ☐ **Re-phrase a negative problem statement into a positive headline**: Pretend the problem has been solved. What did you do to get there? What had to happen?

- ☐ **Admired Expert**: Think of your most esteemed "expert" or famous person that you admire. They are next to you right now. What will they say about the problem/solution?

- ☐ **Build self-esteem and do not be afraid of risk**: The best ideas contain an element of risk. If you have done your due-diligence and followed the steps, give it everything you've got.

- ☐ **Time travel**: What will the situation be in a year? 3 years? 10 years? This opens up possibilities. Write the ideas down.

- ☐ **Paradigms**: Beware the thought patterns that cause us to say, "It won't work here." Paradigms are strong and naturally protective thought filters that can prevent us from considering ALL of the information. Maintain an open mind and question irrational or illogical conclusions. The Five Whys work well in this situation.

- ☐ **Be persistent**: Don't let nay-sayers get you off track or discouraged once you have determined the actions. Successful inventors will tell you that if they allowed others to influence their projects they would not have succeeded. Take failure and risk in stride, knowing you are on the right track.

- ☐ **Start Over**: Consider poor results and failures as a springboard to a better solution. The best problem solving is iterative. Each cycle will bring improvement. In Lean we call this Reflection and Continuous Improvement. Use this cycle to your advantage and don't give up!

NOTES:

Team Leader Tips

As team leader, how can you add the most value to your team? Whether this is your first time as a team leader, or you've served in this role many times, the following tips will help you make the most of your opportunity with the team.

- ☐ **Have a set time and place** for your meetings and stick to it. It's easy to say "We're too busy," but you're always busy!
- ☐ **No reason to meet?** Don't! Use the meeting time to accomplish something more meaningful in your work area, such as 5S or Quick Improvements.
- ☐ **Have a meeting agenda.** Go over the agenda at the beginning of your meeting and ask team members to contribute to it. See Meetings on page 39.
- ☐ **Ask for a volunteer** to take minutes of the meeting and rotate the task every so often. Keep track of your accomplishments and assignments.
- ☐ **Follow the steps to stay on track.** Use the steps process in this booklet.
- ☐ **Use an easel with a pad of paper** to write on so everyone can see it. You can keep it for the presentation and/or write-up and it will assist in documenting your work.

☐ **Ask your coach or facilitator** to help with the steps in the process and demonstrate the tools. They may also assist you when it's time to collect and analyze the data, format it, and present it. With experience, you'll be able to do this yourself.

☐ **Get everyone involved**: Make the meeting comfortable for all members so everyone will feel that they can be heard.

☐ **Invite experts** to your meeting when you need them to find out more about your analysis. Keep their time in the meeting short and to the point.

☐ **Invite your supervisor** or manager in to your meeting for five or ten minutes once in a while to let them know what you are working on. They will appreciate this invitation and the opportunity to see what you are working on.

NOTES:

Meetings

Meetings bring together the company's most valuable resource - people - and represent a large investment. If you are going to gather people together, think out ahead of time how you can maximize output. Use the following tips.

- ☐ Why are you holding the meeting? Could you achieve your purpose another way?
- ☐ If you have to hold a meeting, invite exactly the right people. Inviting people for "information" reasons is wasteful when you can simply update those who need to know after the meeting.
- ☐ Give participants plenty of advance notice of the meeting and send them a simple agenda with the invitation.
- ☐ Assign pre-work if it will reduce the meeting time or make everyone more effective once they assemble.
- ☐ Use a "parking lot" - a large sheet of paper - to list off topic ideas so you don't get off track. You should follow up or assign these for action during and/or after the meeting.
- ☐ Encourage participation and actively manage personalities. Draw out the more quiet individuals to balance the discussion. Express appreciation for contributions.

- ☐ Begin – and end – on time. As much as we already know this rule, we violate it routinely. Participants will appreciate and respect you for maintaining this discipline.
- ☐ End the meeting with a summary and a call to action.
- ☐ Follow up on loose ends, to-do items, and assignments soon after the meeting or they will fall to the side.

NOTES:

A Note About Chronic Problems

A chronic problem is one where people say, "Nothing can be done about it," or "We tried everything and nothing worked."

A chronic problem pops back up right after you think you've solved it. A chronic problem may go dormant, leading you and others to believe that it has been solved, only to terrifyingly reappear in full force when you least expect it.

We don't have time for chronic problems. These types of problems suck all of the energy out of our efforts, leading us to abandon the discipline that we need to solve it once and for all.

Chronic problems, if really solved, will energize our success in supplying quality goods and services to customers, driving profit margins, and maintaining an enjoyable place for people to work.

Follow these tips to solve chronic problems:

- ☐ Acknowledge a chronic problem. Recognize that you've got an issue that will take time and discipline to solve.

- ☐ Commit to using the process and the tools. This guide or something similar (checklist) combined with Lean and Six Sigma tools will speed your analysis.

- ☐ Enlist a facilitator, coach, or sensei to monitor progress and help you. The coach will show you what tools to use and how they work with your particular problem. Chronic problems can be complex and having specialized help and resources will speed your solution.

- ☐ Recognize and acknowledge cultural obstinacy. It is normal for people to fear and fiercely resist change. Don't let this get in your way. Keep asking "Why?" and be careful not to assign blame.
- ☐ Follow up and follow through. 85% of chronic problems reappear because the team thought it was solved and they went on to another issue.
- ☐ **Persevere**. Exercise rigor and discipline in the process and <u>don't give up</u>!

NOTES:

“If you don’t ask the right questions, you don’t get the right answers. A question asked in the right way often points to its own answer. Asking questions is the ABC of diagnosis. Only the inquiring mind solves problems.”

– Edward Hodnett

ABOUT THE AUTHOR

Lisa P. Turner, ScD., has spent 25 years working in business process improvement. She wrote her dissertation on the influence of structured problem solving methods on productivity and quality improvement levels in both administrative and factory personnel in electronics manufacturing.

Lisa holds a BA in English, an AS in Engineering Technology, an MBA in Business, and a Doctor of Science in Management. She obtained certification from the American Society of Quality (ASQ) as a Quality Engineer in 1984 and a Six Sigma Black Belt in 2006, and teaches ASQ preparatory classes for technicians, engineers, and quality auditors.

NOTES:

NOTES:

www.ingramcontent.com/pod-product-compliance
Lightning Source LLC
Chambersburg PA
CBHW022056190326
41520CB00008B/791

Published by Gérard Gremaud
Rte du Pâquier 1, 1723 Marly, Switzerland

ISBN: 978-2-8399-1934-0

Translated from French by Marc Fleury

Original title:

Univers et Matière conjecturés comme un Réseau Tridimensionnel avec des Singularités Topologiques

November 2015
DOI: 10.13140/RG.2.1.2266.5682
http://gerardgremaud.ch/

Universe and Matter conjectured as a 3-dimensional Lattice with Topological Singularities

First part: searching for a new description of the lattice deformation

Second part: searching for a "cosmic lattice"

Gérard Gremaud

Illustration of book cover:

François Gremaud

https://www.2bcompany.ch/index.html

Universe and Matter conjectured as a 3-dimensional Lattice with Topological Singularities

On n'a peut-être pas encore prêté assez d'attention [à] l'utilité dont cette étude [de la Géométrie] peut être pour préparer comme insensiblement les voies à l'esprit philosophique, et pour disposer toute une nation à recevoir la lumière que cet esprit peut y répandre [...]. Bientôt l'étude de la Géométrie conduira [...] à la vraie Philosophie qui par la lumière générale et prompte qu'elle répandra, sera bientôt plus puissante que tous les efforts de la superstition.

Jean le Rond D'Alembert, *article "Géométrie" de L'Encyclopédie, 1772*

The more the universe seems comprehensible, the more it also seems pointless. But if there is no solace in the fruits of our research, there is at least some consolation in the research itself [...] The effort to understand the universe is one of the very few things that lifts human life a little above the level of farce, and gives it some of the grace of tragedy.

Steven Weinberg*, from "The First Three Minutes"*

Imagination is more important than knowledge. For knowledge is limited to all we now know and understand, while imagination embraces the entire world, and all there ever will be to know and understand.

Albert Einstein

Pensons, il en restera toujours quelque chose!
(Think, there will always be something left!)

Snoopy

Acknowledgments

I am grateful to Gianfranco D'Anna, Willy Benoit, Marc Fleury
and Daniele Mari for their valuable comments and discussions,
and should like to thank warmly Marc Fleury for his english translation of the book.

Introduction

One fundamental problem of modern physics is the search for a *theory of everything* able to explain the nature of space-time, what matter is and how matter interacts. There are various propositions, as *Grand Unified Theory, Quantum Gravity, Supersymmetry, String and Superstring Theories, and M-Theory.* However, none of them is able to consistently explain *at the present and same time* electromagnetism, relativity, gravitation, quantum physics and observed elementary particles.

In this book, it is suggested that Universe could be *a massive elastic 3D-lattice*, and that fundamental building blocks of Ordinary Matter could consist of *topological singularities of this lattice*, namely diverse dislocation loops and disclination loops. For an isotropic elastic lattice obeying Newton's law, with specific assumptions on its elastic properties, one shows that the behaviors of this lattice and of its topological defects display "all" known physics, unifying electromagnetism, relativity, gravitation and quantum physics, and resolving some longstanding questions of modern cosmology. Moreover, studying lattices with axial symmetries, represented by *"colored" cubic 3D-lattices*, one can identify a lattice structure whose topological defect loops coincide with the complex zoology of elementary particles, which could open a promising field of research.

This book does not present a *theory of everything* which would be completely elaborated and usable, but it would and could be extremely fruitful to give simple explanations to the modern physics theories which are very difficult, if not impossible, to deeply understand. It could also and above all be useful to define close links and unifying bridges between the diverse theories of modern physics.

In a first part of the book, one summarizes autonomously a first book[1] published in french during year 2013, which lays methodically the foundations of an original approach of the solid lattices deformation using the Euler coordinates, and which introduces in details the concept of tensor dislocation charges and tensor disclination charges within a lattice. This new concept allows one to quantify the topological singularities which can appear at the microscopic scale of a solid lattice. On the basis of this original approach of the solid lattices and their topological singularities, one can deduct a set of fundamental and phenomenological equations allowing to treat rigorously the macroscopic spatiotemporal evolution of a newtonian solid lattice which deforms in the absolute space of an external observer laboratory.

In a second part of the book, one introduces an imaginary lattice, named « *cosmic lattice* » with quite special elastic and structural properties. The Newton equation of this lattice and its topological singularities present then a set of very surprising properties, which will be progressively developed in the course of the chapters. It will appear strong and amazing analogies with all

---

[1] *Théorie eulérienne des milieux déformables, charges de dislocation et de désinclinaison dans les solides, G. Gremaud, Presses polytechniques et universitaires romandes, Lausanne, Suisse, 2013, 750 pages (ISBN 978-2-88074-964-4)*

modern physics theories: Maxwell equations, special relativity, newtonian gravitation, general relativity, modern cosmology, quantum physics and standard model of elementary particles.

The problem of unified field theories

One fundamental problem of modern physics is the search for a *theory of everything* able to explain the nature of space-time, what matter is and how matter interacts. Since the 19th century, physicists have attempted to develop unified field theories, which would consist of a single coherent theoretical framework able to account for several fundamental forces of nature. For instance:

- *Grand Unified Theory* merges electromagnetic, weak and strong interaction forces,
- *Quantum Gravity*, *Loop Quantum Gravity* and *String Theories* attempt to describe the quantum properties of gravity,
- *Supersymmetry* proposes an extension of the space-time symmetry relating the two classes of elementary particles, bosons and fermions,
- *String and Superstring Theories* are theoretical frameworks incorporating gravity in which point-like particles are replaced by one-dimensional strings, whose quantum states describe all types of observed elementary particles,
- *M-Theory* is a unifying theory of five different versions of string theories, with the surprising property that extra dimensions are required for its consistency.

However, none of them is able to consistently explain *at the present and same time* electromagnetism, relativity, gravitation, quantum physics and observed elementary particles. Many physicists believe now that *11-dimensional M-theory* is the theory of everything. However, there is no widespread consensus on this issue and, at present, there is no candidate theory able to calculate the fine structure constant or the mass of the electron. Particle physicists expect that the outcome of the ongoing experiments – search for new particles at the large particle accelerators and search for dark matter – are needed to provide further input for a theory of everything.

In this book, it is suggested that Universe could be *a massive elastic 3D-lattice*, and that fundamental building blocks of Ordinary Matter could consist of *topological singularities of this lattice*, namely diverse dislocation loops and disclination loops. We find, for an isotropic elastic lattice obeying Newton's law, with specific assumptions on its elastic properties, that the behaviours of this lattice and of its topological defects display "all" known physics, unifying electromagnetism, relativity, gravitation and quantum physics, and resolving some longstanding questions of modern cosmology. Moreover, studying lattices with axial symmetries, represented by *"colored" cubic 3D-lattices*, one can identify a lattice structure whose topological defect loops coincide with the complex zoology of elementary particles, which could open a promising field of research.

First part: searching for a new description of the lattice deformation

When one desires to study the solid deformation, one generally uses lagrangian coordinates to describe the evolution of the deformations, and diverse differential geometries to describe the topological defects contained in the solid.

The use of lagrangian coordinates presents a number of inherent difficulties. From the mathematical point of view, the tensors describing the continuous solid deformation are always of order higher than one concerning the spatial derivatives of the displacement field components, which leads to a very complicated mathematical formalism when the solid presents strong distortions (deformations and rotations). To these mathematical difficulties are added physical difficulties when one has to introduce some known properties of solids. Indeed, the lagrangian coordinates become practically unusable, for example when one has to describe the temporal evolution of the microscopic structure of a solid lattice (phase transitions) and of its structural defects (point defects, dislocations, disclinations, boundaries, etc.), or when it is necessary to introduce some physical properties of the medium (thermal, electrical, magnetic or chemical properties) leading to scalar, vectorial or tensorial fields in the real space.

The use of differential geometries in order to introduce topological defects as dislocations in a deformable continuous medium has been initiated by the work of Nye[2] (1953), who showed for the first time the link between the dislocation density tensor and the lattice curvature. On the other hand, Kondo[3] (1952) and Bilby[4] (1954) showed independently that the dislocations can be identified as a crystalline version of the Cartan's concept[5] of torsion of a continuum. This approach was generalized in details by Kröner[6] (1960). However, the use of differential geometries in order to describe the deformable media leads very quickly to difficulties similar to those of the lagrangian coordinates system. A first difficulty arises from the complexity of the mathematical formalism which is similar to the formalism of general relativity, what makes very difficult to handle and to interpret the obtained general field equations. A second difficulty arises with the differential geometries when one has to introduce topological defects other than dislocations. For example, Kröner[7] (1980) has proposed that the existence of extrinsic point defects could be considered as extra-matter and introduced in the same manner that matter in general relativity under the form of Einstein equations, which would lead to a pure riemannian differential geometry in the absence of dislocations. He has also proposed that the intrinsic point defects (vacancies and interstitials) could be approached as a non-metric part of an affine connection. Finally, he has also envisaged introducing other topological defects, as disclinations for example, by using higher order geometries much more complex, as Finsler or Kawaguchi geometries. In fact, the introduction of differential geometries implies generally a heavy mathematical artillery (metric tensor and Christoffel symbols) in order to describe the spatiotemporal evolution in infinitesimal local referentials, as shown for example in the mathematical theory of dislocations of Zorawski[8] (1967).

---

[2] *J.F. Nye, Acta Metall.,vol. 1, p.153, 1953*

[3] *K. Kondo, RAAG Memoirs of the unifying study of the basic problems in physics and engineering science by means of geometry, volume 1. Gakujutsu Bunken Fukyu- Kay, Tokyo, 1952*

[4] *B. A. Bilby , R. Bullough and E. Smith, «Continuous distributions of dislocations: a new application of the methods of non-riemannian geometry», Proc. Roy. Soc. London, Ser. A 231, p. 263–273, 1955*

[5] *E. Cartan, C.R. Akad. Sci., 174, p. 593, 1922 & C.R. Akad. Sci., 174, p.734, 1922*

[6] *E. Kröner, «Allgemeine Kontinuumstheorie der Versetzungen und Eigenspannungen», Arch. Rat. Mech. Anal., 4, p. 273-313, 1960*

[7] *E. Kröner, «Continuum theory of defects», in «physics of defects», ed. by R. Balian et al., Les Houches, Session 35, p. 215–315. North Holland, Amsterdam, 1980.*

[8] *M. Zorawski, «Théorie mathématique des dislocations», Dunod, Paris, 1967.*

Eulerian deformation theory of newtonian lattices

In view of the complexity of calculations in the case of lagrangian coordinates as well as in the case of differential geometries, it seemed to me that it would be better to develop a much simpler approach of deformable solids, but at least equally rigorous, which has been finally published in a first book[1] during year 2013: *la théorie eulérienne des milieux déformables*.

In the first part of the book, one presents a summary of this new and original eulerian approach of the deformation of solids through several sections:

- *a first section (A)* introduces *the eulerian deformation theory of newtonian lattices.* The deformation of a lattice is characterized by *distortions and contortions (chap. 1 to 3).* A vectorial representation of the tensors, presenting undeniable advantages over purely tensorial representation thanks the possibility to use the powerful formalism of the vectorial analysis, allows to obtain *the geometro-compatibility equations* of the lattice which insure its solidity, and *the geometro-kinetics equations* of the lattice, which allow one to describe the deformation kinetics.

One introduces then the physics in this topological context *(chap. 4)*, namely *the newtonian dynamics* and *the eulerian thermo-kinetics* (based on the first and second principles of thermodynamics). With all these ingredients, it becomes possible to describe the particular behaviors of a solid lattice *(chap. 5)*, as *the elasticity, the anelasticity, the plasticity* and *the self-diffusion.* This first section ends with the establishment of the complete set of evolution equations of a lattice in the Euler coordinate system *(chap. 6)*.

- *a second section (B)* is dedicated to *the applications of the eulerian theory (chap. 7).* It presents very succinctly some examples of *phenomenologies of everyday solids.* One shows how to obtain the functions and equations of state of an isotropic solid, what are the elastic and thermal properties which can appear, how waves propagate and why there exist thermoelastic relaxations, what are the mass transport phenomena and why it could appear inertial relaxations, what are the common phenomenologies of anelasticity and plasticity, and finally how it can appear structural transitions of first and second order in a solid lattice.

Dislocation and disclination charges in eulerian lattices

Regarding *the description of defects (topological singularities)* which can appear within a solid, as dislocations and disclinations, it is a domain of physics initiated principally by the idea of macroscopic defects of Volterra[9] (1907). This domain experienced a fulgurant development during the twentieth century, as well illustrated by Hirth[10] (1985). The lattice dislocation theory started up in 1934, when Orowan[11], Polanyi[12] and Taylor[13] published independently papers describing the edge dislocation. In 1939, Burgers[14] described the screw and mixed dislocations. And finally

---

[9] *V. Volterra, «L'équilibre des corps élastiques», Ann. Ec. Norm. (3), XXIV, Paris, 1907*

[10] *J.-P. Hirth, «A Brief History of Dislocation Theory», Metallurgical Transactions A, vol. 16A, p. 2085, 1985*

[11] *E. Orowan, Z. Phys., vol. 89, p. 605,614 et 634, 1934*

[12] *M. Polanyi, Z. Phys., vol.89, p. 660, 1934*

[13] *G. I. Taylor, Proc. Roy. Soc. London, vol. A145, p. 362, 1934*

[14] *J. M. Burgers, Proc. Kon. Ned. Akad. Weten schap., vol.42, p. 293, 378, 1939*

in 1956, Hirsch, Horne et Whelan[15] and Bollmann[16] observed independently dislocations in metals by using electronic microscopes. Concerning the disclinations, it is in 1904 that Lehmann[17] observed them in molecular crystals, and in 1922 that Friedel[18] gave them a physical explanation. From the second part of the century, the physics of lattice defects has grown considerably.

In the first part of this essay, the dislocations and the disclinations are approached by introducing intuitively the concept of dislocation charges by using the famous Volterra pipes[19] (1907) and an analogy with the electrical charges. With Euler coordinates, the concept of dislocation charge density appears then in an equation of geometro-compatibility of the solid, when the concept of flux of charges is introduced in an equation of geometro-kinetics of the solid.

The *rigorous formulation of the charge concept* in the solids makes the essential originality of this approach of the topological singularities. The detailed development of this concept leads to the appearance of tensorial charges of first order, *the dislocation charges*, associated with *the plastic distortions* of the solid (plastic deformations and rotations), and of tensorial charges of second order, *the disclination charges*, associated with *the plastic contortions* of the solid (plastic flexions and torsions). It appears that these topological singularities are quantified in a solid lattice and that they have to appear as *strings (thin tubes)* which can be modelized as unidimensional lines of dislocation or disclination, or as *membranes (thin sheets)* which can be modelized as two-dimensional boundaries of flexion, torsion or accommodation.

The concept of dislocation and disclination charges allows one to find rigorously the main results obtained by the classical dislocation theory. But it allows above all to define a tensor $\vec{\Lambda}_i$ of *linear dislocation charge*, from which one deduces a scalar Λ of *linear rotation charge*, which is associated with the screw part of the dislocation, and a vector $\vec{\Lambda}$ of *linear flexion charge*, which is associated with the edge part of the dislocation. For a given dislocation, both charges Λ and $\vec{\Lambda}$ are perfectly defined without needing a convention at the contrary of the classical definition of a dislocation with its Burger vector! On the other hand, the description of the dislocations in the eulerian coordinate system by the concept of dislocation charges allows one to treat exactly the evolution of the charges and the deformations during very strong volumetric contractions and expansions of a solid medium.

The description of this new approach of the topological defects of a lattice is briefly described by the two following sections of part one of the book:

- *a third section (C)* is dedicated to the introduction of *dislocation charges* and *disclination charges* in the eulerian lattices. After the analytical introduction of the concepts of density and flux of dislocation and disclination charges in the lattices *(chap. 8)*, one presents a detailed review of *the lattice macroscopic and microscopic topological singularities* which can be associated to the dislocation and disclination charges *(chap. 9)*.

Then one discusses the motion of dislocation charges within the lattice by introducing *the dislocation charges flux of the dislocation charges* and *the Orowan relations (chap. 10)*. Finally, one

---

[15] *P. B. Hirsch, R. W. Horne, M. J. Whelan, Phil. Mag., vol. 1, p. 667, 1956*

[16] *W. Bollmann, Phys. Rev., vol. 103, p. 1588, 1956*

[17] *O. Lehmann, «Flussige Kristalle», Engelman, Leibzig, 1904*

[18] *G. Friedel, Ann. Physique, vol. 18, p. 273, 1922*

[19] *V. Volterra, «L'équilibre des corps élastiques», Ann. Ec. Norm. (3), XXIV, Paris, 1907*

deduces the Peach and Koehler force which acts on the dislocations, and one establishes the new set of evolution equations of a lattice in the Euler coordinate system *(chap. 11)*, which takes into account the existence of topological singularities within the lattice.

- *a fourth section (D)* is dedicated to *the applications of the charge concept* within the eulerian solid lattice (chap. 12). It shows the elements of the dislocation theory in the everyday solids. One begins to show that, in the particular case of the deformation of isotropic lattices by pure shears, one can replace the shear strain tensor by the rotation vector, which allows one to find a set of equations, which corresponds strictly *to all the Maxwell equations of electromagnetism*! Then one shows how to calculate the fields and energies of the screw and edge dislocations in an isotropic lattice, just as the interactions, which can occur between dislocations. One finishes this section of applications by presenting *the string model* of dislocations, which is the fundamental model allowing one to explain most of the macroscopic behaviors of anelasticity and plasticity of crystalline solids.

Second part: searching for a "cosmic lattice"

In the first part of the book, it is shown that it is possible to calculate the resting energy E_0 of the dislocations, which corresponds to *the elastic energy stored in the lattice* by their presence, and their kinetic energy E_{cin}, which corresponds to the kinetic energy of the lattice particles mobilized by their movement. This allows to attribute to the dislocations *a virtual inertial mass* M_0 which satisfies relations similar to the famous equation $E_0 = M_0 c^2$ of the Einstein special relativity, but which is obtained here through purely classical calculations, without using relativity principles! Moreover, at high velocity, the dislocation dynamics satisfy also *the special relativity principles* and *the Lorentz transformations*.

It is also shown in the first part that it appears, in the case of isotropic solid media presenting a constant and homogeneous volumetric expansion, a perfect and complete analogy with the Maxwell equations of electromagnetism when the shear stress tensor is replaced by the rotation vector. The existence of an analogy between the electromagnetism and the theory of incompressible continuous media has already been distinguished very long ago by several authors, as shown by Whittaker[20] (1951). However, this analogy is much more complete in my first book[1], because it is not restricted to one of the two Maxwell equation couples in the vacuum, but it is generalized to the two equation couples as well as to *the diverse phenomenologies of dielectric polarization and magnetization of matter*, just as to *the electrical charges and the electrical currents*! The analogy with the Maxwell equations is very surprising on account of the fact that it is initially postulated a solid lattice satisfying a simple and purely newtonian dynamics in the absolute reference frame of the external observer laboratory, which is equipped with absolute orthonormal measuring rods and an absolute clock. At the contrary, the topological singularities within the lattice (dislocations and disclinations) with their respective charges, responsible for the plastic distortions and contortions of the lattice, are submitted to a relativistic dynamics within the lattice, due to the maxwellian equation set governing the shear strains of the massive elastic lattice. From this point of view, the relativistic dynamics of the topological singularities is a direct consequence of the purely classical newtonian dynamics of the elastic lattice in the ab-

[20] *S. E. Whittaker, «A History of the Theory of Aether and Electricity», Dover reprint, vol. 1, p. 142, 1951.*

solute frame of the external observer!

Finally, it also appears in the first part that the tensorial aspect of the distortion fields at short distances of a localized topological singularities cluster formed by one or more dislocation or disclination loops can be easily neglected at great distances of the cluster, because the distortion fields can then be completely described by only two vectorial fields, the vectorial field of rotation by torsion and the vectorial field of curvature by flexion, associated respectively to the only two scalar charges of the cluster, its *scalar rotation charge* Q_λ and its *scalar curvature charge* Q_θ. The rotation charge becomes the perfect analogue of *the electrical charge* in the Maxwell equations, when the curvature charge presents some analogy with *the gravitational mass* in the gravitation theory.

The existence of analogies between the theories of continuum mechanics and solid defects and the theories of electromagnetism, special relativity and gravitation has already been the subject of several publications, from which the more famous are most certainly those of Kröner[4,5]. Excellent reviews in this physics field have also been published, in particular by Whittaker[20] (1951) and Unzicker[21] (2000). But none of these publications has gone as far as the approach published in my first book[1] concerning these highlighted analogies.

The numerous analogies which appear in the first book[1] between the eulerian theory of deformable media and the theories of electromagnetism, gravitation, special relativity, general relativity and even standard model of elementary particles, reinforced by the absence of particles analogue to magnetic monopoles, by a possible solution of the famous paradox of electron field energy and by the existence of a small asymmetry between curvature charges of vacancy or interstitial type, were sufficiently surprising and remarkable to alert any open and curious scientific spirit! But it was also clear that these analogies were, by far, not perfect. It was then tantalizing to analyze much more carefully these analogies and to try to find how to perfect them. That is the reason of this present essay, of which the second part is entirely allotted to the deepening, the improvement and the understanding of these analogies.

The second part of this book is composed of five sections. Progressively, by introducing several judicious conjectures which are summarized in Appendix D, one addresses the problem of the analogies existing between *(i)* the eulerian theory of lattice deformation described in the first part , and applied to a very particular lattice, *the cosmic lattice*, and *(ii)* the modern physics theories of the macrocosm and the microcosm, as the Maxwell equations, the special relativity, the newtonian gravitation, the general relativity, the modern cosmology, the quantum mechanics and the standard model of elementary particles.

The "cosmic lattice" and its Newton's equation

A first section (A) of part two is dedicated to the introduction of the *« cosmic lattice »*. By introducing particular elastic properties for the volumetric expansion, the shear strain and especially the rotation field and by expressing the distortion free energy per volume unit of the lattice, one obtains an imaginary lattice which presents a very particular Newton equation. Indeed, it appears in particular a novel force term directly related to the distortion free energy due to the sin-

[21] *A. Unzicker, «What can Physics learn from Continuum Mechanics?», arXiv:gr-qc/0011064, 2000*

gularities contained in the lattice, which will play subsequently a very important role for the analogies with the gravitation and the quantum physics *(chap. 13).*

Then one shows that the propagation of waves in this cosmic lattice presents interesting particularities *(chap. 14)*: propagation of linear polarization transversal waves is always associated with longitudinal wavelets, and propagation of pure transversal waves can only be done by circularly polarized waves (which will be strongly linked with the photons). On the other hand, when the local value of the lattice volumetric expansion becomes less than a given critical value, propagation of longitudinal waves disappears for the benefit of the appearance of localized longitudinal vibrations modes (which will be strongly linked with the quantum physics).

Afterwards, the calculation of *the curvature of wave rays* in the vicinity of a singularity of the lattice volumetric expansion allows one to find the conditions for which this expansion singularity becomes a real capturing trap for the waves, in other words a *« black hole » (chap. 15)*!

Finally, one shows that such a cosmic lattice, if finite in the absolute space, can present *dynamical volumetric expansion and/or contraction* if it contains some quantity of expansion kinetics energy *(chap. 16).* This phenomenon is perfectly similar to the cosmological expansion of the universe! Following the signs and the values of the lattice elastic modules, several cosmological behaviors of the lattice can appear, some of which presenting phenomena as *big-bang, rapid inflation and acceleration of the expansion velocity*, which can be sometimes followed by *a recontraction of the lattice* driving to *a big-bounce phenomenon*! One deduces that it is *the expansion elastic energy* contained in the lattice which is responsible for these phenomena, and notably for *an expansion velocity increase*, a phenomenon which has been recently discovered by the astrophysicists in the case of the present universe, and which has been attributed to a hypothetical *« black energy »*.

Maxwell's equations and special relativity

A second section (B) is dedicated to *the Maxwell equations* and *the special relativity.* One begins to show that the Newton equation of the cosmic lattice can be separated in a curl part and a divergent part, and that the curl part creates a set of equations for the macroscopic rotation field which is perfectly identical to the set of *the Maxwell equations of the electromagnetism (chap. 17).*

Then one shows that the Newton equation can also be separated in a different manner, in *two partial Newton equations* allowing to calculate on the one hand *the distortion elastic fields* associated with the topological singularities, and on the other hand *the volumetric expansion perturbations* associated with the distortion elastic energies of the topological singularities *(chap. 18).* By using the first partial Newton equation, on can calculate the fields and energies of elastic distortions generated by topological singularities within the cosmic lattice *(chap. 19).* One can then find conditions on the elastic modules of this lattice such as it is possible to attribute in a perfectly conventional manner *an inertial mass* to the topological singularities, which always satisfies the famous Einstein relation $E_0 = M_0 c^2$.

Then one demonstrates that the topological singularities satisfy a typically relativist dynamics when their velocity inside the lattice becomes close to the celerity of the transversal waves *(chap. 20).*

On these foundations, one finishes by discussing the analogy between this theory and *the theory of special relativity (chap. 21)*. One notices that the cosmic lattice acts in fact as *an aether*, in which the topological singularities satisfy exactly the same properties than those of the special relativity concerning *the length contraction, the time dilatation, the Michelson-Morley experiment* and *the Doppler-Fizeau effect*. The existence of the cosmic lattice allows then to explain very simply some obscure sides of the special relativity, as for example *the twin paradox!*

Gravitation, general relativity, weak interaction and cosmology

A third section (C) is dedicated to *the gravitation* and *the cosmology*. Thanks to *the second partial Newton equation*, one begins with the calculation of *the external expansion perturbations*, that is to say *the external scalar gravitation field,* associated with a localized macroscopic topological singularity, knowing either *its distortion elastic energy*, or *its curvature charge*, or *its rotation charge (chap. 22)*.

Immediately afterwards, one describes also *macroscopic vacancy singularities* and *macroscopic interstitial singularities*, which can appear within the lattice in the form of a macroscopic hole in the lattice or an interstitial embedment of a piece of lattice. These singularities will become subsequently the ideal candidates to explain respectively *the black holes* and *the pulsars* of our universe.

By applying the calculations of the external gravitation field of topological singularities to localized microscopic topological singularities, in the form of loops of screw disclination, loops of edge dislocation or loops of mixed dislocation, one deduces the whole of the properties of these loops *(chap. 23)*. It appears then the new concept of « *curvature mass* » of the edge dislocation loops, which corresponds to the equivalent mass associated to the gravitational effects of *the curvature charges of these loops*, and which can be positive (in the case of loops of vacancy type) or negative (in the case of loops of interstitial type). In fact, the curvature charge and the equivalent curvature mass which is associated do not appear in any other physics theory, neither in general relativity, nor in quantum physics, nor in standard model of elementary particles. The appearance of this new curvature charge is *certainly the most important finding of our theory*, because it is precisely that curvature mass which is responsible for *a small asymmetry* between the particles (hypothetically containing edge dislocation loops of interstitial type) and the antiparticles (hypothetically containing edge dislocation loops of vacancy type), which will play a fundamental role concerning the cosmological evolution of the topological singularities within the universe!

By considering the gravitational interactions existing between the topological singularities composed essentially of screw disclination loops, one can deduce the behaviors of *the measuring rods and clocks of local observers* as a function of the local expansion field which takes place within the cosmic lattice *(chap. 24)*. One shows that, for any local observer, and whatever is the value of the local volumetric expansion of the lattice, the Maxwell equations remain always perfectly invariant, so that, for this observer, the transversal wave velocity is a perfect constant, when the transversal wave velocity measured by an observer situated outside the lattice in the absolute space depends strongly on the local expansion of the lattice!

One shows that these gravitational interactions present strong analogies with the Newton's gravitation and with the general relativity, and one discusses in details the perfectly analogue points, as the perfect analogy with the Schwarzschild metric at great distances from massive objects and the curvature of wave rays by massive objects.

But one shows that our eulerian theory of the cosmic lattice provides also new elements to the gravitation theory, notably modifications of the Schwarzschild metric at very short distances from massive objects, and a better understanding of the critical radii associated with black holes: the radii of the photon perturbation sphere and of the point of no return become both equal to the Schwarzschild radius $R_{Schwrzschild} = 2GM / c^2$, and the limit radius for which the time dilatation of a falling observer would stretch to the infinite becomes zero, so that our theory is not limited beyond the Schwarzschild sphere for the description of a black hole.

One establishes next a complete table of *all the gravitational interactions* existing between the diverse topological singularities of the cosmic lattice, and one finds that the gravitational interactions between screw disclination loops is largely dominant *(chap. 25).*

By considering now a topological singularity formed by coupling a screw disclination loop with an edge dislocation loop, called a dispiration loop, it appears *an interaction force similar to a catch potential,* with a very small range, which allows interactions between loops presenting *a perfect analogy with the weak interactions* between elementary particles of the standard model.

On the basis of the cosmological behaviors of a lattice described in section (A), and the gravitational interactions between topological singularities described in section (C), on can imagine *a very plausible scenario for the cosmological evolution of the topological singularities,* leading to the present structure of our universe *(chap. 26).* This scenario allows one to give a very simple explanation of several facts still poorly understood, as *the formation of galaxies, the disappearance of antimatter, the formation of gigantic black holes at the heart of the galaxies,* and even *the famous « dark matter »* that the astrophysicists had to concoct for explaining the gravitational behavior of the galaxies.

In our theory, the *dark matter* would be in fact *a sea of repulsive neutrinos* in which the galaxies would have precipitated and would be immersed. Indeed, in the case of the simplest edge dislocation loops, analogically similar to neutrinos, *the « gravitational curvature mass » dominates the inertial mass,* so that the neutrinos should be the only particles gravitationally repulsive, when the antineutrinos should be gravitationally attractive. It is this surprising particularity which could explain the formation of a repulsive neutrinos sea playing the role of dark matter for the galaxies, due to the compression force exerted by the repulsive neutrinos sea on the galaxies periphery!

Finally, one shows how can be treated *the Hubble constant, the galaxy redshift* and *the evolution of the cosmic microwave background* in the frame of our eulerian theory of cosmic lattice.

Quantum physics, particles spin and photons

A fourth section (D) is dedicated to *the quantum physics* and *the standard model of particles.* One begins by using the second partial Newton equation, in the dynamical case, to show that there exists also *longitudinal gravitational perturbations* associated to moving topological singularities inside the lattice *(chap. 27).* By conjecturing operators similar to those of the quantum mechanics, one shows then that the second partial Newton equation allows one to deduce the gravitational fluctuations associated to a topological singularity moving quasi-freely with relativistic velocities within the lattice.

In the case of non-relativistic topological singularities bonded to a potential, one shows that the second partial Newton equation applied to the longitudinal gravitational fluctuations associated to these singularities leads to *the Schrödinger equation of the quantum physics*, which allows one *for the first time* to give a simple and realistic physical interpretation to the Schrödinger equation and to the quantum wave function: *the quantum wave function deduced from the Schrödinger equation represents the amplitude and the phase lag of longitudinal gravitational vibrations associated to a topological singularity within the cosmic lattice!*

All the consequences of the Schrödinger equation appear now with a simple physical explanation, as for example *the stationary wave equation* of a topological singularity placed inside a static potential, *the Heisenberg uncertainty principle* and *the probability interpretation* of the square of the wave function.

In the case where the gravitational fluctuations of two topological singularities are coupled, it appears also very simply *the concepts of bosons and fermions*, as well as *the Pauli exclusion principle*.

At the heart of a topological singularity loop, one shows that there cannot exist static solutions to the second partial Newton equation for the longitudinal gravitational fluctuations *(chap. 28)*. It becomes then necessary to find a dynamical solution to this equation. The most simplest dynamical solution is to imagine that the loop rotates around one of its diameter. By solving this rotation motion with the second partial Newton equation, which is nothing other than the Schrödinger equation, one obtains a quantified solution for the internal gravitational fluctuations of the loop. This solution is in fact nothing other than *the quantic loop spin*, which can take several different values (1/2, 1, 3/2, ...) and which is perfectly similar to the spin of particles in the standard model! If the loop is composed of a screw disclination loop, it appears also *a magnetic moment of the loop*, proportional to *the famous Bohr magneton*. The notorious argument of the quantum physics pioneers wherein the spin cannot be a real rotation of the particle on itself because the equatorial velocity should become superior to light velocity, is swept out in our theory by the fact that the static expansion at the vicinity of the loop heart is so high that the light velocity becomes much higher that the equatorial rotation velocity of the loop!

In this argumentation about the absolute necessity of a spin of the singularity loops for satisfying the second partial Newton equation, only the exact value of the spin of a loop, namely 1/2 or 1, does not find at the moment a simple explanation!

One finishes by showing how to construct a pure transversal wave packet with a circular polarization *(chap. 29) and why it appears a quantification of the energy of these fluctuations.* These waves packets form quasiparticles which have properties perfectly similar to *the quantum properties of photons: circular polarization, zero mass, non-zero momentum, non-locality, wave-particle duality, quantum entanglement* and *quantum decoherence.*

Standard model of elementary particles and strong interactions

In the second part of this section (D), one searches for the ingredients which have to be added to the cosmic lattice in order to find an analogy between the loops and the diverse particles of the standard model (chap. 30). One shows that, by introducing a cubic lattice with three families of planes (imaginary « colored » in red, green and blue), satisfying some simple rules concerning their successive arrangement and their mutual rotation, one finds topological loops perfectly analogous to all the particles, leptons and quarks, of the first family of elementary particles of the standard model. One finds also topological loops analogous to the W and Z bosons of the standard model. It appears also spontaneously a strong force, in the sense that this force presents an asymptotical behavior, acting between the loops analogous to the quarks of the standard model. This implies that these loops have to group together in triplets to form combinations of three loops analogous to the baryons, or in doublets to form combinations of loop-anti-loop analogous to the mesons. Furthermore, one finds also topological bicolor loops which correspond perfectly to the gluons associated to the strong force in the standard model!

In order to explain the existence of *three families of quarks and leptons in the standard model*, one shows that the introduction of more complicated topological structures of the edge dislocation loops, based on *assembling of pairs of edge disclination loops*, allows one to explain in a satisfactory way the existence of three, or even four, families of particles with very different energies.

Finally, one discusses the interest of this strong analogy between the topological singularities of a cubic « colored » lattice and the elementary particles of the standard model, as well as the numerous questions still pending concerning this analogy.

Vacuum quantum state fluctuations, multiverse cosmological theory and gravitons

A fifth section (E) is dedicated to some very hypothetical consequences concerning *the pure gravitational fluctuations* associated to the perfect cosmic lattice *(chap. 31).* One can imagine the existence of pure longitudinal fluctuations within the cosmic lattice, which are not correlated with the presence of topological singularities, and which can be treated either as *random gravitational fluctuations* that could present some analogy with *the vacuum quantum state fluctuations*, or as *stable gravitational fluctuations* that could lead at the macroscopic scale to *a cosmological theory of multiverse.* At the microscopic scale, stable gravitational fluctuations could also lead to *stable quasiparticles* which could be called *gravitons,* by analogy with the photons, but which have nothing common with the gravitons postulated in the frame of the general relativity.

One finishes this book by a general conclusion in which one shows the central roles played by the Newton equation and by the microscopic structure of the cosmic lattice. One highlights also the numerous positive points, but also the still misunderstood points, which have appeared throughout this essay concerning the analogy between the newtonian cosmic lattice and all the theories of modern physics.

Table of contents

PART I B - Application: examples of phenomenologies of usual solids

PART I C - Dislocation and disclination charges

PART I D - Application: elements of the dislocation theory in the usual solids

PART II A - The "cosmic lattice"

PART II B - Maxwell equations and special relativity

PART I

A

Eulerian theory of newtonian deformable lattices

Distortions and contortions

Geometro-compatibility and geometro-kinetic

Newtonian dynamics and eulerian thermo-kinetics

Elasticity, anelasticity, plasticity and self-diffusion

Evolution equations of a newtonian lattice

Chapter 1

Distortions of a lattice

In this chapter, one introduces a description of the solid deformable lattices in *the Euler coordinate system*. From the definition of the field of average velocity $\vec{\phi}$ of the sites of a lattice in the absolute frame of an observer ***GO***, one deduces *a volumetric expansion scalar* τ and *a rotation vector* $\vec{\omega}$, which satisfy *geometro-kinetic equations* of spatiotemporal evolution.
One shows then these two entities are in fact the trace and the antisymmetric part of a more general tensor, called *the distortion tensor,* which will be represented *in vectorial notation* by $\vec{\beta}_i$, and which satisfies also a *geometro-kinetic equations* in the Euler coordinates. From this tensor, one deduces also *a strain tensor* $\vec{\varepsilon}_i$ and *a shear tensor* $\vec{\alpha}_i$ of the solid lattice.
Later, with some simple examples of velocity fields $\vec{\phi}(\vec{r},t)$ associated with diverse known movements of the medium, one verifies that the topological tensors $\vec{\beta}_i$, $\vec{\varepsilon}_i$, $\vec{\alpha}_i$, $\vec{\omega}$ and τ describe perfectly the distortions, the deformations, the shear strains, the rotations and the volumetric expansions which can appear within a solid lattice.

1.1 - Spatiotemporal evolution of a deformable medium

If an observer, who will be called *the Great Observer **GO***, wants to describe in his laboratory the spatiotemporal evolution of a given continuous medium which moves in the space by translation and rotation, and which can moreover deform in the course of time (figure 1.1), he has first to define the type of kinetic which he has to use. Using as basic axiom that the spatiotemporal evolution of the medium is described by *a galilean kinetic*, and as a consequence by a principle of velocity additivity, the observer ***GO*** can describe this evolution on the basis of the absolute frame of his laboratory. This reference frame is composed of an orthonormal basic frame $Q\xi_1\xi_2\xi_3$, that is *three orthogonal measuring rods of unit length* $(\vec{\mathbf{e}}_1,\vec{\mathbf{e}}_2,\vec{\mathbf{e}}_3)$, and *one universal clock*, insuring that the time t is measured in identical way everywhere in his laboratory (figure 1.1).
In order to describe simply and completely the spatiotemporal evolution of the continuous medium, the observer can use *a lagrangian coordinate system.* First of all, he carries out a marking of the material medium at the initial time $t=0$ by means of a grid of points $\boldsymbol{P}_0$. He can then define a moveless local frame $Ox_1x_2x_3$, situated at the point identified by $\vec{\xi}_O$ in the frame of his laboratory. By equipping this fixed frame $Ox_1x_2x_3$ with measuring rods of unit length $(\vec{e}_1,\vec{e}_2,\vec{e}_3)$, and by positioning it judiciously with regard to the initial position of the medium at time $t=0$, he can measure the positions of all the points $\boldsymbol{P}_0$ of the medium at the initial time $t=0$ by means of vectors $\vec{r}$. At a time $t>0$, a point $\boldsymbol{P}_0$ of the medium will move at $\boldsymbol{P}$, and

the observer can then link up the point P_0 to the point P by using a vector $\vec{u}$, which is called *the displacement vector* of the point P_0. As this vector is dependent on the initial position $\vec{r}$ of the point P_0 and on the time t, the whole of vectors $\vec{u}(\vec{r},t)$ locating all the points of the medium is called *the displacement field* of the medium in lagrangian coordinates.

The concept of continuous medium at the macroscopic scale

The expression of continuous medium is an intuitive concept meaning that the medium does not present during its spatiotemporal evolution and at the macroscopic scale where it is observed, neither the appearance of discontinuous structures, nor the forming of discontinuities such as tears or local breakings, nor the forming of cavities.

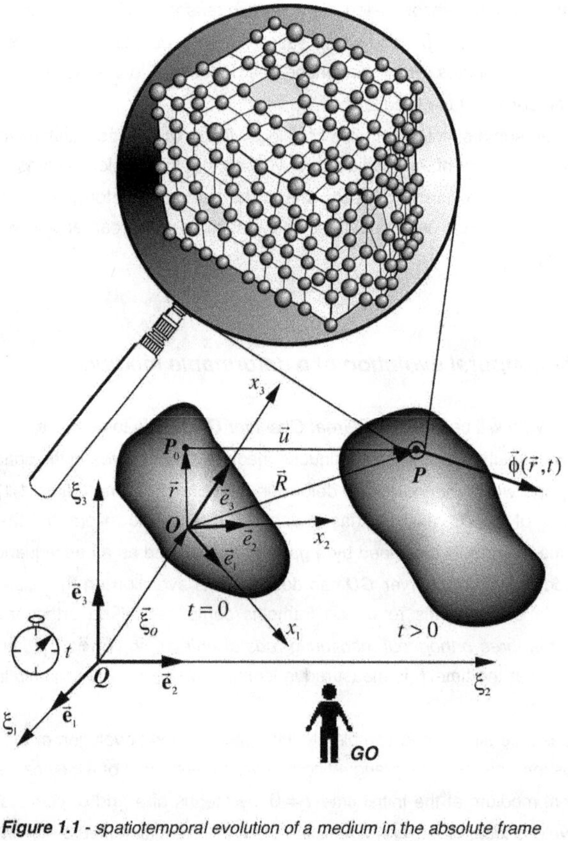

***Figure 1.1** - spatiotemporal evolution of a medium in the absolute frame*

From the macroscopic observation of the behavior of the medium, and in particular of *the continuity of the displacement field* $\vec{u}$, it is possible to attribute some appellations to the observed medium.

If the medium presents a displacement field $\vec{u}$ perfectly continuous during its spatiotemporal evolution, it is qualified of *solid medium.* It owns then a macroscopic shape which is difficult to modify.
On the other hand, if the medium presents a discontinuous displacement field $\vec{u}$ which forms an inextricable tangle in the course of time, it is qualified of *fluid medium.* It owns then the macroscopic property to flow, and has consequently to be maintained in a container from which it takes the shape. In this case, the displacement field $\vec{u}$ lost its original physical sense and meaning, and only *the local velocity* $\vec{\phi}(\vec{\xi},t)$ of the fluid at time t and at the space coordinates $\vec{\xi}$ of the absolute frame maintains a physical sense. In this book, one will be essentially concerned by the case of solid medium.

The concept of continuous medium at the microscopic scale

The concept of continuous solid medium is applicable only when the medium is observed at *a macroscopic scale.* Indeed, an enlargement of the same medium *at a sufficiently microscopic scale* will show a discontinuous collection of objects (fig. 1.1), at which one will afterwards attribute *the generic name of particles* (for example corpuscles, atoms, molecules, etc.). One logically concludes that the global phenomenological properties observed at the macroscopic scale where the medium seems continuous are in fact statistical effects resulting from the great number of particles which interact between themselves at the microscopic scale.
The enlargement of the medium allows also to define some of its important microscopic characteristics, as *its structure*, that is the manner in which its particles are assembled together, and *its chemical composition*, that is the nature of its particles.
A continuous medium will be called *solid* when, at the microscopic scale, it corresponds to a collection of particles such as *the identity of the nearest neighbour particles of a given particle does not change in the course of time.* In other words, each particle is bounded to its nearest neighbour particles by stiff bond which prevent long distance motions. As a consequence, only short distance motions from the nearest neighbour particles are permitted, due to the elasticity of the bonds. By the action of these bonds, the particles of the medium form *a solid lattice.*

The different classes of solid lattices

It is possible to define diverse classes of solid lattices, depending on the arrangement of the particles relative to each other. If the particles presents an arrangement well established, which repeats at long range by translation of an elementary unit cell, one speaks about *a lattice with a crystalline structure.* For example, the two-dimensional lattice reported in figure 1.2(a) and obtained by the translation of a hexagonal unit cell presents a perfect order at long range as well as at short range.
Some solid lattices can present arrangements of their particles without long range order, but only a certain order at short range. One speaks in this case about a lattice with *an amorphous structure.* The two-dimensional example reported in figure 1.2(b) represents an amorphous lattice of particles, obtained by paving the surface with irregular pentagons, hexagons and heptagons with fixed length sides. The short range order of the lattice is reflected by the fact that each particle possesses exactly three closest neighbours.

Solid lattices can also exist with particle arrangement which does not present long range translation order, but a certain order by rotation. One speaks then about *a lattice with a quasicrystalline structure*, represented by the example in figure 1.2(c) which shoes clearly the absence of long range order by translation.

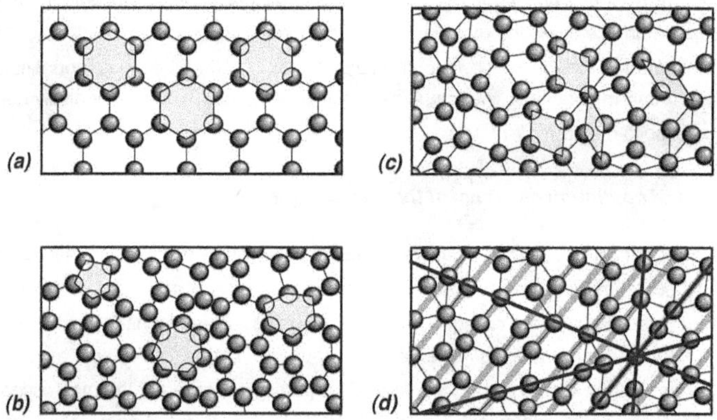

Figure 1.2 *- two-dimensional lattices with crystalline (a), amorphous (b) and quasicrystalline (c-d) arrangements of their particles*

This lattice is obtained by paving the surface with two types of diamond shaped unit cells with different apex angles (Penrose paving). At first sight, this lattice seems amorphous. But a more detailed analysis reported in figure 1.2(d) shows that the particles aligned with parallel straight lines. The distances between these parallel lines are not regular, and there exists in fact five different privileged directions for the orientation of these alignments. This means that this two-dimensional quasicrystalline structure presents a kind of fifth order rotation symmetry, which is not allowed in the case of the crystalline structures obtained by translation of base patterns.

The examples shown in figure 1.3b are two-dimensional representations. It is then necessary to generalize these notions to the three-dimensional space. In three dimensions, the crystalline lattices are constituted by the translation of a three-dimensional elementary volume, called *the unit cell of the lattice* (figure 1.3). The crystalline lattices can be defined by the lengths of the edges of the unit cell and the angles between them, which are called *the lattice parameters*. The crystal structures can be grouped in *seven lattice systems* according to the axial system used to describe their lattice. Considering the possible arrangement of the atoms relative to each other in the unit cell *(lattice centerings P, C, I or F)*, one arrives at *the fourteen Bravais lattices* represented in figure 1.3. Using the symmetry operations of rotation, reflection and inversion that leave at least one point unmoved and the appearance of the crystal structure unchanged, one can still define *32 possible crystallographic point groups*. Using the symmetry operations associated to the translation, one can finally define *230 distinct space groups*.

It is interesting to note that the lattice centerings leads to different values of the number of substitutional sites of the unit cell which can contain a bounded particle. For example in the case of the cubic lattice system, the number of substitutional sites of the unit cell is one for *the simple cubic lattice*, 2 for *the body-centered cubic lattice* and 4 for *the face-centered cubic lattice.*

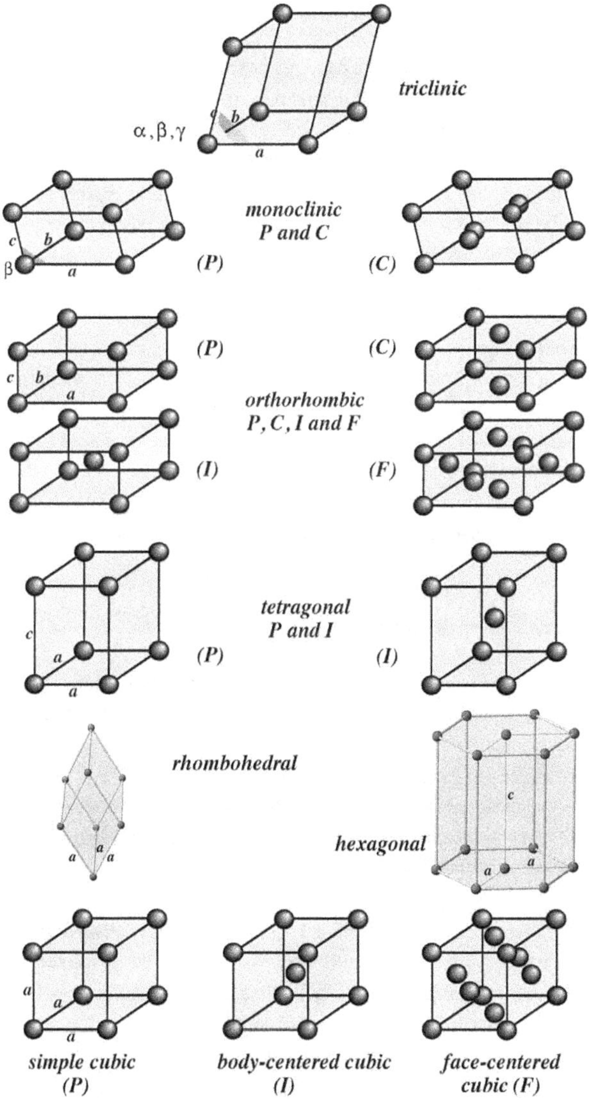

Figure 1.3 - *the seven crystalline systems and the fourteen Bravais lattices*

In the case of unordered solid medium as *the amorphous solids, the quasicrystalline solids* or *the polycrystalline solids with very fine grains*, the notion of lattice unit cell has no more meaning. But the concept of lattice site keeps a physical meaning, in the sense that it corresponds to a substitutional site of the unordered solid lattice.

The concept of topological singularities in a solid lattice

In the case of an ordered solid lattice, it can appear structural defects inside the regular arrangement of the lattice particles. These structural defects have diverse origins as irregularities of the chemical species of the lattice, or topological singularities, that is irregularities in the topological structure of the lattice which can be punctual, linear or planar following their topology. It is also by observations of the dynamics at the microscopic scale, during the macroscopic spatiotemporal evolution of the lattice, that it will be possible to understand the objective reasons of some macroscopic behaviors. For example, one will see that it can exist close links between the macroscopic properties of deformation of ordered lattices and the distorsions of the lattice unit cell induced by the presence of mobile topological singularities within the lattice, as dislocations or disclinations.
As a conclusion, a complete description of the spatiotemporal evolution of a lattice which can be considered as continuous at the macroscopic scale *cannot be obtained without a description of the phenomena which take place at the microscopic scale.* The search of a theory describing the macroscopic spatiotemporal evolution of a continuous deformable lattice has to be based on the definition of average macroscopic fields (scalar, vectorial and tensorial) deduced from a statistical description of the dynamics at the microscopic scale of a multitude of objects interacting with each other.

The lagrangian and the eulerian coordinate systems

In order to describe the spatiotemporal evolution of a deformable continuous medium, there exist two well known coordinate systems: *the lagrangian coordinate system*, used generally to describe the deformation of solids, and *the eulerian coordinate system*, used generally to describe the hydrodynamics of fluids.
The lagrangian coordinate system is based on *the description of the temporal evolution of the previously defined displacement field* $\vec{u}(\vec{r},t)$, knowing the initial coordinates $\vec{r}$ of all the points of the solid in a fixed reference frame $Ox_1x_2x_3$ of the observer laboratory, as illustrated in figure 1.1.
Concerning the eulerian coordinate system, it is based on *the description of the temporal evolution of the velocity field* $\vec{\phi}(\vec{\xi},t)$ of the medium points situated at the space coordinates $\vec{\xi}$ at the time t in the absolute reference frame $Q\xi_1\xi_2\xi_3$ of the observer laboratory.
The lagrangian coordinate system is well adapted to the description of the evolution of solid lattices which deform very weakly, but becomes perfectly unusable to describe strong deformations of a lattice, or to describe a lattice containing topological singularities. The eulerian coordinate system is much more general, because it allows one to describe not only the fluids, but also the solid lattices presenting strong deformations or containing topological singularities. It is the reason why we develop in this essay *an eulerian description of the spatiotemporal evolution of deformable solid lattices.*

1.2 - Definition of local quantities in Euler coordinates

The average local velocity $\vec{\phi}(\vec{r},t)$

In the case of a collection of solid state particles in the space, each particle i has its own velocity $\vec{v}_i$. In order to determine the average local velocity of the particles, one has to fix a small element of volume V_f centered on the space coordinates $\vec{r}$ (figure 1.4), then to measure the velocities $\vec{v}_i$ of all the particles contained in this fixed volume V_f. If the instantaneous number of particles contained in this volume V_f is equal to N, and if N is sufficiently large, *the average velocity* $\vec{\phi}$ at the place $\vec{r}$ and the instant t can be defined by the following expression

$$\vec{\phi} = \vec{\phi}(\vec{r},t) = \frac{1}{N}\sum_{i \in V_f} \vec{v}_i(t) \tag{1.1}$$

If an average velocity $\vec{\phi}$ different of zero is measured, this means also that it is possible to find for each particle a fluctuation $\Delta\vec{v}_i$ to the average velocity $\vec{\phi}$ by the relation

$$\vec{v}_i = \vec{\phi} + \Delta\vec{v}_i \quad \Rightarrow \quad \frac{1}{N}\sum_{i \in V_f} \Delta\vec{v}_i(t) = 0 \tag{1.2}$$

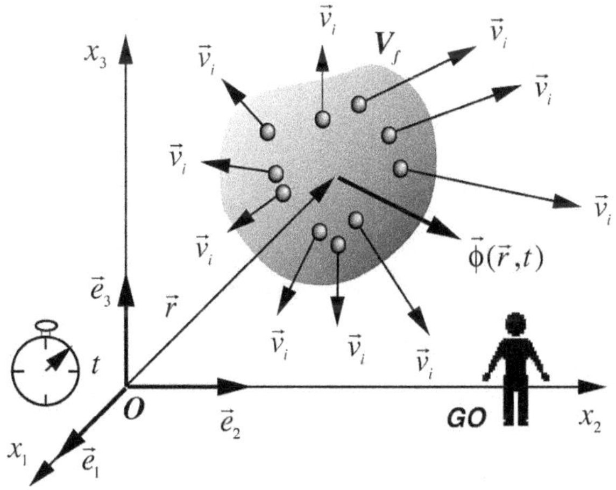

Figure 1.4 *- search of the average local velocity of a lattice in the volume* V_f

In a solid lattice, the existence of an average velocity $\vec{\phi}(\vec{r},t)$ different from zero involves that the solid lattice of particles be submitted to a collective movement. The velocity $\vec{\phi}(\vec{r},t)$ represents then the average local velocity of displacement of the particles bound to the lattice sites, which corresponds to *the average velocity of the lattice sites*, when the $\Delta\vec{v}_i$ are the fluctuations of velocity of the particles bounded to the lattice around each of these sites. For example, in a real solid, such fluctuations are due to the disordered motions of the thermal agitation of the particles associated directly to the solid temperature.

The local density $n(\vec{r},t)$ of elementary substitutional sites of the lattice

Besides the average local velocity $\vec{\phi}(\vec{r},t)$ of the sites of a macroscopically continuous lattice, there is also an other quantity which will play a fundamental role in Euler coordinates: it is *the volume density of elementary substitutional sites of the lattice* which will be written n . This choice will involve to define all the physical quantities characterizing the solid lattice as *average values taken on each site of the lattice.* It is clear also that, in disordered lattice, the quantity n can be related to the volume density of elementary sites of the disordered lattice.

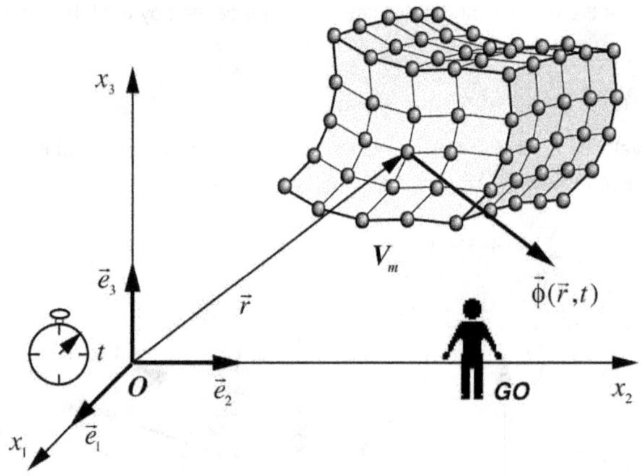

***Figure 1.5** - average velocity of the collective motion of the sites of a solid lattice*

This density is a local average value which can be expressed as a function of $\vec{r}$ and t

$$n = n(\vec{r},t) \tag{1.3}$$

The quantity n has to satisfy a *continuity equation.* In order to express this equation, one has to consider a volume V_m moving with the lattice at the velocity $\vec{\phi}$. The total number N of sites of the lattice contained in the volume V_m is given by

$$N = \iiint_{V_m} n\,dV \tag{1.4}$$

Over times, this total number N can vary only if it exists sources of lattice sites inside the volume V_m , since the volume V_m follows the medium at the velocity $\vec{\phi}$. The continuity equation can be written by equating the temporal variation of N with *a term of source* S_n

$$\frac{dN}{dt} = \frac{d}{dt}\iiint_{V_m} n\,dV = \iiint_{V_m} S_n\,dV \tag{1.5}$$

In this relation S_n is the term of *volume source of sites of lattice,* which is the number of sites of the lattice created of annihilated by unit of volume and unit of time within the lattice under consideration.

With *(A.58)* applied to a mobile volume, expression *(1.5)* becomes

$$\iiint\limits_{V_m}\left[\frac{\partial n}{\partial t}+\operatorname{div}\left(n\vec{\phi}\right)\right]dV=\iiint\limits_{V_m}S_n\,dV \tag{1.6}$$

which holds for any mobile volume V_m we choose, and the equation of local continuity for n can be read off

$$\frac{\partial n}{\partial t}=S_n-\operatorname{div}\left(n\vec{\phi}\right) \tag{1.7}$$

This equation links the local density $n(\vec{r},t)$ of sites of lattice and the average velocity $\vec{\phi}(\vec{r},t)$ of these sites. We will discuss later the possibility of existence of sources of sites S_n within the lattice.

The material derivative in eulerian coordinates

Relation *(1.7)* can be transformed by expanding the term in the divergence

$$\frac{\partial n}{\partial t}=S_n-n\operatorname{div}\vec{\phi}-\vec{\phi}\,\overrightarrow{\operatorname{grad}}\,n \tag{1.8}$$

and then using operator $\vec{\nabla}$, defined in section A.3

$$\frac{\partial n}{\partial t}+(\vec{\phi}\vec{\nabla})n=S_n-n\operatorname{div}\vec{\phi} \tag{1.9}$$

We introduce here an important operator in eulerian coordinates called *the material derivative*, and represented by the symbol (straight derivative instead of curled) d/dt and defined as

$$\frac{d}{dt}=\frac{\partial}{\partial t}+(\vec{\phi}\vec{\nabla}) \tag{1.10}$$

This operator corresponds to the temporal derivative of a quantity observed along the trajectory of the sites of the lattice. It consists in fact in calculating the total derivative of $n(\vec{r},t)$, with respect to time, as shown in the following equalities

$$\frac{dn}{dt}=\frac{\partial n}{\partial t}+\sum_i\frac{\partial n}{\partial x_i}\frac{\partial x_i}{\partial t}=\frac{\partial n}{\partial t}+\sum_i\frac{\partial n}{\partial x_i}\phi_i=\frac{\partial n}{\partial t}+(\vec{\phi}\vec{\nabla})n \tag{1.11}$$

With this operator, the continuity equation n can be presented as a path derivative

$$\frac{dn}{dt}=S_n-n\operatorname{div}\vec{\phi} \tag{1.12}$$

The equation of continuity for density n of sites of lattice is fundamental, because we are going to base all solid lattices in eulerian coordinates.

Non-commutativity of spatial and temporal operators in eulerian coordinates

In eulerian coordinates, the principal operator of time, the material derivative d/dt along the path, does not, in general commute with the operators of space $\overrightarrow{\operatorname{rot}}$, $\overrightarrow{\operatorname{grad}}$ and div. This important property of operators can be verified. Indeed for any vector $\vec{A}$ defined in the medium we have

$$\begin{cases} \overrightarrow{\mathrm{rot}}\left(\dfrac{d\vec{A}}{dt}\right)=\overrightarrow{\mathrm{rot}}\left[\dfrac{\partial\vec{A}}{\partial t}+(\vec{\phi}\vec{\nabla})\vec{A}\right]=\dfrac{\partial}{\partial t}\overrightarrow{\mathrm{rot}}\,\vec{A}+(\vec{\phi}\vec{\nabla})\overrightarrow{\mathrm{rot}}\,\vec{A}+\overrightarrow{\mathrm{grad}}(\vec{\phi}\vec{\nabla})\wedge\vec{A} \\ \overrightarrow{\mathrm{rot}}\left(\dfrac{d\vec{A}}{dt}\right)=\overrightarrow{\mathrm{rot}}\left[\dfrac{\partial\vec{A}}{\partial t}+(\vec{\phi}\vec{\nabla})\vec{A}\right]=\dfrac{d}{dt}\left(\overrightarrow{\mathrm{rot}}\,\vec{A}\right)+\sum_k\overrightarrow{\mathrm{grad}}\,\phi_k\wedge\dfrac{\partial\vec{A}}{\partial x_k} \end{cases} \tag{1.13}$$

We deduce the relation of commutation between the rotational and the material derivative which can be written in the form

$$\overrightarrow{\mathrm{rot}}\left(\frac{d\vec{A}}{dt}\right)-\frac{d}{dt}\left(\overrightarrow{\mathrm{rot}}\,\vec{A}\right)=\sum_k\overrightarrow{\mathrm{grad}}\,\phi_k\wedge\frac{\partial\vec{A}}{\partial x_k} \tag{1.14}$$

Similar expressions can be found for the gradient of a scalar B or the divergence of a vector $\vec{A}$

$$\overrightarrow{\mathrm{grad}}\left(\frac{dB}{dt}\right)-\frac{d}{dt}\left(\overrightarrow{\mathrm{grad}}\,B\right)=\sum_k\overrightarrow{\mathrm{grad}}\,\phi_k\,\frac{\partial B}{\partial x_k} \tag{1.15}$$

$$\mathrm{div}\left(\frac{d\vec{A}}{dt}\right)-\frac{d}{dt}\left(\mathrm{div}\,\vec{A}\right)=\sum_k\overrightarrow{\mathrm{grad}}\,\phi_k\,\frac{\partial\vec{A}}{\partial x_k} \tag{1.16}$$

1.3 - Eulerian geometro-kinetic equations

A definition of the volume expansion of a lattice

The volume density n of sites of the solid lattice is directly linked to the volume expansion of the medium. In fact $n\to 0$ for large expansions and $n\to\infty$ for large contractions. This notion of volume expansion of the lattice can be better captured with a quantity v defined as the inverse of n

$$v=1/n \tag{1.17}$$

Indeed this quantity v has the dimension of a volume. It represents the average volume occupied by an elementary site of the lattice. This volume v translates the intuitive notion of volume expansion of the medium, as $v\to\infty$ for large expansions and $v\to 0$ for intense contractions. But relationship *(1.12)* transforms also in the following way

$$\frac{1}{n}\frac{dn}{dt}=\frac{d}{dt}\left(\ln n+cste\right)=\frac{S_n}{n}-\mathrm{div}\,\vec{\phi} \tag{1.18}$$

In this expression, we can choose the constant to be $-\ln n_0$, and we can then define the dimension-less scalar τ with relation

$$\tau=-\ln\frac{n}{n_0}=\ln\frac{v}{v_0} \tag{1.19}$$

We will call it the *scalar of volume expansion*, which satisfies the following equation of geometro-kinetic

$$\frac{d\tau}{dt}=-\frac{S_n}{n}+\mathrm{div}\,\vec{\phi} \tag{1.20}$$

and which measures perfectly the notion of volume expansion of the lattice as this time $\tau\to\infty$ for large expansions (when $v\to\infty$), and $\tau\to-\infty$ for intense contractions (when $v\to 0$) and

$\tau \to 0$ when $v \to v_0$. Given the construction of the scalar τ, the constants v_0 et n_0 introduced here can be freely adjusted to signify a null expansion of the lattice ($\tau = 0$).

A definition of the global rotation and the contorsions of a lattice

If the field $\vec{\phi}(\vec{r},t)$ is not homogenous in space, then the lattice can have, aside from volume expansions, movements that correspond to a *global rotation* and *local rotations,* which imply contorsions *(flexions and torsions)* of the medium in space. To explain these rotational movements, we must determine the average local angular speed of the lattice. To that end, one must imagine within the lattice a circle of diameter R centered on O' with coordinates $\vec{r}_{O'}$ and oriented perpendicular to axis Ox_k as shown in figure 1.6. The average tangential velocity $\langle\phi\rangle$ along the circle can be written

$$\langle\phi\rangle = \frac{1}{2\pi R}\oint_C \vec{\phi}d\vec{r} \qquad (1.21)$$

If the medium is rotating locally, or globally, about O' around axis Ox_k, the average tangential velocity $\langle\phi\rangle$ thus calculated will be different than zero. The average angular velocity $\langle\dot{\theta}_k\rangle$ of the lattice at O', around the axis Ox_k, is obtained simply by dividing the average tangential velocity $\langle\phi\rangle$ by the circle radius R

$$\langle\dot{\theta}_k\rangle = \frac{\langle\phi\rangle}{R} = \frac{1}{2\pi R^2}\oint_C \vec{\phi}d\vec{r} \qquad (1.22)$$

With this expression of $\langle\dot{\theta}_k\rangle$, the theorem of the rotational *(A.38)* allows one to transform the integral on a circular boundary C to an integral on the surface S enclosed within the boundary

$$\langle\dot{\theta}_k\rangle = \frac{1}{2\pi R^2}\iint_S \overrightarrow{\mathrm{rot}}\,\vec{\phi}\,d\vec{S}_k = \frac{1}{2}\left(\frac{1}{\pi R^2}\iint_S \overrightarrow{\mathrm{rot}}\,\vec{\phi}\,dS_k\right)\vec{e}_k \qquad (1.23)$$

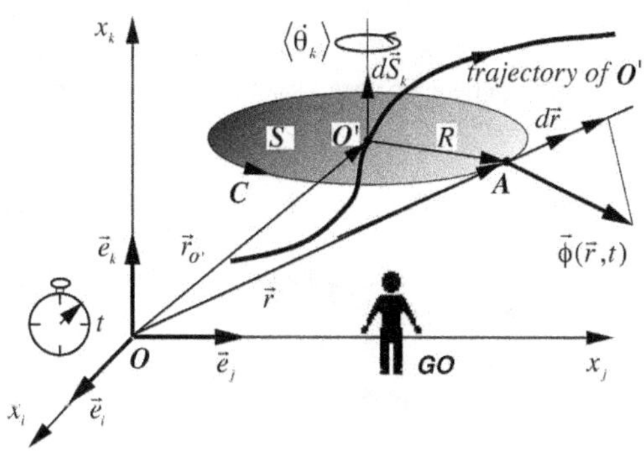

Figure 1.6 *- measure of local angular velocity of rotation of the lattice*

The expression between parenthesis is the average value of $\overrightarrow{\mathrm{rot}}\,\vec{\phi}$ on the disk of radius R, so that when $R \to 0$, there appears the *local angular velocity* $\dot{\theta}_k$ of the medium around the axis Ox_k at the coordinate $\vec{r}$ of space

$$\dot{\theta}_k = \frac{1}{2}\vec{e}_k \lim_{R\to 0}\left(\frac{1}{\pi R^2}\iint_S \overrightarrow{\mathrm{rot}}\,\vec{\phi}\, dS_k\right) = \frac{1}{2}\vec{e}_k\, \overrightarrow{\mathrm{rot}}\,\vec{\phi} \tag{1.24}$$

The vector of local angular velocity $\dot{\vec{\theta}}$ of the medium is easily deduced

$$\dot{\vec{\theta}} = \sum_k \dot{\theta}_k \vec{e}_k = \frac{1}{2}\overrightarrow{\mathrm{rot}}\,\vec{\phi} \tag{1.25}$$

This vector $\dot{\vec{\theta}}$ allows us to define the *vector of rotation* $\vec{\omega}$. Along the path in the medium, the local variation $d\vec{\omega}$ of the angle of rotation is given by $d\vec{\omega} = \dot{\vec{\theta}}dt$, so that it is possible to write a geometro-kinetic equation similar to expression *(1.20)* of the scalar of volume expansion, which involves the material derivative of the vector of rotation $\vec{\omega}$

$$\frac{d\vec{\omega}}{dt} = \dot{\vec{\theta}} = \frac{1}{2}\overrightarrow{\mathrm{rot}}\,\vec{\phi} \tag{1.26}$$

It is interesting to note here that the variation of the scalar of volume expansion τ is due to a divergent part in the field of velocity $\vec{\phi}$ *(1.20)*, whereas the variations in the local rotation vector $\vec{\omega}$ is a consequence of the existence of a rotational part of the same field $\vec{\phi}$ *(1.26)*.

1.4 - Eulerian distortion tensors

In the presence of a velocity field $\vec{\phi}(\vec{r},t)$ which is not homogenous in space, a lattice can exhibit movements, besides the global translation and rotation, corresponding to local rotations (torsions) as well as various types of deformations, such as volume expansion and shear.

To describe these deformations with rotations, which we will generically call *distortions* of the lattice, we have to precisely describe the spatial variations of each of the components of the field of velocities, and we will link them to the temporal variation of a topological measure characterizing the distortions of the lattice. An elegant way of proceeding is to note that the volume expansion τ and the rotation vector $\vec{\omega}$ which we have just defined correspond to the trace and the anti-symmetrical part of a general geometro-kinetic equation based on a *tensor of second order,* β_{ij}*, the gradient of the vectorial velocity field* $\vec{\phi}$.

The vectorial notation and the decomposition properties of second order tensors

For convenience, second order tensor, such as the tensor gradient β_{ij}, can be represented as a *field of 3 vectors.* To that end, we write

$$\begin{pmatrix}\vec{\beta}_1\\ \vec{\beta}_2\\ \vec{\beta}_3\end{pmatrix} = \begin{pmatrix}\beta_{11} & \beta_{12} & \beta_{13}\\ \beta_{21} & \beta_{22} & \beta_{23}\\ \beta_{31} & \beta_{32} & \beta_{33}\end{pmatrix}\begin{pmatrix}\vec{e}_1\\ \vec{e}_2\\ \vec{e}_3\end{pmatrix} \tag{1.27}$$

We will see that this *vectorial representation of tensorial fields* is very powerful and simplifies considerably the physical interpretation of tensorial fields.

From the general tensor $\vec{\beta}_i$ of distortions, we will vectorially describe the deformations of the lattice and we will extract information concerning the different types of deformations we can encounter. The way to proceed is to use the mathematical properties of a tensor $\vec{\beta}_i$, such as separating its symmetrical part and calculating its trace. We will do so now in the next section, in all generality of a second order tensor, without regards to the fact that said tensor $\vec{\beta}_i$ is derived from the field of velocities by a geometro-kinetic equation.

The generic operation of *transposition of a second order tensor* is defined by exchanging the terms of the matrix representation symmetrically to the diagonal. In component view this is achieved by inverting indices. In the vectorial notation of tensors, this operation is simply transcribed as

$$\vec{\beta}_i = \sum_k \beta_{ik}\vec{e}_k \Rightarrow \left[\vec{\beta}_i\right]^{\mathrm{T}} = \sum_k \beta_{ki}\vec{e}_k = \sum_k \vec{e}_k\left(\vec{\beta}_k\vec{e}_i\right) \tag{1.28}$$

By using properties *(A.24)*, the operation of transposition can also be written

$$\left[\vec{\beta}_i\right]^{\mathrm{T}} = \sum_k \vec{e}_k\left(\vec{e}_i\vec{\beta}_k\right) = \vec{\beta}_i + \vec{e}_i \wedge \left(\sum_k \vec{e}_k \wedge \vec{\beta}_k\right) \tag{1.29}$$

This formulation of transposition corresponds to subtracting from the tensor to be transposed the double of its antisymmetric part. In the vectorial notation, *the anti-symmetric part of tensor β_{ij}* can be build as a vector $\vec{\omega}$ by writing

$$\vec{\omega} = -\frac{1}{2}\sum_k \vec{e}_k \wedge \vec{\beta}_k \tag{1.30}$$

To verify that this vector $\vec{\omega}$ indeed represents the anti-symmetric part of tensor β_{ij} as a vector, let's look at the component view where $\vec{\omega}$ is written using the circular permutation on indices ijk

$$\vec{\omega} = -\frac{1}{2}\sum_k \vec{e}_k \wedge \vec{\beta}_k = \sum_k \frac{1}{2}\left(\beta_{ij} - \beta_{ji}\right)\vec{e}_k \tag{1.31}$$

Similarly one can build the *symmetrical part of tensor $\vec{\beta}_i$*, with

$$\left[\vec{\beta}_i\right]^{\mathrm{S}} = \frac{1}{2}\left\{\vec{\beta}_i + \left[\vec{\beta}_i\right]^{\mathrm{T}}\right\} = \vec{\beta}_i + \frac{1}{2}\vec{e}_i \wedge \left(\sum_k \vec{e}_k \wedge \vec{\beta}_k\right) \tag{1.32}$$

Let's call $\vec{\varepsilon}_i$ the symmetrical part of $\vec{\beta}_i$, by using *(1.31)*, we have

$$\vec{\varepsilon}_i = \vec{\beta}_i - \vec{e}_i \wedge \vec{\omega} \tag{1.33}$$

The cyclical permutation on indices ijk to obtain the component view, shows that the tensor $\vec{\varepsilon}_i$ indeed represents the symmetrical part of β_{ij}

$$\vec{\varepsilon}_i = \vec{\beta}_i - \vec{e}_i \wedge \vec{\omega} = \beta_{ii}\vec{e}_i + \frac{1}{2}\left(\beta_{ij} + \beta_{ji}\right)\vec{e}_j + \frac{1}{2}\left(\beta_{ik} + \beta_{ki}\right)\vec{e}_k \tag{1.34}$$

The trace τ of tensors $\vec{\beta}_i$ et $\vec{\varepsilon}_i$ is defined as the sum of the diagonal elements

$$\tau = \sum_k \vec{e}_k\vec{\beta}_k = \sum_k \beta_{kk} = \sum_k \varepsilon_{kk} = \sum_k \vec{e}_k\vec{\varepsilon}_k \tag{1.35}$$

The tensor $\vec{\alpha}_i$ obtained by subtracting the trace from the symmetrical part $\vec{\varepsilon}_i$ is called *the transverse symmetrical part of tensor β_{ij}*

$$\vec{\alpha}_i = \vec{\varepsilon}_i - \frac{1}{3}\tau\vec{e}_i \tag{1.36}$$

This tensor has a null trace, by construction

$$\text{trace}(\vec{\alpha}_i) = \sum_k \vec{e}_k \vec{\alpha}_k = \sum_k \alpha_{kk} \equiv 0 \tag{1.37}$$

Given the definition of the transpose operation, it is clear that the transposition of a *symmetrical* tensor does not change the tensor. This is the case of the symmetrical tensor $\vec{\varepsilon}_i$,

$$\vec{\varepsilon}_i = [\vec{\varepsilon}_i]^{\mathrm{T}} \quad \Rightarrow \quad \vec{\varepsilon}_i = \sum_k \vec{e}_k (\vec{e}_i \vec{\varepsilon}_k) = \vec{\varepsilon}_i + \vec{e}_i \wedge \left(\sum_k \vec{e}_k \wedge \vec{\varepsilon}_k \right) \tag{1.38}$$

from which we deduce that the symmetry of tensors $\vec{\varepsilon}_i$ and $\vec{\alpha}_i$ implies the following properties

$$\vec{\varepsilon}_i = \sum_k \vec{e}_k (\vec{e}_i \vec{\varepsilon}_k) \quad \text{and} \quad \sum_k \vec{e}_k \wedge \vec{\varepsilon}_k \equiv 0 \tag{1.39}$$

$$\vec{\alpha}_i = \sum_k \vec{e}_k (\vec{e}_i \vec{\alpha}_k) \quad \text{and} \quad \sum_k \vec{e}_k \wedge \vec{\alpha}_k \equiv 0 \tag{1.40}$$

The eulerian tensors of distortion, deformation and shear

By using the vectorial notation for second order tensors, we define the *distortion tensor* $\vec{\beta}_i$ with the following geometro-kinetic equation

$$\frac{d\vec{\beta}_i}{dt} = -\frac{S_n}{3n}\vec{e}_i + \overrightarrow{\text{grad}}\,\phi_i \tag{1.41}$$

whose trace gives us the geometro-kinetic equation for the *volume expansion* τ (fig. 1.7)

$$\frac{d\tau}{dt} = -\frac{S_n}{n} + \text{div}\,\vec{\phi} \tag{1.42}$$

and its antisymmetric part represents the geometro-kinetic equation for the *vector of rotation* $\vec{\omega}$ (fig. 1.7)

$$\frac{d\vec{\omega}}{dt} = \frac{1}{2}\overrightarrow{\text{rot}}\,\vec{\phi} \tag{1.43}$$

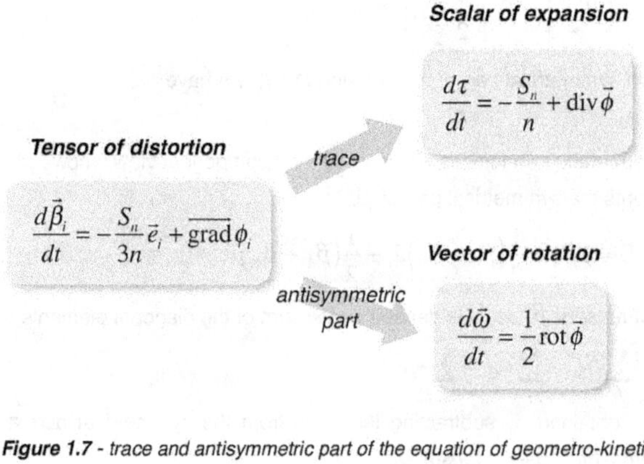

Figure 1.7 *- trace and antisymmetric part of the equation of geometro-kinetic*

Based on the decomposition of tensors of second order, as we have just seen, we can now define generally speaking, the tensor $\vec{\varepsilon}_i$ representing the symmetrical part of tensor $\vec{\beta}_i$. It is called the *tensor of deformation* $\vec{\varepsilon}_i$, since this tensor is equal to the tensor of distortion from which the rotation part (global and local) is subtracted. It satisfies the following geometro-kinetic equation

$$\frac{d\vec{\varepsilon}_i}{dt} = -\frac{S_n}{3n}\vec{e}_i + \overrightarrow{\mathrm{grad}}\,\phi_i - \frac{1}{2}\vec{e}_i \wedge \overrightarrow{\mathrm{rot}}\;\vec{\phi} \tag{1.44}$$

Finally, it is possible to define a tensor $\vec{\alpha}_i$ corresponding to the *transverse symmetrical part* of tensor $\vec{\beta}_i$. As this tensor is obtained from that of deformations $\vec{\varepsilon}_i$ from which the trace representing volume expansion τ is subtracted, it will be called the *shear tensor* $\vec{\alpha}_i$. It satisfies the following equation of geometro-kinetic

$$\frac{d\vec{\alpha}_i}{dt} = \overrightarrow{\mathrm{grad}}\,\phi_i - \frac{1}{2}\vec{e}_i \wedge \overrightarrow{\mathrm{rot}}\;\vec{\phi} - \frac{1}{3}\vec{e}_i\,\mathrm{div}\vec{\phi} \tag{1.45}$$

The complete decomposition of the tensor of distortion follows the schema presented at figure 1.8, it shows the symmetric and anti-symmetric parts, the trace and the transverse symmetric part of the tensor.

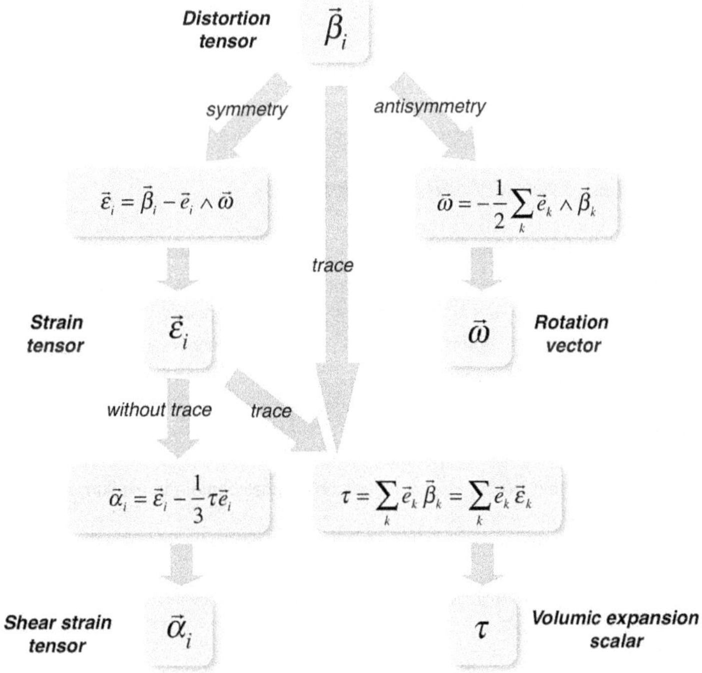

Figure 1.8 *- Decomposition of the eulerian tensor of distortions*

The full set of equations we have obtained is transcribed in table 1.1. With examples we will see later, we will show the adequacy of this system of 'galilean' equations to describe the geometro-kinetic of the topological distortions of any solid lattice in eulerian coordinates.

Table 1.1 - Geometro-kinetics of the distortions of a lattice in eulerian coordinates

Definition of volume expansion

$$\frac{\partial n}{\partial t} = S_n - \operatorname{div}(n\vec{\phi}) \quad (1) \quad \Rightarrow \quad \tau = -\ln\frac{n}{n_0} = \ln\frac{v}{v_0} \quad (2) \quad \Rightarrow$$

Geometro-kinetics of the distortions

$$\begin{cases} \dfrac{d\vec{\beta}_i}{dt} = -\dfrac{S_n}{3n}\vec{e}_i + \overrightarrow{\operatorname{grad}}\,\phi_i & (3) \\ \dfrac{d\vec{\omega}}{dt} = \dfrac{1}{2}\overrightarrow{\operatorname{rot}}\,\vec{\phi} & (4) \\ \dfrac{d\tau}{dt} = -\dfrac{S_n}{n} + \operatorname{div}\vec{\phi} & (5) \end{cases} \qquad \begin{cases} \vec{\omega} = -\dfrac{1}{2}\sum\limits_k \vec{e}_k \wedge \vec{\beta}_k & (6) \\ \tau = \sum\limits_k \vec{e}_k\,\vec{\beta}_k = \sum\limits_k \vec{e}_k\,\vec{\varepsilon}_k & (7) \\ \vec{\varepsilon}_i = \vec{\beta}_i - \vec{e}_i \wedge \vec{\omega} & (8) \\ \vec{\alpha}_i = \vec{\varepsilon}_i - \dfrac{1}{3}\tau\vec{e}_i & (9) \end{cases}$$

1.5 - Examples of velocity and distortion fields

With simple examples of the velocity field $\vec{\phi}(\vec{r},t)$ belonging to known movements of the lattice, this paragraph will show that the topological tensors β_{ij}, ε_{ij}, α_{ij}, $\vec{\omega}$ and τ completely capture the distortions, deformations, shear, rotations and volume expansions that can appear in a lattice, including in the case of large distortions. The values of these tensors simply go to infinity when distortions become very large!

Global translation

Consider the velocity field $\vec{\phi}(\vec{r},t)$ describing a global translation of the medium in the direction $\vec{a}$, where $\vec{a}$ is a constant vector

$$\vec{\phi}(\vec{r},t) = \vec{a}\,g(t) \tag{1.46}$$

With the geometro-kinetic equations, such a field gives us the following particulate derivatives

$$\frac{dn}{dt} = \frac{dv}{dt} = \frac{d\tau}{dt} = \frac{d\vec{\omega}}{dt} = \frac{d\vec{\beta}_i}{dt} = \frac{d\vec{\varepsilon}_i}{dt} = \frac{d\vec{\alpha}_i}{dt} = 0 \tag{1.47}$$

which shows that all topological tensors remain constant along the translation path of the medium.

Figure 1.9(a) represents a velocity field with a global translation of the medium, in the case where the direction $\vec{a}$ of translation, remains in the basis plane defined by $\vec{e}_1$ and $\vec{e}_2$. Figure 1.9(b) shows the temporal evolution between 2 different instants ($t = 0$ and $t > 0$) of a square portion of the medium with this velocity field.

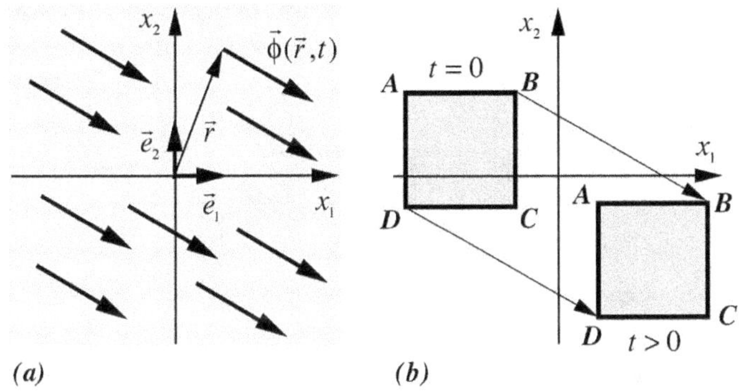

Figure 1.9 *- global translation*

Global rotation

Consider the following velocity field

$$\vec{\phi}(\vec{r},t)=\dot{\vec{\theta}}(t)\wedge\vec{r} \quad \text{with} \quad \dot{\vec{\theta}}(t)=\vec{a}\,g(t) \tag{1.48}$$

where $\vec{a}$ is a constant vector. The component view of this field reads

$$\vec{\phi}(\vec{r},t)=\left[(a_2x_3-a_3x_2)\vec{e}_1+(a_3x_1-a_1x_3)\vec{e}_2+(a_1x_2-a_2x_1)\vec{e}_3\right]g(t) \tag{1.49}$$

Such a velocity field represents a global rotation of the medium around a vector $\vec{a}$. With the equations of geometro-kinetic, we have

$$\frac{dn}{dt}=\frac{dv}{dt}=\frac{d\tau}{dt}=0 \quad \text{and} \quad \frac{d\vec{\varepsilon}_i}{dt}=\frac{d\vec{\alpha}_i}{dt}=0 \tag{1.50}$$

As a consequence, the tensor of deformation $\vec{\varepsilon}_i$, the shear tensor $\vec{\alpha}_i$ and the scalar of volume expansion τ do not change during the evolution of the medium about an axis. The temporal evolution of $\vec{\omega}$ reads

$$\frac{d\vec{\omega}}{dt}=\frac{1}{2}\overrightarrow{\mathrm{rot}}\,\vec{\phi}=\frac{1}{2}\overrightarrow{\mathrm{rot}}\left(\dot{\vec{\theta}}(t)\wedge\vec{r}\right)=\dot{\vec{\theta}}(t) \tag{1.51}$$

Its temporal evolution along the path is equal to the angular velocity $\dot{\vec{\theta}}(t)$, as we deduced in section 1.3.

We can compute the evolution of tensor β_{ij} along the path, and it appears that in the case of a global rotation, the tensor β_{ij} is purely anti-symmetric

$$\frac{d\vec{\beta}_i}{dt}=\vec{e}_i\wedge\frac{d\vec{\omega}}{dt}=\vec{e}_i\wedge\dot{\vec{\theta}}(t) \tag{1.52}$$

Figure 1.10(a) shows the velocity field corresponding to a global rotation of the medium, in the case where direction $\vec{a}$ of the axis of rotation is parallel to the base vector $\vec{e}_3$. Figure 1.10(b) shows the evolution at 2 different times ($t=0$ and $t>0$) of a square portion of the medium.

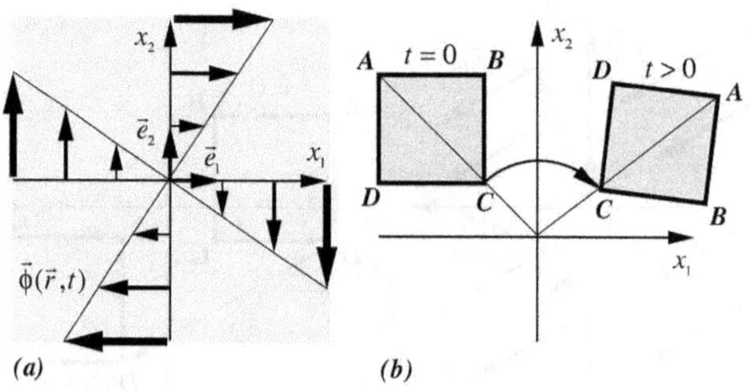

***Figure 1.10** - global rotation*

Elongations and volume expansion

The following velocity field is associated with homogeneous expansions and contractions along the 3 different axis of coordinates

$$\vec{\phi}(\vec{r},t) = \left[a_1 x_1 \vec{e}_1 + a_2 x_2 \vec{e}_2 + a_3 x_3 \vec{e}_3\right] g(t) \qquad (1.53)$$

where the a_i are constants.

With such a field, the medium is elongating or retracting along the 3 coordinate axis depending on the sign of the constants a_i. The equations of geometro-kinetic allow us to deduce β_{ij} and ε_{ij} along the path

$$\frac{d\vec{\beta}_i}{dt} = \frac{d\vec{\alpha}_i}{dt} = a_i g(t) \vec{e}_i \qquad (1.54)$$

Tensors $\vec{\beta}_i$ and $\vec{\varepsilon}_i$ are symmetrical and only have trace components. It is rather easy to compute the evolution of τ, which corresponds to an expansion or a contraction of the medium along the trajectory of points

$$\frac{d\tau}{dt} = (a_1 + a_2 + a_3) g(t) \qquad (1.55)$$

Furthermore it is clear that this type of deformation by dilation is not associated with any local rotation in the medium as we can see from the evolution of vector $\vec{\omega}$

$$\frac{d\vec{\omega}}{dt} = 0 \qquad (1.56)$$

With regards to shear tensor $\vec{\alpha}_i$, it only contains diagonal elements, which take non-null values in this case since, along a give path of a particle of the lattice, we have

$$\frac{d\vec{\alpha}_i}{dt} = \left[a_i - \frac{1}{3}(a_1 + a_2 + a_3)\right] g(t) \vec{e}_i \qquad (1.57)$$

From $\tau(t)$, values for $n(t)$ and $v(t)$ are computed with

$$n(t) = n_0\, \mathrm{e}^{-\tau(t)} \quad \text{and} \quad v(t) = v_0\, \mathrm{e}^{\tau(t)} \qquad (1.58)$$

The scalar τ can be expressed in terms of trajectories $X_i(t)$ of 3 points initially situated at X_i^0 (figure 1.11) on the 3 axis of coordinates.

$$\tau = -\ln\frac{n(t)}{n_0} = \ln\frac{v(t)}{v_0} = \ln\left(1+\frac{\Delta v(t)}{v_0}\right) \tag{1.59}$$

And since the volumes $v(t)$ and v_0 are directly proportional to $\Pi X_i(t)$ and ΠX_i^0, it follows

$$\tau = \ln\left(\prod_i \frac{X_i(t)}{X_i^0}\right) = \sum_i \ln\frac{X_i(t)}{X_i^0} = \sum_i \ln\left(1+\frac{\Delta X_i(t)}{X_i^0}\right) \tag{1.60}$$

By using $\tau = \Sigma\beta_{ii} = \Sigma\varepsilon_{ii}$, we can directly compute β_{ii} et ε_{ii}

$$\beta_{ii} = \varepsilon_{ii} = \ln\frac{X_i(t)}{X_i^0} = \ln\left(1+\frac{\Delta X_i(t)}{X_i^0}\right) \tag{1.61}$$

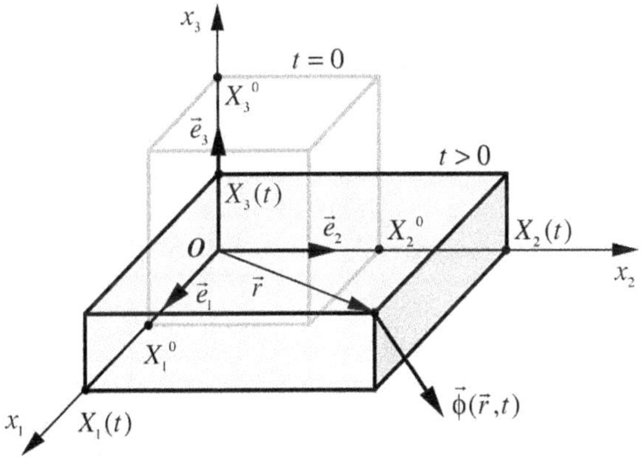

***Figure 1.11** - elongations along the axis of coordinates*

The diagonal elements β_{ii} and ε_{ii} measure the individual expansions and contractions in the 3 directions of space respectively, while the trace τ measures the variation of the volume said expansions and contractions.

Finally it should be noted that in the limit where elongations are small, the scalar τ can be expanded as

$$\tau \cong \frac{\Delta v(t)}{v_0} \quad \text{if} \quad |\Delta v(t)| << v_0 \tag{1.62}$$

The diagonal elements of the tensors of distortion and deformation can also be expanded as such

$$\beta_{ii} = \varepsilon_{ii} \cong \frac{\Delta X_i(t)}{X_i^0} \quad \text{if} \quad |\Delta X_i(t)| << X_i^0 \tag{1.63}$$

Another important remark can be made about the asymptotical behavior of the topological tensors τ, β_{ii} and ε_{ii}. They all go to $\pm\infty$ when the elongations become large

$$\tau, \beta_{ii}, \varepsilon_{ii} \rightarrow +\infty \quad \text{if} \quad \Delta X_i \rightarrow +\infty \tag{1.64}$$

$$\tau, \beta_{ii}, \varepsilon_{ii} \rightarrow -\infty \quad \text{if} \quad \Delta X_i \rightarrow -X_i^0 \tag{1.65}$$

2 extreme cases of evolution of elongation along coordinate axis can now be studied:

- *The case where the volume expansion/contraction is isotropic in space*, meaning where the 3 constants a_i are equal ($a_1 = a_2 = a_3$). We then have

$$\frac{d\tau}{dt} = 3a_i g(t) \quad ; \quad \frac{d\vec{\alpha}_i}{dt} = 0 \quad ; \quad \frac{d\vec{\omega}}{dt} = 0 \tag{1.66}$$

The figure 1.12(a) shows a 2D cut of such a velocity field in the plane defined by vectors $\vec{e}_1$ and $\vec{e}_2$. Figure 1.12(b) shows the evolution between 2 instants ($t = 0$ and $t > 0$) of a square portion of medium under such field. What we see is a volume expansion without shear.

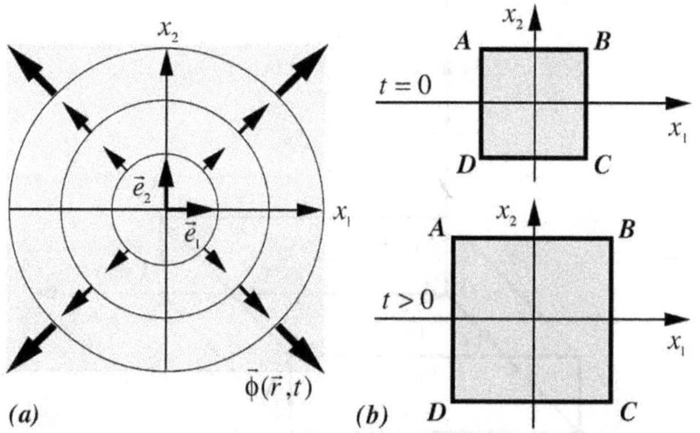

Figure 1.12 *- elongations with volume expansion and no shear*

- *The case where expansions and contractions in the 3 directions of space are such that volume expansion* τ *of the medium is null*, meaning where the sum of the a_i is null ($a_1 + a_2 + a_3 = 0$). It then follows that

$$\frac{d\tau}{dt} = 0 \quad \text{and} \quad \frac{d\vec{\beta}_i}{dt} = \frac{d\vec{\varepsilon}_i}{dt} = \frac{d\vec{\alpha}_i}{dt} = a_i g(t)\vec{e}_i \tag{1.67}$$

Figure 1.13(a) shows the cut of such a field with $a_1 = -a_2 = a$ and $a_3 = 0$.

Figure 1.13(b) shows the spatiotemporal evolution of a square portion of the medium with such a field at 2 different times ($t = 0$ and $t > 0$). By visual inspection we can see there is no volume expansion, the shape changes, but not the volume (as captured by the null trace). We can also see the presence of strong shear in the medium, which can also be seen in the vectorial equations

$$\frac{d\vec{\omega}}{dt} = 0 \quad ; \quad \frac{d\vec{\alpha}_1}{dt} = ag(t)\vec{e}_1 \quad ; \quad \frac{d\vec{\alpha}_2}{dt} = -ag(t)\vec{e}_2 \quad ; \quad \frac{d\vec{\alpha}_3}{dt} = 0 \tag{1.68}$$

Indeed the main difference between the 2 cases we just studied, lies in the isotropic versus anisotropic nature of the elongations along the 3 axis. The different values of expansion gives rise to shear, which is absent in the isotropic case. We conclude that the diagonal components of tensor $\vec{\alpha}_i$ measure shear associated with non-isotropic homogeneous volume expansion of the medium.

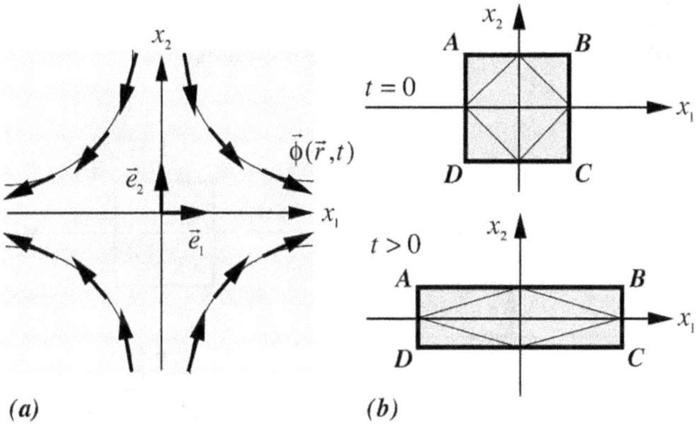

***Figure 1.13** - elongations with null volume expansion (constant volume)*

Shear and local rotations, or torsions

Consider the following velocity field

$$\vec{\phi}(\vec{r},t)=\left[\left(ax_2+bx_3\right)\vec{e}_1+\left(cx_1+dx_3\right)\vec{e}_2+\left(ex_1+fx_2\right)\vec{e}_3\right]g(t) \quad (1.69)$$

where a,b,c,d,e,f are constants.

Given such a field, the equations of geometro-kinetics for tensor β_{ij} gives us the temporal evolution of the β_{ij} components along the path

$$\left(\frac{d\vec{\beta}_i}{dt}\right)=\begin{pmatrix}0 & a & b\\ c & 0 & d\\ e & f & 0\end{pmatrix}g(t) \quad (1.70)$$

It clearly shows that the trace of the tensor of distortion cannot change, and volume expansion τ is a constant

$$\frac{d\tau}{dt}=0 \quad (1.71)$$

We can compute the geometro-kinetic equations satisfied by the rotation vector and the tensors of deformation and shear, those are

$$\begin{cases}\dfrac{d\vec{\omega}}{dt}=\dfrac{1}{2}\left[(f-d)\vec{e}_1+(b-e)\vec{e}_2+(c-a)\vec{e}_3\right]g(t)\\ \left(\dfrac{d\varepsilon_{ij}}{dt}\right)=\left(\dfrac{d\alpha_{ij}}{dt}\right)=\dfrac{1}{2}\begin{pmatrix}0 & a+c & b+e\\ a+c & 0 & d+f\\ b+e & d+f & 0\end{pmatrix}g(t)\end{cases} \quad (1.72)$$

Figure 1.14(a) represents such a velocity field in the case where only the a constant is different from zero ($a\neq 0$ and $b=c=d=e=f=0$).

The spatio-temporal evolution between 2 instants ($t=0$ and $t>0$) of a square portion of medium is shown in figure 1.14(b). The presence of shear in the medium is observed, and there also is a *local rotation* which we will call *torsion* of the medium and we have

$$\frac{d\vec{\omega}}{dt}=-\frac{a}{2}g(t)\vec{e}_3 \quad ; \quad \frac{d\vec{\alpha}_1}{dt}=\frac{a}{2}g(t)\vec{e}_2 \quad ; \quad \frac{d\vec{\alpha}_2}{dt}=\frac{a}{2}g(t)\vec{e}_1 \quad ; \quad \frac{d\vec{\alpha}_3}{dt}=0 \tag{1.73}$$

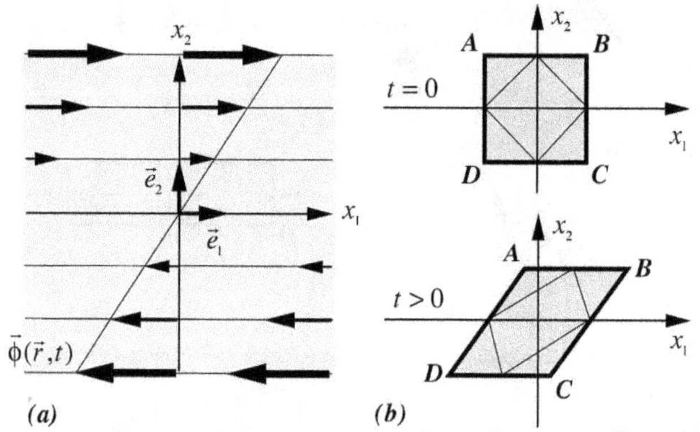

Figure 1.14 *- shear*

In the same way we did for elongations, it is possible to give a topological interpretation to the components of the shear tensor by using the example of distortion presented in figure 1.14. In that case, the velocity field is parallel to $\vec{e}_1$ and is computed as

$$\vec{\phi}(\vec{r},t)=ax_2g(t)\vec{e}_1 \tag{1.74}$$

The displacement $X_1(x_2,t)$ along the $\vec{e}_1$ at a given point A found at coordinate $x_2 \neq 0$ (figure 1.15) can be calculated. We have

$$X_1(x_2,t)=ax_2\int_0^t g(t)dt \tag{1.75}$$

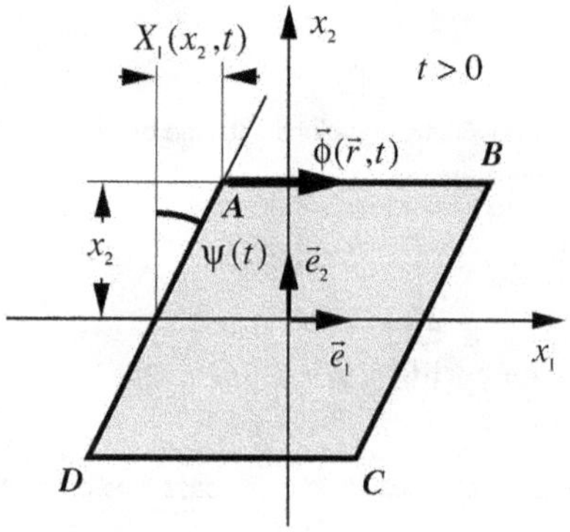

Figure 1.15 *- relation with the tangent of the angle of shear*

We can deduce the temporal evolution of the tangent of shear $\psi(t)$ shown in figure 1.15

$$\operatorname{tg}\psi(t) = \frac{X_1(x_2,t)}{x_2} = a\int_0^t g(t)dt \tag{1.76}$$

The evolution of the vector of rotation and evolution of the tensor of shear along the trajectory can be linked to the value of $\operatorname{tg}\psi(t)$

$$\vec{\omega} = -\frac{a}{2}\vec{e}_3\int_0^t g(t)dt = -\frac{1}{2}\vec{e}_3 \operatorname{tg}\psi(t) \tag{1.77}$$

$$\vec{\alpha}_1 = \frac{a}{2}\vec{e}_2\int_0^t g(t)dt = \frac{1}{2}\vec{e}_2 \operatorname{tg}\psi(t) \tag{1.78}$$

$$\vec{\alpha}_2 = \frac{a}{2}\vec{e}_1\int_0^t g(t)dt = \frac{1}{2}\vec{e}_1 \operatorname{tg}\psi(t) \tag{1.79}$$

The vector of rotation $\vec{\omega}$ and shear tensor $\vec{\alpha}_i$ are linked to half the angle of shear $\psi(t)$.
It is possible to consider now 2 extreme cases of evolution for shear and rotations:

- *The case where tensor β_{ij} is perfectly anti-symmetric,* meaning when $c=-a$, $e=-b$ and $f=-d$

$$\left(\frac{d\beta_{ij}}{dt}\right) = \begin{pmatrix} 0 & a & b \\ -a & 0 & d \\ -b & -d & 0 \end{pmatrix} g(t) \tag{1.80}$$

In this case, the tensor β_{ij} is entirely described by a rotation vector $\vec{\omega}$ and the global rotation of the medium is the one we have already seen with

$$\frac{d\vec{\omega}}{dt} = \dot{\vec{\theta}}(t) = \left(-d\vec{e}_1 + b\vec{e}_2 - a\vec{e}_3\right)g(t) \tag{1.81}$$

- *In the case where β_{ij} is perfectly symmetric,* meaning when $c=a$, $e=b$ and $f=d$. In this case the 3 tensors β_{ij}, ε_{ij} et α_{ij} are equal and the variation of their components along the path is written

$$\left(\frac{d\beta_{ij}}{dt}\right) = \left(\frac{d\varepsilon_{ij}}{dt}\right) = \left(\frac{d\alpha_{ij}}{dt}\right) = \begin{pmatrix} 0 & a & b \\ a & 0 & d \\ b & d & 0 \end{pmatrix} g(t) \tag{1.82}$$

The scalar of volume expansion τ and the vector of rotation $\vec{\omega}$ do not change along the path

$$\frac{d\tau}{dt} = 0 \quad \text{and} \quad \frac{d\vec{\omega}}{dt} = 0 \tag{1.83}$$

The medium does not have volume expansion, nor local or global rotation. As a consequence, only pure shear can appear.

We visually show it in figure 1.16(a), in which we have represented, in the plane of base vectors $\vec{e}_1$ and $\vec{e}_2$, the cut of such a velocity field with $a \neq b$ and $b=c=0$.

Figure 1.16(b) shows the evolution at 2 different times ($t=0$ and $t>0$) of a square portion of medium under such a velocity field. The presence of a head of the medium, without rotation, is seen as

$$\frac{d\vec{\omega}}{dt}=0 \quad ; \quad \frac{d\vec{\alpha}_1}{dt}=ag(t)\vec{e}_2 \quad ; \quad \frac{d\vec{\alpha}_2}{dt}=ag(t)\vec{e}_1 \quad ; \quad \frac{d\vec{\alpha}_3}{dt}=0 \qquad (1.84)$$

The shear is identical to that reported in figure 1.14(b), if a rotation of 45° of the system of coordinates is done, which leads to very similar expressions for the shear tensor in both cases:

$$\left(\frac{d\alpha_{ij}}{dt}\right)=\begin{pmatrix} a & 0 & 0 \\ 0 & -a & 0 \\ 0 & 0 & 0 \end{pmatrix} g(t) \text{ (fig. 1.14) and } \left(\frac{d\alpha_{ij}}{dt}\right)=\begin{pmatrix} 0 & a & 0 \\ a & 0 & 0 \\ 0 & 0 & 0 \end{pmatrix} g(t) \text{ (fig. 1.16)} \qquad (1.85)$$

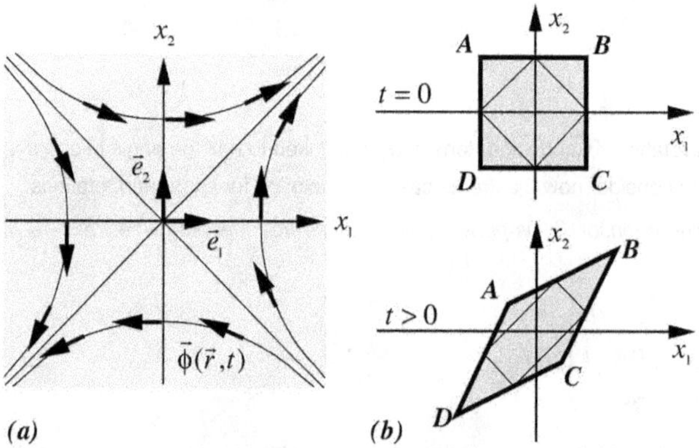

Figure 1.16 *- shear*

Examples of non-uniform distortions

As an example, it is possible to calculate the tensors of distortions with velocity fields that lead to non-uniform distortions of the medium. Consider the following field in figure 1.17(a), *with a cylindrical (axial) symmetry*

$$\vec{\phi}(\vec{r},t)=\left[r\vec{e}_3\wedge\left(x_1\vec{e}_1+x_2\vec{e}_2\right)\right]g(t)=r\left(x_1\vec{e}_2-x_2\vec{e}_1\right)g(t) \quad \text{with} \quad r=\sqrt{x_1^{\,2}+x_2^{\,2}} \qquad (1.86)$$

This velocity field describes a rotation of the medium about the origin of our system of coordinates. The tensor β_{ij} is deduced from the geometro-kinetic equations as

$$\left(\frac{d\beta_{ij}}{dt}\right)=\frac{1}{r}\begin{pmatrix} -x_1x_2 & -\left(r^2+x_2^{\,2}\right) & 0 \\ +\left(r^2+x_1^{\,2}\right) & x_1x_2 & 0 \\ 0 & 0 & 0 \end{pmatrix} g(t) \qquad (1.87)$$

This is a non-uniform rotation, and therefore local, as $\vec{\omega}$ depends on radius r. There is also a volume expansion τ along the path

$$\frac{d\vec{\omega}}{dt}=\frac{3r}{2}g(t)\vec{e}_3 \quad \text{and} \quad \frac{d\tau}{dt}=0 \qquad (1.88)$$

The tensor of deformation ε_{ij} and of shear α_{ij} are identical and show the existence of non-uniform shear

$$\left(\frac{d\varepsilon_{ij}}{dt}\right)=\left(\frac{d\alpha_{ij}}{dt}\right)=\frac{1}{2r}\begin{pmatrix}-2x_1x_2 & x_1^2-x_2^2 & 0\\ x_1^2-x_2^2 & 2x_1x_2 & 0\\ 0 & 0 & 0\end{pmatrix}g(t) \tag{1.89}$$

The second velocity field, represented in figure 1.17(b), has a spherical symmetry

$$\vec{\phi}(\vec{r},t)=r\left[x_1\vec{e}_1+x_2\vec{e}_2+x_3\vec{e}_3\right]g(t) \quad \text{with} \quad r=\sqrt{x_1^{\,2}+x_2^{\,2}+x_3^{\,2}} \tag{1.90}$$

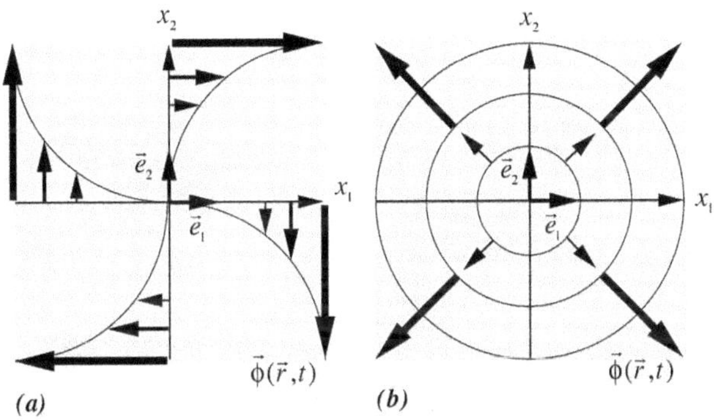

***Figure 1.17** - examples of non-uniform distortions*

This velocity field implies a volume expansion of the medium about the center of system of coordinates. The evolution of tensors β_{ij} and ε_{ij} is the same

$$\left(\frac{d\beta_{ij}}{dt}\right)=\left(\frac{d\varepsilon_{ij}}{dt}\right)=\frac{1}{r}\begin{pmatrix}r^2+x_1^2 & x_1x_2 & x_1x_3\\ x_1x_2 & r^2+x_2^2 & x_2x_3\\ x_1x_3 & x_2x_3 & r^2+x_3^2\end{pmatrix}g(t) \tag{1.91}$$

This means it is an evolution without rotation along the path, but with a non-uniform volume expansion τ which depends on radius r

$$\frac{d\vec{\omega}}{dt}=0 \quad \text{and} \quad \frac{d\tau}{dt}=4r\,g(t) \tag{1.92}$$

Finally, the evolution of shear tensor α_{ij} shows the existence of non-uniform shear as a function of radius r

$$\left(\frac{d\alpha_{ij}}{dt}\right)=\frac{1}{r}\begin{pmatrix}x_1^2-r^2/3 & x_1x_2 & x_1x_3\\ x_1x_2 & x_2^2-r^2/3 & x_2x_3\\ x_1x_3 & x_2x_3 & x_3^2-r^2/3\end{pmatrix}g(t) \tag{1.93}$$

Chapter 2

Local frames and geometro-compatibility of a lattice

In this chapter, we show that eulerian coordinates are well suited to the description of solid lattices. They allow for the description of lattices with large deformations, something that is hard to achieve with lagrangian coordinates. However, we will show that this description of the solid lattice with Euler coordinates requires us to introduce local referential frames. This introduces its own complexity, inherent to the differential geometry description, which makes use of microscopic local frames. To bypass such complexity we will define the concept of *macroscopic local frame.* Finally, in the eulerian coordinate description, we derive the conditions of compatibility that define the lattice as a solid: one where we can define a *continuous* displacement field for the lattice.

2.1 - Definition of local referential frames in solid lattices

The eulerian description of the spatiotemporal evolution of a solid lattice of particles is defined by the *velocity field* $\vec{\phi}(\vec{r},t)$ which describes the average movement of the individual nodes of the lattice relative to an absolute fixed observer (Grand Observer: ***GO***) with a fixed, cartesian, coordinate system $Q\xi_1\xi_2\xi_3$.

From this velocity field we will derive and calculate the *local angular velocity* $d\vec{\omega}/dt$ of the solid rotation as well as the *scalar of local volume expansion* τ of the solid. With these we will describe the time dependent distortions of the solid lattice with the equations of geometro-kinetics we introduced in the first chapter (table 1.1). These equations were based on the 3 following principal equations:

$$\begin{cases} \dfrac{d\vec{\beta}_i}{dt} = -\dfrac{S_n}{3n}\vec{e}_i + \overrightarrow{\text{grad}}\,\phi_i \\ \dfrac{d\vec{\omega}}{dt} = \dfrac{1}{2}\overrightarrow{\text{rot}}\,\vec{\phi} \\ \dfrac{d\tau}{dt} = -\dfrac{S_n}{n} + \text{div}\,\vec{\phi} \end{cases} \qquad (2.1)$$

This method is valid in the case of a perfectly isotropic lattice. However, in the majority of the cases it is not well suited to describe the distortions of a lattice in eulerian coordinates for the following reasons:

- A solid may present an *anisotropic elasticity,* which depends on the direction considered with respect to the *local crystalline directions of the lattice itself.* While the distortion tensors *(2.1)* correctly measure the distortions of a solid, whether it be isotropic or not, the tensors measure and describe the distortions *in the absolute global frame* $Q\xi_1\xi_2\xi_3$ and not along the *local crystalline directions of the lattice*. This complicates the calculation of the free elastic energy of anisotropic solid lattice from the distortion tensors *(2.1)*, specifically if the local rotation is important because the free energy of an anisotropic solid is not built from the invariants of the distortion tensors.

- In the case of a perfect solid lattice with isotropic elasticity (without structural singularities), the free elastic energy can indeed be expressed in terms of the distortion tensors *(2.1)* because it only depends on the invariant quantities of the distortion tensors expressed in said absolute global frame. On the other hand, in the case of an *imperfect* solid lattice with isotropic elasticity, meaning an isotropic lattice which contains anisotropic structural defects and singularities (such as *disclinations* and *dislocations, joints* and other usual crystal defects) the presence of such anisotropic defects requires a description which *uses the local crystalline directions and associated local frame.*

It is imperative to describe the distortions of a lattice with anisotropic elasticity, or an isotropic lattice with defects (non perfect isotropic lattice) with *local frames and local coordinates.* This system of local frames reflects the same average local translation and rotation of the lattice so that the directions of the local system of coordinates maps, on average, to the directions of the local crystalline lattice.

Definition of an orthonormal macroscopic local frame.

In the absolute frame $Q\xi_1\xi_2\xi_3$ of the ***GO*** observer, we can choose to follow a point O of the solid by knowing its instantaneous velocity $\vec{\phi}_O(t)$ expressed in said referential. To this point O we also associate the instantaneous vector of rotation $\vec{\omega}_O(t)$. We can then define at this point O a *macroscopic orthonormal local frame,* $Ox_1x_2x_3$, which follows the solid with a *drag-along velocity* $\vec{\phi}_O(t)$ and which rotates in space with the *drag-along rotational angular velocity* $\vec{\omega}_O(t)$ (figure 2.1).

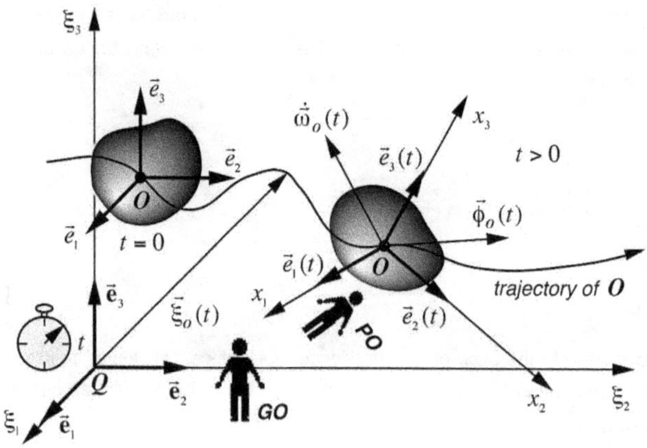

Figure 2.1 *- the orthonormal macroscopic local frame*

The origin O of this new coordinate system can be found with a *location vector* $\vec{\xi}_O(t)$ whose equation of motion, written in the absolute frame $Q\xi_1\xi_2\xi_3$, is simply stated as

$$\frac{d\vec{\xi}_O(t)}{dt} = \vec{\phi}_O(t) \tag{2.2}$$

To this local frame $Ox_1x_2x_3$, we can associate the 3 unitary orthogonal vectors $\vec{e}_i(t)$, the frame. These vectors turn in space with the drag-along rotation speed $\dot{\vec{\omega}}_O(t)$. We can also define the Small Observer ***(PO)*** ("Petit Observateur") attached to this referential and who uses rulers $\vec{e}_i$, *which have the same length as those of **GO***, as well as a clock perfectly synchronized with that of the ***GO*** of the absolute referential.

On the matrix of rotation of the coordinate change

In the absolute referential $Q\xi_1\xi_2\xi_3$ of the ***GO*** observer, the vectors $\vec{e}_i(t)$ of the local coordinate system satisfy the following equation of change, which capture the fact that the vectors $\vec{e}_i(t)$ are rotating with an angular velocity $\dot{\vec{\omega}}_O(t)$ along the trajectory of the O point

$$\frac{d\vec{e}_i(t)}{dt} = \dot{\vec{\omega}}_O(t) \wedge \vec{e}_i(t) \tag{2.3}$$

The local frame vectors $\vec{e}_i(t)$ can be decomposed on the global frame vectors $\vec{\mathbf{e}}_j$ with the matrix of rotation $g_{ij}(t)$, which is then defined as

$$\vec{e}_i(t) = \sum_j g_{ij}(t)\vec{\mathbf{e}}_j \tag{2.4}$$

This relation can be inverted as

$$\vec{\mathbf{e}}_j = \sum_i \left[\vec{e}_i(t)\vec{\mathbf{e}}_j\right]\vec{e}_i(t) = \sum_i g_{ij}(t)\vec{e}_i(t) \tag{2.5}$$

The matrix $g_{ij}(t)$ is in fact the matrix describing the coordinate transformation (map) resulting from the *change of frame* between the two frames, from the global coordinate $Q\xi_1\xi_2\xi_3$ to the local one $Ox_1x_2x_3$. As a rotation, it is an *orthogonal matrix*, and its inverse is its transpose.

$$\left(g_{ij}\right)^{-1} = \left(g_{ij}\right)^{\mathrm{T}} \;\Rightarrow\; \left(g_{ij}\right)\left(g_{ij}\right)^{\mathrm{T}} = \left(g_{ij}\right)\left(g_{ij}\right)^{-1} = \begin{pmatrix} 1 & 0 & 0 \\ 0 & 1 & 0 \\ 0 & 0 & 1 \end{pmatrix} \tag{2.6}$$

As a rotation, its determinant is also equal to 1

$$\left|\left(g_{ij}\right)\left(g_{ij}\right)^{\mathrm{T}}\right| = \left|\left(g_{ij}\right)\left(g_{ij}\right)^{-1}\right| = 1 \;\Rightarrow\; \left|\left(g_{ij}\right)\right| = 1 \tag{2.7}$$

Such a matrix also satisfies the following relations, translating the fact that both sets of vectors $\vec{e}_i$ et $\vec{e}_i$ are indeed unitary

$$\begin{cases} \left|\vec{e}_i\right| = 1 \;\Rightarrow\; \sum_j \left[g_{ij}(t)\right]^2 = 1 \\ \left|\vec{\mathbf{e}}_j\right| = 1 \;\Rightarrow\; \sum_i \left[g_{ij}(t)\right]^2 = 1 \end{cases} \tag{2.8}$$

Given relation *(2.3)*, we deduce the differential equations describing the evolution of the matrix $g_{ij}(t)$ of frame change

$$\sum_j \frac{dg_{ij}(t)}{dt}\vec{\mathbf{e}}_j = \sum_j g_{ij}(t)\left(\dot{\vec{\omega}}_O(t)\wedge\vec{\mathbf{e}}_j\right) \tag{2.9}$$

In the case of a rigid solid (non deformable), the solid only displaces in space by translation or global rotation, i.e. without deformation, and the directions represented by the unit frame vectors $\vec{e}_i$ always correspond exactly to the directions of the *crystalline lattice of the solid.* In the case where the solid under consideration moves and deforms at the same time, as the vectors $\vec{e}_i(t)$ move with a velocity $\vec{\phi}_O(t)$ and rotate with an angular velocity $\dot{\vec{\omega}}_O(t)$, the directions represented by the local unit vectors $\vec{e}_i$ still correspond to the *average directions of the crystal lattice* at the point O.

As a matter of fact, in a vicinity of point O, the directions of the crystalline lattice correspond more or less to the directions of the unit vectors $\vec{e}_i$. The size of this neighborhood depends on how deformed the solid around the point O. We can empirically define this neighborhood as the volume V_m around the point O which satisfies the following criteria: *the volume V_m of lattice around O such that the components of the rotational vector $\vec{\omega}$ and the scalar of volume expansion τ do not vary more than 1% over said volume.*
In figure *2.2,* we represent the crystalline lattice around a point O in the presence of a deformation of the lattice.

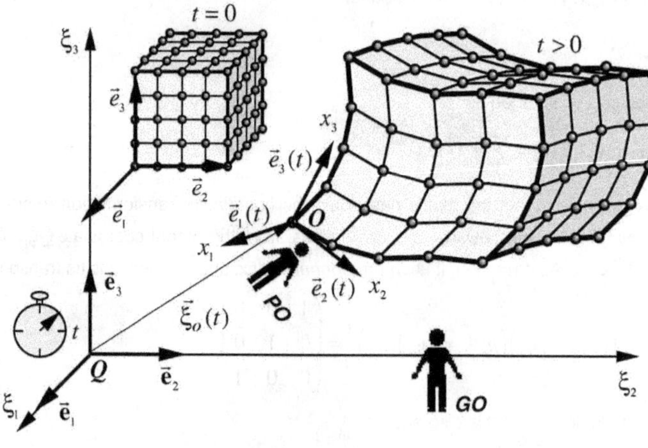

Figure 2.2 *- the solid lattice around the neighborhood of point O with local frame $Ox_1x_2x_3$*

Projection of the fields on the local frame

Every scalar field or vectorial field defined in the absolute space can be projected and decomposed in the global frame $Q\xi_1\xi_2\xi_3$ or locally in a neighborhood of the point O, in the mobile local frame $Ox_1x_2x_3$ (figure 2.5).
For example, the velocity field $\vec{\phi}$ of the solid lattice in the absolute frame is defined at point $\vec{\xi}$ and instant t by its 3 projections ϕ_i on the 3 axes $Q\xi_1$, $Q\xi_2$ and $Q\xi_3$

$$\vec{\phi}(\vec{\xi},t) = \sum_{i=1}^{3} \phi_i(\xi_k,t)\vec{\mathbf{e}}_i \qquad \text{with} \qquad \vec{\xi} = \sum_{k=1}^{3} \xi_k \vec{\mathbf{e}}_k \tag{2.10}$$

In the neighborhood of point $\boldsymbol{O}$ of the local frame, it can also be projected on the 3 axes $\boldsymbol{O}x_1$, $\boldsymbol{O}x_2$ and $\boldsymbol{O}x_3$ (figure 2.3). At point $\boldsymbol{B}$ of the local referential, marked by $\vec{r}$, and at instant t, the velocity field $\vec{\phi}$ can we written in the local frame projected as such

$$\vec{\phi}(\vec{r},t) = \sum_{i=1}^{3} \phi_i(x_k,t)\vec{e}_i \qquad \text{with} \qquad \vec{r} = \sum_{k=1}^{3} x_k \vec{e}_k \tag{2.11}$$

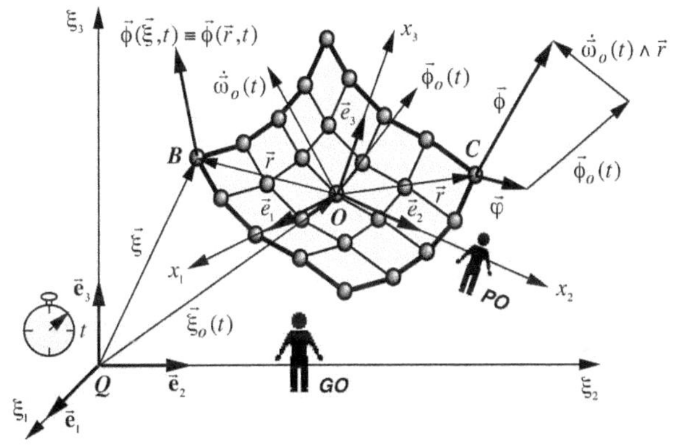

***Figure 2.3** - projection of velocity field $\vec{\phi}$ in the macroscopic local frame*

The expressions *(2.4)* and *(2.5)* giving the local frame vectors $\vec{e}_i$ as functions of the global frame vectors $\vec{\mathbf{e}}_i$ and vice-versa allow us to find a mapping between the ϕ_i components of the velocity field $\vec{\phi}$ in the absolute and local frame.

$$\vec{\phi} = \sum_i \phi_i(\xi_k,t)\vec{\mathbf{e}}_i = \sum_{ij} \phi_i(\xi_k,t) g_{ji}\vec{e}_j = \sum_j \phi_j(x_k,t)\vec{e}_j = \sum_{ij} \phi_j(x_k,t) g_{ji}\vec{\mathbf{e}}_i \quad \Rightarrow$$

$$\phi_i(x_k,t) = \sum_j g_{ij}\,\phi_j(\xi_k,t) \qquad \text{and} \qquad \phi_j(\xi_k,t) = \sum_i g_{ij}\,\phi_i(x_k,t) \tag{2.12}$$

Furthermore, since the position vectors $\vec{r}$ and $\vec{\xi}$ identify respectively the nodes of the lattice in the local and global frame, they are related to each other by the simple relation

$$\vec{r} = \vec{\xi} - \vec{\xi}_O(t) \tag{2.13}$$

The mapping between the components ξ_k et x_k, the position vectors in the absolute frame and the local frame (which identify the same point) are given by

$$\vec{r} = \sum_i x_i\vec{e}_i = \sum_{ij} x_i g_{ij}\vec{\mathbf{e}}_i = \sum_j \xi_j\vec{\mathbf{e}}_j - \vec{\xi}_O(t) = \sum_{ij} \xi_j g_{ij}\vec{e}_i - \vec{\xi}_O(t) \quad \Rightarrow$$

$$x_i = \sum_j g_{ij}\xi_j - \vec{\xi}_O(t)\vec{e}_i \qquad \text{and} \qquad \xi_j = \sum_i g_{ij}x_i + \vec{\xi}_O(t)\vec{\mathbf{e}}_j \tag{2.14}$$

Table 2.1 - Local macroscopic frame $Ox_1x_2x_3$ in a solid lattice

Rotation matrix of frame change

$$\begin{cases} \dfrac{d\vec{e}_i(t)}{dt} = \dot{\vec{\omega}}_O(t) \wedge \vec{e}_i(t) & (1) \\ \sum_j \dfrac{dg_{ij}(t)}{dt}\vec{\mathbf{e}}_j = \sum_j g_{ij}(t)\left(\dot{\vec{\omega}}_O(t) \wedge \vec{\mathbf{e}}_j\right) & (2) \end{cases} \qquad \begin{cases} \left|(g_{ij})\right| = 1 & (3) \\ \sum_j \left[g_{ij}(t)\right]^2 = 1 & (4) \\ \sum_i \left[g_{ij}(t)\right]^2 = 1 & (5) \end{cases}$$

Relations of frame change

$$\begin{cases} \vec{e}_i(t) = \sum_j g_{ij}(t)\vec{\mathbf{e}}_j & (6) \\ \phi_i(x_k,t) = \sum_j g_{ij}\,\phi_j(\xi_k,t) & (7) \\ x_i = \sum_j g_{ij}\xi_j - \vec{\xi}_O(t)\vec{e}_i & (8) \\ \dfrac{\partial}{\partial x_i} = \sum_j g_{ij}\dfrac{\partial}{\partial \xi_j} & (9) \end{cases} \qquad \begin{cases} \vec{\mathbf{e}}_j = \sum_i g_{ij}(t)\vec{e}_i(t) & (10) \\ \phi_j(\xi_k,t) = \sum_i g_{ij}\,\phi_i(x_k,t) & (11) \\ \xi_j = \sum_i g_{ij}x_i + \vec{\xi}_O(t)\vec{\mathbf{e}}_j & (12) \\ \dfrac{\partial}{\partial \xi_j} = \sum_i g_{ij}\dfrac{\partial}{\partial x_i} & (13) \end{cases}$$

Local field of relative velocity

$$\begin{cases} \vec{\varphi}(\vec{r},t) = \vec{\phi}(\vec{r},t) - \vec{\phi}_O(t) - \dot{\vec{\omega}}_O(t) \wedge \vec{r} & (1) \\ \vec{\varphi}(\vec{\xi},t) = \vec{\phi}(\vec{\xi},t) - \vec{\phi}_O(t) - \dot{\vec{\omega}}_O(t) \wedge \left(\vec{\xi} - \vec{\xi}_O(t)\right) & (2) \end{cases}$$

Finally, the relations that exist between the partial derivatives in the absolute frame and the local frame are easily deducted

$$\frac{\partial}{\partial x_i} = \sum_j \frac{\partial}{\partial \xi_j}\frac{\partial \xi_j(y_k)}{\partial x_i} \quad \text{and} \quad \frac{\partial}{\partial \xi_j} = \sum_i \frac{\partial}{\partial x_i}\frac{\partial x_i(x_k)}{\partial \xi_j} \quad \Rightarrow$$

$$\frac{\partial}{\partial x_i} = \sum_j g_{ij}\frac{\partial}{\partial \xi_j} \quad \text{and} \quad \frac{\partial}{\partial \xi_j} = \sum_i g_{ij}\frac{\partial}{\partial x_i} \tag{2.15}$$

The local field of relative velocity

It is possible to define the *local field of relative velocity* $\vec{\varphi}$ of the points of the solid in the neighborhood of point O, in relation to the local mobile frame $Ox_1x_2x_3$. The velocity field is in fact the one measured at point C for example (figure 2.3), by the *Small Observer **(PO)*** linked to the local frame $Ox_1x_2x_3$, who would use rulers $\vec{e}_i$ and a clock synchronized to the clock of the absolute frame to measure the velocities $\vec{\varphi}$ of the nodes of the lattice, while the *Great Observer* ***(GO)*** is linked to the absolute frame and he measures the absolute velocities $\vec{\phi}$ of the nodes of the lattice.

The local field of relative velocity $\vec{\varphi}$ is obtained by subtracting the drag-along velocity $\vec{\phi}_O(t)$ of the local frame $Ox_1x_2x_3$, from the absolute field velocity $\vec{\phi}$ as well as the field of rotational velocity $\dot{\vec{\omega}}_O(t)\wedge\vec{r}$ associated with the rotation of the local frame $Ox_1x_2x_3$ around point O characterized by the angular drag-along velocity $\dot{\vec{\omega}}_O(t)$. We obtain the following equations

$$\vec{\varphi}(\vec{r},t)=\vec{\phi}(\vec{r},t)-\vec{\phi}_O(t)-\dot{\vec{\omega}}_O(t)\wedge\vec{r} \tag{2.16}$$

$$\vec{\varphi}(\vec{\xi},t)=\vec{\phi}(\vec{\xi},t)-\vec{\phi}_O(t)-\dot{\vec{\omega}}_O(t)\wedge\left(\vec{\xi}-\vec{\xi}_O(t)\right) \tag{2.17}$$

On the projection of the invariant operators of space and time in the local frame

The space operators $\overrightarrow{\text{grad}}$, $\overrightarrow{\text{rot}}$ and div applied to scalar or vectorial fields are *invariants,* meaning that they provide a result, scalar or vectorial, that does not depend on the choice of frame in which they are calculated (section A.2.). The same goes for the time operator of *material derivative* d/dt as this measures the temporal variation of quantities along the path. This means that any mathematical relations which invokes the operators $\overrightarrow{\text{grad}}$, $\overrightarrow{\text{rot}}$ and div, as well as d/dt, can be computed in either the absolute frame $Q\xi_1\xi_2\xi_3$ or the local frame $Ox_1x_2x_3$.

In the case of the temporal operator d/dt, the velocity one must introduce in the expression of material derivative is the absolute velocity $\vec{\phi}$ when d/dt is calculated in the absolute frame $Q\xi_1\xi_2\xi_3$ and the relative velocity $\vec{\varphi}$ when d/dt is calculated in the local frame $Ox_1x_2x_3$.

Regarding the formal operator $\vec{\nabla}$, we show, by using relations *(2.4)* et *(2.5)*, that there exists the following link between its formulation in the absolute frame $Q\xi_1\xi_2\xi_3$ and its formulation in the local frame $Ox_1x_2x_3$

$$\vec{\nabla}=\sum_i\vec{e}_i\frac{\partial}{\partial x_i}=\sum_{ij}\vec{e}_i\,g_{ij}\frac{\partial}{\partial\xi_j}=\sum_j\left(\sum_i\vec{e}_i\,g_{ij}\right)\frac{\partial}{\partial\xi_j}=\sum_j\vec{\mathbf{e}}_j\frac{\partial}{\partial\xi_j} \tag{2.18}$$

We can restate the various space and time operators, both in local and absolute frame, in terms of the $\vec{\nabla}$ operator. The following relations allow us to recover the expressions $\overrightarrow{\text{grad}}$, $\overrightarrow{\text{rot}}$ and div, expressions which we show in table 2.2

$$\overrightarrow{\text{grad}}\,f=\vec{\nabla}f \quad\text{and}\quad \overrightarrow{\text{rot}}\,\vec{u}=\vec{\nabla}\wedge\vec{u} \quad\text{and}\quad \text{div}\,\vec{u}=\vec{\nabla}\vec{u} \tag{2.19}$$

We can also define the derivative of a quantity, scalar or vectorial, along a given direction $\vec{n}$, the directional derivative. It is written $(\vec{n}\vec{\nabla})$ and can be expanded in terms of the $\vec{\nabla}$ operator.

An important remark can be made here. In the case of a *perfectly non-deformable* solid lattice, the field of relative velocity $\vec{\varphi}$ defined by the relations *(2.16)* et *(2.17)* is everywhere null in the local frame $Ox_1x_2x_3$. On the other hand, in the case of a *deformable* solid lattice the field of relative velocity $\vec{\varphi}$ only contains velocities associated with the deformations of the lattice as measured in the local frame $Ox_1x_2x_3$, as well as the local rotations associated with these deformations, since the drag-along velocities of the lattice, both of the translational and rotational kind have been subtracted. Notably on the point O relative velocity $\vec{\varphi}$ is null by definition.

Table 2.2 - Expressions of the invariant operators in $Q\xi_1\xi_2\xi_3$ and $Ox_1x_2x_3$

	In absolute frame $Q\xi_1\xi_2\xi_3$ $f = f(\xi_k,t)$ or $\vec{u} = \sum u_i(\xi_k,t)\vec{\mathbf{e}}_i$	In local frame $Ox_1x_2x_3$ $f = f(x_k,t)$ or $\vec{u} = \sum u_i(x_k,t)\,\vec{e}_i$
$\vec{\nabla}$	$\sum_i \vec{\mathbf{e}}_i \dfrac{\partial}{\partial \xi_i}$ (1)	$\sum_i \vec{e}_i \dfrac{\partial}{\partial x_i}$ (2)
$\overrightarrow{\mathrm{grad}}\, f$	$\sum_i \vec{\mathbf{e}}_i \dfrac{\partial f(\xi_k,t)}{\partial \xi_i}$ (3)	$\sum_i \vec{e}_i \dfrac{\partial f(x_k,t)}{\partial x_i}$ (4)
$\overrightarrow{\mathrm{rot}}\,\vec{u}$	$\begin{vmatrix} \vec{\mathbf{e}}_1 & \vec{\mathbf{e}}_2 & \vec{\mathbf{e}}_3 \\ \partial/\partial\xi_1 & \partial/\partial\xi_2 & \partial/\partial\xi_3 \\ u_1(\xi_k,t) & u_2(\xi_k,t) & u_3(\xi_k,t) \end{vmatrix}$ (5)	$\begin{vmatrix} \vec{e}_1 & \vec{e}_2 & \vec{e}_3 \\ \partial/\partial x_1 & \partial/\partial x_2 & \partial/\partial x_3 \\ u_1(x_k,t) & u_2(x_k,t) & u_3(x_k,t) \end{vmatrix}$ (6)
$\mathrm{div}\,\vec{u}$	$\sum_i \dfrac{\partial u_i(\xi_k,t)}{\partial \xi_i}$ (7)	$\sum_i \dfrac{\partial u_i(x_k,t)}{\partial x_i}$ (8)
$(\vec{n}\vec{\nabla})$	$\sum_i (\vec{n}\vec{\mathbf{e}}_i)\dfrac{\partial}{\partial \xi_i}$ (9)	$\sum_i (\vec{n}\vec{e}_i)\dfrac{\partial}{\partial x_i}$ (10)
$\dfrac{d}{dt}$	$\dfrac{\partial}{\partial t} + (\vec{\phi}\vec{\nabla}) = \dfrac{\partial}{\partial t} + \sum_i \phi_i(\vec{\xi},t)\dfrac{\partial}{\partial \xi_i}$ (11)	$\dfrac{\partial}{\partial t} + (\vec{\varphi}\vec{\nabla}) = \dfrac{\partial}{\partial t} + \sum_i \varphi_i(\vec{r},t)\dfrac{\partial}{\partial x_i}$ (12)

We can conclude that in the case of *weakly deformable* solid, meaning when the displacements of the nodes of the lattice as seen in the local frame $Ox_1x_2x_3$ stay small in a neighborhood of point O, it is possible to replace the material derivative d/dt, corresponding normally to the temporal variation along the trajectories of the nodes in the local frame $Ox_1x_2x_3$, by the temporal derivative at a fixed point of the local frame $Ox_1x_2x_3$, meaning the partial derivative with respect to time.

The same goes for a solid which is *deforming very slowly,* since in that case the field of relative velocity $\vec{\varphi}$ is negligible. We derive a remark which will be *very important* for the rest of the treatise:

- *In the vicinity of a point O of a local frame $Ox_1x_2x_3$, if the deformations are small or happen slowly over time, we can use the following approximation for the temporal derivative:*

$$\boxed{\frac{d}{dt} = \frac{\partial}{\partial t} + (\vec{\varphi}\vec{\nabla}) \cong \frac{\partial}{\partial t}} \quad \textit{in the vicinity of } O \qquad (2.20)$$

2.2 - Projection of the geometro-kinetic equations in the local frame

It is now possible to redefine the distortion tensors of a solid by using a formalism similar to that introduced in the first chapter, but in the local frame $Ox_1x_2x_3$ of the small observer ***PO*** (figure 2.3).

Geometro-kinetic equations of distortion in the local frame

In the local frame $Ox_1x_2x_3$, the angular velocity of rotation $d\vec{\omega}/dt$ and the material derivative of the scalar τ of volume expansion of the solid can be computed in a way perfectly similar to that used in the second chapter already. Therefore, the equations of geometro-kinetic introduced above remain valid in the local mobile frame.

$$\frac{d\vec{\omega}}{dt} = \frac{1}{2}\overrightarrow{\text{rot}}\,\vec{\phi} \qquad \text{and} \qquad \frac{d\tau}{dt} = -\frac{S_n}{n} + \text{div}\,\vec{\phi} \tag{2.21}$$

We can now define in the local frame a new *tensor of global distortion* $\vec{\beta}_i$ such that its trace is equal to $d\tau/dt$ and that half of its anti-symmetric part, with an inverted sign, gives us $d\vec{\omega}/dt$. The geometro-kinetic equations which we defined in the second chapter in the following way

$$\frac{d\vec{\beta}_i}{dt} = -\frac{S_n}{3n}\vec{e}_i + \overrightarrow{\text{grad}}\,\phi_i = -\frac{S_n}{3n}\vec{e}_i + \overrightarrow{\text{grad}}\left(\vec{\phi}\vec{e}_i\right) = -\frac{S_n}{3n}\vec{e}_i + \left(\vec{e}_i\vec{\nabla}\right)\vec{\phi} + \vec{e}_i \wedge \overrightarrow{\text{rot}}\,\vec{\phi} \tag{2.22}$$

yield a tensor which satisfy these requirements. Indeed, it follows for the trace of $d\vec{\beta}_i/dt$ and for the half of its anti-symmetric part with inverted sign, the expressions of $d\tau/dt$ and of $d\vec{\omega}/dt$

$$\sum_k \vec{e}_k \frac{d\vec{\beta}_k}{dt} = -\frac{S_n}{n} + \vec{\nabla}\vec{\phi} + \underbrace{\sum_k \overrightarrow{\text{rot}}\,\vec{\phi}\left(\vec{e}_k \wedge \vec{e}_k\right)}_{=0} = -\frac{S_n}{n} + \text{div}\vec{\phi} = \frac{d\tau}{dt} \tag{2.23}$$

$$-\frac{1}{2}\sum_k \vec{e}_k \wedge \frac{d\vec{\beta}_k}{dt} = -\frac{1}{2}\left[\sum_k \vec{e}_k \wedge \left(\vec{e}_k\vec{\nabla}\right)\vec{\phi} + \sum_k \vec{e}_k \wedge \left(\vec{e}_k \wedge \overrightarrow{\text{rot}}\,\vec{\phi}\right)\right] = \frac{1}{2}\overrightarrow{\text{rot}}\,\vec{\phi} = \frac{d\vec{\omega}}{dt} \tag{2.24}$$

The new *deformation tensor* $\vec{\varepsilon}_i$ is obtained by symmetrization of the geometro-kinetic equation $d\vec{\beta}_i/dt$ in the local frame $Ox_1x_2x_3$

$$\frac{d\vec{\varepsilon}_i}{dt} = \frac{d\vec{\beta}_i}{dt}_i - \vec{e}_i \wedge \frac{d\vec{\omega}}{dt} \quad\Rightarrow\quad \frac{d\vec{\varepsilon}_i}{dt} = -\frac{S_n}{3n}\vec{e}_i + \overrightarrow{\text{grad}}\,\phi_i - \frac{1}{2}\vec{e}_i \wedge \overrightarrow{\text{rot}}\,\vec{\phi} \tag{2.25}$$

The new *shear tensor* $\vec{\alpha}_i$, is obtained from the transversal symmetric part of $d\vec{\beta}_i/dt$ in the local frame $Ox_1x_2x_3$

$$\frac{d\vec{\alpha}_i}{dt} = \frac{d\vec{\varepsilon}_i}{dt} - \frac{1}{3}\vec{e}_i\frac{d\tau}{dt} \quad\Rightarrow\quad \frac{d\vec{\alpha}_i}{dt} = \overrightarrow{\text{grad}}\,\phi_i - \frac{1}{2}\vec{e}_i \wedge \overrightarrow{\text{rot}}\,\vec{\phi} - \frac{1}{3}\vec{e}_i\,\text{div}\vec{\phi} \tag{2.26}$$

The tensors of global distortion $\vec{\beta}_i, \vec{\varepsilon}_i, \vec{\alpha}_i$ thus defined depend on the choice of the local frame $Ox_1x_2x_3$, via the presence of the frame vectors $\vec{e}_i$ in their definitions. They are computed from the components of the absolute velocity field $\vec{\phi}$ of the solid projected in the mobile local frame $Ox_1x_2x_3$, in the fashion that the distortion tensors $\vec{\beta}_i, \vec{\varepsilon}_i, \vec{\alpha}_i$ were computed in the second chapter from the projections of the $\vec{\phi}$ field in a fixed frame linked to $Q\xi_1\xi_2\xi_3$.

For the Small Observer ***PO*** in the local mobile frame $Ox_1x_2x_3$, the frame vectors $\vec{e}_i$ appear as constants independent of time, and therefore in $Ox_1x_2x_3$ we have

$$\begin{cases} \dfrac{d\tau}{dt} = \sum_k \vec{e}_k \dfrac{d\vec{\beta}_k}{dt} \Rightarrow \tau = \sum_k \vec{e}_k \vec{\beta}_k \\ \dfrac{d\vec{\omega}}{dt} = -\dfrac{1}{2}\sum_k \vec{e}_k \wedge \dfrac{d\vec{\beta}_k}{dt} \Rightarrow \vec{\omega} = -\dfrac{1}{2}\sum_k \vec{e}_k \wedge \vec{\beta}_k \\ \dfrac{d\vec{\varepsilon}_i}{dt} = \dfrac{d\vec{\beta}_i}{dt} - \vec{e}_i \wedge \dfrac{d\vec{\omega}}{dt} \Rightarrow \vec{\varepsilon}_i = \vec{\beta}_i - \vec{e}_i \wedge \vec{\omega} \\ \dfrac{d\vec{\alpha}_i}{dt} = \dfrac{d\vec{\varepsilon}_i}{dt} - \dfrac{1}{3}\vec{e}_i \dfrac{d\tau}{dt} \Rightarrow \vec{\alpha}_i = \vec{\varepsilon}_i - \dfrac{1}{3}\tau\vec{e}_i \end{cases} \tag{2.27}$$

As a consequence the decompositions of the distortion tensors we saw in figures 1.7 and 1.8 remain valid when we replace the tensors $\vec{\beta}_i, \vec{\varepsilon}_i, \vec{\alpha}_i$ defined in the fixed absolute frame $Q\xi_3\xi_2\xi_3$ by the same tensors $\vec{\beta}_i, \vec{\varepsilon}_i, \vec{\alpha}_i$, this time defined in the mobile local frame $Ox_1x_2x_3$.

Table 2.3 - Geometro-kinetic equations of the distortions of a solid lattice in the local frame $Ox_1x_2x_3$

$$\begin{cases} \dfrac{d\vec{\beta}_i}{dt} = -\dfrac{S_n}{3n}\vec{e}_i + \overrightarrow{\text{grad}}\,\phi_i & (1) \\ \dfrac{d\vec{\omega}}{dt} = \dfrac{1}{2}\overrightarrow{\text{rot}}\,\vec{\phi} & (2) \\ \dfrac{d\tau}{dt} = -\dfrac{S_n}{n} + \text{div}\,\vec{\phi} & (3) \end{cases}$$

$$avec \quad \frac{d}{dt} = \frac{\partial}{\partial t} + \left(\vec{\varphi}\vec{\nabla}\right) \qquad (4)$$

$$\begin{cases} \vec{\omega} = -\dfrac{1}{2}\sum_k \vec{e}_k \wedge \vec{\beta}_k & (5) \\ \tau = \sum_k \vec{e}_k \vec{\beta}_k = \sum_k \vec{e}_k \vec{\varepsilon}_k & (6) \\ \vec{\varepsilon}_i = \vec{\beta}_i - \vec{e}_i \wedge \vec{\omega} & (7) \\ \vec{\alpha}_i = \vec{\varepsilon}_i - \dfrac{1}{3}\tau\vec{e}_i & (8) \end{cases}$$

One should note here that only the vectorial field $\vec{\omega}$ and the scalar field τ (representing respectively the anti-symmetric part and the trace of the distortion tensor $\vec{\beta}_i$) are truly *invariant* quantities, in the sense that we are dealing with a vector and a scalar that do not depend on the choice of frame we use to compute them. The geometro-kinetic equations in the local frame are shown in table 2.2. One should compare them to the geometro-kinetic equations of table 1.1, expressed in the absolute frame: we can observe that they have exactly the same form, but that in these expressions it is *essentially the material derivative that is different* since it depends on $\vec{\phi}$ in the absolute frame $Q\xi_3\xi_2\xi_3$ while it depends on $\vec{\varphi}$ in the mobile frame $Ox_1x_2x_3$.

Local distortions $\vec{\beta}_i^{(\delta)}$ ***and local rotations*** $\vec{\omega}^{(\delta)}$ ***due to pure deformations***

Using the rotational part of the local field of relative velocity $\vec{\varphi}$ expressed in the frame $Ox_1x_2x_3$

$$\overrightarrow{\text{rot}}\,\vec{\varphi} = \overrightarrow{\text{rot}}\left(\vec{\phi} - \vec{\phi}_O(t) - \dot{\vec{\omega}}_O(t) \wedge \vec{r}\right) = \overrightarrow{\text{rot}}\,\vec{\phi} - \dot{\vec{\omega}}_O\,\text{div}\vec{r} + \left(\dot{\vec{\omega}}_O\vec{\nabla}\right)\vec{r} = \overrightarrow{\text{rot}}\,\vec{\phi} - 2\dot{\vec{\omega}}_O \tag{2.28}$$

we can define a local rotation vector $\vec{\omega}^{(\delta)}$ *due to pure deformation* which measures the rotations associated only with the deformations of the solid about the O point, since its material derivative is equal to the global angular velocity $d\vec{\omega}/dt$ from which we subtract the drag-along angular velocity $\dot{\vec{\omega}}_O$ of the frame $Ox_1x_2x_3$

$$\frac{d\vec{\omega}^{(\delta)}}{dt}=\frac{1}{2}\overrightarrow{\mathrm{rot}}\,\vec{\varphi}=\frac{1}{2}\overrightarrow{\mathrm{rot}}\,\vec{\phi}-\dot{\vec{\omega}}_O=\frac{d\vec{\omega}}{dt}-\dot{\vec{\omega}}_O \tag{2.29}$$

It is also possible to build a *local distortion tensor* $\vec{\beta}_i^{(\delta)}$ *due to pure deformation* such that the trace is equal to the scalar of volume expansion τ and that the half of its anti-symmetric part with inverted sign gives us the vector of local rotation $\vec{\omega}^{(\delta)}$. The following geometro-kinetic equation will do

$$\frac{d\vec{\beta}_i^{(\delta)}}{dt}=-\frac{S_n}{3n}\vec{e}_i+\overrightarrow{\mathrm{grad}}\,\varphi_i=-\frac{S_n}{3n}\vec{e}_i+\overrightarrow{\mathrm{grad}}\left(\vec{\varphi}\vec{e}_i\right) \tag{2.30}$$

Indeed, by rewriting the expression for $d\vec{\beta}_i^{(\delta)}/dt$ as

$$\frac{d\vec{\beta}_i^{(\delta)}}{dt}=-\frac{S_n}{3n}\vec{e}_i+\left(\vec{e}_i\vec{\nabla}\right)\vec{\varphi}+\vec{e}_i\wedge\overrightarrow{\mathrm{rot}}\,\vec{\varphi} \tag{2.31}$$

We can verify the 2 following expressions

$$\sum_k\vec{e}_k\frac{d\vec{\beta}_k^{(\delta)}}{dt}=-\frac{S_n}{n}+\mathrm{div}\vec{\varphi}=\frac{d\tau}{dt} \tag{2.32}$$

$$-\frac{1}{2}\sum_k\vec{e}_k\wedge\frac{d\vec{\beta}_k^{(\delta)}}{dt}=\frac{1}{2}\overrightarrow{\mathrm{rot}}\,\vec{\varphi}=\frac{d\vec{\omega}^{(\delta)}}{dt} \tag{2.33}$$

The tensor of local distortion $\vec{\beta}_i^{(\delta)}$ thus defined can be directly linked to the field of absolute velocity . Indeed, the following

$$\begin{aligned}\overrightarrow{\mathrm{grad}}\left(\vec{\phi}\vec{e}_i\right)&=\vec{e}_i\wedge\overrightarrow{\mathrm{rot}}\,\vec{\phi}+\left(\vec{e}_i\vec{\nabla}\right)\vec{\phi}=\vec{e}_i\wedge\overrightarrow{\mathrm{rot}}\,\vec{\varphi}+\vec{e}_i\wedge 2\dot{\vec{\omega}}_O\\&\quad+\left(\vec{e}_i\vec{\nabla}\right)\vec{\varphi}+\left(\vec{e}_i\vec{\nabla}\right)\left(\dot{\vec{\omega}}_O(t)\wedge\vec{r}\right)=\vec{e}_i\wedge\overrightarrow{\mathrm{rot}}\,\vec{\varphi}+\left(\vec{e}_i\vec{\nabla}\right)\vec{\varphi}-\dot{\vec{\omega}}_O\wedge\vec{e}_i\end{aligned} \tag{2.34}$$

shows that

$$\frac{d\vec{\beta}_i^{(\delta)}}{dt}=-\frac{S_n}{3n}\vec{e}_i+\overrightarrow{\mathrm{grad}}\left(\vec{\phi}\vec{e}_i\right)-\vec{e}_i\wedge\dot{\vec{\omega}}_O=\frac{d\vec{\beta}_i}{dt}-\vec{e}_i\wedge\dot{\vec{\omega}}_O \tag{2.35}$$

It is easy to show that the expressions for the deformation field $\vec{\varepsilon}_i$ and shear field $\vec{\alpha}_i$ as a function of the field $\vec{\varphi}$ of relative velocities or as a function of the field $\vec{\phi}$ of absolute velocities are the same, which seems logical since $\vec{\varepsilon}_i$ and $\vec{\alpha}_i$ are only dependent on the pure deformations.

$$\frac{d\vec{\varepsilon}_i}{dt}=-\frac{S_n}{3n}\vec{e}_i+\overrightarrow{\mathrm{grad}}\,\varphi_i-\frac{1}{2}\vec{e}_i\wedge\overrightarrow{\mathrm{rot}}\,\vec{\varphi}=-\frac{S_n}{3n}\vec{e}_i+\left(\vec{e}_i\vec{\nabla}\right)\vec{\varphi}+\frac{1}{2}\vec{e}_i\wedge\overrightarrow{\mathrm{rot}}\,\vec{\varphi} \tag{2.36}$$

$$\frac{d\vec{\alpha}_i}{dt}=\overrightarrow{\mathrm{grad}}\,\varphi_i-\frac{1}{2}\vec{e}_i\wedge\overrightarrow{\mathrm{rot}}\,\vec{\varphi}-\frac{1}{3}\vec{e}_i\,\mathrm{div}\vec{\varphi}=\left(\vec{e}_i\vec{\nabla}\right)\vec{\varphi}+\frac{1}{2}\vec{e}_i\wedge\overrightarrow{\mathrm{rot}}\,\vec{\varphi}-\frac{1}{3}\vec{e}_i\,\mathrm{div}\vec{\varphi} \tag{2.37}$$

Since the frame vectors $\vec{e}_i$ are constants independent of time for the Small Observer ***PO*** in the local mobile frame $Ox_1x_2x_3$, the following relations of decomposition of the tensors of local distortion hold in $Ox_1x_2x_3$

$$\begin{cases} \dfrac{d\tau}{dt} = \sum_k \vec{e}_k \dfrac{d\vec{\beta}_k^{(\delta)}}{dt} & \Rightarrow \quad \tau = \sum_k \vec{e}_k \vec{\beta}_k^{(\delta)} \\ \dfrac{d\vec{\omega}^{(\delta)}}{dt} = -\dfrac{1}{2}\sum_k \vec{e}_k \wedge \dfrac{d\vec{\beta}_k^{(\delta)}}{dt} & \Rightarrow \quad \vec{\omega}^{(\delta)} = -\dfrac{1}{2}\sum_k \vec{e}_k \wedge \vec{\beta}_k^{(\delta)} \\ \dfrac{d\vec{\varepsilon}_i}{dt} = \dfrac{d\vec{\beta}_i^{(\delta)}}{dt} - \vec{e}_i \wedge \dfrac{d\vec{\omega}^{(\delta)}}{dt} & \Rightarrow \quad \vec{\varepsilon}_i = \vec{\beta}_i^{(\delta)} - \vec{e}_i \wedge \vec{\omega}^{(\delta)} \\ \dfrac{d\vec{\alpha}_i}{dt} = \dfrac{d\vec{\varepsilon}_i}{dt} - \dfrac{1}{3}\vec{e}_i \dfrac{d\tau}{dt} & \Rightarrow \quad \vec{\alpha}_i = \vec{\varepsilon}_i - \dfrac{1}{3}\tau \vec{e}_i \end{cases} \tag{2.38}$$

Table 2.4 - Geometro-kinetic equations of local distortions and rotations of a solid lattice in the local frame $Ox_1x_2x_3$

$$\begin{cases} \dfrac{d\vec{\beta}_i^{(\delta)}}{dt} = -\dfrac{S_n}{3n}\vec{e}_i + \overrightarrow{\text{grad}}\,\varphi_i \\ \qquad = \dfrac{d\vec{\beta}_i}{dt} - \vec{e}_i \wedge \dot{\vec{\omega}}_O(t) & (1) \\ \dfrac{d\vec{\omega}^{(\delta)}}{dt} = \dfrac{1}{2}\overrightarrow{\text{rot}}\,\vec{\varphi} \\ \qquad = \dfrac{d\vec{\omega}}{dt} - \dot{\vec{\omega}}_O(t) & (2) \\ \dfrac{d\tau}{dt} = -\dfrac{S_n}{n} + \text{div}\,\vec{\varphi} & (3) \end{cases}$$

$$\begin{cases} \vec{\beta}_i^{(\delta)} = \vec{\beta}_i - \vec{e}_i \wedge \vec{\omega}_O(t) & (4) \\ \vec{\omega}^{(\delta)} = -\dfrac{1}{2}\sum_k \vec{e}_k \wedge \vec{\beta}_k^{(\delta)} = \vec{\omega} - \vec{\omega}_O(t) & (5) \\ \tau = \sum_k \vec{e}_k \vec{\beta}_k^{(\delta)} = \sum_k \vec{e}_k \vec{\beta}_k = \sum_k \vec{e}_k \vec{\varepsilon}_k & (6) \\ \vec{\varepsilon}_i = \vec{\beta}_i^{(\delta)} - \vec{e}_i \wedge \vec{\omega}^{(\delta)} = \vec{\beta}_i - \vec{e}_i \wedge \vec{\omega} & (7) \\ \vec{\alpha}_i = \vec{\varepsilon}_i - \dfrac{1}{3}\tau \vec{e}_i & (8) \end{cases}$$

For the same reason as above, the tensors of local distortion from deformation can be directly linked to the tensors of global distortion in $Ox_1x_2x_3$ by the following relations, in which the rotational vector $\vec{\omega}_O$ only depends on time

$$\frac{d\vec{\beta}_i^{(\delta)}}{dt} = \frac{d\vec{\beta}_i}{dt} - \vec{e}_i \wedge \dot{\vec{\omega}}_O \quad \Rightarrow \quad \vec{\beta}_i^{(\delta)} = \vec{\beta}_i - \vec{e}_i \wedge \vec{\omega}_O(t) \tag{2.39}$$

$$\frac{d\vec{\omega}^{(\delta)}}{dt} = \frac{d\vec{\omega}}{dt} - \dot{\vec{\omega}}_O \quad \Rightarrow \quad \vec{\omega}^{(\delta)} = \vec{\omega} - \vec{\omega}_O(t) \tag{2.40}$$

The local tensors $\vec{\beta}_i^{(\delta)}$ and $\vec{\omega}^{(\delta)}$, associated with local deformations of the lattice in the neighborhood of point O, differ from the global tensors $\vec{\beta}_i$ and $\vec{\omega}$ obtained in the previous paragraph only due to the existence of a drag-along rotation $\vec{\omega}_O(t)$ of the local frame $Ox_1x_2x_3$. In short, the tensors τ, $\vec{\varepsilon}_i$ et $\vec{\alpha}_i$, can be computed indifferently from the field of absolute speed $\vec{\phi}$ or the field of relative speed $\vec{\varphi}$. The equations of geometro-kinetic for the tensors of local distortion due to deformation, deduced from the field $\vec{\varphi}$ in the local mobile frame $Ox_1x_2x_3$, are given in table 2.4.

2.3 - Geometro-compatibility in eulerian coordinates

We have shown section 1.1 that the description of the deformation of compatible solids in lagrangian coordinates is characterized by the existence of a field of displacement. Indeed, in lagrangian coordinates (figure 2.4), the solid is described by a position vector $\vec{r}$ which marks all the points in a frame $Ox_1x_2x_3$ fixed in the absolute referential $Q\xi_1\xi_2\xi_3$ of the ***GO*** observer. The lagrangian vector displacement field $\vec{u}_L(\vec{r},t)$ allows us to find the position of all points of the solid at time t, which were originally at coordinate $\vec{r}$ of the frame $Ox_1x_2x_3$.

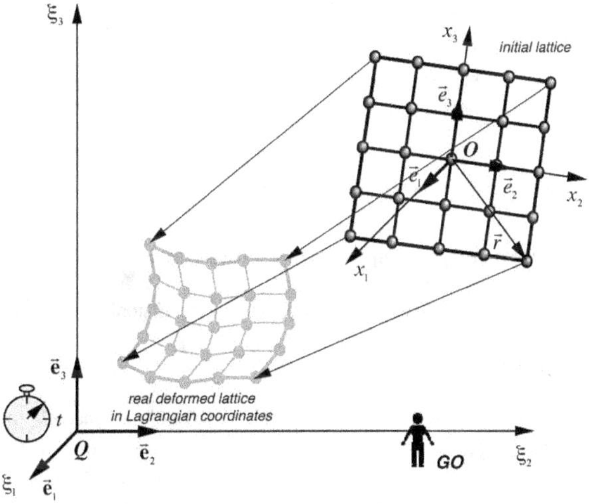

Figure 2.4 *- the displacement field $\vec{u}_L(\vec{r},t)$ in lagrangian coordinates*

It is intuitively clear that the description of the distortions of a solid in eulerian coordinates should allow us to recover the displacement field. Indeed in eulerian coordinates (figure 2.5), the deformed solid is described at a time t in the absolute frame of the observer ***GO***. For a point A of the solid of coordinate $\vec{\xi}$ in this frame we define the eulerian displacement vector $\vec{u}_E(\vec{\xi},t)$ as linking point A to point A' where the A point of the solid was at initial time $t=0$. The same type of construction can be applied in the mobile frame $Ox_1x_2x_3$ of the ***PO*** observer: at a given time t, for each point B of the solide at coordinate $\vec{r}$ in that frame, we define the displacement vector $\vec{u}_E^{(\delta)}(\vec{r},t)$ which links point B to the location B'' of frame $Ox_1x_2x_3$ where point B was at time $t=0$.

The displacement fields $\vec{u}_E(\vec{\xi},t)$ and $\vec{u}_E^{(\delta)}(\vec{r},t)$ thus defined allow us to rebuild the solid as it was at time $t=0$, before the various translations, rotations and deformations it underwent, both in the absolute frame of ***GO*** and in the mobile frame $Ox_1x_2x_3$ of the ***PO*** observer.

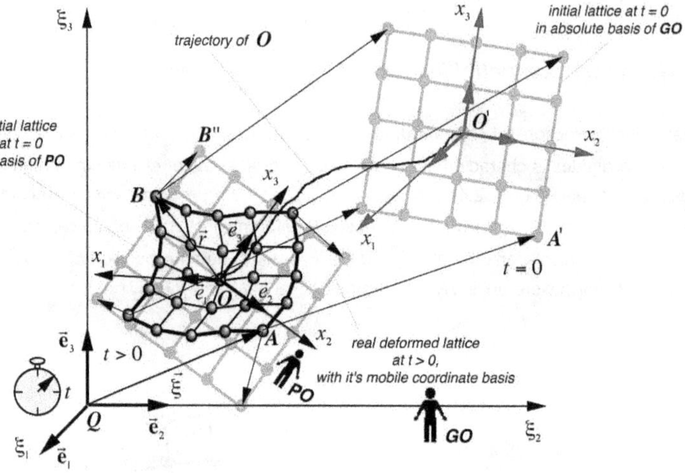

Figure 2.5 *- displacement fields* $\vec{u}_E(\vec{\xi},t)$ *and* $\vec{u}_E^{(\delta)}(\vec{r},t)$ *in eulerian coordinates*

Condition of geometro-compatibility in eulerian coordinates

According to the form of the geometro-kinetic equation for the distortion tensor $\vec{\beta}_i$ we have

$$\frac{d\vec{\beta}_i}{dt} = -\frac{S_n}{3n}\vec{e}_i + \overrightarrow{\mathrm{grad}}\,\phi_i \tag{2.41}$$

It is clear that there is a link between the temporal derivative of the tensor $\vec{\beta}_i$ and the gradient of the field of velocity $\vec{\phi}(\vec{\xi},t)$. Also the velocity field $\vec{\phi}(\vec{\xi},t)$ is related to the time derivative of the displacement field $\vec{u}_E(\vec{\xi},t)$, thus there is link between the tensor $\vec{\beta}_i$ and the gradient of the displacement field $\vec{u}_E(\vec{\xi},t)$. The distortion tensor has to be linked to the gradient of the displacement field.

By decomposing the global distortion $\vec{\beta}_i$ in a local distortion tensor $\vec{\beta}_i^{(\delta)}$, associated to the deformations of the solid in the $Ox_1x_2x_3$ frame, and a second component associated to the drag-along rotation $\vec{\omega}_O(t)$ of said frame we write

$$\vec{\beta}_i = \vec{\beta}_i^{(\delta)} + \vec{e}_i \wedge \vec{\omega}_O(t) \quad \Rightarrow \quad \overrightarrow{\mathrm{rot}}\,\vec{\beta}_i = \overrightarrow{\mathrm{rot}}\,\vec{\beta}_i^{(\delta)} \tag{2.42}$$

Supposing that the tensor of distortion is indeed the gradient of the displacement field we can write *a priori* the following condition of compatibility in eulerian coordinates

Condition of geometro-compatibility: $\overrightarrow{\mathrm{rot}}\,\vec{\beta}_i = \overrightarrow{\mathrm{rot}}\,\vec{\beta}_i^{(\delta)} = 0$ (2.43)

In fact, in section 3.4 we will show that the physical meaning of this relation is, from a topological point of view, that *the displacement field* $\vec{u}_E(\vec{\xi},t)$ *is continuous*, in other words that there are *no discontinuities of the displacement field* $\vec{u}_E(\vec{\xi},t)$, meaning that *there are no dislocations inside the medium.*

We revisit the relations of spatial derivatives of the topological tensors introduced in section 1.4, and we will apply the same manipulations as in 1.4. This condition gives us a complete set of compatibility relations in eulerian coordinates

$$\begin{cases} \overrightarrow{\mathrm{rot}}\,\vec{\beta}_i = \overrightarrow{\mathrm{rot}}\,\vec{\beta}_i^{(\delta)} = 0 \\ \mathrm{div}\,\vec{\omega} = \mathrm{div}\,\vec{\omega}^{(\delta)} = 0 \end{cases} \tag{2.44}$$

$$\begin{cases} \overrightarrow{\mathrm{rot}}\left[\overrightarrow{\mathrm{rot}}\,\vec{\varepsilon}_i\right]^{\mathrm{T}} = 0 \\ \overrightarrow{\mathrm{rot}}\left[\overrightarrow{\mathrm{rot}}\,\vec{\alpha}_i\right]^{\mathrm{T}} - \dfrac{1}{3}\overrightarrow{\mathrm{rot}}\,\overrightarrow{\mathrm{rot}}\left(\vec{e}_i\tau\right) = 0 \end{cases} \tag{2.45}$$

We will show now the consequences of the condition of geometric compatibility in eulerian coordinates:

Consequence 1: existence of potential field of displacement

The condition of geometro-compatibility *(2.43)* implies that $\vec{\beta}_i$ and $\vec{\beta}_i^{(\delta)}$ derive from gradients of scalar fields $\vec{u}_E(\vec{\xi},t)$ and $\vec{u}_E^{(\delta)}(\vec{r},t)$ defined respectively in frame $Q\xi_1\xi_2\xi_3$ and $Ox_1x_2x_3$, which allows us to write without loss of generality

$$\overrightarrow{\mathrm{rot}}\,\vec{\beta}_i = \overrightarrow{\mathrm{rot}}\,\vec{\beta}_i^{(\delta)} = 0 \;\Rightarrow\; \begin{cases} \vec{\beta}_i = -\overrightarrow{\mathrm{grad}}\,u_{Ei}(\vec{\xi},t) = -\sum\limits_k \dfrac{\partial u_{Ei}}{\partial \xi_k}\,\vec{\mathbf{e}}_k \\ \vec{\beta}_i^{(\delta)} = -\overrightarrow{\mathrm{grad}}\,u_{Ei}^{(\delta)}(\vec{r},t) = -\sum\limits_k \dfrac{\partial u_{Ei}^{(\delta)}}{\partial x_k}\,\vec{e}_k \end{cases} \tag{2.46}$$

With the scalar fields $u_{Ei}(\vec{\xi},t)$ and $u_{Ei}^{(\delta)}(\vec{r},t)$ necessary to the description of the tensors $\vec{\beta}_i$ and $\vec{\beta}_i^{(\delta)}$, 2 vectorial fields can be defined $\vec{u}_E(\vec{\xi},t)$ and $\vec{u}_E(\vec{\xi},t)$ such that

$$\vec{u}_E(\vec{\xi},t) = \sum_k u_{Ei}\,(\vec{\xi},t)\,\vec{\mathbf{e}}_k \quad \text{and} \quad \vec{u}_E^{(\delta)}(\vec{r},t) = \sum_k u_{Ei}^{(\delta)}(\vec{r},t)\,\vec{e}_k \tag{2.47}$$

As a consequence, the distortion tensors $\vec{\beta}_i$ and $\vec{\beta}_i^{(\delta)}$ in a geometro-compatible solid lattice, meaning which satisfies the relation of geometric compatibility *(2.44)*, are indeed the *tensor gradients* of *continuous vectorial fields* $\vec{u}_E(\vec{\xi},t)$ and $\vec{u}_E^{(\delta)}(\vec{r},t)$. Dimensional analysis shows us that vectors $\vec{u}_E$ and $\vec{u}_E^{(\delta)}$ have dimensions of displacements. As a consequence, $\vec{u}_E(\vec{\xi},t)$ will be called the *potential field of global displacement of the solid* and $\vec{u}_E^{(\delta)}(\vec{r},t)$ the *potential field of local displacement of the solid.* These fields of potential of displacement correspond, up to a continuous vectorial field $\vec{u}_0(t)$, to the fields of displacement we have already discussed in the eulerian coordinates case (figure 2.5).

It is clear that the field of global displacement $\vec{u}_E(\vec{\xi},t)$ can be expressed in the local frame $Ox_1x_2x_3$ as a function of $\vec{r}$ and t, simply by applying the relation $\vec{\beta}_i\,(\vec{r},t) = -\overrightarrow{\mathrm{grad}}\,u_{Ei}(\vec{r},t)$ in the frame $Ox_1x_2x_3$. Using the decompositions of $\vec{\beta}_i$ and $\vec{\beta}_i^{(\delta)}$ described in tables 2.3 et 2.4, all the tensors of distortion can be directly derived from the vectorial fields of displacement $\vec{u}_E(\vec{\xi},t)$ and $\vec{u}_E^{(\delta)}(\vec{r},t)$ in the frame $Ox_1x_2x_3$

$$\begin{cases} \vec{\beta}_i = -\overrightarrow{\mathrm{grad}}\,u_{Ei} \quad \textit{and} \quad \vec{\beta}_i^{(\delta)} = -\overrightarrow{\mathrm{grad}}\,u_{Ei}^{(\delta)} \\ \vec{\omega} = -\dfrac{1}{2}\overrightarrow{\mathrm{rot}}\,\vec{u}_E \quad \textit{and} \quad \vec{\omega}^{(\delta)} = -\dfrac{1}{2}\overrightarrow{\mathrm{rot}}\,\vec{u}_E^{(\delta)} \\ \tau = -\mathrm{div}\,\vec{u}_E = -\mathrm{div}\,\vec{u}_E^{(\delta)} \end{cases} \tag{2.48a}$$

$$\begin{cases} \vec{\varepsilon}_i = -\overrightarrow{\operatorname{grad}} u_{Ei} + \frac{1}{2}\vec{e}_i \wedge \overrightarrow{\operatorname{rot}} \vec{u}_E = -\overrightarrow{\operatorname{grad}} u_{Ei}^{(\delta)} + \frac{1}{2}\vec{e}_i \wedge \overrightarrow{\operatorname{rot}} \vec{u}_E^{(\delta)} \\ \vec{\alpha}_i = -\overrightarrow{\operatorname{grad}} u_{Ei} + \frac{1}{2}\vec{e}_i \wedge \overrightarrow{\operatorname{rot}} \vec{u}_E + \frac{1}{3}\vec{e}_i \operatorname{div} \vec{u}_E = -\overrightarrow{\operatorname{grad}} u_{Ei}^{(\delta)} + \frac{1}{2}\vec{e}_i \wedge \overrightarrow{\operatorname{rot}} \vec{u}_E^{(\delta)} + \frac{1}{3}\vec{e}_i \operatorname{div} \vec{u}_E^{(\delta)} \end{cases} \tag{2.48b}$$

These relations make use of the fact that we can decompose the displacements $\vec{u}_E(\vec{\xi},t)$ and $\vec{u}_E^{(\delta)}(\vec{r},t)$ in, for example, a purely divergent part, which we will mentally associate with the expansion of the volume, and a purely rotational part, which will associate to global and local rotations, $\vec{\omega}$ and $\vec{\omega}^{(\delta)}$ respectively, by deformation of the solid. It should be noted that the shear tensor of the solid is included in the above.

One should also remark that we have used a negative sign in the defining relations $\vec{\beta}_i = -\overrightarrow{\operatorname{grad}} u_{Ei}$ and $\vec{\beta}_i^{(\delta)} = -\overrightarrow{\operatorname{grad}} u_{Ei}^{(\delta)}$. This motivation becomes clear when one compares figures 2.4 and 2.5. Indeed, we can see that the displacement field written in lagrangian coordinates $\vec{u}_L$ and eulerian coordinates $\vec{u}_E$ are identical but with opposite signs, so that it is necessary to define the relations *(2.46)* in eulerian coordinates. For example imagine a *divergent* field $\vec{u}_L$ in the lagrangian view and coordinates, the divergence is computed as $\tau = \operatorname{div} \vec{u}_L > 0$. The same $\vec{u}_E$ field from the eulerian point of view is *convergent* and $\tau = -\operatorname{div} \vec{u}_E > 0$.

Consequence 2: interpretation of the tensors of local distortion

From the fields of eulerian virtual displacement $\vec{u}^{(\delta)}(\vec{r},t)$, it is easy to give a physical interpretation of each tensor that arises from local distortion. Consider 3 points $A_0^{(i)}$ of the initial lattice undeformed, marked by 3 initial infinitesimal vectors $\vec{a}_0^{(i)} = d\vec{r}^{(i)}$, which are respectively parallel to the 3 axes of coordinates in frame $Ox_1x_2x_3$ (figure 2.6). In this local frame we map the 3 vectors $\vec{a}^{(i)}$ attached to each point $A^{(i)}$ of the real, deformed, lattice. On this visual frame we can see each of the vectors $\vec{a}^{(i)}$ of the deformed solid. It is possible to project on these 3 axes of coordinates the variation of the displacement field under deformation $d\vec{u}^{(\delta)(i)}$. The projections are given by the equations $\vec{a}_0^{(i)} = \vec{a}^{(i)} + d\vec{u}^{(\delta)(i)} = d\vec{r}^{(i)} + d\vec{u}^{(\delta)(i)}$. We can link these 3 projections to an *elongation* $\Delta a_i^{(i)}$ in the direction of the initial vector $\vec{a}_0^{(i)}$ and 2 angles of rotation ψ_{ji} and ψ_{ki} around the 2 remaining axes which are perpendicular to the initial vector $\vec{a}_0^{(i)}$. (In order to lighten figure 2.6, the notation $\vec{u}^{(\delta)(i)}$ is shortened to $\vec{u}^{d(i)}$).

The 3 relative elongations and the 6 relative angles generated by the 3 $d\vec{u}^{(\delta)(i)}$ are linked to the components of the local distortion tensor $\vec{\beta}_i^{(\delta)}$ by deformation

$$\beta_{ii}^{(\delta)} = -\frac{\partial u_i^{(\delta)}}{\partial x_i} = \frac{-du_i^{(\delta)(i)}}{dx_i^{(i)}} = \frac{\Delta a_i^{(i)}}{a_i^{(i)}} = \Delta\left(\ln a_i^{(i)}\right) = \ln a_i^{(i)} - \ln a_{0i}^{(i)} = \ln\left(\frac{a_i^{(i)}}{a_{0i}^{(i)}}\right) \tag{2.49}$$

$$\beta_{ji}^{(\delta)} = -\frac{\partial u_j^{(\delta)}}{\partial x_i} = \frac{-du_j^{(\delta)(i)}}{dx_i^{(i)}} = \tan \psi_{ji} \tag{2.50}$$

From this we deduct that the *diagonal* components $\beta_{ii}^{(\delta)}$ measure the elongation relative to the 3 directions of the coordinate system, while the *non-diagonal* components $\beta_{ji}^{(\delta)}$ measure the tangents of the angles of local rotation. It is essentially the same interpretation we had given for the $\beta_{ii}^{(\delta)}$ and $\beta_{ji}^{(\delta)}$ in section 1.5.

This visual is translated in crisp form if we decompose the local vector of rotation $\vec{\omega}^{(\delta)}$ and the non-diagonal components of the tensor of deformation $\vec{\varepsilon}_i$ and shear $\vec{\alpha}_i$ as functions of the angles of rotation ψ_{jk} et ψ_{kj}, as well as the scalar of expansion τ as a function of the relative elongations $a_i^{(i)} / a_{0i}^{(i)}$, like so

$$\begin{cases} \omega_i^{(\delta)} = \frac{1}{2}\left(\tan\psi_{jk} - \tan\psi_{kj}\right) \\ \varepsilon_{jk} = \alpha_{jk} = \frac{1}{2}\left(\tan\psi_{jk} + \tan\psi_{kj}\right) \\ \tau = \sum_i \ln\left(\frac{a_i^{(i)}}{a_{0i}^{(i)}}\right) = \ln\left(\frac{\prod a_i^{(i)}}{\prod a_{0i}^{(i)}}\right) \end{cases} \tag{2.51}$$

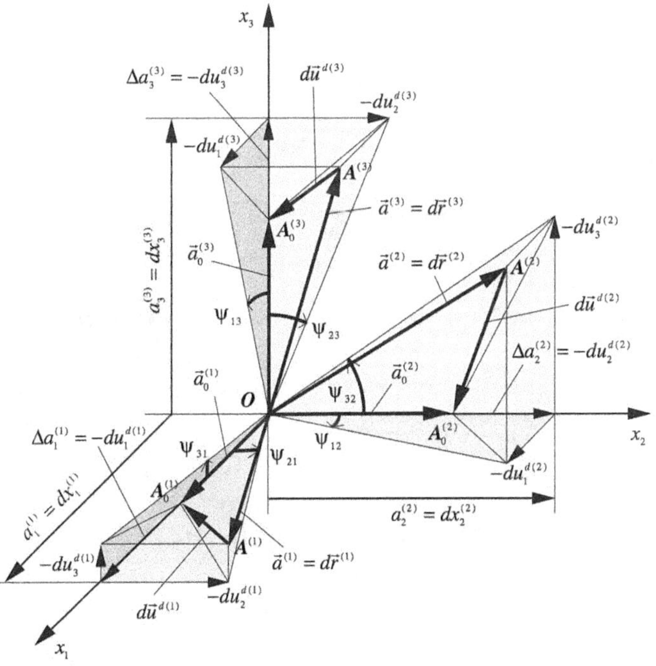

Figure 2.6 - visual decomposition of the displacements in the local frame $Ox_1x_2x_3$

Consequence 3: relation between the displacement field and the velocity field

In the absolute referential $Q\xi_1\xi_2\xi_3$, the global potential field of displacement $\vec{u}_E(\vec{\xi},t)$ must be linked to the field of absolute velocity $\vec{\phi}$. The link between the two can be seen on figure 2.7.

To establish the map between the two, we follow the movement of a point O, a node of a solid lattice, in the absolute frame. Along the path, we associate vector-points $\vec{\xi}_1$ and $\vec{\xi}_2$ corresponding respectively to the positions of point O at time t and at a slightly later moment $t + \delta t$.

The velocity is, by definition, the difference between the 2 vectors of global displacement $\vec{u}_E(\vec{\xi}_1,t)$ and $\vec{u}_E(\vec{\xi}_2,t)$ divided by the time δt, as δt becomes infinitesimally small. This reads

$$\vec{\phi}(\vec{\xi}_1,t)\delta t+\vec{u}_E(\vec{\xi}_2,t+\delta t)=\vec{u}_E(\vec{\xi}_1,t) \tag{2.52}$$

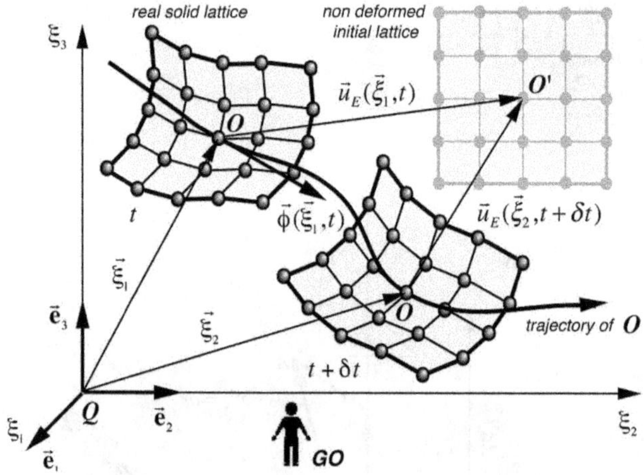

Figure 2.7 *- relation between the field of absolute velocity $\vec{\phi}$ and the field of global displacement $\vec{u}$*

Since $\vec{\xi}_2=\vec{\xi}_1+\vec{\phi}(\vec{\xi}_1,t)\delta t$, the former relation can be transformed in the following way

$$\begin{aligned}\vec{\phi}(\vec{r}_1,t)&=\frac{\vec{u}_E(\vec{\xi}_1,t)-\vec{u}_E(\vec{\xi}_2,t+\delta t)}{\delta t}=\frac{\vec{u}_E(\vec{\xi}_1,t)-\vec{u}_E(\vec{\xi}_1+\vec{\phi}\delta t,t+\delta t)}{\delta t}\\&=\frac{\vec{u}_E(\vec{\xi}_1,t)-\vec{u}_E(\vec{\xi}_1,t+\delta t)-\dfrac{\partial\vec{u}_E}{\partial\xi_i}\phi_i\delta t}{\delta t}=\frac{\vec{u}_E(\vec{\xi}_1,t)-\vec{u}_E(\vec{\xi}_1,t+\delta t)}{\delta t}-\left(\vec{\phi}\vec{\nabla}\right)\vec{u}_E\end{aligned} \tag{2.53}$$

By taking the limit $\delta t\to 0$, the field of absolute velocity $\vec{\phi}$ becomes equal to the material derivative of the field of global displacement $\vec{u}_E$, but with a change of sign

$$\vec{\phi}(\vec{\xi},t)=-\left[\frac{\partial\vec{u}_E}{\partial t}+\left(\vec{\phi}\vec{\nabla}\right)\vec{u}_E\right]=-\frac{d\vec{u}_E}{dt} \tag{2.54}$$

In the local frame $Ox_1x_2x_3$, the field of local displacement $\vec{u}_E^{(\delta)}(\vec{r},t)$ is directly linked to the field of local relative velocity $\vec{\varphi}$. This link can be seen on figure 2.8. Let's consider the movement in a local frame of a point $\boldsymbol{B}$ of a solid lattice. Along the path of this point we identify 2 points $\vec{r}_1$ and $\vec{r}_2$ taken at a time t and a later time $t+\delta t$.
As δt goes to 0 we have

$$\vec{\varphi}(\vec{r}_1,t)\delta t+\vec{u}_E^{(\delta)}(\vec{r}_2,t+\delta t)=\vec{u}_E^{(\delta)}(\vec{r}_1,t) \tag{2.55}$$

$$\vec{\varphi}(\vec{r}_1,t)=\frac{\vec{u}_E^{(\delta)}(\vec{r}_1,t)-\vec{u}_E^{(\delta)}(\vec{r}_2,t+\delta t)}{\delta t}=\frac{\vec{u}_E^{(\delta)}(\vec{r}_1,t)-\vec{u}_E^{(\delta)}(\vec{r}_1+\vec{\varphi}\delta t,t+\delta t)}{\delta t}$$
$$=\frac{\vec{u}_E^{(\delta)}(\vec{r}_1,t)-\vec{u}_E^{(\delta)}(\vec{r}_1,t+\delta t)-\dfrac{\partial\vec{u}_E^{(\delta)}}{\partial x_i}\varphi_i\delta t}{\delta t}=\frac{\vec{u}_E^{(\delta)}(\vec{r}_1,t)-\vec{u}_E^{(\delta)}(\vec{r}_1,t+\delta t)}{\delta t}-\left(\vec{\varphi}\vec{\nabla}\right)\vec{u}_E^{(\delta)} \tag{2.56}$$

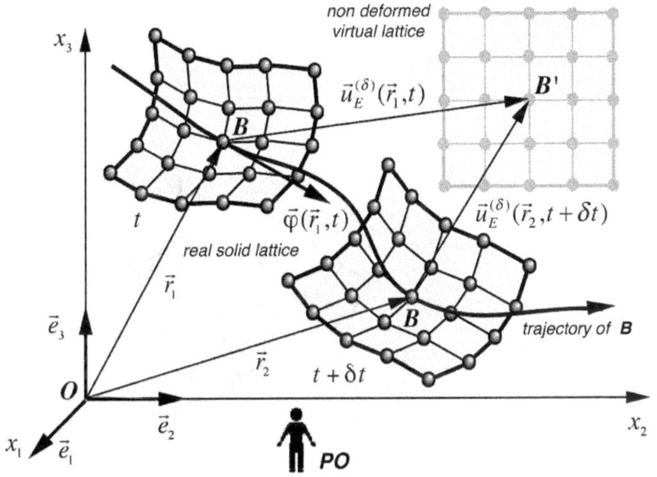

Figure 2.8 *- relation between the local field velocity $\vec{\varphi}$ and the field of local displacement $\vec{u}_E^{(\delta)}$*

By using the relation $\vec{r}_2=\vec{r}_1+\vec{\varphi}(\vec{r}_1,t)\delta t$, and by taking the limit $\delta t\to 0$, the field of local velocity $\vec{\varphi}$ becomes equal to the material derivative of the field of local displacement $\vec{u}_E^{(\delta)}$, but with a sign change

$$\vec{\varphi}(\vec{r},t)=-\left[\frac{\partial\vec{u}_E^{(\delta)}}{\partial t}+\left(\vec{\varphi}\vec{\nabla}\right)\vec{u}_E^{(\delta)}\right]=-\frac{d\vec{u}_E^{(\delta)}}{dt} \tag{2.57}$$

In the local frame $Ox_1x_2x_3$, the partial derivative of $\vec{u}_E^{(\delta)}$ with respect to time also has an interesting physical interpretation which we can discuss with figure 2.9. We will focus this time on the nodes of the real lattice which pass through a fixed point A of the local frame, marked by a position vector $\vec{r}_A$. At a given instant t, it is node (particle) C of the lattice which coincides with fixed (abstract) point A, and then at a later instant $t+\delta t$, it is node B. We define the local vectors of displacement associated with point A at the 2 moments t and $t+\delta t$, denoted $\vec{u}_C^{(\delta)}(\vec{r}_A,t)$ and $\vec{u}_B^{(\delta)}(\vec{r}_A,t+\delta t)$, as pointing respectively to points C' and B' of the initial non-deformed lattice. The end of this vector traces a path in the virtual space of the initial lattice with a velocity we denote $\vec{\psi}(\vec{r}_A,t)$ thus defined by the relation

$$\vec{\psi}(\vec{r}_A,t)=\frac{\vec{u}_B^{(\delta)}(\vec{r}_A,t+\delta t)-\vec{u}_C^{(\delta)}(\vec{r}_A,t)}{\delta t}=\frac{\partial\vec{u}^{(\delta)}(\vec{r}_A,t)}{\partial t} \tag{2.58}$$

We can conclude that the partial derivative with respect to time of the local displacement vector

$\vec{u}_E^{(\delta)}$ represents the velocity with which the extremity of this vector traces its path through the original (not deformed) virtual lattice when it is centered on a fixed point A in the local frame $Ox_1x_2x_3$.

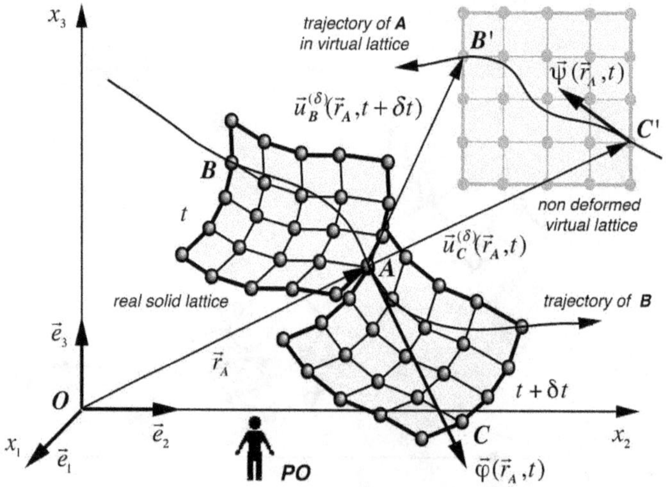

Figure 2.9 *- interpretation of the partial derivative of the local field $\vec{u}_E^{(\delta)}$ with respect to time*

Consequence 4: relation between the 2 potential fields of displacement

Relations *(2.54)* and *(2.57)* can be linked thanks to equation *(2.16)*, which gives us an interesting relation between $\vec{u}_E$ et $\vec{u}_E^{(\delta)}$

$$\frac{d\vec{u}_E}{dt} = -\vec{\phi}(\vec{\xi},t) = -\vec{\phi}(\vec{r},t) = -\vec{\varphi}(\vec{r},t) - \vec{\phi}_O(t) - \dot{\vec{\omega}}_O(t)\wedge\vec{r} = \frac{d\vec{u}_E^{(\delta)}}{dt} - \vec{\phi}_O(t) - \dot{\vec{\omega}}_O(t)\wedge\vec{r} \tag{2.59}$$

This relation holds because displacements $\vec{u}_E$ et $\vec{u}_E^{(\delta)}$ written in the local frame $Ox_1x_2x_3$ have to be related to each other by

$$\vec{u}_E^{(\delta)}(\vec{r},t) = \vec{u}_E(\vec{r},t) + \vec{\omega}_O(t)\wedge\vec{r} + \vec{u}_{E0}(t) \tag{2.60}$$

Chapter 3

Contortions of a lattice

In a solid lattice, the fields of distortion $\vec{\beta}_i$, $\vec{\varepsilon}_i$, $\vec{\alpha}_i$, $\vec{\omega}$, τ represent the distortions, deformations, shears, rotations and volume expansions that happen at every point and throughout the cells of the lattice. If the unit cell of a lattice experiences distortions, which can vary from one point of the lattice to another, effects will appear that are more macroscopic than the flexions and torsions of the solid and which are linked to the continuity of the lattice. These "*curvatures*" of the solid, resulting from a field of distortions are called the *contortions of the lattice*. We will describe this curvature in the first part of the chapter.

In the second part, we interpret the conditions of geometro-compatibility of a lattice by showing that they guarantee that the topology of the lattice is compact and connected.

3.1 - Rotational (curl) and divergence of a tensor

In order to derive the contortion tensors of a solid lattice, we will first find the relations between the first and second spatial derivatives of the various topological tensors. These relations will be *established independently of the fact that* $\vec{\beta}_i$ *can be defined either from the eulerian displacement field* $\vec{u}(\vec{r},t)$ or $\vec{u}^{(\delta)}(\vec{r},t)$ in a geometro-compatible lattice, simply by applying the vectorial operators to the distortion tensors.

The «rotational of a tensor» ("curl of a tensor")

By using the decomposition of the $\vec{\beta}_i$ tensors as documented in figure 1.8, it is possible to calculate the relations that can exist between the spatial derivatives of first order of these tensors. We simply apply the operators of vectorial analysis and their laws of distributivity. Applying the rotational operator to expression *(1.36)* of the symmetric tensor $\vec{\varepsilon}_i$ implies the relation

$$\overrightarrow{\operatorname{rot}}\,\vec{\varepsilon}_i = \overrightarrow{\operatorname{rot}}\,\vec{\alpha}_i - \frac{1}{3}\vec{e}_i \wedge \overrightarrow{\operatorname{grad}}\,\tau = (\vec{e}_i\vec{\nabla})\vec{\omega} + \overrightarrow{\operatorname{rot}}\,\vec{\beta}_i - \vec{e}_i \operatorname{div}\vec{\omega} \qquad (3.1)$$

By using the following transposition rules

$$\left[(\vec{e}_i\vec{\nabla})\vec{\omega}\right]^{\mathrm{T}} = \overrightarrow{\operatorname{grad}}\,\omega_i \quad \text{and} \quad \left[-\vec{e}_i \wedge \overrightarrow{\operatorname{grad}}\,\tau\right]^{\mathrm{T}} = \vec{e}_i \wedge \overrightarrow{\operatorname{grad}}\,\tau \qquad (3.2)$$

the previous expression for $\overrightarrow{\operatorname{rot}}\,\vec{\varepsilon}_i$ can be transposed

$$\left[\overrightarrow{\operatorname{rot}}\,\vec{\varepsilon}_i\right]^{\mathrm{T}} = \left[\overrightarrow{\operatorname{rot}}\,\vec{\alpha}_i\right]^{\mathrm{T}} + \frac{1}{3}\vec{e}_i \wedge \overrightarrow{\operatorname{grad}}\,\tau = \overrightarrow{\operatorname{grad}}\,\omega_i + \left[\overrightarrow{\operatorname{rot}}\,\vec{\beta}_i\right]^{\mathrm{T}} - \vec{e}_i \operatorname{div}\vec{\omega} \qquad (3.3)$$

Relations *(3.1)* and *(3.3)* allow us to find the expression for the symmetrical part of the rotational of the symmetric tensor $\vec{\varepsilon}_i$, and as we can see it only depends on the transverse symmetric tensor $\vec{\alpha}_i$

$$\left[\overrightarrow{\operatorname{rot}}\vec{\varepsilon}_i\right]^S = \left[\overrightarrow{\operatorname{rot}}\vec{\alpha}_i\right]^S = \frac{1}{2}\left(\overrightarrow{\operatorname{grad}}\,\omega_i + (\vec{e}_i\vec{\nabla})\vec{\omega}\right) + \left[\overrightarrow{\operatorname{rot}}\vec{\beta}_i\right]^S - \vec{e}_i \operatorname{div}\vec{\omega} \tag{3.4}$$

The anti-symmetric part of the rotational of a tensor

From the previous relations it is also possible to find direct relations between the anti-symmetric parts of the rotational of a tensor

$$\sum_k \vec{e}_k \wedge \overrightarrow{\operatorname{rot}}\,\vec{\varepsilon}_k = \sum_k \vec{e}_k \wedge \overrightarrow{\operatorname{rot}}\,\vec{\alpha}_k + \frac{2}{3}\overrightarrow{\operatorname{grad}}\,\tau = \sum_k \vec{e}_k \wedge \overrightarrow{\operatorname{rot}}\,\vec{\beta}_k + \overrightarrow{\operatorname{rot}}\,\vec{\omega} \tag{3.5}$$

The trace of the rotational of a tensor and the divergence of its anti-symmetric part

The computation of the divergences of the following expressions, represent respectively the anti-symmetric parts of tensors β_{ij}, ε_{ij} et α_{ij}

$$\sum_k \vec{e}_k \wedge \vec{\beta}_k = -2\vec{\omega} \quad \text{and} \quad \sum_k \vec{e}_k \wedge \vec{\varepsilon}_k \equiv 0 \quad \text{and} \quad \sum_k \vec{e}_k \wedge \vec{\alpha}_k \equiv 0 \tag{3.6}$$

they allow us to find a set of relations regarding the traces of the rotational of these tensors.

$$\sum_k \vec{e}_k \overrightarrow{\operatorname{rot}}\,\vec{\beta}_k = \operatorname{div}\left(-\sum_k \vec{e}_k \wedge \vec{\beta}_k\right) = 2\operatorname{div}\vec{\omega} \tag{3.7}$$

And we can observe, among other things, that the symmetric nature of tensors $\vec{\varepsilon}_i$ et $\vec{\alpha}_i$ implies that the trace of their rotational is null

$$\begin{cases} \sum_k \vec{e}_k \overrightarrow{\operatorname{rot}}\,\vec{\varepsilon}_k = \operatorname{div}\left(-\sum_k \vec{e}_k \wedge \vec{\varepsilon}_k\right) \equiv 0 \\ \sum_k \vec{e}_k \overrightarrow{\operatorname{rot}}\,\vec{\alpha}_k = \operatorname{div}\left(-\sum_k \vec{e}_k \wedge \vec{\alpha}_k\right) \equiv 0 \end{cases} \tag{3.8}$$

The «divergence of a tensor»

We can find relations for the "*divergence*" of these tensors

$$\sum_k \vec{e}_k \operatorname{div}\vec{\varepsilon}_k = \sum_k \vec{e}_k \operatorname{div}\vec{\alpha}_k + \frac{1}{3}\overrightarrow{\operatorname{grad}}\,\tau = \sum_k \vec{e}_k \operatorname{div}\vec{\beta}_k + \overrightarrow{\operatorname{rot}}\,\vec{\omega} \tag{3.9}$$

and we can show that the divergence of a tensor is directly linked to the anti-symmetric part of its rotational

$$\begin{cases} \sum_k \vec{e}_k \operatorname{div}\vec{\beta}_k = -\sum_k \vec{e}_k \wedge \overrightarrow{\operatorname{rot}}\,\vec{\beta}_k + \sum_k \overrightarrow{\operatorname{rot}}\left(\vec{e}_k \wedge \vec{\beta}_k\right) + \sum_k \overrightarrow{\operatorname{grad}}\left(\vec{e}_k \vec{\beta}_k\right) \\ \qquad = -\sum_k \vec{e}_k \wedge \overrightarrow{\operatorname{rot}}\,\vec{\beta}_k - 2\,\overrightarrow{\operatorname{rot}}\,\vec{\omega} + \overrightarrow{\operatorname{grad}}\,\tau \\ \sum_k \vec{e}_k \operatorname{div}\vec{\varepsilon}_k = -\sum_k \vec{e}_k \wedge \overrightarrow{\operatorname{rot}}\,\vec{\varepsilon}_k + \overrightarrow{\operatorname{grad}}\,\tau \\ \sum_k \vec{e}_k \operatorname{div}\vec{\alpha}_k = -\sum_k \vec{e}_k \wedge \overrightarrow{\operatorname{rot}}\,\vec{\alpha}_k \end{cases} \tag{3.10}$$

The contraction operations on the indices

The spatial derivatives of a tensor of order 2, β_{ij}, which we can represent under a vectorial form, have been manipulated thanks to vectorial analysis. These vectorial manipulations of tensorial objects are in fact m*athematical operations of contraction of indices.* From the 27 component of a 3rd order tensor $A_{ijk} = \partial\beta_{ij} / \partial\xi_k$ we can form tensors of order 2, vectors and scalars, as can be seen from relations *(3.3), (3.5)* and *(3.7)* respectively.

It is possible to contract indices on tensors of order larger than 3 to form tensors of order 2, 1 (vectors) and 0 (scalar). For example the rotational in relation *(3.3)* is a tensor of order 2 (9 components) which results from the contraction of a tensor of order 4 (81 components), and which can be calculated, given that $\overrightarrow{\mathrm{rot}}\,\overrightarrow{\mathrm{grad}}\,\omega_i \equiv 0$

$$\overrightarrow{\mathrm{rot}}\left[\overrightarrow{\mathrm{rot}}\,\vec{\varepsilon}_i\right]^{\mathrm{T}} = \overrightarrow{\mathrm{rot}}\left\{\left[\overrightarrow{\mathrm{rot}}\,\vec{\beta}_i\right]^{\mathrm{T}} - \vec{e}_i\,\mathrm{div}\,\vec{\omega}\right\} = \overrightarrow{\mathrm{rot}}\left[\sum_k \vec{e}_k\left(\vec{e}_i\overrightarrow{\mathrm{rot}}\,\vec{\beta}_k\right) - \vec{e}_i\,\mathrm{div}\,\vec{\omega}\right] \tag{3.11}$$

3.2 - Links between the spatial derivatives in the case of geometro-compatibility

The condition of geometro-compatibility $\overrightarrow{\mathrm{rot}}\,\vec{\beta}_i = \overrightarrow{\mathrm{rot}}\,\vec{\beta}_i^{(\delta)} = 0$ implies the following relation

$$\overrightarrow{\mathrm{rot}}\,\vec{\beta}_i^{(\delta)} = \overrightarrow{\mathrm{rot}}\left(\vec{\alpha}_i + \vec{e}_i \wedge \vec{\omega}^{(\delta)} + \frac{1}{3}\tau\vec{e}_i\right) = \overrightarrow{\mathrm{rot}}\,\vec{\alpha}_i + \vec{e}_i\,\mathrm{div}\,\vec{\omega}^{(\delta)} - \left(\vec{e}_i\vec{\nabla}\right)\vec{\omega}^{(\delta)} + \frac{1}{3}\overrightarrow{\mathrm{grad}}\,\tau \wedge \vec{e}_i = 0 \tag{3.12}$$

With $\mathrm{div}\,\vec{\omega}^{(\delta)} = 0$, $\left(\vec{e}_i\vec{\nabla}\right)\vec{\omega}^{(\delta)} = \overrightarrow{\mathrm{grad}}\,\omega_i^{(\delta)} - \vec{e}_i \wedge \overrightarrow{\mathrm{rot}}\,\vec{\omega}^{(\delta)}$ and $\vec{\omega}^{(\delta)} = \vec{\omega} - \vec{\omega}_0(t)$, this relation leads directly to the following relations between the spatial derivatives of the various topological fields

$$\left\{\begin{array}{l} \overrightarrow{\mathrm{rot}}\,\vec{\varepsilon}_i = \overrightarrow{\mathrm{rot}}\,\vec{\alpha}_i - \frac{1}{3}\vec{e}_i \wedge \overrightarrow{\mathrm{grad}}\,\tau = \overrightarrow{\mathrm{grad}}\,\omega_i^{(\delta)} - \vec{e}_i \wedge \overrightarrow{\mathrm{rot}}\,\vec{\omega}^{(\delta)} = \overrightarrow{\mathrm{grad}}\,\omega_i - \vec{e}_i \wedge \overrightarrow{\mathrm{rot}}\,\vec{\omega} \\ -\sum_k \vec{e}_k \wedge \overrightarrow{\mathrm{rot}}\,\vec{\varepsilon}_k = \sum_k \vec{e}_k\,\mathrm{div}\,\vec{\alpha}_k - \frac{2}{3}\overrightarrow{\mathrm{grad}}\,\tau = -\overrightarrow{\mathrm{rot}}\,\vec{\omega}^{(\delta)} = -\overrightarrow{\mathrm{rot}}\,\vec{\omega} \\ \mathrm{div}\,\vec{\omega}^{(\delta)} = \mathrm{div}\,\vec{\omega} = 0 \end{array}\right. \tag{3.13}$$

We will see that these relations are very useful for the description of the evolution of geometro-compatible solid lattices. If, additionally, such a solid has a volume expansion τ independent of the space coordinates, the *spatial derivatives of the rotation vector $\vec{\omega}$ and of the local rotation vector $\vec{\omega}^{(\delta)}$ are only dependent on the spatial derivatives of the shear tensor $\vec{\alpha}_i$*

$$\tau = \tau(t) \neq \tau(\vec{r},t) \;\Rightarrow\; \left\{\begin{array}{l} \overrightarrow{\mathrm{rot}}\,\vec{\alpha}_i = \overrightarrow{\mathrm{grad}}\,\omega_i^{(\delta)} - \vec{e}_i \wedge \overrightarrow{\mathrm{rot}}\,\vec{\omega}^{(\delta)} = \overrightarrow{\mathrm{grad}}\,\omega_i - \vec{e}_i \wedge \overrightarrow{\mathrm{rot}}\,\vec{\omega} \\ \sum_k \vec{e}_k\,\mathrm{div}\,\vec{\alpha}_k = -\overrightarrow{\mathrm{rot}}\,\vec{\omega}^{(\delta)} = -\overrightarrow{\mathrm{rot}}\,\vec{\omega} \\ \mathrm{div}\,\vec{\omega}^{(\delta)} = \mathrm{div}\,\vec{\omega} = 0 \end{array}\right. \tag{3.14}$$

3.3 - Contortion tensors of a solid

The contortion tensor of a solid

Relation *(3.3)* allows us, at least in a compatible lattice, to link directly $\left[\overrightarrow{\mathrm{rot}}\,\vec{\varepsilon}_i\right]^{\mathrm{T}}$ to the gradient of the component of the rotation vector $\vec{\omega}$

$$\overrightarrow{\mathrm{rot}}\,\vec{\beta}_i = 0 \quad \Rightarrow \quad \left[\overrightarrow{\mathrm{rot}}\,\vec{\varepsilon}_i\right]^{\mathrm{T}} = \overrightarrow{\mathrm{grad}}\,\omega_i \tag{3.15}$$

this means that it is possible, in a compatible solid lattice, to find the local variations of the rotational field directly from the field of deformations. To that end, we define a new tensor, *the tensor of contortions* $\vec{\chi}_i$ in the following way:

$$\vec{\chi}_i = \left[\overrightarrow{\mathrm{rot}}\,\vec{\varepsilon}_i\right]^{\mathrm{T}} = \overrightarrow{\mathrm{grad}}\,\omega_i + \left[\overrightarrow{\mathrm{rot}}\,\vec{\beta}_i\right]^{\mathrm{T}} - \vec{e}_i\,\mathrm{div}\,\vec{\omega} \tag{3.16}$$

In a compatible medium, the physical significance of this tensor can be deduced directly from $\overrightarrow{\mathrm{grad}}\,\omega_i$. It is indeed relevant to measure the spatial variations of the rotational vector $\vec{\omega}$, since $\vec{\omega}$ is an invariant of the solid medium, meaning independent of the choice of frame used to calculate it. In a compatible medium, relation *(3.3)* takes the following form

$$\overrightarrow{\mathrm{rot}}\,\vec{\beta}_i = 0 \quad \Rightarrow \quad \vec{\chi}_i = \left[\overrightarrow{\mathrm{rot}}\,\vec{\varepsilon}_i\right]^{\mathrm{T}} = \overrightarrow{\mathrm{grad}}\,\omega_i = \left(\vec{e}_i\vec{\nabla}\right)\vec{\omega} + \vec{e}_i \wedge \overrightarrow{\mathrm{rot}}\,\vec{\omega} \tag{3.17}$$

The components of this tensors are linked to the spatial derivatives of the rotational vector $\vec{\omega}$, as these measure how rotation changes in different points of the solid, such as due to *torsions* or *flexions*. We can define a little more precisely and visually the components of the tensors by using 2 typical examples of spatial rotations $\vec{\omega}$.

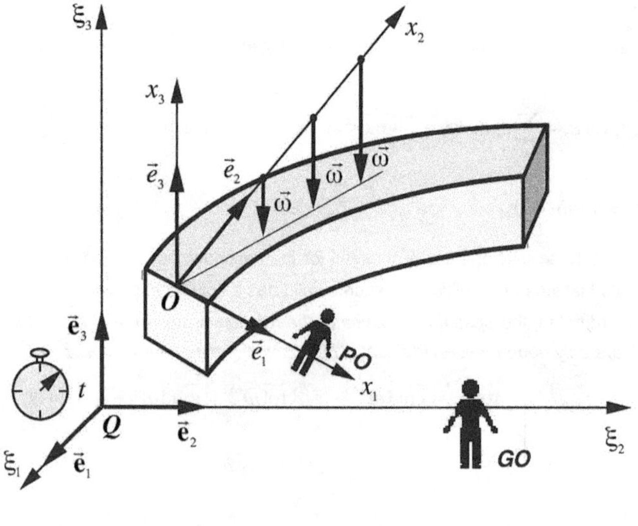

Figure 3.1 - *flexion of a solid medium*

In the first example, a flexed medium shows a rotational vector parallel to Ox_3 (pointing down in figure 3.1), and its amplitude increases along Ox_2 , as shown in figure 3.1. In this case, there exists a *non-diagonal component* of the $\vec{\chi}_i$ tensor, which is not null

$$\chi_{23} = \frac{\partial \omega_3}{\partial x_2} \neq 0 \tag{3.18}$$

This component is associated with the *flexion* of the solid.

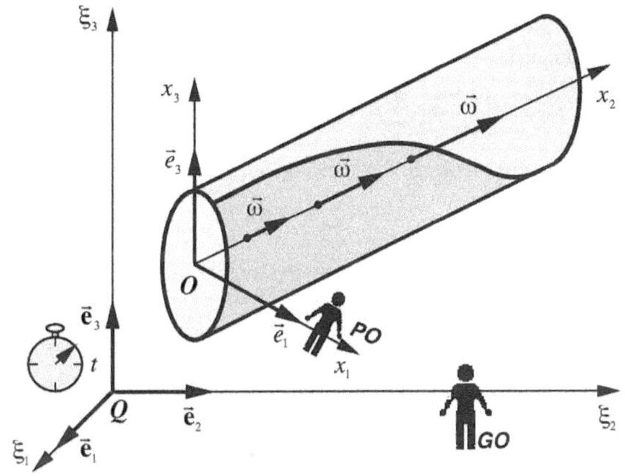

***Figure 3.2** - torsion of a solid medium*

In the second example of figure 3.2, we represent a *torsion* of the solid medium. The rotational vector is along the axis Ox_2 and increases in amplitude along Ox_2 as well. As a result, there is a *diagonal component* of tensor $\vec{\chi}_i$, which is non-null, a fact we translate as

$$\chi_{22} = \frac{\partial \omega_2}{\partial x_2} \neq 0 \tag{3.19}$$

From these two examples, it is clear that the components of the tensor $\vec{\chi}_i$ measure the *flexions* and *torsions* of the solid medium, via non-diagonal and diagonal components respectively. We will call those components the *"contortions" of the medium.* The rotational part of this new *contortion tensor* $\vec{\chi}_i$ satisfies an interesting relationship.

$$\overrightarrow{\mathrm{rot}}\,\vec{\chi}_i = \overrightarrow{\mathrm{rot}}\left[\overrightarrow{\mathrm{rot}}\,\vec{\varepsilon}_i\right]^{\mathrm{T}} = \overrightarrow{\mathrm{rot}}\left\{\left[\overrightarrow{\mathrm{rot}}\,\vec{\beta}_i\right]^{\mathrm{T}} - \vec{e}_i \operatorname{div}\vec{\omega}\right\} \tag{3.20}$$

Indeed, in a compatible medium, the contortion tensor $\vec{\chi}_i$ satisfies a compatibility relation that is equivalent to the compatibility relation of the deformation tensor $\vec{\varepsilon}_i$

$$\overrightarrow{\mathrm{rot}}\,\vec{\beta}_i = 0 \quad \Rightarrow \quad \overrightarrow{\mathrm{rot}}\,\vec{\chi}_i = \overrightarrow{\mathrm{rot}}\left[\overrightarrow{\mathrm{rot}}\,\vec{\varepsilon}_i\right]^{\mathrm{T}} = 0 \tag{3.21}$$

These relations are known as the *equations of De Saint-Venant*. Since the tensor $\vec{\varepsilon}_i$ is symmetrical, these encompass 6 non-trivial relations between the partial derivatives of order 2 of the ε_{ij} .

It should be noted that the trace of $\vec{\chi}_i$ is identically null, meaning that $\vec{\chi}_i$ is a *transverse tensor.* This is due to the simple fact that, as seen in relation *(1.70),* the $\vec{\varepsilon}_i$ tensor is symmetrical

$$\chi = \sum_k \vec{e}_k \vec{\chi}_k = \sum_k \vec{e}_k \left[\overrightarrow{\mathrm{rot}}\, \vec{\varepsilon}_k \right]^{\mathrm{T}} = \sum_k \vec{e}_k \overrightarrow{\mathrm{rot}}\, \vec{\varepsilon}_k \equiv 0 \tag{3.22}$$

The flexion vector of a solid

From the contortion tensor $\vec{\chi}_i$, it is possible to build an invariant vector $\vec{\chi}$ which depends on the anti-symmetric part of the $\vec{\chi}_i$ with the following construct

$$\vec{\chi} = \sum_k \vec{e}_k \wedge \vec{\chi}_k \tag{3.23}$$

Since the anti-symmetric part of a tensor is, by definition, equal to the anti-symmetric part of its transpose, but with an opposite signe, we can deduct the following

$$\vec{\chi} = \sum_k \vec{e}_k \wedge \vec{\chi}_k = \sum_k \vec{e}_k \wedge \left[\overrightarrow{\mathrm{rot}}\, \vec{\varepsilon}_k \right]^{\mathrm{T}} = -\sum_k \vec{e}_k \wedge \overrightarrow{\mathrm{rot}}\, \vec{\varepsilon}_k \tag{3.24}$$

We then use relation *(3.5)*, and we obtain the following relation on the $\vec{\chi}$ vector

$$\vec{\chi} = \sum_k \vec{e}_k \wedge \vec{\chi}_k = -\sum_k \vec{e}_k \wedge \overrightarrow{\mathrm{rot}}\, \vec{\varepsilon}_k = -\overrightarrow{\mathrm{rot}}\, \vec{\omega} - \sum_k \vec{e}_k \wedge \overrightarrow{\mathrm{rot}}\, \vec{\beta}_k \tag{3.25}$$

Thus, in a compatible medium, $\vec{\chi}$ only depends on the rotation vector $\vec{\omega}$

$$\overrightarrow{\mathrm{rot}}\, \vec{\beta}_i = 0 \quad \Rightarrow \quad \vec{\chi} = -\overrightarrow{\mathrm{rot}}\, \vec{\omega} \tag{3.26}$$

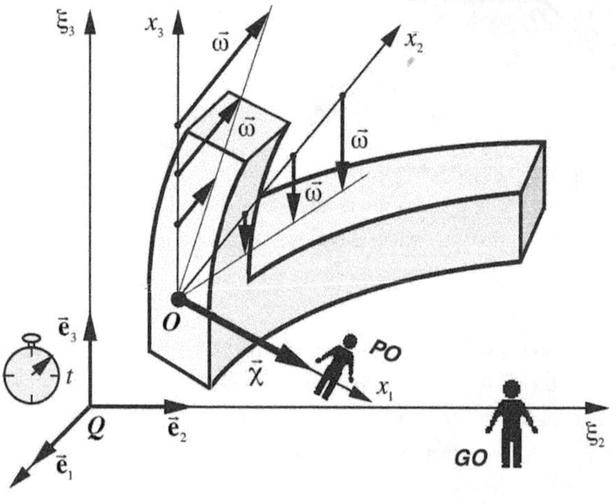

Figure 3.3 *- definition of the flexion vector* $\vec{\chi}$

Since $\vec{\chi}$ is defined as a rotational, it only contains components of the type $\partial\omega_i / \partial x_j$, with $i \neq j$, by definition of a rotational. Thus, it only depends on the non diagonal components of the $\vec{\chi}_i$ tensors, which as we have just seen, are the *flexion components* of the medium. We can verify this assertion on the example of the deformation shown on figure 3.3, in which $\vec{\omega}$ has components along the axis Ox_2 et Ox_3, that grow in the directions Ox_3, and

$$\vec{\omega} = \omega_3(x_2)\vec{e}_3 + \omega_2(x_3)\vec{e}_2 \tag{3.27}$$

which gives us the following tensor $\vec{\chi}_i$

$$\vec{\chi}_1 = 0 \qquad \vec{\chi}_2 = \frac{\partial \omega_2}{\partial x_3}\vec{e}_3 \qquad \vec{\chi}_3 = \frac{\partial \omega_3}{\partial x_2}\vec{e}_2 \tag{3.28}$$

from which we deduct the vector $\vec{\chi}$, with the help of relation *(3.10)*

$$\vec{\chi} = \vec{e}_2 \wedge \frac{\partial \omega_2}{\partial x_3}\vec{e}_3 + \vec{e}_3 \wedge \frac{\partial \omega_3}{\partial x_2}\vec{e}_2 = \left(\frac{\partial \omega_2}{\partial x_3} - \frac{\partial \omega_3}{\partial x_2}\right)\vec{e}_1 \tag{3.29}$$

The $\vec{\chi}$ vector is perpendicular to the general curvature of the medium due to flexion. It is aligned with the direction of the 'radius of curvature'. It's magnitude is inversely proportional to the average radius of curvature in the directions Ox_3 et Ox_2. We will therefore call it the *flexion vector* $\vec{\chi}$ of our solid medium. The $\vec{\chi}$ vector satisfies an interesting relation concerning its divergence. Since the divergence of a rotational is null, we have, following *(3.25)*

$$\begin{aligned} &\operatorname{div}\vec{\chi} = \operatorname{div}\left(\sum_k \vec{e}_k \wedge \vec{\chi}_k\right) = -\operatorname{div}\left(\sum_k \vec{e}_k \wedge \overrightarrow{\operatorname{rot}}\,\vec{\varepsilon}_k\right) = -\operatorname{div}\left(\sum_k \vec{e}_k \wedge \overrightarrow{\operatorname{rot}}\,\vec{\beta}_k\right) \\ &\Rightarrow \quad \operatorname{div}\vec{\chi} = -\sum_k \vec{e}_k \overrightarrow{\operatorname{rot}}\,\vec{\chi}_k = \sum_k \vec{e}_k \overrightarrow{\operatorname{rot}}\,\overrightarrow{\operatorname{rot}}\,\vec{\varepsilon}_k = \sum_k \vec{e}_k \overrightarrow{\operatorname{rot}}\,\overrightarrow{\operatorname{rot}}\,\vec{\beta}_k \end{aligned} \tag{3.30}$$

This relation shows that $\operatorname{div}\vec{\chi}$ is in fact equal to the *trace* of $\overrightarrow{\operatorname{rot}}\,\vec{\chi}_k$ with a changed sign.
In a compatible medium, this relation shows that the flexion vector $\vec{\chi}$ also obeys a compatibility relation which assures that the field of flexion is non divergent:

$$\overrightarrow{\operatorname{rot}}\,\vec{\beta}_i = 0 \quad \Rightarrow \quad \operatorname{div}\vec{\chi} = 0 \tag{3.31}$$

The torsion tensor of a solid: the transverse symmetrical part of the contortion tensor

Since the contortion tensor $\vec{\chi}_i$ is a transverse tensor (of null trace), its symmetrical part $\left[\vec{\chi}_i\right]^{\mathrm{S}}$ is a transverse symmetrical tensor, which cannot be decomposed further. We compute it from the following relations

$$\left[\vec{\chi}_i\right]^{\mathrm{S}} = \vec{\chi}_i + \frac{1}{2}\vec{e}_i \wedge \left(\sum_k \vec{e}_k \wedge \vec{\chi}_k\right) = \vec{\chi}_i + \frac{1}{2}\vec{e}_i \wedge \vec{\chi} \tag{3.32}$$

This transverse symmetrical tensor satisfies the following relations, which are easy to derive using the transpose operation $2\left[\vec{\chi}_i\right]^{\mathrm{S}} = \vec{\chi}_i + \left[\vec{\chi}_i\right]^{\mathrm{T}}$

$$\left[\vec{\chi}_i\right]^{\mathrm{S}} = \left[\overrightarrow{\operatorname{rot}}\,\vec{\varepsilon}_i\right]^{\mathrm{S}} = \left[\overrightarrow{\operatorname{rot}}\,\vec{\alpha}_i\right]^{\mathrm{S}} = \overrightarrow{\operatorname{grad}}\,\omega_i - \frac{1}{2}\vec{e}_i \wedge \overrightarrow{\operatorname{rot}}\,\vec{\omega} + \left[\overrightarrow{\operatorname{rot}}\,\vec{\beta}_i\right]^{\mathrm{S}} - \vec{e}_i \operatorname{div}\vec{\omega} \tag{3.33}$$

The transverse symmetrical part $\left[\vec{\chi}_i\right]^{\mathrm{S}}$ of the contortion tensor therefore represents the torsions of the solid, since it is obtained from the contortion tensor from which we have subtracted the anti-symmetrical part representing the flexions. We thus call $\left[\vec{\chi}_i\right]^{\mathrm{S}}$ the *tensor of torsion* of the solid.
The complete description of the distortions and contortions of a geometro-compatible solid lattice exhibits here a symmetry perfectly described visually in figure 3.3. We also give table 3.1 with the relation which permits a complete topological description of the contortions and the compatibility of a solid lattice. In this table you will find the terms which correspond to the rotation of the distortion tensor and the divergence of the rotation vector, which are by definition null in a compatible medium.
It should be noted here that the transverse symmetrical part of $\vec{\chi}_i$ does not depend directly on the volume expansion τ, but exclusively on the shear tensor $\vec{\alpha}_i$. This means that the *torsions*

$[\vec{\chi}_i]^S$ *of a solid are intimately linked to the pure shear of the solid.* Another important remark can be made about the topological contortions of a medium: since the gradient of the scalar of volume expansion τ does not figure in the anti-symmetric part of the contortion tensor $\vec{\chi}_i$, and furthermore, since this anti-symmetric part is represented by the flexion vector $\vec{\chi}$, it is rather evident that *flexions* $\vec{\chi}$ *of a solid are linked to the volume expansion* τ *of said medium.*

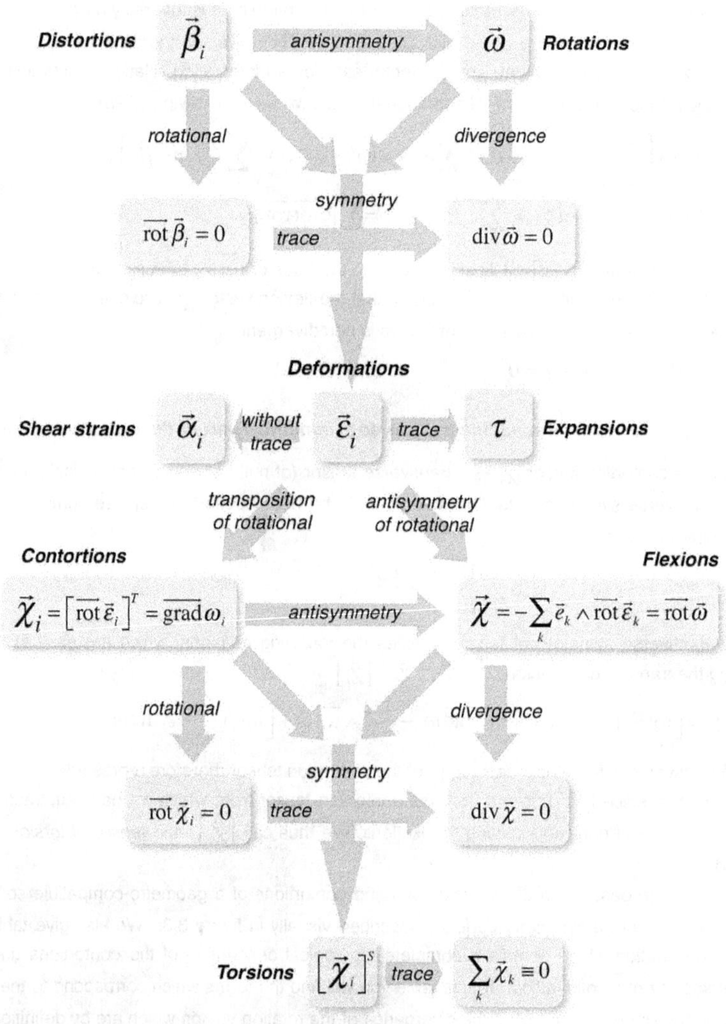

***Figure 3.4** - complete description of the distortions and contortions of a compatible solid lattice*

Table 3.1 - The contortions and compatibility of a solid lattice

Transverse tensor of contortion

$$\vec{\chi}_i = \left[\overrightarrow{\mathrm{rot}}\,\vec{\varepsilon}_i\right]^{\mathrm{T}} = \left[\overrightarrow{\mathrm{rot}}\,\vec{\alpha}_i\right]^{\mathrm{T}} + \frac{1}{3}\vec{e}_i \wedge \overrightarrow{\mathrm{grad}}\,\tau = \overrightarrow{\mathrm{grad}}\,\omega_i \underbrace{\boxed{+\left[\overrightarrow{\mathrm{rot}}\,\vec{\beta}_i\right]^{\mathrm{T}} - \vec{e}_i\,\mathrm{div}\,\vec{\omega}}}_{=0} \quad (1)$$

Flexion vector

$$\vec{\chi} = \sum_k \vec{e}_k \wedge \vec{\chi}_k = -\sum_k \vec{e}_k \wedge \overrightarrow{\mathrm{rot}}\,\vec{\varepsilon}_k = -\sum_k \vec{e}_k \wedge \overrightarrow{\mathrm{rot}}\,\vec{\alpha}_k - \frac{2}{3}\overrightarrow{\mathrm{grad}}\,\tau$$

$$= \sum_k \vec{e}_k\,\mathrm{div}\,\vec{\alpha}_k - \frac{2}{3}\overrightarrow{\mathrm{grad}}\,\tau = -\overrightarrow{\mathrm{rot}}\,\vec{\omega} \underbrace{\boxed{-\sum_k \vec{e}_k \wedge \overrightarrow{\mathrm{rot}}\,\vec{\beta}_k}}_{=0} \quad (2)$$

Null trace of the contortion tensor

$$\chi = \sum_k \vec{e}_k\vec{\chi}_k = \sum_k \vec{e}_k\left[\overrightarrow{\mathrm{rot}}\,\vec{\varepsilon}_k\right]^{\mathrm{T}} = \sum_k \vec{e}_k\overrightarrow{\mathrm{rot}}\,\vec{\varepsilon}_k \equiv 0 \quad (3)$$

Transverse symmetrical part of torsion

$$\left[\vec{\chi}_i\right]^{\mathrm{S}} = \vec{\chi}_i + \frac{1}{2}\vec{e}_i \wedge \vec{\chi} = \left[\overrightarrow{\mathrm{rot}}\,\vec{\varepsilon}_i\right]^{\mathrm{S}} = \left[\overrightarrow{\mathrm{rot}}\,\vec{\alpha}_i\right]^{\mathrm{S}}$$

$$= \overrightarrow{\mathrm{grad}}\,\omega_i - \frac{1}{2}\vec{e}_i \wedge \overrightarrow{\mathrm{rot}}\,\vec{\omega} + \underbrace{\boxed{+\left[\overrightarrow{\mathrm{rot}}\,\vec{\beta}_i\right]^{\mathrm{S}} - \vec{e}_i\,\mathrm{div}\,\vec{\omega}}}_{=0} \quad (4)$$

Geometro-compatibility of distortions and rotations

$$\vec{\beta}_i = \overrightarrow{\mathrm{grad}}\,u_i \quad (5) \quad \Rightarrow \quad \boxed{\overrightarrow{\mathrm{rot}}\,\vec{\beta}_i} = 0 \quad (6)$$

$$\vec{\omega} = \frac{1}{2}\overrightarrow{\mathrm{rot}}\,\vec{u} \quad (7) \quad \Rightarrow \quad \mathrm{div}\,\vec{\omega} = \boxed{\frac{1}{2}\sum_k \vec{e}_k\overrightarrow{\mathrm{rot}}\,\vec{\beta}_k} = 0 \quad (8)$$

Geometro-compatibility of contortions and flexions

$$\overrightarrow{\mathrm{rot}}\,\vec{\chi}_i = \overrightarrow{\mathrm{rot}}\left[\overrightarrow{\mathrm{rot}}\,\vec{\varepsilon}_i\right]^{\mathrm{T}} = \overrightarrow{\mathrm{rot}}\left[\overrightarrow{\mathrm{rot}}\,\vec{\alpha}_i\right]^{\mathrm{T}} - \frac{1}{3}\overrightarrow{\mathrm{rot}}\,\overrightarrow{\mathrm{rot}}\left(\vec{e}_i\tau\right)$$

$$= \boxed{\overrightarrow{\mathrm{rot}}\left\{\left[\overrightarrow{\mathrm{rot}}\,\vec{\beta}_i\right]^{\mathrm{T}} - \vec{e}_i\,\mathrm{div}\,\vec{\omega}\right\}} = 0 \quad (9)$$

$$\mathrm{div}\,\vec{\chi} = -\sum_k \vec{e}_k\overrightarrow{\mathrm{rot}}\,\vec{\chi}_k = \sum_k \vec{e}_k\overrightarrow{\mathrm{rot}}\,\overrightarrow{\mathrm{rot}}\,\vec{\varepsilon}_k = \sum_k \vec{e}_k\overrightarrow{\mathrm{rot}}\,\overrightarrow{\mathrm{rot}}\,\vec{\alpha}_k - \frac{2}{3}\Delta\tau$$

$$= \boxed{\sum_k \vec{e}_k\overrightarrow{\mathrm{rot}}\,\overrightarrow{\mathrm{rot}}\,\vec{\beta}_k} = \boxed{-\mathrm{div}\left(\sum_k \vec{e}_k \wedge \overrightarrow{\mathrm{rot}}\,\vec{\beta}_k\right)} = 0 \quad (10)$$

3.4 - Physical interpretation of the equations of compatibility

The diverse equations of compatibility have a physical meaning we will now explore in this section.

The absence of dislocations and the continuity of the field of displacement.

The compatibility condition $\overrightarrow{\mathrm{rot}}\,\vec{\beta}_i = 0$ for the distortion tensor $\vec{\beta}_i$ implies that the field has nice smoothness properties, in the mathematical sense. To show this, let's calculate $d\vec{u}$ over a closed contour C anywhere inside the lattice

$$\oint_C d\vec{u} = \sum_k \vec{e}_k \oint_C du_k = \sum_k \vec{e}_k \oint_C \overrightarrow{\mathrm{grad}}\, u_k d\vec{r} = -\sum_k \vec{e}_k \oint_C \vec{\beta}_k d\vec{r} \tag{3.34}$$

By using the theorem of the rotational, an integral form of compatibility is obtained over any surface S whose boundary is C

$$\oint_C d\vec{u} = -\sum_k \vec{e}_k \iint_S \overrightarrow{\mathrm{rot}}\vec{\beta}_k d\vec{S} = 0 \quad ; \quad \forall C \tag{3.35}$$

Let's consider the displacement $\vec{u}$ associated with a closed contour C inside the lattice, such as that represented in figure 3.5. For this closed contour from A to B, we have

$$\oint_C d\vec{u} = \vec{u}_B - \vec{u}_A = \vec{B} = 0 \tag{3.36}$$

And thus, if the medium presents a distortion field which satisfies the compatibility condition $\overrightarrow{\mathrm{rot}}\,\vec{\beta}_i = 0$, the closing vector $\vec{B}$, called a *Burgers vector* [1], of contour C' generated by the sum of the $\vec{u}$ vectors is null. This effectively signifies, from a topological point of view, that there are *no discontinuities of displacement, called dislocations* [2] *inside the medium.*

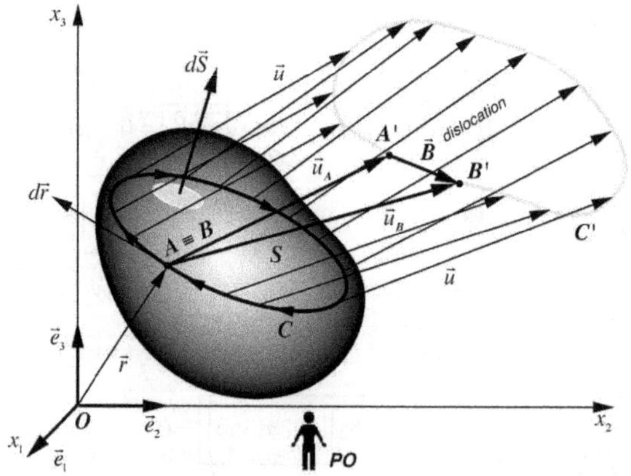

***Figure 3.5** - Interpretation of the condition of compatibility* $\overrightarrow{\mathrm{rot}}\,\vec{\beta}_i = 0$

[1] *J. M. Burgers, Proc. Kon. Ned. Akad. Weten, vol.42, p. 293, 378, 1939*

[2] *A. E. H. Love, «A Treatise on Mathematical Theory of Elasticity», 3rd edition, Cambridge U.P., 1920*

The existence of a displacement field $\vec{u}$ without any discontinuity allows us to say the following:
- The *topology is connected.* This means, from a physical standpoint, that there are no stacking of distortions resulting in a tear and slide of a part,
- The *topology is compact.* This means, from a physical standpoint, that there were no formation of holes inside the medium. In short, the condition $\overrightarrow{\mathrm{rot}}\,\vec{\beta}_i = 0$ *assures the solidity and continuity of the medium.*

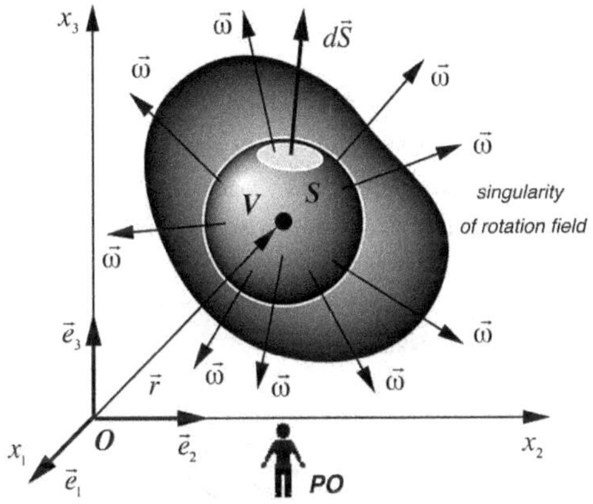

Figure 3.6 *- interpretation of the condition of compatibility* $\mathrm{div}\,\vec{\omega} = 0$

The absence of divergence singularities of the rotation field

To find the significance of the condition of compatibility $\mathrm{div}\,\vec{\omega} = 0$ for the vector of rotation $\vec{\omega}$, the flux of the vector of rotation $\vec{\omega}$, meaning the quantity $\vec{\omega} d\vec{S} = \vec{\omega}\vec{n} dS = \omega_\perp dS$ defined by the scalar product of $\vec{\omega}$ with the element of surface $d\vec{S}$, is integrated on a closed boundary surface S of a volume V of the solid (figure 3.6).
By applying the theorem of divergence we find the integral relations for a medium satisfying the compatibility condition associated with the rotation vector

$$\oiint_S \omega_\perp dS = \oiint_S \vec{\omega} d\vec{S} = \iiint_V \mathrm{div}\,\vec{\omega} dV = 0 \quad ; \quad \forall S \qquad (3.37)$$

Thus, the condition of compatibility for the vector of rotation $\vec{\omega}$, which stipulates that the field of rotations is non-divergent, implies that *there are no rotational singularities of the field of rotation* $\vec{\omega}$ in the solid, so no divergent torsions as seen in figure 3.6.

The absence of disclinations and the continuity of the rotation field by deformation

The condition of compatibility $\overrightarrow{\mathrm{rot}}\,\vec{\chi}_i = 0$ of the tensor of contortion $\vec{\chi}_i$ also presents an interesting interpretation. If $d\vec{\omega}$ is integrated on a closed contour C in a compatible medium, it follows

$$\oint_C d\vec{\omega}^{(\varepsilon)} = \sum_k \vec{e}_k \oint_C d\omega_k^{(\varepsilon)} = \sum_k \vec{e}_k \oint_C \overrightarrow{\text{grad}}\,\omega_k^{(\varepsilon)} d\vec{r} = \sum_k \vec{e}_k \oint_C \vec{\chi}_k d\vec{r} \tag{3.38}$$

In this relation we have introduced a rotational vector $\vec{\omega}^{(\varepsilon)}$ which represents the *rotations directly deduced from the tensor of deformations* $\vec{\varepsilon}_i$, which is identified with the local vector of rotation $\vec{\omega}$ in compatible medium. It follows that

$$\oint_C d\vec{\omega}^{(\varepsilon)} = \sum_k \vec{e}_k \iint_S \overrightarrow{\text{rot}}\vec{\chi}_k\, d\vec{S} = \sum_k \vec{e}_k \iint_S \overrightarrow{\text{rot}}\left[\overrightarrow{\text{rot}}\,\vec{\varepsilon}_k\right]^{\text{T}} d\vec{S} = 0 \quad ; \quad \forall C \tag{3.39}$$

By representing the field of rotation $\vec{\omega}^{(\varepsilon)}$ on a closed contour C in the medium (figure 3.7), we have that, from A to B

$$\oint_C d\vec{\omega}^{(\varepsilon)} = \vec{\omega}_B^{(\varepsilon)} - \vec{\omega}_A^{(\varepsilon)} = \vec{\Omega} = 0 \tag{3.40}$$

In this relation, the condition of compatibility $\overrightarrow{\text{rot}}\,\vec{\chi}_i = 0$ implies that the closing vector $\vec{\Omega}$, called the *Frank vector* [3], is null, which signifies from a topological standpoint that there are no *rotational discontinuities by deformation,* these are called disclinations [4]. Since the compatibility condition for the contortion tensor $\vec{\chi}_i$ is also the compatibility condition for the deformation tensor $\vec{\varepsilon}_i$, we can conclude that the equations of *de Saint-Venant (3.21)* for $\vec{\varepsilon}_i$ are equivalent to the continuity of rotations by deformation.

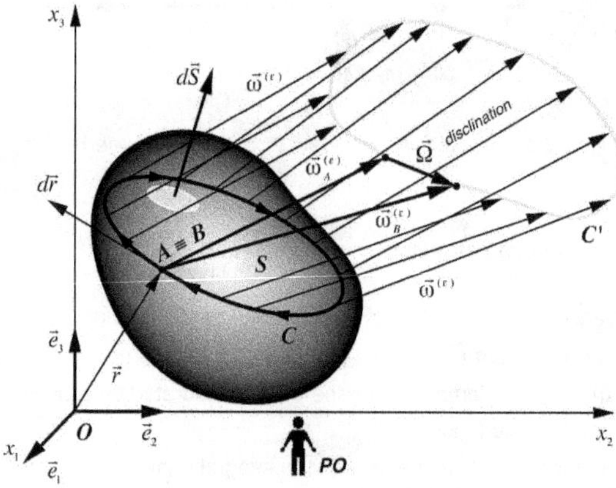

Figure 3.7 *- interpretation of the condition of compatibility* $\overrightarrow{\text{rot}}\,\vec{\chi}_i = 0$

On the absence of divergent singularities within the flexion field

The condition of compatibility $\text{div}\,\vec{\chi} = 0$ of the vector of flexion $\vec{\chi}$ can be interpreted in the following way. By integration of the flux of $\vec{\chi}$ over a closed boundary surface S around a given

---

[3] *R. DeWit, «Theory of disclinations II, III and IV», J. of the Nat. Bureau of Standards A, vol. 77A, p.49-100, p. 359-368, and p. 607-658, 1973*

[4] *F. C. Frank, Disc. Faraday Soc., vol. 25, p. 19, 1958*

volume V of the medium (figure 3.8), and by using the theorem of divergence and the definition of $\vec{\chi}$ from the deformation tensor $\vec{\varepsilon}_i$, it follows that

$$\oiint_S \chi_\perp dS = \oiint_S \vec{\chi} d\vec{S} = \iiint_V \operatorname{div} \vec{\chi} dV = \sum_k \vec{e}_k \iiint_V \overrightarrow{\text{rot}}\,\overrightarrow{\text{rot}}\ \vec{\varepsilon}_k dV = 0 \quad ; \quad \forall S \qquad (3.41)$$

Therefore the compatibility condition on $\vec{\chi}$ implies that the field of flexion is non-divergent, and that there exists no singularity of the field of flexion $\vec{\chi}$ in the medium, such as the one represented on figure 3.8.

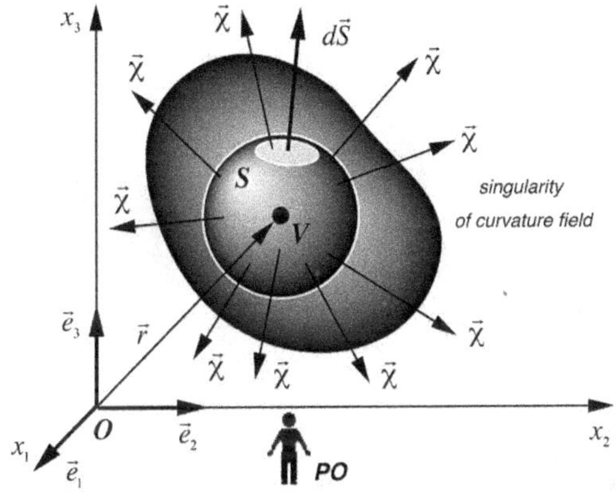

Figure 3.8 *- interpretation of compatibility condition* $\operatorname{div}\vec{\chi} = 0$

One should note that the condition of compatibility for the flexion tensor $\vec{\chi}$ corresponds indeed to the condition of *compatibility for the anti-symmetric part of tensor* $\overrightarrow{\text{rot}}\,\vec{\varepsilon}_i$ since

$$\operatorname{div}\vec{\chi} = -\operatorname{div}\left(\sum_k \vec{e}_k \wedge \overrightarrow{\text{rot}}\,\vec{\varepsilon}_k\right) = \sum_k \vec{e}_k \overrightarrow{\text{rot}}\,\overrightarrow{\text{rot}}\ \vec{\varepsilon}_k \qquad (3.42)$$

3.5 - Pass-through conditions for a compatible interface

The *pass-through conditions* for the topological tensors across a compatible interface between 2 mediums, meaning an interface for which the fields of displacement on each side of the surface are the same, can be deducted from the conditions of compatibility. All one has to do is write down the integration of the condition of compatibility on a really slim volume, which moves with the interface, and where the 2 faces are on each side of the interface (figure 3.9).

The condition of compatibility for the distortion tensor $\vec{\beta}_i$ implies the following relations on a mobile volume V

$$\iiint_V \overrightarrow{\text{rot}}\,\vec{\beta}_i\, dV = \oiint_S d\vec{S} \wedge \vec{\beta}_i = \iint_{S_1} \left(\vec{n} \wedge \vec{\beta}_i^{(1)}\right) dS - \iint_{S_2} \left(\vec{n} \wedge \vec{\beta}_i^{(2)}\right) dS = 0 \qquad (3.43)$$

This means that the tangential components of the $\vec{\beta}_i$ are the same on each side of the interface.

$$\vec{n} \wedge \vec{\beta}_i^{(1)} = \vec{n} \wedge \vec{\beta}_i^{(2)} \qquad (3.44)$$

The topological interpretation of this condition of passage is simple. Indeed, remember that, in a compatible medium, the tensor $\vec{\beta}_i$ derives from the gradient of a field of displacement $\vec{u}$, by the relation $\vec{\beta}_i = \overrightarrow{\mathrm{grad}}\, u_i$, it is possible to re-write the integrals on surfaces S_1 and S_2 under the form

$$\iint_{S_1} \left(\vec{n} \wedge \overrightarrow{\mathrm{grad}}\, u_i^{(1)}\right) dS - \iint_{S_2} \left(\vec{n} \wedge \overrightarrow{\mathrm{grad}}\, u_i^{(2)}\right) dS = 0 \qquad (3.45)$$

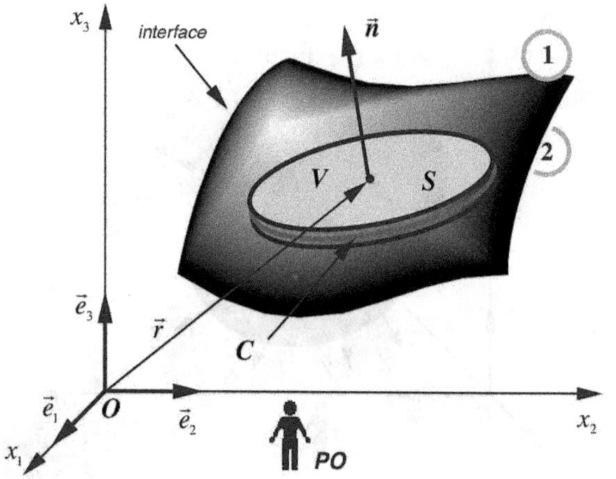

Figure 3.9 *- compatible pass-through conditions between 2 solid medium*

The theorem of the gradient applied to this last relation leads us to

$$\oint_C u_i^{(1)} d\vec{r} - \oint_C u_i^{(2)} d\vec{r} = 0 \qquad (3.46)$$

in which C is the mobile contour situated at the intersection off the mobile surface with the 2 mediums (figure 3.9). For this condition to be satisfied for any contour C we must have

$$u_i^{(1)} = u_i^{(2)} \quad \Rightarrow \quad \vec{u}^{(1)} = \vec{u}^{(2)} \qquad (3.47)$$

In other words the conditions of passage *(3.47)* implies that the displacement $\vec{u}$ on bots sides of the interface are equal, and thus that the 2 mediums in contact cannot 'unglue' nor slip one against the other, defining the "compatible interface".

The relation of compatibility for the vector of rotation $\vec{\omega}$ implies that the following relations

$$\iiint_V \mathrm{div}\,\vec{\omega}\, dV = \oiint_S \vec{\omega}\, d\vec{S} = \iint_{S_1} \vec{\omega}^{(1)} \vec{n}\, dS - \iint_{S_2} \vec{\omega}^{(2)} \vec{n}\, dS = 0 \qquad (3.48)$$

can only be satisfied if

$$\vec{\omega}^{(1)} \vec{n} = \vec{\omega}^{(2)} \vec{n} \qquad (3.49)$$

The interpretation of this pass-through condition is self evident since it assures that the normal component of the field of local rotation is conserved on each side of the interface between 2 mediums; it prevents a sliding at the interface by rotation of one medium with respect to another.

The following integral relations at the interface can be deducted from the condition of compatibility for the tensor of contortion $\vec{\chi}_i$

$$\iiint_V \overrightarrow{\operatorname{rot}}\,\vec{\chi}_i dV = \oiint_S d\vec{S} \wedge \vec{\chi}_i = \iint_{S_1} \left(\vec{n} \wedge \vec{\chi}_i^{(1)}\right) dS - \iint_{S_2} \left(\vec{n} \wedge \vec{\chi}_i^{(2)}\right) dS = 0 \tag{3.50}$$

So that they are satisfied, it must be so that the tangential components of the $\vec{\chi}_i$ be conserved on each side of the interface. This can be written as

$$\vec{n} \wedge \vec{\chi}_i^{(1)} = \vec{n} \wedge \vec{\chi}_i^{(2)} \tag{3.51}$$

To find a topological interpretation for this condition of passage, one has to decompose the surface integral in 2 parts and remember that, in a compatible medium, the tensor $\vec{\chi}_i$ is directly linked to the gradients of the components of the field of rotation $\vec{\omega}^{\varepsilon}$ associated with the deformations $\vec{\varepsilon}_i$. So we have

$$\iint_{S_1} \left(\vec{n} \wedge \overrightarrow{\operatorname{grad}}\,\omega_i^{\varepsilon(1)}\right) dS - \iint_{S_2} \left(\vec{n} \wedge \overrightarrow{\operatorname{grad}}\,\omega_i^{\varepsilon(2)}\right) dS = 0 \tag{3.52}$$

which with the gradient theorem becomes

$$\oint_C \omega_i^{\varepsilon(1)} d\vec{r} - \oint_C \omega_i^{\varepsilon(2)} d\vec{r} = 0 \quad \Rightarrow \quad \omega_i^{\varepsilon(1)} = \omega_i^{\varepsilon(2)} \quad \Rightarrow \quad \vec{\omega}^{\varepsilon(1)} = \vec{\omega}^{\varepsilon(2)} \tag{3.53}$$

In other words the pass-through condition *(3.53)* assures us that the rotations $\vec{\omega}^{\varepsilon}$ implicitly contained in the tensor of deformations $\vec{\varepsilon}_i$ (from which we deduct the tensor of contortions $\vec{\chi}_i$) are equal on each side of the interface between the 2 mediums.

The equation of compatibility for the flexion vector $\vec{\chi}$ implies the following integral relation on the interface

$$\iiint_V \operatorname{div} \vec{\chi}\, dV = \oiint_S \vec{\chi} d\vec{S} = \iint_{S_1} \vec{\chi}^{(1)} \vec{n} dS - \iint_{S_2} \vec{\chi}^{(2)} \vec{n} dS = 0 \tag{3.54}$$

from which we deduct that the normal component of the flexion vector must be conserved on each part of the interface

$$\vec{\chi}^{(1)} \vec{n} = \vec{\chi}^{(2)} \vec{n} \tag{3.55}$$

3.6 - Examples of fields of flexion and torsion

From simple examples of field displacement $\vec{u}^{(\delta)}(\vec{r},t)$ that belong to known deformations of the medium, this paragraph will show that the topological tensors $\vec{\chi}_i$ et $\vec{\chi}$ perfectly describe the contortions, meaning the flexions and torsions that can appear inside the lattice.

The pure flexions of a solid lattice

Imagine the following static field displacement $\vec{u}^{(\delta)}(\vec{r})$ in the neighborhood of the origin of local referential $Ox_1x_2x_3$, which only present one component in the direction of the Ox_3

$$\vec{u}^{(\delta)}(\vec{r}) = \left(R - \sqrt{R^2 + \alpha x_1^2 + \beta x_2^2}\right) \vec{e}_3 \quad \text{for} \quad |x_1| < R \quad \text{and} \quad |x_2| < R \tag{3.56}$$

where the constants α and β can take the values -1, 0 or 3.

If the deformation of the horizontal planes of a solid submitted to such a field of displacement is represented graphically, we observe a *curvature of the planes of flexion* that depends on the value of the parameters α and β, as shown on the 4 examples of figure 3.10.

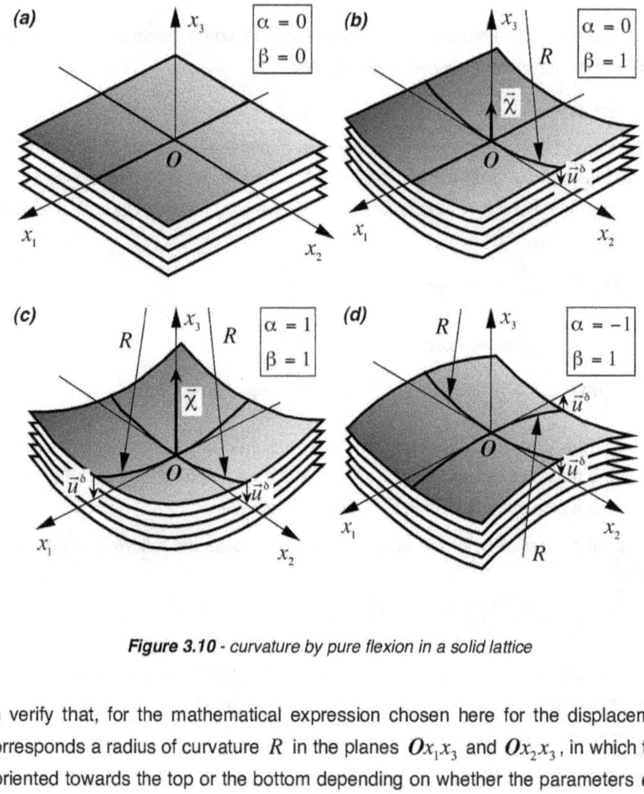

Figure 3.10 *- curvature by pure flexion in a solid lattice*

We can verify that, for the mathematical expression chosen here for the displacement field, there corresponds a radius of curvature R in the planes Ox_1x_3 and Ox_2x_3, in which the curvature is oriented towards the top or the bottom depending on whether the parameters α and β are positive or negative, respectively. If α and β are both null, there is no deformation (figure 3.10a). If $\alpha = 0$ and $\beta = 1$, the horizontal planes are portions of a cylinder of radius R (figure 3.10b). If α and β both are equal to 1, it is a portion of sphere or radius R (figure 3.10c), and if $\alpha = -1$ et $\beta = 1$, the horizontal planes represent a saddle point which is at the origin of co-ordinates (figure 3.10d).

With relations *(2.48)*, the tensors of distortion of the solid can be deduced from the displacement vector $\vec{u}^{(\delta)}(\vec{r})$. The volume expansion τ of the solid is null, because $\tau = -\operatorname{div}\vec{u}^{(\delta)} = 0$, and the local field of rotation $\vec{\omega}^{(\delta)}$ does not present a component along the axis Ox_3

$$\vec{\omega}^{(\delta)} = -\frac{1}{2}\overrightarrow{\operatorname{rot}}\,\vec{u}^{(\delta)} = \frac{\beta x_2}{2\sqrt{R^2 + \alpha x_1^2 + \beta x_2^2}}\vec{e}_1 - \frac{\alpha x_1}{2\sqrt{R^2 + \alpha x_1^2 + \beta x_2^2}}\vec{e}_2 \tag{3.57}$$

One can verify that the expression for field $\vec{\omega}^{(\delta)}$ satisfies the condition of compatibility $\operatorname{div}\vec{\omega}^{(\delta)} = 0$. The tensor of deformation $\vec{\varepsilon}_i$ is equal to the shear tensor $\vec{\alpha}_i$, due to the fact that $\operatorname{div}\vec{u}^{(\delta)} = 0$. As a consequence

$$\left\{\begin{array}{l} \vec{\varepsilon}_1 = \vec{\alpha}_1 = \dfrac{\alpha x_1}{2\sqrt{R^2+\alpha x_1{}^2+\beta x_2{}^2}}\vec{e}_3 \\ \vec{\varepsilon}_2 = \vec{\alpha}_2 = \dfrac{\beta x_2}{2\sqrt{R^2+\alpha x_1{}^2+\beta x_2{}^2}}\vec{e}_3 \\ \vec{\varepsilon}_3 = \vec{\alpha}_3 = \dfrac{\alpha x_1}{2\sqrt{R^2+\alpha x_1{}^2+\beta x_2{}^2}}\vec{e}_1 + \dfrac{\beta x_2}{2\sqrt{R^2+\alpha x_1{}^2+\beta x_2{}^2}}\vec{e}_2 \end{array}\right. \tag{3.58}$$

The tensor of contortions $\vec{\chi}_i$ can also be deducted, and we note that it is of null trace, as we had deduced in *(3.22)*

$$\left\{\begin{array}{l} \vec{\chi}_1 = \overrightarrow{\text{grad}}\,\omega_1{}^{(\delta)} = \dfrac{\alpha\beta x_1\ x_2}{2\left(R^2+\alpha x_1{}^2+\beta x_2{}^2\right)^{3/2}}\vec{e}_1 + \dfrac{\beta\left(R^2+\alpha x_1{}^{22}\right)}{2\left(R^2+\alpha x_1{}^2+\beta x_2{}^2\right)^{3/2}}\vec{e}_2 \\ \vec{\chi}_2 = \overrightarrow{\text{grad}}\,\omega_2{}^{(\delta)} = \dfrac{\alpha\left(R^2+\beta x_2{}^2\right)}{2\left(R^2+\alpha x_1{}^2+\beta x_2{}^2\right)^{3/2}}\vec{e}_1 + \dfrac{\alpha\beta x_1\ x_2}{2\left(R^2+\alpha x_1{}^2+\beta x_2{}^2\right)^{3/2}}\vec{e}_2 \\ \vec{\chi}_3 = \overrightarrow{\text{grad}}\,\omega_3{}^{(\delta)} = 0 \end{array}\right. \tag{3.59}$$

And for the flexion vector $\vec{\chi}$, it follows

$$\vec{\chi} = -\overrightarrow{\text{rot}}\ \vec{\omega}^{(\delta)} = \sum_k \vec{e}_k \wedge \vec{\chi}_k = \frac{(\alpha+\beta)R^2+\alpha\beta\left(x_1^2+x_2^2\right)}{2\left(R^2+\alpha x_1^2+\beta x_2^2\right)^{3/2}}\ \vec{e}_3 \tag{3.60}$$

The norm of the flexion vector $\vec{\chi}$ measured at point O of our system of coordinates is directly linked to the *inverse of the radius of curvature* R, in the following way

$$\left|\vec{\chi}\right|_{x_1=x_2=0} = \left\{\begin{array}{llll} 0 & \text{for } \alpha=0 & \text{and} & \beta=0 \\ 1/2R & \text{for } \alpha=0 & \text{and} & \beta=1 \\ 1/R & \text{for } \alpha=1 & \text{and} & \beta=1 \\ 0 & \text{for } \alpha=-1 & \text{and} & \beta=1 \end{array}\right. \tag{3.61}$$

and in such a way that the norm of the flexion vector in a given point of space is equal to the inverse of radius R of a cylinder (or a sphere) which would be centered at a distance R of this point in the direction of the flexion vector, and which would be tangential at this point to the deformed surface.

The pure torsion of a solid lattice

Consider now the field of local static displacement $\vec{u}^{(\delta)}(\vec{r})$ in the local frame $Ox_1x_2x_3$, where α, β and γ are constants

$$\vec{u}^{(\delta)}(\vec{r}) = \alpha x_2 x_3 \vec{e}_1 + \beta x_1 x_3 \vec{e}_2 + \gamma x_1 x_2 \vec{e}_3 \tag{3.62}$$

It is possible to obtain the vector of torsion (local rotation) $\vec{\omega}^{(\delta)}$ and the scalar of volume expansion τ associated with this field of displacement

$$\begin{cases} \vec{\omega}^{(\delta)} = -\dfrac{1}{2}\overrightarrow{\mathrm{rot}}\,\vec{u}^{(\delta)} = \dfrac{1}{2}\left[(\beta-\gamma)x_1\vec{e}_1 + (\gamma-\alpha)x_2\vec{e}_2 + (\alpha-\beta)x_3\vec{e}_3\right] \\ \tau = -\mathrm{div}\,\vec{u}^{(\delta)} = 0 \end{cases} \tag{3.63}$$

This field of displacement is thus non divergent and it leads to a field of rotation whose components increase or diminish along the coordinate axis, which is characteristic of a *pure torsion of the solid* (which corresponds in fact exactly to the case represented in figure 3.2 if we assume $\beta = 0$ et $\alpha = -\gamma$). Furthermore, we can verify that the vector of local rotation $\vec{\omega}^{(\delta)}$ satisfies the geometro-compatibility condition, which is normal since the divergence of a rotational is null

$$\mathrm{div}\,\vec{\omega}^{(\delta)} = \frac{1}{2}\left[(\beta-\gamma)+(\gamma-\alpha)+(\alpha-\beta)\right] = 0 \tag{3.64}$$

For a tensor of local distortions $\vec{\beta}_i^{(\delta)}$, it follows

$$\begin{cases} \vec{\beta}_1^{(\delta)} = -\overrightarrow{\mathrm{grad}}\,u_1^{(\delta)} = -\alpha x_3\vec{e}_2 - \alpha x_2\vec{e}_3 \\ \vec{\beta}_2^{(\delta)} = -\overrightarrow{\mathrm{grad}}\,u_2^{(\delta)} = -\beta x_3\vec{e}_1 - \beta x_1\vec{e}_3 \\ \vec{\beta}_3^{(\delta)} = -\overrightarrow{\mathrm{grad}}\,u_3^{(\delta)} = -\gamma x_2\vec{e}_1 - \gamma x_1\vec{e}_2 \end{cases} \tag{3.65}$$

and for the tensors of deformation $\vec{\varepsilon}_i$ and shear $\vec{\alpha}_i$

$$\begin{cases} \vec{\varepsilon}_1 = \vec{\alpha}_1 = \vec{\beta}_1^{(\delta)} - \vec{e}_1 \wedge \vec{\omega}^{(\delta)} = -(\alpha+\beta)x_3\vec{e}_2/2 - (\alpha+\gamma)x_2\vec{e}_3/2 \\ \vec{\varepsilon}_2 = \vec{\alpha}_2 = \vec{\beta}_2^{(\delta)} - \vec{e}_2 \wedge \vec{\omega}^{(\delta)} = -(\alpha+\beta)x_3\vec{e}_1/2 - (\beta+\gamma)x_1\vec{e}_3/2 \\ \vec{\varepsilon}_3 = \vec{\alpha}_3 = \vec{\beta}_3^{(\delta)} - \vec{e}_3 \wedge \vec{\omega}^{(\delta)} = -(\alpha+\gamma)x_2\vec{e}_1/2 - (\beta+\gamma)x_1\vec{e}_2/2 \end{cases} \tag{3.66}$$

The tensor of contortions is then written

$$\begin{cases} \vec{\chi}_1 = \overrightarrow{\mathrm{grad}}\,\omega_1^{(\delta)} = (\beta-\gamma)\vec{e}_1/2 \\ \vec{\chi}_2 = \overrightarrow{\mathrm{grad}}\,\omega_2^{(\delta)} = (\gamma-\alpha)\vec{e}_2/2 \\ \vec{\chi}_3 = \overrightarrow{\mathrm{grad}}\,\omega_3^{(\delta)} = (\alpha-\beta)\vec{e}_3/2 \end{cases} \tag{3.67}$$

This tensor is purely diagonal and of null trace. It is therefore equal, in this particular case, to the *tensor of torsion* $\left[\vec{\chi}_i\right]^{S}$. Regarding the vector of flexion, it is in this case null

$$\vec{\chi} = -\overrightarrow{\mathrm{rot}}\;\vec{\omega}^{(\delta)} = \sum_k \vec{e}_k \wedge \vec{\chi}_k = 0 \tag{3.68}$$

Chapter 4

Newtonian dynamics and eulerian thermo-kinetics

If we suppose that the lattice under consideration behaves in a newtonian fashion in the absolute frame of ***GO***, we can introduce in a rigorous manner the dynamics and thermo-kinetics of the lattice in the eulerian coordinates. We will do so on the frame of 3 axioms, which are classical and well-known: the definition of the kinetic energy of a particle and the 2 first principles of thermodynamics. With this rigorous if axiomatic approach, one is led to define average quantities on each site of the lattice, and corresponding sources and flux, which must satisfy 3 *continuity principles that are mandatory in solid newtonian lattices.*

4.1 - The principle of newtonian dynamics

The axiom of newtonian dynamics

It was shown in chapter 1 that a lattice of solid particles in space exhibits a collective movement which corresponds to the global movements of translation, rotation and deformation of the medium in the space of the observer, and which can be described in eulerian coordinates, by a local average velocity $\vec{\phi}(\vec{r},t)$.

We postulate here, axiomatically, that this collective lattice obeys the laws of *newtonian dynamics*. This means that, to the collective movement of particles at velocity $\vec{\phi}$, there corresponds a kinetic energy. For each particle, this kinetic energy can be written, according to newtonian mechanics as:

$$e_{cin} = \frac{1}{2} m \vec{\phi}^2 = \text{kinetic energy per particle} \tag{4.1}$$

This expression of the kinetic energy of a particle in the medium involves a physical invariant scalar that is proper to the particle: its *inertial mass* also called *rest mass,* m.

The equation of local continuity of the inertial mass in the absolute frame of GO

In eulerian coordinates, one has to find an equation that translates the continuity of the mass in the medium in the frame of the ***GO***. To write this equation, we introduce the *density of mass* (mass per unit volume) $\rho(\vec{r},t)$ of the medium, so that the total mass M contained in any given volume V_m, which can be mobile in the medium, is simply obtained by integration of the density of mass over said volume, by definition

$$M = \iiint_{V_m} \rho \, dV \tag{4.2}$$

On this volume V_m, the equation of continuity of mass is written on the frame of the temporal derivative of M, under the simple form

$$\frac{d}{dt}\iiint_{V_m} \rho\, dV = \iiint_{V_m} S_m\, dV - \oiint_{S_m} \vec{J}_m d\vec{S} \tag{4.3}$$

What this equation translates is the fact that, within a given volume V_m, the change in mass must correspond to either a *source density of mass* S_m, representing the creation or annihilation of mass at each point within said volume (the first term), or to the fact that mass leaves the volume. Mass leaving a volume is equivalent to mass passing-through the boundary surface of said volume. This is then captured as a surface exchange term between the inside and outside of the volume and is represented by the *surface flux of mass* $\vec{J}_m$, with respect to the surface S_m (figure 4.1). Equation *(4.3)* can also be transformed by using the derivative of an integral on a mobile volume and the divergence theorem to give us

$$\iiint_{V_m} \left[\frac{\partial \rho}{\partial t} + \operatorname{div}(\rho\vec{\phi})\right] dV = \iiint_{V_m} S_m\, dV - \iiint_{V_m} \operatorname{div}\vec{J}_m\, dV \tag{4.4}$$

We can deduce the local expression for the continuity condition in the $Q\xi_1\xi_2\xi_3$ frame

$$\frac{\partial \rho}{\partial t} = S_m - \operatorname{div}(\rho\vec{\phi} + \vec{J}_m) \tag{4.5}$$

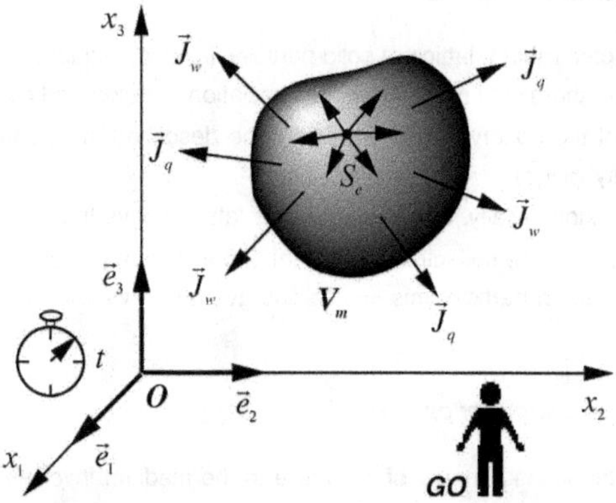

Figure 4.1 *- mass flux $\vec{J}_m$, work flux $\vec{J}_w$, heat flux $\vec{J}_q$, source of mass S_m and source of entropy density S_e on a given mobile volume V_m of medium*

The equation of local continuity of inertial mass in a local frame of PO

We can develop the same calculations we just did but in the local frame $Ox_1x_2x_3$ of ***PO***. With regards to the temporal derivative of ρ, meaning in a fixed point $\vec{r}$ of the local frame, the equation of continuity must involve the local velocity $\vec{\varphi}$ instead of the absolute velocity $\vec{\phi}$, and we obtain the following equation in $Ox_1x_2x_3$, which can also be expressed with the absolute velocity $\vec{\phi}$ by using the relation *(2.16)*

$$\frac{\partial \rho}{\partial t} = S_m - \operatorname{div}(\rho\vec{\varphi} + \vec{J}_m) = S_m - \operatorname{div}\left[\rho\left(\vec{\phi} - \vec{\phi}_O(t) - \dot{\vec{\omega}}_O(t) \wedge \vec{r}\right) + \vec{J}_m\right] \tag{4.6}$$

The eulerian equation of continuity for inertial mass

We can revisit the 2 equations of local continuity of inertial mass in $Q\xi_1\xi_2\xi_3$ and $Ox_1x_2x_3$, and try to derive a continuity equation of the eulerian type, which would only depend on material derivatives, and as a consequence be true both in the absolute frame of ***GO*** and the mobile frame of ***PO***. We rewrite relation (4.5) in the following way

$$\frac{\partial \rho}{\partial t} = S_m - \rho \operatorname{div}\vec{\phi} - \vec{\phi}\,\overrightarrow{\operatorname{grad}}\,\rho - \operatorname{div}\vec{J}_m \tag{4.7}$$

By using the equation of compatibility *(1.42)* for τ, we obtain

$$\frac{\partial \rho}{\partial t} + \vec{\phi}\,\overrightarrow{\operatorname{grad}}\,\rho = S_m - \rho\left(\frac{d\tau}{dt} + \frac{S_n}{n}\right) - \operatorname{div}\vec{J}_m \tag{4.8}$$

The first member of this equation is the material derivative of ρ, so the equation of eulerian continuity for the inertial mass reads

$$\frac{d\rho}{dt} = S_m - \rho\left(\frac{d\tau}{dt} + \frac{S_n}{n}\right) - \operatorname{div}\vec{J}_m \tag{4.9}$$

One would obtain exactly the same results by starting from relation *(4.6)* in the frame $Ox_1x_2x_3$.

The newtonian concept of linear momentum ("quantité de mouvement" in french)

The equation *(4.3)* can be written in a fixed volume V_f of the observer. The temporal derivative of the total mass M contained in a fixed volume V_f is then written

$$\frac{d}{dt}\iiint_{V_f} \rho\, dV = \iiint_{V_f} S_m\, dV - \oiint_{S_f} \vec{P}_m d\vec{S} \tag{4.10}$$

in which the *absolute surface flux of mass* $\vec{P}_m$ with respect to the observer, actually represents an important characteristic measure of newtonian dynamics, the *volume density of linear momentum*. By introducing the *average linear momentum* $\vec{p}$ *of each elementary site of the lattice*, defined as

$$\vec{p} = \vec{P}_m / n \tag{4.11}$$

equation *(4.10)*, which corresponds to the temporal derivative of an integral on a fixed volume in space, can be simply written as

$$\iiint_{V_f} \frac{\partial \rho}{\partial t} dV = \iiint_{V_f} S_m\, dV - \iiint_{V_f} \operatorname{div}(n\vec{p})\, dV \tag{4.12}$$

which leads us to a second expression for the continuity of mass

$$\frac{\partial \rho}{\partial t} = S_m - \operatorname{div}(n\vec{p}) \tag{4.13}$$

Following *(4.5)* et *(4.10)*, the flux of mass can then be represented in 2 equivalent ways

$$\vec{P}_m = n\vec{p} = \rho\vec{\phi} + \vec{J}_m \tag{4.14}$$

which signifies that the absolute transport of mass $\vec{P}_m = n\vec{p}$ is equal to a sum of mass transport $\rho\vec{\phi}$ *by the mobile medium* with velocity $\vec{\phi}$ and an extra term of transport of mass $\vec{J}_m$ via another physical process, such as the self-diffusion in solid lattices which we will introduce later.

4.2 - The principle of energy continuity

To introduce the concept of energy, we must find the total energy contained in a volume V_m of the medium, mobile within it (figure 4.1), meaning having a displacement velocity $\vec{\phi}(\vec{r},t)$. It was shown in section 1.2 that the velocity of each particle i of the medium can simply be written

$$\vec{v}_i = \vec{\phi} + \Delta\vec{v}_i \tag{4.15}$$

where the $\Delta\vec{v}_i$ are the random fluctuations of the velocity of the particle, which we will associate later to thermal agitation.
To the collective movement of particles with velocity $\vec{\phi}$ corresponds, if we admit the postulate of newtonian dynamics, a kinetic energy that can be expressed in the form of a *volume density of kinetic energy* $T(\vec{r},t)$.

The axiom of the first principle of thermodynamics

To the random fluctuations of $\Delta\vec{v}_i$, as well as the interactions that can exist between the particles of the lattice must correspond a kinetic and an internal potential energy. This is precisely the axiom of *first principle of phenomenological thermodynamics*, which postulates, for a given physical system, a state function U, called *internal energy of the system*, which is such that, for all infinitesimal transformation of the system, the following relation holds

$$dU = \delta W + \delta Q \tag{4.16}$$

where δQ represents all exchanges of *heat* between the system and the external world, and δW all exchanges of *work* between the system and the external world.
Along its spatio-temporal evolution, it is certain that the medium will, in general, find itself out of thermodynamical equilibrium, and it is thus necessary to generalize the concept of internal energy by introducing a quantity of local internal energy which will depend on time, under the form of a *volume density of internal energy* $U(\vec{r},t)$.

The eulerian equation of continuity of energy

The total energy $\boldsymbol{E}$ contained in a volume V_m is then equal to the integral on the volume V_m of the sum of kinetic energy density T and internal energy density U, which is to say

$$\boldsymbol{E} = \boldsymbol{T} + \boldsymbol{U} = \iiint_{V_m} (T + U)\,dV \tag{4.17}$$

According to the first principle of thermodynamics, this energy is a conserved quantity inside an isolated V_m. So any variation in said energy can only be due to external influences coming from outside the volume V_m, translating the fact that this volume is in fact *non-isolated.* These variations are due either to work exchange or heat exchange. By definition those exchanges either result from a pass through the boundary surface S_m around the volume V_m, or result from work due to an external work performed, on the internal medium, by an outside field (such as gravity, electro-magnetism, etc). The exchange of work and heat across the surface S_m can be represented as a form of *surface flux of work* $\vec{J}_w$ and of *surface flux of heat* $\vec{J}_q$. As for the volume work due to outside forces, it is introduced in the form of a volume source from external fields, and it is introduced in the form of a *volume source of work of external forces* S_w^{ext}.

The principle of conservation of energy of a given volume V_m can be written as the equality of the temporal derivation of total energy contained in said volume V_m and the different contributions to the volume, which comprises the work volume sources we just discussed and the pass-through contributions which necessarily cross the boundary surface , by definition of a boundary surface S_m

$$\frac{d}{dt}\iiint_{V_m}(T+U)dV = \iiint_{V_m} S_w^{ext}\, dV - \oiint_{S_m}\left(\vec{J}_w + \vec{J}_q\right)d\vec{S} \tag{4.18}$$

Using the theorem of divergence *(A.43)* and the derivation of an integral on a mobile volume *(A. 58)*, this equation can be transformed into the following

$$\iiint_{V_m}\left[\frac{\partial T}{\partial t}+\frac{\partial U}{\partial t}+\operatorname{div}\left(T\vec{\phi}+U\vec{\phi}\right)\right]dV = \iiint_{V_m}\left[S_w^{ext}-\operatorname{div}\left(\vec{J}_w+\vec{J}_q\right)\right]dV \tag{4.19}$$

which under local form, gives us

$$\frac{\partial T}{\partial t}+\frac{\partial U}{\partial t}+\operatorname{div}\left(T\vec{\phi}\right)+\operatorname{div}\left(U\vec{\phi}\right)=S_w^{ext}-\operatorname{div}\vec{J}_w-\operatorname{div}\vec{J}_q \tag{4.20}$$

It will be shown, later in this book, that for convenience reasons, it is judicious to transform this equation, which contains volume densities T and U, by introducing the measures e_{cin} et u, defined respectively as the *medium kinetic energy* and the *medium internal energy*, defined per elementary site of the lattice

$$e_{cin}=\frac{T}{n} \quad \text{and} \quad u=\frac{U}{n} \tag{4.21}$$

Thanks to these definitions, the first part of equation *(4.20)* can be transformed. Indeed, by using the equation of continuity *(1.12)* for n and making explicit the material derivative of u, as well as a term associated to the source S_n of the sites of the lattice, we deduce

$$\frac{\partial U}{\partial t}+\operatorname{div}\left(U\vec{\phi}\right)=n\left[\frac{\partial u}{\partial t}+(\vec{\phi}\vec{\nabla})u\right]+uS_n=n\frac{du}{dt}+uS_n \tag{4.22}$$

Applying transformation *(4.20)*, we obtain the final formulation of the *first principle*, or *principle of continuity of energy* in eulerian coordinates and in its local form

$$n\frac{du}{dt}+n\frac{de_{cin}}{dt}=S_w^{ext}-\operatorname{div}\vec{J}_w-\operatorname{div}\vec{J}_q-uS_n-e_{cin}S_n \tag{4.23}$$

Whether it be written in the absolute referential $Q\xi_1\xi_2\xi_3$ of the ***GO*** or in the local mobile frame $Ox_1x_2x_3$ of the ***PO***, the form of this equation of continuity remains identical to itself. Only the expression of the material derivative (absolute derivative as opposed to partial) changes since it depends on the absolute velocity $\vec{\phi}$ of the lattice in the frame $Q\xi_1\xi_2\xi_3$ and the local relative velocity $\vec{\varphi}$ of the lattice in the mobile frame $Ox_1x_2x_3$.

4.3 - The principle of entropy continuity

The axiom of second principle of thermodynamics

The *second principle of phenomenological thermodynamics* postulates the existence, for a given physical system, of a state function S, called the *entropy* of the system. This state function characterizes the disorder within the system, and is such that all infinitesimal transformation of the system satisfy the following inequality relation

$$dS \geq \frac{\delta Q}{T} \tag{4.24}$$

where δQ represents the sum of heat exchanges between the system and the outside world and T the temperature at which these exchanges happen. During its evolution in space and time, the system is, in general, out of thermodynamical equilibrium, such that it is necessary to generalize the concept of entropy, by introducing the notion of a *local* entropy variable which depends on time, the *volume density of entropy* $S(\vec{r},t)$.

The eulerian equation of continuity of entropy

By using the same mobile volume V_m defined in the previous paragraph, the volume density of entropy $S(\vec{r},t)$ allows us to calculate the total entropy $\boldsymbol{S}$ of a system by relation

$$\boldsymbol{S} = \iiint_{V_m} S\,dV \tag{4.25}$$

The second principle can then be generalized as a *principle of entropy in eulerian coordinates* by writing an integral relation for the temporal derivative of the total entropy $\boldsymbol{S}$ on a mobile volume

$$\frac{d}{dt}\iiint_{V_m} S\,dV = \iiint_{V_m} S_e\,dV - \oiint_{S_m} \frac{\vec{J}_q}{T}\,d\vec{S} \tag{4.26}$$

in which the exchanges of heat with the exterior are introduced via a *surface flux of heat* $\vec{J}_q$ on the boundary of said volume.

The irreversibility of thermodynamic transformations of the medium which is translated through the inequality $dS \geq \delta Q / T$ in phenomenological thermodynamics, was introduced here under the form of a source of volume entropy S_e within the volume V_m. The notion of reversibility or irreversibility of processes of the medium is then described by a macroscopic source term σ of entropy

$$\sigma = \iiint_{V_m} S_e\,dV \tag{4.27}$$

which is such that the evolution of a system will be reversible for $\sigma = 0$ and irreversible for $\sigma > 0$.

The integral relation *(4.26)* can also be written under another form, by using the temporal derivative of an integral on a mobile volume and by making use of the theorem of divergence

$$\iiint_{V_m}\left[\frac{\partial S}{\partial t} + \operatorname{div}\left(S\vec{\phi}\right)\right]dV = \iiint_{V_m} S_e\,dV - \iiint_{V_m} \operatorname{div}\left(\frac{\vec{J}_q}{T}\right)dV \tag{4.28}$$

We can extract a local equation from the second principle of volume density of entropy S

$$\frac{\partial S}{\partial t} + \operatorname{div}\left(S\vec{\phi}\right) = S_e - \operatorname{div}\left(\frac{\vec{J}_q}{T}\right) \tag{4.29}$$

By introducing the value s representing the *average entropy per elementary site of the lattice,* defined by the relation

$$s = \frac{S}{n} \tag{4.30}$$

and by using the transformations described in *(4.22)*, the final formulation of the *second principle*, or *principle of eulerian continuity of entropy* in eulerian coordinates, is written

$$n\frac{ds}{dt} = S_e - \operatorname{div}\left(\frac{\vec{J}_q}{T}\right) - s\,S_n \tag{4.31}$$

Whether it be in the absolute frame $Q\xi_1\xi_2\xi_3$ of the ***GO*** or in the local mobile sphere $Ox_1x_2x_3$ of the ***PO***, this equation of continuity remains identical. Only the expression of the material derivative changes since it depends on the absolute velocity $\vec{\phi}$ of the lattice in the frame $Q\xi_1\xi_2\xi_3$ and on the local relative velocity $\vec{\varphi}$ of the lattice in the mobile frame $Ox_1x_2x_3$.

The 3 fundamental physics principles of continuity of newtonian dynamics and eulerian thermo-kinetics

Therefore, by starting from the axioms of newtonian dynamics and the classic phenomenological thermodynamic, 3 physical principles of continuity were postulated to describe the *"thermo-geometro-dynamic"* in eulerian coordinates, as following

- an eulerian principle of continuity of the density of inertial mass,
- an eulerian principle of continuity of the average energy per site of the lattice,
- an eulerian principle of continuity of the average entropy per site of the lattice.

As we will see during the next chapters, these 3 principles, which are summed up in table 4.1, are *the only physical principles needed for a complete galilean description of the newtonian geometro-dynamic and the phenomenological description of deformable medium in Euler coordinates, whether those are solids or fluids.*

Table 4.1 – Newtonian dynamics and eulerian thermo-kinetics

Absolute flux of mass $\vec{P}_m$ and average linear momentum $\vec{p}$ per site of the lattice

$$\vec{P}_m = n\vec{p} = \rho\vec{\phi} + \vec{J}_m \quad (1)$$

Principle of eulerian continuity for the density of inertial mass

$$\frac{d\rho}{dt} = S_m - \rho\left(\frac{d\tau}{dt} + \frac{S_n}{n}\right) - \operatorname{div}\vec{J}_m \quad (2)$$

Principle of eulerian continuity of the average energy per site of the lattice

$$n\frac{du}{dt} + n\frac{de_{cin}}{dt} = S_w^{ext} - \operatorname{div}\vec{J}_w - \operatorname{div}\vec{J}_q - uS_n - e_{cin}S_n \quad (3)$$

Principle of eulerian continuity of the average entropy per site of the lattice

$$n\frac{ds}{dt} = S_e - \operatorname{div}\left(\frac{\vec{J}_q}{T}\right) - sS_n \quad (4)$$

Chapter 5

Physical properties of a newtonian lattice

In this chapter, the fundamental physical properties of a solid lattice are shown. We will look at the *mechanical properties* and the *properties of mass transport*. Starting from a thermo-kinetic equation of the local solid lattice, we will show how the *elasticity* of the lattice gives rise to thermodynamical potentials called *stress tensors*. We then introduce the notion of *dissipative mechanical phenomenas* in lattices, by decomposing the tensor of distortions in dissipative and non-dissipative components. This leads to 2 phenomenological descriptions of possible dissipations: *anelasticity* and *plasticity*. We describe *self-diffusion*, the transport of mass through the lattice either in the form of *vacancies or self-interstitials*. The newtonian principle of mass conservation implies new fundamental equations for the lattice called *diffusion equations*, as well as new thermodynamic potentials called *chemical potentials*. Finally we derive the quantities associated with the newtonian behavior of the lattice, meaning its *average linear momentum per site*, its *average kinetic energy per site* and *the source of external work*, as in the case where the lattice is submitted to a constant gravity field.

5.1 - Elasticity of a lattice

In the local frame $Ox_1x_2x_3$, we can characterize the state of a solid in the most possible general way, first by giving its global tensor of distortion $\vec{\beta}_i$, and second, with regards to its thermal energy, by the local value of its entropy s. The average internal energy in each site of the lattice is, as defined here, a function of the local distortion and local entropy

$$u = u(\beta_{ij}, s) \tag{5.1}$$

The actual shape of this state function depends on the type of solid lattice we are considering. What we are doing with it is to describe the energy of a given solid phenomenologically, and therefore, we will call it the *phenomenological state function* of the solid.

Expression *(5.1)* of the internal energy is not the usual form found in textbooks to describe common solid latices. Usually only the components of the tensor of deformation are used to compute internal energy, not the β_{ij} components of the tensor of distortion. We will discuss this point further in the book. For now, let's just say that, *à priori*, the description of internal energy using the tensor of distortion $\vec{\beta}_i$ is more general than the one usually obtained from the components of the tensor of deformation.

The distortion stress tensor and the thermo-kinetic equation of elasticity

Since the internal energy u is a function of the scalar of entropy s and the components β_{ij} of the tensor of distortion, its differential form is

$$du = \frac{\partial u}{\partial \beta_{ij}} d\beta_{ij} + \frac{\partial u}{\partial s} ds \tag{5.2}$$

We introduce the *mechanical potential*, the conjugate of the tensor of distortion β_{ij}, and we will call it *distortion stress tensor* Σ_{ij}. We also introduce the *thermal potential*, the conjugate of entropy s, which is simply the *temperature* T. The differential can be written

$$ndu = \Sigma_{ij} d\beta_{ij} + nTds \tag{5.3}$$

The comparison between the 2 previous expressions of the differential of u shows the *phenomenological state equations of the solid*, which is to say the relations that gives us the stress tensor Σ_{ij} and temperature T as functions of the tensor of distortion β_{ij} and the entropy s

$$\begin{cases} \Sigma_{ij} = n\dfrac{\partial u}{\partial \beta_{ij}} = \Sigma_{ij}(\beta_{kl}, s) \\ T = \dfrac{\partial u}{\partial s} = T(\beta_{kl}, s) \end{cases} \tag{5.4}$$

As with the topological tensors, a vectorial notation of the stress tensor Σ_{ij} can be introduced here

$$\vec{\Sigma}_i = \sum_j \Sigma_{ij} \vec{e}_j \tag{5.5}$$

The differential of the internal energy u can then be written with a scalar product in the following form

$$ndu = \vec{\Sigma}_k d\vec{\beta}_k + nTds \tag{5.6}$$

From which we deduce the *thermo-kinetic equation of elasticity* of the lattice

$$n\frac{du}{dt} = \vec{\Sigma}_k \frac{d\vec{\beta}_k}{dt} + nT\frac{ds}{dt} \tag{5.7}$$

The deformation stress tensor and the torque torsor

The equations of thermodynamic obtained in the previous paragraph were written as functions of the components β_{ij} of the tensor of distortion. It could be interesting to rewrite them in terms of the tensor of deformation $\vec{\varepsilon}_i$ and the torsor of rotation $\vec{\omega}$. We recall,

$$\vec{\beta}_i = \vec{\varepsilon}_i + \vec{e}_i \wedge \vec{\omega} \tag{5.8}$$

which we introduce in the differential form of the internal energy u, and we have

$$\begin{aligned} ndu &= \vec{\Sigma}_k d\vec{\beta}_k + nTds = \vec{\Sigma}_k d\vec{\varepsilon}_k + \vec{\Sigma}_k \left(\vec{e}_k \wedge d\vec{\omega}\right) + nTds \\ &= \vec{\Sigma}_k d\vec{\varepsilon}_k + \left(\vec{\Sigma}_k \wedge \vec{e}_k\right) d\vec{\omega} + nTds \end{aligned} \tag{5.9}$$

The mechanical potential which is conjugated to the vector of rotation $\vec{\omega}$ will be called the *torsor of moments* or *torque torsor* $\vec{m}$, which we will then define as

$$\vec{m} = -\sum_k \vec{e}_k \wedge \vec{\Sigma}_k \tag{5.10}$$

It is linked to the anti-symmetric part of the stress tensor $\vec{\Sigma}_i$. In terms of this torque torsor, the differential u takes the form

$$n du = \vec{\Sigma}_k d\vec{\varepsilon}_k + \vec{m} d\vec{\omega} + nT ds = \Sigma_{ij} d\varepsilon_{ij} + m_k d\omega_k + nT ds \tag{5.11}$$

Thus the internal energy u is also a function of the tensor of deformation $\vec{\varepsilon}_i$, of the vector of rotation $\vec{\omega}$ and of entropy s

$$u = u(\varepsilon_{ij}, \omega_k, s) \tag{5.12}$$

Its differential is

$$n du = n\frac{\partial u}{\partial \varepsilon_{ij}} d\varepsilon_{ij} + n\frac{\partial u}{\partial \omega_k} d\omega_k + n\frac{\partial u}{\partial s} ds \tag{5.13}$$

The tensor of deformations $\vec{\varepsilon}_i$ is, by definition, a symmetrical tensor. We will then operate a few transformations on the differential of the internal energy u in order to explicitly show this symmetry which will be very useful later on

$$\begin{aligned} n du &= \frac{1}{2}\Sigma_{ij} d\varepsilon_{ij} + \frac{1}{2}\Sigma_{ji} d\varepsilon_{ji} + m_k d\omega_k + nT ds = \frac{1}{2}\left(\Sigma_{ij} + \Sigma_{ji}\right) d\varepsilon_{ij} + m_k d\omega_k + nT ds \\ &= \frac{n}{2}\left(\frac{\partial u}{\partial \varepsilon_{ij}} + \frac{\partial u}{\partial \varepsilon_{ji}}\right) d\varepsilon_{ij} + n\frac{\partial u}{\partial \omega_k} d\omega_k + n\frac{\partial u}{\partial s} ds \end{aligned} \tag{5.14}$$

We have a symmetric mechanical potential, which is conjugated to the tensor of deformations ε_{ij}, we will call it the *deformation stress tensor* σ_{ij}, and it is defined by

$$\sigma_{ij} = \frac{1}{2}\left(\Sigma_{ij} + \Sigma_{ji}\right) \tag{5.15}$$

This new mechanical potential allows us to write the differential of u simply as

$$n du = \sigma_{ij} d\varepsilon_{ij} + m_k d\omega_k + nT ds \tag{5.16}$$

We define the state equations of the media in the representation $\{\vec{\varepsilon}_i, \vec{\omega}\}$

$$\begin{cases} \sigma_{ij} = \dfrac{n}{2}\left(\dfrac{\partial u}{\partial \varepsilon_{ij}} + \dfrac{\partial u}{\partial \varepsilon_{ji}}\right) = \sigma_{ij}(\varepsilon_{lm}, \omega_n, s) \\ m_k = n\dfrac{\partial u}{\partial \omega_k} = m_k(\varepsilon_{lm}, \omega_n, s) \\ T = \dfrac{\partial u}{\partial s} = T(\varepsilon_{lm}, \omega_n, s) \end{cases} \tag{5.17}$$

The deformation stress tensor σ_{ij} can be written from the tensor Σ_{ij}

$$\vec{\sigma}_i = \sum_j \sigma_{ij}\vec{e}_j = \vec{\Sigma}_i - \frac{1}{2}(\vec{e}_i \wedge \vec{m}) \tag{5.18}$$

The relation $\vec{\varepsilon}_i = \vec{\beta}_i - \vec{e}_i \wedge \vec{\omega}$ allows us to transform the differential of the internal energy u

$$\begin{aligned} n du &= \vec{\Sigma}_k d\vec{\varepsilon}_k + \vec{m} d\vec{\omega} + nT ds = \vec{\sigma}_k d\vec{\varepsilon}_k + \vec{m} d\vec{\omega} + nT ds \\ &= \vec{\sigma}_k d\vec{\beta}_k + \left(\vec{e}_k \wedge \vec{\sigma}_k\right) d\vec{\omega} + \vec{m} d\vec{\omega} + nT ds \end{aligned} \tag{5.19}$$

Since $\vec{\sigma}_i$ is symmetric we have, by definition

$$\sum_k \vec{e}_k \wedge \vec{\sigma}_k = 0 \tag{5.20}$$

In such a way that the differential of the internal energy u can be written as

$$n du = \vec{\sigma}_k d\vec{\beta}_k + \vec{m} d\vec{\omega} + nTds \tag{5.21}$$

This new form of the differential of u depends on the differential $d\vec{\beta}_i$ of the tensor of distortion. This new form was made possible by the introduction of the symmetric stress tensor $\vec{\sigma}_i$. The advantage of this expression is to be able to rewrite the *equation of thermo-kinetic* only using the material derivatives of $\vec{\beta}_i$ and of $\vec{\omega}$, which are linked by the simple geometro-kinetic relations

$$n\frac{du}{dt} = \vec{\sigma}_k \frac{d\vec{\beta}_k}{dt} + \vec{m}\frac{d\vec{\omega}}{dt} + nT\frac{ds}{dt} \tag{5.22}$$

The equation *(5.22)* shows that the temporal variations of the internal energy u along the path is associated with variations of the elastic energy of deformation, the potential energy of rotation and the thermal energy. Furthermore the variations of the elastic energy of deformation can be written in many equivalent ways

$$\vec{\Sigma}_k \frac{d\vec{\varepsilon}_k}{dt} = \vec{\sigma}_k \frac{d\vec{\varepsilon}_k}{dt} = \vec{\sigma}_k \frac{d\vec{\beta}_k}{dt} \tag{5.23}$$

The shear stress tensor, the torque torsor and the scalar of pressure

We use the decomposition of the tensor of deformation $\vec{\varepsilon}_i$ into the shear tensor $\vec{\alpha}_i$ and the scalar of volume expansion τ

$$\vec{\varepsilon}_i = \vec{\alpha}_i + \vec{e}_i \tau / 3 \tag{5.24}$$

to rewrite the differential of u in the following way

$$n du = \vec{\Sigma}_k d\vec{\varepsilon}_k + \vec{m} d\vec{\omega} + nTds = \vec{\Sigma}_k d\vec{\alpha}_k + \frac{1}{3}\vec{e}_k \vec{\Sigma}_k d\tau + \vec{m} d\vec{\omega} + nTds \tag{5.25}$$

This allows us to introduce the conjugate to the scalar of volume expansion τ, which is called the *pressure* p

$$p = -\frac{1}{3}\sum_k \vec{e}_k \vec{\Sigma}_k = -\frac{1}{3}\sum_k \vec{e}_k \vec{\sigma}_k = -\frac{1}{3}\sum_k \Sigma_{kk} = -\frac{1}{3}\sum_k \sigma_{kk} \tag{5.26}$$

So the differential of u is written

$$n du = \vec{\Sigma}_k d\vec{\alpha}_k + \vec{m} d\vec{\omega} - p d\tau + nTds = \Sigma_{ij} d\alpha_{ij} + m_k d\omega_k - p d\tau + nTds \tag{5.27}$$

Internal energy u can be expressed as a function of α_{ij}, ω_k and τ

$$u = u(\alpha_{ij}, \omega_k, \tau, s) \tag{5.28}$$

with the differential

$$n du = n\frac{\partial u}{\partial \alpha_{ij}} d\alpha_{ij} + n\frac{\partial u}{\partial \omega_k} d\omega_k + n\frac{\partial u}{\partial \tau} d\tau + n\frac{\partial u}{\partial s} ds \tag{5.29}$$

The shear tensor $\vec{\alpha}_i$ is symmetric, so

$$n du = \frac{1}{2}\Sigma_{ij} d\alpha_{ij} + \frac{1}{2}\Sigma_{ji} d\alpha_{ji} + m_k d\omega_k - p d\tau + nTds = \frac{1}{2}\left(\Sigma_{ij} + \Sigma_{ji}\right) d\alpha_{ji} + m_k d\omega_k - p d\tau + nTds \tag{5.30}$$

We introduce the *Kronecker symbol* δ_{ij}, such that $\delta_{ij} = 0$ if $i \neq j$ and $\delta_{ij} = 1$ if $i = j$, the sum $\delta_{ij} d\alpha_{ji} = d\alpha_{ii}$ is null, because the shear tensor $\vec{\alpha}_i$ is a null trace tensor, by construction. This allows us to add a term that contains the trace of the stress tensor Σ_{ij} in the differential of u

$$\begin{aligned} ndu &= \frac{1}{2}\left(\Sigma_{ij}+\Sigma_{ji}\right)d\alpha_{ji} - \frac{1}{3}\delta_{ij}d\alpha_{ji}\sum_k \Sigma_{kk} + m_k d\omega_k - pd\tau + nTds \\ &= \left[\frac{1}{2}\left(\Sigma_{ij}+\Sigma_{ji}\right) - \frac{1}{3}\delta_{ij}\sum_k \Sigma_{kk}\right]d\alpha_{ij} + m_k d\omega_k - pd\tau + nTds \end{aligned} \tag{5.31}$$

which can be written as

$$ndu = \left[\frac{n}{2}\left(\frac{\partial u}{\partial \alpha_{ji}} + \frac{\partial u}{\partial \alpha_{ji}}\right) - \frac{n}{3}\delta_{ij}\sum_k \frac{\partial u}{\partial \alpha_{kk}}\right]d\alpha_{ij} + n\frac{\partial u}{\partial \omega_k}d\omega_k + n\frac{\partial u}{\partial \tau}d\tau + n\frac{\partial u}{\partial s}ds \tag{5.32}$$

There appears a new mechanic potential, symmetrical and trace-less, which is conjugated to the shear tensor α_{ij}, which we will call the shear stress tensor s_{ij}, defined by the following relations

$$s_{ij} = \frac{1}{2}\left(\Sigma_{ij}+\Sigma_{ji}\right) - \frac{1}{3}\delta_{ij}\sum_k \Sigma_{kk} = \sigma_{ij} + \delta_{ij}p \tag{5.33}$$

By comparing the differential of u obtained thanks to the potential s_{ij}

$$ndu = s_{ij}d\alpha_{ij} + m_k d\omega_k - pd\tau + nTds \tag{5.34}$$

we can define the following state equations in the representation $\{\vec{\alpha}_i, \vec{\omega}, \tau\}$

$$\left\{ \begin{aligned} & s_{ij} = \frac{n}{2}\left(\frac{\partial u}{\partial \alpha_{ij}} + \frac{\partial u}{\partial \alpha_{ji}}\right) - \frac{n}{3}\delta_{ij}\sum_k \frac{\partial u}{\partial \alpha_{kk}} = s_{ij}(\alpha_{lm}, \omega_n, \tau, s) \\ & m_k = n\frac{\partial u}{\partial \omega_k} = m_k(\alpha_{lm}, \omega_n, \tau, s) \\ & p = -n\frac{\partial u}{\partial \tau} = p(\alpha_{lm}, \omega_n, \tau, s) \\ & T = \frac{\partial u}{\partial s} = T(\alpha_{lm}, \omega_n, \tau, s) \end{aligned} \right. \tag{5.35}$$

The shear stress tensor we just defined can be written in a simple manner in the vectorial form

$$\vec{s}_i = \sum_j s_{ij}\vec{e}_j = \vec{\sigma}_i + p\vec{e}_i \tag{5.36}$$

This tensor is symmetric and with null trace, so the following equations hold

$$\sum_k \vec{e}_k \wedge \vec{s}_k \equiv 0 \quad \text{and} \quad \sum_k \vec{e}_k \vec{s}_k \equiv 0 \tag{5.37}$$

The differential of u is expressed in terms of the shear tensor $\vec{\alpha}_i$

$$ndu = \vec{\Sigma}_k d\vec{\alpha}_k + \vec{m}d\vec{\omega} - pd\tau + nTds = \vec{s}_k d\vec{\alpha}_k + \vec{m}d\vec{\omega} - pd\tau + nTds \tag{5.38}$$

We decompose $\vec{\alpha}_i$

$$\vec{\alpha}_i = \vec{\varepsilon}_i - \frac{1}{3}\tau\vec{e}_i = \vec{\beta}_i - \vec{e}_i \wedge \vec{\omega} - \frac{1}{3}\tau\vec{e}_i \tag{5.39}$$

and exhibit, thanks to *(5.38)*, a direct dependence to $d\vec{\beta}_i$

$$
\begin{aligned}
ndu &= \vec{s}_k d\vec{\beta}_k - \vec{s}_k\left(\vec{e}_k \wedge d\vec{\omega}\right) - \vec{s}_k\left(\frac{1}{3}\vec{e}_k d\tau\right) + \vec{m}d\vec{\omega} - pd\tau + nTds \\
&= \vec{s}_k d\vec{\beta}_k + \left(\sum_k \vec{e}_k \wedge \vec{s}_k\right)d\vec{\omega} - \frac{1}{3}\left(\sum_k \vec{e}_k \vec{s}_k\right)d\tau + \vec{m}d\vec{\omega} - pd\tau + nTds \\
&= \vec{s}_k d\vec{\beta}_k + \vec{m}d\vec{\omega} - pd\tau + nTds
\end{aligned} \tag{5.40}
$$

From this we deduce the thermo-kinetic equation in the representation $\{\vec{\alpha}_i, \vec{\omega}, \tau\}$

$$
n\frac{du}{dt} = \vec{s}_k \frac{d\vec{\beta}_k}{dt} + \vec{m}\frac{d\vec{\omega}}{dt} - p\frac{d\tau}{dt} + nT\frac{ds}{dt} \tag{5.41}
$$

We should note here that each term of the thermo-kinetic equation corresponds respectively to the temporal variations of the energy of elastic shear, global rotation, volume expansion and thermal energy. Furthermore this equation of thermo-kinetic depends on the temporal derivatives of the 3 topological tensors, which we can derive from the velocity field thanks to the geometro-kinetic equations given in table 1.1. Finally the following identities hold

$$
\vec{\Sigma}_k \frac{d\vec{\alpha}_k}{dt} = \vec{\sigma}_k \frac{d\vec{\alpha}_k}{dt} = \vec{s}_k \frac{d\vec{\alpha}_k}{dt} = \vec{s}_k \frac{d\vec{\varepsilon}_k}{dt} = \vec{s}_k \frac{d\vec{\beta}_k}{dt} \tag{5.42}
$$

The decompositions of the stress tensor $\vec{\Sigma}_i$, represented schematically in figure 5.1, follow a procedure completely equivalent, bar a few coefficients, to that used for the decomposition of the tensor of distortion.

Choosing a topological representation

As we have seen, in all generality, the thermo-kinetics of a deformable medium can be described in the three, equivalent, topological representations $\{\vec{\beta}_i\}$, $\{\vec{\varepsilon}_i, \vec{\omega}\}$ or $\{\vec{\alpha}_i, \vec{\omega}, \tau\}$, which each have their conjugate stress potentials, namely $\{\vec{\Sigma}_i\}$, $\{\vec{\sigma}_i, \vec{m}\}$ et $\{\vec{s}_i, \vec{m}, p\}$ respectively. The choice of one representation over another to describe a deformable medium depends on its nature and is a question of convenience as regards to the writing of the internal energy u. Let's give some explicit examples:

- in the case of a fluid, the internal energy u has to be expressed as an explicit function of the volume expansion τ (or of average volume v) and the entropy s, in such a way that the only non-null conservative mechanical potential in a fluid is the pressure

$$
u = u(\tau, s) \quad \text{or} \quad u = u(v, s) \quad \Rightarrow \quad p = -n\frac{\partial u}{\partial \tau} = -\frac{\partial u}{\partial v} \tag{5.43}
$$

which leads us to the following geometro-kinetic equation

$$
u = u(\tau, s) \quad \text{or} \quad u = u(v, s) \quad \Rightarrow \quad n\frac{du}{dt} = -p\frac{d\tau}{dt} + nT\frac{ds}{dt} = -np\frac{dv}{dt} + nT\frac{ds}{dt} \tag{5.44}
$$

- in the case of a usual solid lattice, the potential part of the internal energy depends on the elastic deformations $\vec{\varepsilon}_i$ of the medium, and only on those. It follows that the only non-null mechanical potential for a lattice is generally the symmetrical stress tensor $\vec{\sigma}_i$

$$
u = u\left(\varepsilon_{ij}, s\right) \quad \Rightarrow \quad \sigma_{ij} = \frac{n}{2}\left(\frac{\partial u}{\partial \varepsilon_{ji}} + \frac{\partial u}{\partial \varepsilon_{ij}}\right) \quad \Rightarrow \quad n\frac{du}{dt} = \vec{\sigma}_k \frac{d\vec{\varepsilon}_k}{dt} + nT\frac{ds}{dt} \tag{5.45}
$$

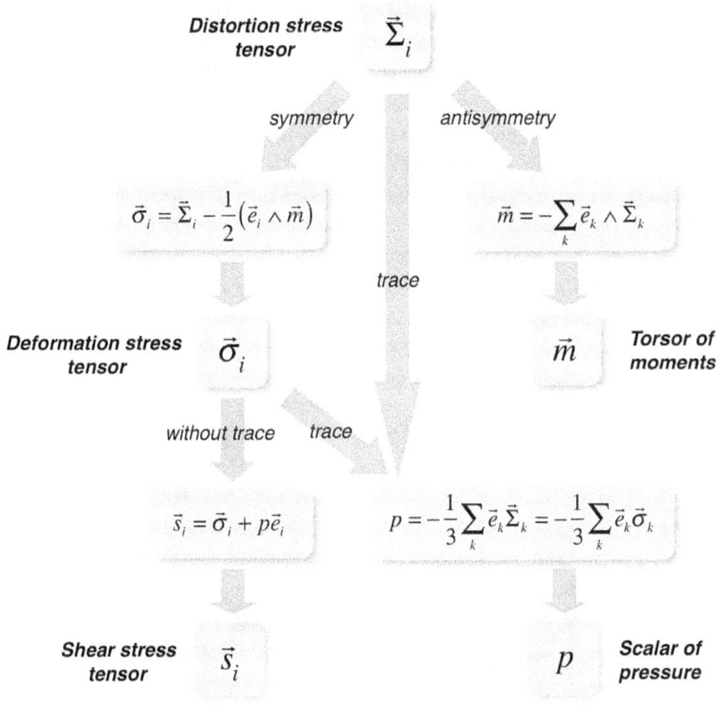

Figure 5.1 *- schematics of the decomposition of the stress tensors*

- in the case of an isotropic lattice, it is preferable to separate the deformations due to shear and the deformations due to volume expansion and to write the internal energy u as a function of the tensor of shear $\vec{\alpha}_i$, the scalar of volume expansion τ and the entropy s. This leads us to the fact that the non-null mechanical potentials are the shear stress tensor $\vec{s}_i$ and the scalar of pressure p

$$u = u(\alpha_{ij}, \tau, s) \quad \Rightarrow \quad s_{ij} = \frac{n}{2}\left(\frac{\partial u}{\partial \alpha_{ji}} + \frac{\partial u}{\partial \alpha_{ij}}\right) - \frac{n}{3}\delta_{ij}\sum_k \frac{\partial u}{\partial \alpha_{kk}} \quad \text{et} \quad p = -n\frac{\partial u}{\partial \tau} \tag{5.46}$$

the thermo-kinetic equation reads

$$u = u(\alpha_{ij}, \tau, s) \quad \Rightarrow \quad n\frac{du}{dt} = \vec{s}_k \frac{d\vec{\alpha}_k}{dt} - p\frac{d\tau}{dt} + nT\frac{ds}{dt} \tag{5.47}$$

Generally there is no direct dependence of the internal energy u on the vector of *global* rotation $\vec{\omega}$, as a global rotation does not deform the medium (merely rotates it). Consequently, the torque torsor $\vec{m}$ is null everywhere. However here we have voluntarily introduced the torque torsor $\vec{m}$ in order to stay general, and to account for the fact that we can easily imagine u as depending on the global vector of rotation $\vec{\omega}$ of the lattice, for example in the case where there

is a coupling of vectorial nature in the lattice having certain directional properties (such as an axial magnetic moment or an electrical axial polarization) and a certain external field defined in an absolute frame (such as a magnetic field or an electric field for example) where a preferred axis would appear and the symmetry of rotation would be broken by the external potential.
While this is true of a *global* rotation, we will show later, that under certain particular conditions, the stress tensor of elastic shear $\vec{s}_i$ can be replaced by a torsor of moments or torque torsor $\vec{m}$ conjugated to the vector of *local* rotation $\vec{\omega}^{(\delta)}$.
Another important remark can be made here. In our description in eulerian coordinates, the mechanic potentials are given by state equations that use the density n of sites of the lattice as a multiplication factor. This density n is directly related to the volume expansion τ of the solid or to the average volume v of the sites of the lattice via relations $n = 1/v = n_0 e^{-\tau}$, which implies that the mechanical potentials will depend in this case non-linearly on the expansion τ of the solid.

5.2 - Anelasticity and plasticity of a lattice

The only way to introduce dissipative processes in a solid lattice of particles is to postulate that the distortions by elastic deformation of the lattice are accompanied by dissipative distortions. It is these dissipative distortions that allow, at the macroscopic scale, relative movement with large distances between particles, by sliding planes of particles one on top of the other. We write in a generic way the total distortions $\vec{\beta}_i$ in the local frame

$$\vec{\beta}_i = \vec{\beta}_i^{(\delta)} + \vec{e}_i \wedge \vec{\omega}_O(t) = \vec{\beta}_i^{\acute{e}l} + \vec{\beta}_i^{dis} + \vec{e}_i \wedge \vec{\omega}_O(t) \tag{5.48}$$

It is clear that the variations of elastic distortions $\vec{\beta}_i^{\acute{e}l}$ contribute to a work of elastic deformation that we will write $\delta W_{\acute{e}l}$,

$$\delta W_{\acute{e}l} = \vec{s}_k d\vec{\beta}_k^{\acute{e}l} + \vec{m} d\vec{\omega}^{\acute{e}l} - p d\tau^{\acute{e}l} \tag{5.49}$$

At the same time the dissipative distortions $\vec{\beta}_i^{dis}$ will be responsible for a variation of work δW_{dis} and a variation of heat δQ_{dis}, which we will thus write

$$\delta W_{dis} + \delta Q_{dis} = \vec{s}_k d\vec{\beta}_k^{dis} + \vec{m} d\vec{\omega}^{dis} - p d\tau^{dis} \tag{5.50}$$

Two possible phenomenologies for dissipation

Relation *(5.50)* can lead to 2 different dissipative phenomenologies, depending on whether the variation of work is null or not:
- if variation of work δW_{dis} is null, then the dissipative distortions do not store potential energy in the network, so *they cannot be recovered.* This thermodynamically irreversible phenomena will be called the *plasticity of the lattice.* The *plastic distortions* satisfy the following relation

$$\delta Q_{pl} = \vec{s}_k d\vec{\beta}_k^{pl} + \vec{m} d\vec{\omega}^{pl} - p d\tau^{pl} \tag{5.51}$$

- if variation of work δW_{dis} is not null, the dissipative distortions do store potential energy in the lattice, so *they are this time recoverable.* However this phenomena is *thermodynamically irre-*

versible since it dissipates heat. It will be called the *anelasticity of the lattice*, and the *anelastic distortions* that are associated with it satisfy the following equation

$$\delta W_{an} + \delta Q_{an} = \vec{s}_k d\vec{\beta}_k^{an} + \vec{m} d\vec{\omega}^{an} - p d\tau^{an} \tag{5.52}$$

Since the anelastic distortions contribute both to variations of work and heat, it is necessary to introduce a way to decompose the previous expression to separate δW_{an} and δQ_{an}. The only logical way to proceed here is to write

$$\begin{cases} \vec{s}_k = \vec{s}_k^{\,cons} + \vec{s}_k^{\,dis} \\ \vec{m} = \vec{m}^{cons} + \vec{m}^{dis} \\ p = p^{cons} + p^{dis} \end{cases} \tag{5.53}$$

and so to decompose the stress tensors in a *conservative part*, which is responsible for the source of work, and a *dissipative part*, which is responsible for the source of heat. This decomposition of stress tensors allows us to write the individual variations of work and heat linked to anelasticity in the following form

$$\begin{cases} \delta W_{an} = \vec{s}_k^{\,cons} d\vec{\beta}_k^{an} + \vec{m}^{cons} d\vec{\omega}^{an} - p^{cons} d\tau^{an} \\ \delta Q_{an} = \vec{s}_k^{\,dis} d\vec{\beta}_k^{an} + \vec{m}^{dis} d\vec{\omega}^{an} - p^{dis} d\tau^{an} \end{cases} \tag{5.54}$$

We can here make a hypothesis of anelasticity, which will allow to simplify the description of dissipative solid medium going forward. Indeed if we suppose that the trace of the anelastic distortion tensor is null, we can immediately conclude that there are no anelastic component of volume expansion, by definition, and therefore that the anelastic deformations are only made of shear

Hypothesis: $$\tau^{an} = \sum_k \vec{\beta}_k^{an} \vec{e}_k \equiv 0 \tag{5.55}$$

The thermo-kinetic equation of an anelastic and plastic lattice

By using relations *(5.49)* and *(5.54)*, and by using the hypothesis *(5.55)*, the variation of work $\delta W_{déf}$ associated with the deformations of a lattice that has elasticity, anelasticity and plasticity is written

$$\delta W_{déf} = \delta W_{él} + \delta W_{an} = \vec{s}_k d\vec{\beta}_k^{él} + \vec{m} d\vec{\omega}^{él} - p d\tau^{él} + \vec{s}_k^{\,cons} d\vec{\beta}_k^{an} + \vec{m}^{cons} d\vec{\omega}^{an} \tag{5.56}$$

The total variation of the work of deformation $\delta W_{déf}$ will be found in the expression for the thermo-kinetic equation for the internal energy u

$$n\frac{du}{dt} = \vec{s}_k \frac{d\vec{\beta}_k^{él}}{dt} + \vec{m}\frac{d\vec{\omega}^{él}}{dt} - p\frac{d\tau^{él}}{dt} + \vec{s}_k^{\,cons} \frac{d\vec{\beta}_k^{an}}{dt} + \vec{m}^{cons} \frac{d\vec{\omega}^{an}}{dt} + nT\frac{ds}{dt} \tag{5.57}$$

This thermo-kinetic equation could be written by replacing the $\vec{\beta}_i$ with $\vec{\alpha}_i$, thanks to property *(5.42)*, so the internal energy function u of a self-diffusive lattice with anelasticity and plasticity is a function of the following state variables

$$u = u\left(\alpha_{ij}^{él}, \alpha_{ij}^{an}, \omega_k^{él}, \omega_k^{an}, \tau^{él}, s\right) \tag{5.58}$$

A first statement can be made about this state function: it does not depend on plastic distortions of the lattice. As a matter of fact, the plastic distortions of the network only appear in the total variation of heat $\delta Q_{déf}$ associated with the dissipative processes of anelastic and plastic deformation

$$\delta Q_{déf} = \delta Q_{an} + \delta Q_{pl} = \vec{s}_k^{\,dis} d\vec{\beta}_k^{an} + \vec{m}^{dis} d\vec{\omega}^{an} + \vec{s}_k d\vec{\beta}_k^{pl} + \vec{m} d\vec{\omega}^{pl} - p d\tau^{pl} \tag{5.59}$$

It is clear that in the description of the anelastic and plastic lattice, this variation of heat will have to appear in the heat equation of the lattice. It is responsible for a source of entropy reflecting the thermodynamical irreversibility of the dissipative processes of anelasticity and plasticity.

The constitutive equations of an anelastic and plastic lattice

Besides the simplifying hypothesis *(5.55)*, there are no restrictions to decomposing the dissipative distortions in relation *(5.48)* in an anelastic part and a plastic part, so that we can write in all generality, the following constitutive equations of a solid lattice that is both anelastic and plastic

$$\begin{cases} \vec{\beta}_i = \vec{\beta}_i^{(\delta)} + \vec{e}_i \wedge \vec{\omega}_0(t) = \vec{\beta}_i^{él} + \vec{\beta}_i^{an} + \vec{\beta}_i^{pl} + \vec{e}_i \wedge \vec{\omega}_0(t) \\ \vec{\omega} = \vec{\omega}^{(\delta)} + \vec{\omega}_0(t) = \vec{\omega}^{él} + \vec{\omega}^{an} + \vec{\omega}^{pl} + \vec{\omega}_0(t) = -\dfrac{1}{2}\sum_k \vec{e}_k \wedge \vec{\beta}_k \\ \tau = \tau^{él} + \tau^{pl} = \sum_k \vec{\beta}_k \vec{e}_k \qquad (\tau^{an} \equiv 0 \textit{ by hypothesis}) \\ \vec{\varepsilon}_i = \vec{\varepsilon}_i^{él} + \vec{\varepsilon}_i^{an} + \vec{\varepsilon}_i^{pl} = \vec{\beta}_i - \vec{e}_i \wedge \vec{\omega} \\ \vec{\alpha}_i = \vec{\alpha}_i^{él} + \vec{\alpha}_i^{an} + \vec{\alpha}_i^{pl} = \vec{\varepsilon}_i - \dfrac{1}{3}\tau \vec{e}_i \end{cases} \tag{5.60}$$

It is clear that the distortions $\vec{\beta}_i$, $\vec{\omega}$ and τ always satisfy, in the local frame, both the equations of geometro-kinetic of table 1.1 that give them in terms of the velocity field $\vec{\phi}$ inside the solid, as well as the equations of geometro-compatibility $\overrightarrow{\mathrm{rot}}\,\vec{\beta}_i = 0$ and $\mathrm{div}\,\vec{\omega} = 0$.

Existence of sources and sinks of sites of lattice in the presence of plasticity

We saw in relation *(1.18)*, that the equation of continuity for n can be represented in the form of a derivative along the trajectory

$$\frac{1}{n}\frac{dn}{dt} = \frac{d}{dt}\left(\ln n + cste\right) = \frac{S_n}{n} - \mathrm{div}\,\vec{\phi} \tag{5.61}$$

By definition there is a link between the volume expansion and the density n of sites of the lattice. But, according to *(5.60)*, the volume expansion contains 2 components since $\tau = \tau^{él} + \tau^{pl}$. One can easily convince oneself that the density n of the lattice sites is not linked directly to the total volume expansion τ, *but only to its elastic part* $\tau^{él}$. This assertion is equivalent to saying

$$\tau^{él} = -\ln\frac{n}{n_0} \quad\Rightarrow\quad n = n_0 \exp(-\tau^{él}) \quad\Rightarrow\quad \frac{1}{n}\frac{dn}{dt} = -\frac{d\tau^{él}}{dt} \tag{5.62}$$

From this assertion we deduce that the equation of continuity *(1.20)* must be written, in presence of elasticity and plasticity as

$$\frac{d\tau^{él}}{dt}=-\frac{S_n}{n}+\operatorname{div}\vec{\phi} \tag{5.63}$$

This relation is very important, as it shows that, in the *presence of plasticity*, the density n of lattice sites cannot obey the same principle of conservation. As a matter of fact, the phenomenas of plastic deformation behave as sources and sinks of lattice sites, which we will justify fully in the rest of this book when we introduce the notion of plastic charges. There appears a *source of lattice sites* S_n different than zero directly linked to the material derivative of the plastic volume expansion

$$S_n=n\frac{d\tau^{pl}}{dt}\quad\Rightarrow\quad\frac{d\tau^{él}}{dt}=-\frac{d\tau^{pl}}{dt}+\operatorname{div}\vec{\phi} \tag{5.64}$$

Thus, in the presence of plasticity the equations of geometro-kinetics will contain, implicitly, a source of lattice sites S_n which is non null and associated to $d\tau^{pl}/dt$. We will write them as

$$\frac{d\vec{\beta}_i}{dt}=\overrightarrow{\operatorname{grad}}\,\phi_i\quad;\quad\frac{d\vec{\omega}}{dt}=\frac{1}{2}\overrightarrow{\operatorname{rot}}\,\vec{\phi}\quad;\quad\frac{d\tau}{dt}=\operatorname{div}\vec{\phi} \tag{5.65}$$

5.3 - Self-diffusion in a lattice

There exists an important consequence to the equation of continuity *(5.61)* for the density n of sites of lattice: the *principle of newtonian conservation of mass* would be violated in the presence of non null site sources! To see this, one only has to multiply the equation of continuity *(1.7)* for n by mass m of the particles of the lattice. Since mass density is equal to $\rho=mn$ and the linear momentum is equal to $\vec{p}=m\vec{\phi}$ in a lattice without point defects, we have the following equation

$$\frac{\partial(mn)}{\partial t}=\frac{\partial\rho}{\partial t}=mS_n-\operatorname{div}\left(mn\vec{\phi}\right)=\rho\frac{d\tau^{pl}}{dt}-\operatorname{div}\left(n\vec{p}\right) \tag{5.66}$$

which has an effective mass source linked to the plasticity of the lattice. It is not possible to admit the non-conservation of mass in a newtonian network, so in a lattice which does not contain point defects, plastic distortions which give us a source of lattice sites S_n different from zero, with $d\tau^{pl}/dt$ non null, are impossible.

On the other hand we can show now that in the case of a network containing self-diffusion due to point defects, plastic distortions lead to a source of lattice sites S_n which is different from zero do not violate the principle of mass conservation and become consequently, possible.

Self-diffusion of particles in a lattice

An intrinsic point defect of *vacancy* type consists of a site of the lattice which does not contain a particle (figure 5.2). It is a *"hole"* in the lattice. An intrinsic point defect of *interstitial* type is a particle which is found in the solid lattice, but which does not occupy a regular site of the lattice (figure 5.2). It is an *"extra"* particle in the lattice.

It is easy to see that the presence of such point defects leads to a mass transport phenomena via the self-diffusion of said defects. We show these 2 mechanisms in a lattice moving with absolute velocity $\vec{\phi}$ (figure 5.2). The movement of a vacancy L (L for *lacuna*, meaning a missing particle in the lattice) with relative velocity $\Delta\vec{\varphi}_L$ with respect to the lattice in a given direction leads to a relative flux of mass in the opposite direction with velocity $-\Delta\vec{\varphi}_L$, while the movement of an interstitial I with relative velocity $\Delta\vec{\varphi}_I$ with respect to the lattice in a given direction, gives rise to a flux of mass in the same direction with velocity $\Delta\vec{\varphi}_I$.

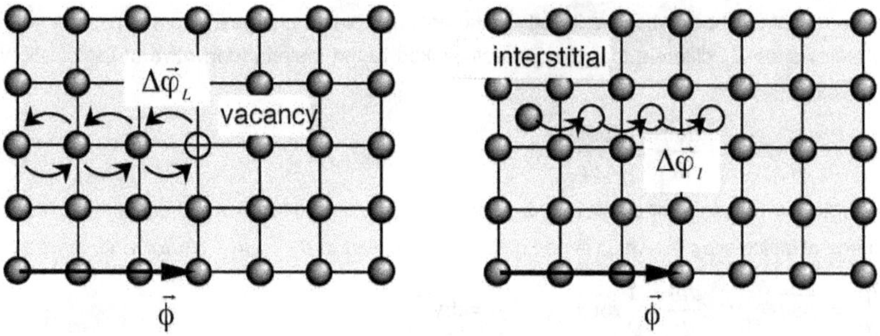

Figure 5.2 *- movement of a hole or interstitial in a solid lattice*

To mathematically transcribe the existence of these punctual defects, one must introduce a notion of volume density of vacancies, n_L, and interstitials, n_I

$$n_L = n_L(\vec{\xi},t) \quad \text{and} \quad n_I = n_I(\vec{\xi},t) \quad \text{in} \quad Q\xi_1\xi_2\xi_3$$
$$n_L = n_L(\vec{r},t) \quad \text{and} \quad n_I = n_I(\vec{r},t) \quad \text{in} \quad Ox_1x_2x_3 \qquad (5.67)$$

From these densities, it is possible to define the notion of *atomic concentrations* of vacancies and self-interstitials with respect to density n of lattice sites with the following

$$C_L = n_L / n \quad \text{and} \quad C_I = n_I / n \qquad (5.68)$$

Contrary to appearances, there is a certain asymmetry between vacancies and self-interstitials, which is seen in the fact that the maximum atomic concentration of vacancies is always limited to 1, when all sites of the lattice are empty, while the atomic concentration of self-interstitials depends on the number of interstitial sites and the number of defects one can put per site.
At point $\vec{\xi}$ of absolute space $Q\xi_1\xi_2\xi_3$ and at instant t, the absolute velocity $\vec{\phi}_L$ of the vacancies and the absolute velocity $\vec{\phi}_I$ of interstitials can be expressed in terms of the relative velocities $\Delta\vec{\varphi}_L$ et $\Delta\vec{\varphi}_I$ of these defects with respect to the lattice as such

$$\vec{\phi}_L = \vec{\phi} + \Delta\vec{\varphi}_L \quad \text{and} \quad \vec{\phi}_I = \vec{\phi} + \Delta\vec{\varphi}_I \qquad (5.69)$$

In the local frame $Ox_1x_2x_3$, velocity $\vec{\varphi}_L$ of the vacancies and velocity $\vec{\varphi}_I$ of the interstitials can be written

$$\vec{\varphi}_L = \vec{\varphi} + \Delta\vec{\varphi}_L \quad \text{and} \quad \vec{\varphi}_I = \vec{\varphi} + \Delta\vec{\varphi}_I \qquad (5.70)$$

If we follow a piece of the lattice during its evolution in space (figure 5.3), the equations of continuity which characterize n, n_L et n_I on this lattice can be found, provided we write them as integrals on this mobile lattice. Indeed on this mobile volume, the number of sites does not change, a fact we translate in a relation valid both in $Q\xi_1\xi_2\xi_3$ as well as $Ox_1x_2x_3$

$$\frac{d}{dt}\iiint_{V_m} n\,dV = 0 \tag{5.71}$$

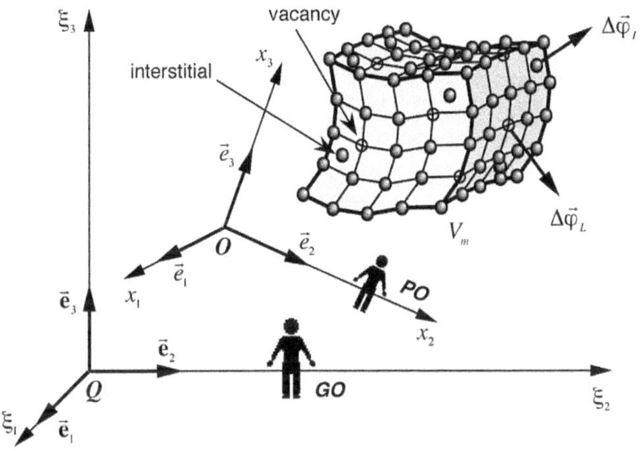

Figure 5.3 - *flux of vacancies and interstitials across the surface of a chunk of lattice*

Regarding the integral equation of evolution describing the number of vacancies or interstitials in a volume V_m, they can be written by taking into account, first the volume sources S_L and S_I associated with the creation and annihilation of vacancies and interstitials in the volume V_m, and second the flux of vacancies and interstitials entering or leaving said volume V_m via its boundary surface with relative velocity $\Delta\vec{\varphi}_L$ or $\Delta\vec{\varphi}_I$

$$\begin{cases} \dfrac{d}{dt}\displaystyle\iiint_{V_m} n_L\,dV = -\oint\!\!\oint_{S_m} n_L\Delta\vec{\varphi}_L\,d\vec{S} + \iiint_{V_m} S_L\,dV \\ \dfrac{d}{dt}\displaystyle\iiint_{V_m} n_I\,dV = -\oint\!\!\oint_{S_m} n_I\Delta\vec{\varphi}_I\,d\vec{S} + \iiint_{V_m} S_I\,dV \end{cases} \tag{5.72}$$

In their local form, in the absolute frame $Q\xi_1\xi_2\xi_3$, and by using the formulas of derivation of an integral on a mobile volume, the equations of continuity for the density of sites of lattice and for the density of vacancies and interstitials can be written as

$$\begin{cases} \dfrac{\partial n}{\partial t} + \operatorname{div}(n\vec{\phi}) = 0 \\ \dfrac{\partial n_L}{\partial t} + \operatorname{div}(n_L\vec{\phi}) = S_L - \operatorname{div}(n_L\Delta\vec{\varphi}_L) \\ \dfrac{\partial n_I}{\partial t} + \operatorname{div}(n_I\vec{\phi}) = S_I - \operatorname{div}(n_I\Delta\vec{\varphi}_I) \end{cases} \tag{5.73}$$

The equations of continuity can be written in their classic form, by making explicit the absolute velocity of vacancies and interstitials thanks to relations *(5.70)*

$$\begin{cases} \dfrac{\partial n}{\partial t} = -\operatorname{div}\left(n\vec{\phi}\right) \\ \dfrac{\partial n_L}{\partial t} = S_L - \operatorname{div}\left(n_L\vec{\phi}_L\right) \\ \dfrac{\partial n_I}{\partial t} = S_I - \operatorname{div}\left(n_I\vec{\phi}_I\right) \end{cases} \tag{5.74}$$

The equation of continuity for the density of lattice sites n does not contain a source term for lattice sites ($S_n = 0$), so that the topological equations of an elastic lattice with self-diffusion remain exactly the same to that established in the case of an elastic lattice. The first equations of continuity *(5.73)* obtained for n_L and n_I can be further transformed by replacing n_L and n_I by nC_L et nC_I and using the material derivative

$$\begin{cases} \dfrac{d(nC_L)}{dt} = S_L - nC_L \operatorname{div}\vec{\phi} - \operatorname{div}\left(n_L\Delta\vec{\varphi}_L\right) \\ \\ \dfrac{d(nC_I)}{dt} = S_I - nC_I \operatorname{div}\vec{\phi} - \operatorname{div}\left(n_I\Delta\vec{\varphi}_I\right) \end{cases} \tag{5.75}$$

Thanks to the following relation, it is easy to verify that for $S_n = 0$

$$\frac{d(nC_L)}{dt} = n\frac{dC_L}{dt} + C_L\frac{dn}{dt} = n\frac{dC_L}{dt} - nC_L \operatorname{div}\vec{\phi} \tag{5.76}$$

Equations that directly describe the variations of the atomic concentrations C_L and C_I along the path, are obtained

$$\begin{cases} n\dfrac{dC_L}{dt} = S_L - \operatorname{div}\left(n_L\Delta\vec{\varphi}_L\right) \\ \\ n\dfrac{dC_I}{dt} = S_I - \operatorname{div}\left(n_I\Delta\vec{\varphi}_I\right) \end{cases} \tag{5.77}$$

We define the *fluxes of diffusion* $\vec{J}_L$ and $\vec{J}_I$ of vacancies and interstitials with respect to the lattice with the following relations

$$\begin{cases} \vec{J}_L = nC_L\Delta\vec{\varphi}_L = nC_L\left(\vec{\phi}_L - \vec{\phi}\right) = nC_L\left(\vec{\varphi}_L - \vec{\varphi}\right) \\ \\ \vec{J}_I = nC_I\Delta\vec{\varphi}_I = nC_I\left(\vec{\phi}_I - \vec{\phi}\right) = nC_I\left(\vec{\varphi}_I - \vec{\varphi}\right) \end{cases} \tag{5.78}$$

We will now rewrite the *equations of diffusion of vacancies and interstitials* in their final form

$$\begin{cases} n\dfrac{dC_L}{dt} = S_L - \operatorname{div}\vec{J}_L \\ \\ n\dfrac{dC_I}{dt} = S_I - \operatorname{div}\vec{J}_I \end{cases} \tag{5.79}$$

These equations, which use the material derivative are valid both in the absolute frame $Q\xi_1\xi_2\xi_3$ as well as the local frame $Ox_1x_2x_3$. Indeed we can verify that the equations *(5.79)* can be deduced from the equations of continuity of integral relations *(5.72)*, but written in the local frame $Ox_1x_2x_3$, namely

$$\begin{cases} \dfrac{\partial n}{\partial t} = -\operatorname{div}(n\vec{\varphi}) \\ \dfrac{\partial n_L}{\partial t} = S_L - \operatorname{div}(n_L\vec{\varphi}_L) \\ \dfrac{\partial n_I}{\partial t} = S_I - \operatorname{div}(n_I\vec{\varphi}_I) \end{cases} \tag{5.80}$$

Chemical potential and thermo-kinetic equation in the presence of self-diffusion

In an elastic, anelastic and plastic lattice, with self-diffusion, it is clear that the atomic concentrations C_L and C_I of vacancies and self-interstitials must also influence the energetic state of the lattice, so that we must complete the state function of internal energy *(5.58)*, and write it now as a function of thermodynamical variables as follows

$$u = u\left(\alpha_{ij}^{él}, \omega_k^{él}, \tau^{él}, \alpha_{ij}^{an}, \omega_k^{an}, C_L, C_I, s\right) \tag{5.81}$$

The differential of the state function u along the path of the sites of lattice can be written

$$du = \frac{\partial u}{\partial \alpha_{ij}^{él}} d\alpha_{ij}^{él} + \frac{\partial u}{\partial \omega_k^{él}} d\omega_k^{él} + \frac{\partial u}{\partial \tau^{él}} d\tau^{él} + \frac{\partial u}{\partial \alpha_{ij}^{an}} d\alpha_{ij}^{an} + \frac{\partial u}{\partial \omega_k^{an}} d\omega_k^{an} + \frac{\partial u}{\partial s} ds + \frac{\partial u}{\partial C_L} dC_L + \frac{\partial u}{\partial C_I} dC_I \tag{5.82}$$

This differential allows us to introduce the mechanical potentials of stress, as well as the thermal potential of temperature $T = \partial u / \partial s$. By analogy, it is now possible to introduce new potentials associated to the partial derivatives $\partial u / \partial C_L$ and $\partial u / \partial C_I$. These potentials will be called the *chemical potentials* μ_L and μ_I of vacancies and interstitials, respectively. Bearing in mind relation *(5.42)*, we can now write the differential

$$n\,du = \vec{s}_k d\vec{\beta}_k^{él} + m_k \omega_k^{él} - p\,d\tau^{él} + \vec{s}_k^{cons} d\vec{\beta}_k^{an} + m_k^{cons} \omega_k^{an} + nT\,ds + n\mu_L dC_L + n\mu_I dC_I \tag{5.83}$$

The thermo-kinetic equation of such a solid lattice can be immediately deduced

$$n\frac{du}{dt} = \vec{s}_k \frac{d\vec{\beta}_k^{él}}{dt} + \vec{m}\frac{d\vec{\omega}^{él}}{dt} - p\frac{d\tau^{él}}{dt} + \vec{s}_k^{cons}\frac{d\vec{\beta}_k^{an}}{dt} + \vec{m}^{cons}\frac{d\vec{\omega}^{an}}{dt} + n\mu_L\frac{dC_L}{dt} + n\mu_I\frac{dC_I}{dt} + nT\frac{ds}{dt} \tag{5.84}$$

Application of the newtonian principle of mass conservation

Indeed, the equations of continuity *(5.74)* and *(5.80)* which were obtained in the case of a self-diffusive lattice can be written taking into account the source S_n of sites of the lattice, due to the plastic component of volume expansion of the solid lattice. The first equation describing n is written in $Q\xi_1\xi_2\xi_3$ and $Ox_1x_2x_3$ respectively

$$\frac{\partial n}{\partial t} = n\frac{d\tau^{pl}}{dt} - \operatorname{div}\left(n\vec{\phi}\right) = n\frac{d\tau^{pl}}{dt} - \operatorname{div}\left(n\vec{\varphi}\right) \tag{5.85}$$

while relations *(5.74)* and *(5.80)* for the density of vacancies n_L and interstitials n_I remain unchanged

$$\begin{cases} \dfrac{\partial n_L}{\partial t} = S_L - \operatorname{div}\left(n_L \vec{\phi}_L\right) = S_L - \operatorname{div}\left(n_L \vec{\varphi}_L\right) \\ \dfrac{\partial n_I}{\partial t} = S_I - \operatorname{div}\left(n_I \vec{\phi}_I\right) = S_I - \operatorname{div}\left(n_I \vec{\varphi}_I\right) \end{cases} \tag{5.86}$$

The density of mass ρ of the solid lattice containing n_L vacancies and n_I interstitials per unit volume can be simply expressed

$$\rho = m\left(n + n_I - n_L\right) = mn\left(1 + C_I - C_L\right) \tag{5.87}$$

We can establish an equation of continuity for the mass density ρ in the absolute frame $Q\xi_1\xi_2\xi_3$ from the equations *(5.85) and (5.86)*

$$\frac{\partial \rho}{\partial t} = \frac{\partial}{\partial t}\left[m\left(n + n_I - n_L\right)\right] = -\operatorname{div}\left[m\left(n\vec{\phi} + n_I\vec{\phi}_I - n_L\vec{\phi}_L\right)\right] + m\left(S_I - S_L\right) + mn\frac{d\tau^{pl}}{dt} \tag{5.88}$$

We can compare this equation with the fundamental equation of continuity of mass obtained from relations *(4.5)* and *(4.14)*

$$\frac{\partial \rho}{\partial t} = S_m - \operatorname{div}\left(n\vec{p}\right) = S_m - \operatorname{div}\left(\rho\vec{\phi} + \vec{J}_m\right) \tag{5.89}$$

From this comparison we have a first relation giving the source of mass S_m per unit volume

$$S_m = m\left(S_I - S_L\right) + mn\frac{d\tau^{pl}}{dt} = m\left(S_I - S_L\right) + mS_n \tag{5.90}$$

If we admit that *principle of conservation of mass* cannot be violated, which we capture with the following hypothesis

Hypothesis: $S_m \equiv 0$ *(5.91)*

then we have the following relation between S_n, S_L et S_I

$$S_n = n\frac{d\tau^{pl}}{dt} = S_L - S_I \tag{5.92}$$

In this expression, the terms of source of vacancies S_L and of interstitial source S_I contain in fact 2 contributions: *the spontaneous creation and annihilation of pairs of 'vacancy-interstitial'*, which we will write as S_{I-L}, and *the creation and annihilation of vacancies and/or interstitials by the process of plastic deformation*, which will be written S_L^{pl} and S_I^{pl}. The total sources S_L and S_I can be decomposed as such

$$S_L = S_{I-L} + S_L^{pl} \quad ; \quad S_I = S_{I-L} + S_I^{pl} \tag{5.93}$$

These values of S_L and S_I can be put in equation *(5.92)*

$$S_n = n\frac{d\tau^{pl}}{dt} = S_L^{pl} - S_I^{pl} \tag{5.94}$$

which translates the fact that plastic distortions with non-null trace, can either be sources or sinks of vacancies and/or interstitials.

Self-diffusion in the presence of sources of lattice sites by plasticity

From the equations of continuity *(5.86)* obtained for n_L and n_I, by replacing n_L and n_I by nC_L and nC_I, we have

$$\begin{cases} d(nC_L)/dt = S_L - nC_L \operatorname{div}\vec{\phi} - \operatorname{div}(n_L\Delta\vec{\varphi}_L) \\ d(nC_I)/dt = S_I - nC_I \operatorname{div}\vec{\phi} - \operatorname{div}(n_I\Delta\vec{\varphi}_I) \end{cases} \tag{5.95}$$

With relation *(5.66)* which contains a non null source S_n of lattice sites, we do the following operations.

$$\begin{cases} \dfrac{d(nC_L)}{dt} = n\dfrac{dC_L}{dt} + C_L\dfrac{dn}{dt} = n\dfrac{dC_L}{dt} + C_L S_n - C_L n \operatorname{div}\vec{\phi} \\ \dfrac{d(nC_I)}{dt} = n\dfrac{dC_I}{dt} + C_I\dfrac{dn}{dt} = n\dfrac{dC_I}{dt} + C_I S_n - C_I n \operatorname{div}\vec{\phi} \end{cases} \tag{5.96}$$

we introduce, again, the fluxes of diffusion $\vec{J}_L$ and $\vec{J}_I$ of vacancies and interstitials, with respect to the lattice, defined in relations *(5.59)*, and the equations of self-diffusion of vacancies and interstitials in the lattice take the following form which is valid both in the global frame $Q\xi_1\xi_2\xi_3$ or the local frame $Ox_1x_2x_3$

$$\begin{cases} n\dfrac{dC_L}{dt} = S_L - C_L S_n - \operatorname{div}\vec{J}_L \\ n\dfrac{dC_I}{dt} = S_I - C_I S_n - \operatorname{div}\vec{J}_I \end{cases} \tag{5.97}$$

Comparatively to the equations of self-diffusion *(5.60)* obtained previously, we have here the extra terms $-C_L S_n$ and $-\,C_I S_n$ associated with source S_n of lattice sites. By using relations *(5.94) and (5.95)*, we explicitly exhibit the sources and sinks of point defects

$$\begin{cases} n\dfrac{dC_L}{dt} = \left(S_{I\text{-}L} + S_L^{pl}\right) - C_L\left(S_L^{pl} - S_I^{pl}\right) - \operatorname{div}\vec{J}_L \\ n\dfrac{dC_I}{dt} = \left(S_{I\text{-}L} + S_I^{pl}\right) - C_I\left(S_L^{pl} - S_I^{pl}\right) - \operatorname{div}\vec{J}_I \end{cases} \tag{5.98}$$

5.4 - Newtonian dynamic of a lattice

If the lattice considered behaves in a newtonian manner in the absolute frame of the ***GO***, we can express its local linear momentum, its local kinetic energy and the source of work of external forces.

The linear momentum

From comparing *(5.88)* and *(5.89)*, there emerges another interesting relation giving the average linear momentum $\vec{p}$ per lattice site

$$\vec{p} = m\left(\vec{\phi} + C_I\vec{\phi}_I - C_L\vec{\phi}_L\right) = \frac{1}{n}\left(\rho\vec{\phi} + \vec{J}_m\right) \tag{5.99}$$

The linear momentum $\vec{p}$ can be written in a slightly different manner by decomposing the velocities $\vec{\phi}_L$ and $\vec{\phi}_I$ thanks to relations *(5.69)*

$$\vec{p} = m\left(\vec{\phi} + C_I - C_L\right)\vec{\phi} + \frac{m}{n}\left(nC_I\Delta\vec{\varphi}_I - nC_L\Delta\vec{\varphi}_L\right) \tag{5.100}$$

We introduce the diffusion fluxes $\vec{J}_L$ and $\vec{J}_I$ of vacancies and interstitials defined in *(5.78), and* we deduce that

$$\vec{p} = m\vec{\phi} + m\left(C_I - C_L\right)\vec{\phi} + \frac{m}{n}\left(\vec{J}_I - \vec{J}_L\right) \tag{5.101}$$

and thus we see that the average linear momentum transported per site of the lattice is the sum of the linear momentum $m\vec{\phi}$ associated with the absolute velocity of the lattice, the linear momentum $m\left(C_I - C_L\right)\vec{\phi}$ due to the transport of vacancies and interstitials by the lattice and finally the linear momentum $m\left(\vec{J}_I - \vec{J}_L\right)/n$ associated to the self-diffusion of vacancies and interstitials by the lattice

We use the mass density ρ *(5.87)*, to write $\vec{p}$ as

$$\vec{p} = \frac{1}{n}\left[\rho\vec{\phi} + m\left(\vec{J}_I - \vec{J}_L\right)\right] \tag{5.102}$$

By comparing this relation with relation *(5.90)*, we obtain

$$\vec{J}_m = m\left(\vec{J}_I - \vec{J}_L\right) \tag{5.103}$$

which also means that the average linear momentum $\vec{p}$ associated to each site of the lattice is the sum of a mass transport $\rho\vec{\phi}/n$ deriving from the local entrained movement of the lattice charged with vacancies and interstitials, and a mass transport $m\left(\vec{J}_I - \vec{J}_L\right)/n = \vec{J}_m/n$ due to the self-diffusion of intrinsic point defects of vacancy and interstitial types.

It is possible to derive the value of $\vec{p}$ by another mean. In the lattice there are n_I interstitial particles with relative velocity $\Delta\vec{\varphi}_I$, that transport the mass in the direction $\Delta\vec{\varphi}_I$. Furthermore, the movement of the n_L vacancies with velocity $\Delta\vec{\varphi}_L$ is associated with a mass flux of n_L particles of the lattice with velocity $-\Delta\vec{\varphi}_L$, that transport mass in the direction opposite to $\Delta\vec{\varphi}_L$. The rest of the particles of the lattice, meaning the $n - n_L$ particles situated on nodes, from which we have subtracted, again, the n_L particles situated on nodes associated with vacancy jumps, move with a null velocity with respect to the lattice, hence the following summary of the dynamic situation of the particles by unit volume of the media:

- There exists $n - 2n_L$ particles with null velocity with respect to the lattice and thus with absolute velocity $\vec{\phi}$,

- There exists n_I particles with relative velocity $\Delta\vec{\varphi}_I$ with respect to the lattice and thus with absolute velocity $\vec{\phi} + \Delta\vec{\varphi}_I = \vec{\phi}_I$,

- There exists n_L particles with relative velocity $-\Delta\vec{\varphi}_L$ with respect to the lattice, and thus with absolute velocity $\vec{\phi} - \Delta\vec{\varphi}_L = 2\vec{\phi} - \vec{\phi}_L$.

The flux of mass, or density of linear momentum $\vec{P}_m$ at point $\vec{\xi}$ and at instant t in the absolute frame $Q\xi_1\xi_2\xi_3$, can be deduced from this dynamical situation of the particles, and we deduce the expressions *(5.99) to (5.102)*

$$\vec{P}_m = m(n-2n_L)\vec{\phi} + mn_L(2\vec{\phi}-\vec{\phi}_L) + mn_I\vec{\phi}_I$$
$$= m(n+n_I-n_L)\vec{\phi} + mn_I\Delta\vec{\varphi}_I - mn_L\Delta\vec{\varphi}_L = \rho\vec{\phi} + m(\vec{J}_I - \vec{J}_L) = n\vec{p} \qquad (5.104)$$

Kinetic energy

We can also deduce the average kinetic energy e_{cin} per site of the lattice. Indeed, the volume density of kinetic energy E_{cin} is simply written from the absolute velocities and the densities of the various flavors of defects

$$E_{cin} = (n-2n_L)\frac{1}{2}m\vec{\phi}^2 + n_I\frac{1}{2}m\vec{\phi}_I^2 + n_L\frac{1}{2}m(2\vec{\phi}-\vec{\phi}_L)^2 \qquad (5.105)$$

We deduce the average kinetic energy e_{cin} per site by division per n

$$e_{cin} = (1-2C_L)\frac{1}{2}m\vec{\phi}^2 + C_I\frac{1}{2}m\vec{\phi}_I^2 + C_L\frac{1}{2}m(2\vec{\phi}-\vec{\phi}_L)^2 \qquad (5.106)$$

The average kinetic energy per site of the lattice is equal to the kinetic energy of a site of the lattice minus a fraction corresponding to the kinetic energy of a couple of particles subtracted from the lattice to have vacancy diffusion, plus a fraction corresponding to the kinetic energy of interstitial and a fraction corresponding to the kinetic energy of vacancy jumps.
The expression e_{cin} can be written by replacing $\vec{\phi}$ by $\vec{\phi}_L - \Delta\vec{\varphi}_L$ in the following term

$$\frac{1}{2}m(2\vec{\phi}-\vec{\phi}_L)^2 - m\vec{\phi}^2 = -\frac{1}{2}m(\vec{\phi}_L^2 - 2\Delta\vec{\varphi}_L^2) \qquad (5.107)$$

which allows us to write

$$e_{cin} = \frac{1}{2}m\left[\vec{\phi}^2 + C_I\vec{\phi}_I^2 - C_L(\vec{\phi}_L^2 - 2\Delta\vec{\varphi}_L^2)\right] \qquad (5.108)$$

The source of work of external forces

With respect to the source of work S_w^{ext} due to a field of external forces, it can be calculated if we know the nature of the field force. For example, if we suppose that the lattice is in a *constant gravity field* $\vec{g}$, we will have

$$S_w^{ext} = m\left[(n-2n_L)\vec{\phi}\vec{g} + n_I\vec{\phi}_I\vec{g} + n_L(2\vec{\phi}-\vec{\phi}_L)\vec{g}\right] = \vec{P}_m\vec{g} = n\vec{p}\vec{g} \qquad (5.109)$$

By introducing the expression *(5.102)* for $\vec{p}$, we have the following expression for the source of work for a constant gravity field

$$S_w^{ext} = \rho\vec{\phi}\vec{g} + (\vec{J}_I - \vec{J}_L)m\vec{g} \qquad (5.110)$$

Chapter 6

Evolution equations for a newtonian lattice

The introduction of the elasticity, of the dissipative mechanical processes of anelasticity and plasticity, of the processes of mass transport and of the newtonian dynamics of the lattice in the fundamental equations of continuity for energy and entropy, allows us to re-derive the entire set of equations that describe the spatiotemporal evolution of said lattice. These behavioral equations are of two kinds: the *fundamental equations* which stay identical to themselves regardless of the lattice under consideration and the *phenomenological equations* which do depend (and describe) the particular nature of the lattice under consideration.

6.1 - Equations of evolution of a lattice

The first part of the equation describing the principle of energy continuity in an elastic, anelastic and plastic solid lattice, in the presence of vacancies and self-interstitials, is written using the relations *(5.108)* for e_{cin} and *(5.84)* for du/dt, all the while assuming here, in all generality, that internal energy u can depend on $\vec{\omega}^{él}$ and $\vec{\omega}^{an}$

$$\begin{aligned} n\frac{de_{cin}}{dt} + n\frac{du}{dt} &= nm\vec{\phi}\frac{d\vec{\phi}}{dt} + n\left[\mu_I + \frac{1}{2}m\vec{\phi}_I^2\right]\frac{dC_I}{dt} + nmC_I\boxed{\vec{\phi}_I}\frac{d\vec{\phi}_I}{dt} \\ &+n\left[\mu_L - \frac{1}{2}m\left(\vec{\phi}_L^2 - 2\Delta\vec{\varphi}_L^2\right)\right]\frac{dC_L}{dt} - nmC_L\boxed{\vec{\phi}_L}\frac{d\vec{\phi}_L}{dt} + 2nmC_L\Delta\vec{\varphi}_L\frac{d\Delta\vec{\varphi}_L}{dt} \\ &+\vec{s}_k\frac{d\vec{\beta}_k^{él}}{dt} + \vec{m}\frac{d\vec{\omega}^{él}}{dt} - p\frac{d\tau^{él}}{dt} + \vec{s}_k^{cons}\frac{d\vec{\beta}_k^{an}}{dt} + \vec{m}^{cons}\frac{d\vec{\omega}^{an}}{dt} + nT\frac{ds}{dt} \end{aligned} \qquad (6.1)$$

The terms in boxes can be decomposed with relations *(5.69)*, and it is possible to introduce the *generalized chemical energies* μ_L^* and μ_I^* of vacancies and interstitials, defined as the sum of chemical potentials and kinetic energies associated with each of the diffusing species

$$\mu_L^* = \mu_L - \frac{1}{2}m\left(\vec{\phi}_L^2 - 2\Delta\vec{\varphi}_L^2\right) = \mu_L + e_{cin}^L \quad \text{and} \quad \mu_I^* = \mu_I + \frac{1}{2}m\vec{\phi}_I^2 = \mu_I + e_{cin}^I \qquad (6.2)$$

Thanks to relation *(5.60)*, the expression

$$\vec{s}_k\frac{d\vec{\beta}_k^{él}}{dt} + \vec{m}\frac{d\vec{\omega}^{él}}{dt} - p\frac{d\tau^{él}}{dt} + \vec{s}_k^{cons}\frac{d\vec{\beta}_k^{an}}{dt} + \vec{m}^{cons}\frac{d\vec{\omega}^{an}}{dt} \qquad (6.3)$$

can be replaced by the following, which has the advantage of explicitly showing $\vec{\beta}_i$, $\vec{\omega}$ and τ

$$\vec{s}_k\frac{d\vec{\beta}_k}{dt} + \vec{m}\frac{d\vec{\omega}}{dt} - p\frac{d\tau}{dt} + \left(-\vec{s}_k^{dis}\frac{d\vec{\beta}_k^{an}}{dt} - \vec{m}^{dis}\frac{d\vec{\omega}^{an}}{dt} - \vec{s}_k\frac{d\vec{\beta}_k^{pl}}{dt} - \vec{m}\frac{d\vec{\omega}^{pl}}{dt} + p\frac{d\tau^{pl}}{dt}\right) \qquad (6.4)$$

We then have

$$n\frac{de_{cin}}{dt}+n\frac{du}{dt}=nm\vec{\phi}\left(\frac{d\vec{\phi}}{dt}+C_I\frac{d\vec{\phi}_I}{dt}-C_L\frac{d\vec{\phi}_L}{dt}\right)+m\vec{J}_I\frac{d\vec{\phi}_I}{dt}-m\vec{J}_L\frac{d}{dt}\left(\vec{\phi}_L-2\Delta\vec{\varphi}_L\right)$$
$$+nT\frac{ds}{dt}+n\mu_I^*\frac{dC_I}{dt}+n\mu_L^*\frac{dC_L}{dt}+\vec{s}_k\frac{d\vec{\beta}_k}{dt}+\vec{m}\frac{d\vec{\omega}}{dt}-p\frac{d\tau}{dt} \tag{6.5}$$
$$+\left(-\vec{s}_k^{\,dis}\frac{d\vec{\beta}_k^{an}}{dt}-\vec{m}^{dis}\frac{d\vec{\omega}^{an}}{dt}-\vec{s}_k\frac{d\vec{\beta}_k^{pl}}{dt}-\vec{m}\frac{d\vec{\omega}^{pl}}{dt}+p\frac{d\tau^{pl}}{dt}\right)$$

In the second line of this expression, the material derivatives of the topological tensors and the atomic concentrations are replaceable using the equations of geometro-kinetic *(5.65)* and the equations of diffusion *(5.98)*. We then have

$$n\frac{de_{cin}}{dt}+n\frac{du}{dt}=\left[nm\frac{d\vec{\phi}}{dt}+nmC_I\frac{d\vec{\phi}_I}{dt}-nmC_L\frac{d\vec{\phi}_L}{dt}-\sum_k\vec{e}_k\operatorname{div}\vec{s}_k+\frac{1}{2}\overrightarrow{\operatorname{rot}}\,\vec{m}+\overrightarrow{\operatorname{grad}}p\right]\vec{\phi}$$
$$-\operatorname{div}\left[\mu_L^*\vec{J}_L+\mu_I^*\vec{J}_I-\phi_k\vec{s}_k-\frac{1}{2}\left(\vec{\phi}\wedge\vec{m}\right)+p\vec{\phi}\right] \tag{6.6}$$
$$+\left[\begin{array}{l}\mu_L^*(S_L-C_LS_n)+\mu_I^*(S_I-C_IS_n)\\ +\vec{J}_I\left(\overrightarrow{\operatorname{grad}}\mu_I^*+m\dfrac{d\vec{\phi}_I}{dt}\right)+\vec{J}_L\left(\overrightarrow{\operatorname{grad}}\mu_L^*-m\dfrac{d\left(\vec{\phi}_L-2\Delta\vec{\varphi}_L\right)}{dt}\right)\\ -\vec{s}_k^{\,dis}\dfrac{d\vec{\beta}_k^{an}}{dt}-\vec{m}^{dis}\dfrac{d\vec{\omega}^{an}}{dt}-\vec{s}_k\dfrac{d\vec{\beta}_k^{pl}}{dt}-\vec{m}\dfrac{d\vec{\omega}^{pl}}{dt}+p\dfrac{d\tau^{pl}}{dt}+nT\dfrac{ds}{dt}\end{array}\right]$$

This expression must be compared to the second part of the first principle *(4.23)*, in which the sources S_n is not null

$$n\frac{de_{cin}}{dt}+n\frac{du}{dt}=S_w^{ext}-\operatorname{div}\vec{J}_w-\operatorname{div}\vec{J}_q-uS_n-e_{cin}S_n \tag{6.7}$$

which can be written, by introducing a source of work *(5.110)* due to a constant gravity field

$$n\frac{de_{cin}}{dt}+n\frac{du}{dt}=\left[\rho\vec{g}\right]\vec{\phi}-\operatorname{div}\left[\vec{J}_w\right]+\left[\left(\vec{J}_I-\vec{J}_L\right)m\vec{g}-\operatorname{div}\vec{J}_q-uS_n-e_{cin}S_n\right] \tag{6.8}$$

From this comparison, we have 3 important equations that must be satisfied at all times in an elastic, anelastic, plastic and self-diffusing lattice.

The lattice's Newton equation

The comparison of the vectorial expressions in brackets, in the scalar product with $\vec{\phi}$, leads to the following equation

$$nm\frac{d\vec{\phi}}{dt}+nmC_I\frac{d\vec{\phi}_I}{dt}-nmC_L\frac{d\vec{\phi}_L}{dt}-\sum_k\vec{e}_k\operatorname{div}\vec{s}_k+\frac{1}{2}\overrightarrow{\operatorname{rot}}\,\vec{m}+\overrightarrow{\operatorname{grad}}\,p=\rho\vec{g} \tag{6.9}$$

By using the material derivative of the average linear momentum $\vec{p}$ per site given by expression *(5.99)*

$$\frac{d\vec{p}}{dt}=m\left(\frac{d\vec{\phi}}{dt}+C_I\frac{d\vec{\phi}_I}{dt}+\vec{\phi}_I\frac{dC_I}{dt}-C_L\frac{d\vec{\phi}_L}{dt}-\vec{\phi}_L\frac{dC_L}{dt}\right) \tag{6.10}$$

equation *(6.9)* becomes a dynamic equation for the linear momentum $\vec{p}$

$$n\frac{d\vec{p}}{dt}=\rho\vec{g}+\sum_k\vec{e}_k\operatorname{div}\vec{s}_k-\frac{1}{2}\overrightarrow{\operatorname{rot}}\ \vec{m}-\overrightarrow{\operatorname{grad}}p+nm\vec{\phi}_I\frac{dC_I}{dt}-nm\vec{\phi}_L\frac{dC_L}{dt} \qquad (6.11)$$

This equation is nothing more than the *Newton equation* applied to the lattice in eulerian coordinates. This equation describes the dynamics of the solid, which depends on gravity $\rho\vec{g}$, volume forces with stress potentials $\vec{s}_i$, $\vec{m}$ and p, and additional linear momentum brought about by increase of the number of point defects in the lattice. Expression *(6.11)* represents in fact the Newton equation we can obtain by using the thermo-kinetic equation *(5.84)*. However, we can verify while working on thermo-kinetic equations *(5.13)* and (5.7), that the same Newton equation can be written under two different forms, which use mechanical stress potentials $\vec{\sigma}_i$ and $\vec{\Sigma}_i$

$$n\frac{d\vec{p}}{dt}=\rho\vec{g}+\sum_k\vec{e}_k\operatorname{div}\vec{\sigma}_k-\frac{1}{2}\overrightarrow{\operatorname{rot}}\,\vec{m}+nm\vec{\phi}_I\frac{dC_I}{dt}-nm\vec{\phi}_L\frac{dC_L}{dt} \qquad (6.12)$$

$$n\frac{d\vec{p}}{dt}=\rho\vec{g}+\sum_k\vec{e}_k\operatorname{div}\vec{\Sigma}_k+nm\vec{\phi}_I\frac{dC_I}{dt}-nm\vec{\phi}_L\frac{dC_L}{dt} \qquad (6.13)$$

The flux of work and the surface force

The comparison of the vectorial terms of the divergence operator, leads to an expression for the *flux of work* within the lattice

$$\vec{J}_w=\mu_L^*\vec{J}_L+\mu_I^*\vec{J}_I-\phi_k\vec{s}_k-\frac{1}{2}(\vec{\phi}\wedge\vec{m})+p\vec{\phi} \qquad (6.14)$$

This flux corresponds to an energy propagation within the solid. We will remark that the contribution of $\vec{m}$ is a vector totally analogous to the *Poynting vector* of electromagnetism. The expression *(6.14)* allows us to write the surface integral that appears in expression *(4.18)*

$$\begin{aligned}-\oint\!\!\!\oint_{S_m}\vec{J}_w d\vec{S}&=-\oint\!\!\!\oint_{S_m}\left(\mu_L^*\vec{J}_L+\mu_I^*\vec{J}_I-\phi_k\vec{s}_k-(\vec{\phi}\wedge\vec{m})/2+p\vec{\phi}\right)d\vec{S}\\&=\oint\!\!\!\oint_{S_m}\vec{\phi}\left[\sum_k\vec{e}_k(\vec{s}_k\vec{n})+(\vec{m}\wedge\vec{n})/2-\vec{n}p\right]dS-\oint\!\!\!\oint_{S_m}\left[\mu_L^*\vec{J}_L+\mu_I^*\vec{J}_I\right]\vec{n}\,dS\end{aligned} \qquad (6.15)$$

The first integral introduces the notion of a *surface force* $\vec{F}_S$, where $\vec{n}$ corresponds to the unit vector normal to the surface under consideration, and the second integral shows the appearance of *energy fluxes associated with the diffusion fluxes of vacancies and interstitials*

$$\begin{cases}\vec{F}_S=\sum_k\vec{e}_k(\vec{s}_k\vec{n})+\dfrac{1}{2}(\vec{m}\wedge\vec{n})-\vec{n}p\\ \vec{J}_w^L=\mu_L^*\vec{J}_L=\left[\mu_L-\dfrac{1}{2}m\left(\vec{\phi}_L^2-2\Delta\vec{\varphi}_L^2\right)\right]nC_L\Delta\vec{\varphi}_L\\ \vec{J}_w^I=\mu_I^*\vec{J}_I=\left[\mu_I+\dfrac{1}{2}m\vec{\phi}_I^2\right]nC_I\Delta\vec{\varphi}_I\end{cases} \qquad (6.16)$$

The last two relations show that the flux of work associated with the flux of diffusion of vacancies and interstitials contain both a flux of potential energy (the chemical potentials μ_L and μ_I

of vacancies and interstitials) and fluxes of kinetic energy.

The first relation could be calculated from the other mechanical potentials and would simply lead to the following expressions

$$\vec{F}_S = \sum_k \vec{e}_k \left(\vec{\sigma}_k \vec{n}\right) + \frac{1}{2}\left(\vec{m} \wedge \vec{n}\right) \qquad (6.17)$$

$$\vec{F}_S = \sum_k \vec{e}_k \left(\vec{\Sigma}_k \vec{n}\right) \qquad (6.18)$$

The three equations for the surface force allow us to specify the *limit conditions* external to the solid. At the limits of the medium, we can link the external forces by unit surface $\vec{F}_S$ to the stress tensors $\vec{\Sigma}_i$, $\vec{\sigma}_i$ or $\vec{s}_i$, the torque vector $\vec{m}$ and the scalar of pressure p.

From these expressions of the surface force $\vec{F}_S$ we easily deduce the physical interpretation, completely classical, of the stress potentials $\vec{\Sigma}_i$, $\vec{\sigma}_i$ or $\vec{s}_i$. Dimensionally they are forces by unit surface: the component Σ_{ij} of tensor $\vec{\Sigma}_i$ is equal to the *force per unit surface applied in direction i on a unit surface perpendicular to direction j.* Schematically, we can represent the whole set of forces and components on a unit cube as in figure 6.1.

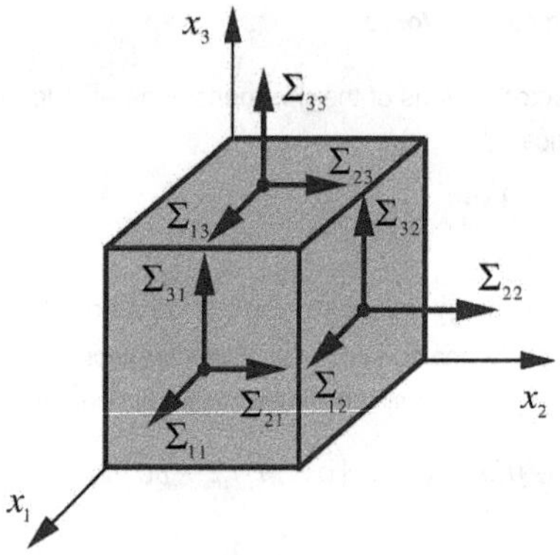

Figure 6.1 *- physical interpretation of the components of the stress tensor* $\vec{\Sigma}_i$

The component of the stress tensor $\vec{\Sigma}_i$ have the dimension of a stress, a force per unit surface, which is why we called $\vec{\Sigma}_i$ the *stress tensor* conjugated to $\vec{\beta}_i$. The same interpretation can be given for tensors $\vec{\sigma}_i$ and $\vec{s}_i$, meaning the stress tensors conjugated to $\vec{\varepsilon}_i$ and $\vec{\alpha}_i$ respectively.

With regards to vector $\vec{m}$, it can be interpreted as in the schematic figure 6.2. The vector $\vec{m}$ represents in fact the local torque (moment) applied to the continuous medium by the surface forces Σ_{ij}. $\vec{m}$ is the torque vector conjugated to the vector $\vec{\omega}$ of rotations, and so we will call $\vec{m}$ the *torsor of moments* or *the torque vector* conjugated to $\vec{\omega}$.

The scalar p is defined as the average of the three stresses of compression Σ_{ii} or σ_{ii}, with a changed sign. This scalar has the dimension of a hydrostatic pressure and we will call it the *scalar of pressure* conjugated to τ.

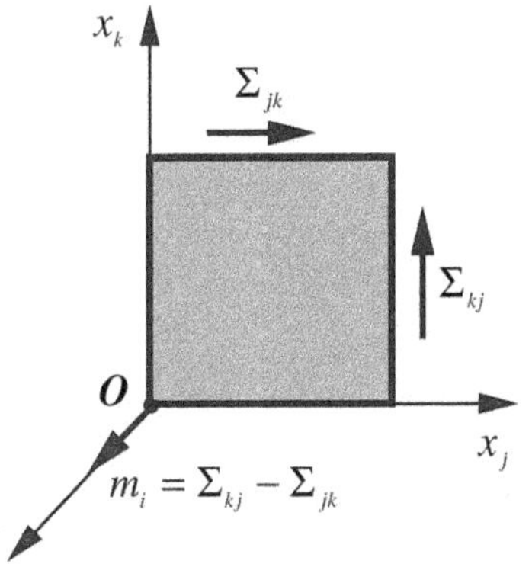

Figure 9.3 *- physical interpretation of the components of the torque vector or torsor of moment* $\vec{m}$

Finally equations *(6.16)* and *(6.18)* allow us to deduce the *pass-through conditions* at the interface of the two medias (see figure 3.9) for the stress potentials $\vec{\Sigma}_i$, $\vec{\sigma}_i$ or $\vec{s}_i$, for the torque vector $\vec{m}$ and for the scalar pressure p, by requiring that the forces be equal across the surface, with the three following equations

$$\sum_k \vec{e}_k\left(\vec{s}_k^{(1)}\vec{n}\right)+\frac{1}{2}\left(\vec{m}^{(1)}\wedge\vec{n}\right)-\vec{n}p^{(1)}=\sum_k \vec{e}_k\left(\vec{s}_k^{(2)}\vec{n}\right)+\frac{1}{2}\left(\vec{m}^{(2)}\wedge\vec{n}\right)-\vec{n}p^{(2)} \qquad (6.19)$$

$$\sum_k \vec{e}_k\left(\vec{\sigma}_k^{(1)}\vec{n}\right)+\frac{1}{2}\left(\vec{m}^{(1)}\wedge\vec{n}\right)=\sum_k \vec{e}_k\left(\vec{\sigma}_k^{(2)}\vec{n}\right)+\frac{1}{2}\left(\vec{m}^{(2)}\wedge\vec{n}\right) \qquad (6.20)$$

$$\sum_k \vec{e}_k\left(\vec{\Sigma}_k^{(1)}\vec{n}\right)=\sum_k \vec{e}_k\left(\vec{\Sigma}_k^{(2)}\vec{n}\right) \qquad (6.21)$$

The equations of heat for the lattice

The comparison of the scalar expressions within brackets leads us to the *equation of heat*, or *thermal equation*, of an elastic, anelastic and plastic media, with self-diffusion, which allows us, in principle, to compute the thermal evolution of the lattice

$$\begin{aligned} nT\frac{ds}{dt} &= -\mu_L^* S_L - \mu_I^* S_I - \left(e_{cin} + u - \mu_L^* C_L - \mu_I^* C_I\right) S_n \\ &\quad -\left(\overrightarrow{\operatorname{grad}}\,\mu_I^* + m\frac{d\vec{\phi}_I}{dt} - m\vec{g}\right)\vec{J}_I - \left(\overrightarrow{\operatorname{grad}}\,\mu_L^* - m\frac{d\left(\vec{\phi}_L - 2\Delta\vec{\varphi}_L\right)}{dt} + m\vec{g}\right)\vec{J}_L \\ &\quad + \vec{s}_k^{\,dis}\frac{d\vec{\beta}_k^{an}}{dt} + \vec{m}^{dis}\frac{d\vec{\omega}^{an}}{dt} + \vec{s}_k\frac{d\vec{\beta}_k^{pl}}{dt} + \vec{m}\frac{d\vec{\omega}^{pl}}{dt} - p\frac{d\tau^{pl}}{dt} - \operatorname{div}\vec{J}_q \end{aligned} \qquad (6.22)$$

As a matter of fact the way *(6.22)* is written is sub-optimal because it does not separate the processes of creation of vacancies and interstitials by creation-annihilation of pairs on the one hand and by plasticity on the other hand. To handle those, it is useful to rewrite this equation by using definition *(5.64)* of $d\tau^{pl}/dt$, and then definitions *(5.93)* for S_L and S_I, as well as the definition *(5.92)* of S_n. We then have an equation of heat that depends solely on S_{I-L}, S_L^{pl} and S_I^{pl}

$$\begin{aligned} nT\frac{ds}{dt} &= -\left(\mu_L^*+\mu_I^*\right)S_{I-L} - \left(\mu_L^*+h^*\right)S_L^{pl} - \left(\mu_I^*-h^*\right)S_I^{pl} \\ &+\left(-\overrightarrow{\text{grad}}\mu_L^* + m\frac{d}{dt}\left(\vec{\phi}_L - 2\Delta\vec{\varphi}_L\right) - m\vec{g}\right)\vec{J}_L + \left(-\overrightarrow{\text{grad}}\mu_I^* - m\frac{d}{dt}\left(\vec{\phi}_I\right) + m\vec{g}\right)\vec{J}_I \\ &+\vec{s}_k^{\,dis}\frac{d\vec{\beta}_k^{an}}{dt} + \vec{m}^{dis}\frac{d\vec{\omega}^{an}}{dt} + \vec{s}_k\frac{d\vec{\beta}_k^{pl}}{dt} + \vec{m}\frac{d\vec{\omega}^{pl}}{dt} - \operatorname{div}\vec{J}_q \end{aligned} \tag{6.23}$$

in which we have introduced energy h^* which is nothing more than the *thermal energy* taken from the lattice for the creation of a lattice site, given by

$$\begin{aligned} h^* &= u + pv + e_{cin} - \mu_L^* C_L - \mu_I^* C_I \\ &= f + Ts + pv + e_{cin} - \mu_L^* C_L - \mu_I^* C_I \end{aligned} \tag{6.24}$$

The denomination h^* given to this energy is simply due to the fact that it takes into account the *enthalpy* $h = u + pv = f + Ts + pv$ per site of lattice. This energy can be written using the definitions *(6.2)*

$$\begin{aligned} h^* &= u + pv + m\vec{\phi}^2/2 - \mu_L C_L - \mu_I C_I \\ &= f + Ts + pv + m\vec{\phi}^2/2 - \mu_L C_L - \mu_I C_I \end{aligned} \tag{6.25}$$

The sources of heat

In heat equation *(6.23)*, we have several *sources of heat:*

- a source of heat $S_q^{formation(I-L)}$ which corresponds to *thermal energy taken from or given to the lattice during the creation or annihilation of a vacancy-interstitial pair*

$$S_q^{formation\ (I-L)} = -\left(\mu_L^* + \mu_I^*\right)S_{I-L} \tag{6.26}$$

- two sources of heat $S_q^{diffusion(L)}$ and $S_q^{diffusion(I)}$ which correspond to the *thermal energies involved in the diffusion of vacancies and interstitials respectively*

$$\begin{cases} S_q^{diffusion\ (L)} = \left(-\overrightarrow{\text{grad}}\mu_L^* + m\dfrac{d}{dt}\left(\vec{\phi}_L - 2\Delta\vec{\varphi}_L\right) - m\vec{g}\right)\vec{J}_L \\ S_q^{diffusion\ (I)} = \left(-\overrightarrow{\text{grad}}\mu_I^* - m\dfrac{d}{dt}\left(\vec{\phi}_I\right) + m\vec{g}\right)\vec{J}_I \end{cases} \tag{6.27}$$

- a source of heat induced by the *dissipative anelastic phenomena*

$$S_q^{anélasticité} = \vec{s}_k^{\,dis}\frac{d\vec{\beta}_k^{an}}{dt} + \vec{m}^{dis}\frac{d\vec{\omega}^{an}}{dt} \tag{6.28}$$

- a source of heat induced by the *dissipative plastic phenomena*, which can be written in two different forms using relation *(5.94)*

$$\begin{aligned} S_q^{plasticité} &= -\left(\mu_L^* + h^*\right)S_L^{pl} - \left(\mu_I^* - h^*\right)S_I^{pl} + \vec{s}_k\frac{d\vec{\beta}_k^{pl}}{dt} + \vec{m}\frac{d\vec{\omega}^{pl}}{dt} \\ &= -\mu_L^* S_L^{pl} - \mu_I^* S_I^{pl} - h^* S_n + \vec{s}_k\frac{d\vec{\beta}_k^{pl}}{dt} + \vec{m}\frac{d\vec{\omega}^{pl}}{dt} \end{aligned} \tag{6.29}$$

which shows that the thermal energies taken from the lattice respectively are worth μ_L^* for the creation of a vacancy, μ_I^* for the creation of a self-interstitial and h^* for the creation of a lattice site.

The source of entropy

If thermal equation *(6.23)* is compared with expression *(4.29)* of the second principle, in which the sources of lattice sites S_n is not null, we have the expression of *volume source of entropy* in an elastic, anelastic, plastic and self-diffusing lattice

$$\begin{aligned} S_e = & -\frac{1}{T}\left(\mu_L^* + \mu_I^*\right)S_{I-L} - \frac{1}{T}\left(\mu_L^* + g^*\right)S_L^{pl} - \frac{1}{T}\left(\mu_I^* - g^*\right)S_I^{pl} \\ & + \frac{1}{T}\left(- \overrightarrow{\operatorname{grad}}\mu_L^* + m\frac{d}{dt}\left(\vec{\phi}_L - 2\Delta\vec{\varphi}_L\right) - m\vec{g}\right)\vec{J}_L + \frac{1}{T}\left(- \overrightarrow{\operatorname{grad}}\mu_I^* - m\frac{d}{dt}\left(\vec{\phi}_I\right) + m\vec{g}\right)\vec{J}_I \\ & + \frac{1}{T}\left(\vec{s}_k^{dis}\frac{d\vec{\beta}_k^{an}}{dt} + \vec{m}^{dis}\frac{d\vec{\omega}^{an}}{dt} + \vec{s}_k\frac{d\vec{\beta}_k^{pl}}{dt} + \vec{m}\frac{d\vec{\omega}^{pl}}{dt}\right) + \vec{J}_q\overrightarrow{\operatorname{grad}}\left(\frac{1}{T}\right) \end{aligned} \qquad (6.30)$$

in which g^* is given by expression

$$\begin{aligned} g^* &= f + pv + e_{cin} - \mu_L^* C_L - \mu_I^* C_I \\ &= f + pv + m\dot{\phi}^2/2 - \mu_L C_L - \mu_I C_I \end{aligned} \qquad (6.31)$$

which contains among others the *free enthalpy* $g = u + pv - Ts = f + pv$ per lattice site.

6.2 - Phenomenological relations of a lattice

The function and equations of state

We already mentioned the state equations in section 5.1. However, the expressions we found there were derived from the internal energy u of the solid and thus depended on entropy s. It is in general a lot more convenient to use the temperature T as an independent variable, instead of entropy s, and we can introduce the state function of *free energy* f per site of lattice by applying the following *Legendre transform*

$$f = u - Ts \quad \Rightarrow \quad df = du - Tds - sdT \qquad (6.32)$$

In this expression of the differential of free energy f, we can introduce the expression for the differential of internal energy u obtained in *(5.83)*

$$ndf = \vec{s}_k d\vec{\alpha}_k^{él} + m_k\omega_k^{él} - pd\tau^{él} + \vec{s}_k^{cons} d\vec{\alpha}_k^{an} + m_k^{cons}\omega_k^{an} + n\mu_L dC_L + n\mu_I dC_I - nsdT \qquad (6.33)$$

The free energy $f = u - Ts$ becomes a function of the following state variables

$$f = f\left(\alpha_{ij}^{él}, \alpha_{ij}^{an}, \omega_k^{él}, \omega_k^{an}, \tau^{él}, C_L, C_I, T\right) \qquad (6.34)$$

An important realization can be made here which concerns this state function, namely that it does not depend on the plastic distortions of the lattice.

As we explained in section 5.1, the functions and state equations can take different forms depending on which of the three topological representations we use $\left\{\vec{\beta}_i\right\}$, $\left\{\vec{\varepsilon}_i, \vec{\omega}\right\}$ or $\left\{\vec{\alpha}_i, \vec{\omega}, \tau\right\}$, which each have their conjugate stress potentials, namely $\{\vec{\Sigma}_i\}$, $\{\vec{\sigma}_i,\vec{m}\}$ and $\{\vec{s}_i,\vec{m},p\}$ respectively. For example, in the expression *(6.34)* for free energy f, we have the various equations of state for a lattice in the topological representation $\{\vec{\alpha}_i,\vec{\omega},\tau\}$ which are directly deduced:

- the *elastic equations of state*:

$$\begin{cases} s_{ij} = \dfrac{n}{2}\left(\dfrac{\partial f}{\partial \alpha_{ij}^{él}} + \dfrac{\partial f}{\partial \alpha_{ji}^{él}}\right) - \dfrac{n}{3}\delta_{ij}\sum_k \dfrac{\partial f}{\partial \alpha_{kk}^{él}} = s_{ij}(\alpha_{lm}^{él},\omega_n^{él},\tau^{él},\alpha_{lm}^{an},\omega_n^{an},C_L,C_I,T) \\ m_k = n\dfrac{\partial f}{\partial \omega_k^{él}} = m_k(\alpha_{lm}^{él},\omega_n^{él},\tau^{él},\alpha_{lm}^{an},\omega_n^{an},C_L,C_I,T) \\ p = -n\dfrac{\partial f}{\partial \tau^{él}} = p(\alpha_{lm}^{él},\omega_n^{él},\tau^{él},\alpha_{lm}^{an},\omega_n^{an},C_L,C_I,T) \end{cases} \tag{6.35}$$

- the *anelastic equations of state*:

$$\begin{cases} s_{ij}^{dis} = \dfrac{n}{2}\left(\dfrac{\partial f}{\partial \alpha_{ij}^{an}} + \dfrac{\partial f}{\partial \alpha_{ji}^{an}}\right) - \dfrac{n}{3}\delta_{ij}\sum_k \dfrac{\partial f}{\partial \alpha_{kk}^{an}} = s_{ij}^{dis}(\alpha_{lm}^{él},\omega_n^{él},\tau^{él},\alpha_{lm}^{an},\omega_n^{an},C_L,C_I,T) \\ m_k^{dis} = n\dfrac{\partial f}{\partial \omega_k^{an}} = m_k^{dis}(\alpha_{lm}^{él},\omega_n^{él},\tau^{él},\alpha_{lm}^{an},\omega_n^{an},C_L,C_I,T) \end{cases} \tag{6.36}$$

- and finally the *equations of state for the entropy and the chemical potentials* for vacancies and interstitials:

$$\begin{cases} s = -\dfrac{\partial f}{\partial T} = s(\alpha_{lm}^{él},\omega_n^{él},\tau^{él},\alpha_{lm}^{an},\omega_n^{an},C_L,C_I,T) \\ \mu_L = \dfrac{\partial f}{\partial C_L} = \mu_L(\alpha_{lm}^{él},\omega_n^{él},\tau^{él},\alpha_{lm}^{an},\omega_n^{an},C_L,C_I,T) \\ \mu_I = \dfrac{\partial f}{\partial C_I} = \mu_I(\alpha_{lm}^{él},\omega_n^{él},\tau^{él},\alpha_{lm}^{an},\omega_n^{an},C_L,C_I,T) \end{cases} \tag{6.37}$$

The dissipative equations of thermo-conduction and self-diffusion

The three terms of the source of entropy *(6.30)* depend on the thermodynamic fluxes and are linked to the following dissipative processes:

- *the process of thermo-conduction:*

$$S_e^{thermoconduction} = \vec{J}_q \, \overrightarrow{\mathrm{grad}}\frac{1}{T} = \vec{J}_q \bullet \vec{X}_q \tag{6.38}$$

- *the process of vacancy diffusion:*

$$S_e^{auto-diffusion\ (L)} = \vec{J}_L \frac{1}{T}\left(-\overrightarrow{\mathrm{grad}}\ \mu_L^* + m\frac{d}{dt}\left(\vec{\phi}_L - 2\Delta\vec{\varphi}_L\right) - m\vec{g}\right) = \vec{J}_L \bullet \vec{X}_L \tag{6.39}$$

- *the process of diffusion of self-interstitials:*

$$S_e^{auto-diffusion\ (I)} = \vec{J}_I \frac{1}{T}\left(-\overrightarrow{\mathrm{grad}}\ \mu_I^* - m\frac{d}{dt}\left(\vec{\phi}_I\right) + m\vec{g}\right) = \vec{J}_I \bullet \vec{X}_I \tag{6.40}$$

These processes can be written as the product of vectorial fluxes by the vectors of thermodynamical forces which are their conjugates. The forces $\vec{X}_L$ and $\vec{X}_I$ conjugated to the flux of diffusion $\vec{J}_L$ and $\vec{J}_I$ contains several terms: a gravitational force, an inertial force and a generalized chemical force, this last one corresponds to the gradient of the chemical potential and the kinetic energy of both species of point defects. According to the *thermodynamics of irreversible processes*, there can exist a coupling between the diverse fluxes and thermodynamical forces which appear in the expression of the source of entropy, as long as these fluxes and forces are of the same tensorial order. In the present case, concerning the *flux of heat by thermo-conduction* $\vec{J}_q$ and the *flux of particles by self-diffusion* $\vec{J}_L$ and $\vec{J}_I$, there must exist *phenomenological dissipation equations of thermo-conduction and self-diffusion*, which depend strongly on the local state of the lattice and therefore on the values of quantities n,T,C_L,C_I

$$\begin{cases} \vec{J}_q = \vec{J}_q(\vec{X}_q,\vec{X}_L,\vec{X}_I,n,T,C_L,C_I,...) \\ \vec{J}_L = \vec{J}_L(\vec{X}_q,\vec{X}_L,\vec{X}_I,n,T,C_L,C_I,...) \\ \vec{J}_I = \vec{J}_I(\vec{X}_q,\vec{X}_L,\vec{X}_I,n,T,C_L,C_I,...) \end{cases} \tag{6.41}$$

To insure the positivity of the sources of entropy, *(6.38) to (6.40)* must depend in a linear fashion on the thermodynamic forces $\vec{X}_q$, $\vec{X}_L$ and $\vec{X}_I$, defined by relations

$$\begin{cases} \vec{X}_q = \overrightarrow{\text{grad}}\dfrac{1}{T} \\ \vec{X}_L = \dfrac{1}{T}\left(-\overrightarrow{\text{grad}}\ \mu_L^* + m\dfrac{d}{dt}\left(\vec{\phi}_L - 2\Delta\vec{\varphi}_L\right) - m\vec{g}\right) \\ \vec{X}_I = \dfrac{1}{T}\left(-\overrightarrow{\text{grad}}\ \mu_I^* - m\dfrac{d}{dt}\left(\vec{\phi}_I\right) + m\vec{g}\right) \end{cases} \tag{6.42}$$

The equation of creation-annihilation of vacancy-interstitial pairs

The first term of the source of entropy uses the product of the source S_{I-L} of pairs of vacancy-interstitial, meaning the number of pairs created or annihilated per unit volume by unit of time inside the lattice, by an entropic term $-(\mu_L^* + \mu_I^*)/T$ which is its conjugate, meaning the *entropy of formation* of vacancy-interstitial pair

$$S_e^{création-annihilation\ (I-L)} = S_{I-L} \bullet \left[-\frac{1}{T}(\mu_L^* + \mu_I^*) \right] \tag{6.43}$$

This entropy of formation is directly linked to the thermal energy (heat) extracted from the lattice to form a vacancy-interstitial pair, as shown by the source of heat $S_q^{formation\ (I-L)}$ obtained in *(6.26)*. *The energy of formation of a vacancy-interstitial pair* is consequently equal to the sum of chemical energy, meaning the chemical potentials μ_L and μ_I, and the kinetic energy of a pair, namely e_{cin}^L and e_{cin}^I as the comparison of relations *(5.108)* and *(6.2)* shows

$$\mu_L^* + \mu_I^* = \mu_L + \mu_I + e_{cin}^L + e_{cin}^I \tag{6.44}$$

The kinetic energies e_{cin}^L and e_{cin}^I need a longer discussion. To that end it is useful to write them using the relations *(6.2)* in which they were defined

$$\begin{cases} e^L_{cin} = -\frac{1}{2} m\left(\vec{\phi}^2_L - 2\Delta\vec{\varphi}^2_L\right) = -\frac{1}{2} m\vec{\phi}^2 - m\vec{\phi}\Delta\vec{\varphi}_L + \frac{1}{2} m\Delta\vec{\varphi}^2_L \\ e^I_{cin} = \frac{1}{2} m\vec{\phi}^2_I = \frac{1}{2} m\vec{\phi}^2 + m\vec{\phi}\Delta\vec{\varphi}_I + \frac{1}{2} m\Delta\vec{\varphi}^2_I \end{cases} \tag{6.45}$$

These energies are composed in a kinetic energy of entrainment proportional to $\vec{\phi}^2$, which is negative since the vacancy it represents is missing kinetic energy at the site under consideration, and an internal kinetic energy (Δe^L_{cin} and Δe^I_{cin}) linked to the relative movement of the point defect with respect to the lattice. The sum of energies e^L_{cin} and e^I_{cin} is then equal to

$$e^L_{cin} + e^I_{cin} = m\vec{\phi}\left(\Delta\vec{\varphi}_I - \Delta\vec{\varphi}_L\right) + \frac{1}{2} m\left(\Delta\vec{\varphi}^2_I + \Delta\vec{\varphi}^2_L\right) = \Delta e^L_{cin} + \Delta e^I_{cin} \tag{6.46}$$

with

$$\begin{cases} \Delta e^L_{cin} = \frac{1}{2} m\Delta\vec{\varphi}^2_L - m\vec{\phi}\Delta\vec{\varphi}_L \\ \Delta e^I_{cin} = \frac{1}{2} m\Delta\vec{\varphi}^2_I + m\vec{\phi}\Delta\vec{\varphi}_I \end{cases} \tag{6.47}$$

The kinetic energy Δe^{I-L}_{cin} necessary for the creation of a pair, is given by the thermal energy of the lattice and is therefore equal to

$$\Delta e^{I-L}_{cin} = \Delta e^L_{cin} + \Delta e^I_{cin} = e^L_{cin} + e^I_{cin} \tag{6.48}$$

so that the *energy of formation of a vacancy-interstitial pair* is composed of the chemical energy $\mu_L + \mu_I$ and the kinetic energy Δe^{I-L}_{cin}

$$\mu^*_L + \mu^*_I = \mu_L + \mu_I + \Delta e^{I-L}_{cin} \tag{6.49}$$

It is clear that the rate S_{I-L} of creation of pairs is strongly dependent on the threshold energy of formation. As a consequence there must exist a *phenomenological equation of creation and annihilation of vacancy-interstitial pair,* which depends on $\mu^*_L + \mu^*_I$. This equation will be written in a symbolic form in which we will include a strong dependence in the intensive local thermodynamic parameters n, T, C_L, C_I

$$S_{I-L} = S_{I-L}\left[\left(\mu^*_L + \mu^*_I\right), n, T, C_L, C_I, \ldots\right] \tag{6.50}$$

The dissipative equations of plasticity

The second and third term for the source of entropy make use of the product of the scalar sources of vacancies and interstitials, due to the creation or annihilation of sites of the lattice during the process of plastic volume expansion, by a term of formation of entropy

$$\begin{aligned} S^{site-L}_e &= S^{pl}_L \bullet \left[-\left(\mu_L^* + g^*\right)/T\right] \\ S^{site-I}_e &= S^{pl}_I \bullet \left[-\left(\mu_I^* - g^*\right)/T\right] \end{aligned} \tag{6.51}$$

This term of entropy of formation is conjugated to the sources of vacancies and interstitials by plasticity and can be written under the following form, by using relation (6.45) for e^L_{cin} and e^I_{cin}, and also using relation (6.31) giving g^*

$$-\frac{1}{T}\Bigg(\underbrace{\mu_L + e^L{}_{cin}}_{\substack{\text{energy of formation}\\ \text{of a vacancy}}} \underbrace{+g + e_{cin} - \mu_L{}^* C_L - \mu_I{}^* C_I}_{\substack{\text{energy of creation}\\ \text{of an additionnal lattice site}}}\Bigg) = -\frac{1}{T}\Bigg(\underbrace{\mu_L + e^L{}_{cin}}_{\substack{\text{energy of formation}\\ \text{of a vacancy}}} \underbrace{+g + \frac{m\vec{\phi}^2}{2} - \mu_L C_L - \mu_I C_I}_{\substack{\text{energy of creation}\\ \text{of an additionnal lattice site}}}\Bigg)$$

$$-\frac{1}{T}\Bigg(\underbrace{\mu_I + e^I{}_{cin}}_{\substack{\text{energy of formation}\\ \text{of an interstitial}}} \underbrace{-g - e_{cin} + \mu_L{}^* C_L + \mu_I{}^* C_I}_{\substack{\text{energy of annihilation}\\ \text{of a lattice site}}}\Bigg) = -\frac{1}{T}\Bigg(\underbrace{\mu_I + e^I{}_{cin}}_{\substack{\text{energy of formation}\\ \text{of an interstitial}}} \underbrace{-g - \frac{m\vec{\phi}^2}{2} + \mu_L C_L + \mu_I C_I}_{\substack{\text{energy of annihilation}\\ \text{of a lattice site}}}\Bigg) \tag{6.52}$$

We can see that this entropies of formation use two very important quantities: *the energy of formation of a vacancy by the creation of an additional site of lattice*, and the *energy of formation of an interstitial by annihilation of a site of lattice*. The creation or annihilation of a lattice site during a plastic deformation will satisfy a kinetic equation that must depend on the energies of formation of vacancies or interstitials, as well as physical conditions such as volume of lattice unit cell v, temperature T, atomic concentrations of vacancies and interstitials, etc. There must exist, as a consequence, *phenomenological equations of creation-annihilation of lattice sites*, which will translate the plasticity of the lattice under volume expansion and which will be symbolically written as such

$$\begin{cases} \dfrac{d\tau^{pl}}{dt} = \dfrac{S_n}{n} = \dfrac{1}{n}\left(S_L{}^{pl} - S_I{}^{pl}\right) \\ S_L{}^{pl} = S_L{}^{pl}\left[\left(\mu_L{}^* + g^*\right), v, T, C_L, C_I, \ldots\right] \\ S_I{}^{pl} = S_I{}^{pl}\left[\left(\mu_I{}^* - g^*\right), v, T, C_L, C_I, \ldots\right] \end{cases} \tag{6.53}$$

As regards the dissipative terms associated with plasticity by shear and rotation, it leads to a source of entropy that we can write with the help of *(6.30) as*

$$S_e^{pl} = \frac{1}{T}\left(\vec{s}_k \frac{d\vec{\beta}_k^{pl}}{dt} + \vec{m}\frac{d\vec{\omega}^{pl}}{dt}\right) \tag{6.54}$$

In this relation, the mechanical potential fields $\vec{s}_i$ and $\vec{m}$ are perfectly known, as it leads to the stress field of elastic shear $\vec{s}_i$ and the torsor of elastic moments $\vec{m}$ deduced from the state equations *(6.35)*. As a consequence the velocity of plastic deformations by shear and rotation must satisfy the *dissipative phenomenological equations of plasticity*

$$\begin{cases} \dfrac{d\vec{\alpha}_i^{pl}}{dt} = \dfrac{d\vec{\alpha}_i^{pl}}{dt}\left(\vec{s}_m, v, T, \ldots\right) \\ \dfrac{d\vec{\omega}^{pl}}{dt} = \dfrac{d\vec{\omega}^{pl}}{dt}\left(\vec{m}, v, T, \ldots\right) \end{cases} \tag{6.55}$$

Writing it in this way reflects the fact that the terms of sources of entropy due to the plastic deformations by shear and rotation can be considered as the product of a *generalized flux* ($d\vec{\alpha}_i^{pl}/dt$ or $d\vec{\omega}^{pl}/dt$) by a *"generalized force"*, which is here represented by the shear stress tensor $\vec{s}_i$ or the torque torsor $\vec{m}$.

The dissipative equations of anelasticity

For the dissipative terms associated with anelasticity by shear and rotation, the source of entropy can be written, thanks to *(6.30)*

$$S_e^{an} = \frac{1}{T}\left(\vec{s}_k^{\,dis} \frac{d\vec{\beta}_k^{an}}{dt} + \vec{m}^{dis} \frac{d\vec{\omega}^{an}}{dt} \right) \tag{6.55}$$

The equations of state *(6.35)* show that the total potential fields $\vec{s}_i$ and $\vec{m}$ are the conjugates of the field of elastic shear deformations $\vec{\alpha}_i^{\acute{e}l}$ and elastic rotation $\vec{\omega}^{\acute{e}l}$ while the conservative potential fields $\vec{s}_i^{\,cons}$ and $\vec{m}^{cons}$ are the conjugates of the fields of anelastic shear deformation $\vec{\alpha}_i^{an}$ and anelastic rotation $\vec{\omega}^{an}$. By writing the decomposition of the energy of anelastic distortion in terms of a source of work and a source of heat, we had to decompose the potential elastic fields $\vec{s}_i$ and $\vec{m}$ in a conservative part and a dissipative part, in the following way

$$\begin{cases} \vec{s}_i = \vec{s}_i^{\,cons} + \vec{s}_i^{\,dis} \\ \vec{m} = \vec{m}^{cons} + \vec{m}^{dis} \end{cases} \tag{6.56}$$

In these relations, the fields $\vec{s}_i$, $\vec{s}_i^{\,cons}$, $\vec{m}$ and $\vec{m}^{cons}$ are given by the state equations. We only have the dissipative components $\vec{s}_i^{\,dis}$ and $\vec{m}^{dis}$ left to discuss.

As a matter of fact, the velocity of anelastic shear distortion $d\vec{\alpha}_i^{an} / dt$ and of anelastic rotation $d\vec{\omega}^{an} / dt$ are responsible for the appearance of intrinsic friction forces which give rise to dissipative stress fields $\vec{s}_i^{\,dis}$ and $\vec{m}^{dis}$. As a consequence, it is reasonable to admit that these dissipative fields are first and foremost function of the velocities of anelastic distortions by writing

$$\begin{cases} \vec{s}_i^{\,dis} = \vec{s}_i^{\,dis}\left(\dfrac{d\vec{\alpha}_m^{an}}{dt}, v, T, ... \right) \\ \vec{m}^{dis} = \vec{m}^{dis}\left(\dfrac{d\vec{\omega}^{an}}{dt}, v, T, ... \right) \end{cases} \tag{6.57}$$

This leads us to write, by taking into account the relations that exist between the various fields of mechanical potential, the *phenomenological dissipative equations of anelasticity*, under the following form

$$\begin{cases} \vec{s}_i = \vec{s}_i^{\,cons}\left(\vec{\alpha}_m^{an}, v, T, ... \right) + \vec{s}_i^{\,dis}\left(\dfrac{d\vec{\alpha}_m^{an}}{dt}, v, T, ... \right) \\ \vec{m} = \vec{m}^{cons}\left(\vec{\omega}^{an}, v, T, ... \right) + \vec{m}^{dis}\left(\dfrac{d\vec{\omega}^{an}}{dt}, v, T, ... \right) \end{cases} \tag{6.58}$$

The dissipative equations we just wrote, written under this form, are nothing more than mathematical differential equations that give us the anelastic responses $\vec{\alpha}_i^{an}$ and $\vec{\omega}^{an}$ of the solid lattice, to the elastic stresses $\vec{s}_i$ et $\vec{m}$.

6.3 - Energetic balance

We carry the following multiplication operations on the three equations of geometro-kinetic *(5.65)* and on the equation of newtonian dynamics *(6.11)*

$$\vec{s}_k \frac{d\vec{\beta}_k}{dt} = \vec{s}_k \overrightarrow{\mathrm{grad}}\, \phi_k \quad \text{and} \quad \vec{m}\frac{d\vec{\omega}}{dt} = \frac{1}{2}\vec{m}\overrightarrow{\mathrm{rot}}\,\vec{\phi} \quad \text{and} \quad -p\frac{d\tau}{dt} = -p\,\mathrm{div}\,\vec{\phi} \tag{6.59}$$

$$n\vec{\phi}\frac{d\vec{p}}{dt} = \rho\vec{g}\vec{\phi} + \phi_k \,\mathrm{div}\,\vec{s}_k - \frac{1}{2}\vec{\phi}\,\overrightarrow{\mathrm{rot}}\,\vec{m} - \vec{\phi}\,\overrightarrow{\mathrm{grad}}\, p + nm\vec{\phi}\vec{\phi}_I \frac{dC_I}{dt} - nm\vec{\phi}\vec{\phi}_L \frac{dC_L}{dt} \tag{6.60}$$

by summing all these equations, we have the *equation of energetic balance* inside the lattice

$$n\vec{\phi}\left(\frac{d\vec{p}}{dt} - m\vec{\phi}_I\frac{dC_I}{dt} + m\vec{\phi}_L\frac{dC_L}{dt}\right) + \vec{s}_k\frac{d\vec{\beta}_k}{dt} + \vec{m}\frac{d\vec{\omega}}{dt} - p\frac{d\tau}{dt} = \rho\vec{g}\vec{\phi} - \mathrm{div}\left[-\phi_k\vec{s}_k - \frac{1}{2}\left(\vec{\phi}\wedge\vec{m}\right) + p\vec{\phi}\right] \tag{6.61}$$

Each term of this equation can be easily interpreted in terms of a variation of energy densities inside the lattice:

$n\vec{\phi}\left(\frac{d\vec{p}}{dt} - m\vec{\phi}_I\frac{dC_I}{dt} + m\vec{\phi}_L\frac{dC_L}{dt}\right)$ is the variation of kinetic energy density,

$\vec{s}_k\; d\vec{\beta}_k / dt$ is the variation in density of shear elastic energy,

$\vec{m}\; d\vec{\omega} / dt$ is the variation in density of potential energy of global rotation of the lattice, if it exists,

$-p\; d\tau / dt$ is the variation of density of elastic energy of volume expansion,

$\rho\vec{g}\vec{\phi}$ is the power per unit volume given by the external field force of gravitation,

$-\phi_k\vec{s}_k - \left(\vec{\phi}\wedge\vec{m}\right)/2 + p\vec{\phi}$ is the flux of work of distorsion $\vec{J}_w$, which is analog of the Poynting vector.

With regards to the equations of energetic balance we can get from the other topological representations of the equation of thermo-kinetic, they are even simpler

$$n\vec{\phi}\left(\frac{d\vec{p}}{dt} - m\vec{\phi}_I\frac{dC_I}{dt} + m\vec{\phi}_L\frac{dC_L}{dt}\right) + \vec{\sigma}_k\frac{d\vec{\beta}_k}{dt} + \vec{m}\frac{d\vec{\omega}}{dt} = \rho\vec{g}\vec{\phi} - \mathrm{div}\left[-\phi_k\vec{\sigma}_k - \frac{1}{2}\left(\vec{\phi}\wedge\vec{m}\right)\right] \tag{6.62}$$

$$n\vec{\phi}\left(\frac{d\vec{p}}{dt} - m\vec{\phi}_I\frac{dC_I}{dt} + m\vec{\phi}_L\frac{dC_L}{dt}\right) + \vec{\Sigma}_k\frac{d\vec{\beta}_k}{dt} = \rho\vec{g}\vec{\phi} - \mathrm{div}\left(-\phi_k\vec{\Sigma}_k\right) \tag{6.63}$$

6.4 - Equations of spatiotemporal evolution of a lattice

It is now possible to combine all the results we have obtained in this chapter and the previous chapters to write the complete equations that describe the spatio-temporal evolution of a solid lattice with self-diffusion and phenomenological elasticity, anelasticity and plasticity. As seen in table 6.1, this system of equations is quite complex, in particular due to the high number of phenomenological equations of state and phenomenological equations of dissipation one needs for a complete description of all the possible behaviors and phenomena of this lattice.

The computation of the evolution of the lattice involves the resolution of this system of *fundamental equations*:

- *the topological equations*, which translate the galilean geometro-kinetic inside the solid, which allows us to calculate the global or local distortion fields from an absolute velocity field $\vec{\phi}$ or a relative velocity field $\vec{\varphi}$ inside the solid, from which we get the volume density n of sites of lattice. Furthermore the conditions of geometro-compatibility based on $\overrightarrow{\mathrm{rot}}\,\vec{\beta}_i = \overrightarrow{\mathrm{rot}}\,\vec{\beta}_i^{(\delta)} = 0$, are evidently valid for a lattice equipped with anelasticity and plasticity, because they assure that the local field $\vec{u}^{(\delta)}$, associated with the elastic, anelastic or plastic deformations, remains continuous. This is a necessary conditions to ensure the solidity of the lattice.
- *the dynamical equations*, translating the newtonian dynamic of the lattice, which lead us to the computation of the linear momentum $\vec{p}$, from which we extract velocities $\vec{\phi}$ and $\vec{\varphi}$, and the calculation of the density of inert mass ρ,
- *the equations of diffusion*, which translate the self-diffusion in the lattice, which lead us to the computation of the densities C_L and C_I of vacancies and self-interstitials,
- *the thermal equation,* which give us entropy s, from which we deduce the temperature T in the solid.

To this fundamental set of equations, which are the same for any type of lattice, we add the set of phenomenological equations which are particular to each type of lattice:

- *the equations of state*, which describe the phenomenological behavior of the lattice with regards to its elastic, anelastic, self-diffusive and thermal properties,
- *the dissipative equations*, which translate the phenomenology of dissipation by thermo-conduction, by self-diffusion, by creation-annihilation of pairs of vacancy-interstitial, by anelasticity and by plasticity of the lattice under consideration.

To the fundamental equations and the phenomenological ones necessary to describe the evolution of the lattice, we can add *additional equations*, such as the equations of mass continuity, flux of work and surface forces, sources of entropy and energetic balance. As a matter of fact, since these equations are deduced from the fundamental equations, they are not necessary to solving the problem of the evolution of the lattice. Nevertheless they shine an interesting light on the problem as far as fluxes of work, surfaces forces and exchanges of energy within the solid are concerned.

Table 6.1 - Fundamental equations of evolution of self-diffusive, elastic, anelastic and plastic solids

Topological equations

$$\begin{cases} \dfrac{d\vec{\beta}_i}{dt} = \overrightarrow{\text{grad}}\,\phi_i & (1) \\ \dfrac{d\vec{\omega}}{dt} = \dfrac{1}{2}\overrightarrow{\text{rot}}\,\vec{\phi} & (2) \\ \dfrac{d\tau}{dt} = \text{div}\,\vec{\phi} & (3) \end{cases}
\qquad
\begin{cases} \overrightarrow{\text{rot}}\,\vec{\beta}_i = 0 & (4) \\ \text{div}\,\vec{\omega} = 0 & (5) \\ d/dt = \partial/\partial t + (\vec{\varphi}\vec{\nabla}) & (6) \\ \vec{\varphi} = \vec{\phi} - \vec{\phi}_O(t) - \dot{\vec{\omega}}_O(t) \wedge \vec{r} & (7) \end{cases}$$

$$\begin{cases} \vec{\beta}_i = \vec{\beta}_i^{(\delta)} + \vec{e}_i \wedge \vec{\omega}_O(t) = \vec{\beta}_i^{\acute{e}l} + \vec{\beta}_i^{an} + \vec{\beta}_i^{pl} + \vec{e}_i \wedge \vec{\omega}_O(t) & (8) \\ \vec{\omega} = -\dfrac{1}{2}\sum_k \vec{e}_k \wedge \vec{\beta}_k = \vec{\omega}^{(\delta)} + \vec{\omega}_0(t) = \vec{\omega}^{\acute{e}l} + \vec{\omega}^{an} + \vec{\omega}^{pl} + \vec{\omega}_0(t) & (9) \\ \tau = \sum_k \vec{\beta}_k \vec{e}_k = \sum_k \vec{\beta}_k^{(\delta)} \vec{e}_k = \tau^{\acute{e}l} + \tau^{pl} \qquad (\tau^{an} \equiv 0 \text{ by hypothesis}) & (10) \\ \vec{\varepsilon}_i = \vec{\beta}_i - \vec{e}_i \wedge \vec{\omega} = \vec{\beta}_i^{(\delta)} - \vec{e}_i \wedge \vec{\omega}^{(\delta)} = \vec{\varepsilon}_i^{\acute{e}l} + \vec{\varepsilon}_i^{an} + \vec{\varepsilon}_i^{pl} & (11) \\ \vec{\alpha}_i = \vec{\varepsilon}_i - \dfrac{1}{3}\tau \vec{e}_i = \vec{\alpha}_i^{\acute{e}l} + \vec{\alpha}_i^{an} + \vec{\alpha}_i^{pl} & (12) \end{cases}$$

Dynamical equations

$$n\frac{d\vec{p}}{dt} = \rho\vec{g} + \sum_k \vec{e}_k \,\text{div}\,\vec{s}_k - \frac{1}{2}\overrightarrow{\text{rot}}\;\vec{m} - \overrightarrow{\text{grad}}p + nm\vec{\phi}_I \frac{dC_I}{dt} - nm\vec{\phi}_L \frac{dC_L}{dt} \qquad (13)$$

$$\begin{cases} n = 1/v = n_0 \exp(-\tau^{\acute{e}l}) & (14) \\ \vec{p} = m\left(\vec{\phi} + C_I\vec{\phi}_I - C_L\vec{\phi}_L\right) = m\vec{\phi} + m\left(C_I - C_L\right)\vec{\phi} + \dfrac{m}{n}\left(\vec{J}_I - \vec{J}_L\right) & \\ \qquad = \left[\rho\vec{\phi} + m\left(\vec{J}_I - \vec{J}_L\right)\right]/n & (15) \\ \rho = mn\left(1 + C_I - C_L\right) & (16) \end{cases}$$

Diffusion equations

$$\begin{cases} n\dfrac{dC_L}{dt} = \left(S_{I-L} + S_L{}^{pl}\right) - C_L\left(S_L{}^{pl} - S_I{}^{pl}\right) - \text{div}\,\vec{J}_L & (17) \\ n\dfrac{dC_I}{dt} = \left(S_{I-L} + S_I{}^{pl}\right) - C_I\left(S_L{}^{pl} - S_I{}^{pl}\right) - \text{div}\,\vec{J}_I & (18) \end{cases}$$

$$\begin{cases} \vec{J}_L = nC_L\Delta\vec{\varphi}_L = nC_L\left(\vec{\phi}_L - \vec{\phi}\right) = nC_L\left(\vec{\varphi}_L - \vec{\varphi}\right) & (19) \\ \vec{J}_I = nC_I\Delta\vec{\varphi}_I = nC_I\left(\vec{\phi}_I - \vec{\phi}\right) = nC_I\left(\vec{\varphi}_I - \vec{\varphi}\right) & (20) \end{cases}$$

Thermal equations

$$nT\frac{ds}{dt} = -\left(\mu_L^* + \mu_I^*\right)S_{I-L} - \left(\mu_L^* + h^*\right)S_L^{pl} - \left(\mu_I^* - h^*\right)S_I^{pl} + T\vec{J}_L\vec{X}_L$$
$$+T\vec{J}_I\vec{X}_I + \vec{s}_k^{\,dis}\frac{d\vec{\beta}_k^{an}}{dt} + \vec{m}^{dis}\frac{d\vec{\omega}^{an}}{dt} + \vec{s}_k\frac{d\vec{\beta}_k^{pl}}{dt} + \vec{m}\frac{d\vec{\omega}^{pl}}{dt} - \operatorname{div}\vec{J}_q \quad (21)$$

$$\mu_L^* = \mu_L - \frac{1}{2}m\left(\vec{\phi}_L^{\,2} - 2\Delta\vec{\varphi}_L^2\right) \quad (22)$$

$$\mu_I^* = \mu_I + \frac{1}{2}m\vec{\phi}_I^{\,2} \quad (23)$$

$$\vec{X}_q = \overrightarrow{\text{grad}}\,\frac{1}{T} \quad (24)$$

$$\vec{X}_L = \frac{1}{T}\left(-\overrightarrow{\text{grad}}\;\mu_L^* + m\frac{d}{dt}\left(\vec{\phi}_L - 2\Delta\vec{\varphi}_L\right) - m\vec{g}\right) \quad (25)$$

$$\vec{X}_I = \frac{1}{T}\left(-\overrightarrow{\text{grad}}\;\mu_I^* - m\frac{d}{dt}\left(\vec{\phi}_I\right) + m\vec{g}\right) \quad (26)$$

$$h^* = f + Ts + pv + \frac{1}{2}m\vec{\phi}^2 - \mu_L C_L - \mu_I C_I \quad (27)$$

Phenomenological equations of evolution of self-diffusive, elastic, anelastic and plastic solids

Functions and equations of state

$$f = f\left(\alpha_{ij}^{\,el}, \alpha_{ij}^{\,an}, \omega_k^{\,el}, \omega_k^{\,an}, \tau^{el}, C_L, C_I, T\right) \quad (28)$$

$$s_{ij} = \frac{n}{2}\left(\frac{\partial f}{\partial \alpha_{ij}^{\,el}} + \frac{\partial f}{\partial \alpha_{ji}^{\,el}}\right) - \frac{n}{3}\delta_{ij}\sum_k \frac{\partial f}{\partial \alpha_{kk}^{\,el}} = s_{ij}(\alpha_{lm}^{\,el}, \omega_n^{\,el}, \tau^{el}, \alpha_{lm}^{\,an}, \omega_n^{\,an}, C_L, C_I, T) \quad (29)$$

$$m_k = n\frac{\partial f}{\partial \omega_k^{\,el}} = m_k(\alpha_{lm}^{\,el}, \omega_n^{\,el}, \tau^{el}, \alpha_{lm}^{\,an}, \omega_n^{\,an}, C_L, C_I, T) \quad (30)$$

$$p = -n\frac{\partial f}{\partial \tau^{el}} = p(\alpha_{lm}^{\,el}, \omega_n^{\,el}, \tau^{el}, \alpha_{lm}^{\,an}, \omega_n^{\,an}, C_L, C_I, T) \quad (31)$$

$$s_{ij}^{dis} = \frac{n}{2}\left(\frac{\partial f}{\partial \alpha_{ij}^{\,an}} + \frac{\partial f}{\partial \alpha_{ji}^{\,an}}\right) - \frac{n}{3}\delta_{ij}\sum_k \frac{\partial f}{\partial \alpha_{kk}^{\,an}} = s_{ij}^{dis}(\alpha_{lm}^{\,el}, \omega_n^{\,el}, \tau^{el}, \alpha_{lm}^{\,an}, \omega_n^{\,an}, C_L, C_I, T) \quad (32)$$

$$m_k^{dis} = n\frac{\partial f}{\partial \omega_k^{\,an}} = m_k^{dis}(\alpha_{lm}^{\,el}, \omega_n^{\,el}, \tau^{el}, \alpha_{lm}^{\,an}, \omega_n^{\,an}, C_L, C_I, T) \quad (33)$$

$$\begin{cases} s = -\dfrac{\partial f}{\partial T} = s(\alpha_{lm}{}^{él}, \omega_n{}^{él}, \tau^{él}, \alpha_{lm}{}^{an}, \omega_n{}^{an}, C_L, C_I, T) & (34) \\ \mu_L = \dfrac{\partial f}{\partial C_L} = \mu_L(\alpha_{lm}{}^{él}, \omega_n{}^{él}, \tau^{él}, \alpha_{lm}{}^{an}, \omega_n{}^{an}, C_L, C_I, T) & (35) \\ \mu_I = \dfrac{\partial f}{\partial C_I} = \mu_I(\alpha_{lm}{}^{él}, \omega_n{}^{él}, \tau^{él}, \alpha_{lm}{}^{an}, \omega_n{}^{an}, C_L, C_I, T) & (36) \end{cases}$$

Equations of dissipation: self-diffusion and creation-annihilation of pairs

$$\begin{cases} \vec{J}_q = \vec{J}_q(\vec{X}_q, \vec{X}_L, \vec{X}_I, n, T, C_L, C_I, ...) & (37) \\ \vec{J}_L = \vec{J}_L(\vec{X}_q, \vec{X}_L, \vec{X}_I, n, T, C_L, C_I, ...) & (38) \\ \vec{J}_I = \vec{J}_I(\vec{X}_q, \vec{X}_L, \vec{X}_I, n, T, C_L, C_I, ...) & (39) \end{cases}$$

$$S_{I-L} = S_{I-L}\left(\mu_L{}^* + \mu_I{}^*, n, T, C_L, C_I, ...\right) \qquad (40)$$

Equations of dissipation: anelasticity

$$\begin{cases} \vec{s}_i = \vec{s}_i^{cons}\left(\vec{\alpha}_m^{an}, v, T, ...\right) + \vec{s}_i^{dis}\left(\dfrac{d\vec{\alpha}_m^{an}}{dt}, v, T, ...\right) & (41) \\ \vec{m} = \vec{m}^{cons}\left(\vec{\omega}^{an}, v, T, ...\right) + \vec{m}^{dis}\left(\dfrac{d\vec{\omega}^{an}}{dt}, v, T, ...\right) & (42) \end{cases}$$

Equations of dissipation: plasticity

$$\begin{cases} \dfrac{d\tau^{pl}}{dt} = \dfrac{S_n}{n} = \dfrac{1}{n}\left(S_L{}^{pl} - S_I{}^{pl}\right) & (43) \\ S_L{}^{pl} = S_L{}^{pl}\left[\left(\mu_L{}^* + g^*\right), v, T, C_L, C_I, ...\right] & (44) \\ S_I{}^{pl} = S_I{}^{pl}\left[\left(\mu_I{}^* - g^*\right), v, T, C_L, C_I, ...\right] & (45) \end{cases}$$

$$g^* = f + pv + m\vec{\phi}^2 / 2 - \mu_L C_L - \mu_I C_I \qquad (46)$$

$$\begin{cases} \dfrac{d\vec{\alpha}_i^{pl}}{dt} = \dfrac{d\vec{\alpha}_i^{pl}}{dt}(\vec{s}_m, v, T, ...) & (47) \\ \dfrac{d\vec{\omega}^{pl}}{dt} = \dfrac{d\vec{\omega}^{pl}}{dt}(\vec{m}, v, T, ...) & (48) \end{cases}$$

Additional equations of evolution

Continuity of mass

$$\frac{\partial \rho}{\partial t} = -\operatorname{div}\left[\rho\vec{\phi} + m\left(\vec{J}_I - \vec{J}_L\right)\right] = -\operatorname{div}\left(n\vec{p}\right) \qquad \text{in } Q\xi_1\xi_2\xi_3 \qquad (49)$$

Flux of work and surface force

$$\left\{\begin{array}{ll} \vec{J}_w = \mu_L^* \vec{J}_L + \mu_I^* \vec{J}_I - \phi_k \vec{s}_k - \frac{1}{2}\left(\vec{\phi} \wedge \vec{m}\right) + p\vec{\phi} & (50) \\ \vec{F}_S = \sum_k \vec{e}_k\left(\vec{s}_k \vec{n}\right) + \frac{1}{2}\left(\vec{m} \wedge \vec{n}\right) - \vec{n}p & (51) \end{array}\right.$$

Source of entropy

$$\begin{aligned} S_e = & -\frac{1}{T}\left(\mu_L^* + \mu_I^*\right)S_{I-L} - \frac{1}{T}\left(\mu_L^* + g^*\right)S_L^{pl} - \frac{1}{T}\left(\mu_I^* - g^*\right)S_I^{pl} \\ & + \vec{J}_L\vec{X}_L + \vec{J}_I\vec{X}_I + \frac{1}{T}\left(\vec{s}_k^{dis}\frac{d\vec{\beta}_k^{an}}{dt} + \vec{m}^{dis}\frac{d\vec{\omega}^{an}}{dt} + \vec{s}_k\frac{d\vec{\beta}_k^{pl}}{dt} + \vec{m}\frac{d\vec{\omega}^{pl}}{dt}\right) \\ & + \vec{J}_q \overrightarrow{\operatorname{grad}}\left(\frac{1}{T}\right) \end{aligned} \qquad (52)$$

Energetic balance

$$\begin{aligned} & n\vec{\phi}\left(\frac{d\vec{p}}{dt} - m\vec{\phi}_I\frac{dC_I}{dt} + m\vec{\phi}_L\frac{dC_L}{dt}\right) + \vec{s}_k\frac{d\vec{\beta}_k}{dt} + \vec{m}\frac{d\vec{\omega}}{dt} - p\frac{d\tau}{dt} \\ & \quad = \rho\vec{g}\vec{\phi} - \operatorname{div}\left[-\phi_k\vec{s}_k - \frac{1}{2}\left(\vec{\phi} \wedge \vec{m}\right) + p\vec{\phi}\right] \end{aligned} \qquad (53)$$

PART I

B

Application: examples of phenomenologies of usual solids

State functions and equations of state of an isotropic solid

Elastic et thermal behaviours

Wave propagations and thermo-elastic relaxations

Transport phenomena and inertial relaxations

Phenomenologies of anelasticity and plasticity

Structural transitions of first-order and second-order

Chapter 7

Phenomenological examples of usual solids

In this chapter, we present some applications of the first part by some simple examples of phenomenologies encountered in common isotropic solids. At first, simplified state functions are introduced to describe the phenomenology of a *'linear' isotropic solid.* From that we deduce the state equations for that type of solid. Then we discuss some phenomenological behaviors associated with elasticity, anelasticity, structural transitions, plasticity and auto-diffusion in the usual solids.

7.1 - Functions and state equations of isotropic solids

For a solid to present isotropic elastic properties (which are independent of direction in space), its free energy f per site of lattice must be a scalar built from the invariants of the tensor of distortion $\vec{\beta}_i$. Given that the operators of divergence and rotational applied to a vector field $\vec{u}$ give us invariant quantities of this field, and hence do not depend on the system of coordinates we choose, the first two scalar invariants of tensor gradient β_{ij} are its trace τ and the square of the norm of its anti-symmetric part $\vec{\omega}$

$$\tau = -\operatorname{div}\vec{u} = \sum_i \vec{e}_i\vec{\beta}_i = \sum_i \vec{e}_i\vec{\varepsilon}_i = \textit{invariant} \tag{7.1}$$

$$\vec{\omega}^2 = \frac{1}{4}\left(\overrightarrow{\operatorname{rot}}\,\vec{u}\right)^2 = \frac{1}{4}\left(\sum_i \vec{e}_i \wedge \vec{\beta}_i\right)^2 = \textit{invariant} \tag{7.2}$$

From the tensor gradient β_{ij}, other invariant scalars can be deduced such as the sum of the minors (concept in linear algebra) of the tensors of distortion of order 2.

$$\sum_i \vec{e}_i\left(\vec{\beta}_j \wedge \vec{\beta}_k\right) = \textit{invariant} \;\; ; \;\; \sum_i \vec{e}_i\left(\vec{\varepsilon}_j \wedge \vec{\varepsilon}_k\right) = \textit{invariant} \;\; ; \;\; \sum_i \vec{e}_i\left(\vec{\alpha}_j \wedge \vec{\alpha}_k\right) = \textit{invariant} \tag{7.3}$$

or the sum of the squares of the components of the tensors of distortion of order 2.

$$\sum_i \vec{\beta}_i^2 = \textit{invariant} \;\; ; \;\; \sum_i \vec{\varepsilon}_i^2 = \textit{invariant} \;\; ; \;\; \sum_i \vec{\alpha}_i^2 = \textit{invariant} \tag{7.4}$$

The state function of elasticity of an isotropic solid

The expression of free energy by lattice site of an isotropic solid must be a function of the following invariants

$$f^{él} = f^{él}\left(\tau, \sum_i \vec{\varepsilon}_i^2, T\right) \quad \text{or} \quad f^{él} = f^{él}\left(\tau, \sum_i \vec{\alpha}_i^2, T\right) \tag{7.5}$$

It must be possible to correctly express it by separating the volume expansion and the shear by a development such as

$$f^{\acute{e}l} = -k_0(T)\tau + k_1(T)\tau^2 + k_2(T)\sum_i \vec{\alpha}_i^2 + k_3(T)\tau\sum_i \vec{\alpha}_i^2 + k_4(T)\tau^3 + ... \tag{7.6}$$

If the development is limited to the terms containing $k_0(T), k_1(T), k_2(T)$, we will speak of a *"linear" isotropic elastic media* which we will henceforth call *perfect solid.* We deduce from that that the free elastic energy $f^{\acute{e}l}$ averaged by site of lattice is

$$f^{\acute{e}l} = -k_0(T)\tau + k_1(T)\tau^2 + k_2(T)\sum_i \vec{\alpha}_i^2 \tag{7.7}$$

The state function associated with the linear anelasticity of a solid

The macroscopic phenomenology of anelasticity of a solid lattice of particles can have many different physical microscopic causes. We can name the short distance defects of the lattice structure (substitutional or interstitial point defects, dislocations or disclinations, grain boundaries in polycrystalline materials or interphase boundaries in polyphase crystals, etc.) and even name some phase transitions. At the macroscopic scale of the continuous description of the lattice, the statistical manifestation of the collective movements of the whole set of structural defects is translated by the existence of a phenomenological equations of dissipative anelasticity.

The most straightforward way to introduce the phenomenological manifestation of anelasticity in a solid lattice is to first limits oneself to an isotropic lattice and suppose that *(i)* the anelasticity of the lattice can be fully described by the shear tensor, which also means we do not have terms from the anelastic rotations in the state function of free energy, *(ii)* that the terms of elasticity do not couple with anelasticity and self-diffusion, and *(iii)* the term of "free energy" by site of lattice linked to the anelasticity is a simple quadratic function of the components of the tensor of anelastic shear, multiplied by a coefficient k_{an} which could in theory depend on temperature T as well as the elastic volume expansion $\tau^{\acute{e}l}$ of the lattice. Under these conditions, the free energy associated with the anelasticity of the lattice can be written as

$$f^{an} = \frac{1}{2}k_{an}(T,\tau^{\acute{e}l})\sum_i (\vec{\alpha}_i^{an})^2 \tag{7.8}$$

As a matter of fact, the anelastic term f^{an} we added to the state function f is nothing more than the average potential energy stored by site of lattice by the restoring forces acting on the structural defects of anelasticity. The quadratic function are equivalent to suppose that these forces are *linear in the tensor of anelastic shear.*

The state function associated with a solid with weak self-diffusion

For a solid lattice of volume density n of sites and containing n_L and n_I , respectively vacancies and interstitials by unit volume, the free energy will be computable if the concentration of point defects is weak enough that we can neglect the interaction of these defects, and that we also neglect the impact of these defects on the elastic and thermic properties of the lattice. We will then talk about a *weakly diffusive solid.* The free energy F by unit of volume of the solid must contain a F^{dp} component due to the presence of *point defects (PD).* If we neglect the interactions between point defects, then the function F^{dp} can be assumed to be proportional to the number n_L et n_I of point defects per unit of volume

$$F^{dp} = n_L \varepsilon_L(\tau^{\acute{e}l}) - T n_L s_L(\tau^{\acute{e}l}) + n_I \varepsilon_I(\tau^{\acute{e}l}) - T n_I s_I(\tau^{\acute{e}l}) - T \Delta S^{conf} \tag{7.9}$$

in which $\varepsilon_L(\tau^{\acute{e}l})$ and $\varepsilon_I(\tau^{\acute{e}l})$ are the *potential energies* of a vacancy and interstitial respectively, which must first depend on the elastic volume expansion $\tau^{\acute{e}l}$ of the lattice. $s_L(\tau^{\acute{e}l})$ and $s_I(\tau^{\acute{e}l})$ are the *entropies* associated with a vacancy or an interstitial and must also depend on $\tau^{\acute{e}l}$, and ΔS^{conf} is the *entropy of configuration*, linked to the disorder introduced inside the lattice by the presence of the n_L and n_I point defects among the n sites per unit volume of the lattice.

The entropy of configuration ΔS^{conf}, by definition, equals $k \ln \Omega^{conf}$, where k is the *constant of Boltzmann* and Ω^{conf} is a number *counting the different microscopic configurations, which are known as 'complexions' of the system, that the n_L and n_I vacancies and interstitial can have.* For a network with a density n of substitutional sites, the density of interstitials is given by $z_I n$, where z_I is the number of interstitial sites associated with each substitutional site, which self evidently depends on the structure of the lattice under consideration. Define that the number of vacancy sites is n_L so that the number of interstitial site that are occupied is given by $n - n_L$ and that the number of unoccupied interstitial sites is $z_I n - n_I$, with these definitions, the number of complexions is written combinatorially

$$\Omega^{conf} = \frac{n!(z_I n)!}{(n - n_L)! n_L! (z_I n - n_I)! n_I!} \tag{7.10}$$

Using the *Stirling's approximation formula*, the entropy of configuration can be written

$$\Delta S^{conf} = k \ln \Omega^{conf} \cong -k \left(n_L \ln \frac{n_L}{n} + n_I \ln \frac{n_I}{z_I n} \right) \tag{7.11}$$

From these relations, the average free energy per site of the lattice due to the presence of point defects can be deduced by dividing F^{dp} by n and by introducing the atomic concentrations of point defects

$$f^{dp} = C_L \left[\varepsilon_L(\tau^{\acute{e}l}) - T s_L(\tau^{\acute{e}l}) \right] + C_I \left[\varepsilon_I(\tau^{\acute{e}l}) - T s_I(\tau^{\acute{e}l}) \right] + kT \left(C_L \ln C_L + C_I \ln \frac{C_I}{z_I} \right) \tag{7.12}$$

The equations of state of a perfect solid

By basing ourselves on the various components of the free energy f, and by introducing another thermic term $f_0(T)$ which contains all *the effects due to the temperature* of the lattice, and by making the following simplifying hypothesis

Hypothesis: *$k_1, k_2, k_{an}, \varepsilon_L, s_L, \varepsilon_I$ and s_I are constants* (7.13)

we have the following expression for the free energy $f \cong f_0(T) + f^{\acute{e}l} + f^{an} + f^{dp}$

$$\begin{aligned} f \cong\ & f_0(T) - k_0(T)\tau^{\acute{e}l} + k_1 \left(\tau^{\acute{e}l}\right)^2 + k_2 \sum \left(\vec{\alpha}_i^{\acute{e}l}\right)^2 + \frac{1}{2} k_{an} \sum_i \left(\vec{\alpha}_i^{an}\right)^2 \\ & + C_L \left[\varepsilon_L - T s_L \right] + C_I \left[\varepsilon_I - T s_I \right] + kT \left(C_L \ln C_L + C_I \ln \frac{C_I}{z_I} \right) \end{aligned} \tag{7.14}$$

The equations of state of the lattice can be written in this case, by recalling that $n = n_0 \exp\left(-\tau^{él}\right)$

$$\left\{\begin{array}{l} \vec{s}_i = n\dfrac{\partial f}{\partial \alpha_{ik}^{él}}\vec{e}_k = 2nk_2\vec{\alpha}_i^{él} \\ p = -n\dfrac{\partial f}{\partial \tau^{él}} = n\left(k_0(T) - 2k_1\tau^{él}\right) \text{ in the perfect solid} \\ \vec{s}_i^{\,cons} = n\dfrac{\partial f}{\partial \alpha_{ij}^{an}}\vec{e}_j = nk_{an}\vec{\alpha}_i^{an} \end{array}\right. \qquad (7.15)$$

$$\left\{ s = -\frac{\partial f}{\partial T} = -\frac{\partial f_0(T)}{\partial T} + \frac{\partial k_0(T)}{\partial T}\tau^{él} + C_L s_L + C_I s_I - kC_L \ln C_L - kC_I \ln\frac{C_I}{z_I} \right. \qquad (7.16)$$

$$\left\{\begin{array}{l} \mu_L = \dfrac{\partial f}{\partial C_L} = \varepsilon_L - Ts_L + kT\left(\ln C_L + 1\right) \\ \mu_I = \dfrac{\partial f}{\partial C_I} = \varepsilon_I - Ts_I + kT\left(\ln\dfrac{C_I}{z_I} + 1\right) \end{array}\right. \qquad (7.17)$$

These state equations are very simple. Amid others, the shear stress $\vec{s}_i$ and $\vec{s}_i^{\,cons}$ depend linearly on the shear tensors $\vec{\alpha}_i^{él}$ and $\vec{\alpha}_i^{an}$. There is another remark to be made: the pressure p in the lattice depends on a term $k_0(T)$ associated with temperature and a term $-2k_1\tau^{él}$ which is a linear function of the volume expansion of the lattice.

7.2 - Elastic moduli of isotropic solids

The Lamé parameters μ and λ at small distortions of the solid

In general, the equations of state of elasticity of an isotropic solid are given in the representation $\left\{\vec{\varepsilon}_i, \vec{\sigma}_i\right\}$. It is very simple to write in this representation by using the relation $\vec{\sigma}_i = \vec{s}_i - p\vec{e}_i$, as well as pressure $p = nk_0(T) - 2nk_1\tau^{él}$

$$\vec{\sigma}_i = n\left(2k_2\vec{\alpha}_i - k_0(T)\vec{e}_i + 2k_1\tau\vec{e}_i\right) \qquad (7.18)$$

With relations $\vec{\alpha}_i = \vec{\varepsilon}_i - \tau\vec{e}_i / 3$ and $\tau = \sum \varepsilon_{kk}$, we also have

$$\vec{\sigma}_i = n\left[2k_2\vec{\varepsilon}_i + 2\left(k_1 - \frac{k_2}{3}\right)\sum_k \varepsilon_{kk}\vec{e}_i - k_0(T)\vec{e}_i\right] \qquad (7.19)$$

We can introduce the new elastic coefficients with the following relations

$$\left\{\begin{array}{l} \mu = nk_2(T) \\ \lambda = 2n\left(k_1(T) - \dfrac{k_2(T)}{3}\right) \end{array}\right. \qquad (7.20)$$

These coefficients μ and λ are called the *Lamé parameters* of the perfect isotropic solid, and they allow us to write the elastic state equations under the form

$$\vec{\sigma}_i = 2\mu\vec{\varepsilon}_i + \vec{e}_i\left(\lambda\sum_k \varepsilon_{kk} - nk_0(T)\right) \qquad (7.21)$$

This form has the distinct advantage of allowing the inversion of the state equation to express tensor $\vec{\varepsilon}_i$ as a function of tensor $\vec{\sigma}_i$

$$\vec{\varepsilon}_i = \frac{1}{2\mu}\vec{\sigma}_i + \vec{e}_i \frac{1}{2\mu+3\lambda}\left(nk_0(T) - \frac{\lambda}{2\mu}\sum_k \sigma_{kk} \right) \qquad (7.22)$$

The Young's modulus E and the Poisson's ratio ν under simple traction

Let's imagine a parallelepipedic solid which is submitted to an *uniaxial tensile testing* (figure 7.1): a normal tensile stress σ_{33}, different than zero, is imposed on the faces perpendicular to the axis Ox_3, while the faces perpendicular to axis Ox_1 and Ox_2 are free of stress ($\sigma_{11}=0$ and $\sigma_{22}=0$). The state equation *(7.22)* allows us to calculate the components ε_{ii} of the tensor of deformation

$$\varepsilon_{11} = \varepsilon_{22} = \frac{1}{2\mu+3\lambda}\left(nk_0(T) - \frac{\lambda}{2\mu}\sigma_{33} \right) \quad \text{et} \quad \varepsilon_{33} = \frac{1}{2\mu+3\lambda}\left(nk_0(T) + \frac{\mu+\lambda}{\mu}\sigma_{33} \right) \qquad (7.23)$$

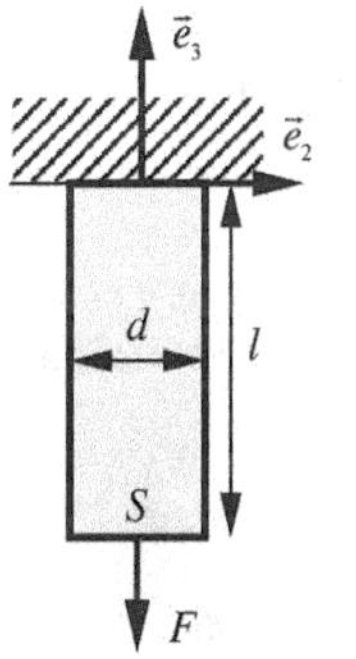

Figure 7.1 *- uniaxial tensile testing*

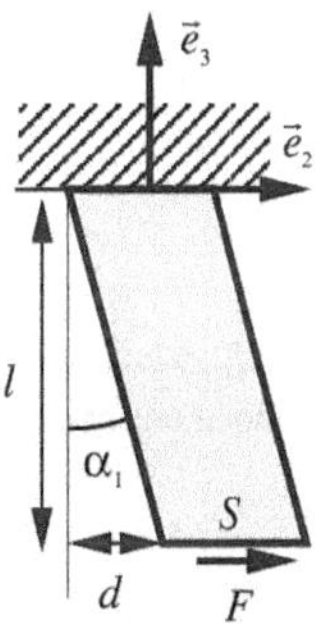

Figure 7.2 *- pure shear strain*

By definition, the *Young's modulus* E is defined by the *uniaxial tensile testing* (figure 7.1) as the ratio of the increase in the stress $\Delta\sigma_{33}$ along the tensile axis and the increase of the deformation $\Delta\varepsilon_{33}$ along the same axis, at a given temperature T

$$E = \frac{\Delta\sigma_{33}}{\Delta\varepsilon_{33}} = \frac{\mu(2\mu+3\lambda)}{\mu+\lambda} = n\frac{18k_1(T)k_2(T)}{6k_1(T)+k_2(T)} \qquad (7.24)$$

With regards to the *Poisson's ratio* ν, it is defined as the ratio between the narrowing $-\Delta\varepsilon_{11}$ or $-\Delta\varepsilon_{22}$ along the transversal axis Ox_1 and Ox_2 and the stretching $\Delta\varepsilon_{33}$ along axis Ox_3 of traction, at a given temperature T

$$\nu = -\frac{\Delta\varepsilon_{11}}{\Delta\varepsilon_{33}} = -\frac{\Delta\varepsilon_{22}}{\Delta\varepsilon_{33}} = \frac{\lambda}{2(\mu+\lambda)} = \frac{3k_1(T)-k_2(T)}{6k_1(T)+k_2(T)} \qquad (7.25)$$

The shear modulus G during pure shear strain

In the case of a deformation of the solid by *pure shear*, a parallelepipedic solid is deformed by shear in the plan i and j by applying tangential stress $\sigma_{ij} = \sigma_{ji}$ on the perpendicular planes to the axes Ox_i and Ox_j (figure 7.2). The state equation *(7.30)* allows us to calculate the components $\varepsilon_{ij} = \varepsilon_{ji}$ of the tensor of deformation as

$$\sigma_{ij} = 2\mu\varepsilon_{ij} = G\varepsilon_{ij} \tag{7.26}$$

The modulus G which appears here is called *shear modulus* or *modulus of rigidity* of the isotropic solid and its value is

$$G = 2\mu = 2nk_2 \tag{7.27}$$

The relations between the elastic moduli

In general moduli E, ν and G are shown in the tables of elastic moduli of isotropic materials. From knowing E, ν and G for a given material, we find the Lamé parameters μ and λ of this material by

$$\mu = \frac{E}{2(1-\nu)} = \frac{G}{2} \quad \text{and} \quad \lambda = \frac{E\nu}{(1+\nu)(1-2\nu)} \tag{7.28}$$

As regards the elastic coefficients k_1 and k_2 of a rigid material, they can be deduced either from the μ and λ parameters, either from the moduli E, ν and G

$$K_1 = \frac{\lambda}{2} + \frac{\mu}{3} = \frac{E}{6(1-2\nu)} \quad \text{et} \quad \lambda = \frac{E\nu}{(1+\nu)(1-2\nu)} \tag{7.29}$$

The Poisson's ratio ν of an isotropic material is very interesting, as it permits us to make the link between the k_1 coefficient associated to the compressibility of the media and the k_2 coefficient, linked to the shear-ability of the media. Indeed

$$\frac{k_1}{k_2} = \frac{1+\nu}{3(1-2\nu)} \tag{7.30}$$

For example, in metals, the Poisson's ratio is in general close to 1/3, which implies that the ratio k_1 / k_2 is approximatively 4/3.

7.3 - Thermal behavior of isotropic solids

In general, experiments show that, in solids at *high temperature*, the coefficient $k_0(T)$ as well as the thermal term $f_0(T)$ of the state function are simple functions of temperature. Indeed, in the usual solids, we observe a *specific heat* c_V measured by site of lattice, which at high temperature is approximatively a constant with value $c_V = 3k$ *(Dulong-Petit law),* where k is the *Boltzmann constant*

$$c_V = T\frac{\partial s}{\partial T} = T\left[-\frac{\partial^2 f_0(T)}{\partial T^2} + \frac{\partial^2 k_0(T)}{\partial T^2}\tau^{él}\right] \cong cste \tag{7.31}$$

This relation can be easily satisfied if we make the hypothesis that the dependence on temperature of the elastic coefficient $k_0(T)$ is linear,

$$T\frac{\partial^2 k_0(T)}{\partial T^2}\tau^{él} \cong 0 \quad \Rightarrow \quad k_0(T) \cong \gamma_0 c_V T \tag{7.32}$$

This hypothesis then implies for $f_0(T)$ that

$$-T\frac{\partial^2 f_0(T)}{\partial T^2} \cong c_V \quad \Rightarrow \quad f_0(T) \cong u_0 - Ts_0 + c_V T(1-\ln T) \tag{7.33}$$

The coefficient $k_0(T)$ depends more or less linearly on T via a constant $\gamma_0 c_V$, or γ_0 which is called the *Grünheisen parameter*, and whose numerical value is between 1,2 and 2,6 for most of the usual isotropic solids. We also observe, in the usual solids, that the coefficients k_1 and k_2 do not depend much on temperature, and can most of the time be treated as constants. These thermal behavior can be obtained from statistical physics, which shows the existence of the Grünheisen parameter γ_0 and that it is linked to the *anharmonic properties of the lattice.*
At constant pressure or with constant volume expansion, as in the case of weak distortions of the lattice, the introduction of *(7.32)* in *(7.15)* shows that the coefficient $k_0(T)$ is directly linked to the *phenomenon of thermic volume expansion* of the lattice

$$\tau^{él}(T)\Big|_{p_{cste}} \cong \frac{1}{2nk_1}\left(\gamma_0 n c_V T - p_{cste}\right) \quad \text{or} \quad p(T)\Big|_{\tau^{él}_{cste}} \cong n\left(\gamma_0 c_V T - 2k_1\tau^{él}_{cste}\right) \tag{7.34}$$

Let's imagine that a parallelepipedic solid is not submitted to any axial tensile stress, so that $\sigma_{ii}=0$, and we change its temperature. Because of coefficient $k_0(T)\cong\gamma_0 c_V T$, the uniaxial strain ε_{ii} (7.23) then becomes dependent on temperature, a phenomenon we call *linear thermal expansion of the solid*

$$\varepsilon_{11}(T)\Big|_{\sigma_{33}=0} = \varepsilon_{22}(T)\Big|_{\sigma_{33}=0} = \varepsilon_{33}(T)\Big|_{\sigma_{33}=0} = \frac{\gamma_0 n c_V T}{2\mu+3\lambda} \tag{7.35}$$

7.4 - Transport phenomena in isotropic solids

Let's consider the phenomenological equations of dissipative thermal conduction and self-diffusion *(6.41)*. In the case where a solid is isotropic and is not too removed from the thermodynamic equilibrium, and additionally if the concentrations of point defects remain small, we can linearize the vectorial dependences of $\vec{J}_q$, $\vec{J}_L$ and $\vec{J}_I$ on the thermodynamic forces $\vec{X}_q$, $\vec{X}_L$ and $\vec{X}_I$ by introducing *kinetic coefficients* of the type $L_{qL}(\tau,T,C_L,C_I)$, which allows us to write the equation of heat transport and the chemical species in the form

$$\begin{cases} \vec{J}_q = L_{qq}\vec{X}_q + L_{qL}\vec{X}_L + L_{qI}\vec{X}_I \\ \vec{J}_L = L_{Lq}\vec{X}_q + L_{LL}\vec{X}_L + L_{LI}\vec{X}_I \\ \vec{J}_I = L_{Iq}\vec{X}_q + L_{IL}\vec{X}_L + L_{II}\vec{X}_I \end{cases} \tag{7.36}$$

The source of entropy $S_e^{transport}$ associated to transport is written

$$S_e^{transport} = \sum_x \vec{X}_x \vec{J}_x \tag{7.37}$$

which we expand as

$$\begin{aligned} S_e^{transport} = & L_{qq}\vec{X}_q^2 + L_{LL}\vec{X}_L^2 + L_{II}\vec{X}_I^2 \\ & + \left(L_{qL}+L_{Lq}\right)\vec{X}_q\vec{X}_L + \left(L_{qI}+L_{Iq}\right)\vec{X}_q\vec{X}_I + \left(L_{LI}+L_{IL}\right)\vec{X}_L\vec{X}_I \end{aligned} \tag{7.38}$$

and which can be transformed by expressing the scalar product in product of the components of the vectors

$$S_e^{transport} = \sum_{k=1}^{3}\left[L_{qq}X_{qk}^2 + L_{LL}X_{Lk}^2 + L_{II}X_{Ik}^2 + \left(L_{qL}+L_{Lq}\right)X_{qk}X_{Lk} + \left(L_{qI}+L_{Iq}\right)X_{qk}X_{Ik} + \left(L_{LI}+L_{IL}\right)X_{Lk}X_{Ik} \right] \tag{7.39}$$

The source of entropy is made up of the sum of three quadratic functions X_{qi}, X_{Li} and X_{Ii}. To insure that whatever the values of X_{qi}, X_{Li} and X_{Ii}, we have the positivity of the source of entropy as required by the second principle, it must be that these three functions are individually positive in the 3D space of X_{qi}, X_{Li} et X_{Ii}, in such a way that the three quadratic functions can be transformed as the sum of square with positive coefficients by an orthogonal transformation of the variables in 3D space. Mathematically speaking, this is only possible if the following matrix with kinetic coefficients, composed of real numbers

$$\begin{pmatrix} L_{qq} & L_{qL} & L_{qI} \\ L_{Lq} & L_{LL} & L_{LI} \\ L_{Iq} & L_{IL} & L_{II} \end{pmatrix} \tag{7.40}$$

is a symmetric matrix, which can be diagonalized with eigenvalues that are positive. We impose then the following conditions, which are known as the *Onsager relations*,

$$L_{qL} = L_{Lq} \quad ; \quad L_{qI} = L_{Iq} \quad ; \quad L_{LI} = L_{IL} \tag{7.41}$$

The coefficients of thermal conductivity and diffusion

In the case of an isotropic solid with no coupling between the flux of particle transport and heat, or if these couplings are negligible, the equations of transport can be rewritten by introducing a *thermal conductivity coefficient* χ *and two coefficients of diffusion* D_L and D_I for the vacancies and the interstitials respectively

$$\begin{cases} \vec{J}_q = L_{qq}\vec{X}_q = L_{qq}\,\overrightarrow{\text{grad}}\dfrac{1}{T} = -\dfrac{L_{qq}}{T^2}\,\overrightarrow{\text{grad}}\,T = -\chi\,\overrightarrow{\text{grad}}\,T \\ \vec{J}_L = L_{LL}\vec{X}_L = nC_L\dfrac{D_L}{k}\vec{X}_L \\ \vec{J}_I = L_{II}\vec{X}_I = nC_I\dfrac{D_I}{k}\vec{X}_I \end{cases} \tag{7.42}$$

If C_L and C_I are small, the previous coefficients do not depend on them, so that we have a dependency on the volume expansion and the temperature (of Boltzmann type for the diffusion coefficients)

$$\chi \cong \chi(\tau,T) \quad ; \quad D_L \cong D_{L0}\exp\left[-E_L^{diff}(\tau)/kT\right] \quad ; \quad D_I \cong D_{I0}\exp\left[-E_I^{diff}(\tau)/kT\right] \tag{7.43}$$

7.5 - Propagation of waves and thermoelastic relaxation

The thermo-plastic equations of the perfect solid

The exact computation of the spatiotemporal evolution of a perfect solid is a rather complex problem, because to all the local deformations of the lattice is associated a local perturbation of the temperature of the lattice. This coupling between deformations and temperature comes from the entropy expression of the solid. By neglecting the effects of the field of gravitation $\rho\vec{g}$, and assuming that there are no point defects, nor anelasticity or plasticity in the lattice, and by using relations *(3.5) and (3.10)*, one obtains for a given compatible solid

$$\begin{cases} \sum_k \vec{e}_k \operatorname{div}\vec{\alpha}_k = -\overrightarrow{\operatorname{rot}}\,\vec{\omega} + \frac{2}{3}\overrightarrow{\operatorname{grad}}\,\tau - \cancel{\sum_k \left(\vec{e}_k \wedge \overrightarrow{\operatorname{rot}}\,\vec{\beta}_k\right)} = -\overrightarrow{\operatorname{rot}}\,\vec{\omega} + \frac{2}{3}\overrightarrow{\operatorname{grad}}\,\tau \\ \sum_k \left(\overrightarrow{\operatorname{grad}}\,\tau\ \vec{\alpha}_k\right)\vec{e}_k = \overrightarrow{\operatorname{grad}}\,\tau \wedge \cancel{\sum_k \left(\vec{e}_k \wedge \vec{\alpha}_k\right)} + \sum_k \left(\vec{e}_k \overrightarrow{\operatorname{grad}}\,\tau\right)\vec{\alpha}_k = \sum_k \left(\vec{e}_k \overrightarrow{\operatorname{grad}}\,\tau\right)\vec{\alpha}_k \end{cases} \tag{7.44}$$

The equation of Newton *(6.11)* can be written by using the state equations *(7.15)* for $\vec{\alpha}_i$ and τ and by using the fact that $k_0(T) \cong \gamma_0 c_V T$ under the form

$$\frac{d\vec{p}}{dt} = -2k_2\overrightarrow{\operatorname{rot}}\,\vec{\omega} - 2k_2\sum_k \left(\vec{e}_k \overrightarrow{\operatorname{grad}}\,\tau\right)\vec{\alpha}_k + \left[\gamma_0 c_V T + \frac{4}{3}k_2 + 2k_1(1-\tau)\right]\overrightarrow{\operatorname{grad}}\,\tau - \gamma_0 c_V \overrightarrow{\operatorname{grad}}\,T \tag{7.45}$$

By using relation *(7.15)* for entropy and by introducing the expressions $k_0(T) \cong \gamma_0 c_V T$ and $f_0(T) \cong u_0 - Ts_0 + c_V T(1-\ln T)$, we obtain the expression of entropy at high temperature inside a perfect lattice

$$s = s_0 + c_V \ln T + \gamma_0 c_V \tau^{el} \tag{7.46}$$

By introducing that expression of entropy as well as the flux of heat $\vec{J}_q = -\chi\,\overrightarrow{\operatorname{grad}}\,T$ in the heat equation *(6.22)*, we obtain the following equation for the *evolution of temperature*

$$\frac{dT}{dt} = -\gamma_0 T\frac{d\tau}{dt} + \frac{\chi}{nc_V}\operatorname{div}\overrightarrow{\operatorname{grad}}\,T = -\gamma_0 T\frac{d\tau}{dt} + \frac{\chi}{nc_V}\Delta T \tag{7.47}$$

The equations *(7.45)* and *(7.47)* are called the *thermo-elastic equations* of the solid.

The propagation of thermo-elastic longitudinal waves

In an isotropic lattice without point defects and without anelasticity, first homogeneous in volume expansion ($\tau = \tau_0$), isothermal ($T = T_0$), non sheared and immobile, we introduce a *longitudinal perturbation* in a local frame $Ox_1x_2x_3$ under the form of a velocity field $\vec{\phi} = \vec{\phi}^{(p)} = \phi_j^{(p)}(x_j,t)\vec{e}_j$ parallel to the axis Ox_j and which changes along the axis Ox_j. Via the geometro-kinetic equations for τ and $\vec{\alpha}_i$, this perturbation of the velocity field implies perturbations $\tau^{(p)}$, $\vec{\omega}^{(p)}$ and $\vec{\alpha}_i^{(p)}$ of distortions along axis Ox_j

$$\begin{cases} \dfrac{d\tau^{(p)}}{dt} = \operatorname{div}\vec{\phi}^{(p)} & \Rightarrow \quad \tau = \tau_0 + \tau^{(p)}(x_j,t) \\ \dfrac{d\vec{\omega}^{(p)}}{dt} = \dfrac{1}{2}\overrightarrow{\operatorname{rot}}\,\vec{\phi}^{(p)} = 0 & \Rightarrow \quad \vec{\omega}^{(p)} = 0 \\ \dfrac{d\vec{\alpha}_i^{(p)}}{dt} = -\dfrac{1}{3}\vec{e}_i \operatorname{div}\vec{\phi}^{(p)} = -\dfrac{1}{3}\dfrac{d\tau^{(p)}}{dt}\vec{e}_i & \Rightarrow \quad \vec{\alpha}_i^{(p)} = -\dfrac{1}{3}\vec{e}_i\tau^{(p)}(x_j,t) \\ \dfrac{d\vec{\alpha}_j^{(p)}}{dt} = \overrightarrow{\operatorname{grad}}\,\phi_j^{(p)} - \dfrac{1}{3}\vec{e}_j \operatorname{div}\vec{\phi}^{(p)} = \dfrac{2}{3}\dfrac{d\tau^{(p)}}{dt}\vec{e}_j & \Rightarrow \quad \vec{\alpha}_j^{(p)} = +\dfrac{2}{3}\vec{e}_j\tau^{(p)}(x_j,t) \\ \dfrac{d\vec{\alpha}_k^{(p)}}{dt} = -\dfrac{1}{3}\vec{e}_k \operatorname{div}\vec{\phi}^{(p)} = -\dfrac{1}{3}\dfrac{d\tau^{(p)}}{dt}\vec{e}_k & \Rightarrow \quad \vec{\alpha}_k^{(p)} = -\dfrac{1}{3}\vec{e}_k\tau^{(p)}(x_j,t) \end{cases} \tag{7.48}$$

By considering the heat equation *(7.56)*, it is clear that there exists a perturbation of the temperature field along the Ox_j axis

$$T = T_0 + T^{(p)}(x_j, t) \tag{7.49}$$

By introducing these perturbations in the equations *(7.45)* and *(7.47)*, and by taking into account the first relation *(7.48)*, we obtain the following set of equations, which we have linearized for weak longitudinal perturbations

$$\begin{cases} \dfrac{\partial^2 \tau^{(p)}}{\partial t^2} \cong \dfrac{1}{m}\left[\gamma_0 c_V T_0 + \dfrac{4}{3}k_2 + 2k_1(1-\tau_0)\right]\Delta\tau^{(p)} - \dfrac{\gamma_0 c_V}{m}\Delta T^{(p)} \\ \dfrac{\partial T^{(p)}}{\partial t} \cong -\gamma_0 T_0 \dfrac{\partial \tau^{(p)}}{\partial t} + \dfrac{\chi}{n c_V}\Delta T^{(p)} \qquad \left(n = n_0 e^{-\tau_0}\right) \end{cases} \tag{7.50}$$

Possible solutions to this set of equations are the *high frequency, isothermal, dampened, longitudinal waves* as well as *the low frequency, adiabatic, non dampened, longitudinal waves*, with respective phase velocities c_{iso} and c_{adia}

$$\begin{cases} c_{iso} = \sqrt{\dfrac{1}{m}\left[\gamma_0 c_V T_0 + \dfrac{4}{3}k_2 + 2k_1(1-\tau_0)\right]} \\ c_{adia} = \sqrt{\dfrac{1}{m}\left[\gamma_0 c_V T_0 + \dfrac{4}{3}k_2 + 2k_1(1-\tau_0)\right] + \dfrac{\gamma_0^2 T_0 c_V}{m}} \end{cases} \tag{7.51}$$

The presence of the term $2k_1(1-\tau_0)$ implies that the velocity of wave propagation depends on the state of volume expansion τ_0 of the lattice.

Logarithmic decrement and the Debye relaxation due to thermo-elasticity

The longitudinal wave described by equations *(7.50)* presents a *spatial logarithmic decrement* Δ_{log}, defined as the logarithm of the ratio of two amplitudes of the wave measured at two points separated by a distance Δx_j which is equal to the wavelength $\lambda = c_\varphi / f$, and evaluates to

$$\Delta_{log} = \delta = \frac{1}{\pi}\ln\frac{\tau^{(p)}(x_j,t)}{\tau^{(p)}(x_j+\lambda,t)} = \Delta_{rel}\,\frac{\omega\tau_{rel}}{1+(\omega\tau_{rel})^2} \tag{7.52}$$

$$\text{with} \quad \Delta_{rel} = \frac{\gamma_0^{\,2} T_0 c_V}{m c_{iso} c_{adia}} \quad \text{et} \quad \tau_{rel} = \frac{\chi}{n c_V c_{iso} c_{adia}} \tag{7.53}$$

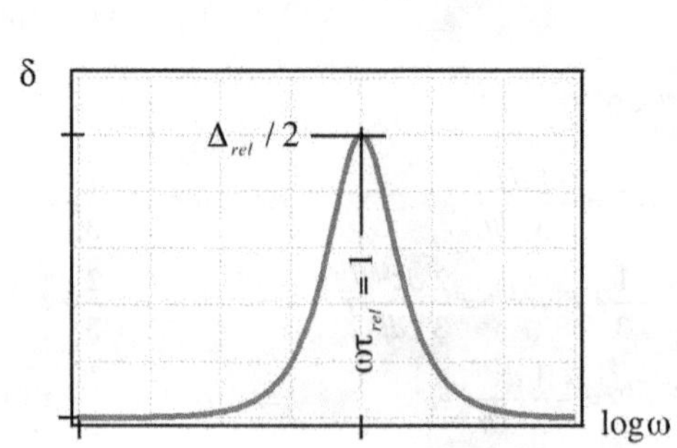

***Figure 7.3** - Loss factor and Debye relaxation due to thermo-elasticity*

The dependency of the *logarithmic decrement* Δ_{log} on the frequency, is also called the *loss factor* δ of the lattice. It is typical of a *relaxation phenomenon.* There appears a characteristic time τ_{rel}, called the *thermo-elastic relaxation time.* The logarithmic decrement Δ_{log} exhibits a *Debye relaxation peak* (figure 7.3) which finds its common origin in the thermal expansion and the process of the thermo-conduction. Indeed, the *relaxation amplitude* $\Delta_{rel} = \gamma_0{}^2 T_0 c_V / (mc_{iso} c_{adia})$ is proportional to γ_0^2 and depends as a consequence directly on the thermal expansion of the solid, while the relaxation time $\tau_{rel} = \chi / (nc_V c_{iso} c_{adia})$ is proportional to χ and depends then essentially to the process of thermo-conduction

The propagation of invariant transversal waves

If we consider deformations of the solid by pure shear, with constant volume expansion τ_0, the equation of Newton *(7.45)* simplifies into the form

$$\frac{d\vec{p}}{dt} = -2k_2 \overrightarrow{\mathrm{rot}}\, \vec{\omega} \tag{7.54}$$

By introducing transversal perturbations of the velocity field $\vec{\phi}^{(p)}$ under the form of a velocity field $\vec{\phi}^{(p)} = \phi_k^{(p)}(x_j, t)\vec{e}_k$ varying along axis Ox_j, we show with the geometro-kinetic equations that this perturbation of the velocity field implies the following perturbations for the volume expansion and the rotational vector along the axis Ox_j

$$\begin{cases} \dfrac{d\tau^{(p)}}{dt} = \mathrm{div}\vec{\phi}^{(p)} = 0 & \Rightarrow \quad \tau = \tau_0 = cste \\[2mm] \dfrac{d\vec{\omega}^{(p)}}{dt} = \dfrac{1}{2}\overrightarrow{\mathrm{rot}}\,\vec{\phi}^{(p)} = \dfrac{1}{2}\dfrac{\partial \phi_k^{(p)}}{\partial x_j}\vec{e}_i & \Rightarrow \quad \vec{\omega}^{(p)} = \omega_i^{(p)}(x_j, t)\vec{e}_i \end{cases} \tag{7.55}$$

it also implies the following perturbations for the shear tensor along the Ox_j axis, which are directly linked to the perturbations of rotation

$$\begin{cases} \dfrac{d\vec{\alpha}_i^{(p)}}{dt} = 0 & \Rightarrow \quad \vec{\alpha}_i^{(p)} = 0 \\[2mm] \dfrac{d\vec{\alpha}_j^{(p)}}{dt} = \dfrac{1}{2}\dfrac{\partial \phi_k^{(p)}}{\partial x_j}\vec{e}_k & \Rightarrow \quad \vec{\alpha}_j^{(p)} = \omega_i^{(p)}(x_j, t)\vec{e}_k \\[2mm] \dfrac{d\vec{\alpha}_k^{(p)}}{dt} = \dfrac{1}{2}\dfrac{\partial \phi_k^{(p)}}{\partial x_j}\vec{e}_j & \Rightarrow \quad \vec{\alpha}_k^{(p)} = \omega_i^{(p)}(x_j, t)\vec{e}_j \end{cases} \tag{7.56}$$

The momentum is linked to the velocity field $\vec{\phi}^{(p)}$ via relation $\vec{p}^{(p)} = m\vec{\phi}^{(p)}$ and the Newton equation can be written

$$\frac{d\vec{p}^{(p)}}{dt} = -2k_2 \overrightarrow{\mathrm{rot}}\,\vec{\omega}^{(p)} \quad \Rightarrow \quad \frac{d\vec{\phi}^{(p)}}{dt} = -2\frac{k_2}{m}\overrightarrow{\mathrm{rot}}\,\vec{\omega}^{(p)} \tag{7.57}$$

Because the perturbations $\vec{\phi}^{(p)}(\vec{r}, t) = \phi_k^{(p)}(x_j, t)\vec{e}_k$ only have one component along the Ox_k axis, and since all the propagating values vary only along the Ox_j axis, the material derivative d/dt can be replaced by the partial derivative $\partial/\partial t$ with respect to time everywhere in the local frame $Ox_1x_2x_3$. This allows us to obtain the following set of scalar differential equations, which control the transversal perturbation in the local frame of the perfect solid

$$\begin{cases} \dfrac{\partial \omega_i^{(p)}}{\partial t} = \dfrac{1}{2} \dfrac{\partial \phi_k^{(p)}}{\partial x_j} \\ \dfrac{\partial \phi_k^{(p)}}{\partial t} = 2 \dfrac{k_2}{m} \dfrac{\partial \omega_i^{(p)}}{\partial x_j} \end{cases} \Rightarrow \begin{cases} \dfrac{\partial^2 \omega_i^{(p)}}{\partial t^2} = \dfrac{k_2}{m} \dfrac{\partial^2 \omega_i^{(p)}}{\partial x_j^2} \\ \dfrac{\partial^2 \phi_k^{(p)}}{\partial t^2} = \dfrac{k_2}{m} \dfrac{\partial^2 \phi_k^{(p)}}{\partial x_j^2} \end{cases} \tag{7.58}$$

Thus, no matter the amplitude of a transversal wave propagating in a perfect solid, it moves about in the local frame $Ox_1x_2x_3$ with a *wave velocity* equal to

$$c_t = \sqrt{\frac{k_2}{m}} \tag{7.59}$$

Therefore, in a perfect isotropic solid, the intrinsic velocity of transversal waves is an «*invariant quantity*», meaning one that does not depend on the volume expansion τ_0 of the lattice, on the condition that it be homogenous and constant. It should be also noted that, if the static volume expansion τ of the solid is not homogenous, meaning if $\overrightarrow{\text{grad}}\,\tau \neq 0$, the propagation of the transverse perturbations is impacted by the presence of a coupling term $-2k_2\sum\left(\vec{e}_k \overrightarrow{\text{grad}}\,\tau\right)\vec{\alpha}_k$ between the fields τ and $\vec{\alpha}_i$ in the Newton equation *(7.45)*. The propagation of transversal waves depends in this case on the gradient of the volume expansion τ, and the resolution of this propagation problem cannot do without the explicit use of the shear field $\vec{\alpha}_i$!

7.6 - Equations of transport and inertial relaxation

The equations of transport for point defects *(7.42)* can be transformed by introducing the fluxes $\vec{J}_L$ et $\vec{J}_I$ *(5.78)* and the forces $\vec{X}_L$ and $\vec{X}_I$ *(6.42)*

$$\begin{cases} \Delta\vec{\varphi}_L = \vec{\phi}_L - \vec{\phi} = \dfrac{D_L(\tau,T)}{kT}\left(-\overrightarrow{\text{grad}}\;\mu_L^* + m\dfrac{d}{dt}\left(\vec{\phi}_L - 2\Delta\vec{\varphi}_L\right) - m\vec{g}\right) \\ \Delta\vec{\varphi}_I = \vec{\phi}_I - \vec{\phi} = \dfrac{D_I(\tau,T)}{kT}\left(-\overrightarrow{\text{grad}}\;\mu_I^* - m\dfrac{d}{dt}\left(\vec{\phi}_I\right) + m\vec{g}\right) \end{cases} \tag{7.60}$$

The new equations link the relative velocities of the point defects with respect to the lattice to the chemical forces and the mechanical forces, via the coefficients D/kT. Such equations bear the name of *Einstein's diffusion relations.* It is possible to transform these equations to explicitly show the *equations of movement* of the vacancies and interstitials with respect to the lattice, along a path of the lattice, by decomposing the absolute velocity of the vacancies and regrouping the terms in a different way. We obtain

$$\begin{cases} m\dfrac{d\Delta\vec{\varphi}_L}{dt} + B_L(\tau,T)\Delta\vec{\varphi}_L = +m\dfrac{d\vec{\phi}}{dt} - m\vec{g} - \overrightarrow{\text{grad}}\;\mu_L^* \\ m\dfrac{d\Delta\vec{\varphi}_I}{dt} + B_I(\tau,T)\Delta\vec{\varphi}_I = -m\dfrac{d\vec{\phi}}{dt} + m\vec{g} - \overrightarrow{\text{grad}}\;\mu_I^* \end{cases} \tag{7.61}$$

in which $B_L(\tau,T)$ and $B_I(\tau,T)$ are the *coefficients of viscous friction* of the vacancies and interstitials respectively, and are linked to the coefficients of diffusion by

$$B_L(\tau,T) = \frac{kT}{D_L(\tau,T)} \quad \textit{and} \quad B_I(\tau,T) = \frac{kT}{D_I(\tau,T)} \tag{7.62}$$

In the equations of movement *(7.61)*, the relative dampened movement of point defects with respect to the lattice is controlled by three different applied forces: a drag-along force due to the acceleration $md\vec{\phi}/dt$ of the lattice, a force $m\vec{g}$ due to gravity, a chemical and kinetic force equal to the gradient of μ_L^* or of μ_I^*.

The processes of inertial relaxation of point defects

If a solid lattice whose point defects are in thermodynamical equilibrium is submitted to a vibration, characterized by an absolute oscillating velocity field $\vec{\phi}$, the movement of the point defects inside the solid is described by the equations of movement *(7.61)*, in which the terms $-\overrightarrow{\text{grad}}\ \mu_L^*$ and $-\overrightarrow{\text{grad}}\ \mu_I^*$ can be considered null since the point defects are at thermodynamical equilibrium. By also neglecting the gravity field, and by imposing in a direction $\vec{n}$ a vibration $\vec{\phi}(t)=\vec{n}\phi_0\cos\omega t$ with angular frequency ω to the isotropic lattice, the field of absolute velocity of the solid can be written, for convenience, by using the formalism of *rotating vectors in the complex plane* (we will underscore complex quantities)

$$\underline{\vec{\phi}}(t)=\vec{n}\phi_0\,\mathrm{e}^{i\omega t} \tag{7.63}$$

According to equations *(7.61)*, the point defects are going to start oscillating but with amplitudes and phase shift angles δ_L and δ_I, which is different for the vacancies and the interstitials

$$\begin{cases}\underline{\Delta\vec{\varphi}_L}(t)=\vec{n}\Delta\varphi_{L0}\phi_0\,\mathrm{e}^{i(\omega t-\delta_L)}=\underline{\alpha_L}\ \underline{\vec{\phi}}(t)\\[2mm] \underline{\Delta\vec{\varphi}_I}(t)=\vec{n}\Delta\varphi_{I0}\phi_0\,\mathrm{e}^{i(\omega t-\delta_I)}=\underline{\alpha_I}\ \underline{\vec{\phi}}(t)\end{cases} \tag{7.64}$$

By introducing the complex coefficients $\underline{\alpha_L}$ and $\underline{\alpha_I}$, the equations of movement *(7.61)* allow us to deduce the values of these coefficients, which are associated to the oscillations of vacancies and interstitials by the following relations

$$\underline{\alpha_L}=-\frac{m\omega}{iB_L-m\omega}=\frac{(\tau_L\omega)^2+i\tau_L\omega}{1+(\tau_L\omega)^2}\quad\text{and}\quad\underline{\alpha_I}=\frac{m\omega}{iB_I-m\omega}=-\frac{(\tau_I\omega)^2+i\tau_I\omega}{1+(\tau_I\omega)^2} \tag{7.65}$$

in which τ_L and τ_I are the *relaxation times for vacancies and interstitials*

$$\tau_L=\frac{m}{B_L}=\frac{mD_L}{kT}\quad\text{and}\quad\tau_I=\frac{m}{B_I}=\frac{mD_I}{kT} \tag{7.66}$$

The real fields of the relative velocities of point defects with respect to the lattice are then deduced as the real part of $\underline{\Delta\vec{\varphi}_L}(t)$ and $\underline{\Delta\vec{\varphi}_I}(t)$. The average momentum $\vec{p}=m(\vec{\phi}-C_L\vec{\phi}_L+C_I\vec{\phi}_I)$ per site of lattice is deduced in the complex representation, which manifests a *complex apparent mass* $\underline{m}$

$$\underline{\vec{p}}(t)=m\left[1-C_L\frac{1+2(\tau_L\omega)^2+i\tau_L\omega}{1+(\tau_L\omega)^2}+C_I\frac{1-i\tau_I\omega}{1+(\tau_I\omega)^2}\right]\underline{\vec{\phi}}(t)=\underline{m}\underline{\vec{\phi}}(t) \tag{7.67}$$

The average momentum $\vec{p}$ per site of lattice in the presence of an oscillatory movement of the lattice then has two contributions: a contribution which is in phase with the movement of the lattice and an other one which is out of phase by $\pi/2$. There appears in fact two *relaxation phenomena* on the average momentum $\vec{p}(t)$ per site of lattice, which are due respectively to the movement of vacancies and interstitials inside the lattice. These relaxation processes depend strongly on the frequency ω of the oscillatory movement and give rise to open cycles when we plot $\vec{pn}$ as a function of $\vec{\phi n}$, as shown in figure 7.4, in which the effects along the vertical axis have been greatly exaggerated

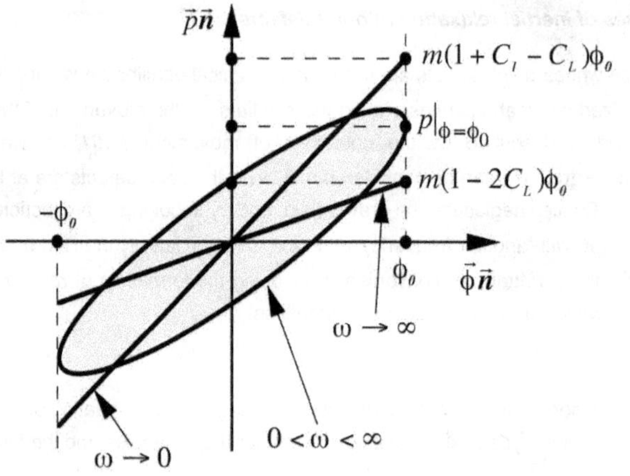

Figure 7.4 *- relaxation cycle of* $\vec{pn}$ as a function of $\vec{\phi n}$

7.7 - Creation-annihilation of a vacancy-interstitial pair

The phenomenological equation *(6.50)* giving the volume source S_{I-L} of pairs of vacancies and interstitials in the lattice can be deduced thanks to a very simplified model of creation-annihilation. Let's assume that the energetic diagram, as a function of distance d separating a vacancy from an interstitial, presents a maximum for a distance which will be called the *activation distance* d^{act} (figure 7.5). For distances d different than zero, but inferior to d^{act}, the vacancy and interstitial do not have a real existence. There are only thermal fluctuations of the position of the particle around the site it occupies. For distances d superior to d^{act}, the vacancy and interstitial become real point defects of the lattice, that have in common an increase in the pair energy $\Delta\varepsilon_0$ with respect to the initial energy of the particle on its initial lattice site. The creation of a pair is due to a thermal fluctuation which momentarily is equal or superior to $\Delta\varepsilon^{act}$. This is why $\Delta\varepsilon^{act}$ will be called *activation energy* of a vacancy-interstitial pair. The annihilation of a vacancy-interstitial pair, meaning the reverse path in the energetic diagram, also needed a thermal fluctuation but of energy superior or equal to $\Delta\varepsilon^{act}-\Delta\varepsilon_0$. The annihilation is thus a thermally activated process which has an *activation energy of annihilation* equal to $\Delta\varepsilon^{act}-\Delta\varepsilon_0$.

The value of the activation energy $\Delta\varepsilon^{act}$ of creation of a pair of vacancy-interstitial is evidently linked to the nature of the lattice under consideration, it must also depend strongly on the hydrostatic pressure p or the local volume expansion τ of the lattice so that

$$\Delta\varepsilon^{act}=\Delta\varepsilon^{act}(\tau) \qquad (7.68)$$

With regards to the activation energy $\Delta\varepsilon^{act}-\Delta\varepsilon_0$ of annihilation of a vacancy-interstitial pair, it depends on $\Delta\varepsilon^{act}(\tau)$, but also on the increase $\Delta\varepsilon_0$ of the energy of the vacancy-interstitial pair with respect to the energy of the particle in its original site. This energy increase $\Delta\varepsilon_0$ is nothing but the sum of *potential energies* $\varepsilon_L(\tau)$ and $\varepsilon_I(\tau)$ of the vacancy and interstitial, introduced previously in the state function Δf^{dp}, and the *increase in kinetic energy* Δe_{cin}^{I-L} defined by the relation *(6.48)*

$$\Delta\varepsilon_0 = \varepsilon_L(\tau) + \varepsilon_I(\tau) + \Delta e_{cin}^{I-L} \tag{7.69}$$

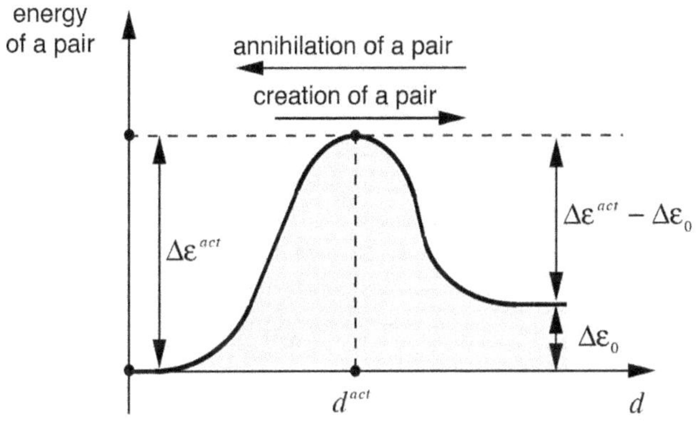

Figure 7.5 *- Energy diagram representing the creation-annihilation of a vacancy-interstitial pair as a function of distance* d

The terms of creation and annihilation of pairs can thus be written

$$S_{I-L}^{création} = n(1 - C_L)v_c \exp\left(-\frac{\Delta\varepsilon^{act}}{kT}\right) \tag{7.70}$$

$$S_{I-L}^{annihilation} = nC_L C_I v_a \exp\left(-\frac{\Delta\varepsilon^{act} - \Delta\varepsilon_0}{kT}\right) \tag{7.71}$$

Indeed, as S_{I-L} represents a volume source term, S_{I-L} must be proportional to the volume density of sites n. In the case of creation, only occupied substitutional sites can contribute to the formation of a pair, hence the proportionality in $(1 - C_L)$. In the case of annihilation, it is clear that the number of annihilations must be proportional to both C_L and C_I of vacancies and interstitials, and thus to the product $C_L C_I$.

In the two expressions we have a frequency of attack $v_c(\tau)$ or $v_a(\tau)$, which corresponds to the frequency at which favorable conditions to the creation or annihilation of a pair occur. This frequency is essentially a function of the local thermal fluctuations and must also depend on τ.

Finally, each of the expressions is multiplied by a *Boltzmann term* corresponding to the probability of the appearance of a thermal fluctuation with an energy equal or superior to the activation energies $\Delta\varepsilon^{act}$ or $\Delta\varepsilon^{act} - \Delta\varepsilon_0$, which allows us to 'jump' above the energy peak shown in figure 7.5. With this elementary model, the term S_{I-L} which is a global source or sink of pairs can be written by remembering that concentration C_L is supposed weak

$$S_{I-L} = S_{I-L}^{création} - S_{I-L}^{annihilation} = n\left[(1-C_L)v_c - C_L C_I v_a \exp\left(\frac{\Delta\varepsilon_0}{kT}\right)\right]\exp\left(-\frac{\Delta\varepsilon^{act}}{kT}\right)$$
$$\cong n\left[v_c - C_L C_I v_a \exp\left(\frac{\varepsilon_L(\tau)+\varepsilon_I(\tau)+\Delta e_{cin}^{I-L}}{kT}\right)\right]\exp\left(-\frac{\Delta\varepsilon^{act}}{kT}\right) \tag{7.72}$$

The global term of source S_{I-L} is null when the argument between brackets is null, meaning when the product $C_L C_I$ is approximatively equal to

$$C_L C_I \cong \frac{v_c}{v_a}\exp\left(-\frac{\Delta\varepsilon_0}{kT}\right) = \frac{v_c}{v_a}\exp\left(-\frac{\varepsilon_L(\tau)+\varepsilon_I(\tau)}{kT}\right)\exp\left(-\frac{\Delta e_{cin}^{I-L}}{kT}\right) \tag{7.73}$$

This condition is nothing more than the *relation of kinetic equilibrium* of the mechanism of creation and annihilation of pairs of vacancy-interstitial. For a temperature T and a volume expansion τ of the solid at thermodynamical equilibrium, it is possible to calculate the *thermodynamical equilibrium concentrations* of vacancies and interstitials from the state function f, simply by looking for the values of C_L and C_I which minimize f. It follows as a consequence the following values for $C_L^{équilibre}$ and $C_I^{équilibre}$

$$\begin{cases} \dfrac{\partial f}{\partial C_L} = \mu_L = 0 \quad \Rightarrow \quad C_L^{équilibre} = \exp\left(\dfrac{s_L(\tau)}{k} - 1\right)\exp\left(-\dfrac{\varepsilon_L(\tau)}{kT}\right) \\ \dfrac{\partial f}{\partial C_I} = \mu_I = 0 \quad \Rightarrow \quad C_I^{équilibre} = z_I \exp\left(\dfrac{s_I(\tau)}{k} - 1\right)\exp\left(-\dfrac{\varepsilon_I(\tau)}{kT}\right) \end{cases} \tag{7.74}$$

These two conditions of thermodynamical equilibrium allows us to calculate the *thermodynamic equilibrium relation* of the mechanism of creation-annihilation of pairs, under the form

$$C_L^{équilibre} C_I^{équilibre} = z_I \exp\left(\frac{s_L(\tau)+s_I(\tau)}{k} - 2\right)\exp\left(-\frac{\varepsilon_L(\tau)+\varepsilon_I(\tau)}{kT}\right) \tag{7.75}$$

As the relations of *kinetic equilibrium (7.73)* and *thermodynamical equilibrium (7.75)* must be equal at thermodynamical equilibrium of the solid, it is necessary that the two following equations be true at thermodynamical equilibrium

$$\begin{cases} \dfrac{v_c}{v_a} = z_I \exp\left(\dfrac{s_L(\tau)+s_I(\tau)}{k} - 2\right) \\ \dfrac{\Delta e_{cin}^{I-L}}{kT} = 0 \end{cases} \tag{7.76}$$

The second relation implies that at thermo-dynamical equilibrium, the kinetic energies of the vacancies and interstitials thermalize, which is expressed by the following equations, with the help of *(6.48)*

$$\Delta e_{cin}^{I-L} = \Delta e_{cin}^{L} + \Delta e_{cin}^{I} = e_{cin}^{L} + e_{cin}^{I} = 0 \tag{7.77}$$

This translates in term of relative velocities of vacancies and interstitials into the relation

$$\Delta e_{cin}^{I-L} = m\vec{\phi}\left(\Delta\vec{\varphi}_I - \Delta\vec{\varphi}_L\right) + \frac{1}{2}m\left(\Delta\vec{\varphi}_I^2 + \Delta\vec{\varphi}_L^2\right) = 0 \quad \textit{(at thermodynamical equilibrium)} \tag{7.78}$$

With regards to the first equation, it implies the existence of a strong relationship between the activation frequencies $\nu_c(\tau)$ and $\nu_a(\tau)$ and the entropies $s_L(\tau)$ and $s_I(\tau)$ of vacancies and interstitials

$$s_L(\tau)+s_I(\tau)=k\ln\left(\frac{e^2\nu_c(\tau)}{z_I\nu_a(\tau)}\right)=k\ln\Omega^{vibration} \qquad (7.79)$$

This shows that the entropies $s_L(\tau)$ and $s_I(\tau)$ measure in fact the variations of $\Omega^{vibration}$, the number of possible *microscopic vibrational configurations of the lattice* in the neighborhood of vacancies and interstitials. From this relation is deduced, thanks to the state equations *(7.15)* for μ_L and μ_I, another relation giving $\nu_a(\tau)$ as a function of $\nu_c(\tau)$ and the chemical potentials

$$\nu_a(\tau)=\frac{\nu_c(\tau)}{z_I}\exp\left(2-\frac{s_L(\tau)+s_I(\tau)}{k}\right)=\frac{\nu_c(\tau)}{C_LC_I}\exp\left(\frac{\mu_L+\mu_I}{kT}\right)\exp\left(-\frac{\varepsilon_L(\tau)+\varepsilon_I(\tau)}{kT}\right) \qquad (7.80)$$

This last relation allows us to express the global term S_{I-L} of sources or sinks of pairs *(7.72)* with the following form

$$S_{I-L}\cong n\nu_c\exp\left(-\frac{\Delta\varepsilon^{act}(\tau)}{kT}\right)\left[1-\exp\left(\frac{\mu_L^*+\mu_I^*}{kT}\right)\right] \qquad (7.81)$$

which corresponds perfectly to the form *(6.50)* previously hypothesized for the phenomenological equation of creation-annihilation of a vacancy-interstitial pair.

7.8 - Phenomenology of anelasticity

The phenomenological equations of anelasticity

The global shear $\vec{\alpha}_i$ of an anelastic solid are given by the sum of the elastic and anelastic shears of a lattice

$$\vec{\alpha}_i=\vec{\alpha}_i^{él}+\vec{\alpha}_i^{an} \qquad (7.82)$$

In this constitutive relation, the tensor of elastic shear and the conservative part of the tensor of anelastic shear are deduced from *(7.15)*

$$\begin{cases}\vec{s}_i=2nk_2(T)\vec{\alpha}_i^{él}=K_{él}\vec{\alpha}_i^{él} & \Rightarrow\quad K_{él}=K_{él}(n,T)=2nk_2(T)\\ \vec{s}_i^{cons}=nk_{an}(T,\tau^{él})\vec{\alpha}_i^{an}=K_{an}\vec{\alpha}_i^{an} & \Rightarrow\quad K_{an}=K_{an}(n,T)=nk_{an}(T,\tau^{él})\end{cases} \qquad (7.83)$$

It is the dissipative part of the phenomenological equation of anelasticity *(6.58)* which must essentially translate the dynamics of the structural defects responsible for anelasticity. It thus takes different forms depending on the obstacles and interaction that counter the movement of defects in the structure. Four typical cases of this dynamic will be reviewed now.

The process of viscous friction and the phenomenon of anelastic relaxation

Since the dissipative part of the phenomenological equation of anelasticity must essentially depend on the velocities of anelastic distortions, the simplest form we can imagine between $\vec{s}_i^{dis}$ and $d\vec{\alpha}_i^{an}/dt$ is simple proportionality. We introduce a *viscous friction coefficient* $B_{an}(n,T)$ which links $\vec{s}_i^{dis}$ with $d\vec{\alpha}_i^{an}/dt$ and which can depend on temperature T and the density of lattice sites

$$\vec{s}_i^{\,dis} = B_{an}(n,T)\frac{d\vec{\alpha}_i^{\,an}}{dt} \tag{7.84}$$

Therefore, the phenomenological equation of dissipative anelasticity *(6.58)* becomes a differential equation of first order in $\vec{\alpha}_i^{\,an}$, which allows us to link the behavior of anelastic shear $\vec{\alpha}_i^{\,an}$ to the spatiotemporal evolution of the field of elastic stress $\vec{s}_i$

$$\vec{s}_i = \vec{s}_i^{\,cons} + \vec{s}_i^{\,dis} = K_{an}(n,T)\,\vec{\alpha}_i^{\,an} + B_{an}(n,T)\,\frac{d\vec{\alpha}_i^{\,an}}{dt} \tag{7.85}$$

The elastic and anelastic shears of the lattice can be represented by a rheological model (figure 7.6a). In fact, the rheological model of the anelastic solid described above corresponds to put in parallel a spring of constant K_{an} with a viscous damper B_{an} in order to represent the anelastic part of the shears described by the equation of anelasticity. To this we add, in series, a spring of constant K_2 to represent the elastic part of the shears described by the equation of state *(15.5)*. In this case, it is clear that the dynamic structural defects responsible for anelasticity is controlled by a mechanism leading to *viscous friction* .

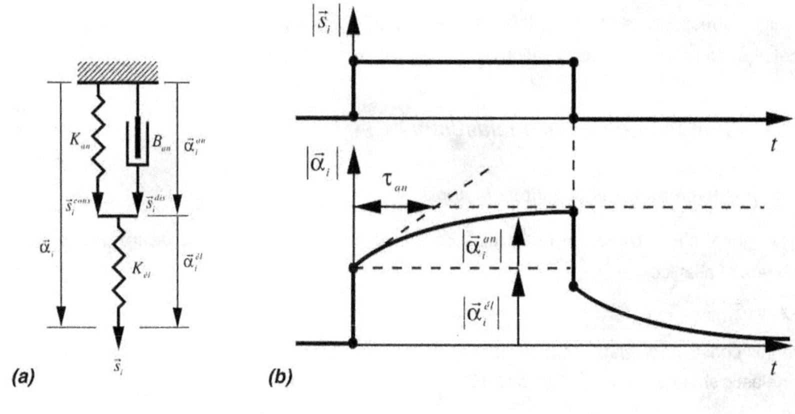

***Figure 7.6** - **(a)** Rheological model of viscous friction*
***(b)** instantaneous elastic response and relaxation*

With such rheological model, the response of the total shear $\vec{\alpha}_i$ to a solicitation by jump of the shear stress $\vec{s}_i$ has an instantaneous elastic part and a delayed anelastic portion (figure 7.6b), along with a *relaxation time* τ_{an} equal to

$$\tau_{an} = \frac{B_{an}}{K_{an}} \tag{7.86}$$

If the lattice is loaded by an oscillatory shear stress field $\vec{s}_i(t) = \vec{n} s_{i0} \cos \omega t$ oscillating in a certain direction $\vec{n}$, the anelastic response will then present a *relaxation phenomenon* at a frequency $\omega = 1/\tau_{an}$ similar to that reported in Figure 7.3. The temporal response of the overall shear field $\vec{\alpha}_i(t)$ to this cyclical shear stress $\vec{s}_i(t)$ itself has an amplitude and opening which correspond to the frequency of oscillation (fig. 7.7).

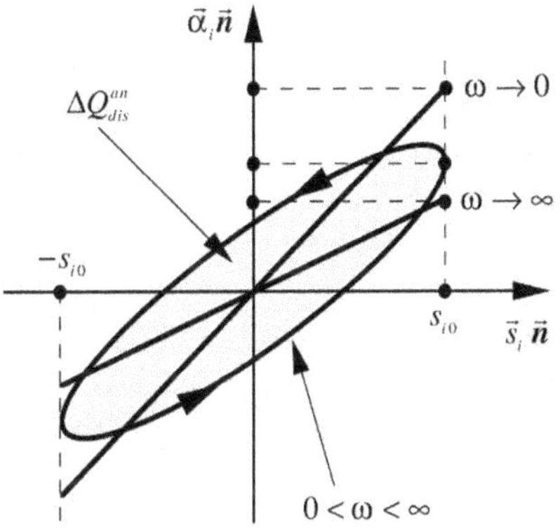

Figure 7.7 *- Hysteresis cycle in presence of viscous friction*

If the temperature T and the volume expansion $\tau^{él}$ of the medium can be considered as constant over time and space, and only in this case, equations *(7.83)* and *(7.85)* can be combined so as to write a single differential equation connecting directly together the shear strain tensor $\vec{\alpha}_i$ and the shear stress tensor $\vec{s}_i$. In this relationship the two components $\vec{\alpha}_i^{él}$ and $\vec{\alpha}_i^{an}$ *do not appear anymore*

$$B_{an}\frac{d\vec{\alpha}_i}{dt} + K_{an}\vec{\alpha}_i = \frac{B_{an}}{K_{él}}\frac{d\vec{s}_i}{dt} + \left(1 + \frac{K_{an}}{K_{él}}\right)\vec{s}_i \tag{7.87}$$

Existence of an inertial term and the phenomenon of anelastic resonance

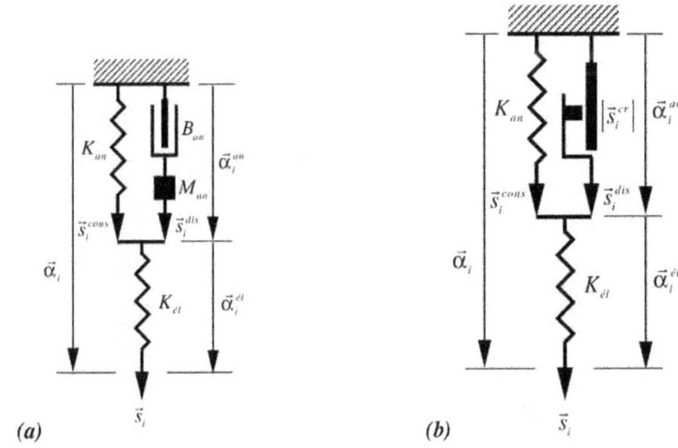

Figure 7.8 *- rheological models* ***(a)*** *with viscous friction and inertial mass*
(b) *with a dry friction pad*

It is still possible that structural defects responsible for anelasticity have their own inertial mass, the effect of which is significant vis-à-vis the viscous friction forces. In this case, the dissipative portion of the phenomenological equation of anelasticity must contain this inertia effect in the form of a term proportional to the second derivative in the tensor of anelastic shears, multiplied by a *coefficient of inertia* $M_{an}(n,T)$ representing the inertial mass involved and which may also depend on the network site density

$$\vec{s}_i^{\,dis} = M_{an}(n,T)\,\frac{d^2\vec{\alpha}_i^{an}}{dt^2} + B_{an}(n,T)\,\frac{d\vec{\alpha}_i^{an}}{dt} \tag{7.88}$$

The presence of the inertial mass transforms the dissipative anelasticity phenomenological equation *(6.58)* in a differential equation of second order in $\vec{\alpha}_i^{an}$, which can be represented by a rheological model of an anelastic solid containing an inertial mass (fig. 7.8a)

$$\vec{s}_i = \vec{s}_i^{\,cons} + \vec{s}_i^{\,dis} = K_{an}(n,T)\,\vec{\alpha}_i^{an} + B_{an}(n,T)\,\frac{d\vec{\alpha}_i^{an}}{dt} + M_{an}(n,T)\,\frac{d^2\vec{\alpha}_i^{an}}{dt^2} \tag{7.89}$$

The equation of anelasticity of the lattice *(7.89)* involves an *anelastic relaxation time* τ_{an} and a *resonance frequency* ω_0, given by the expressions

$$\tau_{an} = \frac{B_{an}}{K_{an}} \quad \text{and} \quad \omega_0 = \sqrt{\frac{K_{an}}{M_{an}}} \tag{7.90}$$

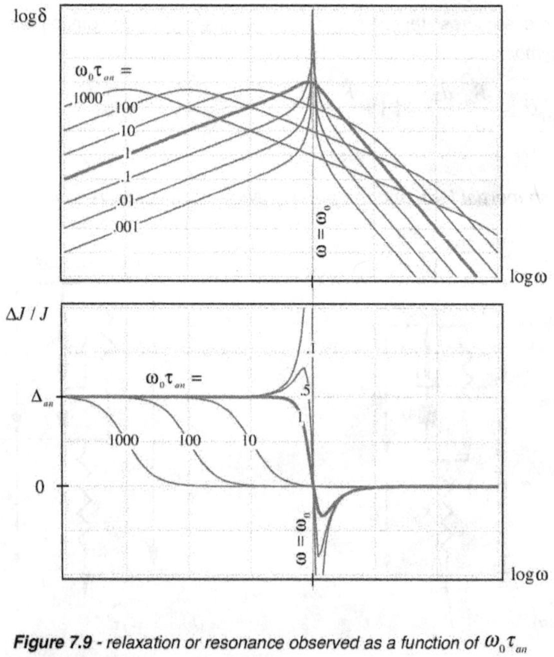

***Figure 7.9** - relaxation or resonance observed as a function of $\omega_0\tau_{an}$*

If the lattice is loaded by an oscillatory shear stress field $\vec{s}_i(t) = \vec{n}s_{i0}\cos\omega t$ in a certain direction $\vec{n}$, the anelastic response may present (fig. 7.9):

- a *phenomenon of relaxation* at frequency $\omega = 1/\tau_{an}$ if $\omega_0\tau_{an} > 1$, because the resonance frequency ω_0 is in this case higher than the relaxation frequency $1/\tau_{an}$,

- a *phenomenon of frequency resonance* at $\omega = \omega_0$ if $\omega_0 \tau_{an} \leq 1$, because the resonant frequency ω_0 is lower than the relaxation frequency $1/\tau_{an}$.

In the case where the temperature T and the volume expansion $\tau^{él}$ of the medium can be considered as constant over time and space, the equations *(7.83)* and *(7.86)* can be combined in a single differential relationship between $\vec{\alpha}_i$ and $\vec{s}_i$

$$M_{an} \frac{d^2\vec{\alpha}_i}{dt^2} + B_{an} \frac{d\vec{\alpha}_i}{dt} + K_{an} \vec{\alpha}_i = \frac{M_{an}}{K_{él}} \frac{d^2\vec{s}_i}{dt^2} + \frac{B_{an}}{K_{él}} \frac{d\vec{s}_i}{dt} + \left(1 + \frac{K_{an}}{K_{él}}\right) \vec{s}_i \tag{7.91}$$

The process of dry friction and the phenomenon of anelastic hysteresis

The two examples of dynamics presented above involve linear phenomena, including a linear coefficient of restoring force K_{an} and a linear viscous damping coefficient B_{an}. The phenomenological relationship of anelasticity is therefore a linear differential equation of first or second order, depending on whether or not there is an inertial mass M_{an}. But nothing prevents *a priori* the existence of strongly nonlinear anelasticity mechanisms. For example, the relationship describing the dissipative term of anelasticity could present a *dry friction* behavior. The easiest way to think of such behavior is to replace in the rheological model in Figure 7.6a, the viscous damper by a *dry friction pad* (fig. 7.8b) whose property is to be blocking while the absolute value $\left|\vec{s}_i^{dis}\right|$ of the stress $\vec{s}_i^{dis}$ is below a certain *critical stress,* $\left|\vec{s}_i^{cr}\right| = s_i^{cr}$, which can depend on the temperature T and the frequency ω of the stress $\vec{s}_i$, and become infinitely flexible when the absolute value $\left|\vec{s}_i^{dis}\right|$ of the stress $\vec{s}_i^{dis}$ becomes equal to the critical stress $\vec{s}_i^{cr}$. Mathematically, the phenomenological equation of dissipative anelasticity *(6.58)* is then replaced by a system of equations and inequalities

$$\begin{cases} \dfrac{d\vec{\alpha}_i^{an}}{dt} = 0 & si \quad \left|\vec{s}_i^{dis}\right| < \left|\vec{s}_i^{cr}\right| \\ K_{an}(n,T)\, \vec{\alpha}_i^{an} = \vec{s}_i - \vec{s}_i^{cr} & si \quad \vec{s}_i^{dis} = \vec{s}_i^{cr} \\ K_{an}(n,T)\, \vec{\alpha}_i^{an} = \vec{s}_i + \vec{s}_i^{cr} & si \quad \vec{s}_i^{dis} = -\vec{s}_i^{cr} \end{cases} \tag{7.93}$$

In this case, the anelastic response to an oscillatory shear field $\vec{s}_i(t) = \vec{n} s_{i0} \cos \omega t$ in the direction $\vec{n}$ does not depend on the frequency ω, but only on the amplitude s_{i0} of the shear stress field. This response can be deduced from equations *(7.93)* and is shown in Figure 7.10. If the amplitude s_{i0} is smaller than the critical stress value of 'undocking', $s_i^{cr} = \left|\vec{s}_i^{cr}\right|$, there is no shear anelastic response ($\vec{\alpha}_i^{an} = 0$). On the other hand, when the amplitude s_{i0} exceeds the critical stress value of undocking, there appears a hysteresis cycle like the one shown in the right figure.

Thus, when the solid's anelasticity is controlled by a dry friction, the behavior of the tensor of anelastic shear $\vec{\alpha}_i^{an}$ explicitly depends only on the magnitude s_{i0} of the shear stress tensor.

One should note that if the shear stress $\vec{s}_i(t)$ is canceled after one cycle whose amplitude s_{i0} has exceeded the critical stress s_i^{cr}, there appears a *non-zero residual value* of the anelastic shear tensor $\vec{\alpha}_i^{an}$. This is sometimes called the phenomenon of "*micro-plasticity*". However, if the dry friction under consideration here is actually an approximation of a highly nonlinear function, which will always be the case if the anelastic process is thermally activated, then this residual value will gradually disappear over time as a result of the restoring anelastic force, so that over the medium term, the anelastic deformation will eventually be fully recovered.

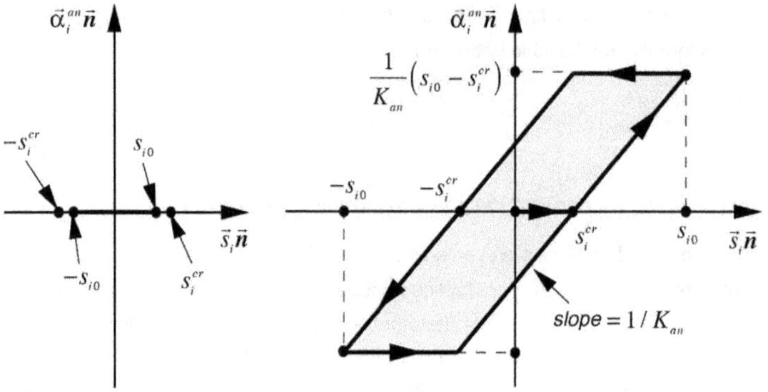

Figure 7.10 *- Hysteresis cycles $\vec{\alpha}_i^{an}$ in the presence of a dry friction*

7.9 - Displacive structural transitions of the first and second order

The existence of phenomena of structural phase transitions in structural nature of the lattice, including *martensitic phase transitions* are sources of anelasticity within the solid. In this section, we will develop for example two simple models for displacive martensitic transformations of the second and first order, meaning, structural transformations without diffusion leading to a homogeneous deformation of the crystal lattice such that it results in a shear deformation on a microscopic scale if the lattice is submitted to an external applied stress.

An example of displacive martensitic transition of second order

We can develop a very simplified imaginary model of martensitic phase transitions, by assuming a lattice with a unit volume slightly distorted by a given local shear angle ψ_0, representing a local shear $\alpha_{ij}^{(0)}$. In a given direction, successive atomic planes can have a positive ψ_0 shear angle and we will call it *variant* ***A***, or a negative shear angle $-\psi_0$ and we will then call it variant **B** (fig. 7.11).

Two successive plans of the lattice in a given direction then have two options of arrangements, namely two planes with the same local shear $\alpha_{ij}^{(0)}$ which we will call ***A-A*** connection, or with $-\alpha_{ij}^{(0)}$ called ***B-B*** connection, or two planes having opposed shears and we will call them ***A-B*** or ***B-A*** connection. Figure 7.11 illustrates such a sequence of successive variations.

If we now consider N successive planes in a given direction of the lattice, we can count the number n_A of layers of type ***A*** and the number n_B of variant ***B*** layers, with

$$n_B = N - n_A \tag{7.94}$$

On the basis of this counting, we can then find the angle ψ of the macroscopic shear

$$\tan\psi = \frac{l}{L} = \frac{\Delta a}{a}\frac{2n_A - N}{N} = \tan\psi_0\left(\frac{2n_A}{N} - 1\right) \qquad (7.95)$$

The macroscopic shear α_{ij}^{an} of the solid associated with the angle ψ can be considered as an anelastic shear since it is not linked to the elasticity of the lattice

$$\alpha_{ij}^{an} = \alpha_{ij}^{(0)}\frac{\tan\psi}{\tan\psi_0} = \alpha_{ij}^{(0)}\left(\frac{2n_A}{N} - 1\right) \qquad (7.96)$$

Suppose that the binding energies between atomic planes are slightly different depending on whether we have a ***A-A***, ***B-B***, ***A-B*** or ***B-A*** type bond. Assume further that ***A-A*** and ***B-B*** bonds are stronger than the bonds ***A-B*** and ***B-A***. In this case, it is clear that the bonding energies per unit cell, which are of course negative, are smaller for the pairs ***A-A*** and ***B-B*** than for the ***A-B*** and ***B-A*** pairs , so that

$$\varepsilon_{A-A} = \varepsilon_{B-B} < \varepsilon_{A-B} = \varepsilon_{B-A} < 0 \qquad (7.97)$$

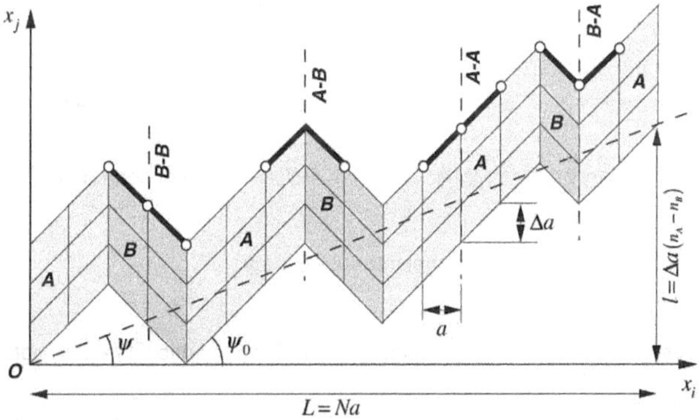

Figure 7.11 *- succession of lattice planes in a given direction with bonds **A-A** , **B-B** and **A-B***

The internal energy $U^{liaison}$ associated to the bonds of the N successive atomic planes in a given direction is written by enumerating the binding pairs

$$U^{liaison} = \varepsilon_{A-A}n_A\frac{n_A}{N} + \varepsilon_{B-B}n_B\frac{n_B}{N} + \varepsilon_{A-B}n_A\frac{n_B}{N} + \varepsilon_{B-A}n_B\frac{n_a}{N} < 0 \qquad (7.98)$$

The configuration entropy $S^{liaison}$ associated with the bonds of the N successive atomic planes in a given direction is deduced from the number of possible configurations binding pairs

$$S^{liaison} = k\ln\Omega = k\ln\frac{N!}{n_A!n_B!} \qquad (7.99)$$

We deduce the binding free energy of the N successive atomic planes with the relationship $F^{liaison} = U^{liaison} - TS^{liaison}$ in which we develop $U^{liaison}$ and $S^{liaison}$

$$F^{liaison} = N\varepsilon_{A-A}\left[\left(\frac{n_A}{N}\right)^2 + \left(\frac{n_B}{N}\right)^2\right] + 2N\varepsilon_{A-B}\left(\frac{n_A}{N}\right)\left(\frac{n_B}{N}\right) - kT\ln\frac{N!}{n_A!n_B!} \qquad (7.100)$$

Using the relation *(7.94)* and Stirling's approximation, we obtain after some calculations the average free energy of the lattice bond $f^{liaison} = F^{liaison} / N$ per unit cell in the direction considered

$$f^{liaison} = 2(\varepsilon_{A-A} - \varepsilon_{A-B})\frac{n_A}{N}\left(\frac{n_A}{N} - 1\right) + \varepsilon_{A-A} + kT\left[\frac{n_A}{N}\ln\left(\frac{n_A}{N}\right) + \left(1 - \frac{n_A}{N}\right)\ln\left(1 - \frac{n_A}{N}\right)\right] \quad (7.101)$$

This average free energy of bonds per unit cell of the lattice can be expressed by introducing the *order parameter* $\alpha_{ij}^{an} / \alpha_{ij}^{(0)}$ associated with the macroscopic shear of the solid lattice, whose values range between -1 and +1. It's expression is then

$$\frac{n_A}{N} = \frac{1}{2}\left(1 + \frac{\alpha_{ij}^{an}}{\alpha_{ij}^{(0)}}\right) \quad \text{and} \quad 1 - \frac{n_A}{N} = \frac{1}{2}\left(1 - \frac{\alpha_{ij}^{an}}{\alpha_{ij}^{(0)}}\right) \quad (7.102)$$

With this order parameter, the average free energy of bonds per unit cell of the lattice can be written as

$$f^{liaison} = \varepsilon_{A-A} + \frac{1}{2}(\varepsilon_{A-A} - \varepsilon_{A-B})\left[\left(\frac{\alpha_{ij}^{an}}{\alpha_{ij}^{(0)}}\right)^2 - 1\right]$$
$$+ kT\left[\left(1 + \frac{\alpha_{ij}^{an}}{\alpha_{ij}^{(0)}}\right)\ln\left(1 + \frac{\alpha_{ij}^{an}}{\alpha_{ij}^{(0)}}\right) + \left(1 - \frac{\alpha_{ij}^{an}}{\alpha_{ij}^{(0)}}\right)\ln\left(1 - \frac{\alpha_{ij}^{an}}{\alpha_{ij}^{(0)}}\right) - 2\ln 2\right] \quad (7.103)$$

At zero temperature, the free energy has a minimum for $\alpha_{ij}^{an} / \alpha_{ij}^{(0)} = \pm 1$. So it is the ***A-A*** or ***B-B*** configurations which are favored at low temperature, and free energy per unit cell is $\varepsilon_{A-A} < 0$

By introducing the energy difference $\Delta\varepsilon$ between ***A-B*** and ***A-A*** configurations

$$\Delta\varepsilon = \varepsilon_{A-B} - \varepsilon_{A-A} > 0 \quad (7.104)$$

it is possible to define the increase in anelastic free energy Δf^{an}

$$\Delta f^{an} = f^{liaison} - \varepsilon_{A-A} \quad (7.105)$$

that is associated with the macroscopic shear state α_{ij}^{an} of the lattice. We then obtain

$$\Delta f^{an} = -\frac{\Delta\varepsilon}{2}\left[\left(\frac{\alpha_{ij}^{an}}{\alpha_{ij}^{(0)}}\right)^2 - 1\right] + \frac{kT}{2}\left[\left(1 + \frac{\alpha_{ij}^{an}}{\alpha_{ij}^{(0)}}\right)\ln\left(1 + \frac{\alpha_{ij}^{an}}{\alpha_{ij}^{(0)}}\right) + \left(1 - \frac{\alpha_{ij}^{an}}{\alpha_{ij}^{(0)}}\right)\ln\left(1 - \frac{\alpha_{ij}^{an}}{\alpha_{ij}^{(0)}}\right) - 2\ln 2\right] \quad (7.106)$$

We can express the state function Δf^{an} as a function of the macroscopic shear α_{ij}^{an} for various temperatures (fig. 7.12), and there is phase transition behavior of the second order, which has a *critical transition temperature* T_c.

For temperatures $T > T_c$ the order parameter is 0 and shearing $\alpha_{ij}^{an} = 0$. The ***A-B*** and ***B-A*** bonds are favored by entropic effect. In other words, the *high-temperature phase* corresponds to a sequence of atomic planes ***A-B-A-B-A-B-...***

For temperatures $T < T_c$, there appears two values $\left(\alpha_{ij}^{an}\right)_1^{st}$ or $\left(\alpha_{ij}^{an}\right)_2^{st}$ of macroscopic shear which correspond to local minima of Δf^{an} and two values of macroscopic shear $\left(\alpha_{ij}^{an}\right)_1^{inst}$ or $\left(\alpha_{ij}^{an}\right)_2^{inst}$ which correspond to inflection points of Δf^{an}. For a given temperature, the alternatives ***A*** and ***B*** tend to cluster, with, in equilibrium, a phase containing an amount of variant ***B*** and having a shear and a phase containing a proportion $\left|\alpha_{ij}^{an}\right|_1^{st} / \alpha_{ij}^{(0)}$ of variant ***A*** and with a shear $\left(\alpha_{ij}^{an}\right)_2^{st} > 0$. Under zero stress, the proportion of these two phases, which are separated by anti-phase walls are equal, so that the overall macroscopic shear can only be zero. Gradually, as the temperature drops, the proportions at equilibrium $\left|\alpha_{ij}^{an}\right|_1^{st} / \alpha_{ij}^{(0)}$ and $\left|\alpha_{ij}^{an}\right|_2^{st} / \alpha_{ij}^{(0)}$ tend to 1, so it should appear at zero temperature *low temperature phase* corresponding to two pure variants ***A-A-A-A-A-...*** and ***B-B-B-B-...*** . separated by anti-phase walls.

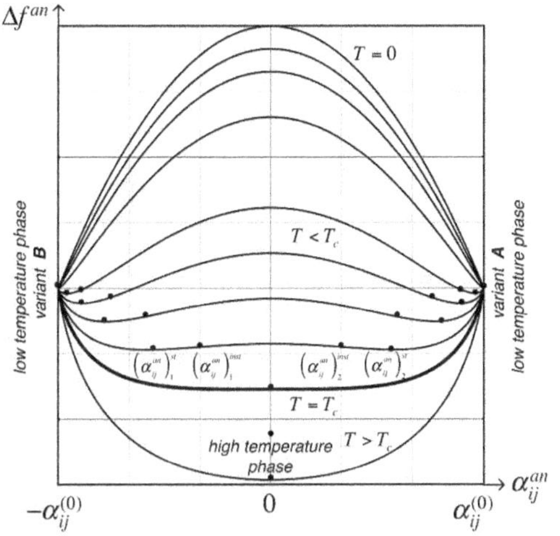

Figure 7.12 *- The anelastic free energy based on anelastic shear for different temperature values*

The *bifurcation diagram* (fig. 7.13) shows the evolution of the various phases with temperature, and the structural aspect of these phases. This second order displacive transformation presents a singular behavior of the specific heat at the critical *transition temperature* T_c . Indeed, the increase Δc_v^{tr} in the specific heat per unit cell due to the phase transition may be deduced from the internal energy $u^{liaison}$ expressed in terms of α_{ij}^{an} using the equation *(B.34)*

$$\Delta c_v^{tr} = \left.\frac{du^{liaison}}{dT}\right|_{v,\alpha_{ij}^{an}} = -\frac{\Delta\varepsilon}{2}\frac{d}{dT}\left(\frac{\alpha_{ij}^{an}}{\alpha_{ij}^{(0)}}\right)^2_{v,\alpha_{ij}^{an}} \tag{7.107}$$

It follows immediately from this relationship that there is a singularity of the specific heat Δc_v^{tr} during the transition as Δc_v^{tr} is proportional to the derivative of the square of α_{ij}^{an} in the bifurcation diagram in figure 7.13.

The effect of shear applied to the second order displacive transformation

From the state function *(7.106)* of anelasticity, we can deduce the equations of state of the lattice for the quasi-static shear stress s_{ij}^{qs} and entropy s^{an}

$$\begin{cases} s_{ij}^{qs} = n\dfrac{\partial \Delta f^{an}}{\partial \alpha_{ij}^{an}} = \dfrac{n}{\alpha_{ij}^{(0)}}\left\{-\Delta\varepsilon\dfrac{\alpha_{ij}^{an}}{\alpha_{ij}^{(0)}} + \dfrac{kT}{2}\left[\ln\left(1+\dfrac{\alpha_{ij}^{an}}{\alpha_{ij}^{(0)}}\right) - \ln\left(1-\dfrac{\alpha_{ij}^{an}}{\alpha_{ij}^{(0)}}\right)\right]\right\} \\ s^{an} = -\dfrac{\partial \Delta f^{an}}{\partial T} = -\dfrac{k}{2}\left[\left(1+\dfrac{\alpha_{ij}^{an}}{\alpha_{ij}^{(0)}}\right)\ln\left(1+\dfrac{\alpha_{ij}^{an}}{\alpha_{ij}^{(0)}}\right) + \left(1-\dfrac{\alpha_{ij}^{an}}{\alpha_{ij}^{(0)}}\right)\ln\left(1-\dfrac{\alpha_{ij}^{an}}{\alpha_{ij}^{(0)}}\right) - 2\ln 2\right] \end{cases} \tag{7.108}$$

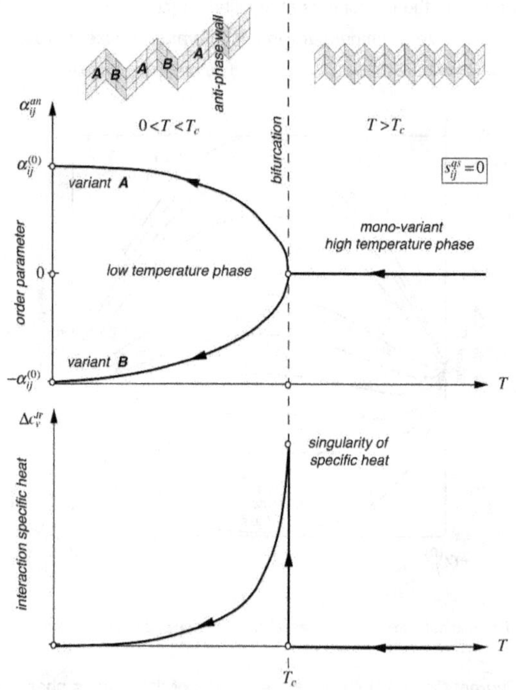

Figure 7.13 *- the bifurcation diagram of the phase transition, and the structural aspect of the phases at different temperatures*

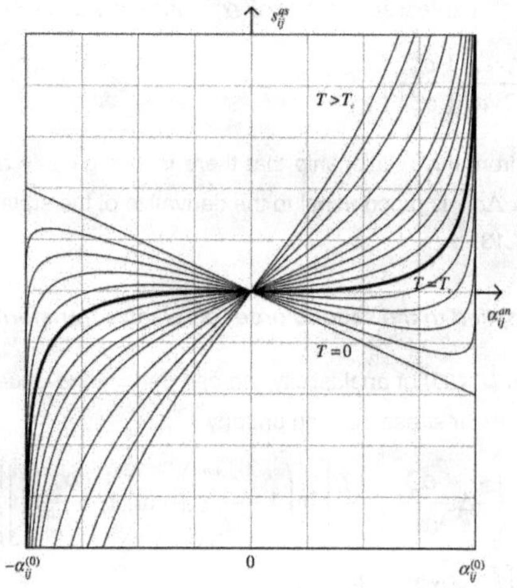

Figure 7.14 *- the anelastic stress-strain* $s_{ij}^{qs}-\alpha_{ij}^{an}$ *diagram*

The stress-strain diagram of anelasticity $s_{ij}^{qs} - \alpha_{ij}^{an}$ can be drawn for different temperatures (fig. 7.14). In this diagram, the shear values in the region $\left[(\alpha_{ij}^{an})_1^{inst}, (\alpha_{ij}^{an})_2^{inst}\right]$ of negative slope correspond to unstable values in the free energy diagram of figure 7.12, and therefore can not be achieved.

This implies that if an increasing or decreasing stress is applied to the solid in the domain $T < T_c$, there will appear a hysteresis in the graph $s_{ij} - \alpha_{ij}^{an}$,, corresponding to an abrupt change of a variant to the other alternative, induced by the external stress effect (fig. 7.15). During the transformation, the applied stress s_{ij} can be separated into quasi-static stress s_{ij}^{qs} derived from the free energy state function, and a dissipative stress s_{ij}^{dis}

$$s_{ij} = s_{ij}^{qs} + s_{ij}^{dis} \tag{7.109}$$

The dissipative stress s_{ij}^{dis} is a source of energy dissipation during the transformation, equal to the area S reported in figure 7.15. This hysteresis effect will cause the appearance of a dissipative phenomenon of internal friction at temperatures lower than T_c, depending on both the temperature and the magnitude of the applied stress s_{ij}.

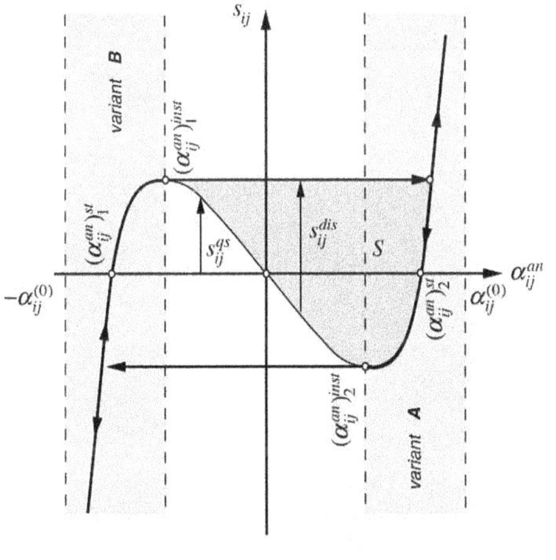

Figure 7.15 *- Hysteresis cycle $s_{ij} - \alpha_{ij}^{an}$ in the domain of temperature $T < T_c$*

Pseudo-plasticity and the irreversible shape memory effect

The combination of the effects of stress and temperature may reveal some surprising behavior of this type of solid.

If the stress is increased to a value $s_{ij(\max)}$ at a given temperature $T < T_c$, variant ***B*** will turn into variant ***A*** and there will appear a nonzero macroscopic shear in the solid (point $(\alpha_{ij}^{an})_1^{inst}$ in figure 7.15). If the stress is then released, the solid will retain a non-zero macroscopic shear, as if he had been plastically deformed (point $(\alpha_{ij}^{an})_2^{st}$ in figure 7.15). If this solid is then heated above the critical temperature, the macroscopic deformation disappears during the formation of the high-temperature phase. If the same solid is then cooled from the high temperature phase without being stressed, after passing the critical temperature, variants ***A*** and ***B*** will appear in equal proportion in the solid, so there will not be macroscopic shear in solid. This is why this low-temperature phenomenon of macroscopic deformation is called *pseudo-plasticity.*

On the other hand, if the solid is cooled from the high temperature phase with a fixed shear stress, the transition will show a macroscopic shear due to the fact that one of the variants will be favored over the other. And this macroscopic shear will again be kept in the low temperature phase if the stress is released. But the initial null macroscopic shear state is again recovered if the solid is heated in its high-temperature phase. This amazing effect is called *irreversible shape memory effect.*

An example of displacive martensitic transformation of first order

Let's consider the highly simplified imaginary model of martensitic phase transition described in the previous section (fig. 7.11), but this time let's assume that the binding energies between atomic planes depend not only on immediate first neighbors, but also the second planes of nearest neighbors. In this case, there are 16 different configurations of four planes of nearest neighbors of a given plane, which are shown in figure 7.16 .

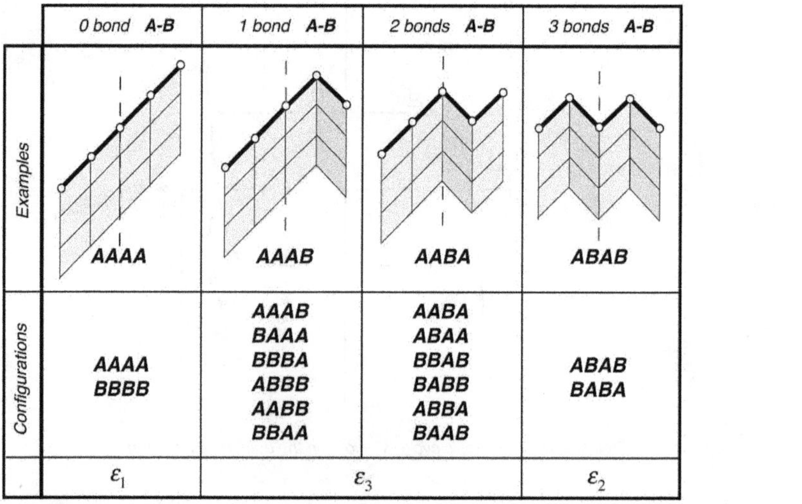
0 bond A-B
1 bond A-B
2 bonds A-B
3 bonds A-B
Examples
AAAA
AAAB
AABA
ABAB
Configurations
AAAA
BBBB
AAAB
BAAA
BBBA
ABBB
AABB
BBAA
AABA
ABAA
BBAB
BABB
ABBA
BAAB
ABAB
BABA
ε1
ε3
ε2

***Figure 7.16** - the 16 possible configurations of the 4 successive atomic planes*

For a set of *N* successive atomic planes, the probability that one of the planes be in one of the possible configurations, for example the probability P_{AABA} of being in the ***AABA*** configuration, is easily calculated

$$P_{AABA} = \frac{n_A}{N}\frac{n_A}{N}\frac{n_B}{N}\frac{n_A}{N} = \frac{n_A^3 n_B}{N^4} \tag{7.110}$$

It is easily verified that the sum of the probabilities of all configurations is worth 1

$$\sum P_{XXX} = \frac{1}{N^4}\left(n_A^4 + n_B^4 + 6n_A^2 n_B^2 + 4n_A n_B^3 + 4n_A^3 n_B\right) = \frac{1}{N^4}\left(n_A + n_B\right)^4 = 1 \tag{7.111}$$

Imagine further that both ***AAAA*** and ***BBBB*** configurations are very stable, with a binding energy $\varepsilon_{AAAA} = \varepsilon_{BBBB} = \varepsilon_1$, and that both ***ABAB*** and ***BABA*** configurations are also very stable, with a binding energy $\varepsilon_{ABAB} = \varepsilon_{BABA} = \varepsilon_2$, and that the 12 other configurations are less stable than the ones we just discussed and that, to simplify the problem, the all share the same binding energy ε_3 .

We then have

$$\varepsilon_1 < \varepsilon_3 < 0 \quad \text{and} \quad \varepsilon_2 < \varepsilon_3 < 0 \tag{7.112}$$

The average internal energy $u^{liaison}$ per unit cell in a given direction is written using the probabilities of the different possible linking configurations. It follows

$$u^{liaison} = \left[\varepsilon_1\left(n_A^4 + n_B^4\right) + \varepsilon_2\left(2n_A^2 n_B^2\right) + \varepsilon_3\left(N^4 - n_A^4 - n_B^4 - 2n_A^2 n_B^2\right)\right] / N^4 \tag{7.113}$$

As to the average entropy $s^{liaison}$ per site of the lattice in a given direction, it is deduced from the number of possible configurations as

$$s^{liaison} = \frac{k}{N}\ln\Omega = \frac{k}{N}\ln\frac{N!}{n_A!n_B!} = -k\left[\frac{n_A}{N}\ln\frac{n_A}{N} + \left(1-\frac{n_A}{N}\right)\ln\left(1-\frac{n_A}{N}\right)\right] \tag{7.114}$$

We deduce the mean free energy per lattice unit in a given direction

$$\begin{aligned} f^{liaison} = u^{liaison} - Ts^{liaison} = \varepsilon_3 + (\varepsilon_1 - \varepsilon_3)\left[\left(\frac{n_A}{N}\right)^4 + \left(\frac{n_B}{N}\right)^4\right] + 2(\varepsilon_2 - \varepsilon_3)\left(\frac{n_A}{N}\right)^2\left(\frac{n_B}{N}\right)^2 \\ + kT\left[\frac{n_A}{N}\ln\frac{n_A}{N} + \left(1-\frac{n_A}{N}\right)\ln\left(1-\frac{n_A}{N}\right)\right] \end{aligned} \tag{7.115}$$

By introducing the following energy differences of the bond energy

$$\Delta\varepsilon_I = \varepsilon_3 - \varepsilon_1 > 0 \quad \text{and} \quad \Delta\varepsilon_{II} = \varepsilon_3 - \varepsilon_2 > 0 \tag{7.116}$$

It is possible to define the increase in anelastic free energy Δf^{an} with the relation

$$\Delta f^{an} = f^{liaison} - \varepsilon_3 \tag{7.117}$$

which is associated to the macroscopic state of shear α_{ij}^{an} via the *order parameter* $\alpha_{ij}^{an} / \alpha_{ij}^{(0)}$ thanks to relations *(7.94)*. We obtain

$$\begin{aligned} \Delta f^{an} = -\frac{\Delta\varepsilon_I}{16}\left[\left(1+\frac{\alpha_{ij}^{an}}{\alpha_{ij}^{(0)}}\right)^4 + \left(1-\frac{\alpha_{ij}^{an}}{\alpha_{ij}^{(0)}}\right)^4\right] - \frac{\Delta\varepsilon_{II}}{8}\left(1+\frac{\alpha_{ij}^{an}}{\alpha_{ij}^{(0)}}\right)^2\left(1-\frac{\alpha_{ij}^{an}}{\alpha_{ij}^{(0)}}\right)^2 \\ + \frac{kT}{2}\left[\left(1+\frac{\alpha_{ij}^{an}}{\alpha_{ij}^{(0)}}\right)\ln\left(1+\frac{\alpha_{ij}^{an}}{\alpha_{ij}^{(0)}}\right) + \left(1-\frac{\alpha_{ij}^{an}}{\alpha_{ij}^{(0)}}\right)\ln\left(1-\frac{\alpha_{ij}^{an}}{\alpha_{ij}^{(0)}}\right) - 2\ln 2\right] \end{aligned} \tag{7.118}$$

One can see the state function Δf^{an} as a function of the macroscopic shear α_{ij}^{an} for different temperatures (fig. 7.17). There is a new behavior with respect to the displacive transition of second order described in the previous section.

At very low temperature, for $T < T_1$, the state function has two minima adjacently located to $\pm\alpha_{ij}^{(0)}$. We will call *phase I* this low temperature phase, which presents a ***A*** variant in the vicinity of $+\alpha_{ij}^{(0)}$ and a ***B*** variant in the vicinity of $-\alpha_{ij}^{(0)}$.

At high temperature $T > T_2$, the state function Δf^{an} no longer has a single minimum for $\alpha_{ij}^{an} = 0$. We will call *phase II* this mono-variant high temperature phase.

In the temperature range $\left[T_1, T_2\right]$, the state function Δf^{an} has two minima located in the vicinity of $\pm\alpha_{ij}^{(0)}$ and a minimum located at $\alpha_{ij}^{an}=0$. This implies that, in this temperature range, there is coexistence of phases I and II.

The phase transition therefore occurs no longer at a fixed temperature as above, but proceeds gradually between the two *critical temperatures* T_1 and T_2, which is a characteristic of a *phase transition of the first order.*

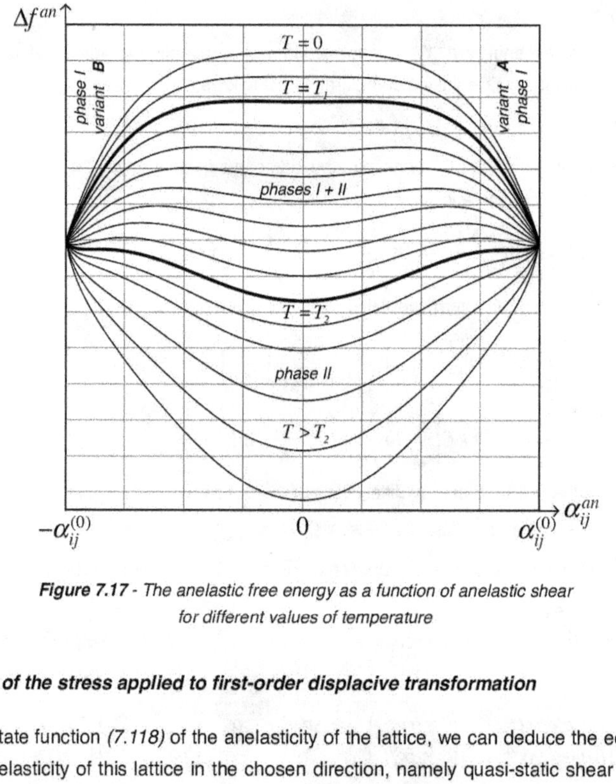

Figure 7.17 *- The anelastic free energy as a function of anelastic shear for different values of temperature*

The effect of the stress applied to first-order displacive transformation

From the state function *(7.118)* of the anelasticity of the lattice, we can deduce the equations of state of anelasticity of this lattice in the chosen direction, namely quasi-static shear stress s_{ij}^{qs} and entropy s^{an}

$$\begin{cases} s_{ij}^{qs}=n\dfrac{\partial\Delta f^{an}}{\partial\alpha_{ij}^{an}}=\dfrac{n}{\alpha_{ij}^{(0)}}\left\{-\dfrac{\Delta\varepsilon_I}{4}\left[\left(1+\dfrac{\alpha_{ij}^{an}}{\alpha_{ij}^{(0)}}\right)^3-\left(1-\dfrac{\alpha_{ij}^{an}}{\alpha_{ij}^{(0)}}\right)^3\right]+\dfrac{\Delta\varepsilon_{II}}{2}\left[1-\left(\dfrac{\alpha_{ij}^{an}}{\alpha_{ij}^{(0)}}\right)^2\right]\dfrac{\alpha_{ij}^{an}}{\alpha_{ij}^{(0)}}\right\} \\ \qquad +\dfrac{kT}{2}\left[\ln\left(1+\dfrac{\alpha_{ij}^{an}}{\alpha_{ij}^{(0)}}\right)-\ln\left(1-\dfrac{\alpha_{ij}^{an}}{\alpha_{ij}^{(0)}}\right)\right] \\ s^{an}=-\dfrac{\partial\Delta f^{an}}{\partial T}=-\dfrac{k}{2}\left[\left(1+\dfrac{\alpha_{ij}^{an}}{\alpha_{ij}^{(0)}}\right)\ln\left(1+\dfrac{\alpha_{ij}^{an}}{\alpha_{ij}^{(0)}}\right)+\left(1-\dfrac{\alpha_{ij}^{an}}{\alpha_{ij}^{(0)}}\right)\ln\left(1-\dfrac{\alpha_{ij}^{an}}{\alpha_{ij}^{(0)}}\right)-2\ln 2\right] \end{cases} \qquad (7.119)$$

The stress-strain diagram of transition anelasticity $s_{ij}^{qs}-\alpha_{ij}^{an}$ can be drawn for different temperatures (fig. 7.18).

In this diagram, the shear values in the negative slope areas correspond to values in the unstable free energy diagram of figure 7.17, and therefore cannot be achieved. By observing this chart in detail, it appears there is actually four characteristic temperatures T_1, T_1', T_2' and T_2, during the phase transformation, as illustrated by the four diagrams shown in the figure 7.19 .

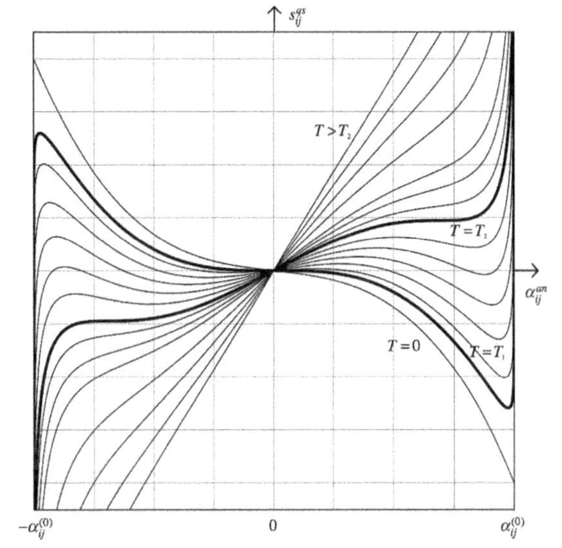

Figure 7.18 *- The stress-strain diagram* $s_{ij}^{qs} - \alpha_{ij}^{an}$ *of the anelasticity of transition with the transition temperatures* T_1 *and* T_2

These four characteristic temperatures, T_1, T_1', T_2' and T_2 are used to define four domains in which the dissipative behavior of anelastic transformation are different if the solid is subjected to an external shear stress s_{ij} (fig. 7.20).
Indeed, if an increasing or decreasing stress is applied to the solid in these different temperature ranges, there appears a phenomena of hysteresis in the diagram $s_{ij} - \alpha_{ij}^{an}$, corresponding to abrupt changes between the two variants of phase I and phase II. In the domain $T < T_1$, only the two variants of the phase I exist, and there are transformations between these two variants. In the domain $[T_1, T_1']$, the two variants in phase I are stable, and phase II is metastable. There are transformations between the two variants of phase I, but also irreversible transformations between the metastable phase II and the two variants of phase I. In the domains $[T_1', T_2']$ and $[T_2', T_2]$, phases I and II are stable, and it is only transformations between the variants of phase I and phase II. During these transformations, the applied stress s_{ij} can be separated into quasi-static strain s_{ij}^{qs} derived from the state function free energy, and dissipative stress s_{ij}^{dis}, according to equation (7.109). The dissipative stress s_{ij}^{dis} is again a source of energy dissipation during transformation, equal to the hatched areas S_1 and S_2 drawn in figure 7.20. Such hysteresis effects result in the onset of dissipative phenomena of internal friction in the different temperature ranges, dependent on both the temperature and the magnitude of the applied

stress s_{ij} .

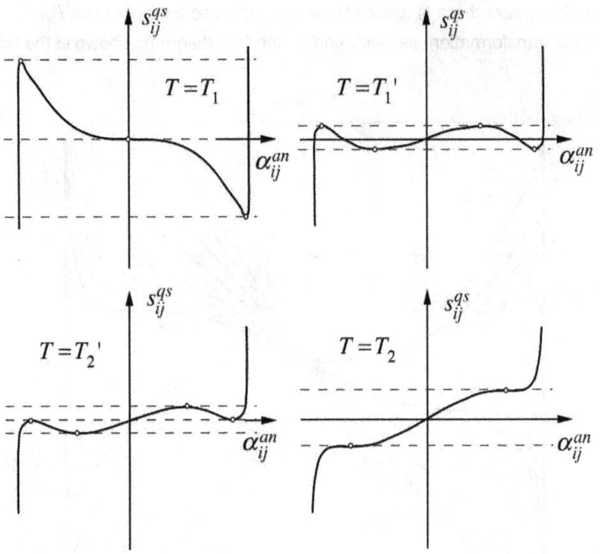

Figure 7.19 - *The diagrams $s_{ij}^{qs}-\alpha_{ij}^{an}$ for the characteristic temperatures T_1, T_1', T_2' and T_2.*

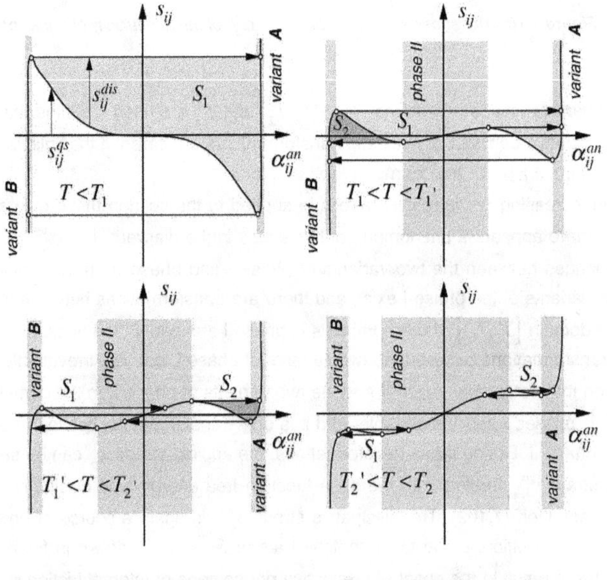

Figure 7.20 - *The cycle of hysteresis $s_{ij}-\alpha_{ij}^{an}$ in the four temperature domains*

Pseudo-plasticity , super-elasticity and the irreversible shape memory effect

In this more complicated case of a solid with first order martensitic transition, the combination of effects of stress and temperature will bring up many quite surprising behaviors of the solid, which we will discuss very briefly.

In the temperature ranges $T < T_2'$, if the stress is increased to a sufficient value $s_{ij(\max)}$, the solid will contain only variant ***A*** and there will appear a nonzero macroscopic shear in the solid. If the stress is then released, the solid will retain a non-zero macroscopic shear, as if he had been plastically deformed. If this solid is then heated above the critical temperature T_2' , the solid turns into phase II and the macroscopic deformation disappears. If the same solid is then cooled without being stressed below the critical temperature, there will appear equal proportions of variants ***A*** and ***B*** of phase I in solid, so that there will no longer be macroscopic shear in the solid. This phenomenon of macroscopic deformation associated with the two variants of the low temperature phase I is called *pseudo-plasticity.*

In the domain $[T_2', T_2]$, the solid is in phase II without applied stress. If one applies sufficient stress, the solid will present a sudden major shear deformation for a small change in the applied stress, as if the solid had deformed plastically, because phase II becomes the variant ***A*** of phase I. And this intense deformation is completely reversible if the stress lowers again. This is called a *super-elastic* phenomenon.

If such solid is cooled from the high temperature stage II under a sufficient shearing stress, the passage of the transition temperature T_2 will bring about a macroscopic shearing due to the fact that one of the phase I variants will be preferred over the other. And macroscopic shear will be retained if the stress is released, provided that $T < T_2'$. The initial zero macroscopic shear state is again recovered if the solid is heated above the temperature T_2' . This phenomena is called the *irreversible shape memory effect.*

Materials with a martensitic phase transition of first order, and phenomenology such as the pseudo-plasticity, super-elasticity and irreversible shape memory effect, as well as other more complex phenomena, but also within the martensitic transformation realm, such as reversible shape memory effect, are well known and well studied. These include amongst the better known, the copper-aluminum-zinc alloys as well as the titanium-nickel alloys.

7.10 - Phenomenology of plasticity

The plasticity of a solid lattice of particles is controlled mainly by the long-distance movements of lattice defects such as dislocations, that move under the influence of a stress or strain field.

At the level of macroscopic description of said continuous solid, the statistical manifestation of movements involving structural defects results from the phenomenological equations of dissipative plasticity *(6.53)* and *(6.55)* .

Phenomenological approach to plasticity

As the dynamics of microscopic structural defects depends essentially on the obstacles they face in the lattice and the nature of the interactions they may have with these obstacles, the shape of the dissipative phenomenological equations of plasticity must therefore reflect this dynamic and can therefore have many different behaviors.

Typically, the plasticity mechanisms are by nature highly nonlinear . The intrinsic non-linearity of plasticity must obviously be captured in the dissipative phenomenological relationship of plasticity *(6.53)* and *(6.55)* .

By considering here, for example, only the plastic behavior of a strong shear, the phenomenological approach to the plasticity of this solid consist in finding an expression of the relation *(6.55)*, so that it best reflects all the plastic shear behavior observed in this solid. In other words, to test the material with the aid of various mechanical tests such as a creep test, a tensile test, a fatigue test, etc. , to select a function $\boldsymbol{F}$ in the following equation

$$\frac{d\vec{\alpha}_i^{pl}}{dt} = \boldsymbol{F}\left(\vec{s}_i, T, n, ...\right) \qquad (7.120)$$

This function $\boldsymbol{F}$ will always be strongly nonlinear and often brutally growing from a certain critical value of the stress $\vec{s}_i^{cr}$ (fig. 7.21a). $\boldsymbol{F}$ will also usually depend heavily on the temperature T , because the plasticity phenomena are generally thermally activated and may also depend on the volume density n of the lattice . Having such a solid plastic behavior can be represented by a rheological model corresponding to a serial setup of a non-linear friction pad, representing plastic deformation and characterized by the function $\boldsymbol{F}\left(\vec{s}_i, T, n, ...\right)$, and a spring representing the elastic deformation and characterized by the elastic modulus K_2 (fig. 7.21b).

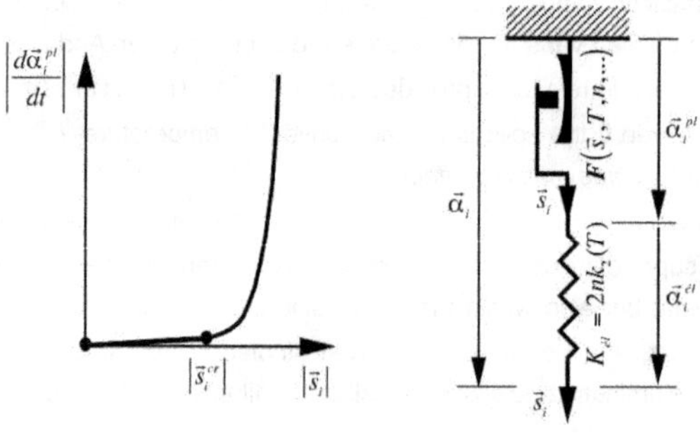

(a) **(b)**

Figure 7.21 *- **(a)** typical behavior of the speed of plastic deformation with shear and **(b)** rheological model for the plastic deformation*

Such an approach may allow a more or less adequate description of the plastic behavior of a solid, including some cases of simple mechanical tests such as creep testing, tensile testing or testing of fatigue.

On the limits of the phenomenological approach to the study of plasticity

The phenomenological approach to the study of continuous plasticity based on the dissipative phenomenological plasticity relations such as relations *(6.53)*, *(6.55)* and *(7.120)* has its limits, as it usually faces several difficulties inherent to the plastic deformation process.

First of all, it is very rare for mechanical tests such as a creep test or a tensile test to be translated properly by a such a simple relationship as the relationships of *(7.120)*. In general, the behavior observed during mechanical testing are *very complex* (nonlinear creep, upper and lower yield points, diverse plateaus, instabilities as Portevin-Le Chatelier effect, hardening effects, etc.) and depend on many parameters, such as the nature of the solid lattice, its crystallinity, its purity, and the history of the thermo-mechanical treatments, etc. These behaviors actually reflect the complexity of the microscopic mechanisms involved during plastic deformation.
On the other hand, the presence of the necessary conditions of geometro-compatibility also imply that the deformations, including plastic deformation, cannot present spatial location, which is not always the case in real strong lattices such as metals, in which it may appear highly localized plastic deformations, such as slip lines for example.
Finally, what essentially distinguishes the phenomenology of solids endowed with anelasticity and plasticity from that of the purely elastic solids is the thermodynamic non-reversibility of anelastic and plastic responses, which results in the existence of sources of entropy associated with anelasticity and plasticity. In the case of the anelasticity, thermodynamic irreversibility leads to a recoverable but *non-immediate* response of anelastic deformation. In other words, the displacement field $\vec{u}$ associated with the anelasticity lags behind the stress field applied to the solid, but it can be recovered if the stress field is canceled. The anelasticity is therefore a Markov process which can relatively easily lead to a phenomenological equation such as the equation *(6.58)*. On the other hand, in the case of plastic deformation, the response of the plastic deformation to a stress field is unrecoverable: after plastic deformation, it inevitably appears a residual plastic displacement field, which can only be canceled by another plastic deformation. This behavior of the plastic deformation, due to the absence of a potential energy stored in the solid lattice by the plastic deformation, often leads to *non-markovian* behaviors of plasticity. In other words, the plastic deformation is a dissipative process which may also depend on the history of the solid, that is to say, of all the previous plastic processes it underwent. As a first approximation, the existence of this "memory" of the solid could be introduced in relation (7.120) by assuming an explicit dependence of the function $\boldsymbol{F}$ on the local state of plastic deformation $\vec{\alpha}_i^{pl}$ such as

$$\boldsymbol{F} = \boldsymbol{F}\left(\vec{s}_i, T, n, \vec{\alpha}_i^{pl}, \ldots\right) \tag{7.121}$$

In fact, the non-markovian behavior of plasticity of solids is extremely difficult, if not impossible, to translate, for the simple reason that it is directly associated to the microscopic structure of the solid, including the behaviors of the defects of this structure (such as dislocations, for example). However, there is no unique, bi-directional, relationship between the local plastic deformation state, characterized by the value of the plastic strain tensor, and the microscopic state of the lattice responsible for that plastic deformation. In other words, there is a multitude of structural defects arrangements leading to *same* local values of a given plastic deformation, which may evolve differently under the effect of a given constraint $\vec{s}_i(t)$.
It is concluded that only a microscopic approach may be able to properly describe the complexity of the plastic behavior of the solid lattice, as it is the microscopic behavior of the topological singularities of this lattice which is responsible for its plasticity. This study will be undertaken in the following chapters.

PART I

C

Dislocation charges and disclination charges in eulerian lattices

Density and flux of dislocation and disclination charges

Topological singularities associated to the charges

Flux dislocation charges and Orowan's relations

Equations of evolution and force of Peach and Koehler

Chapter 8

Density and flux of charges of dislocations and disclinations

The description of the plasticity of solid lattices with a plastic distortion tensor is very limited, especially due to the fact that there is no unique relationship between the local state of plastic deformation $\vec{\beta}_i^{\,pl}$ and the microscopic state of the structural defects of the lattice responsible for this plastic deformation. This is why the way of expressing the presence of distortions in a plastic lattice must be modified so that it is possible to take into account the microscopic state of structural defects of the lattice. A very elegant way to make this change is to introduce the concepts of *dislocation densities and corresponding flow of charges*, which are responsible for the plastic distortions of the solid, and *disclination densities* responsible for the plastic contortions of the solid.

8.1 - On the macroscopic concept of charges of plastic distortions

On the intuitive notion of plastic charges

The concept of *charges of plastic distortions* of the solid, which will now be called simply *dislocation charges*, is intuitively and visually easy to grasp, especially with the help of the famous "*Volterra tubes*"[1] (1907) and an analogy to electrical charges.

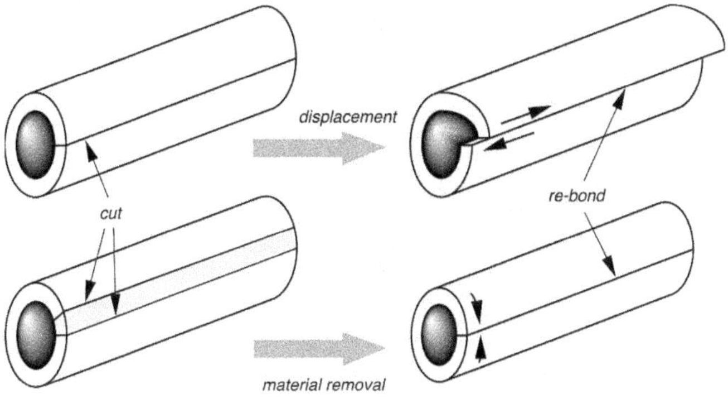

Figure 8.1 - «*Volterra tubes*»

[1] *V. Volterra, L'équilibre des corps élastiques, Ann. Ec. Norm. (3), XXIV, Paris, 1907*

Indeed, just imagine that, either you cut a solid environment and distort it by sliding the jaws, or alternatively you take away a part of the tube and you proceed again to re-bond the tube, as shown by the examples in figure 8.1. In both examples, the deformations undergone by the solid after re-bonding are irreversible and irretrievable, so of the plastic kind. On the other hand, it is intuitively clear that internal forces of stress are developed within the solid after re-bonding. They appeared in the elastic deformation that was imposed on the rest of the solid.
In fact, everything is exactly as if a localized topological discontinuity appeared in the center of the tube after re-bonding. This discontinuity creates an elastic distortion field in the macroscopically continuous medium that makes up the tube. And this field distortion, by its presence, is itself a source of a field-conjugated stresses, which can be called the *field of internal stresses*.
Mathematically, the discontinuity due to re-bonding should translate in terms of a local density of *"plastic charges"*, which are the sources of the elastic distortion field, and thus the internal stress field. In the same way as in electromagnetism the presence a local density of electric charge ρ is responsible for the appearance of an electric displacement field $\vec{D}$, as shown in the Maxwell equation

$$\rho = \operatorname{div}\vec{D} \tag{8.1}$$

and therefore a conjugate electric field $\vec{E}$, as $\vec{D} = \varepsilon_0 \vec{E}$.
The purpose of this chapter will be to mathematically translate the plasticity phenomena inside a solid, not only by introducing densities of plastic charges, but also the flow of these plastic charges, by analogy with the flow of electric charges $\vec{j}$ appearing in the following equation of electromagnetism (wherein $\vec{H}$ represents the magnetic field)

$$\vec{j} = -\frac{\partial \vec{D}}{\partial t} + \overrightarrow{\operatorname{rot}}\,\vec{H} \tag{8.2}$$

On the density and tensorial flow of plastic charges of the solid

By using the definition of the distortion tensor, as it had been obtained in chapter 5, the distortions are the sum of elastic, anelastic and plastic distortions

$$\vec{\beta}_i = \vec{\beta}_i^{\,él} + \vec{\beta}_i^{\,an} + \vec{\beta}_i^{\,pl} \tag{8.3}$$

A new notation for these distortions can be introduced, which allows to separate the contributions of plastic deformation from the contributions of elastic and inelastic deformations

$$\vec{\beta}_i \rightarrow \vec{\beta}_i + \vec{\beta}_i^{\,pl} \qquad \text{with} \qquad \vec{\beta}_i = \vec{\beta}_i^{\,él} + \vec{\beta}_i^{\,an} \tag{8.4}$$

The topological equations for distortions, which were written

$$\frac{d\vec{\beta}_i}{dt} = \overrightarrow{\operatorname{grad}}\,\phi_i \qquad \text{and} \qquad \overrightarrow{\operatorname{rot}}\,\vec{\beta}_i = 0 \tag{8.5}$$

can now be rewritten with the new notation as

$$\frac{d\vec{\beta}_i^{\,pl}}{dt} = -\frac{d\vec{\beta}_i}{dt} + \overrightarrow{\operatorname{grad}}\,\phi_i \qquad \text{and} \qquad -\overrightarrow{\operatorname{rot}}\,\vec{\beta}_i^{\,pl} = \overrightarrow{\operatorname{rot}}\,\vec{\beta}_i \tag{8.6}$$

This allows us to introduce, by analogy with the equations of electromagnetism, the concept of *tensorial charge density* $\vec{\lambda}_i$, the source of plastic distortions, by assuming *a priori* the definition of $\vec{\lambda}_i$ from the plastic distortion tensor $\vec{\beta}_i^{pl}$

$$\vec{\lambda}_i = -\overrightarrow{\text{rot}}\,\vec{\beta}_i^{pl} \tag{8.7}$$

We also introduce the concept of *tensorial flow of charges* $\vec{J}_i$, which is responsible for the temporal variation of the plastic distortion, by assuming *a priori* the following definition deriving from the temporal derivative $d\vec{\beta}_i^{pl} / dt$ of the plastic distortion tensor $\vec{\beta}_i^{pl}$

$$\vec{J}_i = \frac{d\vec{\beta}_i^{pl}}{dt} \tag{8.8}$$

It should be noted here that the concept of charge flow is defined as *a flow with respect to the lattice*, as $\vec{J}_i$ is deduced from the material derivative of $\vec{\beta}_i^{pl}$, that is to say to the temporal derivative taken *along the path of the solid lattice.*

The introduction of these new tensorial densities and charge flows is not gratuitous, they best meet the need to find a way of expressing the presence of plastic distortions. With it, it is possible to take into account the microscopic state of the structural defects of the solid lattice. We will verify *a posteriori*, when interpreting these tensors in the next chapter, that this approach is indeed appropriate. With this approach to the plasticity phenomena with tensors $\vec{J}_i$ and $\vec{\lambda}_i$, the topological equations describing geometro-kinetics and geometro-compatibility of an elastic, anelastic and plastic solid are now written as

$$\begin{cases} \vec{J}_i = -\dfrac{d\vec{\beta}_i}{dt} + \overrightarrow{\text{grad}}\,\phi_i \\ \vec{\lambda}_i = \overrightarrow{\text{rot}}\,\vec{\beta}_i \end{cases} \tag{8.9}$$

in which the distortion tensor $\vec{\beta}_i = \vec{\beta}_i^{él} + \vec{\beta}_i^{an}$ represents both distortions by elasticity and anelasticity.

From this new distortion tensor $\vec{\beta}_i$, we can deduce other distortion tensors always accepting the hypothesis *(5.55)* and decomposition relationships *(5.60)*

$$\begin{cases} \vec{\beta}_i = \vec{\beta}_i^{(\delta)} + \vec{e}_i \wedge \vec{\omega}_O(t) = \vec{\beta}_i^{él} + \vec{\beta}_i^{an} + \vec{e}_i \wedge \vec{\omega}_O(t) \\ \vec{\omega} = -\dfrac{1}{2}\sum_k \vec{e}_k \wedge \vec{\beta}_k = \vec{\omega}^{(\delta)} + \vec{\omega}_0(t) = \vec{\omega}^{él} + \vec{\omega}^{an} + \vec{\omega}_0(t) \\ \tau = \sum_k \vec{\beta}_k \vec{e}_k = \sum_k \vec{\beta}_k^{(\delta)} \vec{e}_k = \tau^{él} \qquad (\tau^{an} \equiv 0 \text{ by hypothesis}) \\ \vec{\varepsilon}_i = \vec{\beta}_i - \vec{e}_i \wedge \vec{\omega} = \vec{\beta}_i^{(\delta)} - \vec{e}_i \wedge \vec{\omega}^{(\delta)} = \vec{\varepsilon}_i^{él} + \vec{\varepsilon}_i^{an} \\ \vec{\alpha}_i = \vec{\varepsilon}_i - \dfrac{1}{3}\tau \vec{e}_i = \vec{\alpha}_i^{él} + \vec{\alpha}_i^{an} \end{cases} \tag{8.10}$$

This new version of the topological distortion tensors and geometro-kinetic equations is in fact nothing else than a change in terminology for the plastic distortions based on an analogy with Maxwell's equations of electromagnetism. To exploit the potential that this formulation of topological equations unleashes will therefore be subject of this chapter.

On the tensorial density of charges of dislocation

Using again relations *(3.35)* and *(3.36)*, in the presence of plastic charges $\vec{\lambda}_i$

$$\vec{B} = \vec{u}_B - \vec{u}_A = \oint_C \delta\vec{u} = -\sum_k \vec{e}_k \iint_S \overrightarrow{\mathrm{rot}}\vec{\beta}_k \, d\vec{S} = -\sum_k \vec{e}_k \iint_S \vec{\lambda}_k \, d\vec{S} \neq 0 \tag{8.11}$$

there appears a closure defect vector $\vec{B} = \vec{u}_B - \vec{u}_A \neq 0$ of the displacement field, around a closed contour C within the medium (fig. 3.5). The differential $\delta\vec{u}$ is no longer a total differential, but a *Pfaff differential form*, which only has meaning under local and differential form, as it represents a basic translation associated with elastic and anelastic deformations and rotations.
The integral on the closed contour C results in a quantity $\vec{B}$ called macroscopic *Burgers vector* [2] defined on the contour C, which corresponds to the macroscopic translation necessary to accommodate the environment in the presence of the charge density $\vec{\lambda}_i$, to ensure the compatibility of total deformations and rotations (the absence of voids and material recoveries in the solid).
The discontinuity $\vec{B}$ is called a *macroscopic dislocation* of the solid in the sense of Volterra, and therefore we will call it *density of dislocation charges,* the tensorial density of charges $\vec{\lambda}_i$ responsible for plastic distortions.

The appearance of macroscopic dislocations

Such macroscopic dislocation is carried out in a continuous solid by locally cutting the solid and moving in parallel the two jaws of the cut, before bonding them back together. This process is illustrated schematically in figure 8.2a using a tube material being cut along the plane $abcd$ and which we glue back together after a parallel shift along the direction of the cut. There appears a *one-dimensional topological singularity* of the localized distortion field on the axis cd. This macroscopic singularity, characterized by a translation vector $\vec{B}$ parallel to the singularity line is called *screw dislocation.*

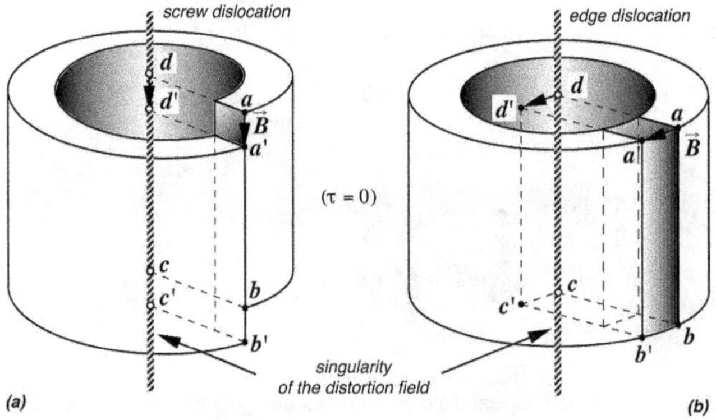

Figure 8.2 - *creation of a screw dislocation **(a)** and an edge dislocation **(b)** by cutting in gluing*

On the other hand, if we glue the two jaws after a parallel translation perpendicular to the plane of the cut, and with the addition or subtraction of a parallelepiped of material (fig. 8.2b), there

[2] *J. M. Burgers, Proc. Kon. Ned. Akad. Weten, vol.42, p. 293, 378, 1939*

appears another one-dimensional topological singularity of the distortion field, located on the axis cd. This macroscopic singularity, characterized by a translation vector $\vec{B}$ perpendicular to the singularity line is called *edge dislocation.* Another approach to achieve an edge dislocation, but without addition or subtraction of material is to glue the two jaws after a parallel translation in the cleavage plane, perpendicular to the direction of the cut, as shown in figure 8.3. Under the *sine qua non* precondition that the elastic volume expansion τ of the medium remained zero during the plastic deformation process, the Burgers vector $\vec{B}$ obtained by the integral *(8.11)* on a contour surrounding the singularity corresponds exactly to the macroscopic translation the jaw $abcd$ underwent.

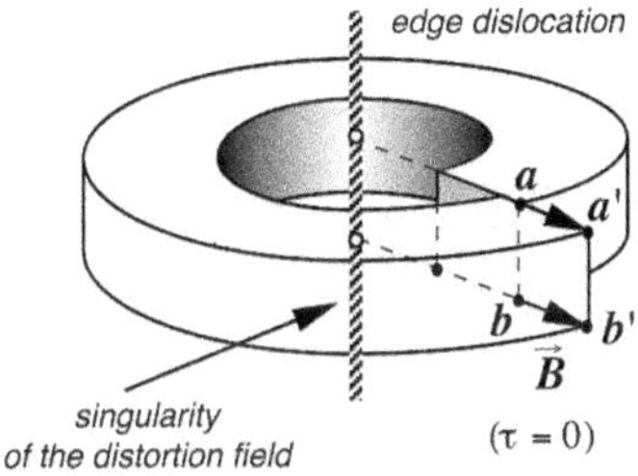

Figure 8.3 *- another creation of an edge dislocation by cutting and gluing*

Let's imagine that within a solid continuum first we cut a torus-shaped hole, as illustrated in the section shown in figure 8.4a, and second we cut the median plane in the center of the torus. The two jaws ab and $a'b'$ so formed can then be displaced with respect to one another, and then reattached.

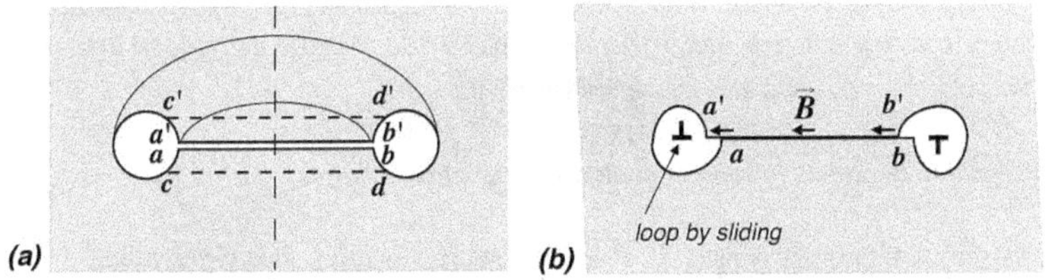

Figure 8.4 *- cut of a torus and the median plane to form loops* ***(a)***
and creation of a dislocation loop by sliding of the jaws ***(b)***

The first possible case is to move the two jaws parallel to the cleavage plane by a distance $\vec{B}$ as shown in figure 8.4b. After rejoining the middle is deformed by shear and the torus contains a macroscopic dislocation of *sliding loop type*, composed of edge, screw and mixed dislocation parts.

One can also insert additional material in the form of a thin disk of thickness $\vec{B}$ between the two jaws and by welding the disk to the two jaws (fig. 8.5A). This results in a deformation of the medium, obviously a curvature of the medium on each side of the torus. As for the core, it is the seat of a macroscopic dislocation of *prismatic loop type*. In this case, the prismatic loop is called interstitial because it contains additional material, and it is composed of a single edge dislocation which closes on itself.

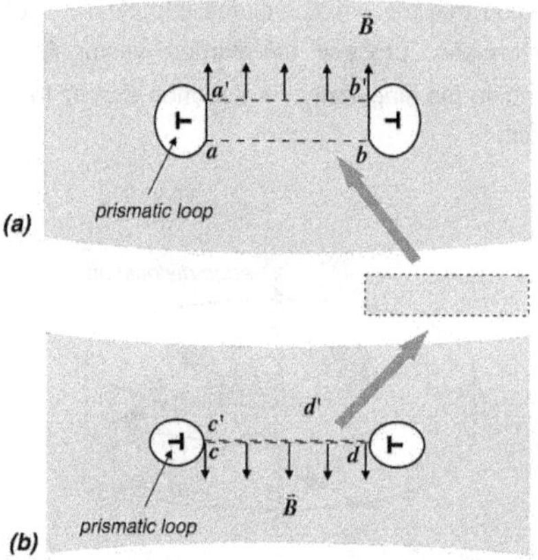

***Figure 8.5** - creation of prismatic loops of dislocation by addition **(a)** or subtraction **(b)** of thin disk of material*

A very similar case is obtained if, instead of adding a disk of material, we subtract a disc of material with thickness $\vec{B}$, as shown in figure 8.5b. We also obtain a macroscopic dislocation of *prismatic loop type*, but the loop is called *vacancy loop* because it lacks a certain amount of material. Within the core, there are also a single edge dislocation that closes on itself.

All resulting singularities are obviously responsible for a distortion field within the solid. Therefore, they require a non-zero energy formation. They are stabilized in the solid by re-bonding of the two jaws of the cut, and thus by the links within the solid.

The equation of conservation of density of dislocation charges

We will note that tensorial density $\vec{\lambda}_i$ of dislocation charges is linked to the rotational of $\vec{\beta}_i$, so that it satisfies the relation

$$\operatorname{div}\vec{\lambda}_i = 0 \tag{8.12}$$

which we will call the *dislocation charge conservation equation*, which will be called later to play a significant role in the topological interpretation of dislocation charges.

The continuity equation connecting the density to the flow of dislocation charges

Topological equations *(8.9)* can be further modified by taking the curl of geometro-kinetic equation and the time derivative of the equation of geometro-compatibility

$$\overrightarrow{\mathrm{rot}}\,\vec{J}_i = -\overrightarrow{\mathrm{rot}}\frac{d\vec{\beta}_i}{dt} \quad \text{and} \quad \frac{d\vec{\lambda}_i}{dt} = \frac{d}{dt}\left(\overrightarrow{\mathrm{rot}}\,\vec{\beta}_i\right) \tag{8.13}$$

From these two relations we obtain

$$\frac{d\vec{\lambda}_i}{dt} = \left[\frac{d}{dt}\left(\overrightarrow{\mathrm{rot}}\,\vec{\beta}_i\right) - \overrightarrow{\mathrm{rot}}\left(\frac{d\vec{\beta}_i}{dt}\right)\right] - \overrightarrow{\mathrm{rot}}\,\vec{J}_i \tag{8.14}$$

which is nothing else than the *continuity equation* for the density of dislocation charges. In fact, in the term in brackets, which is the space-time commutator of $\vec{\beta}_i$, there is a *source term of dislocation charges.* In the event that this commutator is zero, equation *(8.14)* then connects directly the change in the charge density $\vec{\lambda}_i$ along the trajectory to the curl of the flux of charges $\vec{J}_i$ in relation to the lattice, which is characteristic of a continuity equation, in the case of a *tensorial charge density.* It is concluded that the tensorial flux $\vec{J}_i$ is nothing but the flow of dislocation charges.

The scalar density of charges of rotation

Equation *(8.9)* for the compatibility of the distortion tensor $\vec{\beta}_i$, namely $\vec{\lambda}_i = \overrightarrow{\mathrm{rot}}\,\vec{\beta}_i$ allows us to deduce immediately, by calculating the trace, a compatibility equation for the rotation vector $\vec{\omega}$

$$\frac{1}{2}\sum_k \vec{e}_k\vec{\lambda}_k = \frac{1}{2}\sum_k \vec{e}_k\,\overrightarrow{\mathrm{rot}}\,\vec{\beta}_k = \mathrm{div}\,\vec{\omega} \tag{8.15}$$

By introducing a *scalar density charge* λ, which is responsible for the plastic rotation within the solid and defined from the trace of the tensor of density of dislocation charges $\vec{\lambda}_i$

$$\lambda = -\mathrm{div}\,\vec{\omega}^{pl} = \frac{1}{2}\sum_k \vec{e}_k\vec{\lambda}_k \tag{8.16}$$

the compatibility equation of the vector $\vec{\omega}$ for the elastic and anelastic rotations becomes

$$\lambda = \frac{1}{2}\sum_k \vec{e}_k\vec{\lambda}_k = \mathrm{div}\,\vec{\omega} \tag{8.17}$$

It is noted that this scalar charge density allows us to write

$$\oiint_S \omega_\perp\, dS = \oiint_S \vec{\omega}\, d\vec{S} = \iiint_V \mathrm{div}\,\vec{\omega}\, dV = \iiint_V \lambda\, dV \neq 0 \tag{8.18}$$

Thus, the localized presence of a scalar charge density λ gives us a divergence of the rotation field around the charge. That's why we will give to the scalar λ the name of *rotational charge density.*

Note that the equation relating the rotation vector $\vec{\omega}$ to the density of charges of rotation λ is of the same tensorial order and presents a strong analogy with Maxwell's equations $\rho = \mathrm{div}\,\vec{D}$ relating the electric displacement field $\vec{D}$ to the electric charge density ρ.

The vectorial flow of torsion charges

Decomposition properties of the distortion tensor can also be used to derive the geometro-kinetic equation for the rotation vector $\vec{\omega}$. The operation of only taking the anti-symmetric part of the geometro-kinetic equation *(8.9)* leads to the relation

$$-\frac{1}{2}\sum_k \vec{e}_k \wedge \vec{J}_k = -\frac{1}{2}\sum_k \vec{e}_k \wedge \left(-\frac{d\vec{\beta}_k}{dt} + \overrightarrow{\mathrm{grad}}\,\phi_k\right) = -\frac{\partial\vec{\omega}}{\partial t} + \frac{1}{2}\overrightarrow{\mathrm{rot}}\,\vec{\phi} \qquad (8.19)$$

By entering into this relationship a vector flow of charges $\vec{J}$, responsible for the temporal variation of the rotational plastic distortions, or simply a *flux of rotating charges*, defined with respect to the lattice and from the antisymmetric part of the tensor of flow of dislocation charges by the relation

$$\vec{J} = \frac{d\vec{\omega}^{pl}}{dt} = -\frac{1}{2}\sum_k \vec{e}_k \wedge \vec{J}_k \qquad (8.20)$$

the geometro-kinetic equation for the vector of elastic and inelastic rotations $\vec{\omega}$ in the presence of rotational charges becomes

$$\vec{J} = -\frac{d\vec{\omega}}{dt} + \frac{1}{2}\overrightarrow{\mathrm{rot}}\,\vec{\phi} \qquad (8.21)$$

This equation relating the vector $\vec{\omega}$ to the vector of flow of rotation charges $\vec{J}$ is of the same tensorial order and has a strong analogy with Maxwell's equation $\vec{j} = -\partial\vec{D}/\partial t + \overrightarrow{\mathrm{rot}}\,\vec{H}$ relating the electric displacement field $\vec{D}$ to the electric charge flow $\vec{j}$.

The equation of continuity for the charges of rotation

As for the equation of continuity of rotational charges, it is sufficient to combine the divergence of equation *(8.21)* with the material derivative *(8.17)* to get

$$\frac{d\lambda}{dt} = \left[\frac{d}{dt}(\mathrm{div}\,\vec{\omega}) - \mathrm{div}\left(\frac{d\vec{\omega}}{dt}\right)\right] - \mathrm{div}\,\vec{J} \qquad (8.22)$$

wherein, as above, there is a source of rotational charge equal to the spatio-temporal commutator of $\vec{\omega}$.

We can verify that this continuity equation also represents half of the trace of the continuity equation *(8.14)* for the dislocation charges, i.e.

$$\frac{1}{2}\sum_k \vec{e}_k \frac{d\vec{\lambda}_k}{dt} = \frac{1}{2}\sum_k \vec{e}_k \left[\frac{d}{dt}\left(\overrightarrow{\mathrm{rot}}\,\vec{\beta}_k\right) - \overrightarrow{\mathrm{rot}}\left(\frac{d\vec{\beta}_k}{dt}\right)\right] - \frac{1}{2}\sum_k \vec{e}_k \overrightarrow{\mathrm{rot}}\,\vec{J}_k \qquad (8.23)$$

The scalar source of lattice sites

The operation which consists in taking the trace of the equation of geometro-kinetic *(8.9)*

$$\sum_k \vec{e}_k \vec{J}_k = \sum_k \vec{e}_k \left(-\frac{d\vec{\beta}_k}{dt} + \overrightarrow{\mathrm{grad}}\,\phi_k\right) = -\frac{d\tau}{dt} + \mathrm{div}\,\vec{\phi} \qquad (8.24)$$

shows a scalar corresponding to the trace of the *flow of dislocation charges* $\vec{J}_i$.

Using the relation *(5.64)*, it appears as the explicit *scalar volume source of sites of lattice* S_n linked to the trace of the tensorial flow of dislocation charges by the relationship

$$S_n = n\frac{d\tau^{pl}}{dt} = n\sum_k \vec{e}_k \vec{J}_k \qquad (8.25)$$

Table 8.1 - Distortions and charges of dislocation in a solid

Tensors of elastic and anelastic distortions

$$\begin{cases} \vec{\beta}_i = \vec{\beta}_i^{(\delta)} + \vec{e}_i \wedge \vec{\omega}_O(t) = \vec{\beta}_i^{él} + \vec{\beta}_i^{an} + \vec{e}_i \wedge \vec{\omega}_O(t) & (1) \\ \vec{\omega} = -\dfrac{1}{2}\sum_k \vec{e}_k \wedge \vec{\beta}_k = \vec{\omega}^{(\delta)} + \vec{\omega}_0(t) = \vec{\omega}^{él} + \vec{\omega}^{an} + \vec{\omega}_0(t) & (2) \\ \tau = \sum_k \vec{\beta}_k \vec{e}_k = \sum_k \vec{\beta}_k^{(\delta)} \vec{e}_k = \tau^{él} \quad (\tau^{an} \equiv 0 \text{ by hypothesis}) & (3) \\ \vec{\varepsilon}_i = \vec{\beta}_i - \vec{e}_i \wedge \vec{\omega} = \vec{\beta}_i^{(\delta)} - \vec{e}_i \wedge \vec{\omega}^{(\delta)} = \vec{\varepsilon}_i^{él} + \vec{\varepsilon}_i^{an} & (4) \\ \vec{\alpha}_i = \vec{\varepsilon}_i - \dfrac{1}{3}\tau \vec{e}_i = \vec{\alpha}_i^{él} + \vec{\alpha}_i^{an} & (5) \end{cases}$$

Geometro-kinetic of distortions,

flow of dislocation charges, flow of rotational charges, and source of lattice sites

$$\begin{cases} \vec{J}_i = -\dfrac{d\vec{\beta}_i}{dt} + \overrightarrow{\text{grad}}\,\phi_i & (6) \\ \vec{J} = -\dfrac{1}{2}\sum_k \vec{e}_k \wedge \vec{J}_k = -\dfrac{d\vec{\omega}}{dt} + \dfrac{1}{2}\overrightarrow{\text{rot}}\,\vec{\phi} & (7) \\ \dfrac{S_n}{n} = \sum_k \vec{e}_k \vec{J}_k = -\dfrac{d\tau}{dt} + \text{div}\,\vec{\phi} & (8) \end{cases}$$

Geometro-compatibility of distortions

and density of charges of dislocation, of flexion and of rotation

$$\begin{cases} \vec{\lambda}_i = \overrightarrow{\text{rot}}\,\vec{\beta}_i & (9) \\ \vec{\lambda} = -\sum_k \vec{e}_k \wedge \vec{\lambda}_k = -\sum_k \vec{e}_k \wedge \overrightarrow{\text{rot}}\vec{\beta}_k & (10) \\ \lambda = \dfrac{1}{2}\sum_k \vec{e}_k \vec{\lambda}_k = \text{div}\,\vec{\omega} & (11) \end{cases}$$

Conservation and continuity equations of charges of dislocation and of rotation

$$\begin{cases} \text{div}\,\vec{\lambda}_i = 0 & (12) \\ \dfrac{d\vec{\lambda}_i}{dt} = \left[\dfrac{d}{dt}\left(\overrightarrow{\text{rot}}\,\vec{\beta}_i\right) - \overrightarrow{\text{rot}}\left(\dfrac{d\vec{\beta}_i}{dt}\right)\right] - \overrightarrow{\text{rot}}\,\vec{J}_i & (13) \\ \dfrac{d\lambda}{dt} = \left[\dfrac{d}{dt}(\text{div}\,\vec{\omega}) - \text{div}\left(\dfrac{d\vec{\omega}}{dt}\right)\right] - \text{div}\,\vec{J} & (14) \end{cases}$$

Everything happens as if in fact the flow of dislocation $\vec{J}_i$ actually left a *"trace"* of its passage in the solid in the form of creation or annihilation of lattice sites. So we will write the geometro-ki-netic equation for the scalar of elastic volume expansion $\tau = \tau^{el}$ (assuming $\tau^{an} \equiv 0$) in the form

$$\frac{S_n}{n} = -\frac{d\tau}{dt} + \operatorname{div} \vec{\phi}$$

(8.26)

In table 8.1, we wrote all the relations concerning distortions and dislocation densities and flow of charges (or charges of plastic distortion) in a solid. In figure 8.6, we show schematically all the operations which allow us, using the geometro-compatibility and geometro-kinetic of $\vec{\beta}_i$ to deduct this set of equations of compatibility, of continuity and of geometro-kinetic of the disloca-tion charges.

Density of charges of flexion

$$\vec{\lambda} = -\sum_k \vec{e}_k \wedge \vec{\lambda}_k$$

antisymmetry

$$\vec{\lambda} = -\sum_k \vec{e}_k \wedge \overrightarrow{\mathrm{rot}}\vec{\beta}_k$$

Density of charges of dislocation

$$\vec{\lambda}_i = \overrightarrow{\mathrm{rot}}\,\vec{\beta}_i$$

trace

$$\lambda = \frac{1}{2}\sum_k \vec{e}_k \vec{\lambda}_k$$

$$\lambda = \operatorname{div}\vec{\omega}$$ **Density of charges of rotation**

$$\frac{d\vec{\lambda}_i}{dt} + \overrightarrow{\mathrm{rot}}\,\vec{J}_i =$$

$$\left[\frac{d}{dt}\left(\overrightarrow{\mathrm{rot}}\,\vec{\beta}_i\right) - \overrightarrow{\mathrm{rot}}\left(\frac{d\vec{\beta}_i}{dt}\right)\right]$$

Equations of continuity of charges

$$\frac{d\lambda}{dt} + \operatorname{div}\vec{J} =$$

$$\left[\frac{d}{dt}(\operatorname{div}\vec{\omega}) - \operatorname{div}\left(\frac{d\vec{\omega}}{dt}\right)\right]$$

$$\vec{J} = -\frac{1}{2}\sum_k \vec{e}_k \wedge \vec{J}_k$$

antisymmetry

$$\vec{J} = -\frac{d\vec{\omega}}{dt} + \frac{1}{2}\overrightarrow{\mathrm{rot}}\,\vec{\phi}$$

Flux of charges of rotation

$$\vec{J}_i = -\frac{d\vec{\beta}_i}{dt} + \overrightarrow{\mathrm{grad}}\,\phi_i$$

Flux of charges of dislocation

trace

$$S_n = n\sum_k \vec{e}_k \vec{J}_k$$

$$\frac{S_n}{n} = -\frac{d\tau}{dt} + \operatorname{div}\vec{\phi}$$

Source of sites of lattice

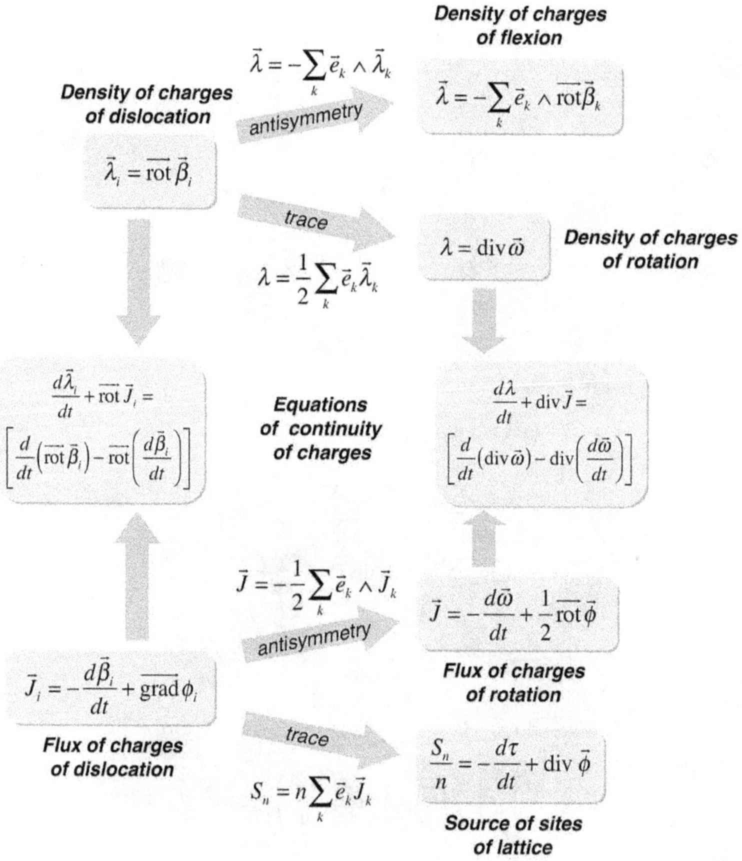

Figure 8.6 - decompositions of the densities and flows of charges
and equations of continuity

8.2 - On the macroscopic concept of charges of plastic contortions

In an elastically and anelastically deformed media, with dislocation charges, we can also express the tensor of elastic and anelastic contortions $\vec{\chi}_i$.

The tensorial density of charges of plastic contortions of a solid

The relationships in table 3.1 show that the tensor expressions for $\vec{\chi}_i$ deriving from the deformation field $\vec{\varepsilon}_i$ or from the shear field $\vec{\alpha}_i$ and the volume expansion τ, are not changed, but that, instead, its expression from the field $\vec{\omega}$ of elastic and anelastic rotations shall make use of the charge density tensors $\vec{\lambda}_i$ and λ. So in the presence of dislocation charges we have for $\vec{\chi}_i$

$$\vec{\chi}_i = \left[\overrightarrow{\mathrm{rot}}\,\vec{\varepsilon}_i\right]^{\mathrm{T}} = \left[\overrightarrow{\mathrm{rot}}\,\vec{\alpha}_i\right]^{\mathrm{T}} + \frac{1}{3}\vec{e}_i \wedge \overrightarrow{\mathrm{grad}}\,\tau = \overrightarrow{\mathrm{grad}}\,\omega_i + \left[\vec{\lambda}_i\right]^{\mathrm{T}} - \vec{e}_i\lambda \tag{8.27}$$

The *tensorial density of charges* $\left[\vec{\lambda}_i\right]^{\mathrm{T}} - \vec{e}_i\lambda$ which appears here is thus responsible for the plastic contortions of the solid. We will therefore call them *density of contortion charges.* They take the following shape

$$\left[\vec{\lambda}_i\right]^{\mathrm{T}} - \vec{e}_i\lambda = \vec{\lambda}_i + \vec{e}_i \wedge \left(\sum_k \vec{e}_k \wedge \vec{\lambda}_k\right) - \vec{e}_i\lambda = \vec{\lambda}_i - \vec{e}_i \wedge \vec{\lambda} - \vec{e}_i\lambda \tag{8.28}$$

The tensor of contortion in a charged solid can thus be written as

$$\vec{\chi}_i = \left[\overrightarrow{\mathrm{rot}}\,\vec{\varepsilon}_i\right]^{\mathrm{T}} = \left[\overrightarrow{\mathrm{rot}}\,\vec{\alpha}_i\right]^{\mathrm{T}} + \frac{1}{3}\vec{e}_i \wedge \overrightarrow{\mathrm{grad}}\,\tau = \overrightarrow{\mathrm{grad}}\,\omega_i + \left(\vec{\lambda}_i - \vec{e}_i \wedge \vec{\lambda} - \vec{e}_i\lambda\right) \tag{8.29}$$

The tensorial density of disclination charges

The compatibility equation in table 8.5 for elastic and anelastic contortions is also modified in the presence of plastic charges, and is written

$$\overrightarrow{\mathrm{rot}}\left(\vec{\lambda}_i - \vec{e}_i \wedge \vec{\lambda} - \vec{e}_i\lambda\right) = \overrightarrow{\mathrm{rot}}\,\vec{\chi}_i = \overrightarrow{\mathrm{rot}}\left[\overrightarrow{\mathrm{rot}}\,\vec{\varepsilon}_i\right]^{\mathrm{T}} = \overrightarrow{\mathrm{rot}}\left[\overrightarrow{\mathrm{rot}}\,\vec{\alpha}_i\right]^{\mathrm{T}} - \frac{1}{3}\overrightarrow{\mathrm{rot}}\,\overrightarrow{\mathrm{rot}}\left(\vec{e}_i\tau\right) \tag{8.30}$$

which is in a form similar to the compatibility of equation obtained for the elastic and anelastic distortions.
This relationship shows a new tensorial density of charges $\vec{\theta}_i$

$$\vec{\theta}_i = \overrightarrow{\mathrm{rot}}\left(\vec{\lambda}_i - \vec{e}_i \wedge \vec{\lambda} - \vec{e}_i\lambda\right) = \overrightarrow{\mathrm{rot}}\,\vec{\lambda}_i - \vec{e}_i\,\mathrm{div}\,\vec{\lambda} + \left(\vec{e}_i\vec{\nabla}\right)\vec{\lambda} + \vec{e}_i \wedge \overrightarrow{\mathrm{grad}}\,\lambda \tag{8.31}$$

Since this density of charges is equal to the rotational of $\vec{\lambda}_i - \vec{e}_i \wedge \vec{\lambda} - \vec{e}_i\lambda$, it satisfies the following relation

$$\mathrm{div}\,\vec{\theta}_i = \mathrm{div}\left[\overrightarrow{\mathrm{rot}}\left(\vec{\lambda}_i - \vec{e}_i \wedge \vec{\lambda} - \vec{e}_i\lambda\right)\right] \equiv 0 \tag{8.32}$$

which we will call the *conservation equation* of the tensorial charges.
Working again on relations *(3.38)* and *(3.40)* in the presence of these charges

$$\vec{\Omega} = \oint_C \delta\vec{\omega}^{\varepsilon} = \sum_k \vec{e}_k \iint_S \overrightarrow{\mathrm{rot}}\,\vec{\chi}_k\, d\vec{S} = \sum_k \vec{e}_k \iint_S \vec{\theta}_i\, d\vec{S} = \sum_k \vec{e}_k \iint_S \overrightarrow{\mathrm{rot}}\left(\vec{\lambda}_k - \vec{e}_k \wedge \vec{\lambda} - \vec{e}_k\lambda\right) d\vec{S} \neq 0 \tag{8.33}$$

there appears a vectorial closure defect $\vec{\Omega} = \vec{\omega}_B^{\varepsilon} - \vec{\omega}_A^{\varepsilon} \neq 0$ of the rotation field by deformation taken on the closed contour C within the medium (fig. 3.7). The differential $\delta\vec{\omega}^{\varepsilon}$ is no longer a total differential, but a *Pfaff differential form*, which only has meaning under local and differential form, where it is associated with an elementary rotation of elastic and anelastic deformations and rotations.

It is still possible to transform the integral over the surface S in an integral on the closed contour C in the form

$$\vec{\Omega} = \oint_C \delta\vec{\omega}^{\varepsilon} = \sum_k \vec{e}_k \iint_S \overrightarrow{\mathrm{rot}}\vec{\chi}_k \, d\vec{S} = \sum_k \vec{e}_k \oint_C \left(\vec{\lambda}_k - \vec{e}_k \wedge \vec{\lambda} - \vec{e}_k \lambda\right) d\vec{r} \neq 0 \tag{8.34}$$

Thus, the integral of the *density* $\left(\vec{\lambda}_i - \vec{e}_i \wedge \vec{\lambda} - \vec{e}_i \lambda\right)$ *of charges of contortion* on a closed contour C results in the existence of a non-zero macroscopic angular quantity $\vec{\Omega}$, called a *macroscopic Frank vector*[3] defined on the contour C, and which corresponds to the macroscopic rotation necessary to accommodate the environment in the presence of the plastic charges. This discontinuity $\vec{\Omega}$ is called a macroscopic disclination of the solid in the sense of macroscopic defects of Volterra. From equation *(18.33)*, it is now possible to interpret the tensorial density of charges $\vec{\theta}_i = \overrightarrow{\mathrm{rot}}\left(\vec{\lambda}_i - \vec{e}_i \wedge \vec{\lambda} - \vec{e}_i \lambda\right)$. Indeed, as the integral over the surface S of the charge density provides the macroscopic Frank vector of a disclination, it will be called the density of disclination charges $\vec{\theta}_i$.

Creation of macroscopic disclinations

It is easy to imagine the realization of a disclination at the macroscopic scale in a solid continuum by locally cutting this solid and rotating a jaw of the cut relative to the other, before gluing them back together. This process is illustrated schematically in figure 8.7,

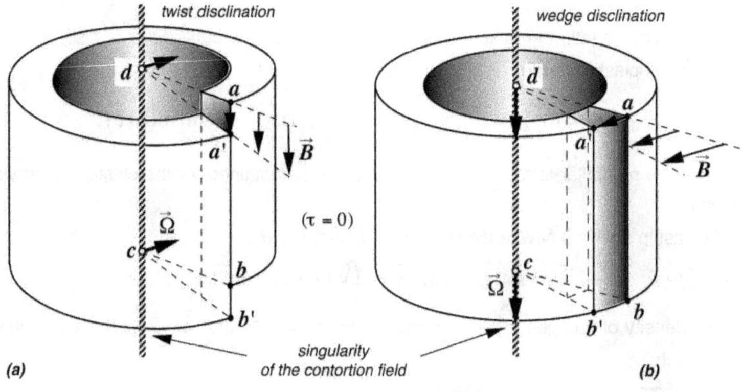

Figure 8.7 *- creation of the twist* ***(a)*** *and wedge* ***(b)*** *disclinations by cutting and welding back*

using a tube $\boldsymbol{abcd}$ material being cut and glued together in two different ways:

- either by shear of the plane $a'b'cd$ without the addition or subtraction of material (fig. 19.3a),

[3] *R. DeWit, «Theory of disclinations II, III and IV», J. of the Nat. Bureau of Standards A, vol. 77A, p. 49-100, p. 359-368, and p. 607-658, 1973*

which leads to a one-dimensional topological singularity located on the axis cd , called *twist disclination*

- by rotation of one of the interfaces around the edge cd with the addition or subtraction of material (fig. 19.3b), which leads to a one-dimensional topological singularity located on the axis, called a *wedge disclination.*

Under the condition that the elastic volume expansion τ of the medium remained zero during the plastic deformation process, the vector $\vec{\Omega}$, obtained by the integral *(8.34)* on a contour C surrounding the singularity corresponds exactly to the macroscopic rotation underwent by the jaw $a'b'cd$. As the vector $\vec{\Omega}$ must remain constant if we vary the diameter of the contour of integration C or if we move this contour vertically, we deduce that disclination charges must be confined to the immediate vicinity of the axis cd of the tube, and that their tensorial density must be constant along the axis.

Topological singularities thus obtained are responsible for a distortion field within the solid. Therefore, they require a non-zero energy formation. They are stabilized in the solid by re-gluing of the two jaws of the cut, and thus by the bonds within the solid.

Comparing figures 8.2 and 8.7, there is a striking resemblance between screw dislocations and twist disclinations and between edge dislocations and wedge disclinations. This similarity is not accidental, since the operations used to generate these discontinuities are very similar. It is interesting to note in particular that the macroscopic disclinations also have a displacement vector $\vec{B}$ going from a to a' (fig. 8.7), as well as the macroscopic dislocation (fig. 8.2). However, this vector $\vec{B}$, in the case of disclinations, increases linearly with the diameter of the integrating loop C used to calculate it. This means that in the presence of a macroscopic disclination associated with a disclination charge distributed along the axis of the pipe cd , there must also be a dislocation charge density, as seen in the last term of equation *(8.34)*. But the latter, instead of being located on the axis of the tube as is the case for a macroscopic dislocation, will be evenly distributed on a surface in the cut plane $abcd$ (fig. 8.7) such that the Burgers vector $\vec{B}$ increases linearly with the diameter of the integration loop C .

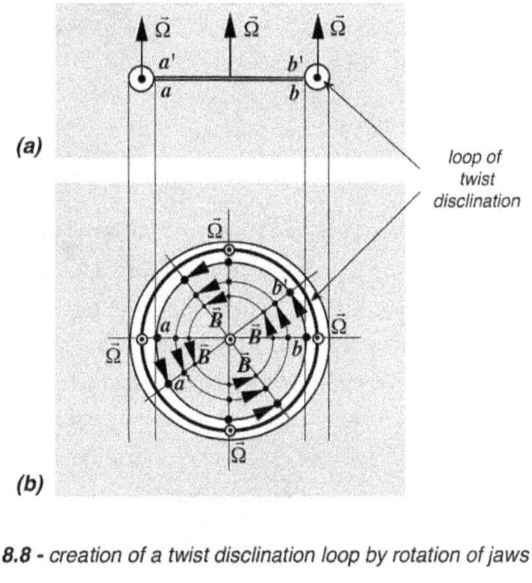

Figure 8.8 *- creation of a twist disclination loop by rotation of jaws*

One can also proceed as in figure 8.4a, and move the two jaws by rotating them relative to each other by a rotational angle $\vec{\Omega}$ in the cleavage plane, as shown in figure 8.8a. After re-bonding, the medium is deformed by the $\vec{\Omega}$ rotation and the torus then contains a *macroscopic twist disclination loop.*

Note here that the field of displacement of the media on each side of the cut plane is tangential to this plane and that the curvilinear displacement motion vector $\vec{B}$ on the cut plane increases from a zero value at the center to a maximum value on the edges of the torus. On the torus, the local displacement field $\vec{B}$ closely resembles the displacement field of a screw dislocation closed in on itself, but it is actually a *pseudo-dislocation* because the curvilinear Burgers vector, tangential to the dislocation line is not preserved [4] (fig. 8.8b).

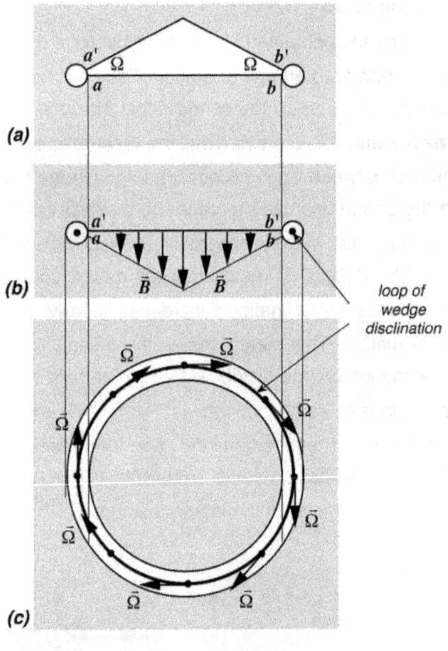

***Figure 8.9** - creation of a wedge disclination loop by adding or removing a part of the media in conical shape*

One could also remove a piece of medium from the center of the torus, with a lenticular or conical shape with angle Ω to the base, as shown in figure 8.9a. In this case, the plane of re-bonding has a local displacement field $\vec{B}$ corresponding to perpendicular Burgers vectors whose lengths have a circular symmetry (fig. 8.9b). On the torus, the required deformation for re-bonding is a rotation $\vec{\Omega}$ tangential to the torus, which would therefore correspond to a *macroscopic wedge disclination loop,* but which is actually a *pseudo-disclination* because the Frank vector, still tangential to the disclination line is not retained along the line (figure 8.9c).

---

[4] *W. Huang, T. Mura, Elastic J. of Applied Physics, vo. 41(13), p. 5175, 1970; A. Unzicker, arXiv:gr-qc/0011064, 2000; A. Unzicker, arXiv:gr-qc/9612061v2, 2010*

The vectorial density of flexion charges

As for the flexion vector $\vec{\chi}$ due to elastic and anelastic bending, it is not modified when expressed as a function of the deformation field $\vec{\varepsilon}_i$ or a function of shear fields $\vec{\alpha}_i$ and volume expansion τ, while its expression from the vector of elastic and anelastic rotations $\vec{\omega}$ depends on a new vector $\vec{\lambda}$ corresponding to the anti-symmetric part of $\vec{\lambda}_i$, the density tensor of dislocation charges

$$\begin{aligned}\vec{\chi} &= \sum_k \vec{e}_k \wedge \vec{\chi}_k = -\sum_k \vec{e}_k \wedge \overrightarrow{\text{rot}}\,\vec{\varepsilon}_k = -\sum_k \vec{e}_k \wedge \overrightarrow{\text{rot}}\,\vec{\alpha}_k - \frac{2}{3}\overrightarrow{\text{grad}}\,\tau \\ &= \sum_k \vec{e}_k \,\text{div}\,\vec{\alpha}_k - \frac{2}{3}\overrightarrow{\text{grad}}\,\tau = -\overrightarrow{\text{rot}}\,\vec{\omega} - \sum_k \vec{e}_k \wedge \vec{\lambda}_k = -\overrightarrow{\text{rot}}\,\vec{\omega} + \vec{\lambda}\end{aligned} \tag{8.35}$$

We deduce the interpretation of this $\vec{\lambda}$ vector which is nothing other than the density of flexion charges, given from the density $\vec{\lambda}_i$ of dislocation charges by the relation

$$\vec{\lambda} = -\sum_k \vec{e}_k \wedge \vec{\lambda}_k = -\sum_k \vec{e}_k \wedge \overrightarrow{\text{rot}}\,\vec{\beta}_k \tag{8.36}$$

From equations *(8.29)* and *(8.35)*, we deduce that the contortions $\vec{\chi}_i$ and flexions $\vec{\chi}$ associated to elastic and anelastic deformation $\vec{\varepsilon}_i$ of the lattice, cannot be deduced directly from the elastic and anelastic rotations $\vec{\omega}$, since it adds terms dependent on the $\vec{\lambda}_i$, $\vec{\lambda}$ and λ density charges.

The scalar density of curvature charges

The compatibility equation in table 3.1 for elastic and anelastic curvature is also modified in the presence of plastic charges, and is written

$$\text{div}\,\vec{\chi} = -\sum_k \vec{e}_k \overrightarrow{\text{rot}}\,\vec{\chi}_k = \sum_k \vec{e}_k \overrightarrow{\text{rot}}\,\overrightarrow{\text{rot}}\,\vec{\varepsilon}_k = \sum_k \vec{e}_k \overrightarrow{\text{rot}}\,\overrightarrow{\text{rot}}\,\vec{\alpha}_k - \frac{2}{3}\Delta\tau = \sum_k \vec{e}_k \overrightarrow{\text{rot}}\,\vec{\lambda}_k \tag{8.37}$$

This relationship shows that it is possible to introduce a new *scalar density of charges* θ, responsible for the plastic contortions of the solid, defined from the divergence of the vector density $\vec{\lambda}$ of flexion charges

$$\theta = \text{div}\,\vec{\lambda} = -\text{div}\left(\sum_k \vec{e}_k \wedge \vec{\lambda}_k\right) = \sum_k \vec{e}_k \overrightarrow{\text{rot}}\,\vec{\lambda}_k = -\sum_k \vec{e}_k \vec{\theta}_k \tag{8.38}$$

Note that this charge density θ can also be connected directly to the trace of the disclination density charge $\vec{\theta}_i = \overrightarrow{\text{rot}}\left(\vec{\lambda}_i - \vec{e}_i \wedge \vec{\lambda} - \vec{e}_i \lambda\right)$, which is easily verified from equation *(8.31)*.
The geometro-compatibility equation *(8.37)* for the flexion vector $\vec{\chi}$ is similar in shape to that obtained for the vector rotation $\vec{\omega}$, and will play later an important role. Now we can write

$$\theta = -\sum_k \vec{e}_k \vec{\theta}_k = \text{div}\,\vec{\lambda} = \text{div}\,\vec{\chi} \tag{8.39}$$

According to figure 3.8, this scalar density θ of plastic charges allows us to write

$$\iiint_V \theta\, dV = \iiint_V \text{div}\,\vec{\lambda}\, dV = \iiint_V \text{div}\,\vec{\chi}\, dV = \oiint_S \vec{\chi}\, d\vec{S} = \oiint_S \chi_\perp\, dS \neq 0 \tag{8.40}$$

Thus, the localized presence of a scalar density of plastic charges $\theta = \text{div}\,\vec{\lambda}$ gives rise to a *divergence of the flexion field* surrounding this charge and therefore *a divergent "curvature"*

around the charge. This is why we call θ the *scalar of curvature charge.*
From equation *(8.35)*, we can also deduce the rotational field of flexion $\vec{\chi}$

$$\overrightarrow{\mathrm{rot}}\vec{\chi} = -\sum_k \overrightarrow{\mathrm{rot}}\left(\vec{e}_k \wedge \overrightarrow{\mathrm{rot}}\,\vec{\varepsilon}_k\right) = -\sum_k \overrightarrow{\mathrm{rot}}\left(\vec{e}_k \wedge \overrightarrow{\mathrm{rot}}\,\vec{\alpha}_k\right) = -\overrightarrow{\mathrm{rot}}\,\overrightarrow{\mathrm{rot}}\,\vec{\omega} + \overrightarrow{\mathrm{rot}}\,\vec{\lambda} \tag{8.41}$$

and it is found to depend on the vectorial density charges $\overrightarrow{\mathrm{rot}}\,\vec{\lambda}$, which represents the *density of rotational flexion charges*, which is put on equal footing with the density of divergent flexion charges, namely the density of charges of curvature. However, this equation *(8.41)* differs from equation *(8.38)* by the fact that it is not a geometro-compatible equation since the rotational field $\vec{\omega}$ appears in it.

8.3 - Complete topological description of charged solids

In figure 8.10, we show all the operations that give all the distortion and contortion tensors governing the topology of the charged solids as well as the compatibility equations that these tensors must satisfy. It shows a beautiful symmetry in the arrangement of relations to derive the tensor of distortion and contortion, as well as in the geometro-compatible equations. It also shows the *torsion tensor* $\left[\vec{\chi}_i\right]^S$, and it is easily shown, using *(18.29)*, that this tensor is expressed from the rotational field $\vec{\omega}$, and a *density of torsion charges* $\vec{\lambda}_i - \left(\vec{e}_i \wedge \vec{\lambda}\right)/2 - \vec{e}_i\lambda$.
The different tensor of density of dislocation and disclination charges are not independent of each other, since all are derived from the initial tensorial density dislocation charge $\vec{\lambda}_i$, as illustrated by figure 8.11.
In fact, *only the two scalar densities* $\lambda = \mathrm{div}\,\vec{\omega}$ and $\theta = \mathrm{div}\,\vec{\lambda}$ are completely independent of one another, because the density of rotational charges λ is deducted from the trace of $\vec{\lambda}_i$, trace that is not involved in the charge curvature θ as it is derived from the anti-symmetric part of θ via the divergence. This independence of the scalar charges λ and θ will play an important role in the future.

Symmetry of the tensor of density of disclination charges

It should also be noted that the *tensorial density* $\vec{\theta}_i$ *of disclination charges is symmetrical.* Indeed, equation *(8.31)* gives the tensor $\vec{\theta}_i$, and using the conservation equation of dislocation charges $\mathrm{div}\,\vec{\lambda}_k \equiv 0$, we show that the anti-symmetric part of $\vec{\theta}_i$ is identically zero

$$\begin{aligned}
&\sum_k \vec{e}_k \wedge \left[\overrightarrow{\mathrm{rot}}\,\vec{\lambda}_k - \vec{e}_k\,\mathrm{div}\,\vec{\lambda} + \left(\vec{e}_k\vec{\nabla}\right)\vec{\lambda} + \vec{e}_k \wedge \overrightarrow{\mathrm{grad}}\,\lambda\right] = \\
&\sum_{k,i}\left\{\vec{e}_k \wedge \left(\vec{\nabla} \wedge \vec{\lambda}_k\right) - \vec{e}_k \wedge \left[\vec{e}_i \wedge \left(\vec{e}_k\vec{\nabla}\right)\vec{\lambda}_i\right] + \frac{1}{2}\vec{e}_k \wedge \left[\vec{e}_k \wedge \vec{\nabla}\left(\vec{e}_i\vec{\lambda}_i\right)\right]\right\} = \\
&\sum_k\left\{\vec{\nabla}\left(\vec{e}_k\vec{\lambda}_k\right) - \left(\vec{e}_k\vec{\nabla}\right)\vec{\lambda}_k - \left(\mathrm{div}\,\vec{\lambda}_k\right)\vec{e}_k + \left(\vec{e}_k\vec{\nabla}\right)\vec{\lambda}_k + \frac{1}{2}\vec{\nabla}\left(\vec{e}_k\vec{\lambda}_k\right) - \frac{3}{2}\vec{\nabla}\left(\vec{e}_k\vec{\lambda}_k\right)\right\} \equiv 0
\end{aligned} \tag{8.42}$$

To finish this chapter, all the relations concerning contortions and the charges of contortions are reported in table 8.2.

Distortions $\vec{\beta}_i$ *anti-symmetry* $\vec{\omega}$ **Rotations**

curl   *divergence*

symmetry

$$\boxed{\vec{\lambda}_i} = \overrightarrow{\mathrm{rot}}\,\vec{\beta}_i \qquad \textit{trace} \qquad \boxed{\lambda} = \mathrm{div}\,\vec{\omega}$$

Charges of dislocation   *Charges of rotation*

Deformations

Shear strains $\vec{\alpha}_i$ *without trace* $\vec{\varepsilon}_i$ *trace* τ **Expansions**

transposition of curl   *anti-symmetry of curl*

Contortions   **Flexions**

$$\vec{\chi}_i = \overline{\mathrm{grad}}\,\omega_i + \boxed{\vec{\lambda}_i - \vec{e}_i \wedge \vec{\lambda} - \vec{e}_i \lambda} \qquad \textit{anti-symmetry} \qquad \vec{\chi} = -\overrightarrow{\mathrm{rot}}\,\vec{\omega} + \boxed{\vec{\lambda}}$$

Charges of contortion   *curl*   *divergence*   *Charges of flexion*

symmetry

$$\boxed{\vec{\theta}_i} = \overrightarrow{\mathrm{rot}}\left(\vec{\lambda}_i - \vec{e}_i \wedge \vec{\lambda} - \vec{e}_i \lambda\right) = \overrightarrow{\mathrm{rot}}\,\vec{\chi}_i \qquad \textit{trace} \qquad \boxed{\theta} = \mathrm{div}\,\vec{\lambda} = \mathrm{div}\,\vec{\chi}$$

Charges of disclination   *Charges of curvature*

Torsions

$$\left[\vec{\chi}_i\right]^S = \overline{\mathrm{grad}}\,\omega_i - \frac{1}{2}\vec{e}_i \wedge \overrightarrow{\mathrm{rot}}\,\vec{\omega} + \boxed{\vec{\lambda}_i - \frac{1}{2}\vec{e}_i \wedge \vec{\lambda} - \vec{e}_i \lambda} \qquad \textit{trace} \qquad \sum_k \vec{\chi}_k \equiv 0$$

Charges of torsion

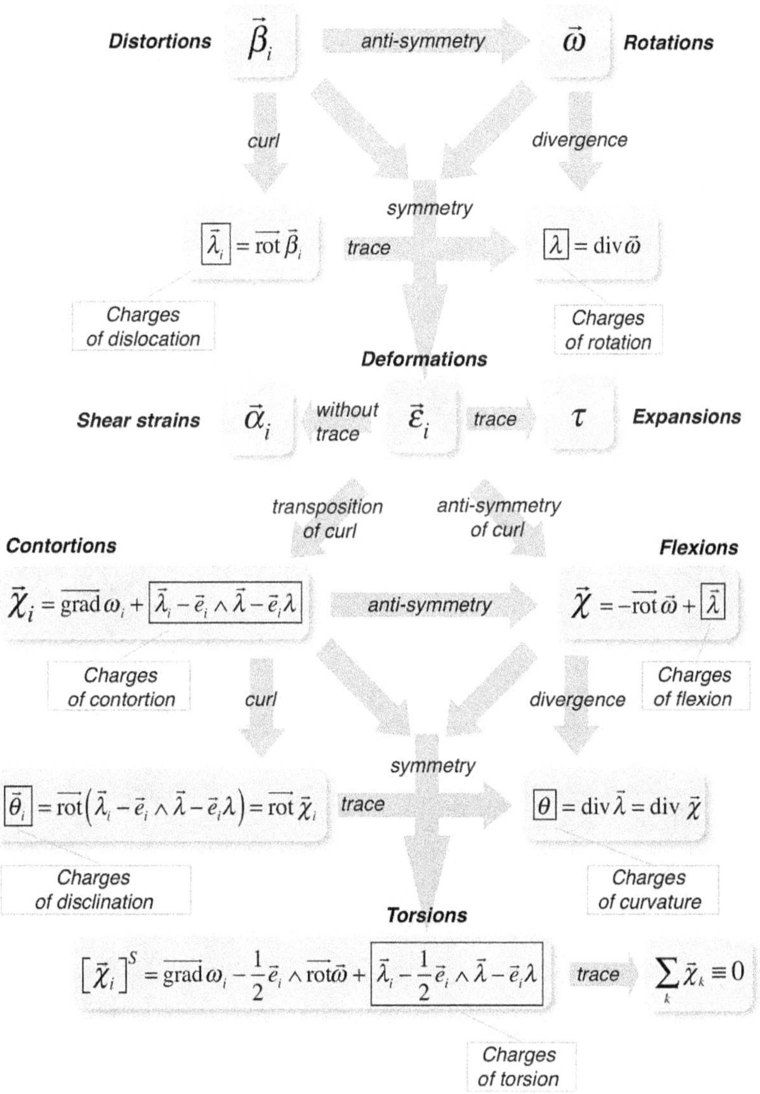

Figure 8.10 - *complete decomposition of the tensors of distortion and contortion.*

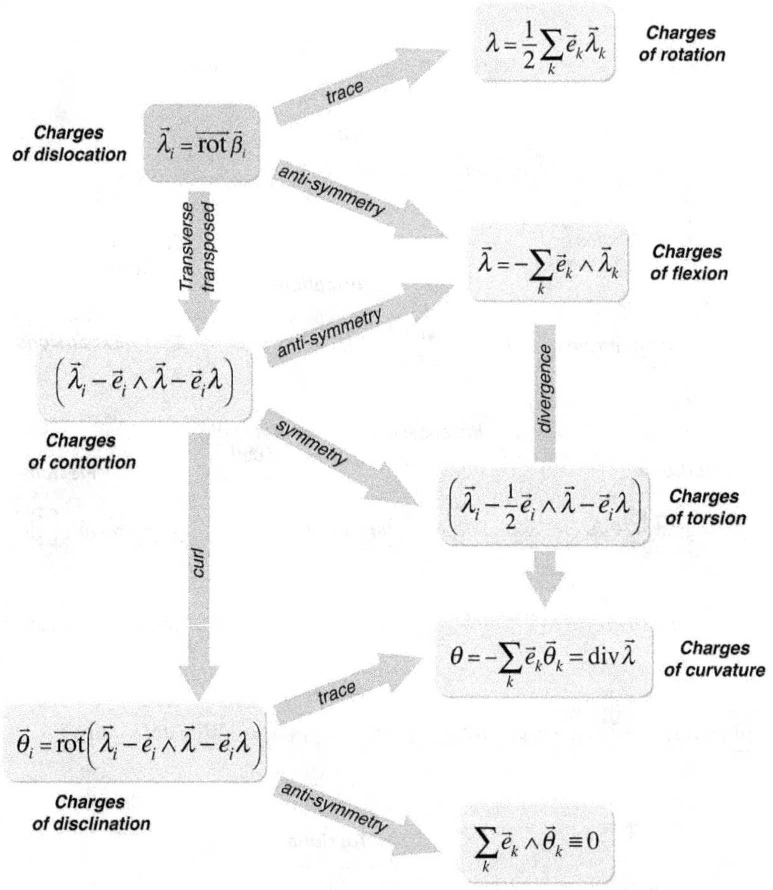
Charges of rotation
$\lambda = \frac{1}{2}\sum_k \vec{e}_k \vec{\lambda}_k$
trace
Charges of dislocation
$\vec{\lambda}_i = \overline{\mathrm{rot}}\,\vec{\beta}_i$
anti-symmetry
Transverse transposed
$\vec{\lambda} = -\sum_k \vec{e}_k \wedge \vec{\lambda}_k$
Charges of flexion
anti-symmetry
$\left(\vec{\lambda}_i - \vec{e}_i \wedge \vec{\lambda} - \vec{e}_i \lambda\right)$
divergence
Charges of contortion
symmetry
$\left(\vec{\lambda}_i - \frac{1}{2}\vec{e}_i \wedge \vec{\lambda} - \vec{e}_i \lambda\right)$
Charges of torsion
curl
$\theta = -\sum_k \vec{e}_k \vec{\theta}_k = \mathrm{div}\,\vec{\lambda}$
Charges of curvature
trace
$\vec{\theta}_i = \overline{\mathrm{rot}}\left(\vec{\lambda}_i - \vec{e}_i \wedge \vec{\lambda} - \vec{e}_i \lambda\right)$
Charges of disclination
anti-symmetry
$\sum_k \vec{e}_k \wedge \vec{\theta}_k \equiv 0$

Figure 8.11 *- Complete decomposition of the tensorial charges*

Table 8.2 - Contortions and charges of disclination in a solid

Transverse tensor of contortion and density charges of contortion

$$\vec{\chi}_i = \overrightarrow{\text{grad}}\,\omega_i + \left(\vec{\lambda}_i - \vec{e}_i \wedge \vec{\lambda} - \vec{e}_i \lambda\right) = \left[\overrightarrow{\text{rot}}\,\vec{\varepsilon}_i\right]^{\text{T}} = \left[\overrightarrow{\text{rot}}\,\vec{\alpha}_i\right]^{\text{T}} + \frac{1}{3}\vec{e}_i \wedge \overrightarrow{\text{grad}}\,\tau \qquad (1)$$

Flexion vector and density of charges of flexion

$$\vec{\chi} = \sum_k \vec{e}_k \wedge \vec{\chi}_k = -\sum_k \vec{e}_k \wedge \overrightarrow{\text{rot}}\,\vec{\varepsilon}_k == -\sum_k \vec{e}_k \wedge \overrightarrow{\text{rot}}\,\vec{\alpha}_k - \frac{2}{3}\overrightarrow{\text{grad}}\,\tau = \sum_k \vec{e}_k\, \text{div}\,\vec{\alpha}_k - \frac{2}{3}\overrightarrow{\text{grad}}\,\tau = -\overrightarrow{\text{rot}}\,\vec{\omega} + \vec{\lambda} \qquad (2)$$

Transversality of the tensor of contortions

$$\sum_k \vec{e}_k \vec{\chi}_k = \sum_k \vec{e}_k \left[\overrightarrow{\text{rot}}\,\vec{\varepsilon}_k\right]^{\text{T}} = \sum_k \vec{e}_k \overrightarrow{\text{rot}}\,\vec{\varepsilon}_k \equiv 0 \qquad (3)$$

Transverse symmetric tensor of torsion and density of torsion charges

$$[\vec{\chi}_i]^{\text{S}} = \vec{\chi}_i + \frac{1}{2}\vec{e}_i \wedge \vec{\chi} = \overrightarrow{\text{grad}}\,\omega_i - \frac{1}{2}\vec{e}_i \wedge \overrightarrow{\text{rot}}\,\vec{\omega} + \left(\vec{\lambda}_i - \frac{1}{2}\vec{e}_i \wedge \vec{\lambda} - \vec{e}_i \lambda\right) = \left[\overrightarrow{\text{rot}}\,\vec{\varepsilon}_i\right]^{\text{S}} = \left[\overrightarrow{\text{rot}}\,\vec{\alpha}_i\right]^{\text{S}} \qquad (4)$$

Geometro-compatibility of contortions and density of disclination charges

$$\vec{\theta}_i = \overrightarrow{\text{rot}}\left(\vec{\lambda}_i - \vec{e}_i \wedge \vec{\lambda} - \vec{e}_i \lambda\right) = \overrightarrow{\text{rot}}\,\vec{\chi}_i = \overrightarrow{\text{rot}}\left[\overrightarrow{\text{rot}}\,\vec{\varepsilon}_i\right]^{\text{T}} = \overrightarrow{\text{rot}}\left[\overrightarrow{\text{rot}}\,\vec{\alpha}_i\right]^{\text{T}} - \frac{1}{3}\overrightarrow{\text{rot}}\,\overrightarrow{\text{rot}}\left(\vec{e}_i \tau\right) \qquad (4)$$

Symmetry of the density tensor of disclination charges

$$\sum_k \vec{e}_k \wedge \vec{\theta}_k = \sum_k \vec{e}_k \wedge \overrightarrow{\text{rot}}\left(\vec{\lambda}_k - \vec{e}_k \wedge \vec{\lambda} - \vec{e}_k \lambda\right) = \sum_k \vec{e}_k \wedge \overrightarrow{\text{rot}}\,\vec{\chi}_k \equiv 0 \qquad (5)$$

Geometro-compatibility of flexions and density of charges of curvature

$$\theta = -\sum_k \vec{e}_k \vec{\theta}_k = \text{div}\,\vec{\lambda} = \text{div}\,\vec{\chi} = \sum_k \vec{e}_k \overrightarrow{\text{rot}}\,\overrightarrow{\text{rot}}\,\vec{\varepsilon}_k = \sum_k \vec{e}_k \overrightarrow{\text{rot}}\,\overrightarrow{\text{rot}}\,\vec{\alpha}_k - \frac{2}{3}\Delta\tau \qquad (6)$$

Equation of conservation of charges of disclination

$$\text{div}\,\vec{\theta}_i = \text{div}\left[\overrightarrow{\text{rot}}\left(\vec{\lambda}_i - \vec{e}_i \wedge \vec{\lambda} - \vec{e}_i \lambda\right)\right] = 0 \qquad (7)$$

Chapter 9

Topological singularities associated with the charges

Since the tensorial density $\vec{\lambda}_i$ of distortion charges must satisfy a conservation equation $\operatorname{div}\vec{\lambda}_i \equiv 0$, it is impossible for it to appear in a punctual way. It must always occupy a field of *nonzero volume* in the solid medium. We therefore treat mathematically the different possible topologies of charged domains, namely *strings, membranes, torus loops* and *clusters*, by illustrating them with quantified topological singularities such as *dislocations, disclinations* and *loops* that appear within a structured lattice.

On the other hand, writing of the tensor density of disclination charges $\vec{\theta}_i = \overrightarrow{\operatorname{rot}}\left(\vec{\lambda}_i - \vec{e}_i \wedge \vec{\lambda} - \vec{e}_i \lambda\right)$ implies that the appearance of disclinations in a place of a solid medium is a direct consequence of the presence in this place of dislocations, so that there cannot be isolated disclinations within a lattice.

Finally, we show that the vector density $\vec{\lambda}$ of flexion charges, the scalar density λ of rotation charges and the scalar density $\theta = \operatorname{div}\vec{\lambda}$ of curvature charges are the three fundamental quantities necessary and sufficient to describe the long-range effects of topological singularities localized within a lattice.

9.1 – Strings and dislocation lines

As the density of charges of distortion $\vec{\lambda}_i$ is a tensorial quantity, the non-zero density areas of charges must be extended in space, and their spatial topology is constrained in order to satisfy the conservation equation $\operatorname{div}\vec{\lambda}_i \equiv 0$. In this section, we will analyze a particular type of field topology of charges, namely tubular areas that can be called *strings.*

The dislocation strings and their charge density $\vec{\lambda}_i$

Suppose a region of a solid lattice, which is approximately in homogeneous volume expansion τ, that is to say wherein the spatial variation of the volume expansion τ is very small. This hypothesis is one that almost always applies to the case of real solids such as metals. But suppose also that significant rotations are possible at great distances in this solid, which is perfectly feasible in the case of real solids. If there exists within this solid a localized area in *string* form, a tubular shape, containing a non-zero density of dislocation charges, shown in figure 9.1 by a shaded area, it is possible to define in the frame $O\xi_1\xi_2\xi_3$ of ***GO*** a volume V partially containing this domain. The integral over the volume V of the conservation equation of tensorial charge density can be transformed into an integral over the surface surrounding the integration volume

$$\iiint_V \operatorname{div}\vec{\lambda}_i \, dV = \oiint_S \vec{\lambda}_i \, d\vec{S} \equiv 0 \qquad (9.1)$$

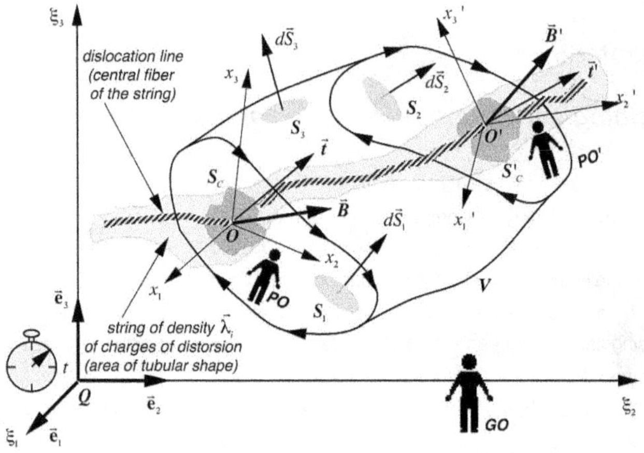

Figure 9.1 *- string (domain of tubular shape) of density $\vec{\lambda}_i$ of charges of distortion and the dislocation line (central fiber of the string)*

With the geometry described in figure 9.1, the surface S may be divided into a side surface S_3, on which the charge density is zero, and therefore does not contribute to the surface integral, and two cutting surfaces S_1 and S_2, areas of non-zero density of charges, so that the curved surface integral can be decomposed by the difference of surface integrals S_1 and S_2

$$\oiint_S \vec{\lambda}_i\, d\vec{S} = \iint_{S_2} \vec{\lambda}_i\, d\vec{S}_2 - \iint_{S_1} \vec{\lambda}_i\, d\vec{S}_1 \equiv 0 \qquad (9.2)$$

The integrals on S_1 and S_2 can be made in the local reference frames $Ox_1x_2x_3$ and $Ox_1'x_2'x_3'$ of observers ***PO*** and ***PO'*** respectively. According to the integral relationship of compatibility *(8.11)*, these integrals then represent the projections of the Burgers vector on the axes of the local frames $Ox_1x_2x_3$ and $Ox_1'x_2'x_3'$, so we have the following consequence of the equation of conservation of the tensorial density $\vec{\lambda}_i$ of charges of dislocation

$$B_i = -\iint_{S_1} \vec{\lambda}_i\, d\vec{S}_1 = -\iint_{S_2} \vec{\lambda}_i\, d\vec{S}_2 = B_i' \qquad (9.3)$$

This relationship has a very important topological result concerning the geometrical forms acceptable for a non-zero density field of charges $\vec{\lambda}_i$. Indeed, as the projections B_i of the Burgers vector $\vec{B}$ on the axes of local frames, projections defined in these local repositories by a transversal surface to the nonzero density of field $\vec{\lambda}_i$, these are invariant scalars throughout the area where density is not null, and thus this area must necessarily be continuous (connected). It cannot end abruptly in the solid as the scalar B_i must be kept on any plane intersecting the domain. Therefore, this area has the form of a *tubular string*, which must necessarily *cross the solid through-and-through*, or be a *loop-shaped ring*.

As for the Burgers vector $\vec{B}$, defined by the relationship $\vec{B} = \sum B_i \vec{e}_i$, *it is not necessarily an invariant vector in the absolute frame* along the dislocation since only its scalar components B_i

are invariant along the string, while the associated base vectors $\vec{e}_i$ associated with local frame may change along the string in the case where the medium undergoes strong rotations from one place to another. That is, if the volume expansion τ remains homogeneous, the vector $\vec{B}$ is of invariant norm in the absolute frame, but its direction can change along the string in exactly the same way as the local frames do along the string in the solid.

On the other hand, the Burgers vector $\vec{B}$ is a real invariant within any local frame $Ox_1x_2x_3$ defined along the dislocation since within the scope of a local frame, τ and $\vec{\omega}$ cannot present significant variations. In other words, *the Burgers vector $\vec{B}$ can be considered as an invariant in local frames*, and this regardless of the direction $\vec{t}$ that the dislocation can take in the local frame , where $\vec{t}$ is the unit vector tangent to the central fiber (figure 9.1).

In relation *(8.11)* giving the Burgers vector $\vec{B}$ of a dislocation from the tensorial density $\vec{\lambda}_i$, it is always possible to integrate on a surface S perpendicular to the direction $\vec{t}$ of the central fiber by writing $d\vec{S} = \vec{t}dS$ (figure 9.1)

$$\vec{B} = -\sum_i \vec{e}_i \left(\vec{t} \iint_S \vec{\lambda}_i \, dS \right) \tag{9.4}$$

Dislocation lines, or dislocations, and their linear charge $\vec{\Lambda}_i$

The field of tubular string of charges $\vec{\lambda}_i$ can be modeled in the form of a *dislocation line*, commonly called *dislocation*, represented by a one-dimensional core fiber in the center of the nonzero density $\vec{\lambda}_i$ of the string of charges (fiber is represented by line hatched in the drawing of figure 9.1). This *dislocation line* must necessarily either pass through the solid from end to end, or form a dislocation loop closed on itself. In the local frame, the vector $\vec{B}$ must remain constant, if ever the diameter C of the contour of integration is varied or this contour along the string is displaced, with the *sine qua non* condition that this contour is always located outside the central area containing the charge density $\vec{\lambda}_i$, and that the volume expansion is uniform in all the local frame. We deduce that if the dislocation string is sufficiently thin (of sufficiently small cross-section), the charge density $\vec{\lambda}_i$ can be represented by a magnitude confined to the immediate vicinity of the central fiber of the string, which will be called *dislocation line*, by introducing the concept of *linear tensorial charge of dislocation* $\vec{\Lambda}_i$, namely a set of three vectors defined on the central fiber by the following relationship

$$\vec{\Lambda}_i = \iint_{S_C} \vec{\lambda}_i \, dS \tag{9.5}$$

where the surface S'_C is the area of the core of the string. These relationships are then used to deduce the Burgers vector $\vec{B}$ from $\vec{\Lambda}_i$, as well as its components B_i (which are invariant along the dislocation line)

$$\vec{B} = \sum_i B_i \vec{e}_i = -\sum_i \left(\vec{t} \vec{\Lambda}_i \right) \vec{e}_i \quad \Rightarrow \quad B_i = -\vec{t} \vec{\Lambda}_i \tag{9.6}$$

To satisfy the relations *(9.6)*, it is necessary and sufficient to state that the vectors $\vec{\Lambda}_i$ representing the linear tensorial charge of the dislocation line have *conservative tangential components* $\vec{\Lambda}_i$ that satisfy the following relationships

$$\vec{\Lambda}_i = -\vec{t} B_i \tag{9.7}$$

Thus, regardless of the direction $\vec{t}$ of the central fiber of a dislocation, the components $\vec{\Lambda}_i$, tangential to the dislocation line are directly deducted from the knowledge of the components $\boldsymbol{B}_i$ of the Burgers vector $\vec{\boldsymbol{B}}$ of the string.

From the *trace* of the linear tensorial charge $\vec{\Lambda}_i$ we deduce linear scalar charge of rotation of the dislocation, using the relations *(8.17)* and *(9.7)*, and note that this scalar charge depends on the scalar dot product of the Burgers vector $\vec{\boldsymbol{B}}$ by the natural vector $\vec{t}$ tangent to the dislocation line

$$\Lambda = \frac{1}{2}\sum_i \vec{e}_i\vec{\Lambda}_i = \frac{1}{2}\sum_i \vec{e}_i\left(-\vec{t}\,\boldsymbol{B}_i\right) = -\frac{1}{2}\vec{\boldsymbol{B}}\vec{t} \tag{9.8}$$

As for the *antisymmetric part* of the linear tensorial charge $\vec{\Lambda}_i$, it can be represented by a *linear vector charge of flexion* of the dislocation, which depends on the cross product of the vector $\vec{\boldsymbol{B}}$ by the natural vector tangent to the dislocation line $\vec{t}$

$$\vec{\Lambda} = -\sum_i \vec{e}_i \wedge \vec{\Lambda}_i = -\sum_i \vec{e}_i \wedge \left(-\vec{t}\,\boldsymbol{B}_i\right) = \vec{\boldsymbol{B}} \wedge \vec{t} \tag{9.9}$$

Thus, the relative orientation of the Burgers vector $\vec{\boldsymbol{B}}$ of the dislocation and of the natural tangent vector $\vec{t}$ to its central fiber plays an important role in the expression of different linear tensorial charges, and actually determines the nature of the dislocation line.

Decomposing the Burgers vector into its parallel part $\vec{\boldsymbol{B}}_{\|}$ and perpendicular part $\vec{\boldsymbol{B}}_{\perp}$ to the tangent vector $\vec{t}$, we have the following relationship

$$\vec{\boldsymbol{B}} = \underbrace{\left(\vec{\boldsymbol{B}}\vec{t}\right)\vec{t}}_{\vec{\boldsymbol{B}}_{\|}} + \underbrace{\vec{t} \wedge \left(\vec{\boldsymbol{B}} \wedge \vec{t}\right)}_{\vec{\boldsymbol{B}}_{\perp}} = -2\Lambda\vec{t} + \vec{t} \wedge \vec{\Lambda} \tag{9.10}$$

which shows that a dislocation can be perfectly and completely described by the data of the *linear charge of rotation* Λ, which is linked to the parallel portion $\vec{\boldsymbol{B}}_{\|}$ of the Burgers vector, and its *linear flexion charge* $\vec{\Lambda}$, which is related to the perpendicular part $\vec{\boldsymbol{B}}_{\perp}$ of the Burgers vector.

On the quantification and taxonomy of dislocations in a lattice

The connected areas of non-zero dislocation density charges can be modeled in the simplest way as thin strings. In this section, we will show that the straight strings appearing in a solid lattice are *quantified* on a microscopic scale (fig. 9.2 and 9.3), and that these strings then represent *elementary plastic distortion singularity fields*, ie "*elementary particles*" of the plastic deformation of the lattice.

If we consider the case of an ordered array of particles on the microscopic scale, one can introduce dislocations by cutting links in a plane of the lattice, parallel movement of the jaws and reconstruction of bonds, as shown in figures 9.9 and 9.10 in the case of a simple cubic lattice.

The Burgers vector $\vec{\boldsymbol{B}}$ of singularities thus obtained is deduced by considering a closed circuit $\boldsymbol{C}$ on the solid, which surrounds the singularity, and seeking the closure $\vec{\boldsymbol{B}}$ of the corresponding open circuit in the undistorted virtual lattice.

With figures 9.2 and 9.3 , we see that the microscopic singularities of the lattice have an essential characteristic: *their Burgers vector is quantized*, that is to say, its components can only be integer multiples of the unit cell length a of the lattice.

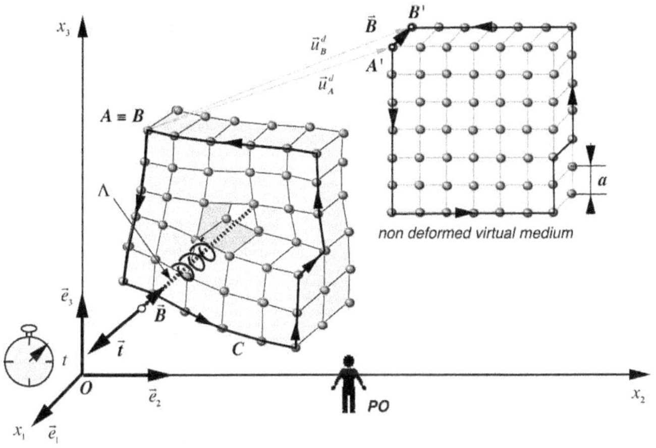

Figure 9.2 *- screw dislocation is quantified in a cubic lattice*

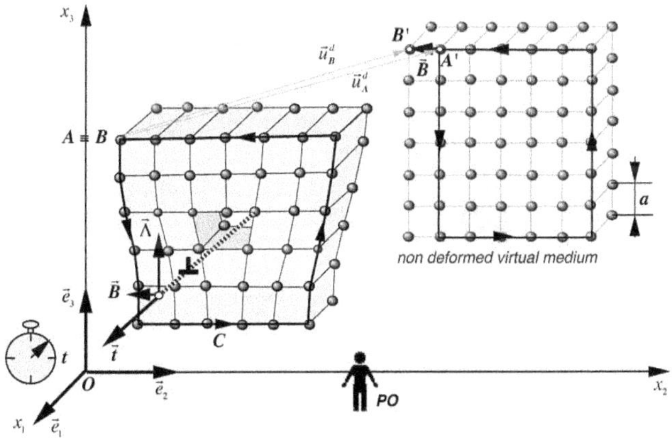

Figure 9.3 *- edge dislocation is quantified in a cubic lattice*

The nature of the microscopic plastic singularity can change according to the respective directions being taken in the local coordinate system, the Burgers vector $\vec{B}$ and the unit vector $\vec{t}$ tangent to the line:

- when $\vec{B}$ is parallel to $\vec{t}$ (fig. 9.2) , the *linear dislocation charge* $\vec{\Lambda}_i$ has a non-zero trace $(\Lambda \neq 0)$, so has rotational charge, and zero antisymmetric part $(\vec{\Lambda} = 0)$. One speaks in this case of *screw dislocations*[1], and of linear charge of rotation Λ of the screw dislocation. We

[1] *J. M. Burgers, Proc. Kon. Ned. Akad. Weten, vol.42, p. 293, 378, 1939*

symbolically represent it by a screw on the dislocation line. As $\Lambda = -\vec{B}\vec{t}\,/\,2$ when the screw dislocation has a *right rotation*, identical to the rotational direction of a normal screw or corkscrew, Λ is positive and vectors $\vec{t}$ and $\vec{B}$ are oriented in opposite directions. On the other hand, if the screw dislocation has a *left rotation*, identical to the rotational direction opposite to that of a normal screw or a corkscrew, Λ is negative and vectors $\vec{t}$ and $\vec{B}$ are oriented in the same direction (fig. 9.4). Note that the choice of a given direction $\vec{t}$ is quite arbitrary as only the sign of Λ is fixed.

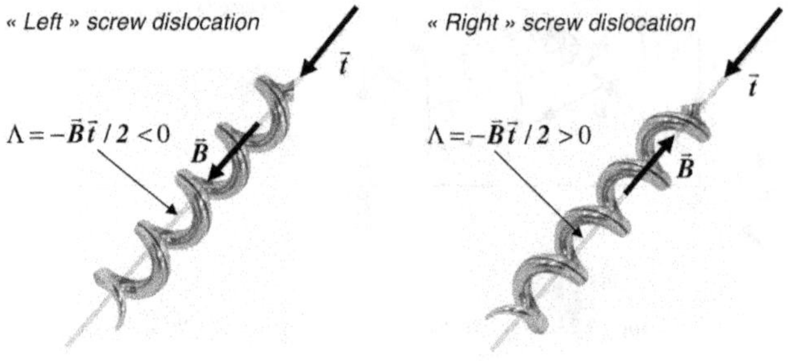
« Left » screw dislocation
$\vec{t}$
$\Lambda = -\vec{B}\vec{t}/2 < 0$
$\vec{B}$
« Right » screw dislocation
$\vec{t}$
$\Lambda = -\vec{B}\vec{t}/2 > 0$
$\vec{B}$

Figure 9.4 *- screw dislocation, left and right respectively*

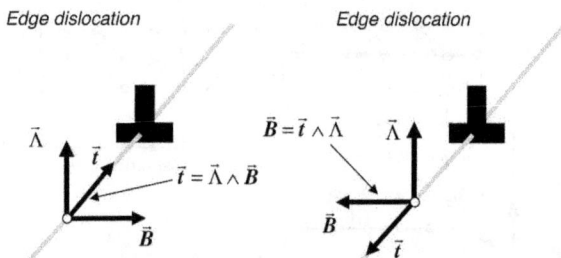

Edge dislocation
$\vec{\Lambda}$
$\vec{t}$
$\vec{t} = \vec{\Lambda} \wedge \vec{B}$
$\vec{B}$
Edge dislocation
$\vec{B} = \vec{t} \wedge \vec{\Lambda}$
$\vec{\Lambda}$
$\vec{B}$
$\vec{t}$

Figure 9.5 *- edge dislocation*

- when $\vec{B}$ is perpendicular to $\vec{t}$ (fig. 9.3), the linear dislocation charge $\vec{\Lambda}_i$ has zero trace ($\Lambda = 0$)), so no charge of rotation, but a non-zero antisymmetric part $(\vec{\Lambda} \neq 0)$. One speaks in this case of *edge dislocations*[2], and of *linear flexion charge* $\vec{\Lambda}$ *of the edge dislocation,* and symbolically we represent the latter by a sign $\perp$ on the dislocation line, oriented so as to represent the additional particle plane. The vector $\vec{\Lambda}$ always has the direction of the additional plane $\perp$ of the edge dislocation. This is the only fixed size, so that $\vec{B} = \vec{t} \wedge \vec{\Lambda}$ if the direction $\vec{t}$ is given, and $\vec{t} = \vec{\Lambda} \wedge \vec{B}$ if the Burgers vector $\vec{B}$ is given (fig. 9.5).
- when $\vec{B}$ is neither parallel nor perpendicular to $\vec{t}$ the *linear charge of dislocation* $\vec{\Lambda}_i$ has a

[2] *E. Orowan, Z. Phys., vol. 89,p.605,614 et 634, 1934; M. Polanyi, Z. Phys., vol.89, p. 660, 1934; G. I. Taylor, Proc. Roy. Soc. London, vol. A145, p. 362, 1934*

non-zero trace $(\Lambda \neq 0)$, but also a non-zero anti-symmetric part $(\vec{\Lambda} \neq 0)$, such that it behaves both as a source of rotations and elastic and anelastic flexion. We speak in this case of *mixed dislocations.*

In a discrete lattice, a dislocation may well change direction. In other words, along the dislocation line, the tangent vector $\vec{t}$ is not necessarily preserved. In this case, as the Burgers vector $\vec{B}$ is conserved in the local frame, this means that the dislocation should change type. For example, figure 9.6 shows a simple cubic lattice model in which a screw dislocation enters the front of left, turns inside the lattice by becoming mixed, and emerges as an edge dislocation on the adjacent face to the right.

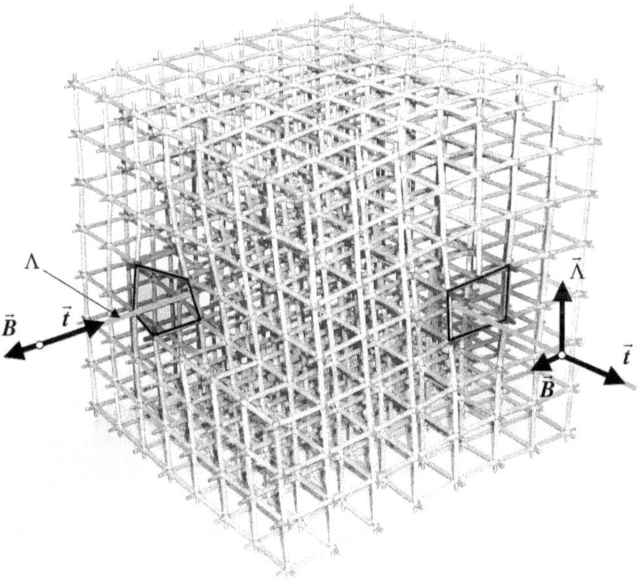

Figure 9.6 - *dislocation changing from a screw type to an edge type in a cubic lattice*

The lattice dislocations, the «elementary particles» of the plastic deformation

Quantized dislocations are the most basic vectors of the plastic deformation of a lattice. In this sense, they might be called the *"elementary particles"* of plastic deformation. Moreover, any dislocation string has its *"anti-string"*. Indeed, it is easy to see that two parallel dislocations in the same direction $\vec{t}$ and Burgers vectors $\vec{B}$ and $-\vec{B}$ respectively, annihilate completely if they come to meet within the lattice.

Using an integration of the compatibility relationship *(8.17)* on a volume V surrounding a screw dislocation on a length L (fig. 9.7) we have

$$\Lambda = \frac{1}{L}\iiint_V \lambda dV = \frac{1}{L}\iiint_{V_0} \lambda dV = \frac{1}{L}\iiint_V \operatorname{div}\vec{\omega} dV = \frac{1}{L}\oiint_S \vec{\omega} d\vec{S} = \frac{1}{L}\iint_{S_3} \vec{\omega} d\vec{S}_3 \qquad (9.11)$$

We notice that screw dislocations, carrying a linear scalar charge $\Lambda \neq 0$, are sources of a *field of divergent local rotations*, which is, as we have seen, the analog of the electric field. Thus, at a distance R of the string, the norm of the rotation field is simply $|\vec{\omega}| = |\Lambda| / 2\pi R$.

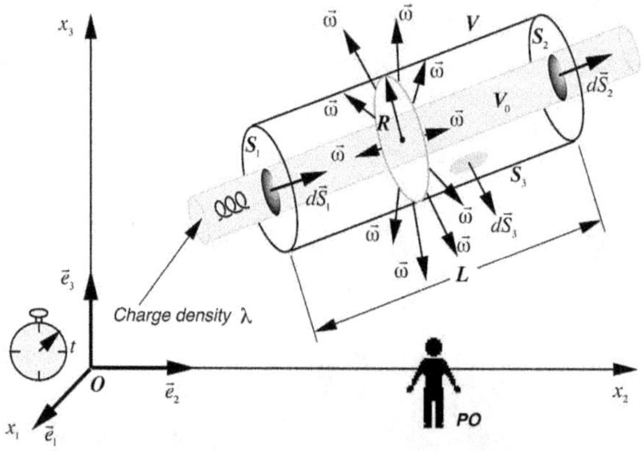

Figure 9.7 *- the radial field of rotation around a screw dislocation*

As to the edge dislocations, which are carriers of a flexion charge $\vec{\Lambda} \neq 0$, they are a source of a *lattice flexion*, and therefore of a local curvature of the lattice in their vicinity as the flexion vector $\vec{\chi}$ satisfies the relationship $\vec{\chi} = -\overrightarrow{\text{rot}}\,\vec{\omega} + \vec{\lambda}$.

On the importance of the microscopic structure of the lattice

The *microscopic structure* of a solid plays a significant role on the nature of dislocations that may appear in this solid, but it is beyond the scope of this book to describe in detail these microscopic aspects. In the excellent book[3] by Hirth and Lothe, the reader will find a detailed description of most of these aspects of lattice dislocations. One can also find an excellent history of the discovery of dislocations in the twentieth century in the article by Hirth[4] (1985).

In fact, it is essentially the structure of the core of the quantized dislocation which will depend on the microscopic structure of the network. A first important effect of the periodic structure of the lattice is that the energy of a line dislocation depends strongly on the direction in which the line points in the lattice. For example, consider an edge dislocation like the one shown in figure 9.3. It seems quite clear that the movement of its position along the axis Ox_2 must change the energy of this dislocation periodically with a spatial frequency which is that of the lattice. The minimum and maximum energy positions are respectively called Peierls valleys and hills, named Peierls[5] who was the first in 1940 to have considered this problem. For example, to move a

[3] *J. P. Hirth and J. Lothe, theory of dislocations, second edition, John Wiley & Sons, New-York, 1982*

[4] *J.-P. Hirth, A Brief History of Dislocation Theory, Metallurgical Transactions A, vol. 16A, p.2085, 1985*

[5] *R. E. Peierls, Proc. Phys. Soc., vol. 52, p. 23, 1940*

dislocation from a Peierls valley to another, we must then apply a force called the *Peierls-Nabarro force*[6].

It is also conceivable that a dislocation is oblique to the Peierls valleys, and that it must therefore overcome these valleys. In this case, it is logical to think that the oblique line dislocation does not necessarily have the lowest energy configuration, and the dislocation could be interested in taking a polygonal shape by the formation of *kinks*[7] that minimize the portions of the dislocation lying on the hills of Peierls, and therefore help to minimize the overall energy (fig.9.8).

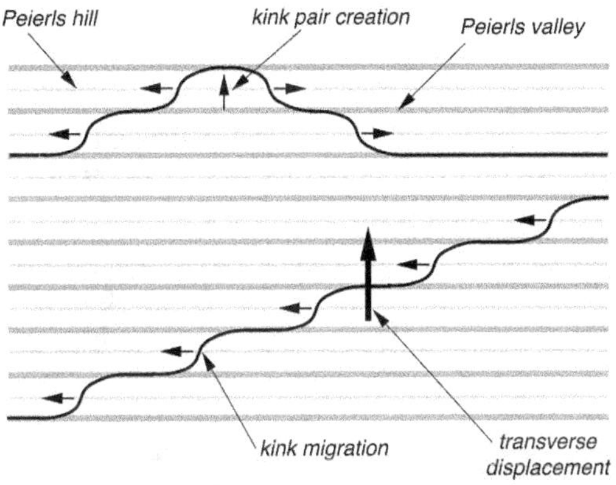

***Figure 9.8** - The kinks and their mobility within the Peierls hills*

The existence of these kinks also has the advantage of allowing the mobility of the dislocation perpendicularly to the valleys Peierls simply by the movement of the kinks along the dislocation line (fig. 9.8). In conventional materials such as metals of different structures we can change the mobility of dislocations through a process of creation and annihilation of pairs of kinks, a process that is usually thermally activated, and is responsible for some of the macroscopic anelastic and plastic properties of these materials. These processes have been extensively discussed in the literature [3].

On the existence of partial dislocation and the 'strong force' that bind them

Dislocations appearing in a more complex structures than the simple cubic lattice, such as face-centered cubic lattices, cubic or hexagonal centered lattices, generally have much more complicated structures within them. Due to energetic reasons and according to the crystal system considered, there will appear a dissociation of the core of the dislocation into two or more *partial dislocations*[8] whose Burgers vectors are individual fractions of the lattice translation vectors.

[6] *F. R. N. Nabarro, Proc. Phys. Soc. London, vol. 59, p. 256, 1947*

[7] *J. Weertman, Phys. Rev., vol. 101, p. 1429, 1956; A. Seeger, Phil. Mag., vol.1, p. 651, 1956*

[8] *J. Frenkel, T. Kontorava, Fiz. Zh., vol.1 , p. 137, 1939; R. D. Heidenreich, W. Shockley, Report of Conf. on Strength of Solids, Phys. Soc., London, p. 57, 1948; F. C. Frank, Proc. Phys. Soc. London, vol. 62A, p. 202, 1949; N. Thompson, Proc. Phys. Soc. London, vol. 66B, P. 481, 1953*

For example, in the face-centered cubic metals (FCCs), the stacking of atoms is characterized by sequences *abc abc abc* ... (fig. 9.9). The Burgers vector $\vec{B}_p$ of a perfect dislocation should in principle be connecting two lattice nodes. But for energetic reasons, the most favorable Burgers vectors are those who have a minimum length because the distortion energy stored in the lattice by a dislocation is proportional to the square of the Burgers vector as we will discuss later. Thus in the case of figure 9.9, the dislocations dissociate on their sliding plane into two partial of Burgers vectors $\vec{B}_I$ and $\vec{B}_2$, so that $\vec{B}_p = \vec{B}_I + \vec{B}_2$. In the case of this dissociation $\vec{B}_1\vec{B}_2 > 0$ so we have $\vec{B}_p^2 = \vec{B}_1^2 + \vec{B}_2^2 + 2\vec{B}_1\vec{B}_1 > \vec{B}_1^2 + \vec{B}_2^2$.

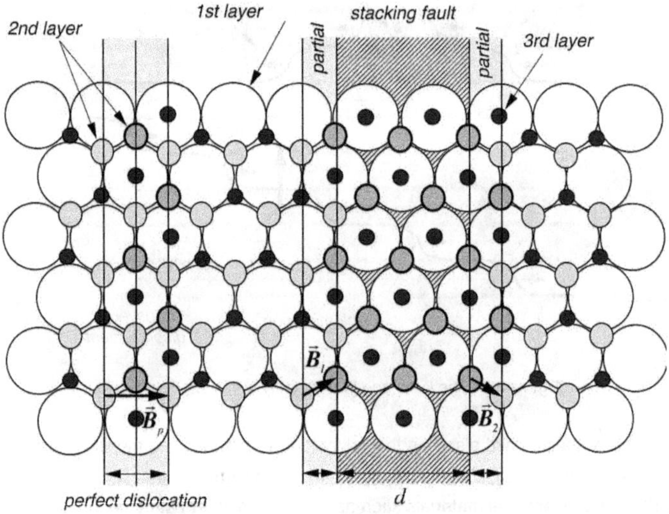

Figure 9.9 *- partial dislocations and ribbon of stacking faults in a FCC*

The two partials obtained in figure 9.9 by the dissociation are called of *Shockley type* [8]. The distance between the two partials is then controlled by a competition between the energy decrease associated with the increased distance between the partials that repel and the energy increase due to the formation of an energetic ribbon of stacking faults (abc abc abc ac ...) between the two partial dislocations, as illustrated by figure 9.10.

As the ribbon has a stacking fault energy γ per unit area, the total energy $E_t(d)$ per unit length of dislocation for a dissociated dislocation over a distance d is written $E_t(d) = \gamma d + E_d(d)$, where $E_d(d)$ is the energy of the two partials depending on the distance that separates them, which is a monotonically decreasing function, decreasing from E_0 for $d = 0$ to $2E_0/3$ pour $d \to \infty$ in the case of the Shockley partials illustrated in figure 9.9.

The energy $E_t(d)$, therefore, presents a minimum for the distance $d = d_0$ (fig. 9.10), which is the equilibrium distance between the two partials, controlled by the competition between the energy decrease associated with the increase in distance between the partials and the energy increase due to the formation of a ribbon of stacking faults between the two partials. This behavior of the energy $E_t(d)$ induces an interaction force between the two partials that could be called a "*strong force*" in the sense that the energy of the pair of partials has a minimum which fixes the position of equilibrium d_0 , but increases if one tries to increase the separation distance beyond it. The "strong force" qualifier proposed here as the behavior of the interaction force between the partials presents an interesting analogy with the strong force acting between quarks in the Standard Model of elementary particles as we shall see later.

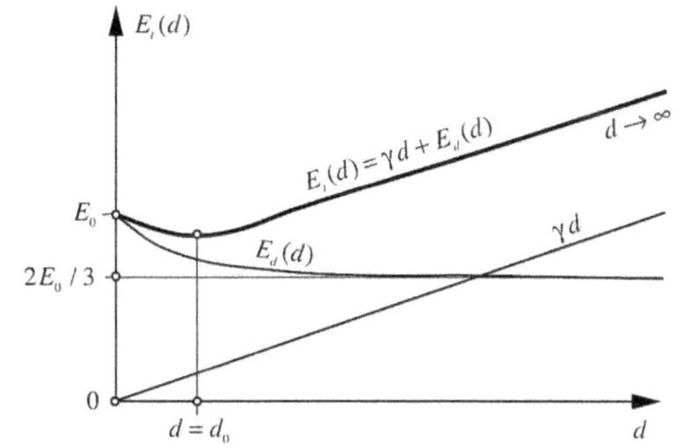

Figure 9.10 *- The energy of two partial dislocations as a function of the distance that separates them*

As an example, consider the illustration (fig. 9.11) of a model of a mixed dislocation (with edge and screw at once) separated into two partial in a face-centered cubic structure. It clearly shows the existence of a stacking fault between the two partials. And since this is a mixed dislocation, both partials present a series of steps (kinks), such as we have described previously. Moreover, one can even observe network flexion induced by the edge portion of the dissociated dislocation.

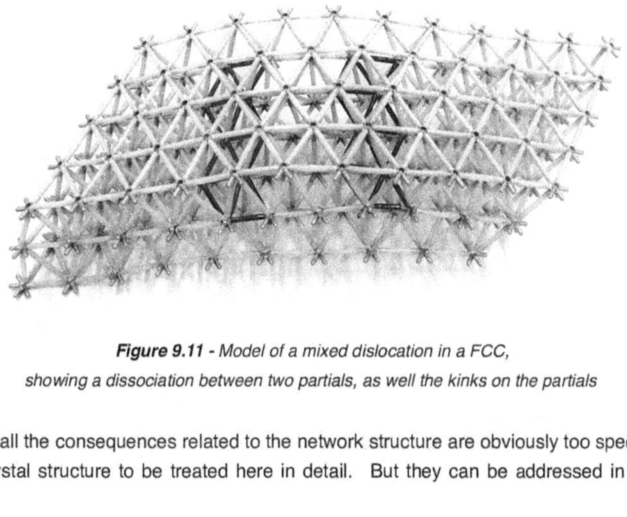

Figure 9.11 *- Model of a mixed dislocation in a FCC,*
showing a dissociation between two partials, as well the kinks on the partials

The set of all the consequences related to the network structure are obviously too specific to the type of crystal structure to be treated here in detail. But they can be addressed in any book dealing with dislocations in the crystal structures.

The effects of a spatial variation of volume expansion

In the case of imaginary medias in which the volume expansion is no longer uniform, say with large variations of τ, it is not possible to represent a dislocation of the lattice by a linear density of charge $\vec{\Lambda}_i$ focused on a line on the center of the dislocation. Indeed, in this case, the integral over a surface S surrounding the center of the dislocation

$$\vec{B} = -\sum_i \vec{e}_i \left(\vec{t} \iint_S \vec{\lambda}_i \, dS \right) \tag{9.12}$$

will not give the same result at different locations along the dislocation line, and does not give the same result according to the size of the selected area S, which implies that the charge density $\vec{\lambda}_i$ is not localized in the center of dislocation, but is now disseminated in a tube more or less bulky around the center, so that the integral (9.12) gives the actual Burgers vector associated with S, the selected surface in the local referential $Ox_1x_2x_3$ we used to measure it (fig. 9.12).

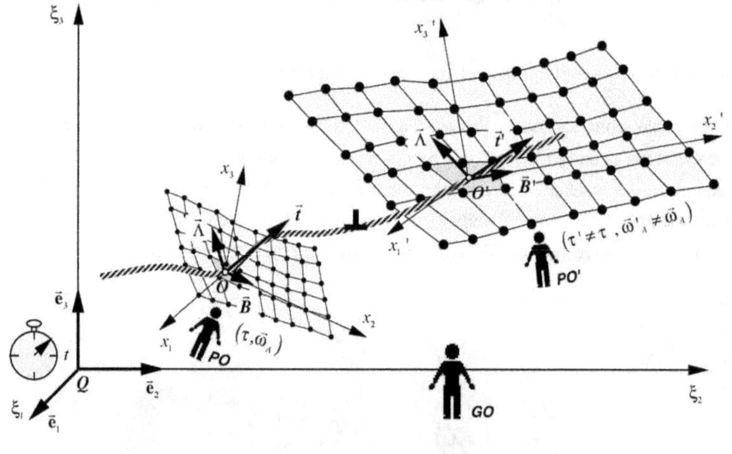

Figure 9.12 *- effects of a spatial variation of the volume expansion on a lattice dislocation*

But it is clear that throughout this tube, the condition (9.2) remains valid if the surfaces S_1 and S_2 are chosen broadly enough to encompass the charge density $\vec{\lambda}_i$ surrounding the core of the lattice dislocation. It goes without saying that this kind of situations where the volume expansion and local rotations are not homogeneous within a large portion of the solid with dislocations, is much more difficult to treat than the case where there is homogeneity of τ and $\vec{\omega}$ over great distances or when the lattice dislocations are confined to small enough regions so that they can be considered homogeneous.

However, it will always be possible to describe locally any dislocation as a localized string in a coordinate system $Ox_1x_2x_3$ as shown in figure 9.12, by using locally relations *(9.8)* to *(9.10)* to calculate the linear densities $\vec{\Lambda}_i$,$\vec{\Lambda}$ and Λ , and by assuming that the volume expansion τ may be regarded as homogeneous in the referential $Ox_1x_2x_3$. In this case, the Burgers vector length is given locally by the relation

$$\left|\vec{B}\right|_{\tau\neq 0} = \left|\vec{B}_0\right|_{\tau=0} \mathrm{e}^{\tau/3} \tag{9.13}$$

so that the linear density in a region where the volume expansion τ is not zero will be connected to the same linear density of the dislocation at $\tau = 0$ by the relations

$$\left|\vec{\Lambda}_i\right|_{\tau\neq 0} = \left|\vec{\Lambda}_{0i}\right|_{\tau=0} \mathrm{e}^{\tau/3} \quad ; \quad \left|\vec{\Lambda}\right|_{\tau\neq 0} = \left|\vec{\Lambda}_0\right|_{\tau=0} \mathrm{e}^{\tau/3} \quad ; \quad \left|\Lambda\right|_{\tau\neq 0} = \left|\Lambda_0\right|_{\tau=0} \mathrm{e}^{\tau/3} \tag{9.14}$$

Since linear density $\vec{\Lambda}_i$ is derived as the product of the volume density $\vec{\lambda}_i$ by a surface, it is deduced that the volume density $\vec{\lambda}_i$ in a region compressed or expanded to a non-zero value of τ is connected to the initial volume density of the same region with $\tau = 0$ by the following relations

$$\left|\vec{\lambda}_i\right|_{\tau\neq 0} = \frac{\left|\vec{\Lambda}_i\right|_{\tau\neq 0}}{\left|S\right|_{\tau\neq 0}} = \frac{\left|\vec{\Lambda}_{0i}\right|_{\tau=0} \mathrm{e}^{\tau/3}}{\left|S_0\right|_{\tau=0} \mathrm{e}^{2\tau/3}} = \left|\vec{\lambda}_{0i}\right|_{\tau=0} \mathrm{e}^{-\tau/3} \tag{9.15}$$

We also deduce that

$$\left|\vec{\lambda}\right|_{\tau\neq 0} = \left|\vec{\lambda}_0\right|_{\tau=0} \mathrm{e}^{-\tau/3} \quad ; \quad \left|\lambda\right|_{\tau\neq 0} = \left|\lambda_0\right|_{\tau=0} \mathrm{e}^{-\tau/3} \tag{9.16}$$

On the impossibility of finding isolated disclination lines

It can be shown that there cannot exist *isolated disclination lines* in a solid, i.e. charged tubes whose overall Frank vector $\vec{\Omega}$ is not zero. Indeed, *as the density of dislocation charges $\vec{\lambda}_i$ and disclination charges $\vec{\theta}_i$ are confined in the same tube*, it is possible to replace $\vec{\theta}_i$ by its expression $\vec{\theta}_i = \overrightarrow{\mathrm{rot}}\left[\vec{\lambda}_i - \vec{e}_i \wedge \vec{\lambda} - \vec{e}_i\lambda\right]$ in equation (8.34). It is then found that the Frank vector $\vec{\Omega}$ may be derived by the integral over the contour C_1 (figure 9.1) of the density $\vec{\lambda}_i - \vec{e}_i \wedge \vec{\lambda} - \vec{e}_i\lambda$, and that this density is zero on this contour on the outside of the tube, so that $\vec{\Omega} = 0$. It therefore follows that there cannot be isolated charged tube whose overall Frank vector $\vec{\Omega}$ is not zero.

9.2 – Membranes of dislocation and torsion, flexion and accommodation boundaries

We call *charged membrane* a thin interface containing charges, which separates two media containing no charges (fig. 9.13). It is clear that these membranes can be any surfaces in space (infinite surfaces, closed spheroidal surfaces or torus, ribbons or hollow tubes, thin plates, etc.), with the only topological condition that on any point of the membrane the dislocation charge conservation equation is satisfied $\mathrm{div}\,\vec{\lambda}_i \equiv 0$ and with the charges of disclination derived from the dislocation charges via the equation $\vec{\theta}_i = \overrightarrow{\mathrm{rot}}\left[\vec{\lambda}_i - \vec{e}_i \wedge \vec{\lambda} - \vec{e}_i\lambda\right]$. *Passthrough conditions* for topological tensor through such a charged membrane are derived from the geometro-compatibility equations and are easy to establish. It is sufficient to write the full compatibility of the equation taking on a thin volume, with the movable membrane, of which each face is located on either side of said membrane (Fig. 9.13).

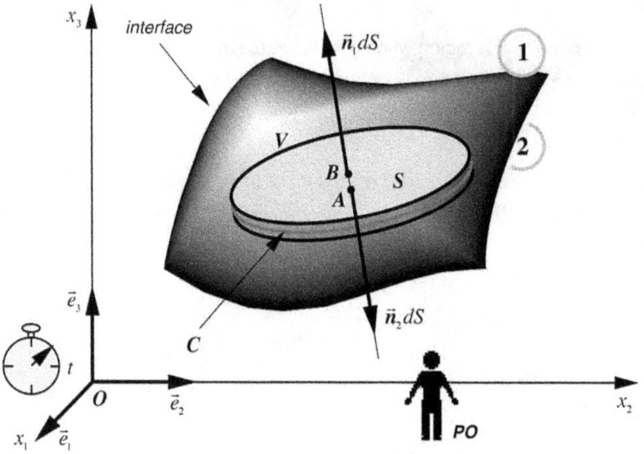

***Figure 9.13** - bi-dimensional charged membrane*

The surface charge $\vec{\Pi}_i$ of dislocation of a joint in the presence of gradients of $\vec{B}$

To establish the passthrough conditions for the distortion tensor, we must integrate its compatibility equation in the volume V of the membrane, which contains the dislocation density $\vec{\lambda}_i$

$$\iiint_V \overrightarrow{\text{rot}}\, \vec{\beta}_i\, dV = \iiint_V \vec{\lambda}_i\, dV = \oiint_S \left(d\vec{S} \wedge \vec{\beta}_i\right) \tag{9.17}$$

Provided that the membrane is sufficiently thin, the surface integral can be written as the sum of the surface integrals on both sides of S and the volume integral can be decomposed as follows

$$\oiint_S \left(d\vec{S} \wedge \vec{\beta}_i\right) = \iint_{S_1} \left(\vec{n}_1 \wedge \vec{\beta}_i^{(1)}\right) dS + \iint_{S_2} \left(\vec{n}_2 \wedge \vec{\beta}_i^{(2)}\right) dS = \iint_S dS \int_A^B \vec{\lambda}_i\, dl \tag{9.18}$$

wherein $\vec{n}_1$ and $\vec{n}_2$ are the unit vectors normal to the surface from the membrane on each side of the membrane.

Thus, if the charged membrane is very thin, it is possible to introduce the concept of *surface charge of dislocation* $\vec{\Pi}_i$ by writing

$$\vec{\Pi}_i = \int_A^B \vec{\lambda}_i\, dl \tag{9.19}$$

which allows us to deduce that

$$\vec{n}_1 \wedge \vec{\beta}_i^{(1)} + \vec{n}_2 \wedge \vec{\beta}_i^{(2)} = \vec{\Pi}_i \tag{9.20}$$

The topological interpretation of this passthrough condition is simple: the existence of a surface charge of dislocation $\vec{\Pi}_i$ in the membrane leads to a discontinuity of the tangential components of the distortion vector $\vec{\beta}_i$ on both sides of it. But on both sides of the membrane, outside the volume, the medium contains no plastic charges, so that the tensor $\vec{\beta}_i$ must derive from the gradient of the components of the displacement field $\vec{u}$, i.e. $\vec{\beta}_i = \overrightarrow{\text{grad}}\, u_i$. It is therefore possible to write

$$\vec{n}_1 \wedge \overrightarrow{\text{grad}}\, u_i^{(1)} + \vec{n}_2 \wedge \overrightarrow{\text{grad}}\, u_i^{(2)} = \vec{\Pi}_i \tag{9.21}$$

In the local framework, this expression implies that the presence of the charged membrane locally induces a dislocation of the displacement field $\vec{u}$ associated with the elastic and anelastic distortions whose Burgers vector $\vec{B}$ is then given by

$$\vec{B} = \int_A^B \delta\vec{u} = \vec{u}^{(1)} - \vec{u}^{(2)} \tag{9.22}$$

so that

$$\vec{\Pi}_i = \vec{n}_1 \wedge \overrightarrow{\text{grad}}\, u_i^{(1)} - \vec{n}_1 \wedge \overrightarrow{\text{grad}}\, u_i^{(2)} = \vec{n}_1 \wedge \overrightarrow{\text{grad}}\, \boldsymbol{B}_i \tag{9.23}$$

According to this relationship, the existence of a nonzero surface charge $\vec{\Pi}_i$ in the membrane is therefore subject to the condition that *there is a gradient of the components of the Burgers vector on the surface of the membrane.* This means among others that if the Burgers vector is constant over the whole membrane, meaning if the membrane is a simple translation of the two surfaces S_1 and S_2 relative to each other, the surface density $\vec{\Pi}_i$ is zero.

Finally, it is noted that the three vectors $\vec{\Pi}_i$ making up the surface tensor charge *must be tangent vectors to the surface of the membrane*, which is in fact a direct result of the dislocation charge conservation equation $\operatorname{div}\vec{\lambda}_i \equiv 0$. Indeed, considering the integral over a volume V in figure 9.13 we have

$$\iiint_V \operatorname{div}\vec{\lambda}_i\, dV = \oiint_S \vec{\lambda}_i\, d\vec{S} = \iint_{S_1} \vec{\lambda}_i\, \vec{n}_1\, dS + \iint_{S_2} \vec{\lambda}_i\, \vec{n}_2\, dS \equiv 0 \tag{9.24}$$

which can be satisfied for any volume V which closely surrounds the charged interface only if the vectors $\vec{\lambda}_i$ are perpendicular to $\vec{n}_1$ and $\vec{n}_2$ on either side of the interface, in other words if they are tangent to the surface of the membrane.

As in the case of one-dimensional dislocation lines, one can extract the trace and the anti-symmetric part of the tensor $\vec{\Pi}_i$ of dislocation surface charge, by writing

$$\begin{cases} \vec{\Pi} = -\sum_i \vec{e}_i \wedge \vec{\Pi}_i = -\sum_i \vec{e}_i \wedge \left(\vec{n}_1 \wedge \overrightarrow{\text{grad}}\, \boldsymbol{B}_i\right) \\ \Pi = \dfrac{1}{2}\sum_i \vec{e}_i\, \vec{\Pi}_i = \dfrac{1}{2}\sum_i \vec{e}_i \left(\vec{n}_1 \wedge \overrightarrow{\text{grad}}\, \boldsymbol{B}_i\right) \end{cases} \tag{9.25}$$

The two-dimensional modeling of a thin membrane obtained by calculating the surface charges is usually called a *joint* or a *boundary*[9]. The boundary is then fully characterized by the given surface tensor $\vec{\Pi}_i$ of charges of dislocation, whose vectors are tangential to the surface. But it can also be given by the anti-symmetric part $\vec{\Pi}$ (the surface charge of a flexion boundary) and the trace Π (the surface charge of a rotation boundary) of the tensor of charges $\vec{\Pi}_i$, as in the

---

[9] *C. Somigliana, Atti. Accad. naz. Lincei Rc., vol. 23, p. 463, 1912, et vol. 24, p.655, 1915; W. T. Read, W. Shockley, Phys. Rev. , vol. 78, p. 275, 1950; F. C. Frank, in Report of the Conf. on Defects in Crystalline Solids, Phys. Soc. London, p. 159, 1955*

case of the one-dimensional lines of dislocation. This point is well illustrated in Figure 9.14, wherein we show three thin membranes whose Burgers vectors linearly increase along the axis Ox_1, and which are respectively oriented along the axes Ox_3 and Ox_1. As these thin membranes of dislocation actually allow us to disorient or to accomodate the solid grains located on either side of the membrane, they are generally called *grain boundaries*.

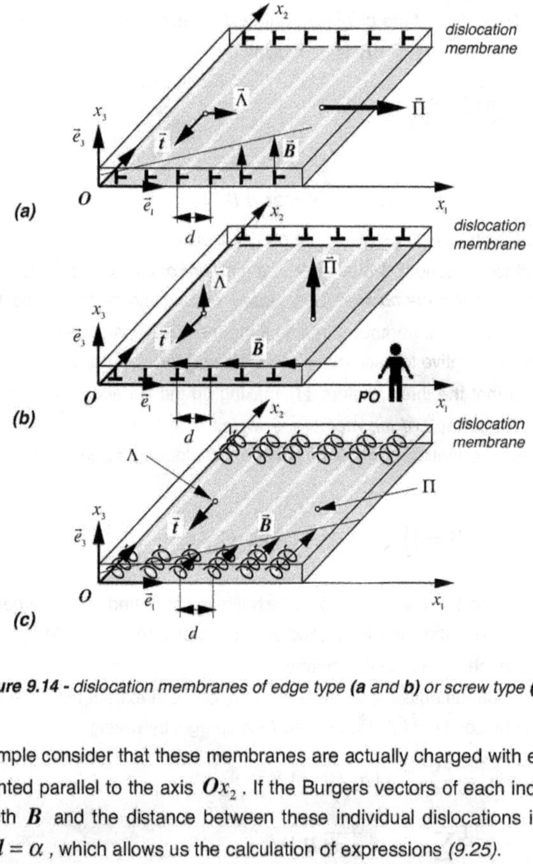

Figure 9.14 *- dislocation membranes of edge type* ***(a** and **b)*** *or screw type* ***(c)***

One can for example consider that these membranes are actually charged with edge and screw dislocations oriented parallel to the axis Ox_2. If the Burgers vectors of each individual dislocation have a length B and the distance between these individual dislocations is d, we have, $|\overrightarrow{\mathrm{grad}}\, B_i| = B / d = \alpha$, which allows us the calculation of expressions *(9.25)*.

But we can also simply represent each individual dislocation by a line charge vector $\vec{\Lambda}$ if we are dealing with an edge dislocation or by a line charge scalar Λ if dealing with a screw dislocation. It then verifies that:

- the edge-type thin membrane with a Burgers vector perpendicular to the surface and increasing along the axis Ox_1 (fig. 9.14a) can be completely characterized by a *vector* $\vec{\Pi}$ *of surface charges of flexion* $\vec{\Pi}$, which is a vector tangential to the plane of the membrane, directed along Ox_1, and worth

$$\vec{\Pi} = \vec{\Lambda} / d \quad ; \quad \vec{\Pi} \,||\, Ox_1 \tag{9.26}$$

Since this type of edge membrane allows us to disorient the solid grains located on either side of the membrane, it is called a disorientation boundary, and in this particular case, as disorientation corresponds to a flexion of the solid, one speaks of a *tilt boundary*.

- the edge-type thin membrane with a Burgers vector parallel to the surface and increasing along the axis Ox_1 (fig. 9.14b) can be fully characterized by a vectorial surface charge of flexion $\vec{\Pi}$, whose vector is perpendicular to the membrane, and which satisfies

$$\vec{\Pi} = \vec{\Lambda} / d \quad ; \quad \vec{\Pi} \parallel Ox_3 \tag{9.27}$$

As this type of edge membrane area actually allows us to change, in direction Ox_1, the density of crystal planes of the solid grains located on either side of the membrane, it can be described as a *misfit boundary*.

- the screw-type thin membrane with a Burgers vector parallel to the membrane and increasing along the axis Ox_1 (fig. 9.14c) is fully characterized by the scalar surface charge of rotation Π, satisfying

$$\Pi = \Lambda / d \tag{9.28}$$

This type of screw membrane also corresponds to a disorientation boundary between the solid grains located on either side of the membrane. In this particular case, as disorientation corresponds to a rotation of the grains with respect to each other, it is called a *twist boundary*.

On the quantification of membranes of dislocation on a lattice

Dislocation membranes also quantify on a microscopic lattice, as shown in figure 9.15 representing a screw dislocation ribbon consisting of three aligned screw dislocations, corresponding to the case of figure 9.14c.

It is of course also possible to imagine ribbons of similar quantified dislocations, but made of edge dislocations, corresponding to the cases of figures 9.14a and 9.14b.

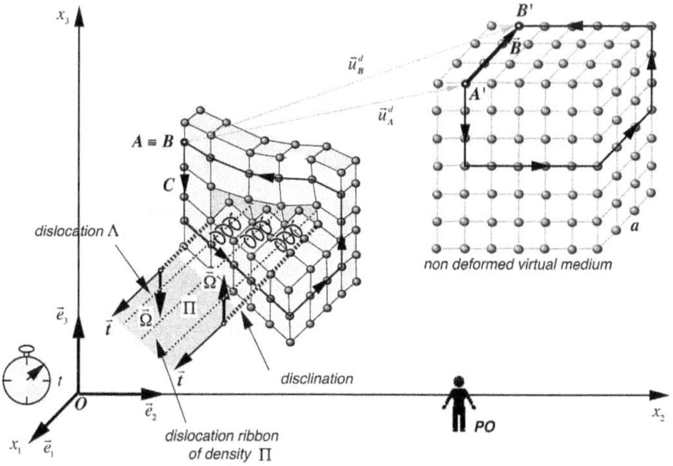

Figure 9.15 - *quantified bi-dimensional dislocation ribbon made up of three dislocations of screw type*

The effects of a volume expansion

In case a medium expands or contracts locally, it is useful to know how the charges of dislocation change. Knowing that the surface charges of flexion $\vec{\Pi}$ and rotation Π of the joint is given by the relationships $\vec{\Lambda}/\boldsymbol{d}$ and $\Lambda/\boldsymbol{d}$ by using equations *(9.16)* and the fact that $\boldsymbol{d}=\boldsymbol{d}_0\,\mathrm{e}^{\tau/3}$, it is found that

$$\left|\vec{\Pi}\right|_{\tau\neq 0}=\frac{\left|\vec{\Lambda}\right|_{\tau\neq 0}}{\boldsymbol{d}\big|_{\tau\neq 0}}=\frac{\left|\vec{\Lambda}_0\right|_{\tau=0}\mathrm{e}^{\tau/3}}{\boldsymbol{d}_0\big|_{\tau=0}\mathrm{e}^{\tau/3}}=\left|\vec{\Pi}_0\right|_{\tau=0} \qquad (9.29)$$

It follows that the surface densities of dislocation charges are conserved when the medium expands or contracts, which leads to the relations

$$\vec{\Pi}_i\Big|_{\tau\neq 0}=\vec{\Pi}_{0i}\Big|_{\tau=0} \quad ; \quad \vec{\Pi}\Big|_{\tau\neq 0}=\vec{\Pi}_0\Big|_{\tau=0} \quad ; \quad \Pi\Big|_{\tau\neq 0}=\Pi_0\Big|_{\tau=0} \qquad (9.30)$$

9.3 – Strings and lines of disclination at the frontiers of membranes of dislocation

While isolated strings of disclinations cannot exist *(19.19)*, it will appear disclinations *in the presence of an extended domain of dislocation charges, such as a dislocation membrane.* One can consider for example the case of a flat membrane of dislocation charges having two regions with different surface densities of charges separated by a boundary (fig. 9.16).

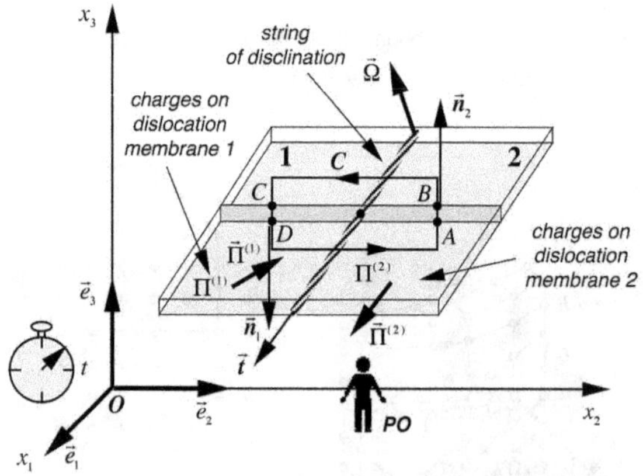

Figure 9.16 *- string of disclination at the frontier of two membranes of dislocations*

Assuming that the surface charges of dislocation $\vec{\Pi}_i^{(1)}$ and $\vec{\Pi}_i^{(2)}$ of the two regions are constant but with different values in each region, it is possible to calculate the Frank vector $\vec{\Omega}$ on a surface S intersecting the membrane surrounding the boundary between the two zones and perpendicular to this border, using the relationship

$$\vec{\Omega} = \sum_i \vec{e}_i \iint_S \vec{\theta}_i d\vec{S} = \sum_i \vec{e}_i \iint_S \overrightarrow{\mathrm{rot}}\left[\vec{\lambda}_i - \vec{e}_i \wedge \vec{\lambda} - \vec{e}_i \lambda\right] d\vec{S} = \sum_i \vec{e}_i \oint_C \left(\vec{\lambda}_i - \vec{e}_i \wedge \vec{\lambda} - \vec{e}_i \lambda\right) d\vec{r} \qquad (9.31)$$

This relation transforms in the following way

$$\vec{\Omega} = \sum_i \vec{e}_i \int_A^B \Big(\underbrace{\vec{\lambda}_i \vec{n}_2}_{=0} - \vec{e}_i \left(\vec{\lambda} \wedge \vec{n}_2\right) - \vec{e}_i \vec{n}_2 \lambda \Big) dl + \sum_i \vec{e}_i \int_C^D \Big(\underbrace{\vec{\lambda}_i \vec{n}_1}_{=0} - \vec{e}_i \left(\vec{\lambda} \wedge \vec{n}_1\right) - \vec{e}_i \vec{n}_1 \lambda \Big) dl \qquad (9.32)$$

In this relationship, the terms $\vec{\lambda}_i \vec{n}_1$ and $\vec{\lambda}_i \vec{n}_2$ are zero since the charge densities $\vec{\lambda}_i$ in a membrane can only be vectors parallel to the membrane *(9.25)*. We have for $\vec{\Omega}$

$$\vec{\Omega} = \int_A^B \left(-\left(\vec{\lambda} \wedge \vec{n}_2\right) - \vec{n}_2 \lambda\right) dl + \int_C^D \left(-\left(\vec{\lambda} \wedge \vec{n}_1\right) - \vec{n}_1 \lambda\right) dl = -\vec{\Pi}^{(1)} \wedge \vec{n}_1 - \Pi^{(1)} \vec{n}_1 - \vec{\Pi}^{(2)} \wedge \vec{n}_2 - \Pi^{(2)} \vec{n}_2 \qquad (9.33)$$

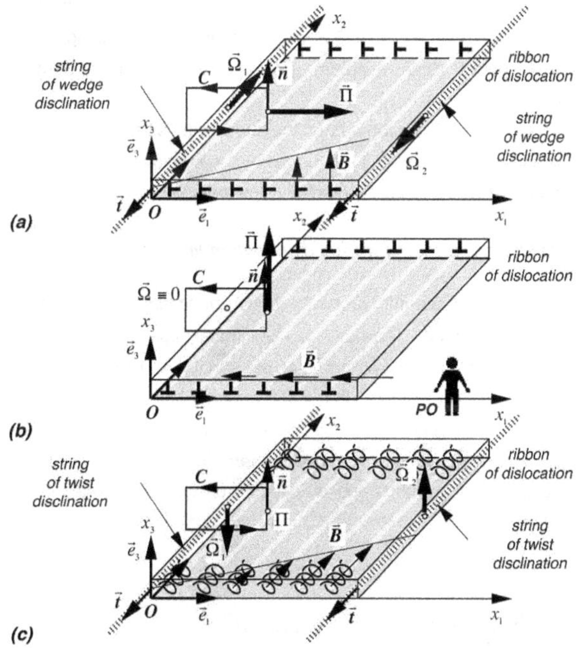

Figure 9.17 *- ribbons of dislocation bordered by disclinations of wedge type* ***(a)*** *and of twist type* ***(c)****. The ribbon of dislocation* ***(b)*** *is not bordered by lines of disclination*

It is found that the value for $\vec{\Omega}$ is independent of the contour chosen as long as this contour surrounds the border between the two regions of the membrane. Thus, for a contour closely surrounding the border, one obtains a line having a constant Frank vector along the line, which can therefore be called *disclination string* or *line of disclination* or simply *disclination*[10]. In figure 9.17, the disclination lines are plotted, and they border the dislocation ribbon similar to those presented in figure 9.14. In this figure, we see the following:

- the ribbon of edge dislocations with a Burgers vector perpendicular to the membrane and increasing along the axis Ox_1 (fig 9.17a) corresponds to a *localized tilt boundary*. It is bordered by two wedge disclinations since $\vec{\Omega}_1 = -\vec{\Omega}_2 = -\vec{\Pi} \wedge \vec{n} = -\left|\vec{\Pi}\right| \vec{t}$,
- the ribbon of edge dislocations with Burgers vector parallel to the membrane and increasing along the axis Ox_1 (fig 9.17b) is a localized accommodation boundary. It is not bordered by disclination since $\vec{\Omega}_1 = \vec{\Omega}_2 \equiv 0$,
- the ribbon of screw dislocations with Burgers vector parallel to the membrane and increasing along the axis Ox_1 (fig 9.17c) corresponds to a *localized twist boundary*. It is bordered by *two twist disclinations* since $\vec{\Omega}_1 = -\vec{\Omega}_2 = -\Pi\vec{n}$.

Note that the disclination of figure 9.17a is the macroscopic disclination shown in figure 8.7b, while the disclination of figure 9.17c corresponds to the macroscopic disclination of figure 8.7a. Furthermore, the quantization of a ribbon of dislocations similar to that of figure 9.17c has already been illustrated in figure 9.15, in which are drawn the two disclinations bordering the quantized dislocation ribbon.

Finally, the relation *(9.33)* justifies *a fortiori* that the surface densities are retained in case of expansion or contraction of the volume, as expressed by the relationship *(9.30)*, since the rotation angles $\vec{\Omega}$ must be an invariant under this kind of transformation.

The scalar charge of curvature of a wedge disclination line

In the previous chapter, we have introduced the notion of scalar density of curvature charges defined by the relation $\theta = \operatorname{div}\vec{\lambda}$. Then one can legitimately ask whether the disinclination obtained at the border of a dislocation membrane as seen in figure 9.16, can be represented by a scalar linear charge Θ, in the same way that a screw dislocation can be fully represented by its linear scalar charge Γ. To answer this question, we need to resume with relationship *(9.31)*, and consider only the terms in $\vec{\lambda}$ and λ since the term in $\vec{\lambda}_i$ is not involved in the case of a disclination on the border of a membrane dislocation, as shown in equation *(9.32)*. We then have

$$\begin{aligned} \vec{\Omega} &= -\sum_i \vec{e}_i \iint_S \overrightarrow{\operatorname{rot}}\left[\vec{e}_i \wedge \vec{\lambda} + \vec{e}_i \lambda\right] d\vec{S} \\ &= -\iint_S \operatorname{div}\vec{\lambda}\, d\vec{S} + \sum_k \iint_S \left(\overrightarrow{\operatorname{grad}}\, \lambda_k\right) dS_k + \iint_S \overrightarrow{\operatorname{grad}}\, \lambda \wedge d\vec{S} \end{aligned} \tag{9.34}$$

In the first term, one can replace $\operatorname{div}\vec{\lambda}$ by θ. The second term is zero by the fact that, in the configuration shown in figure 9.16, there cannot be a gradient of $\vec{\lambda}$ in the direction of the disclination, and the third term can be obtained on the contour surrounding the surface integration, so that

---

[10] *R. DeWit, «Theory of disclinations II, III and IV», J. of the Nat. Bureau of Standards A, vol. 77A, p. 49-100, p. 359-368, and p. 607-658, 1973; E. Kröner, K.-H. Anthony, «Dislocations and Disclinations in Material Structures: The Basic Topological Concepts», Annu. Rev. Mater. Sci., vol. 5, p. 43, 1975*

$$\vec{\Omega} = -\iint_S \theta d\vec{S} - \oint_C \lambda d\vec{r} = \vec{\Omega}\Big|_{\|\vec{t}} + \vec{\Omega}\Big|_{\perp\vec{t}} \tag{9.35}$$

The first term corresponds to the case of a wedge disclination (figure 9.17a) and the second term corresponds to the case of a twist disclination (figure 9.17c). By then introducing the linear scalar curvature charge Θ, we can write for a wedge disclination, in the case of figure 9.16

$$\vec{\Omega}\Big|_{\|\vec{t}} = \vec{\Omega}_{wedge} = -\iint_S \theta d\vec{S} = -\Theta\vec{t} == -\vec{\Pi}^{(1)} \wedge \vec{n}_1 - \vec{\Pi}^{(2)} \wedge \vec{n}_2 \tag{9.36}$$

from which we deduce that it is possible to calculate the *linear charge of curvature* Θ via the expression

$$\Theta = -\vec{\Omega}_{wedge}\vec{t} = \left(\vec{\Pi}^{(1)} \wedge \vec{n}_1\right)\vec{t} + \left(\vec{\Pi}^{(2)} \wedge \vec{n}_2\right)\vec{t} = \vec{\Pi}^{(1)}\left(\vec{n}_1 \wedge \vec{t}\right) + \vec{\Pi}^{(2)}\left(\vec{n}_2 \wedge \vec{t}\right) \tag{9.37}$$

Introducing here the unit vector $\vec{m}$ perpendicular to the disclination line and radial with respect to its center, the following expression is obtained for Θ

$$\Theta = -\vec{\Omega}_{wedge}\vec{t} = \vec{\Pi}^{(1)}\vec{m} + \vec{\Pi}^{(2)}\vec{m} \tag{9.38}$$

The effects of a variation of the volume expansion

In the case where the medium expands or contracts locally, knowing that the surface densities of dislocation of charges $\vec{\Pi}$ on a membrane does not depend on τ *(9.30)*, and using the relation *(9.38)*, we find

$$\Theta\Big|_{\tau\neq 0} = \Theta_0\Big|_{\tau=0} \tag{9.39}$$

The quantification of disclinations of wedge type in a lattice

While isolated disclinations cannot exist, it is possible to imagine a structured solid media which contain line disclinations quantified on the lattice in the case of wedge disclinations, as illustrated by figure 9.18. In this figure, two wedge disclinations are shown with $\Omega = \mp 90°$ in a simple cubic lattice, and there we show the curvature vector $\vec{\chi}$ due to the charge Θ.

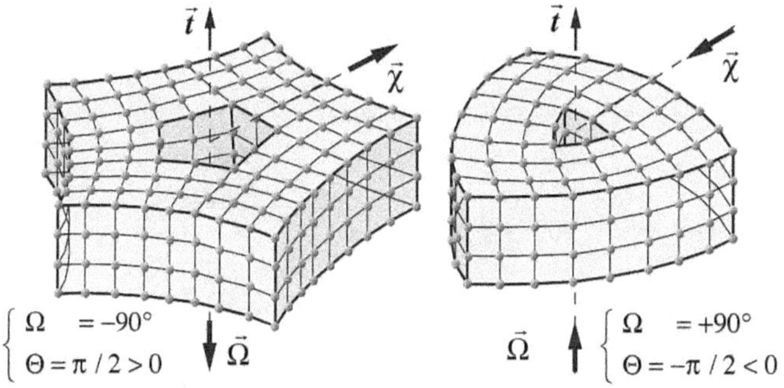

Figure 9.18 *- Examples of quantified wedge disclinations $\Omega = \mp 90°$ on a cubic lattice*

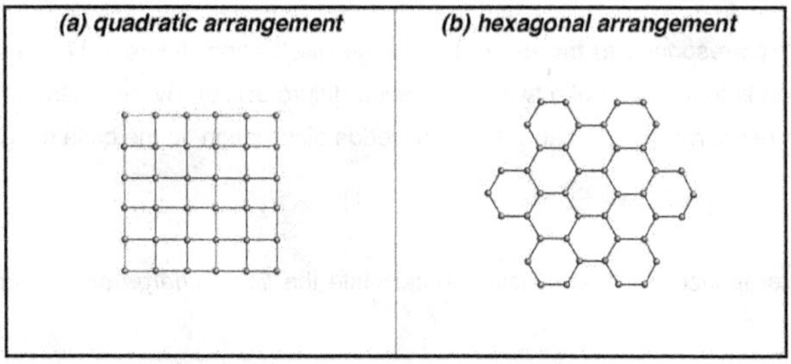

***Figure 9.19** - quadratic or hexagonal arrangements of particles in a plane*

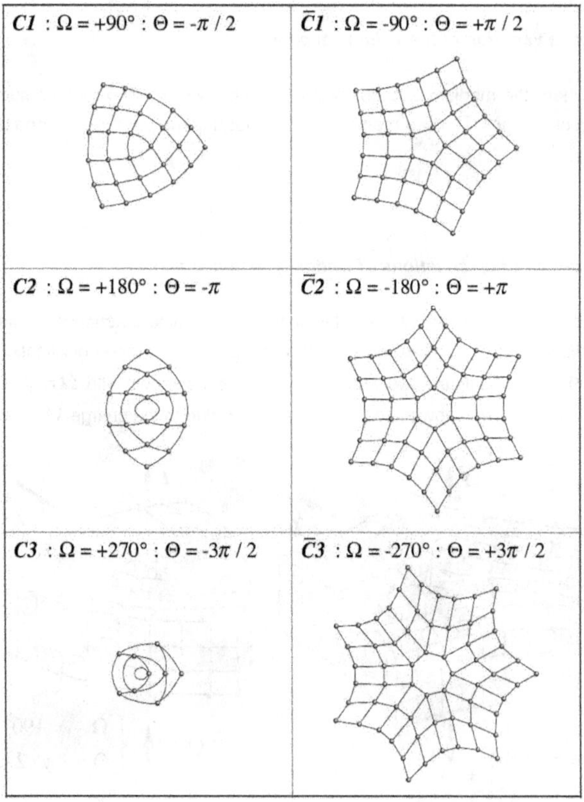

***Figure 9.20** - families of quantized wedge disclinations in a quadratic plane arrangement*

$H1$: $\Omega = +60°$: $\Theta = -\pi / 3$	$\bar{H}1$: $\Omega = -60°$: $\Theta = +\pi / 3$
$H2$: $\Omega = +120°$: $\Theta = -2\pi / 3$	$\bar{H}2$: $\Omega = -120°$: $\Theta = +2\pi / 3$
$H3$: $\Omega = +180°$: $\Theta = -\pi$	$\bar{H}3$: $\Omega = -180°$: $\Theta = +\pi$
$H4$: $\Omega = +240°$: $\Theta = -4\pi / 3$	$\bar{H}4$: $\Omega = -240°$: $\Theta = +4\pi / 3$
$H5$: $\Omega = +300°$: $\Theta = -5\pi / 3$	$\bar{H}5$: $\Omega = -300°$: $\Theta = +5\pi / 3$

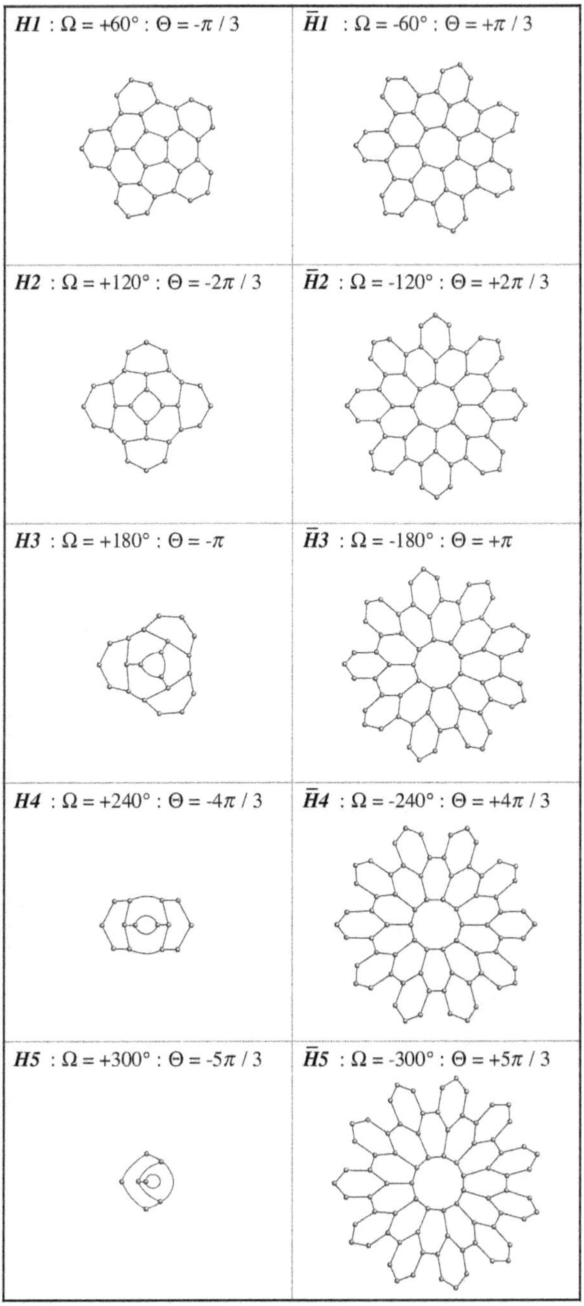

Figure 9.21 - families of quantized wedge disclinations in a hexagonal plane arrangement

One can then imagine that there may be various families of quantized wedge disclinations by considering a solid media with different arrangements of the particles in a secant plane to the disclination line[11]. For example, we will consider here simple arrangements such as quadratic or hexagonal arrangement of the particles in a plane (figure 9.19). But one could obviously be considering more complex arrangements such as three-dimensional centered cubic, hexagonal or cubic face centered structures.

In the case of quadratic arrangement, there may be at most 3 different quantized wedge disclinations , which we will call $C1$, $C2$ and $C3$, with rotation angles Ω of + 90 °, + 180 ° and + 270 ° to which correspond three quantized wedge anti-disclinations $\overline{C1}$, $\overline{C2}$ and $\overline{C3}$, with rotation angles Ω of -90 °, -180 ° and -270 ° (fig. 9.20).

In the case of hexagonal arrangement, there are at most 5 quantized wedge disclinations $H1$, $H2$, $H3$, $H4$ and $H5$, with rotation angles Ω of + 60 °, + 120 °, + 180 °, + 240 ° and + 300 °, to which correspond 5 quantified wedge anti-disclinations $\overline{H1}$, $\overline{H2}$, $\overline{H3}$, $\overline{H4}$ and $\overline{H5}$, with rotation angles Ω of -60 °, -120 °, -180 °, -240 ° and -300 ° (fig. 9.21).

In both cases of figures 9.20 and 9.21, we show the disclinations with a size calculated so that the volume expansion τ is the same in all figures. Note also that the disclinations of + 270 ° in the quadratic arrangement and of +300 ° in the hexagonal arrangement may exist or not exist according to the imaginary medium considered, because their existence is linked to the ability to connect between them two bonds of the same "particle" in the solid medium we consider.

The multiplets of quantized wedge disclinations

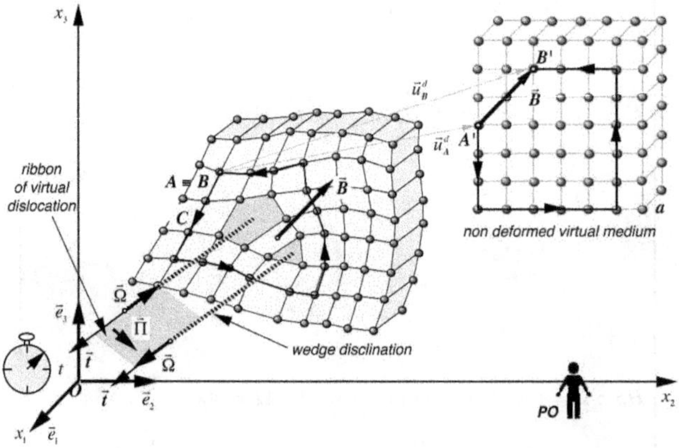

Figure 9.22 *- doublet of quantized wedge disclinations with a ribbon of virtual dislocation.*

It has been shown that there cannot be isolated disclinations with nonzero Frank vector. Therefore, it is necessary to combine several disclinations near each other so that the Frank vector

[11] *The disclinations have been observed essentially in the lattices of flux lines in superconductors of type II, in certain polymers and in the nematic liquid crystals. See: E. Kröner, K.-H. Anthony, Annu. Rev. Mater. Sci., vol. 5, p. 43, 1975; M. Kleman, J. Friedel, Reviews of Modern Physics, vol. 80, p.61-115, 2008*

obtained on a contour surrounding these disclinations is zero.

The example shown in figure 9.22 illustrates this fact: by coupling two quantized wedge disclinations of type $\boldsymbol{C1}$ and $\overline{\boldsymbol{C1}}$ in a simple cubic structure, the Frank vector becomes zero, and there appears a virtual ribbon of edge dislocation between the two disclinations with nonzero overall Burgers vector $\vec{\boldsymbol{B}}$. The ribbon of edge dislocation is similar to the one shown in figure 9.17a and therefore contains a vectorial surface charge of flexion $\vec{\Pi}$. But this flexion charge $\vec{\Pi}$ is not associated with a real quantized dislocation network, but to a virtual ribbon of edge dislocations.

We can find the multiplets of disclinations with global null Frank vector that can be built on the basis of the quantized wedge disclinations that we have described in the case of simple cubic lattice in figure 9.19a. The basic multiplets with zero Frank vector, that is to say, those that cannot be further broken down into two or more multiplets of non null Frank vector, are reported in table 9.1. It is found that, in a simple cubic lattice, there may be three doublets, 4 triplets and 2 quadruplets.

3 doublets
+90° / -90°
+180° / -180°
+270° / -270°

4 triplets	
+90° / +90° / -180°	*-90° / -90° / +180°*
+90° / +180° / -270°	-90° / -180° / +270°

2 quadruplets	
+90° / +90° / +90° / -270°	-90° / -90° / -90° / +270°

Table 9.1 *- the multiplets of quantified wedge disclinations in a simple cubic plane structure*

In the case of the lattice of figure 9.19b, with plans having a hexagonal structure of the particles, there may be 5 doublets, 12 triplets, 8 quadruplets, 4 quintuplets and 2 sextuplets as shown in table 9.2.

In both tables, the multiplets made with disclinations $\boldsymbol{C3}$ (+ 270 °) or $\boldsymbol{H5}$ (+ 300 °) are grayed out because they could not exist, for example in structured media that do not allow a "self connection" of two bonds of a same particle.

5 doublets
+60° / -60°
+120° / -120°
+180° / -180°
+240° / -240°
+300° / -300°

12 triplets	
+60° / +60° / -120°	-60° / -60° / +120°
+60° / +120° / -180°	-60° / -120° / +180°
+60° / +180° / -240°	-60° / -180° / +240°
+120° / +120° / -240°	-120° / -120° / +240°
+60° / +240° / -300°	-60° / -240° / +300°
+120° / +180° / -300°	-120° / -180° / +300°

8 quadruplets	
+60° / +60° / +60° / -180°	-60° / -60° / -60° / +180°
+60° / +60° / +120° / -240°	-60° / -60° / -120° / +240°
+60° / +120° / +120° / -300°	-60° / -120° / -120° / +300°
+60° / +60° / +180° / -300°	-60° / -60° / -180° / +300°

4 quintuplets	
+60°/+60°/+60°/+60°/-240°	-60°/-60°/-60°/-60°/+240°
+60°/+60°/+60°/+120°/-300°	-60°/-60°/-60°/-120°/+300°

2 sextuplets	
60°/+60°/+60°/+60°/+60°/-300°	-60°/-60°/-60°/-60°/-60°/300°

Tableau 9.2 *- the multiplets of quantized wedge disclinations in a hexagonal plane structure.*

In both tables, the multiplets made with disclinations ***C3*** (+ 270 °) or ***H5*** (+ 300 °) are grayed out because they could not exist, for example in structured media that do not allow a "self connection" of two bonds of a same particle.

Examples of quantized multiplets of disclinations in the cubic structure

In this section, as examples, we study some imaginary cases of doublets and triplets of disclinations in the simple cubic structure (fig. 9.23 to 9.28).

For each case shown, we indicate in an inset the number of edge dislocations on each side of the quadrangle **ABCD**. This number of edge dislocations is composed of a fixed number depending on the number of disclinations in play and their respective arrangement, plus a number of edge dislocations due to the inclusion of additional n_i "particle" planes placed between the disclinations to vary the distance between them (these additional plans are shown in red in the figures).

In the figures 9.23 and 9.24, we show the doublets $C1-\bar{C}1$ and $C2-\bar{C}2$, by varying in each case the orientation of the pair relative to the crystal structure (with an angle of 45 ° and 0 ° respectively with respect to the structure). One immediately notices that in the case of doublets, there is always a fixed number of edge dislocations emerging from the quadrangular figure ***ABCD***, and that the addition of insert planes systematically increases this number.

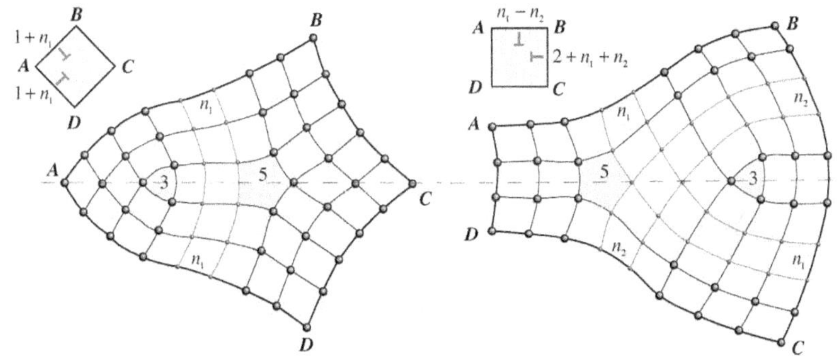

***Figure 9.23** - doublets $C1-\bar{C}1$ of wedge disclinations in a cubic lattice*

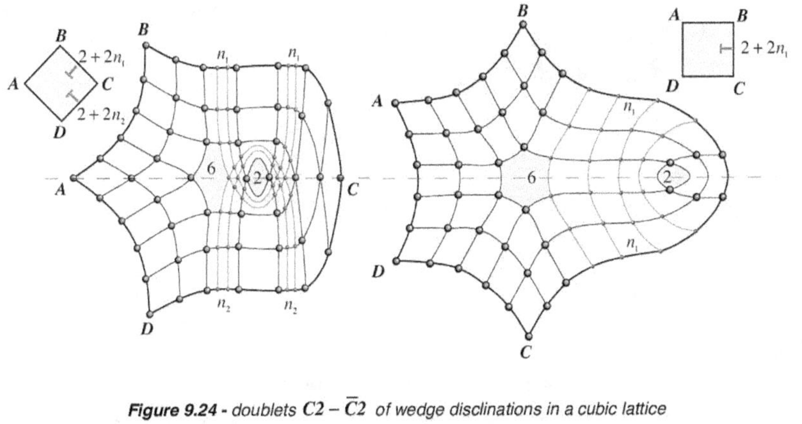

***Figure 9.24** - doublets $C2-\bar{C}2$ of wedge disclinations in a cubic lattice*

In figure 9.25, we plotted triplets $C1-\overline{C}2-C1$, varying the direction of alignment of these triplets with the structure (0 ° and 45 ° respectively), while in figure 9.26, the triplet $C1-\overline{C}2-C1$ and its "antiparticle "$\overline{C}1-C2-\overline{C}1$ are plotted.

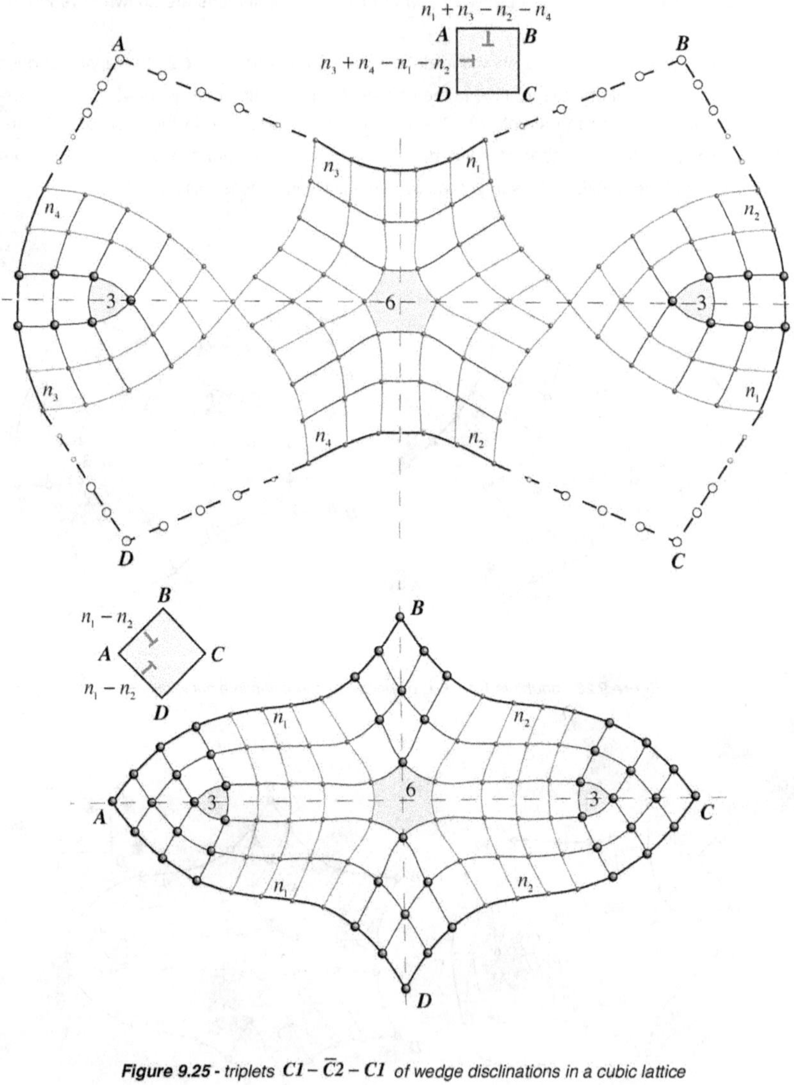

***Figure 9.25** - triplets $C1-\overline{C}2-C1$ of wedge disclinations in a cubic lattice*

In all these cases of triplets, for which disclinations components are aligned on an axis, we see that if the additional insert planes are placed symmetrically, these triplets remain perfectly symmetrical and there is no edge dislocation that emerges in quadrangle ***ABCD***.

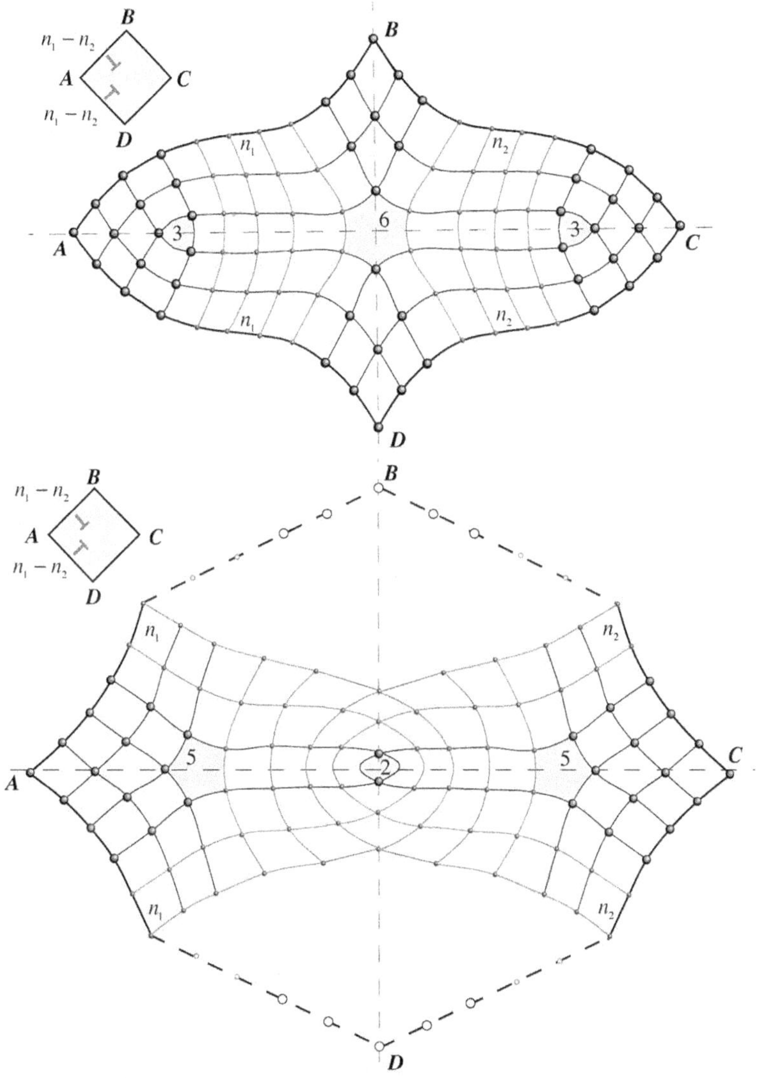

Figure 9.26 *- triplets* $C1-\overline{C}2-C1$ *and* $\overline{C}1-C2-\overline{C}1$ *of wedge disclinations in a cubic lattice*

However, even if no edge dislocation emerges, it is clear that there is strong local deformation of the lattice, which must correspond to a strain energy which is nothing but the energy of the triplet formation. In addition, the introduction of additional infill symmetrical planes has the effect of increasing the distance between the disclinations triplet, and thus increases the energy of the triplet.

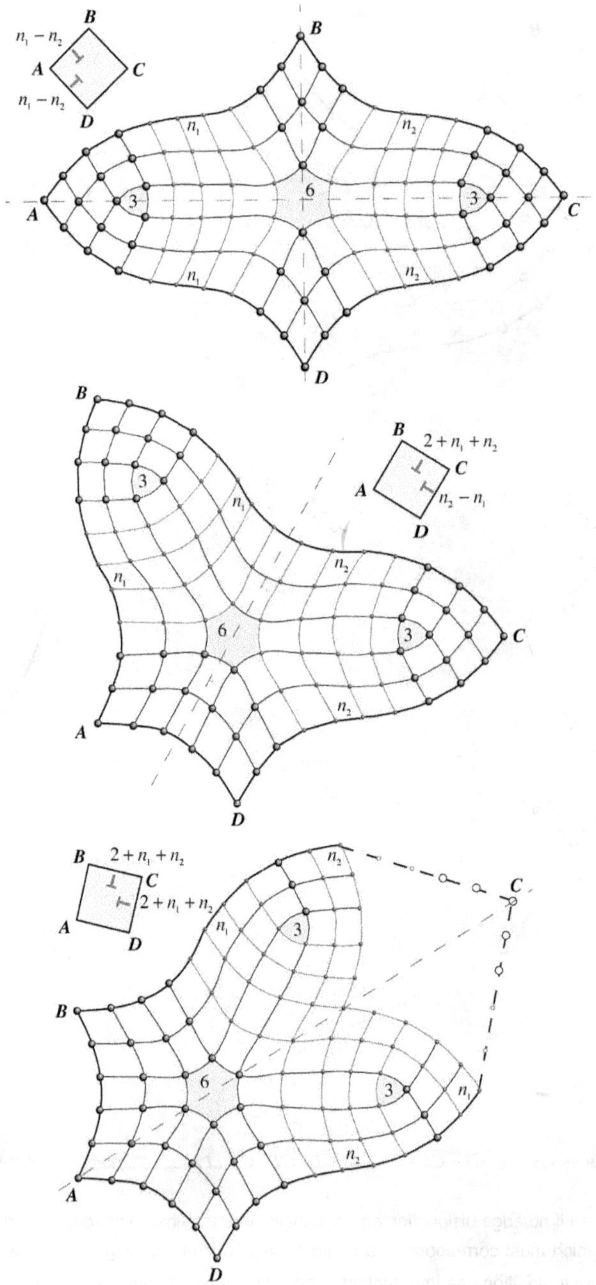

Figure 9.27 - *the different spatial positions of triplets* $C1 - \overline{C}2 - C1$

In figure 9.27, we show triplets $C1-\bar{C}2-C1$, but this time by varying the relative spatial arrangement of disclinations with each other. The angles formed by the straight lines connecting the two disclinations $C1$ and $\bar{C}2$ are 180 °, 120 ° and 60 ° respectively. It is found that the triplet with an asymmetry, in cases 120 ° and 60 °, shows a fixed number of edge dislocations emerging from the quadrangular figure ***ABCD***, and this number increases with the dissymmetry, since it goes from 2 to 4 when the angle goes from 120 ° to 60 °. It is easy to imagine on the basis of these figures that, in all cases, an increase in the distance between the disclinations of the triplet by adding intermediate planes may only increase the energy of the triplet!

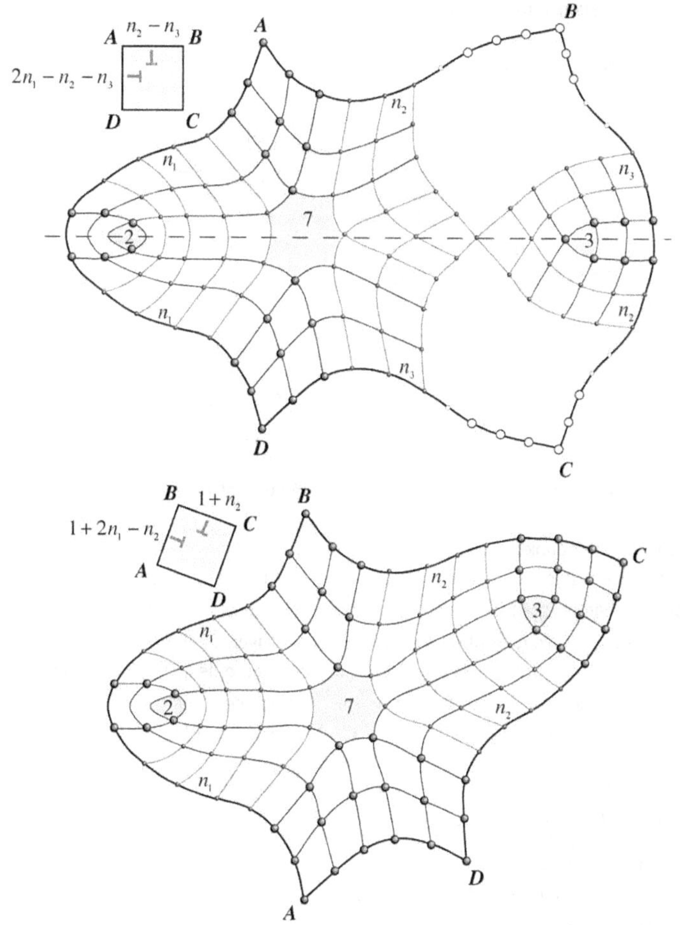

Figure 9.28 *- two spatial configurations of triplet* $C2-\bar{C}3-C1$

Finally, in figure 9.28, we show examples of more "exotic" triplets, by drawing two different spatial configurations of the triplet $C2-\bar{C}3-C1$.

The ribbons of virtual dislocations connecting the disclinations in a multiplet

All the examples we have shown in the previous section show that each multiplet of disclinations can have many different configurations, depending on the number of disclinations comprising the multiplet, the spatial arrangement of disclinations and the number and disposition of infill planes. Moreover, one can well imagine that, for the multiplets obtained in a hexagonal structure, there is a greater multitude of different configurations since the number of singlets, that is to say, quantified disclinations, is greater and the number of possible angles between the arms of the multiplets is also higher. It would therefore be very convenient to find a mathematical way to characterize the various configurations which may occur for a given multiplet in a given structure. We will treat the case of a doublet and a triplet wedge disclinations, the other multiplets can easily be treated similarly.

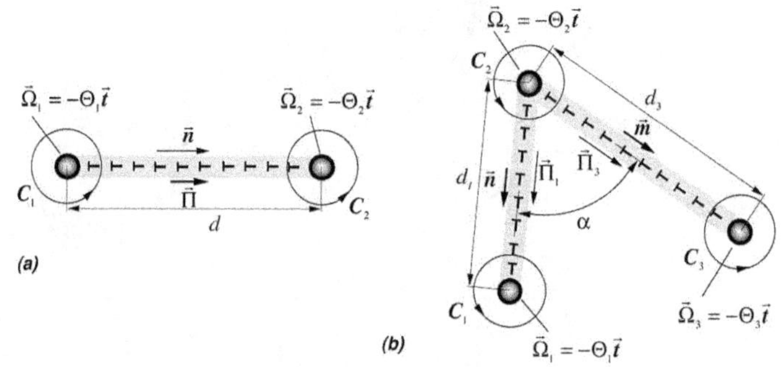

Figure 9.29 *- ribbons of virtual dislocations of a doublet* ***(a)*** *and a triplet* ***(b)*** *of wedge disclinations*

In the case of the doublet shown in figure 9.29a, both disclinations are separated by a distance d, and the direction of the dipole is the one given by the unit vector $\vec{n}$. The disclinations of the doublet are characterized by their Frank vectors $\vec{\Omega}_1 = -\Theta_1 \vec{t}$ and $\vec{\Omega}_2 = -\Theta_2 \vec{t}$. As we have seen previously, the doublet of disclinations must be linked by the existence of virtual dislocations, which we will represent by a dislocation membrane of length d connecting the disclinations and containing a vectorial surface density of dislocation charges $\vec{\Pi}$, with $\vec{\Pi} \parallel \vec{n}$. By applying equation *(9.38)* on the edges C_1 and C_2, the following relations are obtained

$$\begin{cases} \Theta_1 = -\vec{\Omega}_1 \vec{t} = \vec{\Pi} \vec{n} \\ \Theta_2 = -\vec{\Omega}_2 \vec{t} = -\vec{\Pi} \vec{n} \end{cases} \tag{9.40}$$

from which it is deduced that $\Theta_1 = -\Theta_2$.

At large distances, the doublet of disclinations may be considered as an edge dislocation with linear density charge $\vec{\Lambda}$ given by the relation

$$\vec{\Lambda} = d\vec{\Pi} = d\Theta_1 \vec{n} \tag{9.41}$$

This shows that the dislocation equivalent to the doublet of disclinations has a Burgers vector $\vec{B}$ proportional to the scalar charge density Θ_1 and to the distance d between the disclinations

$$\vec{B} = \vec{t} \wedge \vec{\Lambda} = d\Theta_1 \left(\vec{t} \wedge \vec{n} \right) \tag{9.42}$$

As the Burgers vector $\vec{B}$ associated with the charge $\vec{\Lambda}$ must be quantized in a lattice, the minimum value $d_{\min}$ of the distance d obtained when there is no intermediate planes between the disclinations of the doublet, satisfied

$$d_{\min} = \left| \vec{B} \right|_{\min} / \Theta_1 \tag{9.43}$$

Consider then the case of doublets $C1 - \overline{C}1$, $C2 - \overline{C}2$ and $C3 - \overline{C}3$, in a simple cubic lattice, assuming that there are no intermediate planes like those placed between the disclinations of the doublets in figures 9.23 and 9.24. As $\Theta_1 = i\pi / 2$ in the cubic lattice with $i = 1,2,3$ depending on the doublet $Ci - \overline{C}i$, the value $d_{\min}$ satisfies

$$d_{\min} = \left| \vec{B} \right|_{\min} / \Theta_1 = 2 \left| \vec{B} \right|_{\min} / i\pi \tag{9.44}$$

in which $\left| \vec{B} \right|_{\min}$ must be a multiple of the lattice step a, or the diagonal $a\sqrt{2}$ of the base square. Thus we have:

- in the case of figure 19.30a, $\left| \vec{B} \right|_{\min} = a\sqrt{2}$ and $i = 1$, giving $d_{\min} = 2a\sqrt{2} / \pi = 0{,}9a$,
- in the case of figure 19.30b, $\left| \vec{B} \right|_{\min} = 2a$ and $i = 1$, giving $d_{\min} = 4a / \pi = 1{,}28a$,
- in the case of figure 19.31a, $\left| \vec{B} \right|_{\min} = 2a\sqrt{2}$ and $i = 2$, giving $d_{\min} = 4a\sqrt{2} / \pi = 1{,}8a$,
- in the case of figure 19.30a, $\left| \vec{B} \right|_{\min} = 2a$ and $i = 2$, giving $d_{\min} = 2a / \pi = 0{,}64a$.

In the case of a perfect hexagonal lattice[12], the Burgers vector must be quantified as a multiple of the distance b between two hexagons in the perfect lattice. For the doublet $H1 - \overline{H}1$ shown in figure 9.30, $\Theta_1 = \pi / 3$ and it was found that $\left| \vec{B} \right|_{\min} = b$, so we have the relationship $d_{\min} = \left| \vec{B} \right|_{\min} / \Theta_1 = 3b / \pi \cong 0{,}955b$.

In the case of the triplet shown in figure 9.29b, the three disclinations are separated by distances d_1 and d_3 respectively, and the directions of the two branches of the triplet are those given by the unit vector $\vec{n}$ and $\vec{m}$, in such a way that the angle α formed by the two arms is given by the relationship $\cos\alpha = \vec{n}\vec{m}$. The disclinations of the triplet are characterized by Frank vectors $\vec{\Omega}_1 = -\Theta_1 \vec{t}$, $\vec{\Omega}_2 = -\Theta_2 \vec{t}$ and $\vec{\Omega}_3 = -\Theta_3 \vec{t}$. They must be linked by the existence of virtual dislocations which are represented by two membranes of dislocations with length d_1 and d_3, linking the disclinations and containing a vector surface density of dislocation charges $\vec{\Pi}_1$ and $\vec{\Pi}_3$, with $\vec{\Pi}_1 \parallel \vec{n}$ and $\vec{\Pi}_3 \parallel \vec{m}$. By applying equation *(9.38)* on the contours C_1, C_2 and C_3, we have relationships

$$\begin{cases} \Theta_1 = -\vec{\Omega}_1 \vec{t} = -\vec{\Pi}_1 \vec{n} \\ \Theta_2 = -\vec{\Omega}_2 \vec{t} = \vec{\Pi}_1 \vec{n} + \vec{\Pi}_3 \vec{m} \\ \Theta_3 = -\vec{\Omega}_3 \vec{t} = -\vec{\Pi}_3 \vec{m} \end{cases} \tag{9.45}$$

from which we can see that $\sum \Theta_i = 0$.

---

[12] *Multiplets of wedge disclinations leading to edge dislocations can be clearly observed in the hexagonal lattice of the graphen. See: J. C. Meyer, C. Kisielowski, R. Erni, M. D. Rossell, M. F. Crommie, A. Zettl, Nano Lett., vol. 8(11), p. 3582, 2008; M. P. Ariza, M. Ortiz, Journal of the Mechanics and Physics of Solids, vol. 58, p. 710, 2010*

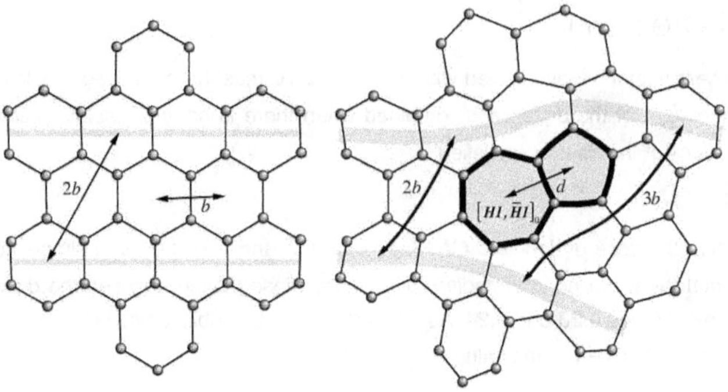

Figure 9.30 *- illustration of doublet* $\boldsymbol{H1}-\bar{\boldsymbol{H}}\boldsymbol{1}$ *and distances* b *and* d

At large distances, the triplet of disclinations may be considered as two parallel strings of edge dislocations with linear charge densities $\vec{\Lambda}_1$ and $\vec{\Lambda}_3$, given by the relations

$$\begin{cases} \vec{\Lambda}_1 = d_1 \vec{\Pi}_{(1)} = -d_1 \Theta_1 \vec{n} \\ \vec{\Lambda}_3 = d_3 \vec{\Pi}_{(3)} = -d_3 \Theta_3 \vec{m} \end{cases} \tag{9.46}$$

These two parallel dislocations can be replaced by a single dislocation, equivalent to the disclination triplet, which has a Burgers vector $\vec{B}$ given by the relation

$$\vec{B} = \vec{t} \wedge \vec{\Lambda} = -d_1 \Theta_1 (\vec{t} \wedge \vec{n}) - d_3 \Theta_3 (\vec{t} \wedge \vec{m}) \tag{9.47}$$

In the case of the symmetric triplet with alignement of three disclinations on a single line (figures 9.29 and 9.30), we have $\vec{m} = -\vec{n}$, and the total Burgers vector is written

$$\vec{B} = \vec{t} \wedge \vec{\Lambda} = (d_3 \Theta_3 - d_1 \Theta_1)(\vec{t} \wedge \vec{n}) \tag{9.48}$$

In this case, if $d_3 \Theta_3 = d_1 \Theta_1$, the triplet has no overall Burgers vector at a distance, but only two local Burgers vector which cancel each other.

According to equations *(9.41)* and *(9.46)*, the energy of a multiplet, associated with the local deformations induced in the lattice, may only increase if the distances d_i separating the disclinations of the multiplet increase. This implies that there is a force of attraction between the disclinations which increases the more one separates these disclinations. By analogy with the naming in the Standard Model of elementary particles, one could speak here of a force of strong nature which links the disclinations of the multiplet, and one could assimilate the virtual dislocation ribbons $\vec{\Pi}_i$ *to kinds of gluons*, since it is these virtual charges that bind together the disclinations, and prohibit the isolation of one of them.

9.4 – Strings of dispiration and solid lattices with axial symmetries

We will call dispiration strings, or lines of dispiration or simply dispirations[13], mixed strings resul-

[13] *W. F. Harris, Philos. Mag. 22, p. 949, 1970; W. F. Harris, Sci. Am. 237(12), p. 130, 1977*

ting from the combination of strings or lines of disclination with strings or lines of dislocation.

On the combination of wedge disclinations and edge dislocations

Figure 9.31 shows two examples of dispirations composed of a wedge disclination and an edge dislocation, represented in the dense plane of a face-centered cubic structure or a hexagonal structure. From the dense plane of a perfect crystal (a), we construct a wedge disclination with Ω =+60° ***(b)***. By then cutting the crystal along the line ***A-B***, and by adding a line ***A'-B'*** of particles, we introduce a linear density $\vec{\Lambda}$ of edge dislocation in the center of the disclination ***(c)***. Similarly, by removing the particles along the line ***C-D*** in figure 9.31 ***(b)***, and by gluing the jaw with the particles of the line ***A-B***, we introduce again a linear density $\vec{\Lambda}$ of edge dislocation in the center of the disclination, as shown in ***(d)*** of figure 9.31.
The absorption and emission of edge dislocations with a quantified wedge disclination is the basic process that explains the movement of a disclination in a lattice[14]. Here we leave the readers to try to illustrate this type of process!

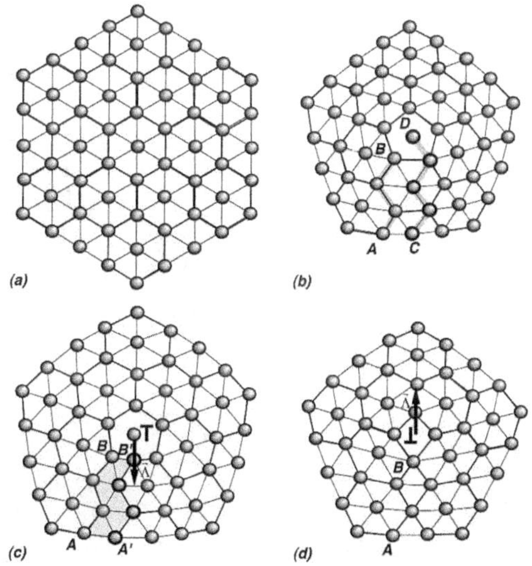

Figure 9.31 *- dispirations **(c)** and **(d)** by combination of a wedge disclination **(b)** with an edge dislocation in the dense plane of a FCC or a hexagonal lattice*

Lattices with axial symmetries

One can imagine lattices which have an axial symmetry in the particles, such as cubic lattices ***(a)*** and ***(b)*** and hexagonal lattices ***(c)*** and ***(d)*** shown in figure 9.32. This axial symmetry of the particles can simply present a preferred orientation of the particles in the plans of the structure,

[14] *W. F. Harris, L. E. Scriven, J. Appl. Phys. 42 , p 3309, 1971; R. deWit, J. Appl. Phys. 42, p. 3304, 1971*

as in the cases ***(a)*** and ***(c)***, or may have an orientation and a preferential direction as in cases ***(b)*** and ***(d)***. This type of lattice thus has an alternating structure of layers ***a, b, c, d, ...***

In addition, the rotation axes of the particles along the vertical axis produces a directed medium, that is referred to as *dextrorotatory* (right-handed) in the case ***(b)*** and *levorotatory* (left-handed) in the cases ***(c)*** and ***(d)***.

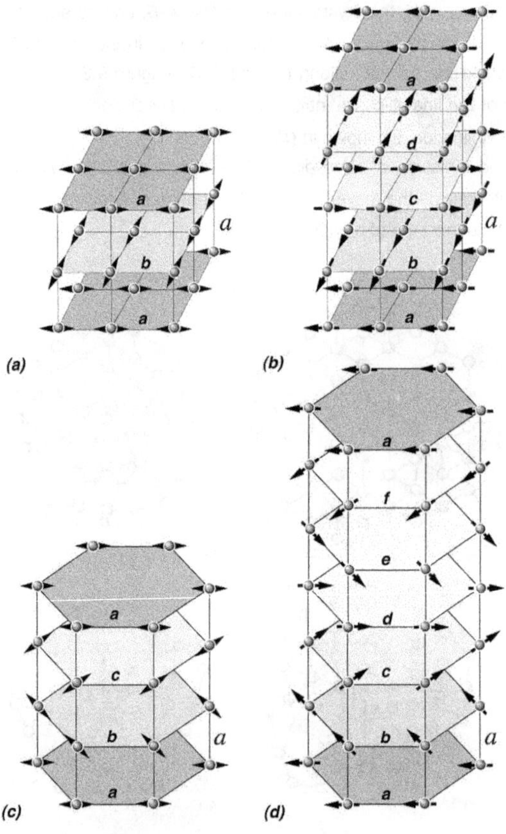

Figure 9.32** - cubic and hexagonal lattices presenting an axial symmetry of the particles of dextrorotatory type (right-handed) in **(b)** and of levorotatory type (left-handed) in **(c)** and **(d)

If it is forbidden to break the axial orientation of the particles in a plane, it is not possible to introduce a vertically oriented screw dislocation with any Burgers vector. Indeed, if the distance between the horizontal planes is a , in order to ensure the continuity of the particle orientation, and also their direction in cases ***(b)*** and ***(d)***, the length of the Burgers vector of the screw dislocation $\vec{B}_{vis}$ has to be equal to $\pm 2a$ in case ***(a)***, $\pm 4a$ in case ***(b)***, $\pm 3a$ in case ***(c)*** and $\pm 6a$ in case ***(d)***.

The dissociation of screw dislocations and the "strong force" linking the partials.

In the axially symmetric environments such as those reported in figure 9.32, screw dislocations must have Burgers vectors $\vec{B}_{vis}$ whose lengths are multiples of the unit length of the lattice a. In this case, the screw dislocations have an incentive to split into partial with Burgers vectors of length $\vec{B}_1$, $\vec{B}_2$ and $\vec{B}_3$, forming respectively 2, 4, 3 or 6 partial in cases ***(a)*** to ***(d)*** respectively. Between the partial dislocations form *ribbons of connection faults* between axial planes **ab, bc, cd,** etc. The separation distance between partials then depends on the energy γ per unit area of the connection fault.

For example, figure 9.33 illustrates the dissociation in the case of axial lattice ***(c)***. In this case, the screw dislocations have an incentive to dissociate into three partial Burgers vectors, $\vec{B}_1$, $\vec{B}_2$ and $\vec{B}_3$, all of length a, so that $\vec{B}_{vis} = \vec{B}_1 + \vec{B}_2 + \vec{B}_3$. In the case of this separation, since the length $\vec{B}_{vis}$ is $3a$, we have $\vec{B}_{vis}^2 = 3\left(\vec{B}_1^2 + \vec{B}_2^2 + \vec{B}_2^2\right)$ and since the energy due to the dislocation is proportional to the square of the Burgers vector, the energy of the three isolated partial is 3 times lower than the energy of the original screw dislocation. If the connection fault ribbon has an energy γ per unit area, the total energy $E_t(d)$ per unit length of dislocation for a screw dislocation dissociated over distances d is written $E_t(d) = 2\gamma d + E_d(d)$ where $E_d(d)$ is the energy of the three partials depending on the distance d separating them. This function is monotonically decreasing from E_0 for $d = 0$ to $E_0 / 3$ for $d \to \infty$ in the case of the partials shown in figure 9.33.

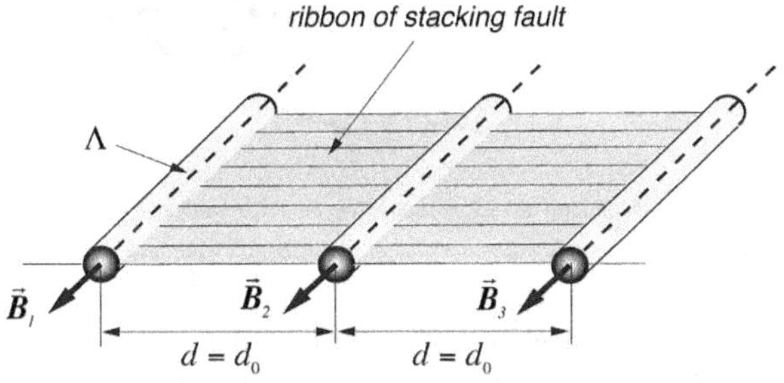

Figure 9.33 - *dissociation of a screw dislocation in three partials in the case of an axial lattice* ***(c)***

The energy $E_t(d)$ thus has a minimum similar to that reported in figure 9.10 for the distance $d = d_0$, which is the equilibrium distance between the three partial, controlled by the competition between energy decrease associated with the increased distance between the partials and increased energy due to the formation of an energy ribbon of connection faults between the three partials.

This behavior of the energy $E_t(d)$ induces and interaction strength between the partials that we qualify as *strong force* in the sense that the energy of the triplet partials will increase if we try to increase the distance separation beyond d_0. This therefore presents a kind of strong force in its behavior and an interesting analogy with the strong force acting between quarks in the Standard Model of elementary particles.

Structural dispirations in a lattice with axial symmetry.

Then we immediately imagine that there must also be connection conditions ensuring the continuity of the axial symmetry if we want to introduce a disclination in such a lattice. In fact, to ensure this continuity, it will be necessary to associate a screw dislocation with the Burgers vector $\vec{B}$ to the disclination. There then appears here a structural need to introduce dispirations in such a media[15].

Figure 9.34 illustrates this perfectly. In order to introduce a disclination $\Omega = +90\,°$ in the medium shown in figure 9.32 **(b)**, it is necessary to add a screw dislocation correctly oriented, with Burgers vector $\vec{B}$ and length a, which ensures the continuity of the axial orientation of particles on the middle of the planes. Note that a screw dislocation with an inverted Burgers vector direction and length $3a$ could also have ensured the continuity of the axial orientation of the particles, so there are two different dispirations with rotation Ω =+90°, both with a linear charge of curvature $\Theta = -\pi/2$, but differing in their Burgers vector $\vec{B}$ associated with a rotation charge per unit length Λ equal to $\Lambda = a$ or $\Lambda = -3a$.

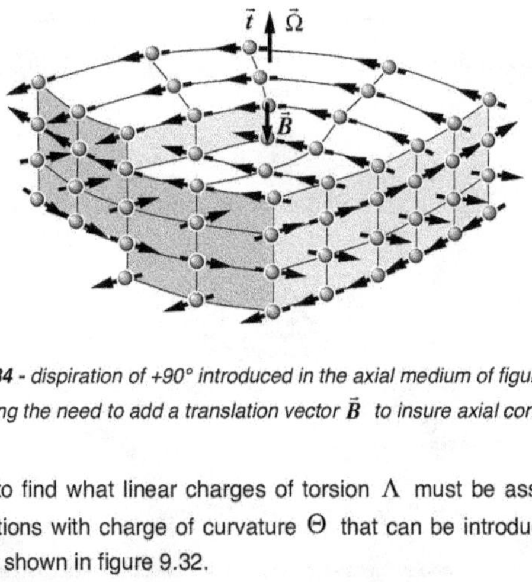

***Figure 9.34** - dispiration of +90° introduced in the axial medium of figure 9.32 **(b)** showing the need to add a translation vector $\vec{B}$ to insure axial continuity*

It is not too difficult to find what linear charges of torsion Λ must be associated with the different wedge dispirations with charge of curvature Θ that can be introduced in the cubic and hexagonal structures shown in figure 9.32.

In table 9.3, we plotted these charges for the cubic structures of figures 9.32 ***(a)*** and ***(b)***. In case ***(a)***, the structure shows no difference between dextrorotatory and levorotatory direction orientation of the lattice particles. On the other hand, there is a difference between these two orientations in case ***(b)***, which implies a change of the sign of the charge Λ between dextrorotatory and levorotatory medias.

In table 9.4, we show the charges Θ and Λ for the hexagonal structures of figures 9.32 ***(c)*** and ***(d)***. Since there is in both cases a difference between the dextrorotatory and levorotatory orientations of the medium, it appears in both cases a sign change of the charge Λ between these two axial directions.

---

15 *W. F. Harris, «Geometry of Desinclinations in Crystals», in Surface and Defect Properties of Solids, volume 3, ed. by M. W. Roberts, J. M. Thomas, The Chemical Society, London, 1974*

	Θ	$\Lambda_{(a)}$	$\Lambda_{(b)}^{(dextrorotatory)}$	$\Lambda_{(b)}^{(levorotatory)}$
vis	0	$\pm 2a$	$\pm 4a$	$\pm 4a$
C1	$-\pi / 2$	$\pm a$	$+a\ ,-3a$	$-a\ ,+3a$
$\bar{C}$1	$+\pi / 2$	$\pm a$	$-a\ ,+3a$	$+a\ ,-3a$
C2	$-\pi$	$0\ ,\pm 2a$	$+2a\ ,-2a$	$+2a\ ,-2a$
$\bar{C}$2	$+\pi$	$0\ ,\pm 2a$	$+2a\ ,-2a$	$+2a\ ,-2a$
C3	$-3\pi / 2$	$\pm a$	$-a\ ,+3a$	$+a\ ,-3a$
$\bar{C}$3	$+3\pi / 2$	$\pm a$	$+a\ ,-3a$	$-a\ ,+3a$

Table 9.3 *- linear charges of curvature Θ and of rotation Λ of the dispirations in the cubic structures of figures 9.39* ***(a)*** *and* ***(b)***

	Θ	$\Lambda_{(c)}^{(dextrorotatory)}$	$\Lambda_{(c)}^{(levorotatory)}$	$\Lambda_{(d)}^{(dextrorotatory)}$	$\Lambda_{(d)}^{(levorotatory)}$
vis	0	$\pm 3a$	$\pm 3a$	$\pm 6a$	$\pm 6a$
H1	$-\pi / 3$	$+a\ ,-2a$	$-a\ ,+2a$	$+a\ ,-5a$	$-a\ ,+5a$
$\bar{H}$1	$+\pi / 3$	$-a\ ,+2a$	$+a\ ,-2a$	$-a\ ,+5a$	$+a\ ,-5a$
H2	$-2\pi / 3$	$-a\ ,+2a$	$+a\ ,-2a$	$+2a\ ,-4a$	$-2a\ ,+4a$
$\bar{H}$2	$+2\pi / 3$	$+a\ ,-2a$	$-a\ ,+2a$	$-2a\ ,+4a$	$+2a\ ,-4a$
H3	$-\pi$	$0\ ,\pm 3a$	$0\ ,\pm 3a$	$\pm 3a$	$\pm 3a$
$\bar{H}$3	$+\pi$	$0\ ,\pm 3a$	$0\ ,\pm 3a$	$\pm 3a$	$\pm 3a$
H4	$-4\pi / 3$	$+a\ ,-2a$	$-a\ ,+2a$	$-2a\ ,+4a$	$+2a\ ,-4a$
$\bar{H}$4	$+4\pi / 3$	$-a\ ,+2a$	$+a\ ,-2a$	$+2a\ ,-4a$	$-2a\ ,+4a$
H5	$-5\pi / 3$	$-a\ ,+2a$	$+a\ ,-2a$	$-a\ ,+5a$	$+a\ ,-5a$
$\bar{H}$5	$+5\pi / 3$	$+a\ ,-2a$	$-a\ ,+2a$	$+a\ ,-5a$	$-a\ ,+5a$

Tableau 9.4 *- linear charges of curvature Θ and rotation Λ of the dispirations in the hexagonal structures of figures 9.35* ***(c)*** *and* ***(d)***

With the dispirations of tables 9.3 and 9.4, it would be possible to build, in each of these four medias, doublets and triplets of dispiration strings having zero overall line charge Θ and some overall line charge Λ which may be zero or not, on the basis of the models shown in figure 9.29.

9.5 – Loops of dislocation and disclination

To satisfy the conservation equation $\operatorname{div}\vec{\lambda}_i = 0$, a line of dislocation or disclination cannot suddenly stop in the middle. However a line closing on itself to form a localized loop still meets the conservation equation. In this section, we will therefore present such loops and their properties in a solid lattice. To simplify the mathematical treatment of loops, it is wise to start by developing the appropriate mathematical tools.

On the vectorial geometry of loops of charged strings

To develop a mathematical tool to simply describe the vectorial geometry of straight or curved strings, we consider a radius of circular string in three-dimensional space (fig. 9.35).
This string is "charged" with vectorial volume densities $\vec{a}$ and $\vec{b}$ and with a scalar density c. We want to write the vector operators acting on the charge densities contained in the string, ie, $c = \operatorname{div}\vec{b}$ et $b = \overrightarrow{\operatorname{rot}}\,\vec{a}$ as operators acting on the linear charges $\vec{A}$, $\vec{B}$ and C, and defined on the central fiber of the string

$$\vec{A} = \iint_{S_1} \vec{a}\, d\vec{S} \quad ; \quad \vec{B} = \iint_{S_1} \vec{b}\, d\vec{S} \quad ; \quad C = \iint_{S_1} c\, dS \tag{9.49}$$

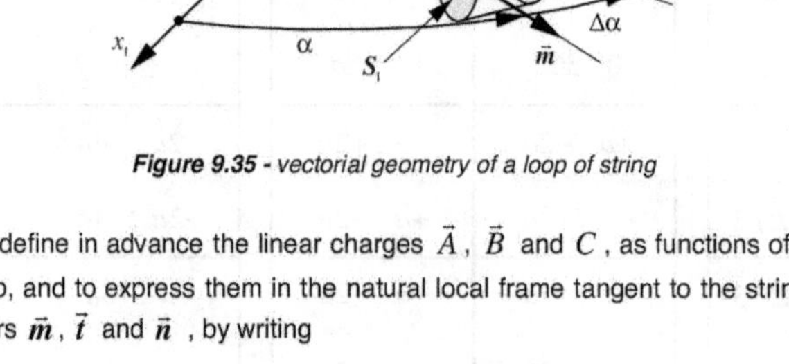

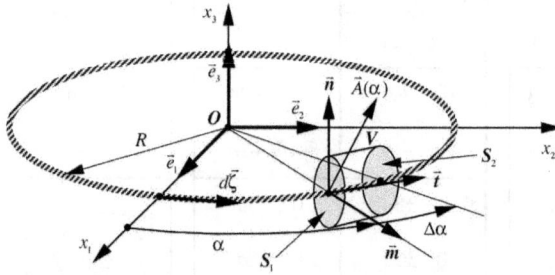

Figure 9.35 *- vectorial geometry of a loop of string*

It is useful to define in advance the linear charges $\vec{A}$, $\vec{B}$ and C, as functions of the angle α along the loop, and to express them in the natural local frame tangent to the string, defined by the unit vectors $\vec{m}$, $\vec{t}$ and $\vec{n}$, by writing

$$\begin{cases} \vec{A} = A_m(\alpha)\vec{m} + A_t(\alpha)\vec{t} + A_n(\alpha)\vec{n} \\ \vec{B} = B_m(\alpha)\vec{m} + B_t(\alpha)\vec{t} + B_n(\alpha)\vec{n} \\ C = C(\alpha) \end{cases} \tag{9.50}$$

In the case of the divergence operator, it is possible to calculate the integral of $c = \operatorname{div}\vec{b}$ on a volume element V contained in the angular region $[\alpha, \alpha + \Delta\alpha]$ (fig. 9.35). With the divergence theorem we have

$$\iiint_V c\,dV = \iiint_V \operatorname{div}\vec{b}\,dV = \oiint_S \vec{b}\,d\vec{S} = \vec{t}\,\vec{B}\Big|_{S_2} - \vec{t}\,\vec{B}\Big|_{S_1} \tag{9.51}$$

which we can write by expressing the integral on the volume as

$$R\Delta\alpha \iint_{S_1} c\,dS = R\Delta\alpha C = \left[\vec{B}(\alpha + \Delta\alpha) - \vec{B}(\alpha)\right]\vec{t} \tag{9.52}$$

so that C is given by the relationship

$$C = \lim_{\Delta\alpha \to 0} \frac{1}{R\Delta\alpha}\left[\vec{B}(\alpha + \Delta\alpha) - \vec{B}(\alpha)\right]\vec{t} \tag{9.53}$$

Taking the limit $\Delta\alpha \to 0$, we have the expression for the divergence operator applied to the linear density $\vec{B}$ which represents the central fibre in the local frame of the string

$$c = \operatorname{div}\vec{b} \quad \Rightarrow \quad C = \frac{1}{R}\frac{\partial \vec{B}}{\partial \alpha}\vec{t} \tag{9.54}$$

In the case of the operator $b = \overrightarrow{\operatorname{rot}}\,\vec{a}$, a similar procedure gives us

$$R\Delta\alpha \iint_{S_1} \vec{b}\,dS = \iiint_V \overrightarrow{\operatorname{rot}}\,\vec{a}\,dV = \oiint_S d\vec{S} \wedge \vec{a} = \vec{t} \wedge \vec{A}\Big|_{S_2} - \vec{t} \wedge \vec{A}\Big|_{S_1} \tag{9.55}$$

which, when we use the fact that $\vec{t} \wedge \vec{A} = (\vec{t} \wedge \vec{m})A_m + (\vec{t} \wedge \vec{n})A_n = A_n\vec{m} - A_m\vec{n}$, can be written

$$\vec{B} = \lim_{\Delta\alpha \to 0} \frac{1}{R\Delta\alpha}\left\{\left[A_n(\alpha + \Delta\alpha) - A_n(\alpha)\right]\vec{m} - \left[A_m(\alpha + \Delta\alpha) - A_m(\alpha)\right]\vec{n}\right\} \tag{9.56}$$

Taking the limit $\Delta\alpha \to 0$, we have the rotational operator expression applied to the linear density $\vec{A}$ of the line representing the central fibre in the local frame of the string

$$b = \overrightarrow{\operatorname{rot}}\,\vec{a} \quad \Rightarrow \quad \vec{B} = \frac{1}{R}\left(\frac{\partial A_n}{\partial \alpha}\vec{m} - \frac{\partial A_m}{\partial \alpha}\vec{n}\right) \tag{9.57}$$

Operators acting on the linear charges of a loop *(9.54)* and *(9.57)* still deserve a few comments:

- choosing the orientation of the loop as shown in figure 9.42, i.e. $\vec{n}$ parallel to $\vec{e}_3$, the natural unit vectors of the natural frame tangent to the string can be expressed in the local frame of the observer ***PO***, and vice versa

$$\begin{cases} \vec{t} = \cos\alpha\,\vec{e}_2 - \sin\alpha\,\vec{e}_1 \\ \vec{m} = \cos\alpha\,\vec{e}_1 + \sin\alpha\,\vec{e}_2 \\ \vec{n} = \vec{e}_3 \end{cases} \quad \text{and} \quad \begin{cases} \vec{e}_1 = -\vec{t}\sin\alpha + \vec{m}\cos\alpha \\ \vec{e}_2 = \vec{t}\cos\alpha + \vec{m}\sin\alpha \\ \vec{e}_3 = \vec{n} \end{cases} \tag{9.58}$$

- the derivatives of the unit vectors $\vec{m}$, $\vec{t}$ and $\vec{n}$ with respect to the angle α can be written

$$\frac{\partial \vec{t}}{\partial \alpha} = -\vec{m} \quad ; \quad \frac{\partial \vec{m}}{\partial \alpha} = \vec{t} \quad ; \quad \frac{\partial \vec{n}}{\partial \alpha} = 0 \tag{9.59}$$

- in the case of a rectilinear string, the operators *(9.54)* and *(9.57)* can be expressed by replacing the derivative in α by a derivative in ζ, *the curvilinear index along the central fibre of the string*, by using the fact that

$$\lim_{R\to 0} \frac{1}{R}\frac{\partial}{\partial \alpha} = \frac{\partial}{\partial \zeta} \qquad (9.60)$$

Glide loops and prismatic loops of dislocations

Now if we want to study the effect of the curvature of a dislocation in the local framework, the easiest method is to consider a circular dislocation loop, and describe it using mathematical tools developed in the previous section. For a circular dislocation loop of radius R , the tensorial linear charge $\vec{\Lambda}_i$ can be connected to the Burgers vector using equation *(9.7)*

$$\vec{\Lambda}_i = -\vec{t}\,\boldsymbol{B}_i \qquad (9.61)$$

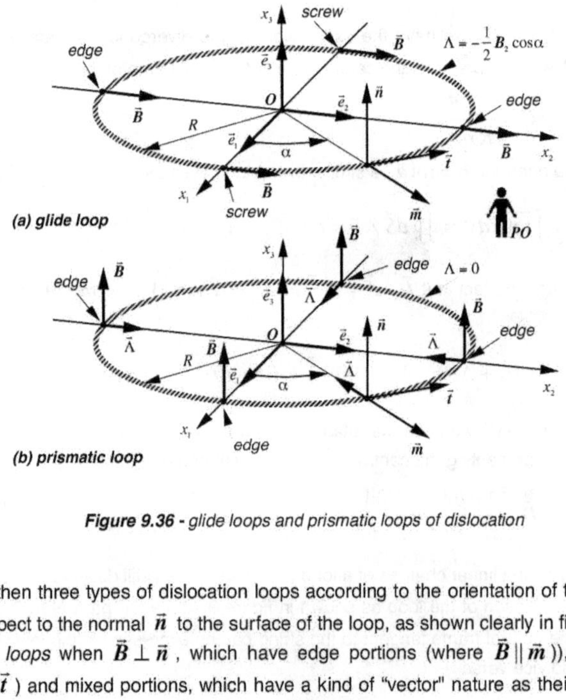

***Figure 9.36** - glide loops and prismatic loops of dislocation*

It appears then three types of dislocation loops according to the orientation of the Burgers vector with respect to the normal $\vec{n}$ to the surface of the loop, as shown clearly in figure 9.36:

- the *glide loops* when $\vec{B} \perp \vec{n}$, which have edge portions (where $\vec{B} \parallel \vec{m}$)), screw portions (where $\vec{B} \parallel \vec{t}$) and mixed portions, which have a kind of "vector" nature as their Burgers vector can take any orientation in the plane perpendicular to $\vec{n}$,
- the *prismatic loops* when $\vec{B} \parallel \vec{n}$, which have a kind of "scalar" nature, since their Burgers vector have a given direction,
- the *mixed loops* when $\vec{B}$ has a component in the direction $\vec{n}$ and a component in the plane of the loop.

We can verify that these dislocation loops satisfy the conservation equation $\operatorname{div}\vec{\lambda}_i = 0$ for the charges of distortion. Indeed using relation *(9.54)*, we have

$$\operatorname{div}\vec{\lambda}_i = 0 \;\Rightarrow\; \frac{1}{R}\frac{\partial \vec{\Lambda}_i}{\partial \alpha}\vec{t} = -\frac{1}{R}\frac{\partial\left(\boldsymbol{B}_i\vec{t}\right)}{\partial \alpha}\vec{t} = -\frac{1}{R}\boldsymbol{B}_i\frac{\partial \vec{t}}{\partial \alpha}\vec{t} - \frac{1}{R}\frac{\partial \boldsymbol{B}_i}{\partial \alpha} = -\frac{1}{R}\frac{\partial \boldsymbol{B}_i}{\partial \alpha} = 0 \qquad (9.62)$$

This relationship implies that $\partial B_i / \partial\alpha \equiv 0$, and therefore that B_i does not depend on the angle α, which corresponds to the expression of the invariant component of the Burgers vector of a dislocation line. We can deduce the linear scalar charge Λ of rotation of the dislocation loops by using relation *(9.8)*

$$\Lambda = -\frac{1}{2}\vec{B}\vec{t} = -\frac{1}{2}\vec{B}\left(\cos\alpha\,\vec{e}_2 - \sin\alpha\,\vec{e}_1\right) = -\frac{1}{2}\left(B_2\cos\alpha - B_1\sin\alpha\right) \tag{9.63}$$

As $\vec{B}$ does not depend on α, it is deduced that the slip dislocation loops have a *dipole moment of rotation charges*. Integrating the linear density Λ of torsional charge on the contour of the loop, it is deduced that the dislocation loops show *no global scalar charge rotation* q_λ $(q_\lambda = 0)$, whichever their nature

$$q_\lambda = \oint_C \Lambda R\,d\alpha = -\frac{1}{2}R\oint_C \left(B_2\cos\alpha - B_1\sin\alpha\right)d\alpha \equiv 0 \tag{9.64}$$

We can still deduce the vectorial linear charge $\vec{\Lambda}$ of flexion of the dislocation loops using the relations *(9.9)* and *(9.58)*. It follows

$$\vec{\Lambda} = \vec{B} \wedge \vec{t} = \vec{B} \wedge \left(\cos\alpha\,\vec{e}_2 - \sin\alpha\,\vec{e}_1\right) \tag{9.65}$$

We then deduce the linear scalar curvature charge Θ of the dislocation through the relationship $\theta = \operatorname{div}\vec{\lambda}$ and the expressions *(9.54)* and *(9.65)*

$$\theta = \operatorname{div}\vec{\lambda} \;\Rightarrow\; \Theta = \frac{1}{R}\frac{\partial\left(\vec{B}\wedge\vec{t}\right)}{\partial\alpha}\vec{t} = \frac{1}{R}\left(\vec{B}\wedge\frac{\partial\vec{t}}{\partial\alpha}\right)\vec{t} = -\frac{1}{R}\left(\vec{B}\wedge\vec{m}\right)\vec{t} \tag{9.66}$$

so that

$$\Theta = -\frac{1}{R}\vec{n}\vec{B} = -\frac{1}{R}B_3 \tag{9.67}$$

As B_3 may only be a constant independent of α, this relationship implies that the *prismatic dislocation loops* have a total scalar curvature charge q_θ given by the integral of the linear density Θ on the contour of the loop

$$q_\theta = \oint_C \Theta R\,d\alpha = -\oint_C \vec{n}\vec{B}\,d\alpha = -2\pi\vec{n}\vec{B} = -2\pi B_3 \tag{9.68}$$

As $\vec{B} = \vec{t} \wedge \vec{\Lambda}$, we also have

$$q_\theta = -2\pi\vec{n}\left(\vec{t}\wedge\vec{\Lambda}\right) = 2\pi\vec{\Lambda}\vec{m} \tag{9.69}$$

The effect of a uniform variation of volume expansion on the curvature charge q_θ

As the scalar charge of curvature q_θ of a prismatic loop does not depend on the radius R of the dislocation loop, but only on the component B_3 of the Burgers vector, the dependency of the curvature charge q_θ on the local volume expansion τ of the medium is easily deduced

$$q_\theta = -2\pi B_3 = -2\pi B_{03}\,e^{\tau/3} = q_{\theta 0}\,e^{\tau/3} \tag{9.70}$$

Quantification of the dislocation loops on a lattice

At the microscopic scale of a solid lattice, the Burgers vectors of dislocation loops are quantized, as shown schematically in figure 9.44 for prismatic loops and glide loops in a cubic lattice. In figure 9.37, there is also clear that:

- the prismatic dislocation loops, which are of a "scalar" nature are obtained by the addition or the deletion of a plane of particles within the loop (translation perpendicular to the plane of the loop), so that the system presents extra material in the loop plane. Note that the overall scalar charge q_θ of diverging curvature of the prismatic loop is directly related to the existence of this "extra material".
- the glide dislocation loops, which are of "vectorial" nature are obtained by slip (translation parallel to the plane of the loop) in the direction of the Burgers vector, so that the lattice does not have "extra-material" in this case. However, the presence of a screw component in regions where $\vec{B} \parallel \vec{t}$ induces a dipolar field of rotation in the vicinity of the glide loop.

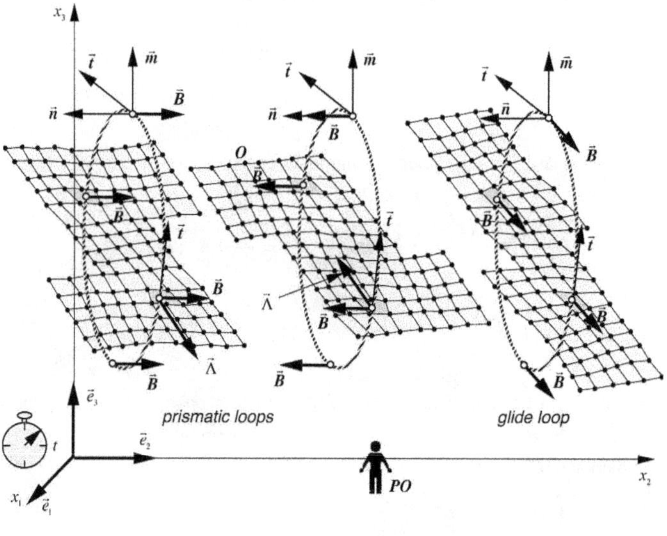

Figure 9.37 *- quantification of the dislocation loops on a lattice*

The loops of wedge disclinations

In figure 8.9, we have shown an embodiment of a macroscopic loop of wedge disclination. Now let's try to express the mathematics of it. For this, consider a loop consisting of a doublet of wedge disclinations bound by a virtual ribbon of dislocation, as shown in figure 9.38. According to the relationship *(9.40)*, the linear density of scalar charge of curvature Θ_1 and Θ_2 of the two disclinations are given by the relations

$$\begin{cases} \Theta_1 = -\vec{\Omega}_1 \vec{t} = -\vec{\Pi} \vec{m} \\ \Theta_2 = -\vec{\Omega}_2 \vec{t} = \vec{\Pi} \vec{m} = -\Theta_1 \end{cases} \tag{9.71}$$

From this we deduce the existence of these two densities Θ_1 and Θ_2 on either side of the dislocation ribbon which generates a dipole field of flexion $\vec{\chi}_{dipolaire}$, located essentially in the vicinity of the two disclinations. This dipole field is shown in figure 9.39 in the case of a doublet of disclinations quantified at $\pm 90^\circ$ in a cubic structure. It is easy to see the positive and negative flexion curvatures surrounding the two disclinations.

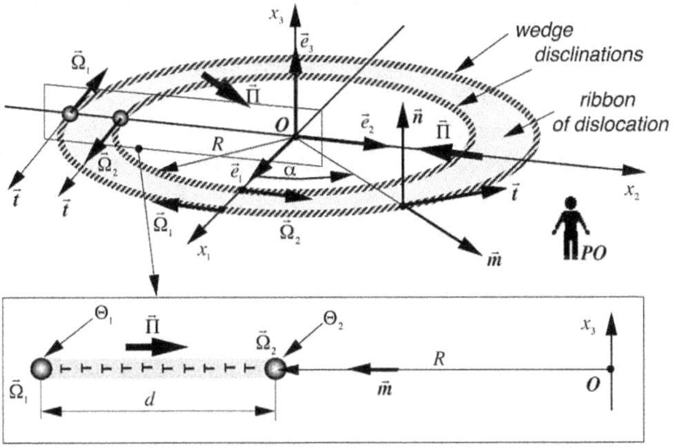

Figure 9.38 *- loop of doublet of wedge disclinations linked by a ribbon of dislocation*

It is interesting to see that the ribbon of dislocation of surface flexion charges $\vec{\Pi}$ can be reduced, by integration over the distance d separating the two disclinations, to a virtual linear charge $\vec{\Lambda}$ of edge dislocation distributed over a radius dislocation loop $R+d/2$. Thus, the disclinations of the doublet loop may be considered similar to a charge of edge dislocation loop with linear charge $\vec{\Lambda}$ such that

$$\vec{\Lambda} = \vec{\Pi} d = -d\Theta_1 \vec{m} \tag{9.72}$$

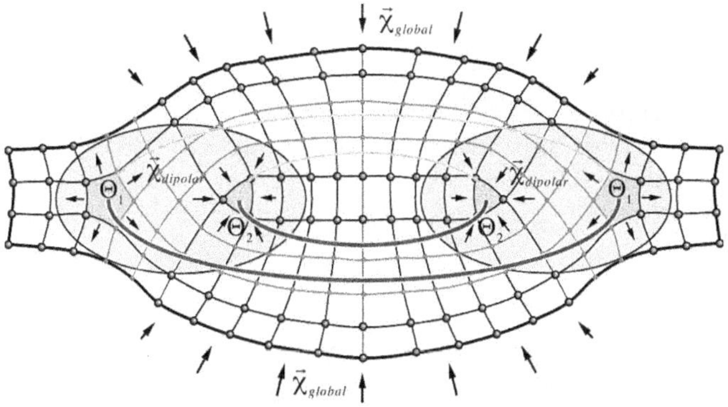

Figure 9.39 *- local dipolar field of flexion due to charges Θ_1 and Θ_2 and divergent flexion field due to the charge Q_θ of a loop of doublet of wedge disclinations*

The relationship *(9.69)* can be used to calculate the overall scalar charge of curvature q_θ of the virtual edge dislocation loop, and we obtain, knowing that $\vec{B} = \vec{t} \wedge \vec{\Lambda}$

$$q_\theta = 2\pi\vec{\Lambda}\vec{m} = 2\pi B_3 = 2\pi d\vec{\Pi}\vec{m} = -2\pi d\Theta_1 \qquad (9.73)$$

One can also imagine that the diameter of the inner loop disclination goes to zero and there remains only the external disclination loop of linear charge Θ_1 and radius d. In this case, the overall charge of curvature q_θ remains unchanged and is always given by the same equation *(9.73)*. The global charge of curvature q_θ is the one due to the dislocation ribbon as a whole, and the one that is seen at large enough distance for the loop to be no longer possible to distinguish this loop from a simple edge dislocation loop. The charge q_θ is then responsible for the overall flexion $\vec{\chi}_{global}$ of the lattice at large scale, as figure 19.46 illustrates in the case of a doublet of quantified disclinations of $\pm 90°$ in a cubic structure.

The loops of twist disclinations

In figure 8.8, we have shown a macroscopic implementation of a loop of twist disclination[16]. We will give their mathematical description here. For this, consider a loop consisting of a twist disclination generated by a rotation $\vec{\Omega}_{twist}$ of the upper plane at an angle α relative to the lower plane, as shown in figure 9.40.

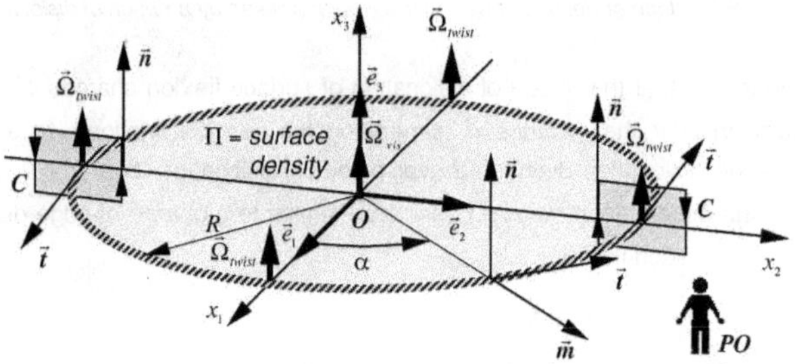

Figure 9.40 *- loop of twist disclination with its charged membrane*

The fact that one glues together two planes which have been moved relative to each other within the loop must create on the plane of the loop a surface charge Π of dislocation. By using relation *(9.35)*, one can write on the contour C surrounding the edge of the loop, and knowing that the surface charge Π comes from a volume density λ distributed in the thickness of the membrane with height h

$$\vec{\Omega}_{twist} = -\oint_C \lambda\, d\vec{r} = -\vec{n}\lambda h = -\vec{n}\Pi \qquad (9.74)$$

in such a way that the angle of rotation α is written

$$\alpha = \Omega_{twist} = -\Pi \qquad (9.75)$$

The surface charge Π of rotation can be integrated on the surface of the loop, giving us the overall charge of rotation q_λ of such a loop

[16] *W. Huang, T. Mura, Elastic J. of Applied Physics, vo. 41(13), p. 5175, 1970; A. Unzicker, arXiv:gr-qc/0011064, 2000; A. Unzicker, arXiv:gr-qc/9612061v2, 2010*

$$q_\lambda = \iint\limits_S \Pi dS = \pi R^2 \Pi \tag{9.76}$$

This global charge q_λ is actually the charge of rotation of the loop of twist disclination as seen at a great distance of the loop. This means that such a loop can act as the source of a divergent field of rotation $\vec{\omega}$ in the solid medium.

Note that it is possible to see a disclination loop somewhat differently. Indeed, the act of carrying out the rotation of the two planes with respect to the other causes a displacement along the string similar to that of a screw dislocation. The Burgers vector and the linear charge of this pseudo-dislocation is then worth

$$\vec{B}_{screw} = R\alpha\vec{t} = -R\Pi\vec{t} \quad \Rightarrow \quad \Lambda_{screw} = -\frac{1}{2}\vec{B}_{screw}\vec{t} = \frac{1}{2}R\Pi \tag{9.77}$$

so that the global charge of this pseudo-loop can be written

$$q_\lambda = \oint \Lambda_{screw}\, ds = 2\pi R \Lambda_{screw} = \pi R^2 \Pi \tag{9.78}$$

We therefore obtain much the same value of the overall charge q_λ that was obtained by considering the surface charge Π, which allows either consider this singularity as a twist disclination loop or as a screw pseudo-dislocation loop!

The effects of a variation of volume expansion on the torsional charge q_λ

Since the scalar charge of torsion q_λ of a twist disclination loop depends on R^2 and Π, and as the surface charge Π is independent of the local volume expansion of the medium τ *(9.30)*, the dependence of the overall charge of rotation q_λ of screw disinclination loops on the local volume expansion τ is easily deduced

$$q_\lambda = \Pi \pi R^2 = \Pi_0 \pi\, R_0^2\, e^{2\tau/3} = q_{\lambda 0}\, e^{2\tau/3} \tag{9.79}$$

On transforming a rectilinear string in a closed loop

A string of dislocation, disclination and/or rectilinear dispiration is fully characterized by its three linear charges Λ, $\vec{\Lambda}$ and Θ. Its linear scalar charge Λ is connected to the screw dislocation portion $\vec{B}_{screw} = -\Lambda\vec{t}$, its linear vectorial charge $\vec{\Lambda}$ is connected to its edge dislocation portion $\vec{B}_{edge} = \vec{t} \wedge \vec{\Lambda}$, and its linear scalar charge Θ is connected to its wedge disclination part $\vec{\Omega}_{wedge} = -\Theta\vec{t}$ (fig. 9.41a). From a straight string satisfying the conditions set out above, it is possible to construct a loop, by bending the string and closing it on itself to form a loop of radius R. For this, there are rules to follow, that we will now discuss.

The edge dislocation part of the string, represented by its Burgers vector $\vec{B}_{edge} = \vec{t} \wedge \vec{\Lambda}$ has to be conserved to meet the conservation equation $\operatorname{div}\vec{\lambda}_i = 0$. To satisfy this conservation necessarily involves bending the rope in the plane perpendicular to the Burgers vector $\vec{B}_{edge}$, which also implies that the vector density $\vec{\Lambda}$ is contained in the plane of the resulting loop (fig. 9.41b). Thus, the loop can only be a *prismatic loop* vis-à-vis its Burgers vector $\vec{B}_{edge}$, so it has a scalar charge of curvature $q_\theta^{disloc} = -2\pi\vec{n}\vec{B}_{edge} = 2\pi\vec{\Lambda}\vec{m}$ according to equation *(9.68)*.

If there is additionally a part of wedge disclinations on the linear string, its scalar charge $\Theta \neq 0$ will generate a virtual dislocation membrane of surface charge $\Theta \neq 0$ within the loop, as shown by the example in figure 9.37. Thus, the portion of the wedge disclination of the linear string leads to a loop of wedge disclination. Using the relation *(9.73)* in which d is replaced by R, one easily obtains the contribution $q_\theta^{disclin} = -2\pi R\Theta = 2\pi R\vec{\Pi}\vec{m}$ of this disclination part to the *scalar charge of curvature.* We deduce that the overall charge of curvature q_θ of the loop due both to the edge dislocation and the wedge disclination of the initial string is worth

$$q_\theta = -2\pi\vec{n}\vec{B}_{edge} - 2\pi R\Theta = 2\pi\vec{\Lambda}\vec{m} + 2\pi R\vec{\Pi}\vec{m} \tag{9.80}$$

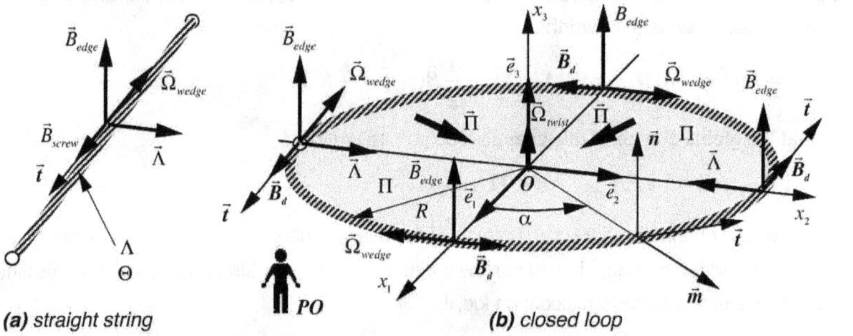

***Figure 9.41** - transformation of a rectilinear string in a closed loop string*

The fact of having closed on itself a string containing a linear charge, associated with a screw Burgers vector $\vec{B}_{screw} = -\Lambda\vec{t}$, implies that this screw Burgers vector $\vec{B}_{screw}$ can be a dislocation vector conserved along the loop, which appears quite paradoxical at first sight. But it is perfectly possible precisely because Λ is a scalar. By closing the string on itself, the screw dislocation $\vec{B}_{screw}$ loses its dislocation identity in favor of a disclination loop of the same type as the one described in figure 9.39. Then former Burgers vector $\vec{B}_{screw}$ becomes a simple displacement $\vec{B}_d$ along the loop, connected with the rotation $\vec{\Omega}_{twist}$ of the upper plane at an angle $\alpha = \Omega_{twist} = -\Pi$ relative to the lower plane of the loop. There must therefore appear a surface charge of rotation Π on the plane of the loop, such that

$$\vec{B}_d = R\alpha\vec{t} = -R\Pi\vec{t} \tag{9.81}$$

But the conservation of the length $\left|\vec{B}_{screw}\right|$ which becomes $\left|\vec{B}_d\right|$ during the process of curving of the string to form a loop, implies that

$$\vec{B}_{screw} = -\Lambda\vec{t} \;\rightarrow\; \vec{B}_d = -R\Pi\vec{t} \;\Rightarrow\; \Pi = \Lambda / R \tag{9.82}$$

The initial linear charge Λ of rotation of the string disappears with the creation of a scalar surface charge Π of rotation of the loop, satisfying the simple transformation relation $\Pi = \Lambda / R$. The loop becomes then a loop of twist disclination vis-a-vis of the ex-vector of Burgers $\vec{B}_{screw}$, which carries then a global scalar charge q_λ *of rotation*, with

$$q_\lambda = \pi R\Lambda = \pi R^2\Pi \tag{9.83}$$

The transformation rules for converting a rectilinear loop in a closed loop are summarized in table 9.5.

Rectilinear string	$\Rightarrow$	*Closed loop*	*Associated charge*
screw dislocation $\Lambda \Rightarrow \vec{B}_{screw} = -\Lambda\vec{t}$	$\Rightarrow$	loop of twist disclination $\Pi = \dfrac{\Lambda}{R} \Rightarrow \vec{\Omega}_{twist} = -\Pi\vec{n} = -\dfrac{\Lambda}{R}\vec{n}$	$q_\lambda = \pi R \Lambda$ $= \pi R^2 \Pi$ $= -\pi R^2 \vec{\Omega}_{twist}\vec{n}$
edge dislocation $\vec{\Lambda} \Rightarrow \vec{B}_{edge} = \vec{t} \wedge \vec{\Lambda}$	$\Rightarrow$	loop of prismatic dislocation $\vec{\Lambda} \Rightarrow \vec{B}_{edge} = \vec{t} \wedge \vec{\Lambda}$	$q_\theta = 2\pi\vec{\Lambda}\vec{m}$ $= -2\pi\vec{n}\vec{B}_{edge}$
wedge disclination $\Theta \Rightarrow \vec{\Omega}_{wedge} = -\Theta\vec{t}$	$\Rightarrow$	loop of wedge disclination $\vec{\Pi} = -\Theta\vec{m} \Rightarrow \vec{\Omega}_{wedge} = (\vec{\Pi}\vec{m})\vec{t} = -\Theta\vec{t}$	$q_\theta = -2\pi R \Theta$ $= 2\pi R \vec{\Pi}\vec{m}$ $= 2\pi R \vec{\Omega}_{wedge}\vec{t}$

Table 9.5 *- rules of transformation of a rectilinear string into a closed loop string*

The quantification of the radius of a localized loop in an ordered lattice

A localized loop can contain, as we have seen, a global scalar charge q_θ of curvature and/or a global scalar charge q_λ of rotation. We can now show that the existence of a non null scalar charge of rotation q_λ may involve, under certain conditions, a quantization of the radius of the loop R if it is in an ordered array. Indeed, if we assume that, during the formation of the loop, the environment in question requires that the upper jaw closes again to the lower jaw in the orientation of the network structure, the angle of rotation $\vec{\Omega}_{twist}$ can then take quantized values $\Omega_{(q)}$ dependent on the lattice structure (for example, $\Omega_{(q)} = \pm n\pi / 2$ in a lattice having a cubic symmetry or $\Omega_{(q)} = \pm n\pi / 3$ in a lattice having a hexagonal symmetry). This implies that the surface charge $\Pi = -\Omega_{twist}$ also takes the quantized values $\Pi_{(q)} = -\Omega_{(q)}$. But since, on the other hand, the length of the Burgers vectors $\vec{B}_{screw}$ of the original string must also take quantized values in an ordered array, the linear charge Λ also has quantized values $\Lambda_{(k)}$.
The result of the quantization of Π and Λ is immediate, and results in a quantization of the radius of the loop R, which can only take eigenvalues $R_{(q,k)}$

$$R_{(q,k)} = \Lambda_{(k)} / \Pi_{(q)} \tag{9.84}$$

There are several interesting consequences of the quantification of the radius of twist disclinations. We will list them here:

- the first and very important result lies in the fact that the quantification of R automatically entails a *quantization of the overall scalar charge of torsion* q_λ of the loop, since in this case

$$q_\lambda = \pi R_{(q,k)}^2 \Pi_{(q)} = \pi R_{(q,k)} \Lambda_{(k)} = \pi \Lambda_{(k)}^2 / \Pi_{(q)} \tag{9.85}$$

- the second interesting result is linked to expressions of the quantized radius and the quantized torsional charge, i.e. $R_{(q,k)} = \Lambda_{(k)} / \Pi_{(q)}$ and $q_\lambda = \pi \Lambda_{(k)}^2 / \Pi_{(q)}$, as they clearly show that the radius and overall charge of rotation of the loop increase with the value of the charge of the string $\Lambda_{(k)}$ from which the loop is formed. This implies among other things that there cannot be a linear relationship between q_λ and $\Lambda_{(k)}$, for a series of given loops, unless the ratio $\Lambda_{(k)} / \Pi_{(q)}$ remains constant, which also implies that the radius does not change from one loop to another.

- the third important consequence is that the value of the global charge of curvature q_θ may depend on the quantification of R, but only its wedge disclination part $2\pi R \vec{\Pi m}$ depends on it, while its edge dislocation part does not depend on it.

- finally, for a loop whose charge of rotation q_λ is zero, there are no conditions requiring quantification of the radius R of the loop in an ordered array, so that the radius can take any value.

9.6 – Clusters of dislocations, disclinations and dispirations

Since the strings of dislocations, disclinations and dispirations containing nonzero tensorial densities of charges can be closed on themselves in loops, it is possible to imagine the existence of small clusters of very localized such loops within a solid medium, such as the cluster shown in the absolute frame of reference in figure 9.42.

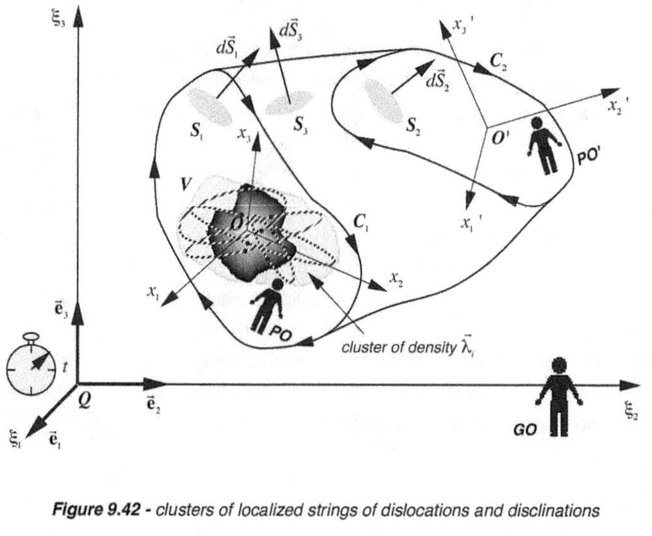

***Figure 9.42** - clusters of localized strings of dislocations and disclinations*

Such clusters are in principle fully characterized by their tensorial density $\vec{\lambda}_i$ of dislocation charges, which are non-zero in the strings, in the area represented by a hatched area in figure 9.42. By then defining a volume V such that its surface S_1 intersects the cluster of charges, the cutoff section is represented by the dark gray area in the figure 9.42, it is possible to break the surface surrounding the volume V into three parts S_1, S_2 and S_3. Then using the relationship *(9.5)*, we can show that, in this case, the components B_i of the Burgers vectors, and consequently also the components Ω_i of the Frank vectors obtained in the local referential $Ox_1x_2x_3$ by integration on the surface S_1 going through the cluster are identically zero

$$\boldsymbol{B}_i = -\iint_{S_1} \vec{\lambda}_i \, d\vec{S}_1 = -\iint_{S_2} \vec{\lambda}_i \, d\vec{S}_2 = \boldsymbol{B}_i \,' \tag{9.86}$$

This is due to the fact that each loop of strings or super-strings of the cluster cuts through the surface S_1 twice, once in one direction and the second time in the opposite direction, so that the contributions associated with each of these crossings exactly offset.

In the local referential $Ox_1x_2x_3$ the lack of vectors $\vec{B}$ and $\vec{\Omega}$ which are nonzero on all the contour surrounding the cluster without crossing it implies that there are no discontinuities in the virtual field of displacement $\vec{u}^{(\delta)}$ and no discontinuity in the rotation field in the solid surrounding the cluster, and that, therefore, the solid *remains perfect outside of the cluster.*

However, the presence of a cluster in the solid must certainly involve an elastic and inelastic distortion field of the *perfect solid surrounding the cluster* and this up to a certain distance thereof. This is analogous to the way the presence of a localized density ρ of electrical charges involves an electric field of displacement $\vec{D}$ at a distance of such charges.

To find this field, it is necessary to introduce here the fact that there is, apart from the tensorial charge conservation equation $\operatorname{div}\vec{\lambda}_i \equiv 0$, no restrictions on the scalar densities λ and θ of charges of rotation and flexion. Therefore, it is to possible, depending on the nature of the charges in the cluster, to have *non-zero global scalar charges* Q_λ *of rotation and* Q_θ *of curvature.* These global scalar charges are defined in the local referential frame by the sums on all closed loops of global charges $q_{\lambda(i)}$ and $q_{\theta(i)}$ as we have defined them for each individual loop in the previous section [17]. These considerations are then used to find the fields of elastic and anelastic distortions at long distances due to the presence of a localized cluster of charges, such as we are going to show in the following sections.

The divergent torsion field associated with a localized cluster of charges

The geometrocompatibility equation $\lambda = \operatorname{div}\vec{\omega}$ for the rotation vector $\vec{\omega}$ gives us the scalar density of torsional charges λ , linked to the trace of the tensorial density $\vec{\lambda}_i$ of distortion charges. It allows us to write a full relationship of compatibility in the following form in the local frame $Ox_1x_2x_3$

$$\oiint_S \vec{\omega} d\vec{S} = \iiint_V \operatorname{div}\vec{\omega} \, dV = \iiint_V \lambda \, dV = Q_\lambda \neq 0 \tag{9.87}$$

This relationship implies that a global scalar charge of rotation Q_λ , localized and non-zero, acts as the source of a divergent field of rotation $\vec{\omega}$ of the perfect solid in the neighborhood of the cluster of charges (fig. 9.43). The rotational field then has a *topological singularity* where is the cluster of charge Q_λ , and its norm $|\vec{\omega}|$ is a decreasing function of $1/R^2$ as a function of the distance R from the cluster.

Everything happens in fact exactly as in electromagnetism, where a local density ρ of electric

---

[17] *One can imagine for example that the curvature charge of a cluster can be due to prismatic dislocation loops of lacunar or interstitial nature (fig. 19.44) and/or to wedge disclination loops (fig. 9.39), and that the rotation charge can be due to twist disclination loops (fig. 9.40).*

charges leads to a localized macroscopic electric charge Q, which behaves like a singularity responsible for an electric displacement field $\vec{D}$ which is divergent in the neighboring space.

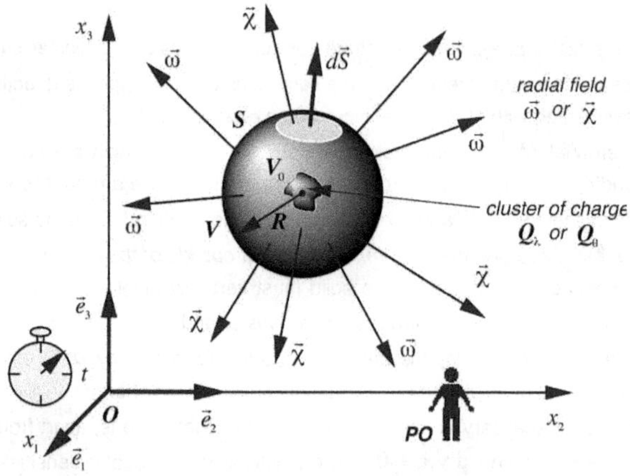

***Figure 9.43** - divergent fields of rotation and flexion in the neighborhood of a cluster of charges*

The divergent flexion field associated with a localized cluster of charges

The equation of geometro-compatibility $\theta = \operatorname{div}\vec{\chi}$ for the vector $\vec{\chi}$ of flexion shows the scalar charge density θ of curvature. It allows us to write a full relationship of compatibility in the following form in the local referential frame $Ox_1x_2x_3$

$$\oiint_S \vec{\chi}d\vec{S} = \iiint_V \operatorname{div}\vec{\chi}\, dV = \iiint_V \theta\, dV = Q_\theta \neq 0 \qquad (9.88)$$

This relationship implies that a macroscopic scalar charge of curvature Q_θ, localized and non-zero, acts as the source of a divergent flexion field $\vec{\chi}$ within the perfect solid surrounding the cluster of charges (fig. 9.43). The flexion field then has a *topological singularity* where is the charge Q_θ, and its norm $|\vec{\chi}|$ also shows a $1/R^2$ decrease as a function of the distance R from the cluster. In other words, in the vicinity of a localized overall charge of curvature Q_θ, the solid bends by flexion with a spherically symmetry around the singularity.

The singularity of the shear strains and the volume expansion in the vicinity of a cluster

As the flexion field $\vec{\chi}$ is directly related to the shear fields $\vec{\alpha}_i$ and the volume expansion τ by *(8.35)*, we can write the relation *(9.88)* in the following form in the local frame $Ox_1x_2x_3$

$$-\oiint_S \left(\sum_k \vec{e}_k \wedge \overrightarrow{\operatorname{rot}}\, \vec{\alpha}_k + \frac{2}{3}\overrightarrow{\operatorname{grad}}\,\tau \right) d\vec{S} = \iiint_V \theta\, dV = Q_\theta \neq 0 \qquad (9.89)$$

which implies that the presence of a localized global scalar charge Q_θ induces radial shear strain and volume expansion fields whose norms $|\vec{\alpha}_i|$ and $|\tau|$ must decrease as $1/R$ in the vicinity of the cluster. However, it is clear that equation *(9.89)* does not allow to determine unambiguously the fields $\vec{\alpha}_i$ and τ, unlike fields $\vec{\omega}$ and $\vec{\chi}$ that are defined unambiguously by the equations *(9.87)* and *(9.88)*. In fact, the exact determination of the distortion fields $\vec{\alpha}_i$ and τ will require an additional equation representing the equilibrium conditions of the elastic forces in the perfect solid surrounding the charges. These equation are necessarily linked to the various elastic moduli of the solid considered.

The simplified description of a solid in the presence of clusters of localized charge

Consider a hypothetical solid wherein the charges are confined in localized clusters, as illustrated for example in figure 9.44, and therefore in which there are *no strings of dislocations and disclinations propagating over large distances* compared to the scale at which the solid is studied. It is clear that, depending on the complexity of the internal structure of these clusters, i.e. the complexity of the intricate curls forming these clusters, the description of the fields of distortion and contortion within clusters can be very complex. But if these clusters have stable internal structures and that they can move individually within the solid, but not sufficiently interact with each other to change their internal structure, it is then possible to greatly simplify the description of the distortion fields prevailing in the solid.

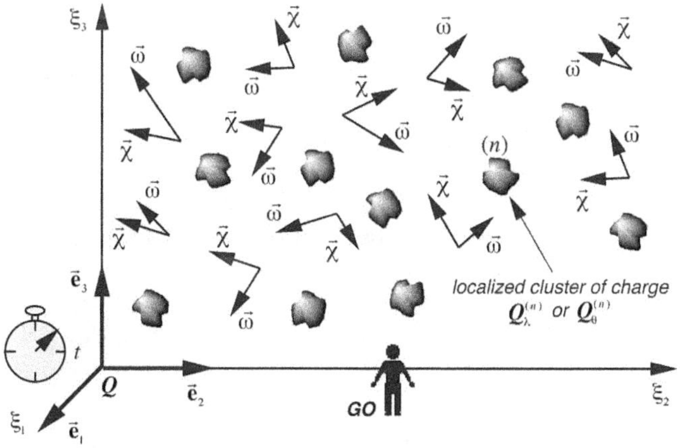

Figure 9.44 *- description of a solid containing localized clusters of charges*

In this case indeed, and so far as we are mainly concerned with describing the fields of distortion and contortion of elastic and anelastic nature in the perfect solid surrounding the clusters, that is to say at a certain distance external to the charged clusters, the problem can be solved much more simply by considering only the scalar charge densities λ and θ within the clusters, which can lead to the existence, at great distance, of two macroscopic scalar charges $Q_{\lambda(n)}$ and $Q_{\theta(n)}$ for each cluster number (n).

In fact, knowledge of the distribution of charge densities λ and θ within a cluster allows us to find purely topological conditions, and therefore independent of the elastic properties of the solid considered, which are imposed on the rotation and flexion fields $\vec{\omega}$ and $\vec{\chi}$ prevailing in the perfect solid outside of the cluster. These conditions are simply expressed by the scalar equations of geometro-compatibility $\lambda = \operatorname{div}\vec{\omega}$ and $\theta = \operatorname{div}\vec{\chi}$.

Depending on the inhomogeneity of the internal distribution of the charge densities λ and θ in the cluster, it may appear dipolar or multipolar fields $\vec{\omega}$ *and* $\vec{\chi}$ at short and medium distance of each cluster.

On the other hand, *at long distances* from the charge clusters, it is essentially the presence of non-null macroscopic scalar charges $Q_{\lambda(n)}$ and $Q_{\theta(n)}$ which will be responsible for the appearance of monopole radial fields of rotation $\vec{\omega}$ and flexion $\vec{\chi}$, as we have already shown in figure 9.43.

Thus, in this particular case of charges located in clusters, both invariant vector fields, namely the rotation $\vec{\omega}$ and flexion $\vec{\chi}$ fields, are affected at great distances from the clusters. And it is remarkable that each of these clusters can be individually and thoroughly characterized, concerning its long-range effects on the fields of distortion and contortion, from only the two macroscopic scalar charges $Q_{\lambda(n)}$ and $Q_{\theta(n)}$, although these clusters may have very complex structures, of tensorial nature and therefore highly dependent on their spatial orientation in the local referential frame.

In the analogy developed with electromagnetism, the rotation field $\vec{\omega}$ is analogous to the electric displacement $\vec{D}$, and the macroscopic charge of rotation $Q_{\lambda(n)}$ is the analog of the macroscopic electric charge Q of a particle in electromagnetism. But is there a similar analogy to the flexion field $\vec{\chi}$ and the overall charge of curvature $Q_{\theta(n)}$? A partially positive answer can be given here. Indeed, the presence of a macroscopic charge cluster $Q_{\theta(n)}$ is responsible for a nonzero vector field $\vec{\chi}$ of flexion, which is divergent in the neighborhood, and thus is also responsible for a spatial curvature of the solid lattice surrounding the cluster, which results in the appearance of nonzero fields of shear strains and volume expansion. Thus, the presence of a cluster of charges $Q_{\theta(n)} \neq 0$ implies vis-à-vis the solid lattice, a result analogous to that stipulated in the general theory of gravitation of Einstein vis-à-vis the spacetime in the presence of matter, i.e. that a matter accumulation (cluster) located in a volume of space is directly responsible for a curvature of the neighboring space-time. We will return later in detail to this analogy.

Chapter 10

Flows of dislocation charges and Orowan's relations

In this chapter, we try to interpret the concept of flow of dislocation charges, and link it to the concept of density of dislocation charges we described in the previous chapter.

We then recover the relationship between the *velocity of the macroscopic plastic deformation* described in section 7.9 and the existence of *flow of dislocation charges* in the solid medium in the form of *Orowan's relationships*.

10.1 – Interpretation of the flow of charges

The macroscopic interpretation of the tensorial density $\vec{\lambda}_i$ of dislocation charges and the conservation equation $\operatorname{div}\vec{\lambda}_i \equiv 0$ which these tensors satisfy, revealed the concepts of strings and loops of dislocation, disclination and dispiration. It was also shown that, at great distance from the clusters of plastic charges, it is essentially the two invariant vector fields, ie the fields of rotation $\vec{\omega}$ and of curvature $\vec{\chi}$, which are affected by the scalar components λ and θ of charge density. We still have to make the connection between these quantities and the flows of charges $\vec{J}_i$ and $\vec{J}$ introduced in chapter 9.

The continuity equation of the dislocation charges

Consider a tube filled with a dislocation density $\vec{\lambda}_i$, which moves with a *relative velocity* $\vec{\boldsymbol{v}}$ *with regard to the lattice*, which itself moves at velocity $\vec{\phi}$ in the local frame $Ox_1x_2x_3$ of a ***PO***. Define a mobile contour C surrounding the charge tube, with a movable surface S, which moves along with the tube at velocity $\vec{\boldsymbol{v}}+\vec{\phi}$ with respect to $Ox_1x_2x_3$, as depicted in figure 10.1.

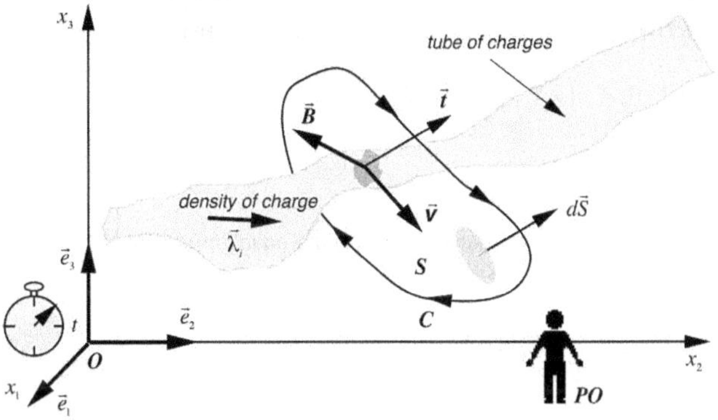

Figure 10.1 *- tube of charges $\vec{\lambda}_i$ moving with velocity $\vec{\boldsymbol{v}}+\vec{\phi}$ in the frame $Ox_1x_2x_3$*

According to relation *(9.3)*, the global Burgers vector $\vec{B}$ defined by this contour is given by

$$\vec{B} = \sum_i B_i \vec{e}_i = -\sum_i \vec{e}_i \iint_S \vec{\lambda}_i d\vec{S} \tag{10.1}$$

so that the temporal variation of the Burgers vector on the tube of charges in movement can be written

$$\frac{d\vec{B}}{dt} = \sum_i \frac{dB_i}{dt} \vec{e}_i = -\sum_i \vec{e}_i \frac{d}{dt} \iint_S \vec{\lambda}_i \, d\vec{S} \tag{10.2}$$

It is possible to apply the formula of the derivative of an integral on a mobile surface *(see appendix F)* to relation *(10.2)* and we obtain

$$\frac{d\vec{B}}{dt} = -\sum_i \vec{e}_i \iint_S \left[\frac{\partial \vec{\lambda}_i}{\partial t} + (\vec{v} + \vec{\varphi}) \operatorname{div} \vec{\lambda}_i - \overrightarrow{\operatorname{rot}} \left[(\vec{v} + \vec{\varphi}) \wedge \vec{\lambda}_i \right] \right] d\vec{S} \tag{10.3}$$

As the components of the Burgers vector are locally conserved in the referential $Ox_1x_2x_3$, in order to be a change in the overall Burgers vector $\vec{B}$ defined by the movable contour, it is necessary that there is a non-zero source of charges $\vec{S}_i^{(\vec{\lambda}_i)}$ in the lattice, so that

$$\frac{d\vec{B}}{dt} = \sum_i \frac{dB_i}{dt} \vec{e}_i = -\sum_i \vec{e}_i \iint_S \vec{S}_i^{(\vec{\lambda}_i)} \, d\vec{S} \tag{10.4}$$

For the equality between relations *(10.3)* and *(10.4)* to be satisfied for any velocity $\vec{v} + \vec{\varphi}$, it is necessary that

$$\frac{\partial \vec{\lambda}_i}{\partial t} + (\vec{v} + \vec{\varphi}) \operatorname{div} \vec{\lambda}_i - \overrightarrow{\operatorname{rot}} \left[(\vec{v} + \vec{\varphi}) \wedge \vec{\lambda}_i \right] = \vec{S}_i^{(\vec{\lambda}_i)} \tag{10.5}$$

The condition of geometro-compatibility imposes that $\operatorname{div} \vec{\lambda}_i \equiv 0$, and we obtain the following relation, defined on a mobile lattice with velocity $\vec{\varphi} \neq 0$

$$\begin{aligned} \frac{\partial \vec{\lambda}_i}{\partial t} &= \vec{S}_i^{(\vec{\lambda}_i)} + \overrightarrow{\operatorname{rot}} \left[(\vec{v} + \vec{\varphi}) \wedge \vec{\lambda}_i \right] = \vec{S}_i^{(\vec{\lambda}_i)} + \overrightarrow{\operatorname{rot}} \left[\vec{v} \wedge \vec{\lambda}_i \right] + \overrightarrow{\operatorname{rot}} \left[\vec{\varphi} \wedge \vec{\lambda}_i \right] \\ &= \vec{S}_i^{(\vec{\lambda}_i)} + \overrightarrow{\operatorname{rot}} \left[\vec{v} \wedge \vec{\lambda}_i \right] + \left[\underbrace{\vec{\varphi} \operatorname{div} \vec{\lambda}_i}_{\equiv 0} - \vec{\lambda}_i \operatorname{div} \vec{\varphi} + (\vec{\lambda}_i \vec{\nabla}) \vec{\varphi} - (\vec{\varphi} \vec{\nabla}) \vec{\lambda}_i \right] \end{aligned} \tag{10.6}$$

It then follows that the equation of continuity describing the material derivative for the density of charges in a lattice which moves at velocity $\vec{\varphi} \neq 0$ in the referential frame $Ox_1x_2x_3$

$$\frac{d\vec{\lambda}_i}{dt} = \left[\vec{S}_i^{(\vec{\lambda}_i)} - \vec{\lambda}_i \operatorname{div} \vec{\varphi} + (\vec{\lambda}_i \vec{\nabla}) \vec{\varphi} \right] + \overrightarrow{\operatorname{rot}} \left[\vec{v} \wedge \vec{\lambda}_i \right] \tag{10.7}$$

The term of sources of dislocation charges

This continuity equation can now be compared to the equation of continuity for the charges $\vec{\lambda}_i$ *(8.14)* which states that

$$\frac{d\vec{\lambda}_i}{dt} = \left[\frac{d}{dt} \left(\overrightarrow{\operatorname{rot}} \, \vec{\beta}_i \right) - \overrightarrow{\operatorname{rot}} \left(\frac{d\vec{\beta}_i}{dt} \right) \right] - \overrightarrow{\operatorname{rot}} \, \vec{J}_i \tag{10.8}$$

Thus, in the case of a charge tube moving with velocity $\vec{v}$ relative to the lattice, the term of non-commutativity between brackets, which expresses the creation of local non-compatibility due to the charges may be connected to the existence of a non-zero source of charges $\vec{S}_i^{(\vec{\lambda}_i)}$ and the terms $\vec{\lambda}_i \operatorname{div}\vec{\varphi}$ and $\left(\vec{\lambda}_i\vec{\nabla}\right)\vec{\varphi}$

$$\frac{d}{dt}\left(\overrightarrow{\operatorname{rot}}\,\vec{\beta}_i\right)-\overrightarrow{\operatorname{rot}}\left(\frac{d\vec{\beta}_i}{dt}\right)=\vec{S}_i^{(\vec{\lambda}_i)}+\left(\vec{\lambda}_i\vec{\nabla}\right)\vec{\varphi}-\vec{\lambda}_i \operatorname{div}\vec{\varphi} \qquad (10.9)$$

The term of charge sources $\vec{S}_i^{(\vec{\lambda}_i)}$ can only be due to the existence of a process of creating charges in the lattice, for example the *production of dislocations by a Frank-Read type mechanism* (see section 12.9).

The source term $\vec{\lambda}_i \operatorname{div}\vec{\varphi}$ is a decrease or an increase of the charge density $\vec{\lambda}_i$ due respectively to an *increase or a decrease of the local volume expansion* $\operatorname{div}\vec{\varphi}$. In the case where the local volume expansion *varies homogeneously*, this term ensures that charge density $\vec{\lambda}_i$ indeed changes according to the equation *(9.16)*.

Concerning the source term $\left(\vec{\lambda}_i\vec{\nabla}\right)\vec{\varphi}$, it provides an increase or a decrease of the density of charges $\vec{\lambda}_i$ due respectively to an *elongation or a shrinkage of the medium in the direction of the dislocation string* since the vectors $\vec{\lambda}_i$ are oriented in the direction of the string. In the event that the medium elongates only in the direction of the dislocation line, we can verify that this term $\left(\vec{\lambda}_i\vec{\nabla}\right)\vec{\varphi}$ compensates exactly the term $\vec{\lambda}_i \operatorname{div}\vec{\varphi}$, and thus ensures that the charge density $\vec{\lambda}_i$ does not change in this case.

But near the origin of the frame $Ox_1x_2x_3$, the velocity is, by definition, very low, and if we assume that $\vec{\boldsymbol{v}} > \vec{\varphi}$, the charge continuity equation can be written simply as follows, as we will adopt in the later parts of this chapter

$$\frac{\partial\vec{\lambda}_i}{\partial t}\cong\frac{d\vec{\lambda}_i}{dt}\cong\vec{S}_i^{(\vec{\lambda}_i)}+\overrightarrow{\operatorname{rot}}\left[\vec{\boldsymbol{v}}\wedge\vec{\lambda}_i\right] \qquad (10.10)$$

The relations between flows and densities of charges

The comparison of equations *(10.7)* and *(10.8)* allows us to write a simple correspondence between the curl operators applied to $\vec{\boldsymbol{v}}\wedge\vec{\lambda}_i$ and $-\vec{J}_i$

$$\overrightarrow{\operatorname{rot}}\left(\vec{\boldsymbol{v}}\wedge\vec{\lambda}_i\right)=-\overrightarrow{\operatorname{rot}}\,\vec{J}_i \qquad (10.11)$$

From which we deduce the relation that must exist between the charge density $\vec{\lambda}_i$ moving with velocity $\vec{\boldsymbol{v}}$ with respect to the lattice and flow of charges $\vec{J}_i$ associated with this movement

$$\vec{J}_i=\vec{\lambda}_i\wedge\vec{\boldsymbol{v}}+\overrightarrow{\operatorname{grad}}\,\boldsymbol{A}_i \qquad (10.12)$$

In this relationship, we introduced the term $\overrightarrow{\operatorname{grad}}\,\boldsymbol{A}_i$, which is a vector field, gradient of an arbitrary scalar $\boldsymbol{A}_i$, which perfectly meets the relation *(10.10)*.

We can then deduce the relationship between the vector flow of charge $\vec{J}$ and the charge density $\vec{\lambda}_i$ by using equation *(8.20)*

$$\vec{J}=-\frac{1}{2}\sum_k\vec{e}_k\wedge\vec{J}_k=-\frac{1}{2}\sum_k\vec{e}_k\wedge\left(\vec{\lambda}_k\wedge\vec{\boldsymbol{v}}+\overrightarrow{\operatorname{grad}}\,\boldsymbol{A}_k\right)=\frac{1}{2}\sum_k\left(\vec{e}_k\vec{\lambda}_k\right)\vec{\boldsymbol{v}}-\frac{1}{2}\sum_k\left(\vec{e}_k\vec{\boldsymbol{v}}\right)\vec{\lambda}_k-\frac{1}{2}\sum_k\vec{e}_k\wedge\overrightarrow{\operatorname{grad}}\,\boldsymbol{A}_k \qquad (10.13)$$

The first term of this relationship can be expressed as a function of the scalar torsion charge λ thanks to the relation *(8.16)*, and the third term is none other than the curl of the vector $\vec{\boldsymbol{A}}$ constructed with components $\boldsymbol{A}_i$

$$\vec{J} = \lambda\vec{v} - \frac{1}{2}\sum_k (\vec{e}_k \vec{v})\vec{\lambda}_k + \frac{1}{2}\overrightarrow{\mathrm{rot}}\,\vec{A} \tag{10.14}$$

With regards to the second term we can use the following relation

$$\sum_k (\vec{e}_k \vec{v})\vec{\lambda}_k = \sum_k \underbrace{(\vec{\lambda}_k \vec{v})}_{=0}\vec{e}_k - \sum_k \vec{v} \wedge (\vec{e}_k \wedge \vec{\lambda}_k) = -\vec{v} \wedge \sum_k \vec{e}_k \wedge \vec{\lambda}_k = \vec{v} \wedge \vec{\lambda} \tag{10.15}$$

in which the term $\vec{\lambda}_k\vec{v}$ is zero since $\vec{\lambda}_k$ is parallel to $\vec{t}$, while $\vec{v}$ is perpendicular to $\vec{t}$, and in which the sum of terms $\vec{e}_k \wedge \vec{\lambda}_k$ can be replaced by the relation *(8.36)*. The vectorial flow of charges $\vec{J}$ is therefore composed of two components $\vec{J}^{(\lambda)}$ and $\vec{J}^{(\vec{\lambda})}$ which depend, respectively, on the scalar charge density λ and the vectorial charge density $\vec{\lambda}$

$$\vec{J} = \vec{J}^{(\lambda)} + \vec{J}^{(\vec{\lambda})} + \frac{1}{2}\overrightarrow{\mathrm{rot}}\,\vec{A} = \lambda\vec{v} + \frac{1}{2}(\vec{\lambda} \wedge \vec{v}) + \frac{1}{2}\overrightarrow{\mathrm{rot}}\,\vec{A} \tag{10.16}$$

Finally we can calculate the trace of the vectorial flow $\vec{J}_i$ by using relation *(8.25)*

$$\frac{S_n}{n} = \sum_k \vec{e}_k\vec{J}_k = \sum_k \vec{e}_k\left(\vec{\lambda}_k \wedge \vec{v} + \overrightarrow{\mathrm{grad}}\,A_k\right) = \sum_k \vec{v}\left(\vec{e}_k \wedge \vec{\lambda}_k\right) + \sum_k \vec{e}_k\,\overrightarrow{\mathrm{grad}}\,A_k \tag{10.17}$$

By using once again relation *(8.36)*, and by remarking that the second term is the divergence of the vector $\vec{A}$ built with components A_i, we have

$$\frac{S_n}{n} = -\vec{\lambda}\vec{v} + \mathrm{div}\,\vec{A} \tag{10.18}$$

The three main relationships we have obtained between the charge density tensors and the tensors of charge flows are given in table 10.1. As we do no have, for the moment, a direct interpretation for the vector $\vec{A}$ appearing in the continuity relations, other than the fact that this vector is the origin of flows $\vec{J}_i$ and $\vec{J}$ and source S_n / n which are not directly related to charges $\vec{\lambda}_i$, $\vec{\lambda}$ and λ, and since these flows and this source disappear in the continuity equation *(10.10)* of charges, we neglect them in table 10.1.

The continuity equations of the densities of dislocation charges

For the movements of charges at homogeneous velocity $\vec{v}$ such that $\vec{v} \gg \vec{\varphi}$, it is easy to show thanks to the equation of conservation *(8.12)*, that

$$\overrightarrow{\mathrm{rot}}\left(\vec{v} \wedge \vec{\lambda}_i\right) = \vec{v}\underbrace{\mathrm{div}\,\vec{\lambda}_i}_{\equiv 0} - (\vec{v}\vec{\nabla})\vec{\lambda}_i = -(\vec{v}\vec{\nabla})\vec{\lambda}_i \tag{10.19}$$

which allows to write the equation of continuity *(10.10)* for the tensorial density $\vec{\lambda}_i$ of charges in the following equivalent forms

$$\frac{d\vec{\lambda}_i}{dt} \underset{\vec{v}\gg\vec{\varphi}}{\cong} \vec{S}_i^{(\vec{\lambda}_i)} + \overrightarrow{\mathrm{rot}}\left(\vec{v} \wedge \vec{\lambda}_i\right) = \vec{S}_i^{(\vec{\lambda}_i)} - (\vec{v}\vec{\nabla})\vec{\lambda}_i = \vec{S}_i^{(\vec{\lambda}_i)} - \overrightarrow{\mathrm{rot}}\,\vec{J}_i \tag{10.20}$$

By applying relation *(8.36)* to this last relation, we have the equation of continuity for the density $\vec{\lambda}$ of charges of flexion

$$\frac{d\vec{\lambda}}{dt} \cong -\sum_k \vec{e}_k \wedge \frac{d\vec{\lambda}_k}{dt} \underset{\vec{v}\gg\vec{\phi}}{\cong} -\sum_k \vec{e}_k \wedge \vec{S}_k^{(\vec{\lambda}_k)} + \sum_k \vec{e}_k \wedge (\vec{v}\vec{\nabla})\vec{\lambda}_k = \vec{S}^{(\vec{\lambda})} - (\vec{v}\vec{\nabla})\vec{\lambda} \tag{10.21}$$

in which we have introduced the source $\vec{S}^{(\vec{\lambda})}$ of charges $\vec{\lambda}$, defined as

$$\vec{S}^{(\vec{\lambda})} = -\sum_k \vec{e}_k \wedge \vec{S}_k^{(\vec{\lambda}_i)} \tag{10.22}$$

By using again the following relation, with the density of charges of curvature $\theta = \operatorname{div}\vec{\lambda}$

$$\overrightarrow{\operatorname{rot}}\left(\vec{\boldsymbol{v}} \wedge \vec{\lambda}\right) = -2\ \overrightarrow{\operatorname{rot}}\,\vec{J}^{(\vec{\lambda})} = \vec{\boldsymbol{v}}\operatorname{div}\vec{\lambda} - \left(\vec{\boldsymbol{v}}\vec{\nabla}\right)\vec{\lambda} = \vec{\boldsymbol{v}}\theta - \left(\vec{\boldsymbol{v}}\vec{\nabla}\right)\vec{\lambda} \tag{10.23}$$

which calls upon the density of charges of curvature $\theta = \operatorname{div}\vec{\lambda}$, the vectorial flow of charges of flexion $\vec{J}^{(\vec{\lambda})} = (\vec{\lambda} \wedge \vec{\boldsymbol{v}})/2$ and the vectorial flow of charges of curvature $\theta\vec{\boldsymbol{v}}$, the equation of continuity for the charges $\vec{\lambda}$ can be written under the following forms

$$\frac{d\vec{\lambda}}{dt} \cong \vec{S}^{(\vec{\lambda})} - \left(\vec{\boldsymbol{v}}\vec{\nabla}\right)\vec{\lambda} = \vec{S}^{(\vec{\lambda})} + \overrightarrow{\operatorname{rot}}\left(\vec{\boldsymbol{v}} \wedge \vec{\lambda}\right) - \vec{\boldsymbol{v}}\operatorname{div}\vec{\lambda} = \vec{S}^{(\vec{\lambda})} - 2\ \overrightarrow{\operatorname{rot}}\,\vec{J}^{(\vec{\lambda})} - \vec{\boldsymbol{v}}\theta \tag{10.24}$$

Table 10.1 - Relations between tensors of charge densities and tensorial flows of charges

Relations between flows and densities
for charges at velocity $\vec{\boldsymbol{v}}$ *with respect to the lattice*

$$\begin{cases} \vec{J}_i = \vec{\lambda}_i \wedge \vec{\boldsymbol{v}} & (1) \\ \vec{J} = \vec{J}^{(\lambda)} + \vec{J}^{(\vec{\lambda})} = \lambda\vec{\boldsymbol{v}} + \dfrac{1}{2}\left(\vec{\lambda} \wedge \vec{\boldsymbol{v}}\right) & (2) \\ \dfrac{S_n}{n} = -\vec{\lambda}\vec{\boldsymbol{v}} & (3) \end{cases}$$

Continuity equations for the charges in the case $\vec{\boldsymbol{v}} >> \vec{\varphi}$

$$\begin{cases} \dfrac{d\vec{\lambda}_i}{dt} \cong \vec{S}_i^{(\vec{\lambda}_i)} - \left(\vec{\boldsymbol{v}}\vec{\nabla}\right)\vec{\lambda}_i = \vec{S}_i^{(\vec{\lambda}_i)} + \overrightarrow{\operatorname{rot}}\left(\vec{\boldsymbol{v}} \wedge \vec{\lambda}_i\right) = \vec{S}_i^{(\vec{\lambda}_i)} - \overrightarrow{\operatorname{rot}}\,\vec{J}_i & (4) \\ \dfrac{d\vec{\lambda}}{dt} \cong \vec{S}^{(\vec{\lambda})} - \left(\vec{\boldsymbol{v}}\vec{\nabla}\right)\vec{\lambda} = \vec{S}^{(\vec{\lambda})} + \overrightarrow{\operatorname{rot}}\left(\vec{\boldsymbol{v}} \wedge \vec{\lambda}\right) - \vec{\boldsymbol{v}}\operatorname{div}\vec{\lambda} = \vec{S}^{(\vec{\lambda})} - 2\ \overrightarrow{\operatorname{rot}}\,\vec{J}^{(\vec{\lambda})} - \vec{\boldsymbol{v}}\theta & (5) \\ \dfrac{d\lambda}{dt} \cong S^{(\lambda)} - \left(\vec{\boldsymbol{v}}\vec{\nabla}\right)\lambda = S^{(\lambda)} - \operatorname{div}\left(\lambda\vec{\boldsymbol{v}}\right) = S^{(\lambda)} - \operatorname{div}\vec{J}^{(\lambda)} & (6) \end{cases}$$

with $$\begin{cases} \vec{S}^{(\vec{\lambda})} = -\sum_k \vec{e}_k \wedge \vec{S}_k^{(\vec{\lambda}_i)} & (7) \\ S^{(\lambda)} = \dfrac{1}{2}\sum_k \vec{e}_k \vec{S}_k^{(\vec{\lambda}_i)} & (8) \end{cases}$$

By applying relation *(8.17)* to relation *(10.20)*, we obtain the continuity equation for the density λ of charges of rotation

$$\frac{d\lambda}{dt} = \frac{1}{2}\sum_k \vec{e}_k \frac{d\vec{\lambda}_k}{dt} \underset{\vec{\boldsymbol{v}}>>\vec{\varphi}}{\cong} \frac{1}{2}\sum_k \vec{e}_k \vec{S}_k^{(\vec{\lambda}_i)} - \frac{1}{2}\sum_k \vec{e}_k\left(\vec{\boldsymbol{v}}\vec{\nabla}\right)\vec{\lambda}_k = S^{(\lambda)} - \left(\vec{\boldsymbol{v}}\vec{\nabla}\right)\lambda \tag{10.25}$$

in which we have introduced the scalar source $S^{(\lambda)}$ of charges λ, defined as

$$S^{(\lambda)} = \frac{1}{2}\sum_k \vec{e}_k \vec{S}_k^{(\vec{\lambda}_i)} \tag{10.26}$$

By using again the following relation

$$\operatorname{div}(\lambda\vec{v}) = \vec{v}\,\overrightarrow{\operatorname{grad}}\,\lambda = (\vec{v}\vec{\nabla})\lambda \tag{10.27}$$

which calls upon the vectorial flow $\vec{J}^{(\lambda)} = \lambda\vec{v}$, the equation of continuity for the scalar charges λ can be written under the following forms

$$\frac{d\lambda}{dt} \cong S^{(\lambda)} - (\vec{v}\vec{\nabla})\lambda = S^{(\lambda)} - \operatorname{div}(\lambda\vec{v}) = S^{(\lambda)} - \operatorname{div}\vec{J}^{(\lambda)} \tag{10.28}$$

The equations of continuity for the density of charges $\vec{\lambda}_i$, $\vec{\lambda}$ and λ are also shown in table 10.1, with the definitions of the sources of charges $\vec{S}^{(\vec{\lambda})}$ and $S^{(\lambda)}$.

10.2 – Charges and linear flows for lines of dislocations

To interpret the relationships of table 10.1, it is useful to apply them to the case of dislocation lines. So consider a line like that reported in figure 10.2, which moves relative to the lattice with velocity $\vec{v}$.

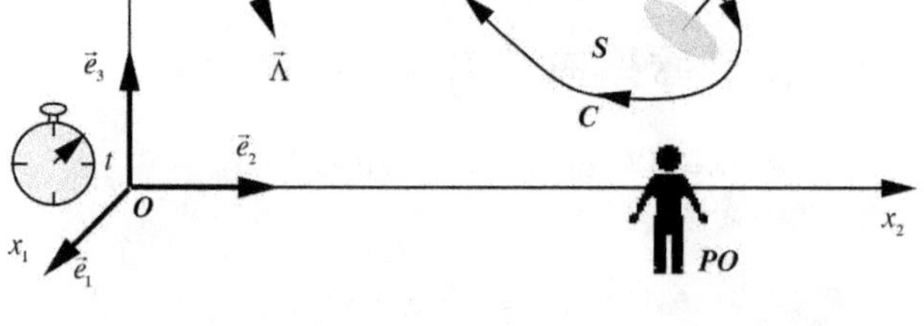

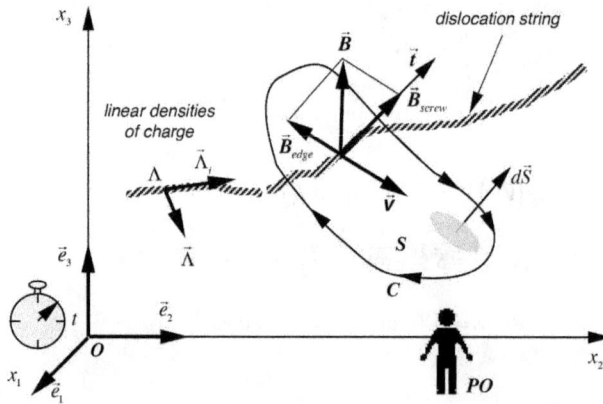

Figure 10.2 *- lines of dislocation moving with velocity $\vec{v}$ in the referential $Ox_1x_2x_3$*

Clearly, in the case of a line, the velocity $\vec{v}$ can only be perpendicular at all points to the direction $\vec{t}$ of the line. In the case of a line, one can integrate relationships of table 10.1 on the mobile contour C. Integrations of the contour's surface of charge flows $\vec{J}_i$ and $\vec{J}$ will give the *linear flow associated with the movable dislocation,* which we represent by the symbols $\vec{\Upsilon}_i$ and $\vec{\Upsilon}$. As flows $\vec{J}_i$ and $\vec{J}$ have dimension of the inverse of a time (1/s), the linear flows $\vec{\Upsilon}_i$ and $\vec{\Upsilon}$ have as dimension a surface per time unit (m²/s). As for the source of lattice sites S_n / n, its integral also represents a surface per unit of time (m²/s), and we will write it Υ. We have, taking into account relationships *(9.5)*, *(9.8)* and *(9.9)* for the linear charges $\vec{\Lambda}_i$, $\vec{\Lambda}$ and Λ of the dislocation

$$\begin{cases} \vec{\Upsilon}_i = \iint_S \vec{J}_i\, dS = \iint_S \vec{\lambda}_i\, dS \wedge \vec{\boldsymbol{v}} = \vec{\Lambda}_i \wedge \vec{\boldsymbol{v}} \\ \vec{\Upsilon} = \iint_S \vec{J}\, dS = \Lambda \vec{\boldsymbol{v}} + \dfrac{1}{2}\left(\vec{\Lambda} \wedge \vec{\boldsymbol{v}}\right) \\ \Upsilon = \iint_S \dfrac{S_n}{n}\, dS = -\vec{\Lambda}\vec{\boldsymbol{v}} \end{cases} \tag{10.30}$$

We may now introduce the relations which link the linear charge of a dislocation with he Burgers vector of said dislocation, namely

$$\begin{cases} \vec{\boldsymbol{B}} = -\sum_i \left(\vec{t}\vec{\Lambda}_i\right)\vec{e}_i & \Rightarrow \quad \vec{\Lambda}_i = -\vec{t}\boldsymbol{B}_i \\ \vec{\boldsymbol{B}}_{screw} = \left(\vec{\boldsymbol{B}}\vec{t}\right)\vec{t} = -2\,\Lambda\vec{t} & \Rightarrow \quad \Lambda = -\vec{\boldsymbol{B}}\vec{t}\,/\,2 \\ \vec{\boldsymbol{B}}_{edge} = \vec{t} \wedge \left(\vec{\boldsymbol{B}} \wedge \vec{t}\right)t = \vec{t} \wedge \vec{\Lambda} & \Rightarrow \quad \vec{\Lambda} = \vec{\boldsymbol{B}} \wedge \vec{t} \end{cases} \tag{10.31}$$

in relations *(10.30)* and we obtain the following expressions for the linear flows

$$\begin{cases} \vec{\Upsilon}_i = \vec{\Lambda}_i \wedge \vec{\boldsymbol{v}} = -\left(\vec{t} \wedge \vec{\boldsymbol{v}}\right)\boldsymbol{B}_i = \left(\vec{\boldsymbol{v}} \wedge \vec{t}\right)\left(\vec{\boldsymbol{B}}\vec{e}_i\right) \\ \vec{\Upsilon} = \Lambda\vec{\boldsymbol{v}} + \dfrac{1}{2}\left(\vec{\Lambda} \wedge \vec{\boldsymbol{v}}\right) = -\dfrac{1}{2}\left(\vec{\boldsymbol{B}}\vec{t}\right)\vec{\boldsymbol{v}} - \dfrac{1}{2}\left[\vec{\boldsymbol{v}} \wedge \left(\vec{\boldsymbol{B}} \wedge \vec{t}\right)\right] = -\dfrac{1}{2}\left(\vec{\boldsymbol{B}}\vec{t}\right)\vec{\boldsymbol{v}} + \dfrac{1}{2}\left(\vec{\boldsymbol{v}}\vec{\boldsymbol{B}}\right)\vec{t} \\ \Upsilon = -\vec{\Lambda}\vec{\boldsymbol{v}} = \left(\vec{t} \wedge \vec{\boldsymbol{B}}\right)\vec{\boldsymbol{v}} \end{cases} \tag{10.32}$$

Thanks to these relations, we will be able to interpret the different terms of the flow of charges associated with the movement of dislocations.

The flow associated with the slipping of a screw dislocation

If the dislocation contains only a screw component, the flux $\vec{\Upsilon}$ will be given by relation

$$\vec{\Upsilon}_{screw} = \Lambda_{screw}\vec{\boldsymbol{v}} = -\frac{1}{2}\left(\vec{\boldsymbol{B}}_{screw}\vec{t}\right)\vec{\boldsymbol{v}} \tag{10.33}$$

This relationship shows that, as the velocity $\vec{\boldsymbol{v}}$ is always perpendicular to the direction $\vec{t}$ of the line, the pure screw dislocations can move in all directions perpendicular to the direction $\vec{t}$. One speaks in this case of a slipping movement of screw dislocations, and the planes on which the screw dislocation moves are called *slip planes*.

The flows associated with the slipping or the climbing of an edge dislocation

In the case of a pure edge dislocation, the flows $\vec{\Upsilon}$ and Υ may appear, since the edge dislocation is completely represented by the charge $\vec{\Lambda}$, so that

$$\begin{cases} \vec{\Upsilon}_{edge}^{(slip)} = \dfrac{1}{2}\left(\vec{\Lambda}_{edge} \wedge \vec{\boldsymbol{v}}\right) = \dfrac{1}{2}\left[\left(\vec{\boldsymbol{B}}_{edge} \wedge \vec{t}\right) \wedge \vec{\boldsymbol{v}}\right] = \dfrac{1}{2}\left(\vec{\boldsymbol{v}}\,\vec{\boldsymbol{B}}_{edge}\right)\vec{t} \\ \Upsilon_{edge}^{(climb)} = -\vec{\boldsymbol{v}}\vec{\Lambda}_{edge} = \left(\vec{t} \wedge \vec{\boldsymbol{B}}_{edge}\right)\vec{\boldsymbol{v}} \end{cases} \tag{10.34}$$

With relations *(10.33)* and *(10.34)*, one understands the dimensions (m^2/s) of the linear flows Υ and $\vec{\Upsilon}$ since these flows are the product of the length of the Burgers vector with the relative speed of the line relative to lattice.

According to *(10.34)*, there appears two types of movements of the edge dislocation:

- the movement for which $\vec{\boldsymbol{v}}$ is perpendicular to $\vec{\Lambda}$, and which is responsible for a vectorial flow of charges $\vec{\Upsilon}^{(slip)}_{edge}$. This movement is shown in figure 10.3a. It corresponds to a conservative *slipping* movement of the edge dislocation on its *slip plane*, which is defined as the plane perpendicular to $\vec{\Lambda}$, so the plane that contains both the Burgers vector $\vec{\boldsymbol{B}}_{edge}$, the direction $\vec{\boldsymbol{t}}$ of the line and the velocity vector $\vec{\boldsymbol{v}}$.

- the movement for which $\vec{\boldsymbol{v}}$ is parallel to $\vec{\Lambda}$ and which is responsible for a scalar flow of charges $\Upsilon^{(climb)}_{edge}$. This movement is shown in figure 10.3b. It corresponds to a non-conservative upward movement of the edge dislocation perpendicular to its slip plane. The dislocation goes "up" in the lattice, it climbs, creating or destroying a plane of the lattice. This movement is not conservative in the sense that it destroys or builds the lattice, and it is this movement that is responsible for the existence of a source of lattice sites S_n in the equations of geometro-kinetic volume expansion *(8.25)*, and which is given by

$$S_n = -n\vec{\boldsymbol{v}}\vec{\Lambda}_{edge} = n\left(\vec{\boldsymbol{t}} \wedge \vec{\boldsymbol{B}}_{edge}\right)\vec{\boldsymbol{v}} \qquad (10.35)$$

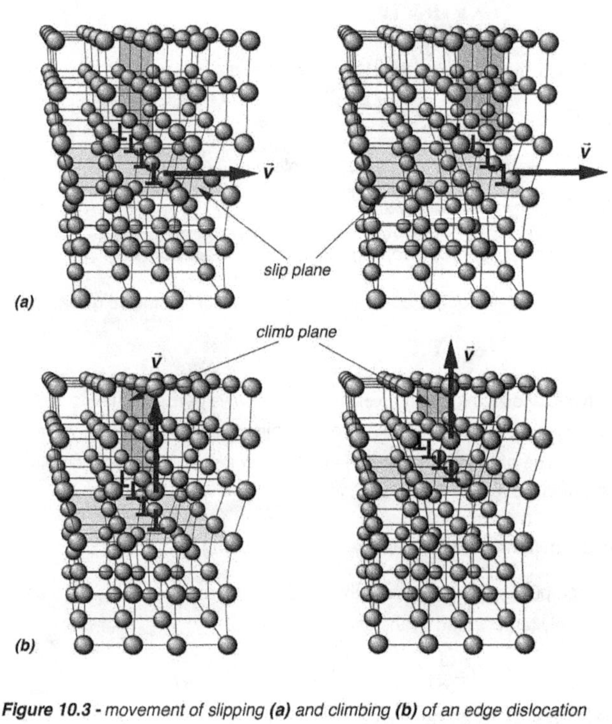

***Figure 10.3** - movement of slipping **(a)** and climbing **(b)** of an edge dislocation*

The various flows associated with the slipping and climbing movements of dislocations are summarized in table 10.2.

Table 10.2 - Relations between tensors of density and flow for a line dislocation

$\vec{v}$ = *relative velocity of the string with respect to the lattice* $\Rightarrow$

$$\begin{cases} \vec{\Upsilon}_i = \vec{\Lambda}_i \wedge \vec{v} = -(\vec{t} \wedge \vec{v}) B_i = (\vec{v} \wedge \vec{t})(\vec{B}\vec{e}_i) & (1) \\ \vec{\Upsilon} = \vec{\Upsilon}_{screw} + \vec{\Upsilon}^{(slip)}_{edge} = \Lambda_{screw}\vec{v} + \frac{1}{2}(\vec{\Lambda}_{edge} \wedge \vec{v}) = -\frac{1}{2}(\vec{B}_{screw}\vec{t})\vec{v} + \frac{1}{2}(\vec{v}\vec{B}_{edge})\vec{t} & (2) \\ \Upsilon = \Upsilon^{(climb)}_{edge} = -\vec{v}\vec{\Lambda}_{edge} = (\vec{t} \wedge \vec{B}_{edge})\vec{v} & (3) \end{cases}$$

The flow associated with the movement of a local charge of rotation

If we consider a localized charge of rotation Q_λ moving with velocity $\vec{v}$ relative to the solid medium, we obtain the relation between charge and flow by integrating the equation *(10.13)* on the volume of the charge, by asking that $\vec{\lambda} = 0$. It then comes, by introducing a current $\vec{I}$ which has a dimension of volume per time unit (m³/s)

$$\vec{I} = \iiint_V \vec{J}\, dV = \iiint_V \lambda\vec{v}\, dV = Q_\lambda \vec{v} \qquad (10.36)$$

This relationship implies that a localized scalar charge Q_λ, whose analogy is a localized electric charge becomes, as it moves in the solid medium, the source of a current $\vec{I}$ which is an analog current to the electric current.

10.3 – Relations of Orowan

From the knowledge of the flows $\vec{J}_i$ of dislocation charges, it is possible to retrieve the *macroscopic plastic distortion of the solid* $\vec{\beta}_i^{pl}$, as defined in chapter 8, thanks to the relation *(8.8)*. Total derivatives $\dot{\vec{\beta}}_i^{pl}$, $\dot{\vec{\varepsilon}}_i^{pl}$, $\dot{\vec{\alpha}}_i^{pl}$, $\dot{\vec{\omega}}^{pl}$ and $\dot{\tau}^{pl}$ can be written

$$\begin{cases} \dot{\vec{\beta}}_i^{pl} = \vec{J}_i = \vec{\lambda}_i \wedge \vec{v} \\ \dot{\vec{\omega}}^{pl} = \vec{J} = \lambda\vec{v} + \frac{1}{2}(\vec{\lambda} \wedge \vec{v}) \\ \dot{\vec{\varepsilon}}_i^{pl} = \vec{J}_i - \vec{e}_i \wedge \vec{J} = \vec{\lambda}_i \wedge \vec{v} - (\vec{e}_i \wedge \vec{v})\lambda - \frac{1}{2}\vec{e}_i \wedge (\vec{\lambda} \wedge \vec{v}) \\ \dot{\tau}^{pl} = \frac{S_n}{n} = -\vec{\lambda}\vec{v} \\ \dot{\vec{\alpha}}_i^{pl} = \vec{J}_i - \vec{e}_i \wedge \vec{J} - \frac{1}{3}\vec{e}_i \frac{S_n}{n} = \vec{\lambda}_i \wedge \vec{v} - (\vec{e}_i \wedge \vec{v})\lambda - \frac{1}{2}\vec{e}_i \wedge (\vec{\lambda} \wedge \vec{v}) + \frac{1}{3}\vec{e}_i(\vec{\lambda}\vec{v}) \end{cases} \qquad (10.37)$$

Similar relationships can be established in the case of many dislocations moving with velocity $\vec{v}$ relative to the solid medium. For example, imagine the same dislocations, all parallel to one direction and moving all at the same velocity $\vec{v}$. The integral $\vec{\beta}_i^{pl}$ on a surface S perpendicular to the direction of the lines is written

$$\iint_S \dot{\vec{\beta}}_i^{pl}\, dS = \dot{\vec{\beta}}_i^{pl} S = \iint_S \left(\vec{\lambda}_i \wedge \vec{\boldsymbol{v}}\right) dS = \iint_S \vec{\lambda}_i\, dS \wedge \vec{\boldsymbol{v}} = N\left(\vec{\Lambda}_i \wedge \vec{\boldsymbol{v}}\right) \qquad (10.38)$$

where N is the number of lines which cut perpendicularly the surface S.
We obtain

$$\dot{\vec{\beta}}_i^{pl} = \frac{N}{S}\left(\vec{\Lambda}_i \wedge \vec{\boldsymbol{v}}\right) = \Lambda\left(\vec{\Lambda}_i \wedge \vec{\boldsymbol{v}}\right) \qquad (10.39)$$

wherein $\Lambda = N/S$ is the density of *active dislocations,* that is to say the number of dislocation lines crossing perpendicularly the surface unit. It can be ensured that the dimension of the total derivative $\dot{\vec{\beta}}_i^{pl}$ corresponds to the inverse of a time (1/s), because the density Λ has dimension of the inverse of a surface (1/m^2) and the product $\vec{\Lambda}_i \wedge \vec{\boldsymbol{v}}$ has dimension (m^2/s). The quantity Λ can also be interpreted as *the length of lines of active dislocations per unit volume of lattice* since its size can also be written as (m/m^3).
By applying the same treatment to all relations *(10.36)*, we obtain the following equations set, in which we can introduce the linear flow of charges $\vec{\Upsilon}_i$, $\vec{\Upsilon}$ and Υ associated with the movement of the dislocation lines

$$\begin{cases} \dot{\vec{\beta}}_i^{pl} = \Lambda\left[\vec{\Lambda}_i \wedge \vec{\boldsymbol{v}}\right] = \Lambda\vec{\Upsilon}_i \\ \dot{\vec{\omega}}^{pl} = \Lambda\left[\Lambda_{screw}\vec{\boldsymbol{v}} + \frac{1}{2}\left(\vec{\Lambda}_{edge} \wedge \vec{\boldsymbol{v}}\right)\right] = \Lambda\vec{\Upsilon} \\ \dot{\vec{\varepsilon}}_i^{pl} = \Lambda\left[\vec{\Lambda}_i \wedge \vec{\boldsymbol{v}} + \left(\vec{\boldsymbol{v}} \wedge \vec{e}_i\right)\Lambda_{screw} + \frac{1}{2}\vec{e}_i \wedge \left(\vec{\boldsymbol{v}} \wedge \vec{\Lambda}_{edge}\right)\right] = \Lambda\left(\vec{\Upsilon}_i - \vec{e}_i \wedge \vec{\Upsilon}\right) \\ \dot{\tau}^{pl} = \Lambda\left[-\vec{\boldsymbol{v}}\vec{\Lambda}_{edge}\right] = \Lambda\Upsilon \\ \dot{\vec{\alpha}}_i^{pl} = \Lambda\left[\vec{\Lambda}_i \wedge \vec{\boldsymbol{v}} + \left(\vec{\boldsymbol{v}} \wedge \vec{e}_i\right)\Lambda_{screw} + \frac{1}{2}\vec{e}_i \wedge \left(\vec{\boldsymbol{v}} \wedge \vec{\Lambda}_{edge}\right) + \frac{1}{3}\vec{e}_i\left(\vec{\boldsymbol{v}}\vec{\Lambda}_{edge}\right)\right] = \Lambda\left(\vec{\Upsilon}_i - \vec{e}_i \wedge \vec{\Upsilon} - \vec{e}_i\Upsilon/3\right) \end{cases}$$

(10.40)

These relationships between the macroscopic deformation velocity and the dislocation movement velocities at the microscopic level are usually called *relations of Orowan*[1]. We can also express these relationships of Orowan as functions of the Burgers vectors of the dislocations instead of the linear charge densities of the dislocation lines

$$\begin{aligned} \dot{\vec{\beta}}_i^{pl} &= \Lambda\left[\left(\vec{\boldsymbol{v}} \wedge \vec{t}\right)\left(\vec{\boldsymbol{B}}\vec{e}_i\right)\right] \\ \dot{\vec{\omega}}^{pl} &= \Lambda\left[-\frac{1}{2}\left(\vec{\boldsymbol{B}}_{vis}\vec{t}\right)\vec{\boldsymbol{v}} + \frac{1}{2}\left(\vec{\boldsymbol{v}}\vec{\boldsymbol{B}}_{coin}\right)\vec{t}\right] \\ \dot{\vec{\varepsilon}}_i^{pl} &= \Lambda\left[\left(\vec{\boldsymbol{v}} \wedge \vec{t}\right)\left(\vec{\boldsymbol{B}}\vec{e}_i\right) + \frac{1}{2}\left(\vec{e}_i \wedge \vec{\boldsymbol{v}}\right)\left(\vec{\boldsymbol{B}}_{vis}\vec{t}\right) - \frac{1}{2}\left(\vec{e}_i \wedge \vec{t}\right)\left(\vec{\boldsymbol{v}}\vec{\boldsymbol{B}}_{coin}\right)\right] \\ \dot{\tau}^{pl} &= \Lambda\left[\left(\vec{t} \wedge \vec{\boldsymbol{B}}_{coin}\right)\vec{\boldsymbol{v}}\right] \\ \dot{\vec{\alpha}}_i^{pl} &= \Lambda\left[\left(\vec{\boldsymbol{v}} \wedge \vec{t}\right)\left(\vec{\boldsymbol{B}}\vec{e}_i\right) + \frac{1}{2}\left(\vec{e}_i \wedge \vec{\boldsymbol{v}}\right)\left(\vec{\boldsymbol{B}}_{vis}\vec{t}\right) - \frac{1}{2}\left(\vec{e}_i \wedge \vec{t}\right)\left(\vec{\boldsymbol{v}}\vec{\boldsymbol{B}}_{coin}\right) - \frac{1}{3}\vec{e}_i\left[\left(\vec{t} \wedge \vec{\boldsymbol{B}}_{coin}\right)\vec{\boldsymbol{v}}\right.\right. \end{aligned} \qquad (10.41)$$

[1] *E. Orowan, Z. Phys., vol. 89, p. 605, 614, 634, 1934*

Chapter 11

Evolution equations of a charged lattice and the force of Peach and Koehler

The tensor of density $\vec{\lambda}_i$ and the tensor of flow $\vec{J}_i$ of dislocation charges, defined in the previous chapters, allow us to find in this chapter all the fundamental and phenomenological equations of spatiotemporal evolution to be met by a lattice presenting elastic, anelastic and self-diffusive properties and containing dislocation charges. Furthermore, from the energy balance equation of charged solids, it is possible to find the forces acting on the dislocation charges in the presence of non-zero stress fields. These forces are called *forces of Peach and Koehler.*

11.1 - Replacement of the tensor of plastic distortion

To write the spatiotemporal evolution equations of an elastic and anelastic solid containing dislocation charges, we will simply consider all evolution equations obtained in chapter 6, and reported in table 6.1, and we will replace in them all the terms showing the tensor of plastic distortions $\vec{\beta}_i^{pl}$ by the concepts of density and flow of dislocation charges that we developed in the previous chapters. In other words, we perform the following systematic replacements of the plastic distortion tensors

$$\begin{cases} \vec{\beta}_i \rightarrow \vec{\beta}_i + \vec{\beta}_i^{pl} \\ d\vec{\beta}_i^{pl} / dt \rightarrow \vec{J}_i \\ \overrightarrow{\mathrm{rot}}\, \vec{\beta}_i^{pl} \rightarrow -\vec{\lambda}_i \end{cases} \qquad \begin{cases} \vec{\omega} \rightarrow \vec{\omega} + \vec{\omega}^{pl} \\ d\vec{\omega}^{pl} / dt \rightarrow \vec{J} \\ \mathrm{div}\, \vec{\omega}^{pl} \rightarrow -\vec{\lambda} \end{cases} \qquad \begin{cases} \tau \rightarrow \tau + \tau^{pl} \\ d\tau^{pl} / dt \rightarrow S_n / n \end{cases} \tag{11.1}$$

In reality, the number of transformations to be performed is relatively low. Indeed, we find that only the topological equations, the heat equations and the energy balance equation are affected by these replacements. While the transformation of topological equations are immediate since we established them in detail in the previous chapter, we need to discuss in more detail in this chapter the changes in the thermal equations and the energy balance equation.

The equation of heat and the source of entropy

Considering again equations *(6.23)* and *(6.30)*, and doing all the necessary changes, we have

$$\begin{aligned} nT\frac{ds}{dt} &= -\left(\mu_L^* + \mu_I^*\right)S_{I-L} - \left(\mu_L^* + h^*\right)S_L^{pl} - \left(\mu_I^* - h^*\right)S_I^{pl} \\ &+\left(-\overrightarrow{\mathrm{grad}}\mu_L^* + m\frac{d}{dt}\left(\vec{\phi}_L - 2\Delta\vec{\varphi}_L\right) - m\vec{g}\right)\vec{J}_L + \left(-\overrightarrow{\mathrm{grad}}\mu_I^* - m\frac{d}{dt}\left(\vec{\phi}_I\right) + m\vec{g}\right)\vec{J}_I \\ &+\vec{s}_k^{dis}\frac{d\vec{\beta}_k^{an}}{dt} + \vec{m}^{dis}\frac{d\vec{\omega}^{an}}{dt} + \vec{s}_k\vec{J}_k + \vec{m}\vec{J} - \mathrm{div}\,\vec{J}_q \end{aligned} \tag{11.2}$$

$$S_e = -\frac{1}{T}\left(\mu_L^* + \mu_I^*\right)S_{I-L} - \frac{1}{T}\left(\mu_L^* + g^*\right)S_L^{pl} - \frac{1}{T}\left(\mu_I^* - g^*\right)S_I^{pl}$$
$$+\frac{\vec{J}_L}{T}\left(-\overrightarrow{\text{grad}}\mu_L^* + m\frac{d}{dt}\left(\vec{\phi}_L - 2\Delta\vec{\varphi}_L\right) - m\vec{g}\right) + \frac{\vec{J}_I}{T}\left(-\overrightarrow{\text{grad}}\mu_I^* - m\frac{d}{dt}\left(\vec{\phi}_I\right) + m\vec{g}\right) \quad (11.3)$$
$$+\frac{1}{T}\left(\vec{s}_k^{dis}\frac{d\vec{\beta}_k^{an}}{dt} + \vec{m}^{dis}\frac{d\vec{\omega}^{an}}{dt} + \vec{s}_k\vec{J}_k + \vec{m}\vec{J}\right) + \vec{J}_q\overrightarrow{\text{grad}}\left(\frac{1}{T}\right)$$

The dissipative equations of plasticity

The terms of source of entropy associated with plastic deformation generated by the movement of the plastic charges in equation *(11.3)* are the following

$$S_e^{pl} = -\frac{1}{T}\left(\mu_L^* + g^*\right)S_L^{pl} - \frac{1}{T}\left(\mu_I^* - g^*\right)S_I^{pl} + \frac{1}{T}\left(\vec{s}_k\vec{J}_k + \vec{m}\vec{J}\right) \quad (11.4)$$

From these sources of entropy, it is possible to define the dissipative equations of plasticity that characterize the tensorial flow of dislocation charges. The first of these concerns the trace of $\vec{J}_k$, which is associated with the source of creation or annihilation of lattice sites $S_n / n = -\vec{\lambda}\vec{\boldsymbol{v}}$ via *the phenomenon of climbing of edge dislocations* with velocity $\vec{\boldsymbol{V}}$. This climbing is controlled by the source of vacancies and interstitials, and depends essentially on creation-annihilation energies $\mu_L^* + g^*$ and $\mu_I^* - g^*$, but certainly also on the local pressure within the medium and the local dislocation density charge $\vec{\lambda}$ of edge dislocation which are able to leave a *"trace"* of their passage

$$\begin{cases} \dfrac{S_n}{n} = \sum\limits_k \vec{e}_k\vec{J}_k = -\vec{\lambda}\vec{\boldsymbol{v}} = \dfrac{1}{n}\left(S_L^{pl} - S_I^{pl}\right) \\ S_L^{pl} = S_L^{pl}\left[\left(\mu_L^* + g^*\right), p, \vec{\lambda}, v, T, C_L, C_I, \ldots\right] \\ S_I^{pl} = S_I^{pl}\left[\left(\mu_I^* - g^*\right), p, \vec{\lambda}, v, T, C_L, C_I, \ldots\right] \end{cases} \quad (11.5)$$

The other two relationships are deduced from the product of charge flows $\vec{J}_k$ and $\vec{J}$ and thermodynamic forces $\vec{s}_k / T$ and $\vec{m} / T$. These products show that the tensorial flow of charges $\vec{J}_k$ must depend essentially on the tensor of shear stress $\vec{s}_i$, while the vector flow of charges $\vec{J}$ should depend on the torsor of moment $\vec{m}$, so that

$$\begin{cases} \vec{J}_k = \vec{\lambda}_k \wedge \vec{\boldsymbol{v}} = \vec{J}_k\left(\vec{s}_m, \vec{\lambda}_n, v, T, \ldots\right) \\ \vec{J} = -\dfrac{1}{2}\sum\limits_k \vec{e}_k \wedge \vec{J}_k = \lambda\vec{\boldsymbol{v}} + \dfrac{1}{2}\left(\vec{\lambda} \wedge \vec{\boldsymbol{v}}\right) = \vec{J}\left(\vec{m}, \lambda, v, T, \ldots\right) \end{cases} \quad (11.6)$$

These two relationships are the *dynamic equations of motion of charges* $\vec{\lambda}_k$ and λ within the medium. Indeed, we will see in section 11.2 that forces dependent on $\vec{s}_k$ and $\vec{m}$ apply to the charges of dislocation $\vec{\lambda}_k$ and λ. In this way, the associated flows $\vec{J}_k = \vec{\lambda}_k \wedge \vec{\boldsymbol{V}}$ and $\vec{J} = \lambda\vec{\boldsymbol{v}} + \left(\vec{\lambda} \wedge \vec{\boldsymbol{v}}\right)/2$ depend on $\vec{s}_k$ and $\vec{m}$ via a dynamic equation which should allow us to calculate the velocity $\vec{\boldsymbol{V}}$ of the charges by knowing the forces that depend on $\vec{s}_k$ and $\vec{m}$. We will return later to the problem of the dynamic equation of dislocation movement, which will lead us to *the string model of the dislocation.*

The dissipative equations of sources of plastic charges

By taking the continuity equations for dislocation charges of table 10.1, there appears the sources of charges $\vec{S}_i^{\lambda_i}$, $\vec{S}^{(\vec{\lambda})}$ and S^{λ}, which are partially responsible for the non-commutative nature of the operators of time and space on distortion tensors $\vec{\beta}_i$ and $\vec{\omega}$. The expression of these sources $\vec{S}_i^{\vec{\lambda}_i}$, $\vec{S}^{(\vec{\lambda})}$ and S^{λ} will certainly depend on the variables $\vec{s}_m, \vec{\lambda}_n, v, T, \ldots$, but can be obtained through the description of the dynamics of dislocation charges at the level of the lattice itself. We cannot fully understand the nature of this until we develop the *string model of dislocation*, the dynamic equation of motion of the dislocations, which will allow us to describe the existence of dislocation sources (see section 12.9).

The equation of energy balance

The main change to the energy balance equation is the emergence of new dissipative terms associated with the flows $\vec{J}_i$ and $\vec{J}$, as well as the source of lattice sites S_n

$$n\vec{\phi}\left(\frac{d\vec{p}}{dt} - m\vec{\phi}_I \frac{dC_I}{dt} + m\vec{\phi}_L \frac{dC_L}{dt}\right) + \left(\vec{s}_k \frac{d\vec{\beta}_k}{dt} + \vec{m}\frac{d\vec{\omega}}{dt} - p\frac{d\tau}{dt}\right)$$
$$= -\left(\vec{s}_k \vec{J}_k + \vec{m}\vec{J} - p\frac{S_n}{n}\right) + \rho\vec{g}\vec{\phi} - \operatorname{div}\left[-\phi_k \vec{s}_k - \frac{1}{2}(\vec{\phi} \wedge \vec{m}) + p\vec{\phi}\right] \tag{11.7}$$

These terms $\vec{s}_k \vec{J}_k + \vec{m}\vec{J} - pS_n / n$ will allow us to find, in the following section, the forces acting on the dislocation charges in the presence of non-zero stress fields.

11.2 - Force of Peach and Koehler acting on the dislocation charges

In the energy balance equations, the power $\boldsymbol{P}_{ch}$ supplied to the dislocation charges by the stress fields corresponds to the balance of the term containing the flow of charges

$$\boldsymbol{P}_{ch} = \vec{s}_k \vec{J}_k + \vec{m}\vec{J} - p\frac{S_n}{n} \tag{11.8}$$

In this term of power, it is possible to replace flows $\vec{J}_k$, $\vec{J}$ and S_n / n by their expression as a function of the velocity $\vec{\boldsymbol{v}}$ of the charges, from table 10.1. One obtains an expression of $\boldsymbol{P}_{ch}$ containing the charge densities $\vec{\lambda}_k$, $\vec{\lambda}$ and λ

$$\boldsymbol{P}_{ch} = \sum_k \vec{s}_k \left(\vec{\lambda}_k \wedge \vec{\boldsymbol{v}}\right) + \vec{m}\lambda\vec{\boldsymbol{v}} + \frac{1}{2}\vec{m}\left(\vec{\lambda} \wedge \vec{\boldsymbol{v}}\right) + p\vec{\lambda}\vec{\boldsymbol{v}} = \left[\sum_k \left(\vec{s}_k \wedge \vec{\lambda}_k\right) + \lambda\vec{m} + \frac{1}{2}\left(\vec{m} \wedge \vec{\lambda}\right) + \vec{\lambda}p\right]\vec{\boldsymbol{v}} \tag{11.9}$$

The power $\boldsymbol{P}_{ch}$ given to the charges is thus the product of a velocity by a term that is a force $\vec{\boldsymbol{f}}_{PK}$ acting by unit of volume on the charge densities $\vec{\lambda}_k$, $\vec{\lambda}$, λ

$$\boldsymbol{P}_{ch} = \vec{\boldsymbol{f}}_{PK}\vec{\boldsymbol{v}} \quad \Rightarrow \quad \vec{\boldsymbol{f}}_{PK} = \sum_k \left(\vec{s}_k \wedge \vec{\lambda}_k\right) + \lambda\vec{m} + \frac{1}{2}\left(\vec{m} \wedge \vec{\lambda}\right) + \vec{\lambda}p + \vec{\boldsymbol{v}} \wedge \vec{\boldsymbol{A}} \tag{11.10}$$

This force depends on the stress tensors $\vec{s}_k$, $\vec{m}$ and/or p , and is usually called the *force of Peach and Koehler*. The use of the stress tensors $\vec{s}_k$, $\vec{m}$ and/or p depends of course, as we saw in chapter 5, on the choice of how we write the free energy state function of the solid medium considered.

As the dimension is an inverse of a length (1/m) and the dimension of the torsor of moments is a moment per unit volume, namely (Nm/ m³), the force $\vec{f}_{PK}$ has the dimension of a force per unit volume (N/m³).
The last term containing a vector $\vec{A}$ is added here because it is a term for a possible force which would not produce work, so that it does not appear in the expression *(11.9)* of power. We will also later see that this term actually corresponds to a relativistic force similar to the *Lorentz force* in electromagnetism.

The force of Peach and Kohler acting on a dislocation line

The force of Peach and Koehler can also be written in the case of a dislocation, integrating equation *(11.10)* on the surface of a section of the string. It then comes, neglecting the term containing the vector $\vec{A}$

$$\vec{F}_{PK} = \iint_S \vec{f}_{PK}\, dS = \iint_S \left[\sum_k \left(\vec{s}_k \wedge \vec{\lambda}_k \right) + \lambda \vec{m} + \frac{1}{2}\left(\vec{m} \wedge \vec{\lambda} \right) + \vec{\lambda} p \right] dS \tag{11.11}$$

The integral of $\vec{\lambda}_k$, $\vec{\lambda}$, λ on the section of a dislocation string gives linear densities, so that the force of Peach and Koehler acting on the dislocation line is given by

$$\vec{F}_{PK} = \sum_k \left(\vec{s}_k \wedge \vec{\Lambda}_k \right) + \Lambda \vec{m} + \frac{1}{2}\left(\vec{m} \wedge \vec{\Lambda} \right) + \vec{\Lambda} p \tag{11.12}$$

The dimension of the force $\vec{F}_{PK}$ acting on the dislocation is a force per unit length (N/m). This is in fact *the force per unit length of the string* in the presence of the stress fields $\vec{s}_k, \vec{m}, p$. Writing the force of Peach and Koehler using the constraint tensors $\vec{s}_k, \vec{m}, p$ is in fact very interesting because it gives a much clearer picture than by using the notation $\vec{F}_{PK} = \sum \left(\vec{\sigma}_k \wedge \vec{\Lambda}_k \right)$ usually used with the symmetric stress tensor $\vec{\sigma}_k$. Indeed, imagine a solid in which the volume expansion is zero, so that one would have a negligible pressure p. In this case, it is already known that we can replace the shear tensor $\vec{\alpha}_k$ by the vector of rotation $\vec{\omega}$, so that the force becomes a *sliding force*, which is written $\vec{F}_{PK} = \Lambda \vec{m} + \left(\vec{m} \wedge \vec{\Lambda} \right)/2$, wherein the term $\Lambda \vec{m}$ is the force acting on the screw component Λ of the dislocation and the term $\left(\vec{m} \wedge \vec{\Lambda} \right)/2$ is the force acting on the edge component $\vec{\Lambda}$ of the dislocation. As the component m_k of torque is associated with the shear components σ_{ij} and σ_{ji} of the stress tensor, it can give a very clear representation of the forces acting on a dislocation. The same here goes for the pressure force $\vec{\Lambda} p$, which acts on the edge component $\vec{\Lambda}$ of dislocations and which matches, given its direction (figure 10.3b), a *climbing force for the dislocations*.
We can again use relations *(10.31)* to express the Peach and Koehler force directly from the Burgers vector of the dislocation

$$\vec{F}_{PK} = \sum_k \left(\vec{t} \wedge \vec{s}_k \right)\left(\vec{B} \vec{e}_k \right) - \frac{1}{2}\left(\vec{B} \vec{t} \right) \vec{m} - \frac{1}{2}\left[\left(\vec{B} \wedge \vec{t} \right) \wedge \vec{m} \right] + \left(\vec{B} \wedge \vec{t} \right) p \tag{11.13}$$

The Peach and Kohler force acting on a localized charge of rotation

In the case of a localized charge of rotation Q_λ, we obtain the force of Peach and Koehler by integrating equation *(11.10)* on the volume of the localized charge. It gives

$$\vec{F}_{PK} = \iiint_V \vec{f}_{PK}\, dV = \vec{m} \iiint_V \lambda\, dV = Q_\lambda \vec{m} \tag{11.14}$$

In this case, the dimension of $\vec{F}_{PK}$ is that of a pure force, expressed in (N), and which is analogous to the electrical force acting on a localized electrical charge ($\vec{F}_{electrical} = q\vec{E}$).
The possible expressions of the strength of Peach and Koehler are shown in Table 11.1.

Table 11.1 - Forces of Peach et Koehler

Volume force on volume density of dislocation charges

$$\vec{f}_{PK} = \sum_k \left(\vec{s}_k \wedge \vec{\lambda}_k\right) + \lambda\vec{m} + \frac{1}{2}\left(\vec{m} \wedge \vec{\lambda}\right) + \vec{\lambda} p + \vec{\boldsymbol{v}} \wedge \vec{\boldsymbol{A}} \qquad (1)$$

Linear forces on a line of dislocation

$$\vec{F}_{PK} = \sum_k \left(\vec{s}_k \wedge \vec{\Lambda}_k\right) + \Lambda\vec{m} + \frac{1}{2}\left(\vec{m} \wedge \vec{\Lambda}\right) + \vec{\Lambda} p \qquad (2)$$

$$\vec{F}_{PK} = \sum_k \left(\vec{t} \wedge \vec{s}_k\right)\left(\vec{\boldsymbol{B}}\vec{e}_k\right) - \frac{1}{2}\left(\vec{\boldsymbol{B}}\vec{t}\right)\vec{m} - \frac{1}{2}\left[\left(\vec{\boldsymbol{B}} \wedge \vec{t}\right) \wedge \vec{m}\right] + \left(\vec{\boldsymbol{B}} \wedge \vec{t}\right) p \qquad (3)$$

Global force on a localized charge of rotation

$$\vec{F}_{PK} = \iiint_V \vec{f}_{PK}\, dV = \vec{m} \iiint_V \lambda\, dV = \boldsymbol{Q}_\lambda \vec{m} \qquad (4)$$

11.3 - Equations of spatio-temporal evolution in charged solids

It is now easy to combine all of the results obtained in this chapter with the equations of tables 6.1 and 10.1 to obtain the complete equations of spatio-temporal evolution of a solid self-diffusive lattice, with phenomenological behaviors of elasticity and anelasticity, and containing densities and flows of dislocation charges. As shown in Table 11.2, this system of equations is quite complex, especially in the high number of phenomenological equations of state and phenomenological equations of dissipation required for a complete description of all possible phenomena in such an environment .
We can already imagine that the new concepts of density and flow of dislocation charges are expected to describe the plasticity phenomena and the anelasticity at the microscopic scale in a discrete solid lattice by introducing a scalable microstructure of plastics charges which should help us to encode the non-Markovian behavior of plasticity. Moreover, the approach by plastic charges of a microscopically discrete solid lattice should also permit, in principle, to find exact local expressions for the dissipative equations *(11.5)* to *(11.9)* associated with the plastic charges.

Table 11.2 - Fundamental equations of evolution of self-diffusive solids with elasticity, anelasticity and charges of dislocations

Topological equations

$$\begin{cases} \vec{J}_i = -\dfrac{d\vec{\beta}_i}{dt} + \overrightarrow{\mathrm{grad}}\,\phi_i & (1) \\ \vec{J} = -\dfrac{1}{2}\sum_k \vec{e}_k \wedge \vec{J}_k = -\dfrac{d\vec{\omega}}{dt} + \dfrac{1}{2}\overrightarrow{\mathrm{rot}}\,\vec{\phi} & (2) \\ \dfrac{S_n}{n} = \sum_k \vec{e}_k \vec{J}_k = -\dfrac{d\tau}{dt} + \mathrm{div}\,\vec{\phi} & (3) \end{cases}$$

$$\begin{cases} \vec{\lambda}_i = \overrightarrow{\mathrm{rot}}\,\vec{\beta}_i \quad avec \quad \mathrm{div}\,\vec{\lambda}_i = 0 & (4) \\ \vec{\lambda} = -\sum_k \vec{e}_k \wedge \vec{\lambda}_k = -\sum_k \vec{e}_k \wedge \overrightarrow{\mathrm{rot}}\,\vec{\beta}_k & (5) \\ \lambda = \dfrac{1}{2}\sum_k \vec{e}_k \vec{\lambda}_k = \mathrm{div}\,\vec{\omega} & (6) \end{cases}$$

$$\begin{cases} \vec{\beta}_i = \vec{\beta}_i^{(\delta)} + \vec{e}_i \wedge \vec{\omega}_O(t) = \vec{\beta}_i^{\acute{e}l} + \vec{\beta}_i^{an} + \vec{e}_i \wedge \vec{\omega}_O(t) & (7) \\ \vec{\omega} = -\dfrac{1}{2}\sum_k \vec{e}_k \wedge \vec{\beta}_k = \vec{\omega}^{(\delta)} + \vec{\omega}_0(t) = \vec{\omega}^{\acute{e}l} + \vec{\omega}^{an} + \vec{\omega}_0(t) & (8) \\ \tau = \sum_k \vec{\beta}_k \vec{e}_k = \sum_k \vec{\beta}_k^{(\delta)} \vec{e}_k = \tau^{\acute{e}l} \quad (\tau^{an} \equiv 0 \textit{ par hypothèse}) & (9) \\ \vec{\varepsilon}_i = \vec{\beta}_i - \vec{e}_i \wedge \vec{\omega} = \vec{\beta}_i^{(\delta)} - \vec{e}_i \wedge \vec{\omega}^{(\delta)} = \vec{\varepsilon}_i^{\acute{e}l} + \vec{\varepsilon}_i^{an} & (10) \\ \vec{\alpha}_i = \vec{\varepsilon}_i - \dfrac{1}{3}\tau \vec{e}_i = \vec{\alpha}_i^{\acute{e}l} + \vec{\alpha}_i^{an} & (11) \end{cases}$$

$$\begin{cases} d/dt = \partial/\partial t + (\vec{\varphi}\vec{\nabla}) & (12) \\ \vec{\varphi} = \vec{\phi} - \vec{\phi}_O(t) - \dot{\vec{\omega}}_O(t) \wedge \vec{r} & (13) \end{cases}$$

Dynamic equations

$$n\frac{d\vec{p}}{dt} = \rho\vec{g} + \sum_k \vec{e}_k\,\mathrm{div}\,\vec{s}_k - \frac{1}{2}\overrightarrow{\mathrm{rot}}\;\vec{m} - \overrightarrow{\mathrm{grad}}\,p + nm\vec{\phi}_I \frac{dC_I}{dt} - nm\vec{\phi}_L \frac{dC_L}{dt} \quad (14)$$

$$\begin{cases} n = 1/v = n_0 e^{-\tau} & (15) \\ \vec{p} = m\left(\vec{\phi} + C_I\vec{\phi}_I - C_L\vec{\phi}_L\right) = m\vec{\phi} + m\left(C_I - C_L\right)\vec{\phi} + \dfrac{m}{n}\left(\vec{J}_I - \vec{J}_L\right) = \dfrac{1}{n}\left[\rho\vec{\phi} + m\left(\vec{J}_I - \vec{J}_L\right)\right] & (16) \\ \rho = mn\left(1 + C_I - C_L\right) & (17) \end{cases}$$

Diffusion equations

$$\begin{cases} n\dfrac{dC_L}{dt} = \left(S_{I-L} + S_L{}^{pl}\right) - C_L\left(S_L{}^{pl} - S_I{}^{pl}\right) - \mathrm{div}\,\vec{J}_L & (18) \\ n\dfrac{dC_I}{dt} = \left(S_{I-L} + S_I{}^{pl}\right) - C_I\left(S_L{}^{pl} - S_I{}^{pl}\right) - \mathrm{div}\,\vec{J}_I & (19) \end{cases}$$

$$\begin{cases} \vec{J}_L = nC_L\Delta\vec{\varphi}_L = nC_L\left(\vec{\phi}_L - \vec{\phi}\right) = nC_L\left(\vec{\varphi}_L - \vec{\varphi}\right) & (20) \\ \vec{J}_I = nC_I\Delta\vec{\varphi}_I = nC_I\left(\vec{\phi}_I - \vec{\phi}\right) = nC_I\left(\vec{\varphi}_I - \vec{\varphi}\right) & (21) \end{cases}$$

Thermal equations

$$nT\frac{ds}{dt}=-\left(\mu_L^*+\mu_I^*\right)S_{I-L}-\left(\mu_L^*+h^*\right)S_L^{pl}-\left(\mu_I^*-h^*\right)S_I^{pl}+T\vec{J}_L\vec{X}_L$$
$$+T\vec{J}_I\vec{X}_I+\vec{s}_k^{dis}\frac{d\vec{\beta}_k^{an}}{dt}+\vec{m}^{dis}\frac{d\vec{\omega}^{an}}{dt}+\vec{s}_k\vec{J}_k+\vec{m}\vec{J}-\operatorname{div}\vec{J}_q \qquad (22)$$

$$\left\{\begin{array}{ll}\mu_L^*=\mu_L-\dfrac{1}{2}m\left(\vec{\phi}_L^2-2\Delta\vec{\varphi}_L^2\right) & (23)\\ \mu_I^*=\mu_I+\dfrac{1}{2}m\vec{\phi}_I^2 & (24)\end{array}\right.$$

$$\left\{\begin{array}{ll}\vec{X}_q=\overrightarrow{\operatorname{grad}}\dfrac{1}{T} & (25)\\ \vec{X}_L=\dfrac{1}{T}\left(-\overrightarrow{\operatorname{grad}}\ \mu_L^*+m\dfrac{d}{dt}\left(\vec{\phi}_L-2\Delta\vec{\varphi}_L\right)-m\vec{g}\right) & (26)\\ \vec{X}_I=\dfrac{1}{T}\left(-\overrightarrow{\operatorname{grad}}\ \mu_I^*-m\dfrac{d}{dt}\left(\vec{\phi}_I\right)+m\vec{g}\right) & (27)\end{array}\right.$$

$$h^*=f+Ts+pv+\frac{1}{2}m\vec{\phi}^2-\mu_LC_L-\mu_IC_I \qquad (28)$$

Charge equations

$$\left\{\begin{array}{ll}\dfrac{d\vec{\lambda}_i}{dt}\cong\vec{S}_i^{(\vec{\lambda}_i)}-\left(\vec{\boldsymbol{v}}\vec{\nabla}\right)\vec{\lambda}_i=\vec{S}_i^{(\vec{\lambda}_i)}+\overrightarrow{\operatorname{rot}}\left(\vec{\boldsymbol{v}}\wedge\vec{\lambda}_i\right)=\vec{S}_i^{(\vec{\lambda}_i)}-\overrightarrow{\operatorname{rot}}\vec{J}_i & (29)\\ \dfrac{d\vec{\lambda}}{dt}\cong\vec{S}^{(\vec{\lambda})}-\left(\vec{\boldsymbol{v}}\vec{\nabla}\right)\vec{\lambda}=\vec{S}^{(\vec{\lambda})}+\overrightarrow{\operatorname{rot}}\left(\vec{\boldsymbol{v}}\wedge\vec{\lambda}\right)-\vec{\boldsymbol{v}}\operatorname{div}\vec{\lambda}=\vec{S}^{(\vec{\lambda})}-2\ \overrightarrow{\operatorname{rot}}\vec{J}^{(\vec{\lambda})}-\vec{\boldsymbol{v}}\theta & (30)\\ \dfrac{d\lambda}{dt}\cong S^{(\lambda)}-\left(\vec{\boldsymbol{v}}\vec{\nabla}\right)\lambda=S^{(\lambda)}-\operatorname{div}\left(\lambda\vec{\boldsymbol{v}}\right)=S^{(\lambda)}-\operatorname{div}\vec{J}^{(\lambda)} & (31)\end{array}\right.$$

$$\left\{\begin{array}{ll}\vec{J}_i=\vec{\lambda}_i\wedge\vec{\boldsymbol{v}} & (32)\\ \vec{J}=\vec{J}^{(\lambda)}+\vec{J}^{(\vec{\lambda})}=\lambda\vec{\boldsymbol{v}}+\left(\vec{\lambda}\wedge\vec{\boldsymbol{v}}\right)/2 & (33)\\ S_n/n=-\vec{\lambda}\vec{\boldsymbol{v}} & (34)\end{array}\right.$$

$$\left\{\begin{array}{ll}\vec{S}^{(\vec{\lambda})}=-\displaystyle\sum_k\vec{e}_k\wedge\vec{S}_k^{(\vec{\lambda}_i)} & (35)\\ S^{(\lambda)}=\dfrac{1}{2}\displaystyle\sum_k\vec{e}_k\vec{S}_k^{(\vec{\lambda}_i)} & (36)\end{array}\right.$$

$$\vec{f}_{PK}=\sum_k\left(\vec{s}_k\wedge\vec{\lambda}_k\right)+\lambda\vec{m}+\frac{1}{2}\left(\vec{m}\wedge\vec{\lambda}\right)+\vec{\lambda}p+\vec{\boldsymbol{v}}\wedge\vec{\boldsymbol{A}} \qquad (37)$$

Phenomenological equations of evolution of self-diffusive solids with elasticity, anelasticity and charges of dislocations

Functions and equations of state

$$f = f\left(\alpha_{ij}^{él}, \alpha_{ij}^{an}, \omega_k^{él}, \omega_k^{an}, \tau^{él}, C_L, C_I, T\right) \qquad (38)$$

$$\begin{cases} s_{ij} = \dfrac{n}{2}\left(\dfrac{\partial f}{\partial \alpha_{ij}^{él}} + \dfrac{\partial f}{\partial \alpha_{ji}^{él}}\right) - \dfrac{n}{3}\delta_{ij}\sum_k \dfrac{\partial f}{\partial \alpha_{kk}^{él}} = s_{ij}(\alpha_{lm}^{él}, \omega_n^{él}, \tau^{él}, \alpha_{lm}^{an}, \omega_n^{an}, C_L, C_I, T) & (39) \\ m_k = n\dfrac{\partial f}{\partial \omega_k^{él}} = m_k(\alpha_{lm}^{él}, \omega_n^{él}, \tau^{él}, \alpha_{lm}^{an}, \omega_n^{an}, C_L, C_I, T) & (40) \\ p = -n\dfrac{\partial f}{\partial \tau^{él}} = p(\alpha_{lm}^{él}, \omega_n^{él}, \tau^{él}, \alpha_{lm}^{an}, \omega_n^{an}, C_L, C_I, T) & (41) \end{cases}$$

$$\begin{cases} s_{ij}^{dis} = \dfrac{n}{2}\left(\dfrac{\partial f}{\partial \alpha_{ij}^{an}} + \dfrac{\partial f}{\partial \alpha_{ji}^{an}}\right) - \dfrac{n}{3}\delta_{ij}\sum_k \dfrac{\partial f}{\partial \alpha_{kk}^{an}} = s_{ij}^{dis}(\alpha_{lm}^{él}, \omega_n^{él}, \tau^{él}, \alpha_{lm}^{an}, \omega_n^{an}, C_L, C_I, T) & (42) \\ m_k^{dis} = n\dfrac{\partial f}{\partial \omega_k^{an}} = m_k^{dis}(\alpha_{lm}^{él}, \omega_n^{él}, \tau^{él}, \alpha_{lm}^{an}, \omega_n^{an}, C_L, C_I, T) & (43) \end{cases}$$

$$\begin{cases} s = -\dfrac{\partial f}{\partial T} = s(\alpha_{lm}^{él}, \omega_n^{él}, \tau^{él}, \alpha_{lm}^{an}, \omega_n^{an}, C_L, C_I, T) & (44) \\ \mu_L = \dfrac{\partial f}{\partial C_L} = \mu_L(\alpha_{lm}^{él}, \omega_n^{él}, \tau^{él}, \alpha_{lm}^{an}, \omega_n^{an}, C_L, C_I, T) & (45) \\ \mu_I = \dfrac{\partial f}{\partial C_I} = \mu_I(\alpha_{lm}^{él}, \omega_n^{él}, \tau^{él}, \alpha_{lm}^{an}, \omega_n^{an}, C_L, C_I, T) & (46) \end{cases}$$

Dissipation equations: self-diffusion and creation-annihilation of pairs

$$\begin{cases} \vec{J}_q = \vec{J}_q(\vec{X}_q, \vec{X}_L, \vec{X}_I, n, T, C_L, C_I, \ldots) & (47) \\ \vec{J}_L = \vec{J}_L(\vec{X}_q, \vec{X}_L, \vec{X}_I, n, T, C_L, C_I, \ldots) & (48) \\ \vec{J}_I = \vec{J}_I(\vec{X}_q, \vec{X}_L, \vec{X}_I, n, T, C_L, C_I, \ldots) & (49) \end{cases}$$

$$S_{I-L} = S_{I-L}\left(\mu_L^* + \mu_I^*, n, T, C_L, C_I, \ldots\right) \qquad (50)$$

Dissipation equations: anelasticity

$$\begin{cases} \vec{s}_i = \vec{s}_i^{\,cons}\left(\vec{\alpha}_m^{an}, v, T, \ldots\right) + \vec{s}_i^{\,dis}\left(\dfrac{d\vec{\alpha}_m^{an}}{dt}, v, T, \ldots\right) & (51) \\ \vec{m} = \vec{m}^{cons}\left(\vec{\omega}^{an}, v, T, \ldots\right) + \vec{m}^{dis}\left(\dfrac{d\vec{\omega}^{an}}{dt}, v, T, \ldots\right) & (52) \end{cases}$$

Dissipation equations: flows of dislocation charges

$$\begin{cases} \vec{J}_k = \vec{\lambda}_k \wedge \vec{\mathbf{v}} = \vec{J}_k\left(\vec{s}_m, \vec{\lambda}_n, v, T, ...\right) & (53) \\ \vec{J} = -\dfrac{1}{2}\sum_k \vec{e}_k \wedge \vec{J}_k = \lambda\vec{\mathbf{v}} + \dfrac{1}{2}\left(\vec{\lambda} \wedge \vec{\mathbf{v}}\right) = \vec{J}\left(\vec{m}, \lambda, v, T, ...\right) & (54) \\ \dfrac{S_n}{n} = \sum_k \vec{e}_k \vec{J}_k = -\vec{\lambda}\vec{\mathbf{v}} = \dfrac{1}{n}\left(S_L^{pl} - S_I^{pl}\right) & (55) \end{cases}$$

$$\begin{cases} S_L^{pl} = S_L^{pl}\left[\left(\mu_L^* + g^*\right), p, \vec{\lambda}, v, T, C_L, C_I, ...\right] & (56) \\ S_I^{pl} = S_I^{pl}\left[\left(\mu_I^* - g^*\right), p, \vec{\lambda}, v, T, C_L, C_I, ...\right] & (57) \end{cases}$$

$$g^* = f + pv + m\vec{\phi}^2 / 2 - \mu_L C_L - \mu_I C_I \qquad (58)$$

Dissipation equations: sources of dislocation charges

$$\vec{S}_i^{\vec{\lambda}_i} = \vec{S}_i^{\vec{\lambda}_i}(\vec{s}_m, \vec{\lambda}_n, v, T, ...) \quad (59) \quad \Rightarrow \quad \begin{cases} \vec{S}^{(\vec{\lambda})} = -\sum_k \vec{e}_k \wedge \vec{S}_k^{(\vec{\lambda}_i)} = \vec{S}^{(\vec{\lambda})}(\vec{s}_m, \vec{\lambda}_n, v, T, ...) & (60) \\ S^{(\lambda)} = \dfrac{1}{2}\sum_k \vec{e}_k \vec{S}_k^{(\vec{\lambda}_i)} = S^{(\lambda)}(\vec{s}_m, \vec{\lambda}_n, v, T, ...) & (61) \end{cases}$$

Additional equations of evolution

Mass continuity

$$\frac{\partial \rho}{\partial t} = -\operatorname{div}\left[\rho\vec{\phi} + m\left(\vec{J}_I - \vec{J}_L\right)\right] = -\operatorname{div}\left(n\vec{p}\right) \quad \text{dans } Q\xi_1\xi_2\xi_3 \qquad (62)$$

Flux of work and surface forces

$$\vec{J}_w = \mu_L^* \vec{J}_L + \mu_I^* \vec{J}_I - \phi_k \vec{s}_k - \frac{1}{2}\left(\vec{\phi} \wedge \vec{m}\right) + p\vec{\phi} \qquad (63)$$

$$\vec{F}_S = \sum_k \vec{e}_k \left(\vec{s}_k \vec{\mathbf{n}}\right) + \frac{1}{2}\left(\vec{m} \wedge \vec{\mathbf{n}}\right) - \vec{\mathbf{n}} p \qquad (64)$$

Entropy source

$$\begin{aligned} S_e = & -\frac{1}{T}\left(\mu_L^* + \mu_I^*\right)S_{I-L} - \frac{1}{T}\left(\mu_L^* + g^*\right)S_L^{pl} - \frac{1}{T}\left(\mu_I^* - g^*\right)S_I^{pl} + \vec{J}_L \vec{X}_L \\ & + \vec{J}_I \vec{X}_I + \frac{1}{T}\left(\vec{s}_k^{dis} \frac{d\vec{\beta}_k^{an}}{dt} + \vec{m}^{dis} \frac{d\vec{\omega}^{an}}{dt} + \vec{s}_k \vec{J}_k + \vec{m}\vec{J}\right) + \vec{J}_q \overrightarrow{\operatorname{grad}}\left(\frac{1}{T}\right) \end{aligned} \qquad (65)$$

Energy balance

$$\begin{aligned} & n\vec{\phi}\left(\frac{d\vec{p}}{dt} - m\vec{\phi}_I \frac{dC_I}{dt} + m\vec{\phi}_L \frac{dC_L}{dt}\right) + \left(\vec{s}_k \frac{d\vec{\beta}_k}{dt} + \vec{m}\frac{d\vec{\omega}}{dt} - p\frac{d\tau}{dt}\right) \\ & + \left(\vec{s}_k \vec{J}_k + \vec{m}\vec{J} - p\frac{S_n}{n}\right) = \rho\vec{g}\vec{\phi} - \operatorname{div}\left[-\phi_k \vec{s}_k - \frac{1}{2}\left(\vec{\phi} \wedge \vec{m}\right) + p\vec{\phi}\right] \end{aligned} \qquad (66)$$

PART I

D

Application: elements of the dislocation theory in the usual solids

Maxwell's equations at constant volume expansion

Fields and energies of the screw and edge dislocations

Interactions between dislocations

String model of dislocation

Chapter 12

Elements of dislocation theory in usual solids

In this chapter, we introduce the simplest solid that can be considered, namely the perfect isotropic solid, and discuss Newton's equation of this solid. We can then show that this solid is perfectly described by equations analogous to the Maxwell equations when the volume expansion is homogeneous in the solid.

Then we calculate the distortion fields, energies and interactions of dislocations in that perfect solid. In the case of immobile dislocations in the solid lattice, the *static lattice distortions* induced by these dislocation store elastic energy within the lattice. This stored energy can then be considered as *the rest energy* of these immobile dislocations. In the case where dislocations are mobile within the lattice, *the movement of lattice itself* induced by the movement of dislocations are associated with a *kinetic energy.* At low speed, the kinetic energy is directly linked to the rest energy of the dislocations through relationships similar to the famous expression of Einstein $E_0 = M_0 c^2$. This allows us to introduce, in a *completely classical way* the notion of *inertial mass of dislocations.*

From the distortion fields induced by the dislocations and the force of Peach and Koehler, we then describe some interactions that may occur between dislocations. Finally, we introduce *the string model of dislocation*, which can process the dynamics of a dislocation line that moves and deforms in the lattice.

12.1 – The perfect solid and its equation of Newton

We will call *perfect lattice*, the isotropic lattice as it has been defined in section 7.1, with the state function *(7.7)*. It is further assumed that, very generally, only the modulus k_2 is strictly positive in this state function, while the moduli k_1 and k_0 can be either positive, negative or zero. Furthermore we will assume that these moduli can be temperature dependent (section 7.3). The purely elastic part of the state function of the perfect lattice depends only on the values of τ, τ^2 and $(\vec{\alpha}_i^{\,el})^2$

$$f^{el} = -k_0(T)\tau + k_1\tau^2 + k_2\sum_i \vec{\alpha}_i^{\,2} \qquad (12.1)$$

so that the equations of state which completely characterize the elasticity of the *perfect lattice* are limited to the scalar pressure p and the symmetric tensor of transverse shear stresses $\vec{s}_i$

$$\vec{s}_i = n\frac{\partial f}{\partial \alpha_{ik}^{el}}\vec{e}_k = 2nk_2\vec{\alpha}_i^{el} \quad \text{and} \quad p = -n\frac{\partial f}{\partial \tau^{el}} = n\left(k_0(T) - 2k_1\tau^{el}\right) \qquad (12.2)$$

It is found that the pressure p depends on scalar modulus $k_0(T)$ which may contain the effects of temperature, and the volume expansion τ through the modulus k_1.

The equation of Newton of the perfect solid

By neglecting the force of gravity, knowing that the torsor of moment $\vec{m}$ is null, and by using state equations *(12.2)*, the equation of Newton *(6.11)* is written as

$$\begin{aligned} n\frac{d\vec{p}}{dt} &= \sum_k \vec{e}_k \operatorname{div}\vec{s}_k - \overrightarrow{\operatorname{grad}}\, p + nm\vec{\phi}_I \frac{dC_I}{dt} - nm\vec{\phi}_L \frac{dC_L}{dt} \\ &= 2k_2 \sum_k \vec{e}_k \operatorname{div}\left(n\vec{\alpha}_i^{el}\right) - \overrightarrow{\operatorname{grad}}\left(nk_0(T) - 2nk_1\tau^{el}\right) + nm\vec{\phi}_I \frac{dC_I}{dt} - nm\vec{\phi}_L \frac{dC_L}{dt} \end{aligned} \tag{12.3}$$

By taking into account the fact that $\overrightarrow{\operatorname{grad}}\, n = -n\, \overrightarrow{\operatorname{grad}}\, \tau$, we obtain the following equation of Newton

$$\frac{d\vec{p}}{dt} = 2k_2 \sum_k \vec{e}_k \operatorname{div}\vec{\alpha}_k - 2k_2 \sum_k \left(\vec{e}_k \overrightarrow{\operatorname{grad}}\, \tau\right)\vec{\alpha}_k + \overrightarrow{\operatorname{grad}}\left[\left(2k_1(1-\tau) + k_0\right)\tau\right] + m\vec{\phi}_I \frac{dC_I}{dt} - m\vec{\phi}_L \frac{dC_L}{dt} \tag{12.4}$$

Written in this form, the Newton's equation does not allow to separate the effects leading to a "rotational" quantity of movement, which would be related to pure shears without volume expansion of the lattice, from those leading to a "divergent" quantity of movement, which would be linked to the local volume expansions of the lattice.
But it is possible to find another formulation of the Newton's equation to separate these effects. Indeed, one can use the equations giving the *vector of flexion* to connect the space derivatives of the distortion tensors

$$\vec{\chi} = -\sum_k \vec{e}_k \wedge \overrightarrow{\operatorname{rot}}\, \vec{\alpha}_k - \frac{2}{3}\overrightarrow{\operatorname{grad}}\, \tau = \sum_k \vec{e}_k \operatorname{div}\vec{\alpha}_k - \frac{2}{3}\overrightarrow{\operatorname{grad}}\, \tau = -\overrightarrow{\operatorname{rot}}\, \vec{\omega} + \vec{\lambda} \tag{12.5}$$

If we assume the existence of elasticity in the lattice but *only by shear,* the deformations of the lattice are entirely characterized by $\vec{\alpha}_i$ and τ so that

Hypothesis 1: $\quad \vec{\alpha}_i = \vec{\alpha}_i^{el} + \vec{\alpha}_i^{an}$ *and* $\quad \tau = \tau^{el}$ *(12.6)*

Expression *(12.5)* can be written

$$\sum_k \vec{e}_k \operatorname{div}\left(\vec{\alpha}_k^{el} + \vec{\alpha}_k^{an}\right) = \vec{\lambda} - \overrightarrow{\operatorname{rot}}\left(\vec{\omega}^{el} + \vec{\omega}^{an}\right) + \frac{2}{3}\overrightarrow{\operatorname{grad}}\, \tau^{el} \tag{12.7}$$

This relationship combines elasticity effects with anelasticity effects, which cannot be separated, given the presence of the density $\vec{\lambda}$ of flexion charges, which greatly limits the applicability of this relationship. But we can assume that the presence of a vector density $\vec{\lambda}$ of flexion charges in the lattice should primarily be related to the behavior of elastic distortions, so that we will make the simplifying assumption that the relation *(12.7)* can be split into two separate relationships, one for the elastic part and the other for the anelastic part of the distortions in the following form

Hypothesis 2:

$$\begin{cases} \sum_k \vec{e}_k \operatorname{div}\vec{\alpha}_k^{el} = \vec{\lambda} - \overrightarrow{\operatorname{rot}}\, \vec{\omega}^{el} + \frac{2}{3}\overrightarrow{\operatorname{grad}}\, \tau^{el} \\ \sum_k \vec{e}_k \operatorname{div}\vec{\alpha}_k^{an} = -\overrightarrow{\operatorname{rot}}\, \vec{\omega}^{an} \end{cases} \tag{12.8}$$

Using then the first relation *(12.8)*, the equation of Newton *(12.4)* can also be written

$$\frac{d\vec{p}}{dt}=-2k_2\overrightarrow{\text{rot}}\,\vec{\omega}-2k_2\sum_k\left(\vec{e}_k\overrightarrow{\text{grad}}\,\tau\right)\vec{\alpha}_k+\overrightarrow{\text{grad}}\left[\left(\frac{4}{3}k_2+2k_1(1-\tau)+k_0\right)\tau\right]$$
$$+m\vec{\phi}_I\frac{dC_I}{dt}-m\vec{\phi}_L\frac{dC_L}{dt}+2k_2\vec{\lambda} \qquad (12.9)$$

In this new form, Newton's equation becomes really interesting because it is now possible to separate the effects of pure shear in the lattice without volume expansion from the effects of pure volume expansion of the lattice. The only term of this equation which is not linear is the coupling term $2k_2\sum\left(\vec{e}_k\overrightarrow{\text{grad}}\,\tau\right)\vec{\alpha}_k$. For small deformations, as in the common solids, this second order term may be usually neglected.

12.2 - Analogy with Maxwell's equations under homogeneous expansion

By supposing a solid in which the volume expansion is homogeneous, and which therefore satisfies the following

Hypothesis 3: $\tau=\tau(t)\neq\tau(\vec{r},t)$ *(12.10)*

and by assuming further that the concentrations of vacancies and interstitials are independent of time

Hypothesis 4: $\dfrac{dC_I}{dt}=\dfrac{dC_L}{dt}=0$ *(12.11)*

the Newton equation becomes

$$\frac{d\vec{p}}{dt}=-2k_2\overrightarrow{\text{rot}}\,\vec{\omega}+2k_2\vec{\lambda} \qquad (12.12)$$

In this case, it is possible to introduce in equation *(12.12)* a new quantity, namely the torsor of moment $\vec{m}$ copied after the torsor of moment which appears in the equation of Newton *(6.11)*, by writing

$$\frac{d\vec{p}}{dt}=-2k_2\overrightarrow{\text{rot}}\,\vec{\omega}+2k_2\vec{\lambda}=-\frac{1}{2n}\overrightarrow{\text{rot}}\,\vec{m}+2k_2\vec{\lambda} \qquad (12.13)$$

Thus, in the *perfect lattice* with homogeneous volume expansion, the torsor of moment $\vec{m}$ is linked to the vector of elastic rotation $\vec{\omega}^{el}$ by a very simple relation

$$\vec{m}=4nk_2\vec{\omega}^{el} \qquad (12.14)$$

As a consequence, if we imagine that the torsor of moment $\vec{m}$ derives from a state equation, we obtain the *volume density of free energy of elastic rotation* by lattice site, at homogenous expansion, under the form

$$m_k=n\frac{\partial f^{el}_{rotation}}{\partial\omega_k^{el}}=4nk_2\omega^{el}\Rightarrow f^{el}_{rotation}=2k_2\left(\vec{\omega}^{el}\right)^2 \qquad (12.15)$$

Thus, the *volume density of free energy of elastic rotation*, linked to the deformation by pure elastic rotations at homogenous volume expansion is equal to the elastic energy of deformation by pure shear at homogenous volume expansion

$$f^{el}_{rotation}\Big|_{\tau=\tau_0}=2k_2\left(\vec{\omega}^{el}\right)^2=k_2\sum_i(\vec{\alpha}_i^{el})^2=f^{el}_{shear}\Big|_{\tau=\tau_0} \qquad (12.16)$$

Finally, by using the second relation of hypothesis *(12.8)*, namely that there exists a one-to-one relationship $\sum \vec{e}_k \operatorname{div}\vec{\alpha}_k^{an} = -\overrightarrow{\operatorname{rot}}\,\vec{\omega}^{an}$ between spatial derivatives of the anelastic shears and anelastic rotations, we deduce that the anelasticity of the lattice by pure anelastic shear without changes in the volume expansion can also be reduced, at homogenous volume expansion, to a simple *lattice anelasticity by pure anelastic rotation.* This anelasticity can then be represented by a global rotation vector of the lattice $\vec{\omega}$, by writing the following relationship

$$\vec{\omega} = \vec{\omega}^{el} + \vec{\omega}^{an} + \vec{\omega}_0(t) = \frac{\vec{m}}{4nk_2} + \vec{\omega}^{an} + \vec{\omega}_0(t) \tag{12.17}$$

where $\vec{\omega}_0(t)$ represents the uniform global rotation of the lattice.
It is concluded that, in the lattice where the volume expansion may be regarded as homogeneous, it is possible to treat univocally the elastic and anelastic pure lattice shear using only rotation vectors $\vec{\omega}^{el}$ and $\vec{\omega}^{an}$, which is very useful for dealing with the problem of the fields, energies and interactions of dislocations in a perfect lattice.
As the density $\vec{\lambda}$ of flexion charges associated with the edge dislocation charges, which are likely to bring locally a solid volume expansion variation, it will be assumed a-priori here a solid in which this type of charge is negligible, posing

Hypothesis 5: $\vec{\lambda} \approx 0$ (12.18)

As n is a constant in this case, we can rewrite the equations of Newton *(12.13)* as

$$\frac{d\vec{p}}{dt} = -2k_2\overrightarrow{\operatorname{rot}}\,\vec{\omega} = -\frac{1}{2n}\overrightarrow{\operatorname{rot}}\,\vec{m} \tag{12.19}$$

The equations needed for a complete description of the pure shear of a perfect solid still need to incorporate the topological equations for the rotation vector $\vec{\omega}$, i.e. the geometrokinetic equations and the equation of geometrocompatibility in the presence of dislocation charges

$$\vec{J} = -\frac{1}{2}\sum_k \vec{e}_k \wedge \vec{J}_k = -\frac{d\vec{\omega}}{dt} + \frac{1}{2}\overrightarrow{\operatorname{rot}}\,\vec{\phi} \quad \text{and} \quad \lambda = \frac{1}{2}\sum_k \vec{e}_k\vec{\lambda}_k = \operatorname{div}\vec{\omega} \tag{12.20}$$

With hypothesis 4, we have that $C_L = cste$ and $C_I = cste$, so that the density of mass ρ becomes a constant

$$\rho = m(n + n_I - n_L) = mn(1 + C_I - C_L) = cste \tag{12.21}$$

We also have that the equation of evolution of this density in the local referential $Ox_1x_2x_3$ allows us to deduce that the divergence of $n\vec{p}$ is null

$$\partial\rho/\partial t = 0 = -\operatorname{div}(n\vec{p}) \quad \Rightarrow \quad \operatorname{div}\vec{p} = 0 \tag{12.22}$$

The quantity $\vec{p}$ can then be deduced thanks to the following relationships

$$\vec{p} = \frac{1}{n}\left(\rho\vec{\phi} + \vec{J}_m\right) = \frac{1}{n}\left[\rho\vec{\phi} + m\left(\vec{J}_I - \vec{J}_L\right)\right] = m\left[\vec{\phi} + (C_I - C_L)\vec{\phi} + \frac{1}{n}\left(\vec{J}_I - \vec{J}_L\right)\right] \tag{12.23}$$

Furthermore, by imagining that there are no sources S^λ of charges of rotation

Hypothesis 6: $S^\lambda = 0$ (12.24)

we also obtain an equation for the continuity of charges of rotation

$$\frac{d\lambda}{dt} = -\operatorname{div}\vec{J} \tag{12.25}$$

Finally, we can establish a continuity equation of the energy from equations *(12.19)* and *(12.20)*

$$-\vec{m}\vec{J} = \vec{m}\frac{d\vec{\omega}}{dt} + \vec{\phi}\frac{d(n\vec{p})}{dt} - \operatorname{div}\left(\frac{1}{2}\vec{\phi}\wedge\vec{m}\right) \qquad (12.26)$$

The relations thus obtained for the pure shears of a perfect solid in the local framework $Ox_1x_2x_3$ in translation $\vec{\phi}_O(t)$ and in rotation $\dot{\vec{\omega}}_O(t)$ in the absolute referential frame are written in table 12.1, and compared to the *equations of Maxwell for electromagnetism*

Table 12.1 - "Maxwellian" formulation of the equations of a perfect solid presenting homogeneous expansion τ in the local frame $Ox_1x_2x_3$

$\begin{cases} -\dfrac{d(2\vec{\omega})}{dt} + \overrightarrow{\operatorname{rot}}\vec{\phi} = (2\vec{J}) \\ \operatorname{div}(2\vec{\omega}) = (2\lambda) \end{cases}$	$\Leftrightarrow$	$\begin{cases} -\dfrac{\partial\vec{D}}{\partial t} + \overrightarrow{\operatorname{rot}}\vec{H} = \vec{j} \\ \operatorname{div}\vec{D} = \rho \end{cases}$
$\begin{cases} \dfrac{d(n\vec{p})}{dt} = -\overrightarrow{\operatorname{rot}}\left(\dfrac{\vec{m}}{2}\right) \\ \operatorname{div}(n\vec{p}) = 0 \end{cases}$	$\Leftrightarrow$	$\begin{cases} \dfrac{\partial\vec{B}}{\partial t} = -\overrightarrow{\operatorname{rot}}\vec{E} \\ \operatorname{div}\vec{B} = 0 \end{cases}$
$\begin{cases} (2\vec{\omega}) = \left(\dfrac{1}{nk_2}\right)\left(\dfrac{\vec{m}}{2}\right) + (2\vec{\omega}^{an}) + \left(2\vec{\omega}_0(t)\right) \\ (n\vec{p}) = (nm)\left[\vec{\phi} + (C_l - C_L)\vec{\phi} + \left(\dfrac{1}{n}(\vec{J}_l - \vec{J}_L)\right)\right] \end{cases}$	$\Leftrightarrow$	$\begin{cases} \vec{D} = \varepsilon_0\vec{E} + \vec{P} + \vec{P}_0(t) \\ \vec{B} = \mu_0\left[\vec{H} + (\chi^{para} + \chi^{dia})\vec{H} + \vec{M}\right] \end{cases}$
$\left\{ \dfrac{d(2\lambda)}{dt} = -\operatorname{div}(2\vec{J}) \right.$	$\Leftrightarrow$	$\left\{ \dfrac{\partial\rho}{\partial t} = -\operatorname{div}\vec{j} \right.$
$\begin{cases} -\left(\dfrac{\vec{m}}{2}\right)(2\vec{J}) = \\ \vec{\phi}\dfrac{d(n\vec{p})}{dt} + \left(\dfrac{\vec{m}}{2}\right)\dfrac{d(2\vec{\omega})}{dt} - \operatorname{div}\left(\vec{\phi}\wedge\left(\dfrac{\vec{m}}{2}\right)\right) \end{cases}$	$\Leftrightarrow$	$\begin{cases} -\vec{E}\vec{j} = \\ \vec{H}\dfrac{\partial\vec{B}}{\partial t} + \vec{E}\dfrac{\partial\vec{D}}{\partial t} - \operatorname{div}\left(\vec{H}\wedge\vec{E}\right) \end{cases}$
$\left\{ c_t = \sqrt{\dfrac{nk_2}{nm}} = \sqrt{\dfrac{k_2}{m}} \right.$	$\Leftrightarrow$	$\left\{ c = \sqrt{\dfrac{1}{\varepsilon_0\mu_0}} \right.$

It is noted that there is a remarkable and absolutely complete analogy between these two sets of equations, except for the fact that the evolution equations involve the total derivative, while the Maxwell equations only involve the partial derivative with respect to time. However, it is recalled here that the material derivative in the local frame $Ox_1x_2x_3$ can be replaced without problems by the partial derivative with respect to time if the strains are sufficiently small and/or slow in the vicinity of the origin of the local frame $Ox_1x_2x_3$!

This analogy will be discussed in detail in the third part of the book. For now, let's enumerate the following analogies:

- the *rotational field* $2\vec{\omega}$ is analogous to the *electrical field of displacement* $\vec{D}$,
- the *quantity of movement and the flow of mass* $n\vec{p}$ is analogous of the *magnetic induction field* $\vec{B}$,
- the *torsor of moment* $\vec{m}/2$ is analogous to the *electric field* $\vec{E}$,
- the *velocity field* $\vec{\phi}$ is analogous to the *magnetic field* $\vec{H}$,
- the *density* 2λ *of rotational charges* is analogous to the *density* ρ *of electric charges*
- the *flux* $2\vec{J}$ *of rotational charges* is analogous to the *density* $\vec{j}$ *of electric current*
- the anelasticity $2\vec{\omega}^{an}$ is analogous to the *dielectric polarisation* $\vec{P}$ *of matter*,
- the *mass transport* $nm\vec{\phi}$ *by the movement of the lattice is analogous to the term* $\mu_0\vec{H}$ of *magnetic induction of vacuum*,
- the *mass transport* $nm(C_I - C_L)\vec{\phi}$ *by the drag-along movement of point defects of the lattice* is analogous to the term $\mu_0(\chi^{para} + \chi^{dia})\vec{H}$ *of paramagnetism and diamagnetism of matter*,
- the *mass transport* $m(\vec{J}_I - \vec{J}_L)$ *due to self-diffusion of vacancies and interstitials* is analogous to the *magnetization* $\vec{M}$ *of matter*,
- the *inverse of elastic modulus* $1/K_2$ is analogous to the *dielectric permittivity of vacuum* ε_0,
- the *density of mass* $nm\vec{\phi}$ is analogous to the *magnetic permeability of vacuum* μ_0,
- the *speed of transversal waves* c_t is analogous to the *speed of light* c,
- the *flow of elastic energy* $\vec{\phi}\wedge\vec{m}/2$ is analogous to the *vector of Poynting* $\vec{H}\wedge\vec{E}$.

12.3 – Fields and energies of a screw dislocation

The rotational field of a screw dislocation

Let's consider a *screw dislocation string*, in the form of a straight cylinder of infinite length and radius, containing a density λ of rotational charges (figure 12.1).

Apply the compatibility relationship $\operatorname{div}\vec{\omega}^{él} = \lambda$ and integrate on a cylindrical volume of radius r and length unit. We have

$$\iiint_V \operatorname{div}\vec{\omega}^{screw}\, dV = \oiint_S \vec{\omega}^{screw}\vec{n}\, dS = \vec{\omega}^{screw}\vec{n}\, 2\pi r = \iiint_V \lambda\, dV \qquad (12.27)$$

in which $\vec{n}$ is the vector normal to the surface cylinder.

Outside the charge, meaning for $r > R$, this relation gives us the following $\vec{\omega}$ field

$$\vec{\omega}_{ext}^{screw}\vec{n} = \frac{1}{2\pi r}\left(\lambda\pi R^2\right) = \frac{\Lambda}{2\pi r} \quad \Rightarrow \quad \vec{\omega}_{ext}^{screw} = \frac{\Lambda}{2\pi}\frac{\vec{r}}{r^2} \quad (r > R) \qquad (12.28)$$

where Λ is the *linear charge of rotation of the string* given by the integration of density λ on the unit length of the string

$$\Lambda = \pi R^2 \lambda \qquad (12.29)$$

Inside the charge, meaning for $r < R$, we then have

$$\vec{\omega}_{int}^{screw}\vec{n} = \frac{1}{2\pi r}\left(\pi\lambda r^2\right) = \frac{\Lambda}{2\pi R^2} r \quad \Rightarrow \quad \vec{\omega}_{int}^{screw} = \frac{\Lambda}{2\pi R^2}\vec{r} \quad (r < R) \qquad (12.30)$$

The behavior of the module of $\vec{\omega}^{screw}$ as a function of distance r from the center of the cylinder is shown in figure 12.2.

The displacement field by rotation of a screw dislocation

Knowing the field of external rotations $\vec{\omega}_{ext}^{screw}$ outside the string, it is possible to derive easily the elastic displacement field $\vec{u}_{ext}^{screw}$ associated with the rotation due to the presence of the charged string since outside of the string λ is zero. We have

$$\vec{\omega}_{ext}^{screw} = -\frac{1}{2}\overrightarrow{\mathrm{rot}}\,\vec{u}_{ext}^{screw} = \frac{\Lambda}{2\pi}\frac{\vec{r}}{r^2} = \frac{\Lambda}{2\pi}\frac{x_1\vec{e}_1 + x_3\vec{e}_3}{x_1^{\,2} + x_3^{\,2}} \qquad (r > R) \tag{12.31}$$

It is clear that the field $\vec{u}_{ext}^{screw}$ must be parallel to $\vec{e}_2$ to satisfy this relationship, so that

$$\vec{u}_{ext}^{screw} = f\vec{e}_2 \quad \Rightarrow \quad \overrightarrow{\mathrm{rot}}\,\vec{u}_{ext}^{screw} = -\frac{\partial f}{\partial x_3}\vec{e}_1 + \frac{\partial f}{\partial x_1}\vec{e}_3 \tag{12.32}$$

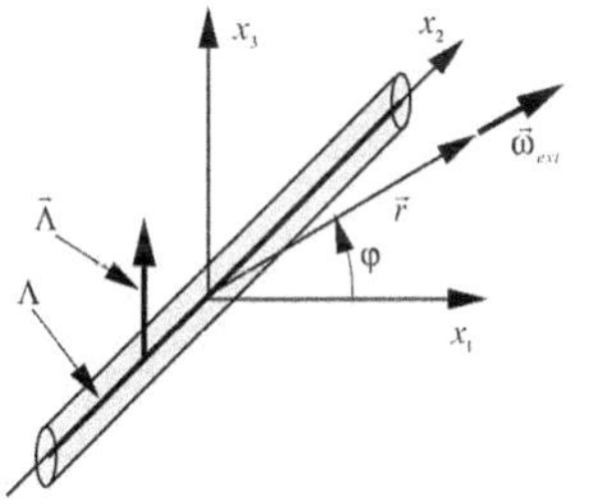

Figure 12.1 *- cylindrical charge of radius R and density λ*

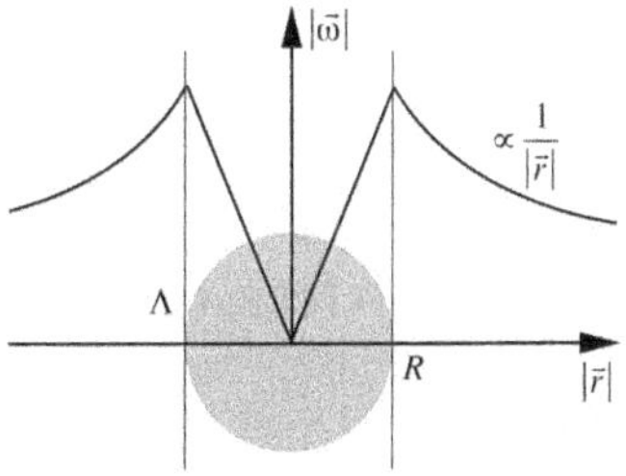

Figure 12.2 *- the norm of the field $\vec{\omega}$ inside and outside of the linearly charged string Λ*

so that we obtain the following system of partial differential equations

$$\frac{\partial f}{\partial x_3} = +\frac{\Lambda}{\pi}\frac{x_1}{x_1^2 + x_3^2} \quad \textit{et} \quad \frac{\partial f}{\partial x_1} = -\frac{\Lambda}{\pi}\frac{x_3}{x_1^2 + x_3^2} \tag{12.33}$$

for which the solution is

$$\vec{u}_{ext}^{screw} = -\vec{e}_2\frac{\Lambda}{\pi}\arctan\frac{x_1}{x_3} \tag{12.34}$$

by introducing the polar coordinates $x_1 = r\cos\varphi$ and $x_2 = r\sin\varphi$ in the plane of the cylinder, the displacement field can be written

$$\vec{u}_{ext}^{screw} = \vec{e}_2\frac{\Lambda}{\pi}\left(\varphi - \frac{\pi}{2}\right) \tag{12.35}$$

In polar coordinates, this field does not depend on r and increases with angle φ. It does indeed correspond to a displacement of type 'screw' along the axis Ox_2.

The rotational rest energy of a screw dislocation in a perfect lattice

As the rotational plane of the screw dislocation is not dependent on the volume expansion field, we can use without problem equation *(12.15)*, and write the stored elastic energy of rotation stored outside of the string per unit length in a "quasi-infinite" environment, as such

$$E_0^{screw\,ext} = 2nk_2 \int_R^{R_\infty} \frac{\Lambda^2}{4\pi^2 r^2} 2\pi r\, dr = \frac{nk_2\Lambda^2}{\pi} \ln\frac{R_\infty}{R} \tag{12.36}$$

and the energy stored inside the string as

$$E_0^{screw\,int} = 2nk_2 \int_0^{R} \frac{\Lambda^2 r^2}{4\pi^2 R^4} 2\pi r\, dr = \frac{nk_2\Lambda^2}{4\pi} \tag{12.37}$$

It is remarkable that the energy of the interior of the dislocation $E_0^{screw\,int}$ thus obtained does not depend on the radius of the string R.
The sum of the two energies stored in the lattice can be considered as the rest energy E_0^{screw} of the dislocation string, expressed by unit length of the string or radius R, and thus is worth

$$E_0^{screw} = E_0^{screw\,int} + E_0^{screw\,ext} = \frac{nk_2\Lambda^2}{\pi}\left(\frac{1}{4} + \ln\frac{R_\infty}{R}\right) \tag{12.38}$$

We notice that the internal energy of the string, which does not depend on radius R, is a lot smaller than the energy stored outside the string, which depends on the radius R of the string and the size of the lattice $R_\infty >> R$. Since the internal energy by unit of length does not depend on R, we can neglect it compared to $E_0^{screw\,ext}$ and write approximatively the rest energy of dislocation under the following forms, by using the relation $\Lambda = -\vec{\boldsymbol{B}}_{screw}\vec{t}\,/\,2$ and the *Lamé coefficient* $\mu = K_2$

$$E_0^{screw} \cong \frac{nk_2\Lambda^2}{\pi} \ln\frac{R_\infty}{R} = \frac{nk_2\vec{\boldsymbol{B}}_{screw}^2}{4\pi} \ln\frac{R_\infty}{R} = \frac{\mu\vec{\boldsymbol{B}}_{screw}^2}{4\pi} \ln\frac{R_\infty}{R} \tag{12.39}$$

In the expressions of the rest energy of a screw dislocation appears the external dimension of the lattice in question. Thus, the rest energy of a single line of dislocation $R_\infty >> R$ in an infinite network should be infinite. However, apart from the fact that a real solid is not of infinite size and the presence of a single dislocation in a solid is an exceptional fact, the size R_∞ is generally limited to smaller values than the actual size R_∞ of the solid . Indeed, if there exists in the lattice a large number of dislocations separated by an average distance d, and these dislocations have random orientations in the solid, we have a screening effect of the individual fields of displacement of each lines by the opposite sign lines located in the immediate vicinity, so that long-distance energy stored by an individual line becomes negligible (one can show that it actually decreases as $\sim 1/r^2$ for distances r greater than d). In this case, a fairly good estimate of the rest energy of the dislocation is obtained by choosing, for R_∞, the average distance d between two adjacent lines of opposite sign. Note that R_∞ appears in a logarithm, which greatly mitigates the effects of large variations of d.
It should also be noted that the *energy associated with a line segment of length* L *of screw dislocation does not depend on the volume expansion of the solid lattice*. Indeed, it is easy to verify that

$$\left.E_0^{screw}\right|_L = L\frac{nk_2\vec{\boldsymbol{B}}_{screw}^2}{4\pi} \ln\frac{R_\infty}{R} = L_0\frac{n_0k_2\vec{\boldsymbol{B}}_{0\,screw}^2}{4\pi} \ln\frac{R_{\infty 0}}{R_0} = \left.E_{00}^{screw}\right|_{L_0} \tag{12.40}$$

The kinetic energy of rotation at low velocity of a screw dislocation

In case a screw dislocation is moving in the direction Ox_1 at low velocity $\vec{v}$ compared to the speed of transverse waves c_t, the dynamic displacement field associated with rotations is obtained by substituting x_1 by $x_1(t) = x_1 - \mathbf{v}t$ in the expression *(12.34)*

$$\vec{u}_{ext}^{screw} = -\vec{e}_2 \frac{\Lambda}{\pi} \arctan \frac{x_1 - \mathbf{v}t}{x_3} \qquad (12.41)$$

The velocity field is then obtain by temporal derivation of this expression

$$\vec{\phi}_{ext}^{screw} = -\frac{d\vec{u}}{dt} = \vec{e}_2 \frac{\Lambda}{\pi} \frac{\partial}{\partial t}\left(\arctan \frac{x_1 - \mathbf{v}t}{x_3} \right) = \vec{e}_2 \frac{\Lambda}{\pi} \frac{-x_3}{x_3^{\,2} + \left(x_1 - \mathbf{v}t\right)^2} \mathbf{v} \qquad (12.42)$$

The kinetic energy per lattice particle can then be deduced. For this, it is convenient to express it in the referential frame $Ox_1'x_2'x_3'$, which is in translation with velocity $\vec{v}$ with the dislocation line and in which $x_1 - \mathbf{v}t$ is x_1', and then switch to polar coordinates in this referential frame with $x_1' = r'\cos\varphi'$ and $x_3' = r'\sin\varphi'$ in the plane of the moving cylinder. We have

$$e_{kin}^{screw} = \frac{1}{2} m \left(\vec{\phi}_{ext}^{screw}\right)^2 = \frac{1}{2} m \left(\frac{\Lambda}{\pi}\right)^2 \left(\frac{x_3'}{x_3'^2 + x_1'^2}\right)^2 \mathbf{v}^2 = \frac{1}{2} m \left(\frac{\Lambda}{\pi}\right)^2 \left(\frac{\sin^2\varphi'}{r'}\right)^2 \mathbf{v}^2 \qquad (12.43)$$

The kinetic energy stored in the lattice by the dynamic field of rotation, expressed in unit of length of the screw dislocation that is in movement with velocity $\vec{v}$ is obtained by the integration of energy density $e_{kin}^{\vec{\omega}}$ on a cylinder of radius R_∞ and of unit length

$$E_{kin}^{screw} = \frac{1}{2} m \int_R^{R_\infty} \int_0^{2\pi} n \left(\frac{\Lambda}{\pi}\right)^2 \frac{\sin^2\varphi'}{r'^2} \mathbf{v}^2 r' d\varphi' dr' = \frac{1}{2} mn\mathbf{v}^2 \frac{\Lambda^2}{\pi} \ln \frac{R_\infty}{R} \qquad (12.44)$$

Using the fact that $\Lambda = -\vec{B}_{screw}\vec{t}\,/\,2$ and that the specific mass of the medium is $\rho = mn$, we can write the kinetic energy associated with the movement at low velocity of the screw dislocation under the following forms

$$E_{kin}^{screw} = \frac{mn\Lambda^2}{2\pi}\left(\ln \frac{R_\infty}{R}\right)\mathbf{v}^2 = \frac{mn\,\vec{B}_{screw}^2}{8\pi}\left(\ln \frac{R_\infty}{R}\right)\mathbf{v}^2 = \frac{\rho\,\vec{B}_{screw}^2}{8\pi}\left(\ln \frac{R_\infty}{R}\right)\mathbf{v}^2 \qquad (12.45)$$

By comparing the kinetic energy E_{kin}^{screw} stored in the medium by the movement of the string with the elastic potential energy E_0^{screw} stored in the medium by the presence of this line, using the expression *(12.39)* of the rest energy of the screw dislocation and the relationship $c_t^2 = k_2 / m$ of the speed of the transverse waves in the medium *(7.59)*, we find the following relationship between the rest energy and the kinetic energy of a screw dislocation

$$E_{kin}^{screw} = \frac{1}{2}\frac{m}{k_2}\frac{nk_2\Lambda^2}{\pi}\ln\frac{R_\infty}{R}\mathbf{v}^2 = \frac{1}{2}\frac{E_0^{screw}}{c_t^2}\mathbf{v}^2 \qquad (12.50)$$

This kinetic energy E_{kin}^{screw} stored in the solid lattice by the mobile dislocation can be considered as the kinetic energy of the dislocation in motion, and we can therefore introduce an *inertial mass M_0^{screw} of the dislocation* by writing relations

$$E_{kin}^{screw} = \frac{1}{2} M_0^{screw} \mathbf{v}^2 \quad \Rightarrow \quad M_0^{screw} = \frac{1}{c_t^2} E_0^{screw} \quad \Rightarrow \quad E_0^{screw} = M_0^{screw} c_t^2 \qquad (12.51)$$

We thus find for the dislocation line the famous *expression of Einstein* connecting the inertial energy to the rest energy via the speed of transverse waves. But here, this relationship is found without *any appeal to a relativistic dynamics of the string*, because it is due to the fact that the rest energy and the kinetic energy of the line are the elastic potential energy and the kinetic energy stored in the lattice by the dynamic distortions imposed on the lattice by the presence of the moving screw dislocation!

12.4 – Fields and energies of an edge dislocation

The rotational field and the volume expansion field of an edge dislocation

Consider now an *edge dislocation string* in the form of a straight cylinder of infinite length and radius R, as shown in figure 12.1, containing this time a vector density of flexion charges $\vec{\lambda}$ perpendicular to the direction of the string, which points in the direction of the axis Ox_3. In conventional solid media, the volume expansion is still very low, and the modulus k_0 associated with thermal effects is negligible vis-à-vis the moduli k_1 and k_2. So we can make three simplifying assumptions: the volume expansion τ is still very small ($\tau_0 \cong 0$) and can therefore be treated as a disturbance $\tau^{(p)} << 1$, and the modulus k_0 can be neglected

Hypothesis: $\tau = \tau^{(p)} << 1$ and $k_0 \cong 0$ and $\tau_0 \cong 0$ (12.52)

One can then use the linearized equation of Newton *(12.9)*, expressed in terms of $\tau^{(p)}$ and $\vec{\omega}^{el}$, to find the static fields of elastic torsion $\vec{\omega}^{el}$ and perturbation of volume expansion $\tau^{(p)}$ associated with the presence of the vector charge density $\vec{\lambda}$. We thus get the following equilibrium equation around the charge

$$\overrightarrow{\text{rot}}\,\vec{\omega}^{el} - \frac{2k_2+3k_1}{3k_2}\,\overrightarrow{\text{grad}}\,\tau^{(p)} \cong \begin{cases} \vec{\lambda} & (r<R) \\ 0 & (r>R) \end{cases} \tag{12.53}$$

To avoid manipulating expressions containing the elastic moduli k_1 and k_2 we can use *the Poisson's modulus (7.25 and 7.30)* of the solid, and write the equilibrium equation

$$\overrightarrow{\text{rot}}\,\vec{\omega}^{el} - \frac{1-\nu}{1-2\nu}\,\overrightarrow{\text{grad}}\,\tau^{(p)} \cong \begin{cases} \vec{\lambda} & (r<R) \\ 0 & (r>R) \end{cases} \tag{12.54}$$

Outside of the dislocation string $(r > R)$, the vector of rotation $\vec{\omega}^{el}$ should have a single-axis component along Ox_2 which must depend on $\cos\varphi$, while the expansion scalar $\tau^{(p)}$ should depend on $\sin\varphi$. On the other hand, the two variables $\vec{\omega}^{el}$ and $\tau^{(p)}$ should decrease with the distance as $1/r$ in the same way as for the screw dislocation string treated before, so the solution must surely be written in the following form

$$\begin{cases} \vec{\omega}^{el} = A\dfrac{\cos\varphi}{r}\vec{e}_2 = A\dfrac{x_1}{x_1^2+x_3^2}\vec{e}_2 \\ \qquad\qquad\qquad\qquad\qquad\qquad (r>R) \\ \tau^{(p)} = B\dfrac{\sin\varphi}{r} = B\dfrac{x_3}{x_1^2+x_3^2} \end{cases} \tag{12.55}$$

where A and B are constants of integration.

We can then calculate $\overrightarrow{\mathrm{rot}}\,\vec{\omega}^{el}$ and $\overrightarrow{\mathrm{grad}}\,\tau^{(p)}$

$$\begin{cases} \overrightarrow{\mathrm{rot}}\,\vec{\omega}^{el} = A\left[\dfrac{2x_1x_3}{\left(x_1^2+x_3^2\right)^2}\vec{e}_1 + \dfrac{x_3^2-x_1^2}{\left(x_1^2+x_3^2\right)^2}\vec{e}_3\right] \\ \overrightarrow{\mathrm{grad}}\,\tau^{(p)} = -B\left[\dfrac{2x_1x_3}{\left(x_1^2+x_3^2\right)^2}\vec{e}_1 + \dfrac{x_3^2-x_1^2}{\left(x_1^2+x_3^2\right)^2}\vec{e}_3\right] \end{cases} \tag{12.56}$$

which, when introduced in the equation of equilibrium, give us the relation existing between A and B

$$A + \frac{1-\nu}{1-2\nu}B = 0 \tag{12.57}$$

Inside the string, we can integrate the equilibrium equation on a cylinder containing the density of charge $\vec{\lambda}$, of radius R and of unit length. We then have the integral relation

$$\iiint_V \overrightarrow{\mathrm{rot}}\,\vec{\omega}^{el}\,dV - \frac{1-\nu}{1-2\nu}\iiint_V \overrightarrow{\mathrm{grad}}\,\tau^{(p)}\,dV \cong \iiint_V \vec{\lambda}\,dV \tag{12.58}$$

As the integral on $\vec{\lambda}$ on the cylinder of unit length gives us the linear charge $\vec{\Lambda}$ of the string, the previous relation is easily transformed in

$$\oiint_{cylinder} d\vec{S}\wedge\vec{\omega}^{el} - \frac{1-\nu}{1-2\nu}\oiint_{cylinder} \tau^{(p)}\,d\vec{S} \cong \vec{\Lambda} \tag{12.59}$$

By symmetry, the integrals on the two sections (lateral faces) of the cylinder cancel each other, so that with $d\vec{S} = R\left(\vec{e}_1\cos\varphi + \vec{e}_3\sin\varphi\right)d\varphi$

$$\begin{cases} \iint_S d\vec{S}\wedge\vec{\omega}^{el} = \iint_S A\dfrac{\cos\varphi}{R}\left(d\vec{S}\wedge\vec{e}_2\right) = \displaystyle\int_0^{2\pi} A\frac{\cos\varphi}{R}R\left(\vec{e}_3\cos\varphi - \vec{e}_1\sin\varphi\right)d\varphi = A\pi\vec{e}_3 \\ \iint_S \tau^{(p)}\,d\vec{S} = \iint_S B\dfrac{\sin\varphi}{R}d\vec{S} = \displaystyle\int_0^{2\pi} B\frac{\sin\varphi}{R}R\left(\vec{e}_1\cos\varphi + \vec{e}_3\sin\varphi\right)d\varphi = B\pi\vec{e}_3 \end{cases} \tag{12.60}$$

We then have as a result of the equation of equilibrium inside the string

$$A\pi\vec{e}_3 - \frac{1-\nu}{1-2\nu}B\pi\vec{e}_3 \cong \vec{\Lambda} \quad\Rightarrow\quad A - \frac{1-\nu}{1-2\nu}B \cong \frac{\vec{\Lambda}\vec{e}_3}{\pi} \tag{12.61a}$$

From relations *(12.57)* and *(12.61)*, we deduce the constants A and B

$$A = \frac{\vec{\Lambda}\vec{e}_3}{2\pi} \quad\text{and}\quad B = -\left(\frac{1-2\nu}{1-\nu}\right)\frac{\vec{\Lambda}\vec{e}_3}{2\pi} \tag{12.61b}$$

With the help of relations *(12.55)*, and recalling that the linear charge $\vec{\Lambda}$ is worth $\vec{\Lambda} = \vec{\boldsymbol{B}}_{edge}\wedge\vec{\boldsymbol{t}}$, the fields $\vec{\omega}^{el}$ and $\tau^{(p)}$ outside the dislocation charge are then completely known

$$\begin{cases} \vec{\omega}^{el} = \dfrac{\vec{\Lambda}\vec{e}_3}{2\pi}\dfrac{\cos\varphi}{r}\vec{e}_2 = \dfrac{\vec{\Lambda}\vec{e}_3}{2\pi}\dfrac{x_1}{x_1^2+x_3^2}\vec{e}_2 \\ \tau^{(p)} = -\dfrac{\vec{\Lambda}\vec{e}_3}{2\pi}\dfrac{1-2\nu}{1-\nu}\dfrac{\sin\varphi}{r} = -\dfrac{\vec{\Lambda}\vec{e}_3}{2\pi}\dfrac{1-2\nu}{1-\nu}\dfrac{x_3}{x_1^2+x_3^2} \end{cases} \quad (r > R) \tag{12.62}$$

The displacement field of an edge dislocation

It is possible to find the displacement field $\vec{u}$ outside an edge dislocation string. Indeed, as the field $\vec{u}$ can only have components along Ox_1 and Ox_3, and that $\vec{u}$ must not depend on x_2, we have

$$\vec{u} = u_1(x_1,x_3)\vec{e}_1 + u_3(x_1,x_3)\vec{e}_3 \tag{12.63}$$

So that we have two differential equations

$$\begin{cases} \vec{\omega}^{el} = -\dfrac{1}{2}\overrightarrow{\mathrm{rot}}\,\vec{u} = \dfrac{1}{2}\left(\dfrac{\partial u_3}{\partial x_1} - \dfrac{\partial u_1}{\partial x_3}\right)\vec{e}_2 \;\Rightarrow\; \dfrac{\partial u_3}{\partial x_1} - \dfrac{\partial u_1}{\partial x_3} = \dfrac{\vec{\Lambda}\vec{e}_3}{\pi}\dfrac{x_1}{x_1^2+x_3^2} \\ \tau^{(p)} = -\mathrm{div}\,\vec{u} = -\left(\dfrac{\partial u_1}{\partial x_1} + \dfrac{\partial u_3}{\partial x_3}\right) \;\Rightarrow\; \dfrac{\partial u_1}{\partial x_1} + \dfrac{\partial u_3}{\partial x_3} = \dfrac{\vec{\Lambda}\vec{e}_3}{2\pi}\dfrac{1-2\nu}{1-\nu}\dfrac{x_3}{x_1^2+x_3^2} \end{cases} \tag{12.64}$$

We can try the following classic solution[1] for the displacement field $\vec{u}$

$$\begin{cases} u_1(x_1,x_3) = \alpha\arctan\left(\dfrac{x_3}{x_1}\right) + \beta\dfrac{x_1x_3}{x_1^2+x_3^2} \\ u_3(x_1,x_3) = \gamma\ln\left(x_1^2+x_3^2\right) + \delta\dfrac{x_1^2-x_3^2}{x_1^2+x_3^2} \end{cases} \tag{12.65}$$

where $\alpha,\beta,\delta,\gamma$ are constants to be determined. The fields $\vec{\omega}^{el}$ and $\tau^{(p)}$ are then written

$$\begin{cases} \vec{\omega}^{el} = -\dfrac{1}{2}\overrightarrow{\mathrm{rot}}\,\vec{u} = \left[-\dfrac{1}{2}(\alpha+\beta-2\gamma)\dfrac{x_1}{x_1^2+x_3^2} + (\beta+2\delta)\dfrac{x_1x_3^2}{\left(x_1^2+x_3^2\right)^2}\right]\vec{e}_2 \\ \tau^{(p)} = -\mathrm{div}\,\vec{u} = \left[(\alpha-\beta-2\gamma)\dfrac{x_3}{x_1^2+x_3^2} + 2(\beta+2\delta)\dfrac{x_1^2x_3}{\left(x_1^2+x_3^2\right)^2}\right] \end{cases} \tag{12.66}$$

Comparing these relations with *(12.62)*, we deduce three relations

$$\alpha-2\gamma+\beta = -\frac{\vec{\Lambda}\vec{e}_3}{\pi} \quad ; \quad \alpha-2\gamma-\beta = -\frac{\vec{\Lambda}\vec{e}_3}{2\pi}\frac{1-2\nu}{1-\nu} \quad ; \quad \beta+2\delta = 0 \tag{12.67}$$

whose solutions are

$$\alpha-2\gamma = -\frac{\vec{\Lambda}\vec{e}_3}{2\pi}\frac{3-4\nu}{2(1-\nu)} \quad ; \quad \beta = -\frac{\vec{\Lambda}\vec{e}_3}{2\pi}\frac{1}{2(1-\nu)} \quad ; \quad \delta = \frac{\vec{\Lambda}\vec{e}_3}{2\pi}\frac{1}{4(1-\nu)} \tag{12.68}$$

With this system, the constants α and γ are still undetermined and there exists an infinite number of possible equilibrium solutions. The determination of these constants is actually a balance energy minimization problem in the case of a solid free of all external constraints, or with boundary conditions in the case of a solid subjected to restrictive external constraints. We can solve for the moment the general problem without making a prior choice as to the respective values of α and γ. Therefore introduce *an adjustable parameter* ς such that we can write

[1] *Hirth and Lothe, theory of dislocations, second edition, John Wiley and Sons, 1982, p.78*

$$\begin{cases} \alpha = -\dfrac{\vec{\Lambda}\vec{e}_3}{2\pi}\varsigma \\ \gamma = \dfrac{\vec{\Lambda}\vec{e}_3}{2\pi}\left[\dfrac{3-4\nu}{4(1-\nu)} - \dfrac{\varsigma}{2}\right] \end{cases} \quad et \quad \begin{cases} \beta = -\dfrac{\vec{\Lambda}\vec{e}_3}{2\pi}\dfrac{1}{2(1-\nu)} \\ \delta = \dfrac{\vec{\Lambda}\vec{e}_3}{2\pi}\dfrac{1}{4(1-\nu)} \end{cases} \qquad (12.69)$$

The displacement field is then written

$$\begin{cases} u_1(x_1,x_3) = -\dfrac{\vec{\Lambda}\vec{e}_3}{2\pi}\varsigma \arctan\left(\dfrac{x_3}{x_1}\right) - \dfrac{\vec{\Lambda}\vec{e}_3}{2\pi}\dfrac{1}{2(1-\nu)}\dfrac{x_1 x_3}{x_1^2+x_3^2} \\ u_3(x_1,x_3) = \dfrac{\vec{\Lambda}\vec{e}_3}{2\pi}\left[\dfrac{3-4\nu}{4(1-\nu)} - \dfrac{\varsigma}{2}\right]\ln\left(x_1^2+x_3^2\right) + \dfrac{\vec{\Lambda}\vec{e}_3}{2\pi}\dfrac{1}{4(1-\nu)}\dfrac{x_1^2-x_3^2}{x_1^2+x_3^2} \end{cases} \qquad (12.70)$$

How we got here the displacement field associated with an edge dislocation is new. Indeed, it is based on finding the equilibrium of the solid conditions on the basis of Newton's equation *(12.9)* containing a vector density $\vec{\lambda}$ of flexion charges, whereas in the classical approach[2], the field displacement is obtained from a particular solution of the differential equation for the *Airy function* of plane stress, which is much more complicated, and involves a strong a priori of not having restrictive external constraints!

The deformation and shear fields of an edge dislocation

From these expressions, using the relationship *(2.48)*, we can deduce the elastic deformation tensor $\vec{\varepsilon}_i^{\,él}$ and the tensor of elastic shear strains $\vec{\alpha}_i^{\,el}$ outside of the dislocation line

$$\begin{cases} \vec{\varepsilon}_1^{\,el} = -\dfrac{\partial u_1}{\partial x_1}\vec{e}_1 - \dfrac{1}{2}\left(\dfrac{\partial u_1}{\partial x_3} + \dfrac{\partial u_3}{\partial x_1}\right)\vec{e}_3 \\ \vec{\varepsilon}_2^{\,el} = 0 \\ \vec{\varepsilon}_3^{\,el} = -\dfrac{1}{2}\left(\dfrac{\partial u_1}{\partial x_3} + \dfrac{\partial u_3}{\partial x_1}\right)\vec{e}_1 - \dfrac{\partial u_3}{\partial x_3}\vec{e}_3 \end{cases} \qquad (12.71)$$

$$\begin{cases} \vec{\alpha}_1^{\,el} = \dfrac{1}{3}\left(\dfrac{\partial u_3}{\partial x_3} - 2\dfrac{\partial u_1}{\partial x_1}\right)\vec{e}_1 - \dfrac{1}{2}\left(\dfrac{\partial u_1}{\partial x_3} + \dfrac{\partial u_3}{\partial x_1}\right)\vec{e}_3 \\ \vec{\alpha}_2^{\,el} = \dfrac{1}{3}\left(\dfrac{\partial u_1}{\partial x_1} + \dfrac{\partial u_3}{\partial x_3}\right)\vec{e}_2 \\ \vec{\alpha}_3^{\,el} = -\dfrac{1}{2}\left(\dfrac{\partial u_1}{\partial x_3} + \dfrac{\partial u_3}{\partial x_1}\right)\vec{e}_1 + \dfrac{1}{3}\left(\dfrac{\partial u_1}{\partial x_1} - 2\dfrac{\partial u_3}{\partial x_3}\right)\vec{e}_3 \end{cases} \qquad (12.72)$$

The derivatives of the components of the displacement field that appear in these relations are deduced from *(23.56)*

$$\begin{cases} \dfrac{\partial u_1}{\partial x_1} = \dfrac{\vec{\Lambda}\vec{e}_3}{2\pi}\dfrac{1}{2(1-\nu)}\left\{\left[2(1-\nu)\varsigma - 1\right]\dfrac{x_3}{x_1^2+x_3^2} + \dfrac{2x_1^2 x_3}{\left(x_1^2+x_3^2\right)^2}\right\} \\ \dfrac{\partial u_1}{\partial x_3} = \dfrac{\vec{\Lambda}\vec{e}_3}{2\pi}\dfrac{1}{2(1-\nu)}\left\{-\left[2(1-\nu)\varsigma + 1\right]\dfrac{x_1}{x_1^2+x_3^2} + \dfrac{2x_1 x_3^2}{\left(x_1^2+x_3^2\right)^2}\right\} \end{cases} \qquad (12.73a)$$

[2] *Hirth and Lothe, theory of dislocations, second edition, John Wiley and Sons, 1982, p.75*

$$\begin{cases} \dfrac{\partial u_3}{\partial x_1} = \dfrac{\vec{\Lambda}\vec{e}_3}{2\pi} \dfrac{1}{2(1-\nu)} \left\{ \left[4(1-\nu) - 2(1-\nu)\varsigma \right] \dfrac{x_1}{x_1^2 + x_3^2} + \dfrac{x_1 x_3^2 - x_1^3}{\left(x_1^2 + x_3^2\right)^2} \right\} \\ \dfrac{\partial u_3}{\partial x_3} = \dfrac{\vec{\Lambda}\vec{e}_3}{2\pi} \dfrac{1}{2(1-\nu)} \left\{ \left[2(1-2\nu) - 2(1-\nu)\varsigma \right] \dfrac{x_3}{x_1^2 + x_3^2} + \dfrac{x_3^3 - x_1^2 x_3}{\left(x_1^2 + x_3^2\right)^2} \right\} \end{cases} \quad (12.73b)$$

As will be preferable thereafter, for the sake of simplicity, to work in polar coordinates $x_1 = r\cos\varphi$ and $x_2 = r\sin\varphi$ in the plane of the cylinder (fig. 12.1), the relationships

$$\frac{\cos\varphi}{r} = \frac{x_1}{x_1^2 + x_3^2} \Rightarrow x_1 = r\cos\varphi \quad \text{and} \quad \frac{\sin\varphi}{r} = \frac{x_3}{x_1^2 + x_3^2} \Rightarrow x_3 = r\sin\varphi \quad (12.74)$$

allow us to obtain the derivatives as function of coordinates $[r,\varphi]$

$$\begin{cases} \dfrac{\partial u_1}{\partial x_1} = \dfrac{\vec{\Lambda}\vec{e}_3}{2\pi} \dfrac{1}{2(1-\nu)} \left\{ \left[2(1-\nu)\varsigma - 1 \right] + 2\cos^2\varphi \right\} \dfrac{\sin\varphi}{r} \\ \dfrac{\partial u_1}{\partial x_3} = \dfrac{\vec{\Lambda}\vec{e}_3}{2\pi} \dfrac{1}{2(1-\nu)} \left\{ -\left[2(1-\nu)\varsigma + 1 \right] + 2\sin^2\varphi \right\} \dfrac{\cos\varphi}{r} \\ \dfrac{\partial u_3}{\partial x_1} = \dfrac{\vec{\Lambda}\vec{e}_3}{2\pi} \dfrac{1}{2(1-\nu)} \left\{ \left[4(1-\nu) - 2(1-\nu)\varsigma \right] - \cos 2\varphi \right\} \dfrac{\cos\varphi}{r} \\ \dfrac{\partial u_3}{\partial x_3} = \dfrac{\vec{\Lambda}\vec{e}_3}{2\pi} \dfrac{1}{2(1-\nu)} \left\{ \left[2(1-2\nu) - 2(1-\nu)\varsigma \right] - \cos 2\varphi \right\} \dfrac{\sin\varphi}{r} \end{cases} \quad (12.75)$$

The tensor of elastic deformation $\vec{\varepsilon}_i^{\,el}$ and the tensor of elastic shear $\vec{\alpha}_i^{\,el}$ can then be expressed in polar coordinates by introducing these derivatives in *(23.58)* and *(23.59)*

$$\begin{cases} \vec{\varepsilon}_1^{\,el} = -\dfrac{\vec{\Lambda}\vec{e}_3}{4\pi(1-\nu)} \left\{ \left[2(1-\nu)\varsigma + \cos 2\varphi \right] \dfrac{\sin\varphi}{r} \vec{e}_1 + \left[2(1-\nu)(1-\varsigma) - \cos 2\varphi \right] \dfrac{\cos\varphi}{r} \vec{e}_3 \right\} \\ \vec{\varepsilon}_2^{\,el} = 0 \\ \vec{\varepsilon}_3^{\,el} = -\dfrac{\vec{\Lambda}\vec{e}_3}{4\pi(1-\nu)} \left\{ \left[2(1-\nu)(1-\varsigma) - \cos 2\varphi \right] \dfrac{\cos\varphi}{r} \vec{e}_1 + \left[2(1-2\nu) - 2(1-\nu)\varsigma - \cos 2\varphi \right] \dfrac{\sin\varphi}{r} \vec{e}_3 \right\} \end{cases}$$

(12.76)

$$\begin{cases} \vec{\alpha}_1^{\,el} = -\dfrac{\vec{\Lambda}\vec{e}_3}{4\pi(1-\nu)} \left\{ \left[2(1-\nu)\varsigma - \dfrac{2}{3}(1-2\nu) + \cos 2\varphi \right] \dfrac{\sin\varphi}{r} \vec{e}_1 + \left[2(1-\nu)(1-\varsigma) - \cos 2\varphi \right] \dfrac{\cos\varphi}{r} \vec{e}_3 \right\} \\ \vec{\alpha}_2^{\,el} = \dfrac{\vec{\Lambda}\vec{e}_3}{4\pi(1-\nu)} \left[\dfrac{2}{3}(1-2\nu) \right] \dfrac{\sin\varphi}{r} \vec{e}_2 \\ \vec{\alpha}_3^{\,el} = -\dfrac{\vec{\Lambda}\vec{e}_3}{4\pi(1-\nu)} \left\{ \left[2(1-\nu)(1-\varsigma) - \cos 2\varphi \right] \dfrac{\cos\varphi}{r} \vec{e}_1 + \left[-2(1-\nu)\varsigma + \dfrac{4}{3}(1-2\nu) - \cos 2\varphi \right] \dfrac{\sin\varphi}{r} \vec{e}_3 \right\} \end{cases}$$

(12.77)

The rest energy of an edge dislocation

The rest energy of an edge dislocation is obtained by integrating the elastic potential energy stored outside of the string by the lattice

$$e_0 = k_2 \sum_i \left(\vec{\alpha}_i^{el}\right)^2 + k_1 \left(\tau^{(p)}\right)^2 \tag{12.78}$$

We deduce with the following integration

$$E_0^{edge} = \int_R^{R_\infty} \int_0^{2\pi} n e_0(r,\varphi)\, r\, d\varphi\, dr = \int_R^{R_\infty} \int_0^{2\pi} n \left[k_2 \sum_i \left(\vec{\alpha}_i^{el}\right)^2 + k_1 \left(\tau^{(p)}\right)^2 \right] r\, d\varphi\, dr \tag{12.79}$$

the rest energy E_0^{edge} by unit of length of the edge dislocation by using relations *(12.77)* and *(12.62)*

$$E_0^{edge} = \frac{n\vec{\Lambda}^2}{4\pi} \left\{ k_2 \left[4\varsigma^2 - 8\varsigma + \frac{2}{3}\left(\frac{7v^2 - 13v + 7}{(1-v)^2} \right) \right] + k_1 \left(\frac{1-2v}{1-v} \right)^2 \right\} \ln \frac{R_\infty}{R} \tag{12.80}$$

It is found that this rest energy contains a term dependent on modulus k_2 associated to shear deformations, and a term dependent on modulus k_1 associated to deformation by volume expansion of the medium. The rest energy per unit length of the edge dislocation line can also be expressed more simply by using the relation *(7.30)*

$$E_0^{edge} = \frac{nk_2\vec{\Lambda}^2}{4\pi} \left[4\varsigma^2 - 8\varsigma + \frac{4v^2 - 9v + 5}{(1-v)^2} \right] \ln \frac{R_\infty}{R} \tag{12.81}$$

It should be noted here that *the energy associated to a certain segment of length L of a line of edge dislocation does not depend on the state of volume expansion of the solid lattice*. Indeed it can easily be verified that

$$E_0^{edge}\Big|_L = L_0 \frac{n_0 k_2 \left(\vec{\Lambda}_0\right)^2}{4\pi} \left[4\varsigma^2 - 8\varsigma + \frac{4v^2 - 9v + 5}{(1-v)^2} \right] \ln \frac{R_{\infty 0}}{R_0} = E_{00}^{edge}\Big|_{L_0} \tag{12.82}$$

To compare the value of the rest energy of an edge dislocation with that of a screw dislocation, it is enough to remember that the relationship between the linear charge $\vec{\Lambda}$ and the Burgers vector $\vec{\boldsymbol{B}}_{edge}$ is simply written $\vec{\Lambda}^2 = \vec{\boldsymbol{B}}_{edge}^2$ and the modulus k_2 is connected to the *Lamé coefficient* by the relationship $\mu = nk_2$. It is then found that for Burgers vectors of equal length, the edge dislocation has a different rest energy than the screw dislocation by a factor that depends on both the tunable parameter ς and the Poisson's ratio v of the material

$$\left\{ \begin{aligned} E_0^{screws} &= \frac{\mu \vec{\boldsymbol{B}}_{screws}^2}{4\pi} \ln \frac{R_\infty}{R} \\ E_0^{edge} &= \frac{\mu \vec{\boldsymbol{B}}_{edge}^2}{4\pi} \ln \frac{R_\infty}{R} \left[4\varsigma^2 - 8\varsigma + \frac{4v^2 - 9v + 5}{(1-v)^2} \right] \end{aligned} \right. \tag{12.83}$$

The kinetic energy of a low velocity edge dislocation

In the case where a string of edge dislocation is moving in direction Ox_1, with velocity $\vec{\boldsymbol{v}}$ which is small compared with the speed of transversal waves c_t, the components of the velocity field are obtained by replacing x_1 by $x_1(t) = x_1 - \boldsymbol{v}t$ in expressions *(12.70)* of the components of the field of displacement, and then by taking the derivative $\vec{\phi} = -d\vec{u}\,/\,dt$

$$\begin{cases} \phi_1(x_1,x_3,t) = +\dfrac{\vec{\Lambda}\vec{e}_3}{2\pi}\dfrac{\partial}{\partial t}\left\{\varsigma \arctan\left(\dfrac{x_3}{x_1 - \mathbf{v}t}\right) + \dfrac{1}{2(1-\nu)}\dfrac{(x_1-\mathbf{v}t)x_3}{(x_1-\mathbf{v}t)^2 + x_3^2}\right\} \\ \phi_3(x_1,x_3,t) = -\dfrac{\vec{\Lambda}\vec{e}_3}{2\pi}\dfrac{\partial}{\partial t}\left\{\left(\dfrac{3-4\nu}{4(1-\nu)} - \dfrac{\varsigma}{2}\right)\ln\left[(x_1-\mathbf{v}t)^2 + x_3^2\right] + \dfrac{1}{4(1-\nu)}\dfrac{(x_1-\mathbf{v}t)^2 - x_3^2}{(x_1-\mathbf{v}t)^2 + x_3^2}\right\} \end{cases} \tag{12.84}$$

After derivation, it is useful to jump into the coordinate system $Ox_1'x_2'x_3'$ which is in translation with velocity $\vec{\mathbf{v}}$ with the line of dislocation, in which $x_1 - \mathbf{v}t$ becomes x_1'

$$\begin{cases} \phi_1(x_1',x_3') = +\dfrac{\vec{\Lambda}\vec{e}_3}{2\pi}\left[\left(\varsigma + \dfrac{1}{2(1-\nu)}\right)\dfrac{x_1'^2 x_3'}{(x_1'^2 + x_3'^2)^2} + \left(\varsigma - \dfrac{1}{2(1-\nu)}\right)\dfrac{x_3'^3}{(x_1'^2 + x_3'^2)^2}\right]\mathbf{v} \\ \phi_3(x_1',x_3') = -\dfrac{\vec{\Lambda}\vec{e}_3}{2\pi}\left[\left(\varsigma - \dfrac{5-4\nu}{2(1-\nu)}\right)\dfrac{x_3'^2 x_1'}{(x_1'^2 + x_3'^2)^2} + \left(\varsigma - \dfrac{3-4\nu}{2(1-\nu)}\right)\dfrac{x_1'^3}{(x_1'^2 + x_3'^2)^2}\right]\mathbf{v} \end{cases} \tag{12.85}$$

and to use in $Ox_1'x_2'x_3'$ the polar coordinates $x_1' = r'\cos\varphi'$ and $x_3' = r'\sin\varphi'$ expressed in the plane of the moving cylinder

$$\begin{cases} \phi_1(r',\varphi') = +\dfrac{\vec{\Lambda}\vec{e}_3}{2\pi}\dfrac{1}{r'}\left[\left(\varsigma + \dfrac{1}{2(1-\nu)}\right)\cos^2\varphi'\sin\varphi' + \left(\varsigma - \dfrac{1}{2(1-\nu)}\right)\sin^3\varphi'\right]\mathbf{v} \\ \phi_3(r',\varphi') = -\dfrac{\vec{\Lambda}\vec{e}_3}{2\pi}\dfrac{1}{r'}\left[\left(\varsigma - \dfrac{5-4\nu}{2(1-\nu)}\right)\sin^2\varphi'\cos\varphi' + \left(\varsigma - \dfrac{3-4\nu}{2(1-\nu)}\right)\cos^3\varphi'\right]\mathbf{v} \end{cases} \tag{12.86}$$

The kinetic energy of the edge dislocation is then obtain by operating the following integral

$$E_{kin}^{edge} = \frac{1}{2}m\int_R^{R_\infty}\int_0^{2\pi} n\vec{\phi}^2(r',\varphi')\, r'd\varphi'\, dr' \tag{12.87}$$

It is a rather long computation that leads finally to the kinetic energy of an edge dislocation. This energy is given by unit of length, and it depends on the parameter ς and the Poisson modulus ν, just as with the rest energy of a screw dislocation *(12.81)*

$$E_{kin}^{edge} = \frac{mn\vec{\Lambda}^2}{64\pi}\ln\frac{R_\infty}{R}\left[16\varsigma^2 - 32\varsigma + \frac{26 - 56\nu + 32\nu^2}{(1-\nu)^2}\right]\mathbf{v} \tag{12.88}$$

12.5 – Effects of the boundary conditions and of the nature of the lattice

The edge dislocation in a usual, finite and free medium

In the case of a *finite and free* lattice, that is to say not subjected to any boundary stress, the presence of an edge dislocation may bend the solid, as illustrated by the model in figure 9.11, which is taken into account by a non-zero value of the constant γ in the relations *(12.65)*. To determine the constants α and γ completely, look for the value of the adjustable parameter ς, so it minimizes the rest energy of the edge dislocation. From equation *(12.81)* we have

$$\frac{\partial E_0^{edge}}{\partial \varsigma}=\frac{nk_2\vec{\Lambda}^2}{\pi}2(\varsigma-1)\ln\frac{R_\infty}{R}=0 \quad \Rightarrow \quad \varsigma=1 \tag{12.89}$$

Thus, the value of ς that minimizes the rest energy of the edge dislocation string is 1. It can be verified that if the string moves relative to the lattice with velocity $\vec{\mathbf{v}}$, the value that minimizes the sum of potential energy and kinetic energy is always equal to 1. Indeed, as we have, from the relation *(12.88)*

$$\partial E_{kin}^{edge} / \partial \varsigma = 0 \quad \Rightarrow \quad \varsigma = 1 \tag{12.90}$$

Thus in the case of a *usual medium, free and finite,* the energies of the edge dislocation are written

$$\begin{cases} E_0^{edge}=\dfrac{nk_2\vec{\Lambda}^2}{4\pi(1-v)}\ln\dfrac{R_\infty}{R}=\left[\dfrac{4(1-v)}{8v^2-12v+5}\right]M_0^{edge}c_t^2 \\ E_{kin}^{edge}=\dfrac{mn\vec{\Lambda}^2}{8\pi}\left[\dfrac{8v^2-12v+5}{4(1-v)^2}\right]\left(\ln\dfrac{R_\infty}{R}\right)\mathbf{v}^2=\dfrac{1}{2}M_0^{edge}\mathbf{v}^2 \end{cases} \tag{12.91}$$

The rest energy then corresponds to the classic value obtained in the literature[3].
Since the usual solids have a Poisson modulus close to 1/3, the Einstein relation is modified and is approximatively worth $E_0^{edge} \cong 1.41\, M_0^{edge} c_t^2$.
It is interesting to realize that the field of displacement is in this case written as

$$\begin{cases} u_1^{edge}(x_1,x_3)=\dfrac{\vec{\Lambda}\vec{e}_3}{2\pi}\left[-\arctan\dfrac{x_3}{x_1}-\dfrac{1}{2(1-v)}\dfrac{x_1x_3}{x_1^2+x_3^2}\right] \\ u_3^{edge}(x_1,x_3)=\dfrac{\vec{\Lambda}\vec{e}_3}{2\pi}\left[\dfrac{1-2v}{4(1-v)}\ln\left(x_1^2+x_3^2\right)+\dfrac{1}{4(1-v)}\dfrac{x_1^2-x_3^2}{x_1^2+x_3^2}\right] \end{cases} \tag{12.92}$$

and that it corresponds exactly with what is found by the classical theories of dislocations[3]. Only the sign of the expressions $u_1^{edge}(x_1,x_3)$ and $u_3^{edge}(x_1,x_3)$ are inverted, which is a result of the fact that the displacement field expressed in Euler coordinates has a sign opposite to the displacement field expressed in Lagrange coordinates. The tensor of elastic deformation $\vec{\varepsilon}_i^{\,edge}$ can be written

$$\begin{cases} \vec{\varepsilon}_1^{\,edge}=\dfrac{\vec{\Lambda}\vec{e}_3}{4\pi(1-v)}\dfrac{1}{r}\left[-(2-2v+\cos 2\varphi)\sin\varphi\,\vec{e}_1+\cos 2\varphi\cos\varphi\,\vec{e}_3\right] \\ \vec{\varepsilon}_2^{\,edge}=0 \\ \vec{\varepsilon}_3^{\,edge}=\dfrac{\vec{\Lambda}\vec{e}_3}{4\pi(1-v)}\dfrac{1}{r}\left[\cos 2\varphi\cos\varphi\,\vec{e}_1+(2v+\cos 2\varphi)\sin\varphi\,\vec{e}_3\right] \end{cases} \tag{12.93}$$

and the elastic shear strain $\vec{\alpha}_i^{\,edge}$

---

[3] *Hirth and Lothe, theory of dislocations, second edition, John Wiley and Sons, 1982, p.78*

$$\begin{cases} \vec{\alpha}_1^{\,edge} = \dfrac{\vec{\Lambda}\vec{e}_3}{4\pi(1-\nu)}\dfrac{1}{r}\left\{\left[-2(2-\nu)/3-\cos 2\varphi\right]\sin\varphi\,\vec{e}_1+\cos 2\varphi\cos\varphi\,\vec{e}_3\right\} \\ \vec{\alpha}_2^{\,edge} = \dfrac{\vec{\Lambda}\vec{e}_3}{4\pi(1-\nu)}\dfrac{1}{r}\left\{2(1-2\nu)\sin\varphi\,\vec{e}_2/3\right\} \\ \vec{\alpha}_3^{\,edge} = \dfrac{\vec{\Lambda}\vec{e}_3}{4\pi(1-\nu)}\dfrac{1}{r}\left\{\left[2(1+\nu)/3+\cos 2\varphi\right]\sin\varphi\,\vec{e}_3+\cos 2\varphi\cos\varphi\,\vec{e}_1\right\} \end{cases} \tag{12.94}$$

The edge dislocation in a usual medium which is prevented from bending

If the solid medium is subjected to boundary conditions, it is the adjustable parameter ς which will be chosen so as to satisfy said boundary conditions. The simplest case we can consider, for example, is to assume a rectangular solid in which an edge dislocation is introduced in the direction Ox_2 with a line charge $\vec{\Lambda}$ pointing in the direction Ox_3. In case the solid is completely free, we know that it bends in the direction Ox_1x_3 because of the term $\ln\left(x_1^2+x_3^2\right)$ of the component $u_3^{edge}(x_1,x_3)$ in the relations *(12.65)*. But if, due to the boundary conditions imposed on the solid, it is unable to bend, the term $\ln\left(x_1^2+x_3^2\right)$ in the relations *(12.65)* must be zero, which is possible since there is the free parameter ς to adjust $u_1^{edge}(x_1,x_3)$ and $u_3^{edge}(x_1,x_3)$. We then have

$$\frac{3-4\nu}{4(1-\nu)}-\frac{\varsigma}{2}=0 \;\Rightarrow\; \varsigma=\frac{3-4\nu}{2(1-\nu)} \tag{12.95}$$

The energies of the dislocation in this constrained lattice are deduced from *(12.81) and (12.88)*

$$\begin{cases} E_0^{edge}=\dfrac{nk_2\vec{\Lambda}^2}{4\pi}\left[\dfrac{16\nu^2-20\nu+8}{4(1-\nu)^2}\right]\ln\dfrac{R_\infty}{R}=\left[\dfrac{16\nu^2-20\nu+8}{16\nu^2-20\nu+7}\right]M_0^{edge}c_t^2 \\ E_{kin}^{edge}=\dfrac{mn\vec{\Lambda}^2}{8\pi}\left[\dfrac{16\nu^2-20\nu+7}{4(1-\nu)^2}\right]\left(\ln\dfrac{R_\infty}{R}\right)\mathbf{v}^2=\dfrac{1}{2}M_0^{edge}\mathbf{v}^2 \end{cases} \tag{12.96}$$

Since solid media usually have a Poisson modulus on the order of 1/3, the Einstein relation is modified to $E_0^{edge}=1.47\,M_0^{edge}c_t^2$.

The dislocation in an auxetic media, finite and free

In a more exotic environment where the modulus k_1 would be a negligible modulus vis-à-vis the k_2 modulus, so a medium for which the Poisson modulus would be close to -1, and which is called an *auxetic solid medium*, the energy of the edge dislocation is written

$$\begin{cases} E_0^{edge}=\dfrac{nk_2\vec{\Lambda}^2}{8\pi}\ln\dfrac{R_\infty}{R}=\left[\dfrac{8}{25}\right]M_0^{edge}c_t^2 \\ E_{kin}^{edge}=\dfrac{\rho\vec{\Lambda}^2}{8\pi}\left[\dfrac{25}{16}\right]\left(\ln\dfrac{R_\infty}{R}\right)\mathbf{v}^2=\dfrac{1}{2}M_0^{edge}\mathbf{v}^2 \end{cases} \tag{12.97}$$

In the auxetic medium, finite and free, the Einstein relation reads $E_0^{edge}\cong 0.32\,M_0^{edge}c_t^2$.

The edge dislocation in an auxetic media, which is prevented from bending

In the case of an auxetic medium prevented from bending, the energy of an edge dislocation reads

$$\begin{aligned} E_0^{edge} &= \frac{K_2\vec{\Lambda}^2}{4\pi}\left[\frac{11}{4}\right]\ln\frac{R_\infty}{R} = \left[\frac{44}{43}\right]M_0^{edge}c_t^2 \\ E_{cin}^{edge} &= \frac{mn\vec{\Lambda}^2}{8\pi}\left[\frac{43}{16}\right]\left(\ln\frac{R_\infty}{R}\right)\mathbf{v}^2 = \frac{1}{2}M_0^{edge}\mathbf{v}^2 \end{aligned} \qquad (12.98)$$

In an auxetic media which is prevented from bending, something quite remarkable happens, namely that the Einstein equation is very close to an exact Einstein equation, since $E_0^{edge} = 1.02\, M_0^{edge}c_t^2$.

12.6 – Interactions between dislocations

Distortion fields associated with dislocations allow us to calculate the interaction forces acting between dislocations through the force of Peach and Koehler. In this section, we will present some cases of these interactions, limiting ourselves to the dislocations usually found in conventional, finite and free solids, that is to say, including edge dislocations satisfying the "traditional" relationships *(12.91) to (12.94)*.

The interactions between two dislocations

In the case of screw and edge dislocations, the force of Peach and Koehler acting per unit length of a dislocation *(2)* by the effect of a dislocation *(1)* is given by the expression *(11.12)*, namely

$$\vec{F}_{PK(2)} = \sum_k\left(\vec{s}_{k(1)}\wedge\vec{\Lambda}_{k(2)}\right) + \Lambda_{(2)}\vec{m}_{(1)} + \frac{1}{2}\left(\vec{m}_{(1)}\wedge\vec{\Lambda}_{(2)}\right) + \vec{\Lambda}_{(2)}p_{(1)} \qquad (12.99)$$

in which the linear charges $\vec{\Lambda}_{k(2)},\vec{\Lambda}_{(2)}$ and $\Lambda_{(2)}$ are obtained from the vector of Burgers $\vec{\boldsymbol{B}}_{(2)}$ of the dislocation *(2)* and the tangent vector $\vec{t}$ to the dislocation *(2)* by the following relationships

$$\begin{cases} \vec{\Lambda}_{k(2)} = -\vec{t}_{(2)}\boldsymbol{B}_{k(2)} \\ \vec{\Lambda}_{(2)} = \vec{\boldsymbol{B}}_{(2)}\wedge\vec{t}_{(2)} \\ \Lambda_{(2)} = -\vec{\boldsymbol{B}}_{(2)}\vec{t}_{(2)}/2 \end{cases} \qquad (12.100)$$

With regards to the stress fields $\vec{m}_{(1)}$, $p_{(1)}$ and $\vec{s}_{k(1)}$ which appear in the force of Peach and Koehler, they are obtained from the elastic distortion fields generated by dislocation *(1)*, via relations *(12.2) and (12.14)*, and are written

$$\begin{cases} \vec{s}_{i(1)} = 2nk_2\vec{\alpha}_{i(1)}^{él} \\ \vec{m}_{(1)} = 4nk_2\vec{\omega}_{(1)}^{él} \\ p_{(1)} = n\left(k_0 - 2k_1\tau_{(1)}^{él}\right) \end{cases} \qquad (12.101)$$

For example, we will calculate in the following sections the interaction forces that occur between two parallel dislocations, for two screw strings, an edge and a screw string, and two edge strings.

The interaction between two parallel screw dislocations

A screw dislocation with charge $\Lambda_{(1)} = -\vec{B}_{screw(1)}\vec{e}_2 / 2$ in the direction Ox_2, with $\vec{t} = \vec{e}_2$, located at coordinates $x_1 = x_3 = 0$, is the source of a field of moments obtained from expression *(12.31)*, which we can recast in polar coordinate in the plane Ox_1x_3

$$\vec{m}_{(1)} = 4nk_2\vec{\omega}_{(1)} = \frac{2nk_2\Lambda_{(1)}}{\pi}\frac{\vec{r}}{r^2} = \frac{2nk_2\Lambda_{(1)}}{\pi}\frac{1}{r}\left(\cos\varphi\,\vec{e}_1 + \sin\varphi\,\vec{e}_3\right) \quad (r > R) \tag{12.102}$$

If a second string parallel to the first one with a charge $\Lambda_{(2)} = -\vec{B}_{screw(2)}\vec{e}_2 / 2$, is situated at a distance r from it, with polar coordinates r and φ in the plane Ox_1x_3, it undergoes a force by unit length which is given by the force of Peach and Koehler $\vec{F}_{PK(2)}$

$$\vec{F}_{PK(2)} = \Lambda_{(2)}\vec{m}_{screw(1)} = \frac{2nk_2}{\pi}\frac{\Lambda_{(1)}\Lambda_{(2)}}{r}\frac{\vec{r}}{r} = \frac{nk_2}{2\pi}\frac{\vec{B}_{screw(1)}\vec{B}_{screw(2)}}{r}\left(\cos\varphi\,\vec{e}_1 + \sin\varphi\,\vec{e}_3\right) \tag{12.103}$$

The interaction force between two parallel screw dislocation is therefore repulsive if $\Lambda_{(1)}\Lambda_{(2)} > 0$ and attractive if $\Lambda_{(1)}\Lambda_{(2)} < 0$.

The interaction between an edge dislocation and a parallel screw dislocation

An edge dislocation in the direction Ox_2 ($\vec{t} = \vec{e}_2$), located at coordinates $x_1 = 0$ and $x_3 = 0$, of Burgers vector in the direction Ox_1, has a charge $\vec{\Lambda}_{(1)} = \vec{B}_{edge(1)}\vec{e}_3$. It is then the source of stress fields $\vec{m}_{(1)}$, $p_{(1)}$ et $\vec{s}_{k(1)}$ obtained from expressions *(12.62) and (12.94)*

$$\begin{cases} \vec{m}_{(1)} = 4nk_2\vec{\omega}_{(1)} = \dfrac{2nk_2\left(\vec{\Lambda}_{(1)}\vec{e}_3\right)}{\pi}\dfrac{1}{r}\cos\varphi\,\vec{e}_2 \\ p_{(1)} = -2nk_1\tau_{(1)} = \dfrac{nk_1\left(\vec{\Lambda}_{(1)}\vec{e}_3\right)}{\pi}\dfrac{1}{r}\left(\dfrac{1-2\nu}{1-\nu}\right)\sin\varphi \end{cases} \quad (r > R) \tag{12.104}$$

$$\begin{cases} \vec{s}_{1(1)} = \dfrac{nk_2\left(\vec{\Lambda}_{(1)}\vec{e}_3\right)}{2\pi(1-\nu)}\dfrac{1}{r}\left\{\sin\varphi\left[-\dfrac{2}{3}(2-\nu) - \cos 2\varphi\right]\vec{e}_1 + \cos\varphi\cos 2\varphi\,\vec{e}_3\right\} \\ \vec{s}_{2(1)} = \dfrac{nk_2\left(\vec{\Lambda}_{(1)}\vec{e}_3\right)}{2\pi(1-\nu)}\dfrac{1}{r}\left\{\sin\varphi\left[\dfrac{2}{3}(1-2\nu)\right]\vec{e}_2\right\} \\ \vec{s}_{3(1)} = \dfrac{nk_2\left(\vec{\Lambda}_{(1)}\vec{e}_3\right)}{2\pi(1-\nu)}\dfrac{1}{r}\left\{\sin\varphi\left[\dfrac{2}{3}(1+\nu) + \cos 2\varphi\right]\vec{e}_3 + \cos\varphi\cos 2\varphi\,\vec{e}_1\right\} \end{cases} \tag{12.105}$$

If a second line parallel to the first one, but of screw type, is at a distance r, with charge $\Lambda_{(2)} = -\vec{B}_{screw(2)}\vec{e}_2 / 2$, it will undergo a force per unit length given by the force of Peach and Koehler $\vec{F}_{PK(2)}$

$$\vec{F}_{PK(2)} = \vec{s}_{2(1)} \wedge \vec{\Lambda}_{2(2)} = \frac{nk_2\left(\vec{\Lambda}_{(1)}\vec{e}_3\right)}{2\pi(1-\nu)}\frac{1}{r}\left[\left(\frac{2}{3} - \frac{4}{3}\nu\right)\sin\varphi\right]\vec{B}_{2(2)}\left(\vec{e}_2 \wedge \vec{e}_2\right) = 0 \tag{12.106}$$

We conclude that the interaction force between an edge dislocation and a parallel screw dislocation are always null, whatever their respective positions.

The interaction between two parallel edge dislocations

Given an edge dislocation in direction Ox_2 $(\vec{t} = \vec{e}_2)$, located at $x_1 = 0$ and $x_3 = 0$, with a Burgers vector in the direction Ox_1, with a charge $\vec{\Lambda}_{(1)} = \boldsymbol{B}_{edge(1)}\vec{e}_3$, which interacts with a parallel edge dislocation at polar coordinates $[r,\varphi]$ in the plane Ox_1x_3 (fig. 12.3), and with charges $\vec{\Lambda}_{i(2)}$ or $\vec{\Lambda}_{(2)}$ are given by

$$\begin{cases} \vec{\Lambda}_{1(2)} = -\vec{t}_{(2)}\boldsymbol{B}_{1(2)} = -\vec{e}_2\boldsymbol{B}_{edge(2)}\cos\theta \\ \vec{\Lambda}_{3(2)} = -\vec{t}_{(2)}\boldsymbol{B}_{3(2)} = -\vec{e}_2\boldsymbol{B}_{edge(2)}\sin\theta \\ \vec{\Lambda}_{(2)} = \vec{\boldsymbol{B}}_{coin(2)}\wedge\vec{t}_{(2)} = \boldsymbol{B}_{edge(2)}\cos\theta\vec{e}_3 - \boldsymbol{B}_{edge(2)}\sin\theta\vec{e}_1 \end{cases} \tag{12.107}$$

Th first line *(1)* is the source of fields of stress $\vec{m}_{(1)}$, $p_{(1)}$ and $\vec{s}_{k(1)}$ given by *(12.104)* and *(12.105)*. In those fields, the line *(2)* undergoes the following force of Peach and Koehler by unit length

$$\vec{\boldsymbol{F}}_{PK(2)} = \sum_k\left(\vec{s}_{k(1)}\wedge\vec{\Lambda}_{k(2)}\right) + \vec{\Lambda}_{(2)}p_{(1)} \tag{12.108}$$

which is written

$$\vec{\boldsymbol{F}}_{PK(2)} = -\vec{\Lambda}_{1(2)}\wedge\vec{s}_{1(1)} - \vec{\Lambda}_{3(2)}\wedge\vec{s}_{3(1)} + \left(\boldsymbol{B}_{edge(2)}\cos\theta\vec{e}_3 - \boldsymbol{B}_{edge(2)}\sin\theta\vec{e}_1\right)p_{(1)} \tag{12.109}$$

and finally

$$\begin{aligned} \vec{\boldsymbol{F}}_{PK(2)} &= \vec{e}_2\boldsymbol{B}_{edge(2)}\cos\theta\wedge\frac{nk_2\left(\vec{\Lambda}_{(1)}\vec{e}_3\right)}{2\pi(1-\nu)}\frac{1}{r}\left\{\sin\varphi\left[\frac{2}{3}\nu-\frac{4}{3}\nu-\cos2\varphi\right]\vec{e}_1 + \cos\varphi\cos2\varphi\vec{e}_3\right\} \\ &+\vec{e}_2\boldsymbol{B}_{edge(2)}\sin\theta\wedge\frac{nk_2\left(\vec{\Lambda}_{(1)}\vec{e}_3\right)}{2\pi(1-\nu)}\frac{1}{r}\left\{\sin\varphi\left[\frac{2}{3}+\frac{2}{3}\nu+\cos2\varphi\right]\vec{e}_3 + \cos\varphi\cos2\varphi\vec{e}_1\right\} \\ &+\left(\boldsymbol{B}_{edge(2)}\cos\theta\vec{e}_3 - \boldsymbol{B}_{edge(2)}\sin\theta\vec{e}_1\right)\frac{nk_1\left(\vec{\Lambda}_{(1)}\vec{e}_3\right)}{\pi}\frac{1}{r}\left(\frac{1-2\nu}{1-\nu}\right)\sin\varphi \end{aligned} \tag{12.110}$$

By further transforming this expression, we obtain the classical expression of the interaction between two parallel edge dislocations

$$\begin{aligned} \vec{\boldsymbol{F}}_{PK(2)} &= \frac{nk_2\left(\vec{\Lambda}_{(1)}\vec{e}_3\right)}{2\pi(1-\nu)}\frac{1}{r}\left\{\left[\boldsymbol{B}_{edge(2)}\cos\theta\right]\cos\varphi\cos2\varphi + \left[\boldsymbol{B}_{edge(2)}\sin\theta\right]\sin\varphi\cos2\varphi\right\}\vec{e}_1 \\ &+\frac{nk_2\left(\vec{\Lambda}_{(1)}\vec{e}_3\right)}{2\pi(1-\nu)}\frac{1}{r}\left\{\left[\boldsymbol{B}_{edge(2)}\cos\theta\right]\sin\varphi(2+\cos2\varphi) - \left[\boldsymbol{B}_{edge(2)}\sin\theta\right]\cos\varphi\cos2\varphi\right\}\vec{e}_3 \end{aligned} \tag{12.111}$$

which can also be written by reintroducing the vector of charge $\vec{\Lambda}^{(2)}$ under the form

$$\begin{aligned} \vec{\boldsymbol{F}}_{PK(2)} &= \frac{nk_2\left(\vec{\Lambda}_{(1)}\vec{e}_3\right)}{2\pi(1-\nu)}\frac{1}{r}\left\{\left(\vec{\Lambda}_{(2)}\vec{e}_3\right)\cos\varphi\cos2\varphi - \left(\vec{\Lambda}_{(2)}\vec{e}_1\right)\sin\varphi\cos2\varphi\right\}\vec{e}_1 \\ &+\frac{nk_2\left(\vec{\Lambda}_{(1)}\vec{e}_3\right)}{2\pi(1-\nu)}\frac{1}{r}\left\{\left(\vec{\Lambda}_{(2)}\vec{e}_3\right)\sin\varphi(2+\cos2\varphi) + \left(\vec{\Lambda}_{(2)}\vec{e}_1\right)\cos\varphi\cos2\varphi\right\}\vec{e}_3 \end{aligned} \tag{12.112}$$

In figure 12.4, we have shown the direction of the components of the force of Peach and Koehler in the case where charge $\vec{\Lambda}_{(2)}$ of line *(2)* points along the axis Ox_3.

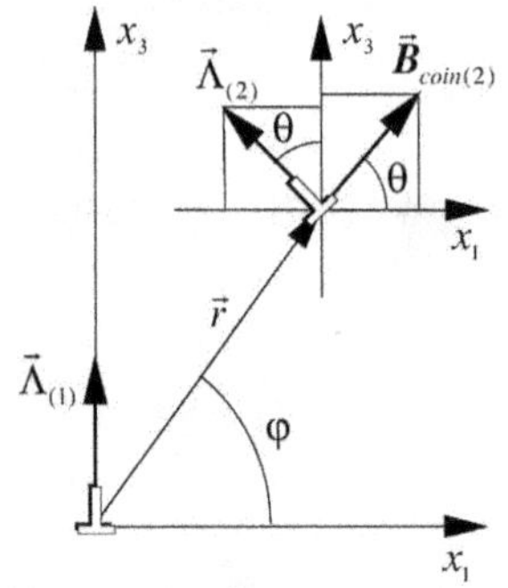

Figure 12.3 *- interaction between two parallel edge dislocations along* Ox_2 *with charges* $\vec{\Lambda}_{(1)}$ *and* $\vec{\Lambda}_{(2)}$

One interesting consequence of this figure is related to the fact that, as an edge dislocation can only easily move by sliding, it is the component of the force of Peach Koehler along the axis Ox_1 which is important in this particular case. The second interesting consequence of this figure is that the force along the axis Ox_1 tends to separate the two lines if the angle φ is less than 45 °, but it tends to superimpose them, if the angle φ is greater than 45 °. The latter case explains why there is formation of a stable ribbon of aligned edge dislocations, with their charges $\vec{\Lambda}$ more or less in the plane of the ribbon, and can thus form a dislocation membrane bounded by two lines of disclination, these are called *flexion joints* as were shown in fig. 9.17a.

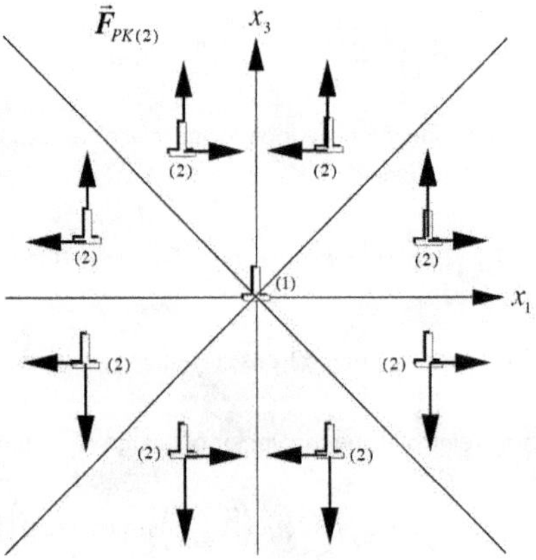

Figure 12.4 *- components of the force of Peach and Koehler due to the interaction between two edge dislocations parallel to axis* Ox_2 *and with charges* $\vec{\Lambda}_{(1)}$ *and* $\vec{\Lambda}_{(2)}$ *along the axis* Ox_3

12.7 – The string model of curved dislocation

The dynamic behavior of a dislocation that curves in a conventional solid, finite and free, can be addressed by introducing the concept of line tension of the dislocation, which allows us to write Newton's equation for an infinitesimal segment of string.

The line tension of a dislocation

Imagine a dislocation which passes through two fixed points $\boldsymbol{A}$ and $\boldsymbol{B}$ separated by a distance L, which is smoothly curved into an arc of radius of curvature r between the two points (fig. 12.5). At great distance from the string, it can be seen as a straight dislocation, but in the vicinity of the string, it has a length equal to $L+\Delta L$ because of its curvature. If we introduce the quantity l corresponding to the distance where we go from the short distance to the long distance, and which roughly corresponds to several times the length L, we can calculate the energy $\boldsymbol{E}_0$ of the dislocation segment of length L using the formulas *(12.43)* for a screw dislocation or *(12.99)* for an edge dislocation, the mathematical expressions are used to separate the energies of the string at short and long distance. We have

$$E_0 = L\frac{\alpha n k_2 \vec{B}^2}{4\pi}\ln\frac{R_\infty}{l} + (L+\Delta L)\frac{\alpha n k_2 \vec{B}^2}{4\pi}\ln\frac{l}{R} = L\frac{\alpha n k_2 \vec{B}^2}{4\pi}\ln\frac{R_\infty}{R} + \Delta L\frac{\alpha n k K_2 \vec{B}^2}{4\pi}\ln\frac{l}{R} \qquad (12.113)$$

where $1 \le \alpha \le 1/(1-\nu)$, since α is equal to 1 in the case of the screw dislocation and $1/(1-\nu)$ in the case of an edge dislocation.

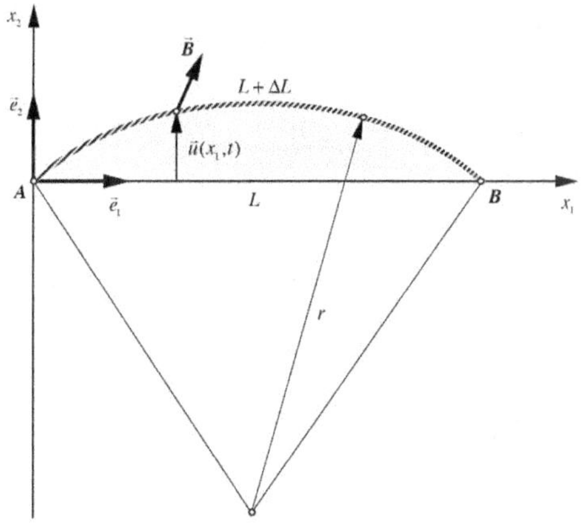

Figure 12.5 *- curved dislocation between two points **A** and **B** separated by a distance* L

The increase in energy ΔE_0 due to the curvature of the string reads

$$\Delta E_0 = \Delta L\frac{\alpha n k_2 \vec{B}^2}{4\pi}\ln\frac{l}{R} \quad \textit{with} \quad l \approx L \qquad (12.114)$$

The fact that the energy of a curved string is greater than the energy of the straight string implies that a curved string will try to become straight like a rubber band stretched between two points. In the case of an elastic, the elongation of a length equal to ΔL requires an energy equal to $\tau\Delta L$ where τ is the tension force of the elastic. By analogy with an elastic, we can introduce a concept of *line tension* τ *of the dislocation*, corresponding to a fictitious force tangential to the string, which is therefore

$$\Delta E_0 = \Delta L\tau = \Delta L\frac{\alpha n k_2 \vec{B}^2}{4\pi}\ln\frac{l}{R} \quad\Rightarrow\quad \tau = \frac{\alpha n k_2 \vec{B}^2}{4\pi}\ln\frac{l}{R} \tag{12.115}$$

In fact, the analogy between the line tension and an elastic is certainly not very rigorous and somewhat unclear[4]. But despite this, it plays a significant role in dislocation theory, for it is that representation that is used for solving dislocation dynamics problems.

The force of Peach and Koehler acting on a dislocation

Imagine a curved dislocation situated in its slip plane, that is to say in the plane containing both its Burgers vector $\vec{B}$ and $\vec{t}$, *the* vector tangential to the line. Suppose further that this dislocation is subjected to a shear stress tangential to the slip plane, which can be fully characterized by a torsor of moment $\vec{m}$ belonging to the slip plane and perpendicular to the plane of shear. The Peach and Koehler force acting on unit length of this dislocation can then be deduced from the relation *(11.13)*, taking only the terms containing the torsor moment

$$\vec{F}_{PK} = -\frac{1}{2}\left(\vec{B}\vec{t}\right)\vec{m} - \frac{1}{2}\left[\left(\vec{B}\wedge\vec{t}\right)\wedge\vec{m}\right] \tag{12.116}$$

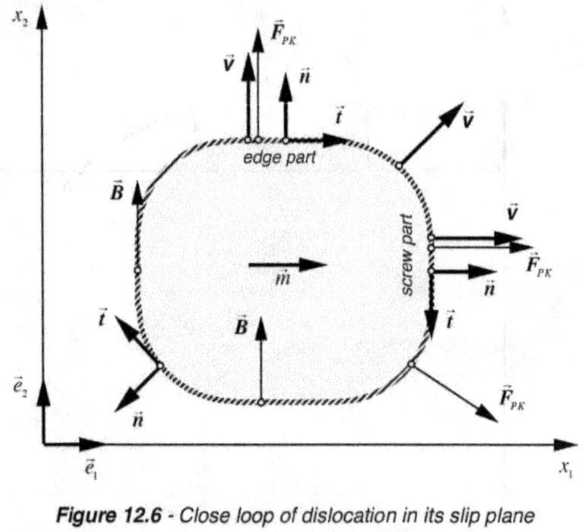

***Figure 12.6** - Close loop of dislocation in its slip plane*

We represent in figure 12.6 a closed loop dislocation in its slip plane Ox_1x_2, with a tangential vector $\vec{t}$ going clockwise and with Burgers vector $\vec{B} = B\vec{e}_2$ in the direction Ox_2. This dislocation is subjected to pure shear stress s_{23} applied in the plane Ox_2x_3. This shear stress can be perfectly represented by a torsor of moment $\vec{m}$ headed in the direction Ox_1, with $\vec{m} = -2s_{23}\vec{e}_1$, so that the strength of the Peach and Koehler force $\vec{F}_{PK}$ can be written as

4 *J. P. Hirth, T. Jossang, J. Lothe, J. Appl. Phys., vol. 37, p.110, 1966*

$$\vec{F}_{PK} = \left[(\vec{e}_2 \vec{t}) \vec{e}_1 \right] B s_{23} + \left[(\vec{e}_2 \wedge \vec{t}) \wedge \vec{e}_1 \right] B s_{23} \tag{12.117}$$

We then check that the Peach and Koehler force is always directed along the normal vector $\vec{n}$ of the dislocation, contained in the slip plane Ox_1x_2, whatever the direction $\vec{t}$, so that

$$\vec{F}_{PK} = \vec{n}\, B s_{23} \tag{12.118}$$

The dynamic equation of a non-relativistic string

Now consider a similar dislocation to that described in the previous section, but with any shape in its slip plane Ox_1x_2, and subjected to pure shear stress $s_{23}(t)$ which may be time dependent (fig. 12.7). Describe its shape by a displacement vector $\vec{u}(x_1,t)$ in the slip plane, oriented in the direction Ox_2.

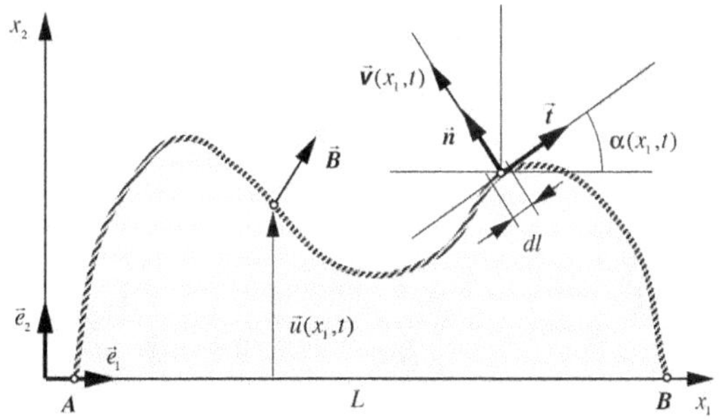

Figure 12.7 *- definition of displacement $\vec{u}(x_1,t)$ of a dislocation in its slip plane*

The goal is then to find the dynamic equation which manages the evolution of the shape $\vec{u}(x_1,t)$ of the dislocation subjected to shear stress. For this, we will try to write Newton's equation of an infinitesimal segment of string of length dl, which has an inclination angle α relative to the axis Ox_1, which is obviously defined by

$$\tan\alpha(x_1,t) = \frac{\partial u(x_1,t)}{\partial x_1} \tag{12.119}$$

If the dislocation moves in the sliding plane, the local velocity $\vec{v}(x_1,t)$ can only be perpendicular to the string, and its projection on the axis Ox_2 must be equal to the time derivative $\dot{u}(x_1,t)$ of the displacement. The local velocity is thus written

$$\vec{v}(x_1,t) = \vec{n}\, v(x_1,t) = \vec{n}\, \frac{\dot{u}(x_1,t)}{\cos\alpha(x_1,t)} \tag{12.120}$$

As for the local acceleration $\vec{a}(x_1,t)$ of the dislocation, it can also only be perpendicular to the string, and such that its projection onto the axis Ox_2 is equal to the second time derivative of the displacement $\ddot{u}(x_1,t)$. The local acceleration is written therefore

$$\vec{a}(x_1,t) = \dot{\vec{v}}(x_1,t) = \vec{n}\, a(x_1,t) = \vec{n}\, \frac{\ddot{u}(x_1,t)}{\cos\alpha(x_1,t)} \tag{12.121}$$

Newton's equation of non-relativistic dislocation segment dl then involves the inertial mass at rest of the segment $M_0 dl$, and is written thus

$$M_0 dl\, \vec{\boldsymbol{a}} = \sum \vec{f}_i \tag{12.122}$$

where the $\vec{f}_i$ represent the diverse forces acting on the segment dl .

We can now find the set of forces that act on this infinitesimal segment (fig. 12.8), namely:

- *the force of Peach and Koehler* due to the shear stress $s_{23}(t)$

$$\vec{f}_{PK} = \vec{\boldsymbol{n}} \boldsymbol{B} s_{23} dl \tag{12.123}$$

- *the tension forces acting tangentially on the two extremities of the segment*

$$\vec{f}_\tau^{(1)} = -\tau \vec{t}\,\Big|_{x_1+dx/2} \quad \text{and} \quad \vec{f}_\tau^{(2)} = \tau \vec{t}\,\Big|_{x_1-dx/2} \tag{12.124}$$

- *a braking force* related to the possible existence of a viscous friction of the moving dislocation within the solid (this is for example the case of braking of dislocations by their interaction with phonons and electrons in metals). If it exists, the braking force is proportional at the first order to the speed of the segment, and can be written by introducing a *coefficient of viscous friction* B_f *by unit of length of the string*

$$\vec{f}_{freinage} = -B_f dl\, \vec{\boldsymbol{v}}(x_1,t) \tag{12.125}$$

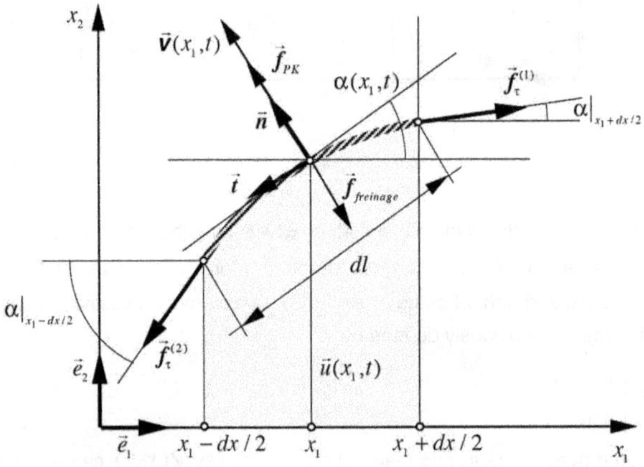

Figure 12.8 *- the forces acting on an infinitesimal segment of string in the sliding plane*

With these forces, the non-relativistic vectorial equation of Newton for the segment dl is written

$$M_0 dl\, \vec{\boldsymbol{a}} = \vec{\boldsymbol{n}} \boldsymbol{B} s_{23} dl - \tau \vec{t}\,\Big|_{x_1+dx/2} + \tau \vec{t}\,\Big|_{x_1-dx/2} - B_f dl\, \vec{\boldsymbol{v}} \tag{12.126}$$

which becomes, by introducing the values of $\vec{\boldsymbol{v}}(x_1,t)$ and $\vec{\boldsymbol{a}}(x_1,t)$

$$M_0 dl \,\frac{\ddot{u}}{\cos\alpha}\, \vec{\boldsymbol{n}} = \boldsymbol{B} s_{23} dl\, \vec{\boldsymbol{n}} - \tau \vec{t}\,\Big|_{x_1+dx/2} + \tau \vec{t}\,\Big|_{x_1-dx/2} - B_f dl\, \frac{\dot{u}}{\cos\alpha}\, \vec{\boldsymbol{n}} \tag{12.127}$$

And taking into account the following relations

$$\begin{cases} dl = dx / \cos\alpha \\ \textit{projection of } \vec{n} \textit{ on } \boldsymbol{Ox_2} = \cos\alpha \\ \textit{projection of } -\tau\vec{t}\big|_{x_1+dx/2} \textit{ on } \boldsymbol{Ox_2} = \tau\sin\alpha\big|_{x_1+dx/2} \\ \textit{projection of } +\tau\vec{t}\big|_{x_1-dx/2} \textit{ on } \boldsymbol{Ox_2} = -\tau\sin\alpha\big|_{x_1-dx/2} \end{cases} \tag{12.128}$$

we can project equation *(12.135)* on axis $\boldsymbol{Ox_2}$, and we have

$$M_0 \frac{\ddot{u}}{\cos\alpha} + B_f \frac{\dot{u}}{\cos\alpha} = \boldsymbol{B}s_{23} + \frac{\left(\tau\sin\alpha\big|_{x_1+dx/2} - \tau\sin\alpha\big|_{x_1-dx/2}\right)}{dx} \tag{12.129}$$

The ratio in the last term is the infinitesimal expression of the partial derivative of $\tau\sin\alpha$ with respect to x_1, so that

$$M_0 \frac{\ddot{u}}{\cos\alpha} + B_f \frac{\dot{u}}{\cos\alpha} - \frac{\partial}{\partial x_1}(\tau\sin\alpha) = \boldsymbol{B}s_{23} \tag{12.130}$$

All we have to do is to use *(12.127)* in order to express $\cos\alpha$ and $\sin\alpha$ as a function of $\vec{u}(x_1,t)$

$$\tan\alpha = \frac{\partial u}{\partial x_1} \quad \Rightarrow \quad \begin{cases} \sin\alpha = \dfrac{\tan\alpha}{\sqrt{1+\tan^2\alpha}} = \dfrac{\partial u}{\partial x_1} \Big/ \sqrt{1+(\partial u/\partial x_1)^2} \\ \cos\alpha = \dfrac{1}{\sqrt{1+\tan^2\alpha}} = 1 \Big/ \sqrt{1+(\partial u/\partial x_1)^2} \end{cases} \tag{12.131}$$

and we obtain the following differential equation, which is called the *"string model of dislocation"*

$$\left(M_0 \frac{\partial^2 u}{\partial t^2} + B_f \frac{\partial u}{\partial t}\right)\sqrt{1+(\partial u/\partial x_1)^2} - \frac{\partial}{\partial x_1}\left[\tau \frac{\partial u}{\partial x_1} \Big/ \sqrt{1+(\partial u/\partial x_1)^2}\right] = \boldsymbol{B}s_{23} \tag{12.132}$$

that calculates the dynamics $\vec{u}(x_1,t)$ of a non-straight dislocation on its slip plane, in a solid subjected to shear stress $s_{23}(t)$.

In this equation, the line tension τ may depend on the orientation of the string in the sliding surface since the line tension is different for a screw dislocation and an edge dislocation, in which case we must introduce τ as a function of the angle $\alpha(x_1,t)$ with the form $\tau = \tau(\alpha(x_1,t)) = \tau(\arctan(\partial u(x_1,t)/\partial x_1))$. But, generally, we introduce an approximate constant value of τ , which already allows us to obtain satisfactory results.

12.8 – Applications of the string model

The string model is extremely useful and effective for treating plasticity and anelasticity problems due to dislocation motion and involving conventional solids, such as metals for example. But it is beyond the scope of this book to detail these phenomena, which can be read about in many books on this particular subject. However, we will outline how this problem can be addressed on the basis of the string model.

The plasticity and anelasticity due to the dynamics of the dislocations

The dislocation motion interacting with obstacles is one of the main phenomena responsible for the plasticity and the anelasticity of conventional materials[5]. Plasticity is then associated with irreversible long-distance movement of dislocations under the effect of an external stress, while anelasticity is associated with short, but recoverable, dissipative movements of dislocations strongly anchored in the solid and subject to a cyclical external constraint. If these microscopic movements at long or short distance can be calculated for a dislocation from the string model, then it is easy to deduce the speed of plastic $\dot{\vec{\varepsilon}}_i^{pl}$ or anelastic $\dot{\vec{\varepsilon}}_i^{an}$ macroscopic deformation of the solid due to a set of dislocation density Λ through the relationships of Orowan *(10.37)* *(10.40)* or *(10.41)*.

The control of mobility of dislocations by obstacles

In conventional materials, such as metals for example, the movement of dislocations is typically controlled by interactions at short or long range, with obstacles more or less localized, as phonons and electrons, other dislocations, point defects, precipitates, grain boundaries, crystal lattice itself, etc.

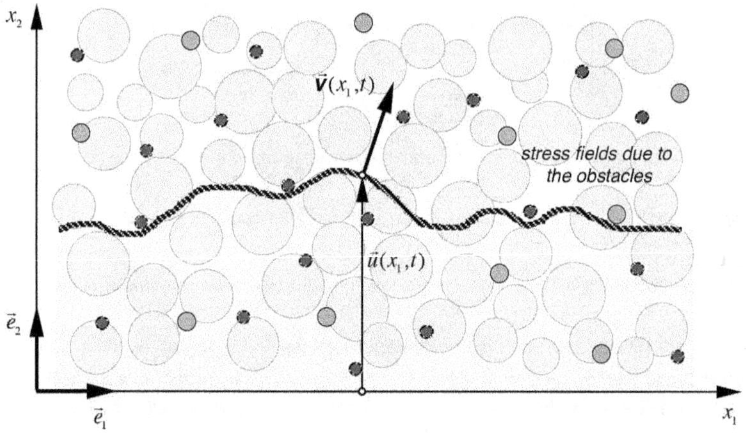

Figure 12.9 *- The stress fields due to obstacles represented in the slip plane*

To introduce this type of interaction in the string model, we must know the spatial distribution of the obstacles in the solid, and the internal stress fields generated by them. These stress fields due to obstacles can be expressed and displayed in the dislocation slip plane (fig. 12.9), where they become responsible for a Peach and Koehler force acting on the dislocation (relations of table 11.1).

In this way, one can add in the equation of the string *(12.132)* all the forces $f_n\left(x_1,u(x_1,t)\right)$ due to the N obstacles surrounding a dislocation

[5] *J. P. Hirth, J. Lothe, «Theory of dislocations», second edition, John Wiley and Sons, New York, 1982; V. L. Indenbom, J. Lothe, «Elastic strain fields and dislocation mobility», Elsevier, North-Holland, 1992*

$$\left(M_0\frac{\partial^2 u}{\partial t^2}+B_f\frac{\partial u}{\partial t}\right)\sqrt{1+(\partial u/\partial x_1)^2}-\frac{\partial}{\partial x_1}\left[\tau\frac{\partial u}{\partial x_1}\Big/\sqrt{1+(\partial u/\partial x_1)^2}\right]=\boldsymbol{B}s_{23}+\sum_{n=1}^{N}f_n\left(x_1,u(x_1,t)\right) \tag{12.133}$$

But it is clear that such an approach to dislocation dynamics quickly proves very complex. In general, we address these problems of dislocation interactions with obstacles in a much more pragmatic way, by developing simple models based on the string model and judiciously adapted to the problem at hand. To further study this, we find many examples of dislocation interaction mechanisms with obstacles, illustrated by experimental results and theoretical models in many books on dislocations, or in review articles as "*dislocation point defect interactions*" [6] and "*dislocation-lattice interaction*" [7] .

The role of thermal activation during interactions with obstacles

The string equation *(12.133)* in the presence of interactions with obstacles is a purely mechanical equation, which cannot take into account the effects of temperature, such as migration of obstacles by diffusion or obstacle crossing by thermal activation. Introducing the effects of temperature in the string equation is theoretically possible by developing an image of "*Brownian*" dislocation, that is to say by introducing a thermal fluctuation of the local force $\boldsymbol{F}_{fluctuation}\left(x_1,u(x_1,t),t\right)$ due to local temperature in the equation of the string, modeled after the term of thermal fluctuations in the model of the *Langevin equation* [8]

$$\left(M_0\frac{\partial^2 u}{\partial t^2}+B_f\frac{\partial u}{\partial t}\right)\sqrt{1+(\partial u/\partial x_1)^2}-\frac{\partial}{\partial x_1}\left[\tau\frac{\partial u}{\partial x_1}\Big/\sqrt{1+(\partial u/\partial x_1)^2}\right]=\boldsymbol{B}s_{23}+\sum_{n=1}^{N}f_n\left(x_1,u(x_1,t)\right)+\boldsymbol{F}_{fluctuation}\left(x_1,u(x_1,t),t\right) \tag{12.134}$$

Such an approach of thermal fluctuation phenomenon proves again very complex, so that, in general, we also address the more pragmatic problems by developing models of the thermal activation suitably adapted to the problem to be treated.

For interested readers, typical examples of this type of thermal activation approach is developed by the author in the articles *"overview on dislocation-point defect interaction: the brownian picture of dislocation motion"* [9] and *"theory of plasticity and anelasticity due to dislocation creep through a multi-scale hierarchy of obstacles"* [10]. In the latter, it is shown among other things that the thermally activated motion of a dislocation moving in a hierarchy of different barriers is of a fractal appearance (fig. 12.10), with interesting effects on the plasticity and anelasticity of the

[6] *G. Gremaud, chapter 3.3, in «Mechanical spectroscopy», Trans Tech Publications, Zürich, 2001, p. 178-246*

[7] *W. Benoit, chapter 3.2, in «Mechanical spectroscopy», Trans Tech Publications, Zürich, 2001, p.158-177*

[8] *Isihara, statistical physics, Academic Press, N.-Y., 1971, chapitre 7*

[9] *G. Gremaud, Materials Science and Engineering, A 370, p. 191-198, 2004*

[10] *G. Gremaud, Materials Science and Engineering, A 521-522, p. 12-17, 2009*

macroscopic solid.

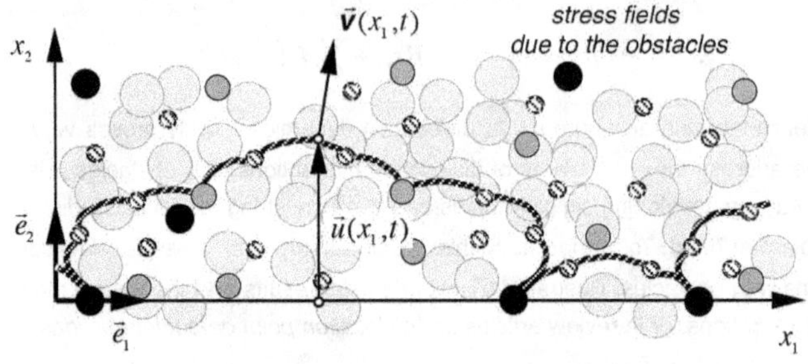

Figure 12.10 *- the fractal behavior of a dislocation moving amid many types of obstacles*

12.9 – The effects of a strong static stress on a pinned dislocation

Suppose a dislocation segment of length L anchored on its sliding surface at two pinning points $\boldsymbol{A}$ and $\boldsymbol{B}$ located on the axis Ox_1, $x_1 = 0$ and $x_1 = L$ as shown in figure 12.7. Such pinning of the dislocation in the medium may be due to the existence of strong localized interactions between the dislocation and obstacles (other dislocations, precipitates, etc.).

The static deformation of a pinned dislocation

If this segment is subjected to a constant shear stress s_{23}, it will take a curved static form which we will try to determine. We make the simplifying assumption that one can use an approximately constant line tension, which is independent of the orientation of the segment in the sliding plane. The static equation of the string model *(12.140)* can then be written

$$\frac{\partial}{\partial x_1}\left[\frac{\partial u}{\partial x_1}\Big/\sqrt{1+\left(\partial u/\partial x_1\right)^2}\right] = -\frac{\boldsymbol{B}s_{23}}{\tau} \qquad (12.135)$$

This equation can easily be integrated a first time with respect to x_1. We have, by introducing an integration constant C, a new differential equation

$$\frac{\partial u}{\partial x_1}\Big/\sqrt{1+\left(\partial u/\partial x_1\right)^2} = -\frac{\boldsymbol{B}s_{23}}{\tau}x_1 + C \qquad (12.136)$$

This equation can also be modified to extract a value of the derivative $\partial u/\partial x_1$, and we obtain the following differential equation of first order

$$\frac{\partial u}{\partial x_1} = \frac{C - x_1/r}{\sqrt{1-\left(C - x_1/r\right)^2}} \quad \text{with} \quad r = \frac{\tau}{\boldsymbol{B}s_{23}} \qquad (12.137)$$

The solution is simply

$$u(x_1) = r\sqrt{1-\left(C - x_1/r\right)^2} + K \qquad (12.138)$$

where K is a second integration constant. The two integration constants C and K are obtained from the boundary conditions of the string, namely the pinning points for which $u(0)=0$ and $u(L)=0$

$$C=\frac{L}{2r} \quad et \quad K=-r\sqrt{1-C^2}=-\sqrt{r^2-(L/2)^2} \tag{12.139}$$

With these values, the deformation of the string segment is described by

$$u(x_1)=\sqrt{r^2-(L/2-x_1)^2}-\sqrt{r^2-(L/2)^2} \tag{12.140}$$

This deformation is an arc of circle which goes through the pinning points $\boldsymbol{A}$ and $\boldsymbol{B}$, whose *radius of curvature* corresponds to $r=\tau/(\boldsymbol{B}s_{23})$, as illustrated in figure 12.5.

The limit of the static deformation and the source of dislocations of Frank-Read

The radius of curvature of the segment $r=\tau/(\boldsymbol{B}s_{23})$ is inversely proportional to the static shear stress s_{23}, which means that it decreases when s_{23} increases. However, it is clear that there is a minimum limit for the radius of curvature, which occurs for a critical stress such that the radius of curvature $s_{23}\big|_{cr}$ becomes equal to $L/2$, therefore

$$s_{23}\big|_{cr}=\frac{2\tau}{\boldsymbol{B}L} \tag{12.141}$$

For any value of s_{23} greater than $s_{23}\big|_{cr}$, there cannot be static solutions for the deformation of the string. What appears instead is a complex dynamic solution of equation *(12.132)*, which corresponds to a *mechanism of source of dislocations of Frank-Read* [11] (fig. 12.11).

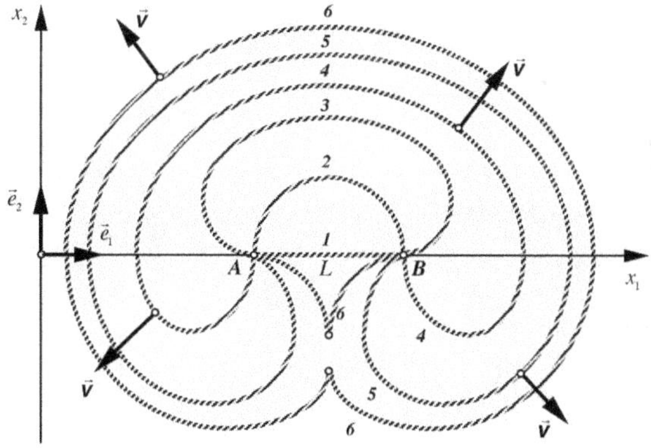

Figure 12.11 *- the mechanism of source of dislocations of Frank-Read*

The initial rectilinear segment represented by *(1)* in figure 25.7 bends between the two pinning points to form a half-circle *(2)*. Then it continues to extend beyond the pinning points, steps *(3)*, *(4)* and *(5)*, until the strand segment leaving $\boldsymbol{A}$ join the segment strand starting from $\boldsymbol{B}$ *(6)*. Here, as the two strands have the same Burgers vector, they bind to each other forming, first of

[11] *F. C. Frank, W. T. Read, Phys. Rev., vol.79, p.722, 1950*

all a new segment *(1)* growing between the pinning points $\boldsymbol{A}$ and $\boldsymbol{B}$, and secondly a closed loop *(6)* that will not stop growing. This mechanism is equivalent to an uninterrupted source of dislocation loops. It is essentially this phenomenon, which is very well observed by electron microscopy, which makes it possible to perform large plastic deformations of certain solids such as metals. And it is this phenomenon that may be responsible for the existence of a non-zero source of dislocation charges in the continuity equation *(10.9)* of the density of dislocation charges.

12.10 – The effects of a dynamical stress on a pinned dislocation

Assume again a dislocation segment pinned in its sliding plane at two points $\boldsymbol{A}$ and $\boldsymbol{B}$ located on the axis Ox_1, at $x_1 = 0$ and $x_1 = L$, as shown in figure 12.7. And suppose further that this segment is subject to an external dynamical stress $s_{23}(t)$ such that $s_{23}(t) << s_{23}\big|_{cr}$. To solve the problem of this moving pinned dislocation, it is clear that the string model *(12.134)* is far too sophisticated. In the case of low stress, it is useful to develop simplified models.

The classical model of the string for weak constraints [12]

In the case where $s_{23}(t) << s_{23}\big|_{cr}$, the angle α that the string forms with respect to axis Ox_1 is always weak so that

$$\tan\alpha = \partial u / \partial x_1 << 1 \tag{12.142}$$

We rewrite the string model by supposing that $\partial u / \partial x_1 \cong 0$, and one obtains the well-known *'classic model'* of the string

$$M_0 \frac{\partial^2 u}{\partial t^2} + B_f \frac{\partial u}{\partial t} - \tau \frac{\partial^2 u}{\partial x_1{}^2} = \boldsymbol{B} s_{23} \tag{12.143}$$

In this model we show that the static deformation obtained in the case of a stress s_{23} is approximated by a parabola of equation

$$u(x_1) = \frac{\boldsymbol{B} s_{23}}{2\tau}(L - x_1)\, x_1 \tag{12.144}$$

which maximum is worth

$$\max(u) = u(x_1 = L/2) = \frac{\boldsymbol{B} L^2}{8\tau} s_{23} \tag{12.145}$$

The «rigid rod model» of a pinned dislocation submitted to weak stresses

The model of the classic string can further be simplified by looking for an approximate equation for the mean displacement $\bar{u}$ of the dislocation, defined by relation

$$\bar{u} = \frac{1}{L}\int_0^L u(x_1)dx_1 = \frac{1}{L}\int_0^L \frac{\boldsymbol{B} s_{23}}{2\tau}(L - x_1)x_1 dx_1 = \frac{\boldsymbol{B} L^2}{12\tau} s_{23} \tag{12.146}$$

---

[12] *A. Granato and K. Lücke, J. Appl. Phys., vol. 27, p. 583, 1956; A. Granato and K. Lücke, J. Appl. Phys., vol. 27, p. 789, 1956*

For a constant stress s_{23}, the average statical displacement $\bar{u}$ satisfies the following equation, by introducing a *restoring coefficient* K_u *due to the line tension* τ of the string

$$K_u \bar{u} = \boldsymbol{B} s_{23} \quad \text{with} \quad K_u = \frac{12\tau}{L^2} \tag{12.147}$$

We can then find a dynamical model which is approximative for the pinned dislocation, by writing that the average displacement $\bar{u}(t)$ satisfies the following equation

$$M_0\, \ddot{\bar{u}}(t) + B_f\, \dot{\bar{u}}(t) + K_u\, \bar{u}(t) = \boldsymbol{B} s_{23} \tag{12.148}$$

This equation is called the *"rigid rod model"* because it describes the average displacement of the dislocation, as if it were a rigid bar subjected to a restoring force. It is in fact nothing but the equation of a *damped harmonic oscillator*, and is obviously a rough approximation of reality, since it presupposes that the dynamic deformation of the dislocation remains a parabola which is clearly not the case if the speed of variation of s_{23} becomes very large!

The «phonon relaxation» of dislocations in metals

In a metal, there is always a substantial density Λ of dislocations. They generally form a three-dimensional network known as the *Frank network* [13], consisting of nodes to which dislocations converge. These nodes are to be considered as anchor points for lattice dislocations, so that we can consider that, for low applied stresses, dislocations are represented by segments, of average length $\bar{L}$ that bend under the effect of applied stresses. If such a metal is subjected to an ultrasonic dynamic stress $s_{23}(t) = s_0 \sin\omega t$ of low amplitude but high frequency, these segments will begin to vibrate.

In the equation of the rigid rod model *(12.148)*, we use the rest mass M_0 per unit length of dislocation, and the viscous friction coefficient B_f per unit length of string. In the absence of other interactions, the viscous friction coefficient B_f of a metal is essentially due to the interaction of mobile dislocations with phonons of the lattice [14]. The friction coefficient is controlled by two mechanisms: the *"phonon wind"* which appears with a linear temperature dependence of B_f, and the *"slow phonon relaxation"*, which appears with a parabolic behavior of B_f while temperature is below the Debye temperature of the metal, and with a constant behavior of B_f when the temperature is higher than the Debye temperature of the metal. The behavior of the dislocation segments under the effect of ultrasound then depends primarily on the relative values of M_0 and B_f.

To study this behavior, we can show that the short distance microscopic movements of the dislocation segments will be perceived as a macroscopic anelastic shear deformation α_{23}^{an}. Indeed, applying the relationships of Orowan *(10.40)* to the case of the string shown in figure 12.6, assuming that the dislocation loop extends at a constant velocity $\boldsymbol{V}$ in all directions, we can verify that we always have the following relationships, throughout the string portions

---

[13] *F. C. Frank, in Report of the Conference on Defects in Crystalline Solids, Phys. Soc., London, p. 159, 1954*

[14] *V. I. Alshits, the Phonon-dislocation interaction and its rôle in dislocation dragging and thermal resistivity, chapter 11 in «Elastic strain fields and dislocation mobility», Elsevier, North-Holland, 1992*

$$\begin{cases} \dot{\vec{\alpha}}_1{}^{pl} = 0 \\ \dot{\vec{\alpha}}_2{}^{pl} = -\Lambda \boldsymbol{B} \mathbf{v} \vec{e}_3 / 2 \\ \dot{\vec{\alpha}}_3{}^{pl} = -\Lambda \boldsymbol{B} \mathbf{v} \vec{e}_2 / 2 \end{cases} \tag{12.149}$$

which imply that

$$\dot{\alpha}_{23}^{pl} = \dot{\alpha}_{32}^{pl} = \frac{1}{2} \Lambda \boldsymbol{B} \mathbf{v} = \frac{1}{2} \Lambda \boldsymbol{B} \dot{\vec{u}} \tag{12.150}$$

Thus, at the macroscopic scale of the metal, the microscopic motion represented by equation *(12.148)* can be put in the form of an equation of anelasticity

$$\dot{\vec{u}} = \frac{2}{\Lambda \boldsymbol{B}} \dot{\alpha}_{23}^{an} \quad \Rightarrow \quad \frac{2M_0}{\Lambda \boldsymbol{B}^2} \ddot{\alpha}_{23}^{an} + \frac{2B_f}{\Lambda \boldsymbol{B}^2} \dot{\alpha}_{23}^{an} + \frac{2K_u}{\Lambda \boldsymbol{B}^2} \alpha_{23}^{an} = s_{23} \tag{12.151}$$

which corresponds exactly to the phenomenological equation of anelasticity *(7.89)* which we introduced in chapter 7, namely

$$M_{an} \ddot{\alpha}_{23}^{an} + B_{an} \dot{\alpha}_{23}^{an} + K_{an} \alpha_{23}^{an} = s_{23} \tag{12.152}$$

In the presence of such an equation, we saw in chapter 7 that the response to a solicitation $s_{23}(t) = s_0 \sin \omega t$ might be a resonance at frequency $\omega_0 = \sqrt{K_{an} / M_{an}} = \sqrt{K_u / M_0}$, or a relaxation with a relaxation time $\tau_{an} = B_{an} / K_{an} = B_f / K_u$, depending on whether the product $\omega_0 \tau_{an}$ is greater or lesser than 1, so whether $B_f > \sqrt{M_0 K_u}$ or $B_f < \sqrt{M_0 K_u}$.
Experimentally, there is a relaxation around 20 to 150 MHz in many metals, which could probably be attributed to the interaction dislocation-phonon [15]. Therefore, it was concluded that in general the coefficient of friction B_f due to phonon outweighs the inertial mass M_0 of dislocations, and therefore we can neglect the term in M_0 in the string equation when dealing with dynamic problems of dislocations in metals.
There are several interesting consequences to this experimental observation:

- the fact that the dislocation segments have a relaxation in the field of MHz in the metal implies that the dislocations become perfectly still for frequencies well above the relaxation frequency, so that this field of very high frequency is ideal for measuring pure elastic properties of the lattice without interference due to dislocations,

- this same fact implies that if we study dislocations by mechanical spectroscopy in a frequency range well below the frequency of the phonon relaxation, it is perfectly possible to neglect the term B_f, and consequently the term M_0 in the string model, meaning that the pinned dislocation segments are always on the mechanical balance with the low frequency applied stress,

- as the coefficient of friction B_f due to the phonons prevails over the inertial mass M_0 of the dislocations in the metals, the dislocations are over-amortized, so that pinned dislocation segments cannot present vibrations like we could observe on a tight rope. This means among other things that, under the effect of thermal disturbances of the network, a dislocation segment has to behave as a "Brownian segment" which would deform continuously under the effect of local thermal perturbations of stress.

---

[15] *A. Granato, K. Lücke, J. Appl. Phys., vol. 27, p. 789, 1956*

It is on the basis of these very important observations that we can quite easily develop many models of interactions between dislocations and obstacles, such as the reader may consult in numerous books on dislocations, and in the articles [2,3,5,6] cited above.

PART II

A

The "cosmic lattice"

Equation of Newton

Waves propagation modes
and localized vibrations

Curvature of the wave rays
and "black holes"

Cosmological expansion of a finite lattice
and "black energy"

Chapter 13

Perfect lattices and equations of Newton

In this chapter, we introduce a new *perfect isotropic lattice*. It is purely imaginary, and we will call it the *cosmic lattice*, the development of the free energy of deformation is expressed per unit volume, which depends linearly on volume expansion and quadratically on volume expansion, shear strain and torsional rotation deformations. It is then compared with the usual perfect isotropic solid, namely the one we used in chapters 7 and 12 of the applications of the theory to the usual isotropic solids. We also compare the Newton's equations of these two types of perfect lattices.

13.1 – The «perfect solid» and its equation of Newton

We have seen in sections 7.1 and 12.1 that the elastic part of the free energy of the usual isotropic solids depends only on the volume expansion τ and shear strains $\vec{\alpha}_i^{él}$ and $\vec{\alpha}_i^{an}$, and that we can calculate the *free energy per lattice site* based on these quantities. The free energy function of the simplest deformation that we can imagine is that of perfect isotropic lattice, which we name the *"perfect solid"*, and which is written

$$f^{déf} = -k_0\tau + k_1\tau^2 + k_2\sum_i(\vec{\alpha}_i^{él})^2 + k_{an}\sum_i(\vec{\alpha}_i^{an})^2 \tag{13.1}$$

In the absence of anelastic expansion, the state equations of the deformation is limited to the scalar pressure p and the transverse symmetrical tensor of shear stresses $\vec{s}_i$ and $\vec{s}_i^{cons}$ as shown by the relationship *(7.15)*

$$\begin{cases} p = -n\dfrac{\partial f^{él}}{\partial \tau} = nk_0 - 2nk_1\tau \\ \vec{s}_i = n\dfrac{\partial f^{él}}{\partial \alpha_{ik}^{él}}\vec{e}_k = 2nk_2\vec{\alpha}_i^{él} \end{cases} \quad \text{and} \quad \left\{ \vec{s}_i^{cons} = n\dfrac{\partial f^{él}}{\partial \alpha_{ik}^{an}}\vec{e}_k = 2nk_{an}\vec{\alpha}_i^{an} \right. \tag{13.2}$$

We also saw in section 12.1 that, in the perfect solid, pure shear in the presence of a homogeneous volume expansion can be treated using the elastic $\vec{\omega}^{él}$ and anelastic $\vec{\omega}^{an}$ rotational vectors, with the help of the torsor moment $\vec{m}$. With these equations of state, it was shown in section 12.1 that Newton's equation can be written in two ways, namely in terms of $\vec{\alpha}_k^{él}$ *(12.4)* or $\vec{\omega}^{él}$ according to *(12.9)*

$$\frac{d\vec{p}}{dt} = 2k_2\sum_k \vec{e}_k \operatorname{div}\vec{\alpha}_k^{él} - 2k_2\sum_k\left(\vec{e}_k\overrightarrow{\operatorname{grad}}\,\tau\right)\vec{\alpha}_k^{él} + \overrightarrow{\operatorname{grad}}\left[\left(2k_1(1-\tau)+k_0\right)\tau\right] + m\vec{\phi}_I\frac{dC_I}{dt} - m\vec{\phi}_L\frac{dC_L}{dt} \tag{13.3}$$

$$\frac{d\vec{p}}{dt} = -2k_2\overrightarrow{\operatorname{rot}}\,\vec{\omega}^{él} - 2k_2\sum_k\left(\vec{e}_k\overrightarrow{\operatorname{grad}}\,\tau\right)\vec{\alpha}_k^{él} + \overrightarrow{\operatorname{grad}}\left[\left(\frac{4}{3}k_2 + 2k_1(1-\tau)+k_0\right)\tau\right] + m\vec{\phi}_I\frac{dC_I}{dt} - m\vec{\phi}_L\frac{dC_L}{dt} + 2k_2\vec{\lambda} \tag{13.4}$$

Newton's equation *(13.4)* of the usual perfect solid presents the advantage of directly depending on the elastic rotation vector, such that the shear tensor only appears once in the coupling term with the gradient field of volume expansion.
This Newton equation, coupled with the heat propagation equation, allowed us to calculate the propagation modes of thermoelastic waves in the usual *isotropic solids* (section 7.3). It also allowed us to calculate the fields at equilibrium of an edge dislocation (section 12.4) and to show that the spatial and temporal evolution of a homogeneous isotropic solid in volume expansion is described by a set of equations perfectly analogous to Maxwell's equations of electromagnetism.

13.2 – The «cosmic lattice» and its equation of Newton

We introduce here a new imaginary lattice, which we will arbitrarily call *"cosmic lattice"*. It's *free energy per unit volume* becomes a development in τ , $\vec{\alpha}_i^{él}$ and $\vec{\alpha}_i^{an}$, but also depends directly on the rotational vectors $\vec{\omega}^{él}$ and $\vec{\omega}^{an}$ by elastic and anelastic rotational deformations. Our initial conjecture is *a priori* the following

Conjecture 0: *the free energy of the "cosmic lattice" per unit volume of lattice is*

$$F^{déf} = F^{déf}\left[\tau, \tau^2, (\vec{\alpha}_i^{él})^2, (\vec{\omega}^{él})^2, (\vec{\alpha}_i^{an})^2, (\vec{\omega}^{an})^2\right] \tag{13.5}$$

Such a lattice is actually the most generic perfect isotropic lattice imaginable if we assume that its energy depends both linearly on volume expansion and quadratically on volume expansions, shear strains and torsional rotation deformation. Always in a spirit of simplification, we can still assume that there is no anelasticity by volume expansion in the lattice. The state function per unit volume of the cosmic lattice is written consequently

$$F^{déf} = -K_0\tau + K_1\tau^2 + K_2\sum_i(\vec{\alpha}_i^{él})^2 + 2K_3(\vec{\omega}^{él})^2 + K_1^{an}\sum_i(\vec{\alpha}_i^{an})^2 + 2K_2^{an}(\vec{\omega}^{an})^2 \tag{13.6}$$

from which we deduce directly the free energy per site of lattice

$$f^{déf} = \frac{F^{déf}}{n} = \frac{1}{n}\left(-K_0\tau + K_1\tau^2 + K_2\sum_i(\vec{\alpha}_i^{él})^2 + 2K_3(\vec{\omega}^{él})^2 + K_1^{an}\sum_i(\vec{\alpha}_i^{an})^2 + 2K_2^{an}(\vec{\omega}^{an})^2\right) \tag{13.7}$$

Comparing this expression of $f^{déf}$ with the expression *(13.1)* of the *usual perfect solid*, we see immediately that it is the presence of the factor $1/n$ and the terms $\vec{\omega}^{él}$ and $\vec{\omega}^{an}$ which primarily differentiates the two terms of free energy. From *(13.7)*, we deduce the five equations of state of elasticity and anelasticity respectively for the pressure scalar p, the transverse symmetric tensor of shear stresses $\vec{s}_i$ and $\vec{s}_i^{cons}$, as we as the torsor of moments $\vec{m}$ and $\vec{m}^{cons}$

$$\begin{cases} p = -n\dfrac{\partial f^{déf}}{\partial \tau} = K_0 - 2K_1\tau - F^{déf} \\ \vec{s}_i = n\dfrac{\partial f^{déf}}{\partial \alpha_{ik}^{él}}\vec{e}_k = 2K_2\,\vec{\alpha}_i^{él} \\ \vec{m} = n\dfrac{\partial f^{déf}}{\partial \omega_i^{él}}\vec{e}_i = 4K_3\,\vec{\omega}^{él} \end{cases} \quad \text{and} \quad \begin{cases} \vec{s}_i^{cons} = n\dfrac{\partial f^{déf}}{\partial \alpha_{ik}^{an}}\vec{e}_k = 2K_1^{an}\,\vec{\alpha}_i^{an} \\ \vec{m}^{cons} = n\dfrac{\partial f^{déf}}{\partial \omega_i^{an}}\vec{e}_i = 4K_2^{an}\,\vec{\omega}_i^{an} \end{cases} \tag{13.8}$$

deformations	velocity field	components of the energy of deformation
homogenous expansion 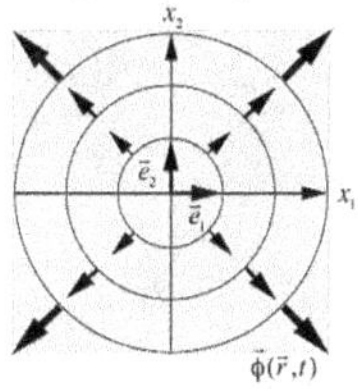	$\vec{\phi} = a\left[x_1\vec{e}_1 + x_2\vec{e}_2 + x_3\vec{e}_3\right]\dot{g}(t)$	$F^{\tau} = 3ag(t)\left[3K_1 ag(t) - K_0\right]$ $\boxed{F^{\vec{\alpha}_i^{el}} = 0}$ $\boxed{F^{\vec{\omega}} = 0}$
non-homogenous expansion 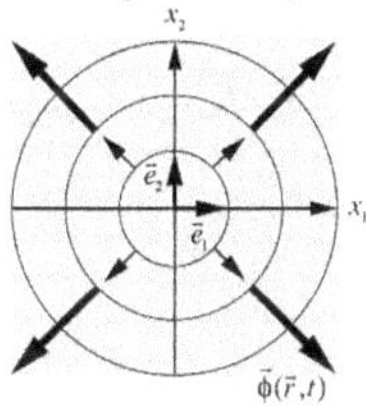	$\vec{\phi} = \dot{g}(t) r \sum_i x_i\vec{e}_i$ $où\ \ r = \sqrt{x_1^2 + x_2^2 + x_3^2}$	$F^{\tau} = 4rg(t)\left[4K_1 rg(t) - K_0\right]$ $F^{\vec{\alpha}_i^{el}} = 6K_2 r^2 g^2(t)$ $\boxed{F^{\vec{\omega}} = 0}$
elongation-shortening 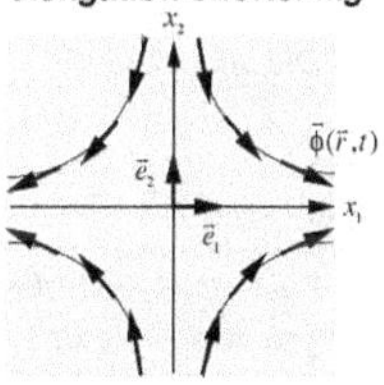	$\vec{\phi} = \left[a_1 x_1\vec{e}_1 - a_2 x_2\vec{e}_2\right]\dot{g}(t)$	$F^{\tau} = K_1\left(a_1 - a_2\right)^2 g^2(t)$ $- K_0\left(a_1 - a_2\right)g(t)$ $F^{\vec{\alpha}_i^{el}} = K_2\left(a_1^2 + a_2^2\right)g^2(t)$ $-\dfrac{K_2}{3}\left(a_1 - a_2\right)^2 g^2(t)$ $\boxed{F^{\vec{\omega}} = 0}$
shear 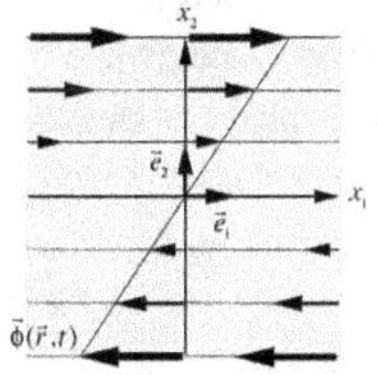	$\vec{\phi} = a x_2 \dot{g}(t)\vec{e}_1$	$\boxed{F^{\tau} = 0}$ $F^{\vec{\alpha}_i^{el}} = K_2 a^2 g^2(t)/2$ $F^{\vec{\omega}^{el}} = K_3 a^2 g^2(t)/2$
non-homogenous rotation 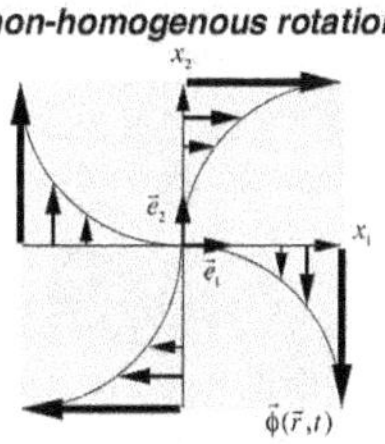	$\vec{\phi} = r\left(x_1\vec{e}_2 - x_2\vec{e}_1\right)g(t)$ $où\ \ r = \sqrt{x_1^2 + x_2^2}$	$\boxed{F^{\tau} = 0}$ $F^{\vec{\alpha}_i^{el}} = K_2 r^2 g^2(t)/2$ $F^{\vec{\omega}^{el}} = 9K_3 r^2 g^2(t)/2$

***Table 13.1** - examples of deformations and energies of expansion, shear and torsion*

deformations	*velocity field*	*components of energy of deformation*
torsion 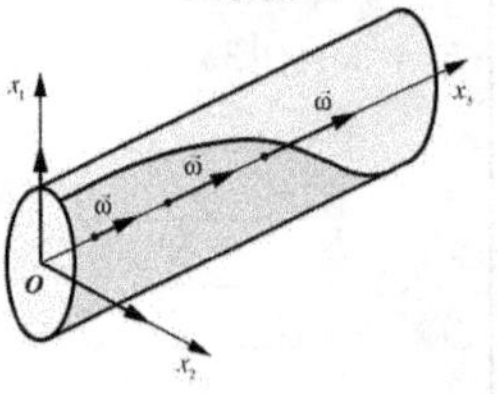	$\vec{\phi} = (x_1\vec{e}_2 - x_2\vec{e}_1)x_3\,\dot{g}(t)$	$\boxed{F^{\tau} = 0}$ $F^{\vec{\alpha}_i^{\,el}} = K_2(x_1^2 + x_2^2)g^2(t)/2$ $F^{\vec{\omega}} = K_3(4x_3^2 + x_1^2 + x_2^2)g^2(t)/2$
flexion 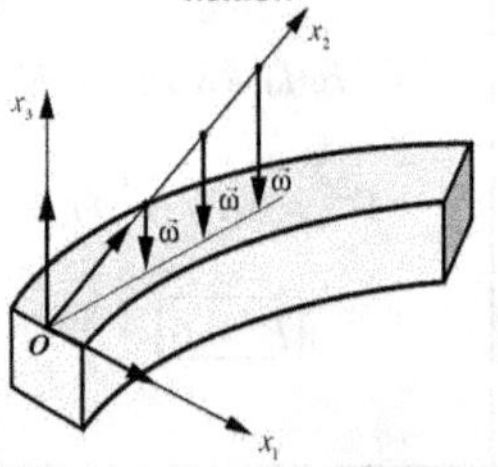	$\vec{\phi} = 2(x_2\vec{e}_1 + x_1\vec{e}_2)x_2\,\dot{g}(t)$	$F^{\tau} = 2x_1 g(t)[2K_1 x_1 g(t) - K_0]$ $F^{\vec{\alpha}_i^{\,el}} = 2K_2(9x_2^2 + 4x_1^2/3)g^2(t)$ $F^{\vec{\omega}} = 2K_3 x_2^2 g^2(t)$

Table 13.2 *- pure torsion and flexion, and energies of expansion, shear and rotation*

With regards to the state equations, one can also wonder what differentiates the *usual perfect solid* from the *cosmic lattice*. The first obvious difference, and one that *will play a major role* thereafter, is the expression of the pressure $p = K_0 - 2K_1\tau - F^{déf}$ within the cosmic lattice, which directly depends on the free energy $F^{déf}$ of the local elastic deformation, while it is simply written $p = n(k_0 - 2k_1\tau)$ in the usual perfect solid.

In table 13.1, we show different types of deformation and energy terms resulting in the cosmic lattice. If the homogeneous volume expansion does not change the shear energy and the elastic rotation energy, the inhomogeneous deformation of volume by elongation or shrinkage in a given direction changes the shear energy but do not affect the energy of elastic rotation. In fact, only the pure shear or non-uniform rotation, that could be called torsional deformation of the lattice affect both the energies of elastic shear and rotation, but in different ways: in the case of pure shear, the energy of shear and elastic rotation are equal, whereas in the case of non-homogeneous rotation, shear energies are different from elastic rotation energy. Moreover, the torsion of the pure medium, as described in figure 3.2, as well as pure bending, as described in figure 3.1, also affect both the shear tensor and the elastic rotational vector. Indeed, suppose the following velocity fields representing respectively a torsion of the medium along Ox_3 and a flexion of the medium along Ox_3

$$\vec{\phi}_{torsion} = (x_1\vec{e}_2 - x_2\vec{e}_1)x_3\,\dot{g}(t) \quad \text{and} \quad \vec{\phi}_{flexion} = 2(x_2\vec{e}_1 + x_1\vec{e}_2)x_2\,\dot{g}(t) \qquad (13.9)$$

With these velocity fields, we can calculate the temporal evolution of the distortion tensors by using the geometrokinetic relations of table 2.3. In the case of torsion we have

$$\begin{cases} \tau = 0 \\ \left(\alpha_{ij}^{él}\right) = \begin{pmatrix} 0 & 0 & -x_2 \\ 0 & 0 & x_1 \\ -x_2 & x_1 & 0 \end{pmatrix} \dfrac{g(t)}{2} \\ \vec{\omega}^{él} = \left(x_3\vec{e}_3 - x_1\vec{e}_1 / 2 - x_2\vec{e}_2 / 2\right) g(t) \end{cases} \tag{13.10}$$

and in the case of flexion

$$\begin{cases} \tau = 2x_1 g(t) \\ \left(\alpha_{ij}^{él}\right) = \begin{pmatrix} -2x_1/3 & 3x_2 & 0 \\ 3x_2 & 4x_1/3 & 0 \\ 0 & 0 & -2x_1/3 \end{pmatrix} g(t) \\ \vec{\omega}^{él} = -x_2 g(t)\vec{e}_3 \end{cases} \tag{13.11}$$

We deduce from these expressions the expansion energies, the elastic shear and rotation energies and we show them in table 13.2. It is found that pure bending and twisting both contribute to energy of shear and elastic rotation, but that they have different expressions.

The *cosmic lattice Newton's equation* is derived from the expression *(6.11)*, in which the term $\rho\vec{g}$ is neglected and we introduced state equations *(13.8)*

$$\begin{aligned} n\frac{d\vec{p}}{dt} &= 2K_2 \sum_i \vec{e}_i \operatorname{div}\vec{\alpha}_i^{él} - 2K_3 \overrightarrow{\operatorname{rot}}\,\vec{\omega}^{él} + 2K_1 \overrightarrow{\operatorname{grad}}\,\tau + nm\vec{\phi}_I \frac{dC_I}{dt} - nm\vec{\phi}_L \frac{dC_L}{dt} \\ &+ \overrightarrow{\operatorname{grad}} \underbrace{\left(K_2 \sum_i (\vec{\alpha}_i^{él})^2 + K_1^{an} \sum_i (\vec{\alpha}_i^{an})^2 + 2K_3 (\vec{\omega}^{él})^2 + 2K_2^{an} (\vec{\omega}^{an})^2 + K_1\tau^2 - K_0\tau \right)}_{F^{déf}} \end{aligned} \tag{13.12}$$

By using again relation *(12.8)*, which was deducted directly from expression *(12.5)* of the flexion vector, namely

$$\sum_k \vec{e}_k \operatorname{div}\vec{\alpha}_k^{él} = \vec{\lambda} - \overrightarrow{\operatorname{rot}}\,\vec{\omega}^{él} + \frac{2}{3}\overrightarrow{\operatorname{grad}}\,\tau^{él} \quad \text{and} \quad \sum_k \vec{e}_k \operatorname{div}\vec{\alpha}_k^{an} = -\overrightarrow{\operatorname{rot}}\,\vec{\omega}^{an} \tag{13.13}$$

the *equation of Newton of the cosmic lattice* can also be written as

$$\begin{aligned} n\frac{d\vec{p}}{dt} &= -2\left(K_2 + K_3\right)\overrightarrow{\operatorname{rot}}\,\vec{\omega}^{él} + \left(\frac{4}{3}K_2 + 2K_1\right)\overrightarrow{\operatorname{grad}}\,\tau + 2K_2\vec{\lambda} + nm\vec{\phi}_I \frac{dC_I}{dt} - nm\vec{\phi}_L \frac{dC_L}{dt} \\ &+ \overrightarrow{\operatorname{grad}} \underbrace{\left(K_2 \sum_i (\vec{\alpha}_i^{él})^2 + K_1^{an} \sum_i (\vec{\alpha}_i^{an})^2 + 2K_3 (\vec{\omega}^{él})^2 + 2K_2^{an} (\vec{\omega}^{an})^2 + K_1\tau^2 - K_0\tau \right)}_{F^{déf}} \end{aligned} \tag{13.14}$$

We will see later that Newton's equation will play an absolutely central role in the behavior of the cosmic lattice. The equation is fairly complex, especially due to the presence of the flexion charge density $\vec{\lambda}$, of the terms related to the diffusion of intrinsic point defects and especially of the term dependent on volume density of free energy of deformation $F^{déf}$.

13.3 – The «hidden face» of the cosmic lattice

The fact of introducing rotational energy with terms $\vec{\omega}^{\acute{e}l}$ and $\vec{\omega}^{an}$ in expression *(13.6)* of the free energy of the cosmic lattice, as well as the fact that we develop its free energy per unit volume and not by lattice site are not at all trivial to understand, and really make this lattice a perfectly imaginary lattice for which we have absolutely no equivalent among the usual solids.

On the search of analogies with the great theories of physics

Rather than engaging in a superfluous interpretation of the *"hidden side"* of this imaginary lattice, it seems preferable to begin by exploring in detail the consequences that this *hidden face* entails in behaviors that may be present in the cosmic lattice. This is what we will do for the rest of the book. To do this, we will show that this Newton's equation we deduce has spectacular properties and is central to many analogies that we will develop in the third part with the great theories of physics ,i.e. the Electromagnetism of Maxwell, the Gravitation of Newton, the Einstein's general relativity, the Lorentz Transformation and the Special Relativity of Einstein, and even the quantum physics and the Standard Model of elementary particles.

Chapter 14

Modes of wave propagations and of localized expansion vibrations in the cosmic lattice

In this chapter, we are interested in the *propagation of waves* in the cosmic lattice. There appears quite surprising phenomena, such as the appearance of a longitudinal mode coupled to the propagation of transverse waves that are polarized linearly, which disappears for circularly polarized transverse waves. There is also the possibility of propagation of *longitudinal waves*. But under certain conditions dependent on elastic moduli, the *longitudinal propagation mode* disappears in favor of *localized vibration modes of the expansion.*

14.1 – Propagation of transversal waves

In this section, we will proceed with a detailed study of the propagation of transversal waves in the cosmic lattice. Let's assume the following

Hypothesis 1:
$$\begin{cases} \vec{\alpha}_i^{an} = 0 \quad \& \quad \vec{\omega}^{an} = 0 \\ C_I = C_L \cong 0 \\ \tau = \tau_0 + \tau^{(p)} \\ \vec{\lambda} = 0 \end{cases} \tag{14.1}$$

that is to say, if it is assumed that there is no anelasticity, that the concentrations of vacancies and interstitials are negligible, that the background of the volume expansion is a constant τ_0 and that there is no density of flexion charges, then the transverse propagation of disturbances can be calculated easily.

Newton's equation *(13.14)* can be written through the relationship

$$\vec{p}^{(p)} = m\vec{\phi}^{(p)} \tag{14.2}$$

which is obtained in the absence of vacancies and interstitials, under the form

$$mn\frac{d\vec{\phi}^{(p)}}{dt} = -2\left(K_2 + K_3\right)\overrightarrow{\mathrm{rot}}\,\vec{\omega}^{(p)} + \left(\frac{4}{3}K_2 + 2K_1\right)\overrightarrow{\mathrm{grad}}\,\tau + \overrightarrow{\mathrm{grad}}\left(K_2\sum_i(\vec{\alpha}_i^{(p)})^2 + 2K_3(\vec{\omega}^{(p)})^2 + K_1\tau^2 - K_0\tau\right) \tag{14.3}$$

The coupling with longitudinal wavelets

If in a local referential $Ox_1x_2x_3$, let's consider a transversal perturbation of the velocity field $\vec{\phi}^{(p)}$ in the form $\phi_k^{(p)}(x_j,t)\vec{e}_k$, parallel to Ox_k and varying along Ox_j. The argument of the second gradient will vary along the axis Ox_j under the effect of the square $(\vec{\alpha}_i^{(p)})^2$ and $(\vec{\omega}^{(p)})^2$ which are non-zero, and necessarily associated with $\vec{\omega}^{(p)}$. As a result there appears a force in the direction Ox_j, and therefore a velocity $\phi_j^{(p)}(x_j,t)\vec{e}_j$ along the axis Ox_j, as well as a disturbance of the volume expansion $\tau^{(p)}(x_j,t)$. So we will write in this case a velocity perturbation in the form

$$\vec{\phi}^{(p)} = \phi_k^{(p)}(x_j,t)\vec{e}_k + \phi_j^{(p)}(x_j,t)\vec{e}_j \tag{14.4}$$

Such a perturbation $\vec{\phi}^{(p)}$ implies the following perturbations $\tau^{(p)}$ and $\vec{\omega}^{(p)}$ of the expansion field and the rotation field along the axis Ox_j, which we deduce via the geometro-kinetic equations for τ and $\vec{\omega}^{él}$

$$\begin{cases} \dfrac{d\tau^{(p)}}{dt} = \operatorname{div}\vec{\phi}^{(p)} = \dfrac{\partial \phi_j^{(p)}}{\partial x_j} & \Rightarrow \quad \tau = \tau_0 + \tau^{(p)}(x_j,t) \\ \dfrac{d\vec{\omega}^{(p)}}{dt} = \dfrac{1}{2}\overrightarrow{\operatorname{rot}}\,\vec{\phi}^{(p)} & \Rightarrow \quad \vec{\omega}^{(p)} = \omega_i^{(p)}(x_j,t)\vec{e}_i \end{cases} \tag{14.5}$$

And the perturbation $\vec{\phi}^{(p)}$ implies also a perturbation of shear $\vec{\alpha}_i^{(p)}$ along the axis Ox_j, which we deduce via the geometro-kinetic equations *(2.26)* for $\vec{\alpha}_i$

$$\begin{cases} \vec{\alpha}_i^{(p)} = -\dfrac{1}{3}\vec{e}_i\tau^{(p)}(x_j,t) \\ \vec{\alpha}_j^{(p)} = +\dfrac{2}{3}\vec{e}_j\tau^{(p)}(x_j,t) + \omega_i^{(p)}(x_j,t)\vec{e}_k \\ \vec{\alpha}_k^{(p)} = -\dfrac{1}{3}\vec{e}_k\tau^{(p)}(x_j,t) + \omega_i^{(p)}(x_j,t)\vec{e}_j \end{cases} \tag{14.6}$$

Equation *(14.3)* can then be fully expressed in terms of $\phi_k^{(p)}$, $\phi_j^{(p)}$, $\omega_i^{(p)}$, and $\tau_0 + \tau^{(p)}$, by taking into account the fact that the density n depends on τ by the relation $n = n_0\,\mathrm{e}^{-\tau}$. The transverse component of this equation is written along the Ox_k axis as

$$\frac{d\phi_k^{(p)}}{dt} = 2\frac{K_2+K_3}{m}\frac{\partial \omega_i^{(p)}}{\partial x_j} = 2\frac{K_2+K_3}{mn_0}\mathrm{e}^{\tau_0+\tau^{(p)}}\frac{\partial \omega_i^{(p)}}{\partial x_j} \tag{14.7}$$

It's longitudinal component along axis Ox_j

$$\frac{d\phi_j^{(p)}}{dt} = \frac{1}{mn_0}\mathrm{e}^{\tau_0+\tau^{(p)}}\frac{\partial}{\partial x_j}\left[\left(\frac{4}{3}K_2 + 2K_1(1+\tau_0) - K_0\right)\tau^{(p)} + \left(K_1 + \frac{2}{3}K_2\right)(\tau^{(p)})^2 + 2(K_2+K_3)(\omega_i^{(p)})^2\right] \tag{14.8}$$

The transversal component can be associated with the geometro-kinetic equation for $\vec{\omega}^{(p)}$

$$\frac{d\vec{\omega}^{(p)}}{dt} = \frac{1}{2}\overrightarrow{\operatorname{rot}}\,\vec{\phi}^{(p)} \quad \Rightarrow \quad \frac{d\omega_i^{(p)}}{dt} = \frac{1}{2}\frac{\partial \phi_k^{(p)}}{\partial x_j} \tag{14.9}$$

For these perturbations, the material derivative can be written, if the local referential $Ox_1x_2x_3$ is in translation $\vec{\phi}_O(t)$ and in rotation $\dot{\vec{\omega}}_O(t)$ in the absolute referential as

$$\frac{d}{dt} = \frac{\partial}{\partial t} + \vec{\varphi}\vec{\nabla} = \frac{\partial}{\partial t} + \left(\vec{\phi}^{(p)}(\vec{r},t) - \vec{\phi}_O(t) - \dot{\vec{\omega}}_O(t)\wedge\vec{r}\right)\vec{\nabla} \tag{14.10}$$

Let's assume that the local referential is not in rotation

Hypothesis 2: *The local referential is not in rotation* $\Rightarrow \quad \dot{\vec{\omega}}_O(t) = 0$ *(14.11)*

Then as the perturbations $\vec{\phi}^{(p)}(\vec{r},t) = \phi_k^{(p)}(x_j,t)\vec{e}_k$ only have a component along the axis Ox_k, the relative velocity field $\vec{\varphi}^{(p)}(\vec{r},t) = \vec{\phi}^{(p)}(\vec{r},t) - \vec{\phi}_0(t)$ only has one component along Ox_k, worth $\vec{\varphi}^{(p)}(\vec{r},t) = \left[\phi_k^{(p)}(x_j,t) - \phi_k^{(p)}(x_j=0,t)\right]\vec{e}_k$, so that

$$\frac{d}{dt}=\frac{\partial}{\partial t}+(\vec{\varphi}\vec{\nabla})=\frac{\partial}{\partial t}+(\varphi_k\frac{\partial}{\partial x_k}) \tag{14.12}$$

But since all the propagation values vary only along Ox_j, the partial derivative along x_k does not feature in the equation, and the material derivative d/dt can be replaced by the partial derivative $\partial/\partial t$ with respect to time in whole of the local framework $Ox_1x_2x_3$.

The transversal perturbation are then governed by the following equations

$$\begin{cases}\dfrac{\partial\omega_i^{(p)}}{\partial t}=\dfrac{1}{2}\dfrac{\partial\phi_k^{(p)}}{\partial x_j}\\[2ex] \dfrac{\partial\phi_k^{(p)}}{\partial t}=2\dfrac{K_2+K_3}{mn_0}\mathrm{e}^{\tau_0+\tau^{(p)}}\dfrac{\partial\omega_i^{(p)}}{\partial x_j}\end{cases} \tag{14.13}$$

In this couple of equations, the value of $n=n_0\,\mathrm{e}^{-\tau}$ slightly depends on t and x_j since the propagation of the transversal wave is coupled to a longitudinal wavelet expressed as $\tau=\tau_0+\tau^{(p)}(x_j,t)$. By neglecting this dependence, assuming in first order that $\tau\cong\tau_0$, the weak transversal perturbations $\omega_i^{(p)}$ in the perfect solid with shear will satisfy the following linearized wave equation

$$\frac{\partial^2\omega_i^{(p)}}{\partial t^2}\cong\frac{K_2+K_3}{mn_0}\mathrm{e}^{\tau_0}\frac{\partial^2\omega_i^{(p)}}{\partial x_j^2}=\frac{K_2+K_3}{mn}\frac{\partial^2\omega_i^{(p)}}{\partial x_j^2} \tag{14.14}$$

which implies progressive waves

$$\omega_i^{(p)}(x_j,t)\cong\omega_{i0}^{(p)}\exp\left[i\left(k_t x_j-\omega t\right)\right]\quad\text{with}\quad k_t=\sqrt{\frac{mn}{K_2+K_3}}\omega=\mathrm{e}^{-\tau_0/2}\sqrt{\frac{mn_0}{K_2+K_3}}\omega \tag{14.15}$$

which propagates with a phase velocity approximately equal to

$$c_t=\frac{\omega}{k_t}\cong\sqrt{\frac{K_2+K_3}{mn}}=\mathrm{e}^{\tau_0/2}\sqrt{\frac{K_2+K_3}{mn_0}} \tag{14.16}$$

The longitudinal component associated with the propagation of the transversal perturbation is then obtained by using relation *(14.8)* and the following equation of geometro-kinetic

$$\frac{d\tau^{(p)}}{dt}=\operatorname{div}\vec{\phi}^{(p)}=\operatorname{div}\left(\phi_j^{(p)}\vec{e}_j\right)\quad\Rightarrow\quad\frac{d\tau^{(p)}}{dt}=\frac{\partial\phi_j^{(p)}}{\partial x_j} \tag{14.17}$$

This system of equations is evidently non-linear, but if we suppose weak amplitudes to $\tau^{(p)}$, we can obtain a couple of linearized equations in $\tau^{(p)}$

$$\begin{cases}\dfrac{\partial\tau^{(p)}}{\partial t}\cong\dfrac{\partial\phi_j^{(p)}}{\partial x_j}\\[2ex] mn\dfrac{\partial\phi_j^{(p)}}{\partial t}\cong\left(\dfrac{4}{3}K_2+2K_1\left(1+\tau_0\right)-K_0\right)\dfrac{\partial\tau^{(p)}}{\partial x_j}+2\left(K_2+K_3\right)\omega_i^{(p)}\dfrac{\partial\omega_i^{(p)}}{\partial x_j}\end{cases} \tag{14.18}$$

which reduces to a single differential equation of second order in $\tau^{(p)}$

$$mn\frac{\partial^2\tau^{(p)}}{\partial t^2}\cong\left(\frac{4}{3}K_2+2K_1\left(1+\tau_0\right)-K_0\right)\frac{\partial^2\tau^{(p)}}{\partial x_j^2}+2\left(K_2+K_3\right)\left[\left(\frac{\partial\omega_i^{(p)}}{\partial x_j}\right)^2+\omega_i^{(p)}\frac{\partial^2\omega_i^{(p)}}{\partial x_j^2}\right] \tag{14.19}$$

By expressing the term in $\omega_i^{(p)}$ from the transversal solution *(14.15)*, we obtain a differential equation for $\tau^{(p)}$ with a second member

$$mn\frac{\partial^2\tau^{(p)}}{\partial t^2}-\left(\frac{4}{3}K_2+2K_1(1+\tau_0)-K_0\right)\frac{\partial^2\tau^{(p)}}{\partial x_j^2}\cong -4(K_2+K_3)k_i^2\left(\omega_{i0}^{(p)}\right)^2\exp\left[i2(k_ix_j-\omega t)\right] \quad (14.20)$$

By introducing the following solution for $\tau^{(p)}$

$$\tau^{(p)}(x_j,t)\cong\tau_0^{(p)}\exp\left[i2(k_ix_j-\omega t)\right] \quad (14.21)$$

in the wave equation *(14.19)*, we have the relation between the amplitude $\tau_0{}^{(p)}$ of the longitudinal wavelet $\tau^{(p)}$ coupled to the transversal waves, as a function of amplitude $\omega_{i0}^{(p)}$ of this wave

$$\tau_0^{(p)}\cong-\frac{K_2+K_3}{K_2/3+2K_1(1+\tau_0)-K_0-K_3}\left(\omega_{i0}^{(p)}\right)^2 \quad (14.22)$$

Thus, the propagation of a linearly polarized transverse perturbation in the cosmic lattice satisfies a wave equation *(14.14)* which is completely conventional provided that its amplitude is not too strong. But it is always accompanied by a longitudinal wavelet which propagates in the same direction at the same velocity as the transversal disturbance. The frequency of this longitudinal wavelet is twice the frequency of the transverse perturbation and its magnitude is proportional to the square of the amplitude of the transverse perturbations. Note also that the velocity of propagation of transverse perturbation strongly depends on the background volume expansion τ_0 of the lattice as shown in equation *(14.16)*.

On the necessity to introduce a circular polarization to obtain a pure transversal propagation mode

Suppose now that we couple to this rotational wave polarized along Ox_i a rotational wave polarized along Ox_k , but out of phase by $\pm\pi/2$ with the same amplitude and so

$$\vec{\omega}(x_j,t)=\omega_{i0}^{(p)}\exp\left[i(k_ix_j-\omega t)\right]\vec{e}_i\pm i\omega_{i0}^{(p)}\exp\left[i(k_ix_j-\omega t)\right]\vec{e}_k \quad (14.23)$$

We will say that such a wave presents a left or right *circular polarization* depending on the sign $\pm$. We talk of *positive or negative helicity.* In this case, it is interesting to see what becomes of the longitudinal wavelets. Let's consider again the coupling term appearing in *(14.18)* and calculate it. We have

$$\omega_i^{(p)}\frac{\partial\omega_i^{(p)}}{\partial x_j}+\omega_k^{(p)}\frac{\partial\omega_k^{(p)}}{\partial x_j}=\left\{\begin{matrix}+ik_i\left(\omega_{i0}^{(p)}\right)^2\exp\left[i2(k_ix_j-\omega t)\right]\\-ik_i\left(\omega_{i0}^{(p)}\right)^2\exp\left[i2(k_ix_j-\omega t)\right]\end{matrix}\right\}=0 \quad (14.24)$$

We deduce that the *circularly polarized transversal waves are pure*, meaning that they are not coupled to longitudinal wavelets.

14.2 – Propagation of longitudinal waves

In a cosmic lattice, initially homogenous in volume expansion $(\tau=\tau_0)$ and without shear strains, satisfying the hypothesis *(14.11)*, we introduce a *perturbation* in the local referential $Ox_1x_2x_3$ under the form of a velocity field $\vec{\phi}=\vec{\phi}^{(p)}=\phi_j^{(p)}(x_j,t)\vec{e}_j$ parallel to the axis Ox_j and varying along the Ox_j axis. In the presence of this longitudinal perturbation, there does not exist a coupling term in equation *(14.3)* which could give rise to a transversal perturbation $\phi_k^{(p)}(x_j,t)\vec{e}_k$, so that the equations which govern the purely longitudinal perturbation is equation *(14.8)*, in which we remove the terms dependent on $\omega_i^{(p)}$, and equation *(14.17)* gives

$$\begin{cases} \dfrac{d\phi_j^{(p)}}{dt} = \dfrac{1}{mn_0} \mathrm{e}^{\tau_0+\tau^{(p)}} \dfrac{\partial}{\partial x_j}\left[\left(\dfrac{4}{3}K_2 + 2K_1(1+\tau_0) - K_0\right)\tau^{(p)} + \left(K_1 + \dfrac{2}{3}K_2\right)(\tau^{(p)})^2\right] \\ \dfrac{d\tau^{(p)}}{dt} = \dfrac{\partial \phi_j^{(p)}}{\partial x_j} \end{cases} \tag{14.25}$$

The fact that this system of equations is not linear in the variables $\phi_j^{(p)}$ and $\tau^{(p)}$ implies that the propagation of longitudinal waves in the perfect lattice is a complex process which, in addition, strongly depends on the background expansion state τ_0 of the lattice. One can hypothesize that if the disturbances are sufficiently low, we can neglect second order terms disturbances that appear in these equations directly, and indirectly through material derivatives. In this case, the small perturbations $\tau^{(p)}$ must obey the following linearized differential equation of second order

$$\frac{\partial^2 \tau^{(p)}}{\partial t^2} \cong \frac{1}{mn}\left[\frac{4}{3}K_2 + 2K_1(1+\tau_0) - K_0\right]\Delta\tau^{(p)} \tag{14.26}$$

We immediately deduce that the longitudinal waves only exist in a cosmic lattice if and only if the term in brackets is positive. In this case, for weak amplitudes, the longitudinal waves propagate with the following phase velocity

$$c_l \cong \sqrt{\frac{1}{mn}\left[\frac{4}{3}K_2 + 2K_1(1+\tau_0) - K_0\right]} = \mathrm{e}^{\tau_0/2}\sqrt{\frac{1}{mn_0}\left[\frac{4}{3}K_2 + 2K_1(1+\tau_0) - K_0\right]} \tag{14.27}$$

For larger amplitudes we cannot linearize the differential equations, so that the longitudinal waves become strongly non-linear and depend strongly on the amplitude of perturbations $\tau^{(p)}$.

14.3 – Localized longitudinal vibrational modes

In the cosmic lattice, if the phase velocity of longitudinal waves becomes an imaginary number, there is no longer propagation of longitudinal waves. In this case, we can rewrite the solution of complex disturbances in the form

$$\underline{\tau}^{(p)} = \tau_0^{(p)} \exp\left[-x_j/\delta\right]\exp\left[-i\omega t\right] \quad \text{si} \quad 4K_2/3 + 2K_1(1+\tau_0) - K_0 < 0 \tag{14.28}$$

in which the *spatial range δ of perturbations* is given by

$$\delta = \frac{1}{\omega}\sqrt{\frac{1}{mn}\left|\frac{4}{3}K_2 + 2K_1(1+\tau_0) - K_0\right|} = \frac{\mathrm{e}^{\tau_0/2}}{\omega}\sqrt{\frac{1}{mn_0}\left|\frac{4}{3}K_2 + 2K_1(1+\tau_0) - K_0\right|} \tag{14.29}$$

Here we have a surprising phenomenon, namely the appearance of localized eigenmodes of longitudinal vibrations, which do not propagate over long distances, but are instead confined over distance range of the order of δ. For large amplitudes, these localized modes of longitudinal disturbances will become non-linear and strongly dependent on the amplitude of the disturbances $\tau^{(p)}$.

14.4 – Analogy with Einstein gravitation and Quantum Mechanics

The various properties of the cosmic lattice as introduced in the previous chapter and this one are summarized in table 14.2. They can be compared with the similar properties obtained in a perfect usual isotropic solid, which are reported in Table 14.1.

The comparison of these two tables is quite telling, especially concerning the wave propagation speed in these two lattices. In fact, the propagation of shear waves is invariant (independent of lattice expansion τ) in the usual perfect solid while indirectly depends on the local volume expansion τ in the cosmic lattice, via the presence of the density of sites n in the expression of the velocity c_t . The same goes for the speed of the longitudinal waves, although there is a dependency to the multiplicative terms k_1 et K_1.

On the analogy with the helicity of photons

Another key difference between the two types of lattices resides in the fact that the linearly polarized transverse waves are perfectly pure and invariant in usual perfect solid whereas they are necessarily coupled to longitudinal wavelets in the cosmic lattice. In the case of cosmic lattice, the only transverse waves which are pure, not coupled to longitudinal wavelets are then right or left circularly polarized waves, that is to say the transverse waves of positive or negative helicity. Strangely, we have already found a property of photons in the real universe, namely that photons are necessarily of non-zero helicity! Since photons are quantum objects, here we find a surprising feature which will be discussed later.

The cosmic lattice introduced in the previous chapter is, as we have already said, a purely imaginary construct, in the sense that it is difficult to find a match with an existing real solid. But this lattice becomes very interesting when we revisit more thoroughly the analogies we have already seen emerge between the eulerian deformation theory and other theories of physics, such as the analogies with the Maxwell equations (table 12.1) or the Einstein equation $E_0^{vis} = M_0^{vis} c_t^2$ in the case of screw dislocations (section 12.3). In fact, the cosmic lattice will serve us primarily to pursue further our search for analogies with the great theories of physics and to issue different conjectures to narrow down these analogies.

On the analogy with the lack of longitudinal waves in Einstein gravitation

The existence of domains of volume expansion of the cosmic lattice in which the propagation of longitudinal waves is not possible, for $4K_2/3+2K_1(1+\tau_0)-K_0<0$, is a good analogy to the fact that there is no propagation of longitudinal waves in the theory of General Relativity of Einstein. Indeed, in the latter, gravitational waves are always transverse waves, defined as the spread of perturbations of the space-time metric. These waves have a tensorial symmetry, with two independent polarizations perpendicular to the direction of propagation, unlike the longitudinal disturbances that have a scalar symmetry.

The condition $4K_2/3+2K_1(1+\tau_0)-K_0<0$ that there is no longitudinal waves implies the existence of a *critical expansion* τ_{0cr} *of the lattice* between the domains where there is and there is not longitudinal waves. Hence the following conjecture, which of course admits the existence of pure transverse waves of circular polarization

Conjecture 1: *for the cosmic lattice to present analogies with Einstein Gravitation, with electromagnetism and with the photons of quantum mechanics, the following must hold:*

- $\exists$ *pure transverse waves circularly polarized* $\Leftrightarrow K_2 + K_3 > 0$
- $\nexists$ *longitudinal waves* $\Leftrightarrow \begin{cases} \tau_0 < \tau_{0cr} = \dfrac{K_0}{2K_1} - \dfrac{2K_2}{3K_1} - 1 & (K_1 > 0) \\ \tau_0 > \tau_{0cr} = \dfrac{K_0}{2K_1} - \dfrac{2K_2}{3K_1} - 1 & (K_1 < 0) \end{cases}$ *(14.30)*

This conjecture implies that it is only in the domain where the background volume expansion τ_0 and the moduli are such that relations *(14.30)* are satisfied that we can find analogies to General Relativity, Electromagnetism and Quantum Physics.

On the analogy with quantum gravity and the quantum fluctuation of the vacuum

In the absence of longitudinal waves, the cosmic lattice presents localized eigenmodes of longitudinal disturbances, and so local variations of the scalar τ of volume expansion. Such modes immediately remind us of the ideas of quantum fluctuations of gravity at very small scale since they affect the scalar τ which undeniably has a link to the gravitational field. But these localized disturbances of the scalar volume expansion are also reminiscent of the quantum vacuum fluctuations described by quantum physics. We can therefore, based on this analogy between the gravitational field and the expansion field τ, ask the question: "Is it necessary to quantify gravity at small scale, or rather, is it gravitation which is, at very small scale, responsible for quantum physics? " We try to bring in the following chapters a few answers to this relevant topic.

Table 14.1 - Waves and eigenmodes in a perfect solid

Function and equations of state

$$f^{él} = -k_0\tau + k_1\tau^2 + k_2\sum_i (\vec{\alpha}_i^{\,él})^2$$

$$\Rightarrow \quad \begin{cases} p = -n\dfrac{\partial f^{él}}{\partial \tau} = nk_0 - 2nk_1\tau \\ \vec{s}_i = n\dfrac{\partial f^{él}}{\partial \alpha_{ik}^{él}}\vec{e}_k = 2nk_2\vec{\alpha}_i^{\,él} \end{cases}$$

Newton's equation

$$\frac{d\vec{p}}{dt} = -2k_2\overrightarrow{\mathrm{rot}}\,\vec{\omega}^{él} + \left[\frac{4}{3}k_2 + 2k_1(1-\tau) + k_0\right]\overrightarrow{\mathrm{grad}}\,\tau \underbrace{- 2k_2\sum_k\left(\vec{e}_k\overrightarrow{\mathrm{grad}}\,\tau\right)\vec{\alpha}_k^{\,él}}_{\text{couplage entre } \vec{\alpha}_k^{\,él} \text{ et } \overrightarrow{\mathrm{grad}}\,\tau} + 2k_2\vec{\lambda}$$

Propagation of pure 'invariant' transversal waves

$$\omega_i^{(p)}(x_j,t) \cong \omega_{i0}^{(p)}\exp\left[i\left(k_t x_j - \omega t\right)\right]$$

$$c_t = \frac{\omega}{k_t} = \sqrt{\frac{k_2}{m}}$$

Propagation of pure longitudinal waves

$$\underline{\tau}^{(p)} = \tau_0^{(p)}\exp\left[i\left(k_l x_j - \omega t\right)\right] \quad \textit{si} \quad \frac{4}{3}k_2 + 2k_1(1-\tau_0) + k_0 > 0$$

$$c_l \cong \frac{\omega}{k_l} \cong \sqrt{\frac{1}{m}\left[\frac{4}{3}k_2 + 2k_1(1-\tau_0) + k_0\right]}$$

Localized longitudinal eigenmodes

$$\underline{\tau}^{(p)} = \tau_0^{(p)}\exp\left[-x_j/\delta\right]\exp\left[-i\omega t\right] \quad \textit{if} \quad \frac{4}{3}k_2 + 2k_1(1-\tau_0) + k_0 < 0$$

$$\delta = \frac{1}{\omega}\sqrt{\frac{1}{m}\left|4k_2/3 + 2k_1(1-\tau_0) + k_0\right|}$$

Table 14.2 - Waves and eigenmodes in the cosmic lattice

Function and equations of state

$$F^{déf} = -K_0\tau + K_1\tau^2 + K_2\sum_i(\vec{\alpha}_i^{él})^2 + 2K_3(\vec{\omega}^{él})^2 + K_1^{an}\sum_i(\vec{\alpha}_i^{an})^2 + 2K_2^{an}(\vec{\omega}^{an})^2 \Rightarrow f^{déf} = \frac{F^{déf}}{n}$$

$$\Rightarrow \begin{cases} p = -n\dfrac{\partial f^{déf}}{\partial \tau} = K_0 - 2K_1\tau - F^{déf} \\ \vec{s}_i = n\dfrac{\partial f^{déf}}{\partial \alpha_{ik}^{él}}\vec{e}_k = 2K_2\vec{\alpha}_i^{él} \\ \vec{m} = n\dfrac{\partial f^{déf}}{\partial \omega_i^{él}}\vec{e}_i = 4K_3\vec{\omega}^{él} \end{cases} \quad \text{et} \quad \begin{cases} \vec{s}_i^{cons} = n\dfrac{\partial f^{déf}}{\partial \alpha_{ik}^{an}}\vec{e}_k = 2K_1^{an}\vec{\alpha}_i^{an} \\ \vec{m}^{cons} = n\dfrac{\partial f^{déf}}{\partial \omega_i^{an}}\vec{e}_i = 4K_2^{an}\vec{\omega}_i^{an} \end{cases}$$

Newton's equation

$$n\frac{d\vec{p}}{dt} = -2(K_2+K_3)\overrightarrow{\mathrm{rot}}\,\vec{\omega}^{él} + \left(\frac{4}{3}K_2 + 2K_1\right)\overrightarrow{\mathrm{grad}}\,\tau + 2K_2\vec{\lambda} + nm\vec{\phi}_I\frac{dC_I}{dt} - nm\vec{\phi}_L\frac{dC_L}{dt}$$
$$+\ \overrightarrow{\mathrm{grad}}\underbrace{\left(K_2\sum_i(\vec{\alpha}_i^{él})^2 + K_1^{an}\sum_i(\vec{\alpha}_i^{an})^2 + 2K_3(\vec{\omega}^{él})^2 + 2K_2^{an}(\vec{\omega}^{an})^2 + K_1\tau^2 - K_0\tau\right)}_{F^{déf}}$$

Propagation of transversal waves coupled to longitudinal wavelets

$$\begin{cases} \omega_i^{(p)}(x_j,t) \cong \omega_{i0}^{(p)}\exp\left[i\left(k_t x_j - \omega t\right)\right] \\ \tau^{(p)}(x_j,t) \cong \tau_0^{(p)}\exp\left[i\left(2k_t x_j - 2\omega t\right)\right] \end{cases}$$

$$c_t \cong \sqrt{\frac{K_2+K_3}{mn}} = e^{\tau_0/2}\sqrt{\frac{K_2+K_3}{mn_0}} \quad \text{et}$$

$$\tau_0^{(p)} \cong -\frac{K_2+K_3}{K_2/3 + 2K_1(1+\tau_0) - K_0 - K_3}\left(\omega_{i0}^{(p)}\right)^2$$

Propagation of pure longitudinal waves

$$\underline{\tau}^{(p)} = \tau_0^{(p)}\exp\left[i\left(k_l x_j - \omega t\right)\right] \quad \textit{if} \quad 4K_2/3 + 2K_1(1+\tau_0) - K_0 > 0$$

$$c_l \cong \sqrt{\frac{1}{mn}\left[\frac{4}{3}K_2 + 2K_1(1+\tau_0) - K_0\right]} = e^{\tau_0/2}\sqrt{\frac{1}{mn_0}\left[\frac{4}{3}K_2 + 2K_1(1+\tau_0) - K_0\right]}$$

Localized longitudinal eigenmodes

$$\underline{\tau}^{(p)} = \tau_0^{(p)}\exp\left[-x_j/\delta\right]\exp\left[-i\omega t\right] \quad \textit{if} \quad 4K_2/3 + 2K_1(1+\tau_0) - K_0 < 0$$

$$\delta = \frac{1}{\omega}\sqrt{\frac{1}{mn}\left|\frac{4}{3}K_2 + 2K_1(1+\tau_0) - K_0\right|} = \frac{e^{\tau_0/2}}{\omega}\sqrt{\frac{1}{mn_0}\left|\frac{4}{3}K_2 + 2K_1(1+\tau_0) - K_0\right|}$$

Chapter 15

Curvature of wave rays by a singularity of expansion and black holes

Among the surprising behavior that may be present in a cosmic lattice is the curvature of wave rays by a volume expansion gradient due to the presence of a strong topological singularity of expansion τ. This curvature can lead to the formation of *"black holes"* absorbing all waves passing in its vicinity, or impenetrable *"white holes"* pushing all the waves away from its vicinity.

15.1 – Non-dispersive curvature of wave rays

In a cosmic lattice, the presence of a non-null gradient of volume expansion will give rise to a non-dispersive curvature in the propagation of wave-rays which we will now calculate.

On the curvature of waves in the presence of a singularity of volume expansion

The fact that transverse *(14.16)* and longitudinal *(14.25)* waves phase velocities increase non-linearly with the value of the static volume expansion τ via the value of the site density n will lead to a curvature in the propagation of these waves if they pass in the direct neighborhood of a singularity of volume expansion within the lattice, as shown in figure 15.1.

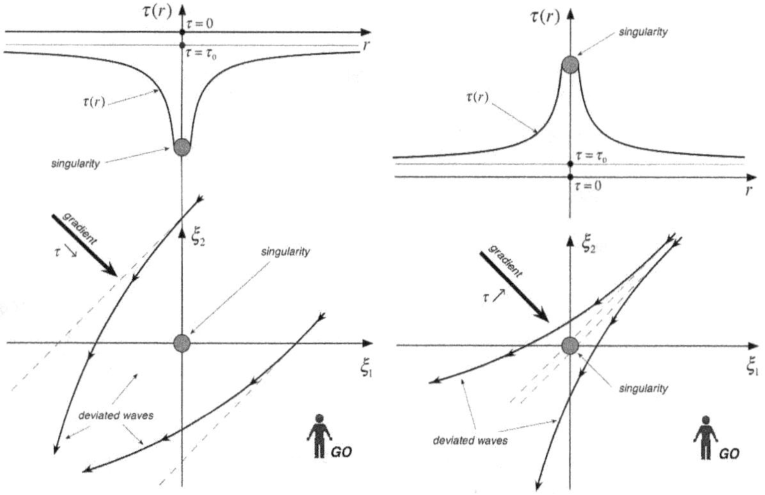

Figure 15.1 - *curvature of wave rays in the vicinity of a singularity of volume expansion* τ *with spherical symmetry*

Indeed, imagine a stationary cosmic lattice in the absolute reference frame of the observer ***GO*** containing a volume expansion singularity of spherical symmetry located in the center of the coordinate system $O\xi_1\xi_2\xi_3$. Consider also a longitudinal or transverse wave, initially plane, arriving on this singularity along the $O\xi_1$ axis. The propagation velocity increases or decreases, approaching the singularity, under the effect of the lattice density $n = n_0\, e^{-\tau}$. Depending on whether the singularity is "positive" (reaches a maximum at the origin) or "negative" (passes through a minimum at the origin), there will appear a curvature of the wave rays, and the rays seem repulsed by a "positive singularity" or attracted by a "negative singularity".
This phenomenon does not depend on the shape of the field around the singularity, but only on its gradient. For a plane wave incident on the singularity, this phenomenon of acceleration or braking of the wave will then produce a deformation of the plane wave similar to the effect of a negative lens in the case of a "positive singularity" or a converging lens in the case of a "negative singularity". Furthermore, as this phenomenon does not depend on the frequency of the wave, the singularity behaves like a *non-dispersive*, converging or diverging lens in the cosmic lattice.

15.2 – Perturbation sphere and 'black holes'

Now imagine that in a stationary cosmic lattice in the absolute reference frame of the observer ***GO*** and containing a "negative singularity" of the volume expansion of spherical symmetry, located in the center of the coordinate system $O\xi_1\xi_2\xi_3$, we have a transverse ($c_i = c_t$) or longitudinal ($c_i = c_l$) wave in the vicinity of the singularity at a distance $r = r_{cr}$ from the origin of the singularity such that the following relation is satisfied

$$\frac{c_i(r_{cr}+dr)}{c_i(r_{cr})} = \frac{r_{cr}+dr}{r_{cr}} \tag{15.1}$$

In this case, the wave planes adjacent to r_{cr} will always be parallel to a line passing through the origin, so that the radius of the transverse or longitudinal wave is in fact $r = r_{cr}$, a circle centered on the origin!
The condition *(15.1)* can be explained. Indeed, it is easy to show that this condition entails that

$$\left.\frac{\partial c_i}{\partial r}\right|_{r_{cr}} = \frac{c_i(r_{cr})}{r_{cr}} \tag{15.2}$$

Thus, if a transverse or longitudinal wave passes at a distance $r \le r_{cr}$ satisfying this relationship, it becomes impossible for it to escape from the virtual sphere radius r_{cr}. If the field of the singularity is an increasing monotone gradient from its origin, wave rays of curvature located within this critical area will be further enhanced, so that all these waves will be permanently trapped by the singularity. By analogy with the *"photon sphere"* around a *black hole* in general relativity, we will call *"sphere of transverse and longitudinal perturbations"* the layer at a distance $r = r_{cr}$ from the heart of the singularity. It is clear that the existence of such a *sphere of perturbations* is subject to the condition that it is located outside of the *"object"* responsible for the negative singularity of the expansion field. If the radius of this *"object"* is R, with the speed of transverse waves given by *(14.16)* or longitudinal waves given by *(14.25)*, we deduce the conditions of existence of a *"black hole"* using *(15.2)*

Existence conditions of a "black hole": $\left.\frac{\partial \tau(r)}{\partial r}\right|_{r_{cr}} = \frac{2}{r_{cr}}$ and $r_{cr} > R$ (15.3)

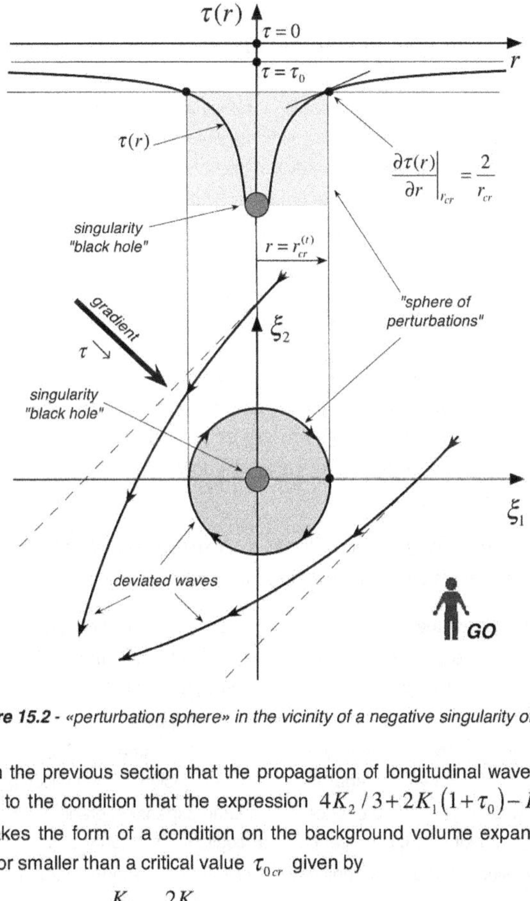

Figure 15.2 *- «perturbation sphere» in the vicinity of a negative singularity of τ*

We have seen in the previous section that the propagation of longitudinal waves in the perfect lattice is subject to the condition that the expression $4K_2/3+2K_1(1+\tau_0)-K_0$ be positive. This condition takes the form of a condition on the background volume expansion τ_0, which must be greater or smaller than a critical value τ_{0cr} given by

$$\begin{cases} K_1 > 0 \quad \Rightarrow \quad \tau_0 > \tau_{0cr} = \dfrac{K_0}{2K_1} - \dfrac{2K_2}{3K_1} - 1 \\[2ex] K_1 < 0 \quad \Rightarrow \quad \tau_0 < \tau_{0cr} = -\dfrac{K_0}{2|K_1|} + \dfrac{3K_2}{3|K_1|} - 1 \end{cases} \qquad (15.4)$$

If propagation of longitudinal waves is possible in the lattice, that is to say if these relations are satisfied, then the longitudinal waves will also suffer the trapping phenomenon at the boundary $r = r_{cr}$.

In the case where $K_1 > 0$ there still occurs another phenomena. Indeed, if the singularity is of increasing monotone gradient from its origin, there may be a surrounding radius $r = r_{cr2}^{(l)} < r_{cr}$ beyond which the value $\tau(\vec{r})$ falls below τ_{0cr}, so any longitudinal wave initially trapped in the

limit $r = r_{cr}$, will then reach the second limit $r = r_{cr2}^{(l)} < r_{cr}$ beyond which it can no longer propagate, it will then increase the longitudinal vibration eigenmodes localized within this volume.
In the case where $K_1 < 0$, the same phenomenon does not exist since the existence of a propagation implies that $\tau(r) < \tau_0 < \tau_{0cr}$.

15.3 – Analogy with Einstein's gravitation

The cosmic lattice presents a very interesting analogy with the General Theory of Gravitation of Einstein. In the vicinity of singularities of volume expansion, the *spheres of disturbances* are very similar to the *photon sphere* surrounding a black hole. We therefore deduce from this *non-dispersive effect of curvature of rays by gradients of the volume expansion that the scalar volume expansion undoubtedly has a strong analog relationship with the gravitational field* in General Relativity.

On the analogy with black holes in general relativity

It is also interesting to note that only a negative singularity of τ has properties similar to that of a "black hole", namely catching all the waves passing by its vicinity, while a positive singularity would behave like a "white hole", that is to say as an entity that would repel the waves and which could not be penetrated by waves. Hence the following conjecture has to be satisfied for our analogy with Gravitation

Conjecture 2: *the usual singularities of the field of expansion must be 'negative' for them to correspond to the usual gravitational field* *(15.5)*

It is also remarkable that the curvature of the waves by a gradient of volume expansion and the existence of a localized sphere of perturbartions around a singularity of volume expansion is exclusively due to *the development of the free energy per unit volume* that we have used for the cosmic lattice. Indeed, if we look more closely at what happens in the case of conventional perfect solid, for which it is the free energy per lattice site that is developed, we find that the speed of transverse waves is *"invariant"* regardless of the expansion of the lattice, which does not lead to a bending of wave rays in the presence of a volume expansion gradient, or the appearance of a sphere of perturbations in the presence of a localized singularity of volume expansion. This analogy thus justifies a fortiori the conjecture 0 *(13.5)* that we formulated in section 13.2, since such analogy cannot appear within the usual perfect solid.

Chapter 16

Cosmological evolution of a finite perfect lattice

Considering a finite imaginary sphere of a perfect solid or a cosmic lattice, we can introduce the concept of *"cosmological evolution"* of the lattice, assuming that one injects a certain amount of kinetic energy inside the lattice. In this case, the lattice has strong temporal variations of its volume expansion, that can be modeled very simplistically assuming that volume expansion remains perfectly homogeneous throughout the lattice during its evolution.

16.1 – Cosmological behavior of a sphere of perfect solid

Let's imagine that in an absolute referential $O\xi_1\xi_2\xi_3$, the ***GO*** observes a solid, of spherical form, of radius R_U, made of a lattice of N nodes (fig. 16.1). Let's assume this solid possesses a homogeneous volume expansion of the background with depends on time in the form

$$\tau(t) = \tau_0(t) \neq \tau_0(\vec{\xi},t) \tag{16.1}$$

In this case, the ***GO*** will observe that the radius R_U of the sphere will depend on time

$$R_U = R_U(t) \tag{16.2}$$

and that this sphere will expand or contract. This behavior will be described as a *"cosmological behaviour"* by analogy to the *theories of the cosmological expansion of the Universe.* We assume that the total energy $\boldsymbol{E}$ of the solid is a constant. It is made up of the elastic energy $\boldsymbol{F}^{él}(\tau)$ and the kinetic energy $\boldsymbol{T}(\tau)$ of expansion.

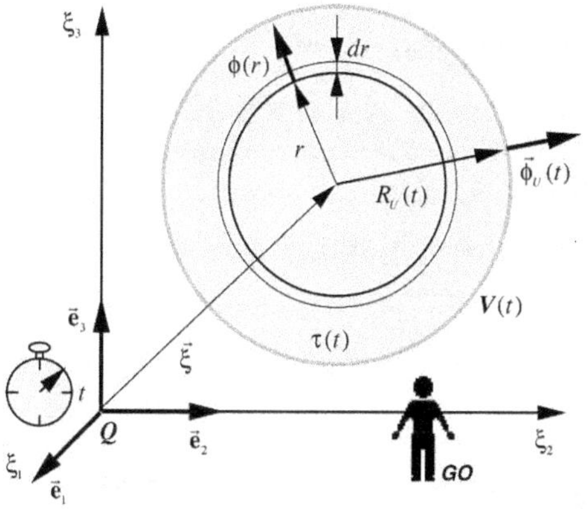

Figure 16.1 *- "Cosmological" volume expansion $\tau(t)$ of an imaginary solid sphere*

The total kinetic energy $\boldsymbol{T}$ of volume expansion is linked to the speed of expansion which we can characterize by the velocity $\vec{\phi}_U(t)$ of the surface of the sphere (fig. 16.1). The kinetic energy $\boldsymbol{T}$ can then be obtained by integrating on all the sphere and considering the energy located in the lattice contained in the volume between radius r and $r+dr$. The velocity of expansion of the surface is

$$\phi(r)=\phi_U\,\frac{r}{R_U} \tag{16.3}$$

since the volume expansion τ was supposed homogeneous. And using the fact that the density of sites n is given by

$$n=n_0\,\mathrm{e}^{-\tau}=\frac{N}{V_U}=\frac{3N}{4\pi R_U^3} \tag{16.4}$$

we have for the kinetic energy

$$\boldsymbol{T}=\int_0^{R_U}\frac{1}{2}mn\phi^2(r)4\pi r^2dr=\frac{3Nm\phi_U^2}{2R_U^5}\int_0^{R_U}r^4\,dr=\frac{3}{10}Nm\phi_U^2 \tag{16.5}$$

We deduce that the velocity of expansion ϕ_U is proportional to $\sqrt{\boldsymbol{T}}=\sqrt{\boldsymbol{E}-\boldsymbol{F}^{él}}$

$$\phi_U(\tau)=\sqrt{\frac{10}{3Nm}\boldsymbol{T}(\tau)}=\sqrt{\frac{10}{3Nm}\left(\boldsymbol{E}-\boldsymbol{F}^{él}(\tau)\right)} \tag{16.6}$$

We also deduce that

$$\frac{d\phi_U(\tau)}{d\tau}=-\frac{5}{3Nm}\frac{d\boldsymbol{F}^{él}(\tau)/d\tau}{\phi_U(\tau)} \tag{16.7}$$

and so, as a consequence, the derivative $d\phi_U/d\tau$ tends towards $\pm\infty$ if ϕ_U goes to zero and $d\boldsymbol{F}^{él}/d\tau$ is finite and not null. Let's study the cosmological behavior of perfect solids and cosmic lattices that we have previously defined.

16.2 – Cosmological evolution of a perfect solid

Cosmological evolution of a perfect solid with $k_0=0$ ***and*** $k_1>0$

To simplify the problem, hypothesize that there is no shear strains, which is perfectly plausible since the volume expansion is assumed to be homogeneous, so that, for a perfect solid, one can calculate the energy $\boldsymbol{F}^{él}(\tau)$ as a function of expansion τ from equation *(13.1)*

$$\boldsymbol{F}^{él}=Nf^{él}=\underbrace{-Nk_0\tau}_{k_0=0}+Nk_1\tau^2+\underbrace{Nk_2\sum_i(\vec{\alpha}_i^{él})^2}_{\vec{\alpha}_i^{él}=0}=Nk_1\tau^2 \tag{16.8}$$

If no phenomenon has dissipated total energy $\boldsymbol{E}$, for example as heat, energy will be a conserved quantity.

If the modulus k_1 of the perfect solid is positive, it will only be able to oscillate between a minimum volume expansion $\tau_{\min}$ and a maximum volume expansion $\tau_{\max}$, as illustrated in figure 16.2.

If we show in the diagrams of $\boldsymbol{F}^{él}(\tau)$ and $\vec{\phi}_U(\tau)$, the critical value $\tau_{0cr}=1+2k_2/3k_1>1$ *above which longitudinal waves exist in the perfect lattice*, we notice that during it's 'cosmological evolution', the solid will transition between a domain ($\tau \leq \tau_{0cr}$) where there are both longitudinal and transverse waves and a domain ($\tau \geq \tau_{0cr}$) where there are only transverse waves but localized longitudinal vibration eigenmodes.

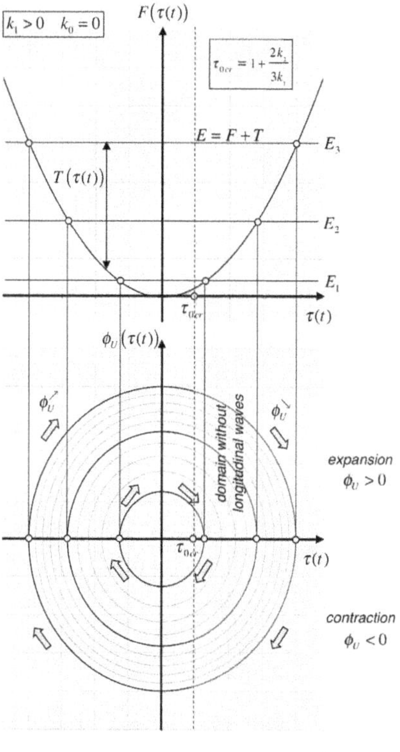

Figure 16.2 *- "cosmological" behavior of elastic energy $F^{él}(\tau)$ of expansion and velocity $\vec{\phi}_U(\tau)$ of expansion of a perfect imaginary solid with $k_1 > 0$*

Cosmological evolution of a perfect solid with $k_0 = 0$ and $k_1 < 0$

For a perfect imaginary solid where modulus k_1 would be negative, depending on the value of the total energy E, this solid could have many different *"cosmological behaviors"* as illustrated in figure 16.3:

- if $E \leq 0$, it can contract and expand in a oscillating fashion between $\tau \to -\infty$ and τ_1 or expand indefinitely from a value τ_2. It should be noted that it is hard to imagine a solid which would evolve by contracting from $\tau = \infty$, which is why we classify these behavior in a 'grey zone',
- if $E \geq 0$, it can dilate indefinitely from $\tau \to -\infty$. In this case also, the longitudinal waves disappear as soon as $\tau \geq \tau_{0cr} = 1 - 2k_2 / 3|k_1| < 1$.

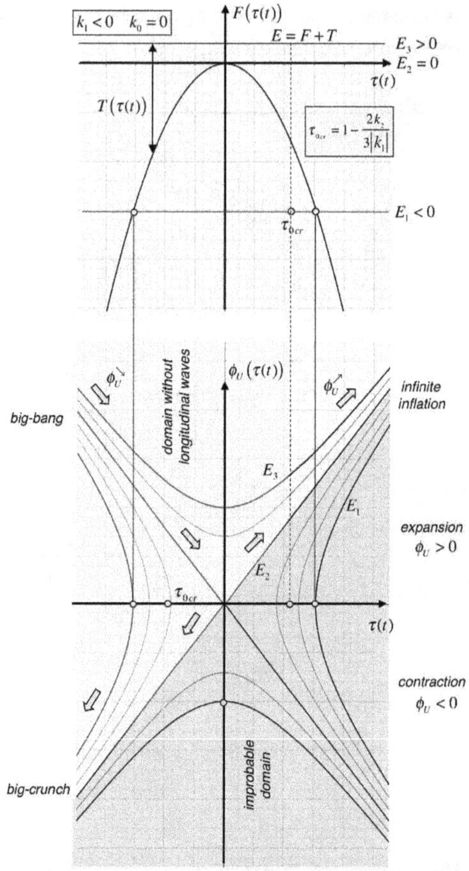

Figure 16.3 *- "cosmological" behavior of elastic energy $F^{él}(\tau)$ of expansion and velocity $\vec{\phi}_U(\tau)$ of expansion of a perfect imaginary solid with $k_1 < 0$*

On the analogy with the cosmic evolution of our Universe

The various *"cosmological behaviors"* deduced for a perfect solid can be compared with the cosmological behavior that is assigned to our real Universe. Indeed, in the case of the real universe, we have a system that does not have longitudinal waves, as in the theory of general relativity, and which follows a cosmological evolution that is in several stages: a "big bang" from a singularity of space-time, followed by a period of very rapid inflation, a slowdown in inflation, followed, according to very recent observations, by an expansion whose speed seems to increase over time. This last point is one that would correspond to the present state of our Universe.

Amongst the "cosmological behaviors" derived for the perfect solid, only the perfect solid with $k_1 < 0$ presents some analogy with the cosmological behavior of the real Universe. Indeed, the perfect solid with $k_1 < 0$, if E is greater than zero (fig. 16.3), goes through all the stage, the big bang, inflation, slowing of inflation and growing at steady velocity in the domain where there are no longitudinal waves. But for this solid, the stage of increasing speed expansion inevitably continues towards $\tau \to +\infty$.
Also note that the elastic energy $E^{él}$ in the solid lattice has a very interesting analogy with the concept of *"dark energy"* of astrophysicists used to explain the increase in the rate of expansion of the Universe, since it is this elastic energy that is responsible for the expansion of the solid by increasing speed through the the modulus $k_1 < 0$ of the solid.

16.3 – Cosmological evolution of a cosmic lattice

The cosmological evolution of a cosmic lattice with $K_0 = 0$ *and* $K_1 > 0$

Imagine that in the absolute framework $O\xi_1\xi_2\xi_3$, the ***GO*** observes a *cosmic lattice* of spherical radius R_U , consisting of a lattice with N meshes (fig. 16.1). The elastic energy $\boldsymbol{F}^{él}$ is written, by using *(13.7)*

$$\boldsymbol{F}^{él} = Nf^{él} = \underbrace{-\frac{NK_0}{n}\tau}_{K_0=0} + \frac{NK_1}{n}\tau^2 + \underbrace{\frac{NK_2}{n}\sum_i(\vec{\alpha}_i^{él})^2}_{\vec{\alpha}_i^{él}=0} + \underbrace{\frac{NK_3}{n}(\vec{\omega}^{él})^2}_{\vec{\omega}^{él}=0} = \frac{NK_1}{n}\tau^2 = \frac{NK_1}{n_0}\tau^2 e^{\tau} \qquad (16.9)$$

To plot the behavior of $\boldsymbol{F}^{él}$ as a function of τ for this lattice, we must look for the extremas of $\boldsymbol{F}^{él}(\tau)$

$$\frac{d\boldsymbol{F}^{él}}{d\tau} = \frac{NK_1}{n_0}(2+\tau)\tau e^{\tau} = 0 \quad \Rightarrow \quad \tau = 0, -2 \ et\ -\infty \qquad (16.10)$$

If the modulus K_1 *of the lattice is positive,* then for $\tau = 0$ and $\tau \to -\infty$, the value of $\boldsymbol{F}^{él}$ tends towards two minimas equal to zero, while for $\tau = -2$, the value of $\boldsymbol{F}^{él}$ goes through a minimum equal to $\boldsymbol{F}^{él}_{max} = 4e^{-2}NK_1/n_0 \cong 0{,}54NK_1/n_0$. The graph of $\boldsymbol{F}^{él}(\tau)$ is shown in figure 16.4a, and we can see that it is very different from the case of a lattice of usual perfect solid.
There are three modes of oscillation depending on the value of $\boldsymbol{E}$, as illustrated in figure 16.4:
- if $\boldsymbol{E} \le \boldsymbol{F}^{él}_{max} = 4e^{-2}NK_1/n_0$, there are two modes of oscillation possible, a first mode between $\tau \to -\infty$ and $\tau_1 < 0$ and a second mode between $\tau_2 < 0$ and $\tau_3 > 0$,
- if $\boldsymbol{E} \ge \boldsymbol{F}^{él}_{max} = 4e^{-2}NK_1/n_0$, there is a third mode of oscillation possible, between $\tau \to -\infty$ and $\tau_4 > 0$.

In the graphs of figure 16.4, we can show the limit $\tau_{0cr} = -1 - 2K_2/3K_1 < -1$ taken from relation *(14.26)*. We show a value of τ_{0cr} close to -1, corresponding to the case where $K_1 >> K_2$.
There are again domains of different behaviors of the solid: a domain where there coexists transverse and longitudinal waves (for $\tau \ge \tau_{0cr}$) and a domain where there are only transverse waves and localized vibrational eigenmodes (for $\tau \le \tau_{0cr}$) . But unlike the perfect solid, in the *cosmic lattice*, the position of these domains is reversed along the τ axis!

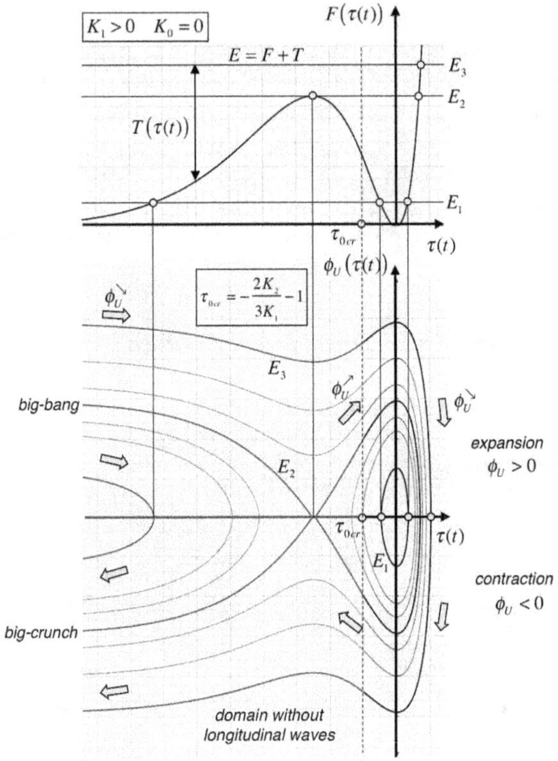

Figure 16.4 *- "cosmological" behavior of elastic energy $F^{él}(\tau)$ of expansion and velocity $\vec{\phi}_U(\tau)$ of expansion of an imaginary cosmic lattice with $K_1 > 0$*

The cosmological evolution of a cosmic lattice with $K_0 = 0$ and $K_1 < 0$

If the modulus K_1 of the lattice is negative, the plastic energy $F^{él}(\tau)$ presents two maximas for $\tau \to -\infty$ and $\tau = 0$, and a minima for $\tau = -2$. Furthermore, $F^{él}(\tau) \to -\infty$ for $\tau \to \infty$ (fig. 16.5). We can consider here three different cases following the value of total energy E :

- if $E < 0$, there exists a *"cosmological solution"* around the value $\tau = -2$, for which the lattice contacts and expands indefinitely between values $\tau_1 < 0$ and $\tau_2 < 0$, and a second solution for which the lattice dilates indefinitely, at constant velocity, from the value $\tau_3 > 0$,

- if $E = 0$, there exists a solution for values inferior to $\tau = 0$, and a solution for the superior values. The lattice can dilate from $\tau \to -\infty$, and can afterwards, either contract towards $\tau \to -\infty$ and start the cycle again, or dilate indefinitely towards $\tau \to +\infty$,

- if $E > 0$, there exists a unique solution for which the lattice dilates one time only from

$\tau \to -\infty$ to $\tau \to +\infty$. The symmetric solution which would consist in the lattice contracting form $\tau = +\infty$, where the lattice possesses a phenomenal kinetic energy of contraction, to $\tau \to -\infty$ is not prohibited but it is seems strongly improbable!

In the case of this lattice, we also notice the existence of domains of different behavior: for $\tau > \tau_{0cr} = -1 + 2K_2 / 3|K_1| > -1$ a domain where there are both transverse and longitudinal waves, and for $\tau < \tau_{0cr} = -1 + 2K_2 / 3|K_1| > -1$ a domain where there are only transverse waves and localized longitudinal eigenmodes of vibration.

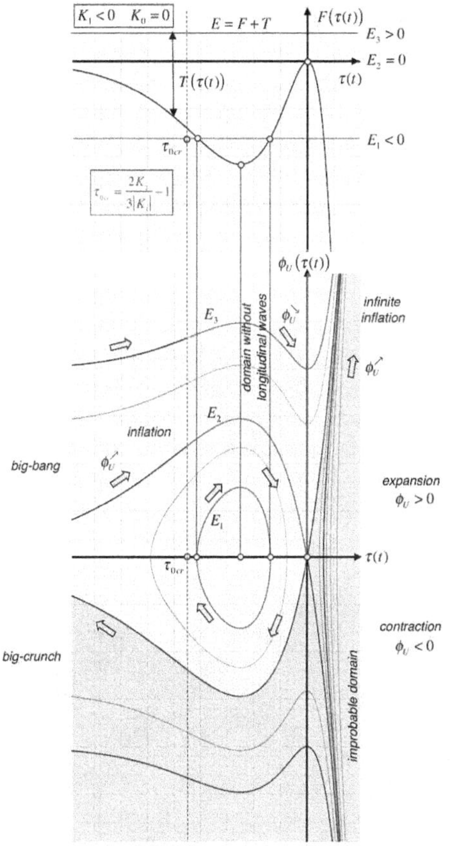

***Figure 16.5** - "cosmological behavior" of elastic energy $F^{él}(\tau)$ of expansion and velocity $\bar{\phi}_U(\tau)$ of expansion of an imaginary cosmic lattice with $K_1 < 0$*

The cosmological evolution of a cosmic lattice with $K_1 = 0$ and $K_0 < 0$

If, in the absolute referential $O\xi_1\xi_2\xi_3$ of the **GO** we consider a cosmic lattice with $K_1 = 0$ and $K_0 < 0$, the elastic energy of expansion $F^{él}(\tau)$ is written

$$F^{él} = Nf^{él} = -\frac{NK_0}{n}\tau + \underbrace{\frac{NK_1}{n}\tau^2}_{K_1=0} + \underbrace{\frac{NK_2}{n}\sum_i(\vec{\alpha}_i^{él})^2}_{\vec{\alpha}_i^{él}=0} + \underbrace{\frac{NK_3}{n}(\vec{\omega}^{él})^2}_{\vec{\omega}^{él}=0} = -\frac{NK_0}{n}\tau = -\frac{NK_0}{n_0}e^{\tau}\tau \quad (16.11)$$

To plot the behavior of $F^{él}(\tau)$ as a function of τ for this solid, we must seek the extremes of $F^{él}(\tau)$

$$\frac{dF^{él}(\tau)}{d\tau} = -\frac{NK_0}{n_0}(1+\tau)e^{\tau} = 0 \quad \Rightarrow \quad \tau = -1 \ \ et \ \ \tau \to -\infty \quad (16.12)$$

If the modulus K_0 of the solid is negative, the energy $F^{él}(\tau)$ as a function of τ presents a minimum for $\tau = -1$ as illustrated in figure 16.6.

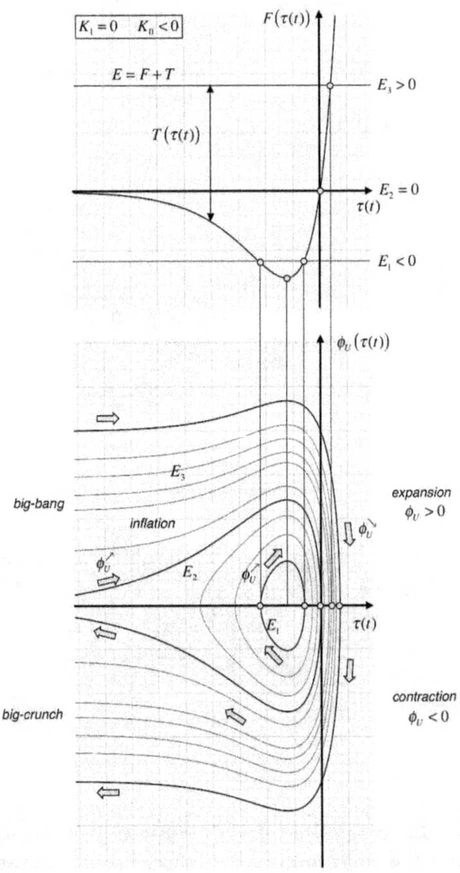

***Figure 16.6** - "cosmological behavior" of elastic energy $F^{él}(\tau)$ of expansion and velocity $\vec{\phi}_U(\tau)$ of expansion of an imaginary cosmic lattice with $K_0 < 0$*

We then deduce the "cosmological behavior" of this type of lattice and show it in figure 16.6:

- if $E<0$, the lattice oscillates indefinitely on a closed trajectory between minimum τ_{min} and maximum τ_{max},
- if $E \geq 0$, the lattice oscillates between $\tau \to -\infty$ and a maximum τ_{max}.

In this case the lattice still presents longitudinal waves as since $K_0<0$, we also have $K_0<4K_2/3$.

The cosmological evolution of a cosmic lattice with $K_1=0$ and $K_0>0$

If the K_0 modulus of a cosmic lattice is positive, the energy $E^{él}(\tau)$ as a function of τ presents a maximum for $\tau=-1$ as illustrated in figure 16.7.

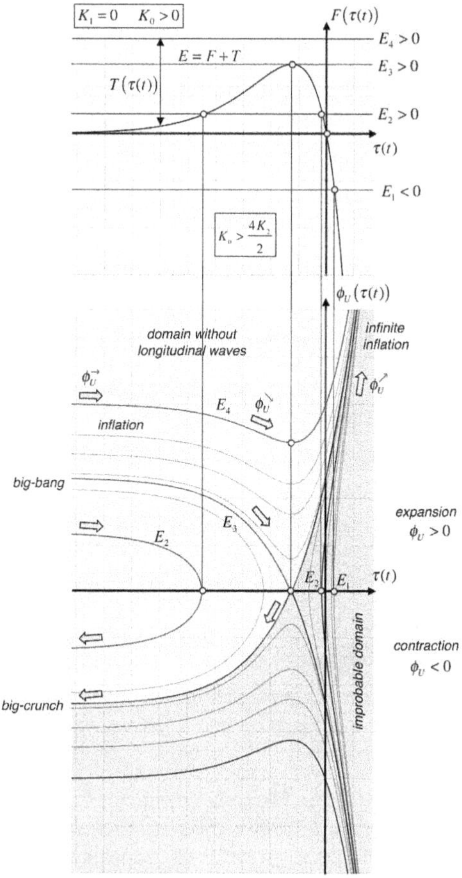

***Figure 16.7** - "cosmological behavior" of elastic energy $F^{él}(\tau)$ of expansion and velocity $\vec{\phi}_U(\tau)$ of expansion of an imaginary cosmic lattice with $K_0>0$*

We deduce the "cosmological behavior" of this type of lattice as shown in figure 16.7:

- if $0 < E < F^{él}_{max} = NK_0 / e n_0$, the lattice presents two possible trajectories, one that oscillates indefinitely between $\tau \to -\infty$ and a maximal value τ_{max} , and one which corresponds to an irreversible expansion, at constant velocity, from an initial value with $\vec{\phi}_U = 0$,

- if $E \geq F^{él}_{max} = NK_0 / e n_0$, the lattice presents an irreversible expansion from $\tau = -\infty$ to $\tau \to +\infty$, with a first decreasing velocity and then increasing velocity,

- if $E \leq 0$, the evolution of the lattice only has one trajectory presenting an irreversible expansion from value $\tau_{min} > 0$ to $\tau \to +\infty$, with an increasing velocity.

In this case, the lattice presents longitudinal waves if $0 < K_0 < 4K_2 / 3$, but does not present them if $K_0 > 4K_2 / 3$.

The cosmological evolution of a cosmic lattice with $K_0 > 0$ and $K_1 > 0$

The elastic free energy of this lattice is written

$$F^{él}(\tau) = Nf^{él} = N\left(-\frac{K_0}{n}\tau + \frac{K_1}{n}\tau^2 \right) = \frac{N}{n_0}\left(K_1\tau - K_0\right)\tau\, e^{\tau} \tag{16.13}$$

This function is represented at the top of figure 16.8. It has zeroes for

$$F^{él}(\tau) = 0 \quad \Leftrightarrow \quad \tau = \begin{cases} 0 \\ K_0 / K_1 \\ \to -\infty \end{cases} \tag{16.14}$$

as well as a maximum in the domain $\tau < 0$ and a minimum in the domain $\tau > 0$. According to relation *(16.6)*, these extremas correspond respectively to the minimum and maximum of the velocity of expansion $\phi_U(\tau)$ of the lattice, so that we have

$$\frac{dF^{él}(\tau)}{d\tau} = 0 \Rightarrow \begin{cases} \tau_{F^{él}\max} = \tau_{\phi_U \min} = \left(\dfrac{K_0}{2K_1} - 1\right) - \sqrt{\dfrac{K_0}{2K_1}\left(\dfrac{K_0}{2K_1} + 1\right)} \underset{\frac{K_0}{K_1} >> 1}{\longrightarrow} -\dfrac{3}{2} \\ \tau_{F^{él}\min} = \tau_{\phi_U \max} = \left(\dfrac{K_0}{2K_1} - 1\right) + \sqrt{\dfrac{K_0}{2K_1}\left(\dfrac{K_0}{2K_1} + 1\right)} \underset{\frac{K_0}{K_1} >> 1}{\longrightarrow} \dfrac{K_0}{K_1} - \dfrac{1}{2} \end{cases} \tag{16.15}$$

We deduce the "cosmological behavior" of this type of lattice, as shown in figure 16.8:

- if $E < 0$, the lattice presents only one possible trajectory, entirely in the domain $\tau > 0$, and which corresponds to a contraction and an expansion that keeps on going between two extreme values of τ ,

- if $0 < E < F^{él}_{max}$, the lattice presents two possible trajectories: the first one is an expansion/contraction that goes on indefinitely between a positive and a negative value of τ , and the second corresponds to an indefinite oscillation between a negative value of τ and an expansion going to $\tau \to -\infty$,

- if $E > F^{él}_{max}$, the lattice presents only one trajectory which is rather interesting. We oscillate indefinitely between a *big-bang* and a *big crunch*! The big-bang is followed by an expansion phase which is very fast, then a slowdown, and then again an expansion with increasing velocity, and suddenly an inversion of the velocity of expansion, so it contracts by retracing all the steps followed during the expansion phase. The contraction finishes with a big crunch, which can only be followed by a big-bang since the lattice has accumulated a total kinetic energy T equal to E, this phenomena is called *"big bounce"!*

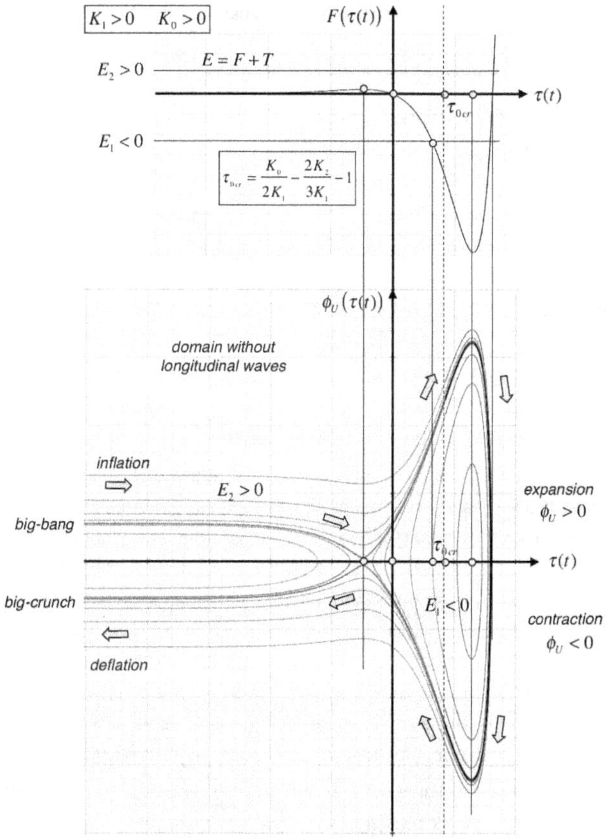

***Figure 16.8** - «cosmological behavior» of elastic energy $F^{él}(\tau)$ of expansion and velocity $\vec{\phi}_U(\tau)$ of expansion of an imaginary cosmic lattice with $K_0>0$ and $K_1>0$*

In the case of this lattice, we notice too the existence of domains of volume expansion that present different behaviors with regards to longitudinal waves: a domain where we have transverse and longitudinal waves for $\tau>\tau_{0cr}=K_0/2K_1-2K_2/3K_1-1$, and a domain for $\tau<\tau_{0cr}=K_0/2K_1-2K_2/3K_1-1$ where there are only transverse waves and localized vibrational eigenmodes. The domain where there are no longitudinal waves corresponds precisely to the domain of the big-bang, the inflation, the slowdown of inflation, finally followed by an acceleration of the expansion!

16.4 – Analogy with the cosmological evolution of our universe and origin of the 'dark energy'

In figure 16.9, we show eight different behaviors that can be obtained with a cosmic lattice, depending on the values that the moduli K_0 and K_1 can take. It is also shown in this figure the domains of expansion in which the longitudinal waves cannot exist.

On the analogy with the cosmology of the real universe

It is noted that there are four different *"cosmological behavior"*, three of which have convincing analogies with the cosmology of the real universe:

- cosmological lattices with $K_1 < 0$ which are reported in figures 16.9 (a), (c) and (d). These three types of lattices all have a big bang followed by high speed inflation, a slowdown in inflation and ultimately an expansion at increasing velocity towards $\tau \to +\infty$. All the stages follow in perfect order. The disappearance of the longitudinal waves takes place in these networks to higher expansions than a critical value τ_{0cr}, which depends on the value of the shear modulus $K_2 > 0$,

- the cosmic lattice of figure 16.9 (b), with $K_1 = 0$ and $K_0 > 0$ for which there are never longitudinal waves provided that $K_0 > 4K_2 / 3$, making it a very simple and very interesting case to describe the cosmological behavior of the real universe,

- the cosmic lattice with $K_1 > 0$ or $K_1 = 0$ and $K_0 < 0$ are shown in figures 16.9 (e), (g) and (h). These three types of lattice go through the four stages of the cosmology of the real universe, in the absence of longitudinal waves (a "big bang" from a singularity of space-time, followed by a period of very rapid inflation and a slowdown in inflation, followed by an expansion whose speed seems to increase over time), before entering an expansion phase during which the longitudinal waves appear, and precede a symmetrical contraction phase back to the singularity state $\tau \to -\infty$ ("big crunch"). In this case, there is a region of the diagram for which $\tau < \tau_{0cr}$ where there are no longitudinal waves, and wherein the lattice is expanding with increasing velocity. Note that the lattice of figure 16.9 (g) could be an excellent candidate to describe the cosmological behavior of the real universe, because all its elastic moduli are positive,

- finally, the cosmic lattice of figure 16.9 (f), with $K_1 = 0$ and $K_0 < 0$, does not present the stages corresponding to the cosmology of the real universe, and it always has longitudinal waves. It is clearly not suitable to describe the cosmological behavior of our universe.

The "cosmological behavior" of a cosmic lattice can be illustrated more clearly by plotting the velocity of volume expansion $d\tau / dt$ as a function of the volume expansion τ, as shown in the cases (c) and (d) with $K_1 < 0$ in figure 16.10 and for the case (g) and (h) with $K_1 > 0$ in figure 16.11. To find these behaviors, we retrieve the value R_U of the expression *(16.4)*

$$R_U = \left(\frac{3N}{4\pi n_0} \right)^{1/3} e^{\tau/3} \tag{16.16}$$

and we deduce the velocity of expansion $\phi_U(\tau)$

$$\phi_U(\tau) = \frac{dR_U}{dt} = \frac{1}{3} \left(\frac{3N}{4\pi n_0} \right)^{1/3} e^{\tau/3} \frac{d\tau}{dt} \tag{16.17}$$

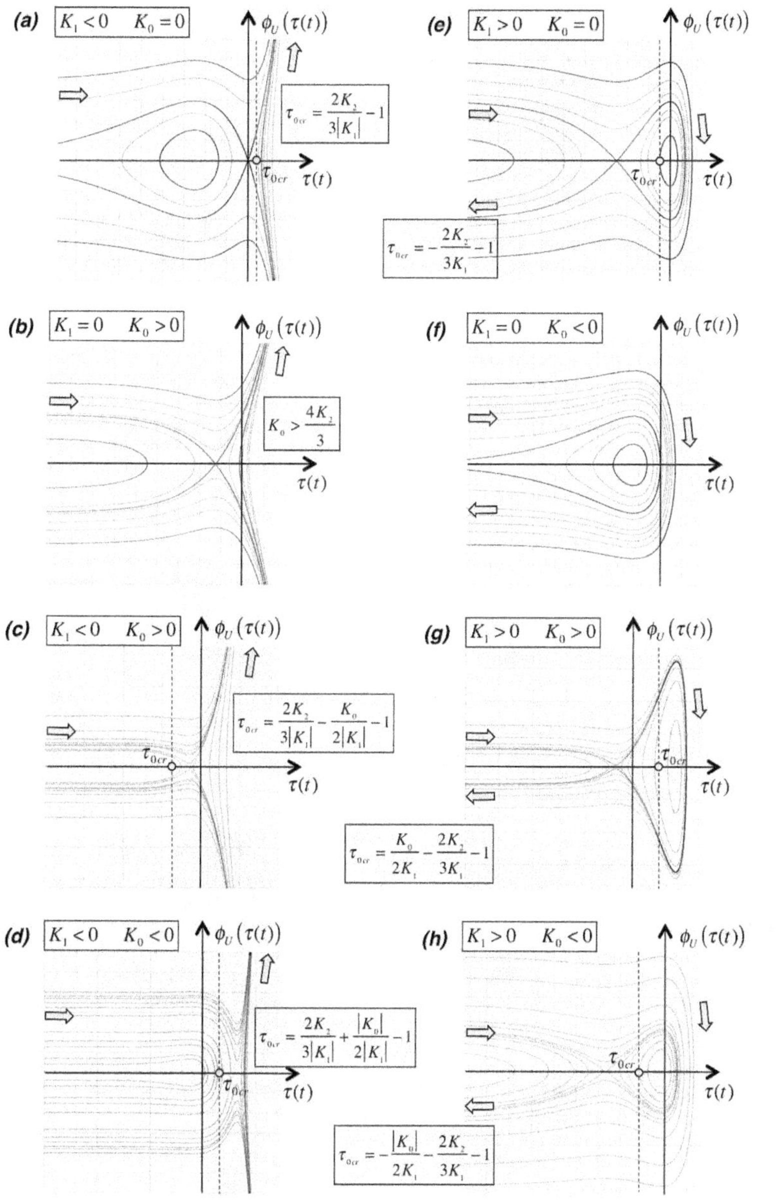

Figure 16.9 *- all the "cosmological behaviors" that are possible for cosmic lattices, depending on values* K_0 *and* K_1*:* ***(a)*** *through* ***(d)*** *the lattices with infinite accelerating expansion,* ***(e)*** *through* ***(h)*** *the lattices oscillating from big-bang to big-crunch*

which, compared to expression *(16.6)* of $\phi_U(\tau)$ allows us to write

$$\frac{d\tau}{dt}=3\left(\frac{4\pi n_0}{3N}\right)^{1/3}\mathrm{e}^{-\tau/3}\phi_U(\tau)=3\left(\frac{4\pi n_0}{3N}\right)^{1/3}\mathrm{e}^{-\tau/3}\sqrt{\frac{10}{3Nm}\left(\boldsymbol{E}-\boldsymbol{F}^{él}(\tau)\right)} \qquad (16.18)$$

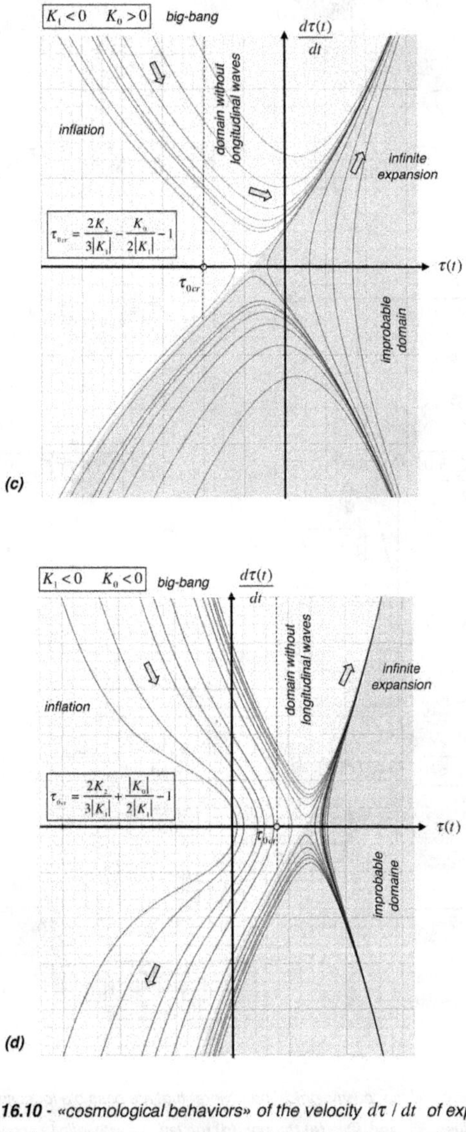

Figure 16.10 *- «cosmological behaviors» of the velocity $d\tau / dt$ of expansion as a function of expansion τ of two imaginary cosmic lattices with $K_1 < 0$*

The behavior of the rate of volume expansion $d\tau / dt$ as a function of τ can then be deduced from the knowledge of $F^{él}(\tau)$, which allows us to do the plots of figures 16.10 and 16.11.

(g)

(h)

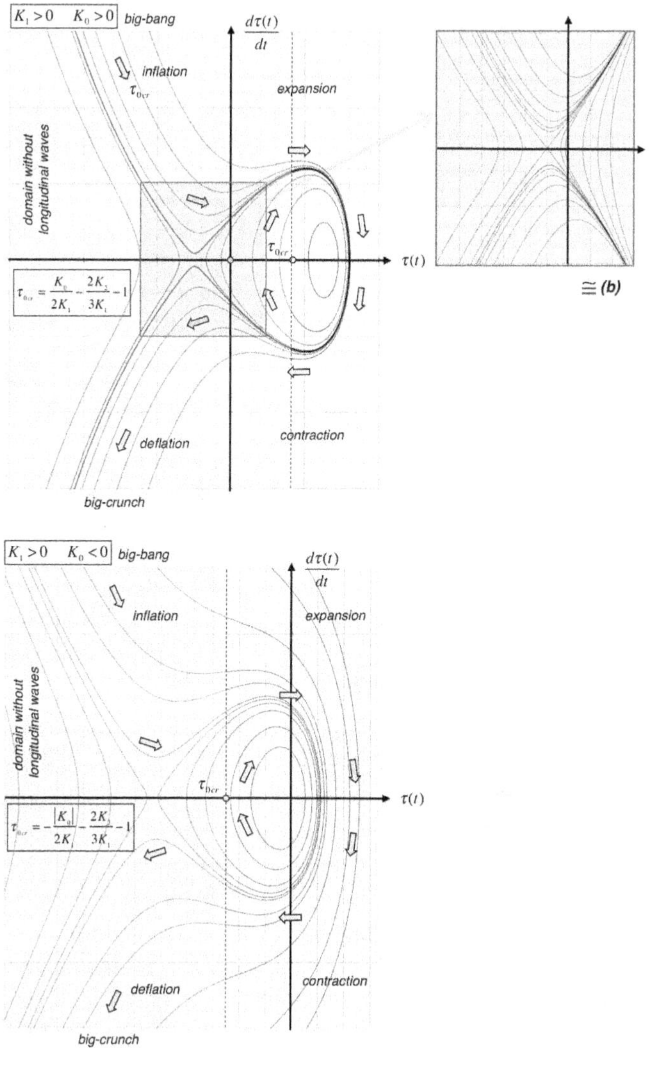

Figure 16.11 *- «cosmological behaviors» of the velocity $d\tau / dt$ of expansion as a function of expansion τ of two imaginary cosmic lattices with $K_1 > 0$*

The figures 16.10 and 16.11 are very interesting because they clearly show the existence of an extremely fast initial stage of inflation of the volume expansion in cosmic lattices since

$d\tau / dt \to \pm\infty$ for $\tau \to -\infty$ just after the big bang stage or just before the big crunch, and the rate of expansion or contraction of the volume is at a minimum before accelerating again, just after the stage of inflation or just after the stage of re-contraction.

On the limits of our model

It goes without saying that the modeling used in this chapter to describe the "cosmological behaviors" of imaginary lattices is extremely simple, if not simplistic. It is essentially the initial assumption of a homogeneous volume expansion throughout the lattice that can be questioned, because with this hypothesis was evaded the two major problems that would lead in principle to much more complicated models: the fact that the solid is subjected to Newtonian dynamics in the absolute space of ***GO***, and the fact that we should have put a condition on the validity of the pressure at the outer edge of the solid sphere. But despite the extreme simplifications of our modeling, the overall predicted behaviors in figures 16.9 to 16.11 should still remain close enough to the behaviors which could have been obtained by a more realistic treatment of the problem!

On the 'reasonable' choice of a cosmic lattice to describe the real universe

Among the various lattices proposed in this chapter, it is clear that the cosmic lattices have more interesting features than the perfect solids to describe the experimental observations of cosmologists. It is obviously not possible here to choose the cosmic lattice which is close to most of the known cosmological evolution of the real universe. But from a philosophical point of view and from the point of view of common sense, the cosmic lattices (fig. 16.9 (e) to (h)) which have a big bang followed by a big crunch, and thus ultimately a big-bounce, are much more satisfying for a Cartesian mind than cosmic lattices presenting a single and infinite expansion (fig. 16.9 (a) to (d)). One can then emit here a conjecture of 'philosophical nature'

Conjecture 3: *It seems more 'reasonable' to imagine cosmic lattices with* $K_1 > 0$ *,*
so as to have a finite expansion *(16.19)*

As for value of K_0 , nothing allows us for the moment to propose a positive, zero or negative value, as the cases illustrated in figure 16.11 are both very interesting!

On the origin of 'dark energy"

It should finally be noted that the elastic energy $F^{el}(\tau)$ contained in the cosmic lattice could very well correspond to the 'dark energy' which astrophysicists introduce to explain the acceleration of the velocity of expansion of the universe which was recently observed experimentally, since it is that elastic energy which is fully responsible for an increase of the rate of volume expansion via relation *(16.18)*.

PART II

B

Maxwell's equations and special relativity

Separability of the Newton's equation

Maxwell's equations of the field of rotation

*Fields and energies of elastic distortions
due to topological singularities*

Inertial mass

Relativistic dynamics of the topological singularities

Special relativity of Einstein

Chapter 17

Maxwell's equations of evolution of the field of rotation of a cosmic lattice

In this chapter, we start by showing we can separate the field of volume expansion from the other fields in the Newton equation of a cosmic lattice in the case where the concentration of point defects are constant. Then we use these results to obtain the Maxwell's equations of evolution of a lattice in the case where the volume expansion can be treated as constant.

17.1 – Separability of Newton's equation of a cosmic lattice in a 'rotational' part and a 'divergent' part

Assume that the field of volume expansion in the cosmic lattice is represented by a homogenous background field τ_0 on which we superpose a field of elastic expansion $\tau^{él}$

Hypothesis 1: $\tau = \left(\tau_0 + \tau^{él}\right)$ *(17.1)*

Introduce this field in the equation of Newton *(13.9)*. We have

$$n\frac{d\vec{p}}{dt} = -2\left(K_2+K_3\right)\overrightarrow{\mathrm{rot}}\,\vec{\omega}^{él} + \overrightarrow{\mathrm{grad}}\left[\left(\frac{4}{3}K_2+2K_1\left(1+\tau_0\right)-K_0\right)\tau^{él}+K_1\left(\tau^{él}\right)^2+F^{rot}\right]+2K_2\vec{\lambda}+nm\vec{\phi}_I\frac{dC_I}{dt}-nm\vec{\phi}_L\frac{dC_L}{dt} \qquad (17.2)$$

in which F^{rot} represents the density of energy of deformation by elastic and anelastic shear strains and rotations, and is worth

$$F^{rot} = K_2\sum_i(\vec{\alpha}_i^{él})^2 + K_1^{an}\sum_i(\vec{\alpha}_i^{an})^2 + 2K_3(\vec{\omega}^{él})^2 + 2K_2^{an}(\vec{\omega}^{an})^2 \qquad (17.3)$$

Suppose further than the atomic concentrations of vacancies and interstitials are homogeneous constants in the lattice such that

Hypothesis 2: $\dfrac{dC_I}{dt} = \dfrac{dC_L}{dt} = 0$ *(17.4)*

In this case the equation of Newton simplifies into

$$n\frac{d\vec{p}}{dt} = -2\left(K_2+K_3\right)\overrightarrow{\mathrm{rot}}\,\vec{\omega}^{él} + \overrightarrow{\mathrm{grad}}\left[\left(\frac{4}{3}K_2+2K_1\left(1+\tau_0\right)-K_0\right)\tau^{él}+K_1\left(\tau^{él}\right)^2+F^{rot}\right]+2K_2\vec{\lambda} \qquad (17.5)$$

in which the quantity of movement can be written according to *(5.101) and (5.78)*

$$\begin{cases} \vec{p} = m\vec{\phi} + m\left(C_I - C_L\right)\vec{\phi} + m\left(\vec{J}_I - \vec{J}_L\right)/n \\ \vec{J}_L = nC_L\left(\vec{\phi}_L - \vec{\phi}\right) \\ \vec{J}_I = nC_I\left(\vec{\phi}_I - \vec{\phi}\right) \end{cases} \qquad (17.6)$$

Thanks to the second hypothesis, the linearity of equations *(17.6)* with respect to the various velocities, means that it is possible to separate the equations in two different sets by separating the velocities $\vec{\phi}$, $\vec{\phi}_L$ and $\vec{\phi}_I$ in a component indexes *«rot»*, associated with the deformations by shear and rotation on one hand, and a component indexed by *«div»*, associated with the deformations by volume expansion on the other hand. We write

$$\vec{\phi} = \vec{\phi}^{rot} + \vec{\phi}^{div} \quad ; \quad \vec{\phi}_L = \vec{\phi}_L^{rot} + \vec{\phi}_L^{div} \quad ; \quad \vec{\phi}_I = \vec{\phi}_I^{rot} + \vec{\phi}_I^{div} \tag{17.7}$$

We also have two contributions to the equation of Newton:

- *a contribution which pilots the elastic fields of shear and rotation,* via the vectorial field of rotation $\vec{\omega}^{él}$. This contribution only depends on volume expansion τ via the presence of the density of sites $n = n_0\,\mathrm{e}^{-(\tau_0+\tau^{él})}$, and it is written

$$n\frac{d\vec{p}^{rot}}{dt} = -2\left(K_2 + K_3\right)\overrightarrow{\mathrm{rot}}\,\vec{\omega}^{él} + 2K_2\vec{\lambda}^{rot} \tag{17.8}$$

$$\text{with} \quad \begin{cases} \vec{p}^{rot} = m\vec{\phi}^{rot} + m\left(C_I - C_L\right)\vec{\phi}^{rot} + m\left(\vec{J}_I^{rot} - \vec{J}_L^{rot}\right)/\,n \\ \vec{J}_L^{rot} = nC_L\left(\vec{\phi}_L^{rot} - \vec{\phi}^{rot}\right) = nC_L\Delta\vec{\varphi}_L^{rot} \\ \vec{J}_I^{rot} = nC_I\left(\vec{\phi}_I^{rot} - \vec{\phi}^{rot}\right) = nC_I\Delta\vec{\varphi}_I^{rot} \end{cases} \tag{17.9}$$

- *a contribution which pilots the field of perturbation of volume expansion,* and which depends on the previous solution via the density of energy F^{rot} of deformation by elastic and anelastic shear strains and rotations, and which is written

$$n\frac{d\vec{p}^{div}}{dt} = \overrightarrow{\mathrm{grad}}\left[\left(\frac{4}{3}K_2 + 2K_1\left(1+\tau_0\right) - K_0\right)\tau^{él} + K_1\left(\tau^{él}\right)^2 + F^{rot}\right] + 2K_2\vec{\lambda}^{div} \tag{17.10}$$

$$\text{with} \quad \begin{cases} F^{rot} = K_2\sum_i(\vec{\alpha}_i^{él})^2 + K_1^{an}\sum_i(\vec{\alpha}_i^{an})^2 + 2K_3(\vec{\omega}^{él})^2 + 2K_2^{an}(\vec{\omega}^{an})^2 \\ \vec{p}^{div} = m\vec{\phi}^{div} + m\left(C_I - C_L\right)\vec{\phi}^{div} + m\left(\vec{J}_I^{div} - \vec{J}_L^{div}\right)/\,n \\ \vec{J}_L^{div} = nC_L\left(\vec{\phi}_L^{div} - \vec{\phi}^{div}\right) = nC_L\Delta\vec{\varphi}_L^{div} \\ \vec{J}_I^{div} = nC_I\left(\vec{\phi}_I^{div} - \vec{\phi}^{div}\right) = nC_I\Delta\vec{\varphi}_I^{div} \end{cases} \tag{17.11}$$

The density of flexion charges was also separated in two parts: the *charges of rotational flexion* and the *charges of divergent flexion,* which satisfy the following relations

$$\vec{\lambda} = \vec{\lambda}^{rot} + \vec{\lambda}^{div} \quad \text{such that} \quad \overrightarrow{\mathrm{rot}}\,\vec{\lambda}^{rot} \neq 0 \quad \text{and} \quad \mathrm{div}\,\vec{\lambda}^{div} = \theta \tag{17.12}$$

They connect the Newton's equation for expansion $\tau^{él}$ *(17.10)* to the density of charge of curvature θ within the lattice.

This split of Newton's equation, in the case where concentrations of interstitials and vacancies are homogeneous constants allows us, with equations of table 11.2, to resolve the spatiotemporal evolution problems of the generalized perfect lattice, separating the solving of fields of elastic shear and rotation from the solving of the volume expansion of the lattice. With additional simplifying assumptions, it is possible to solve completely these two sets of equations. This is what we will show in the next section, considering the particular case where the volume expansion field can be considered almost constant.

17.2 – Maxwellian behavior of the rotational part

Now make the assumption that the average value of the volume expansion $\langle\tau\rangle=\tau_0+\langle\tau^{él}\rangle$ in the cosmological lattice can be considered in first approximation as a homogeneous constant, so that the site density n may also be regarded on average as a constant

Hypothesis 3: $\langle\tau\rangle=\tau_0+\langle\tau^{él}\rangle\cong cste \quad\Rightarrow\quad n\cong\langle n\rangle\cong n_0\,e^{\tau_0+\langle\tau^{él}\rangle}\cong cste$ (17.13)

With these hypothesis, we can re-write the equations of Newton *(17.8)* by introducing a *vectorial moment* $\vec{m}$ conjugated to rotations $\vec{\omega}^{él}$, under the form

$$\frac{d(n\vec{p}^{rot})}{dt}=-2(K_2+K_3)\overrightarrow{\mathrm{rot}}\,\vec{\omega}^{él}+2K_2\vec{\lambda}^{rot}=-\frac{1}{2}\overrightarrow{\mathrm{rot}}\,\vec{m}+2K_2\vec{\lambda}^{rot} \tag{17.15}$$

By hypothesis, the *anelasticity of the lattice* manifests itself purely by shear and/or rotation, so that it can be represented here by a vector of anelastic rotation $\vec{\omega}^{an}$, by writing the relation *(2.40)* under the form

$$\vec{\omega}^{\delta}=\vec{\omega}+\vec{\omega}_0(t)=\vec{\omega}^{él}+\vec{\omega}^{an}+\vec{\omega}_0(t)=\frac{1}{4(K_2+K_3)}\vec{m}+\vec{\omega}^{an}+\vec{\omega}_0(t) \tag{17.16}$$

Note that you can imagine in this case that the torsor of moments $\vec{m}$ derives from a virtual state equation. This results in a virtual free energy density of elastic rotation per lattice site in the form

$$m_k=n\frac{\partial f^{él}_{rotation}}{\partial\omega^{él}_k}=4(K_2+K_3)\vec{\omega}^{él}\;\Rightarrow\; f^{él}_{rotation}=\frac{2(K_2+K_3)}{n}\left(\vec{\omega}^{él}\right)^2 \tag{17.17}$$

so that the *volume density of virtual free energy of elastic rotation*, linked to the deformations by shear strains and pure elastic rotations, without volume expansion can be written

$$F^{él}_{rotation}\left(\vec{\omega}^{él}\right)=2(K_2+K_3)\left(\vec{\omega}^{él}\right)^2 \tag{17.18}$$

The equations needed for the description of elastic shear and rotation of the cosmic lattice have yet to incorporate the topological equations for the elastic rotation vector $\vec{\omega}^{él}$, i.e. the geometro-kinetic equation and the equation of geometro-compatibility in the presence of dislocation charges

$$\vec{J}=-\frac{1}{2}\sum_k\vec{e}_k\wedge\vec{J}_k=-\frac{d\vec{\omega}^{él}}{dt}+\frac{1}{2}\overrightarrow{\mathrm{rot}}\,\vec{\phi}^{rot}\quad\text{and}\quad\lambda=\frac{1}{2}\sum_k\vec{e}_k\vec{\lambda}_k=\mathrm{div}\,\vec{\omega}^{él} \tag{17.19}$$

With regards to density ρ of inertial mass of lattice, the hypothesis 2 and 3 allow to insure it is a constant

$$\rho=m(n+n_I-n_L)=mn(1+C_I-C_L)=cste \tag{17.20}$$

so that the evolution equation of this density in the local referential $Ox_1x_2x_3$ allow us to deduce that the divergence of $n\vec{p}^{rot}$ is null

$$\frac{\partial\rho}{\partial t}=0=-\mathrm{div}\left(n\vec{p}^{rot}\right)\quad\Rightarrow\quad\mathrm{div}\left(n\vec{p}^{rot}\right)=0 \tag{17.21}$$

This quantity $n\vec{p}^{rot}$ is directly deduced from *(17.9)* and can be written under the following form

$$n\vec{p}^{rot} = mn\left[\vec{\phi}^{rot} + (C_I - C_L)\vec{\phi}^{rot} + \frac{1}{n}\left(\vec{J}_I^{rot} - \vec{J}_L^{rot}\right)\right] = \rho\vec{\phi}^{rot} + m\left(\vec{J}_I^{rot} - \vec{J}_L^{rot}\right) \tag{17.22}$$

From relations *(8.22)* and *(10.28)*, we can suppose that there are no sources of charges of rotation S^{λ} in the lattice

Hypothesis 4: $$S^{\lambda} = \left[\frac{d}{dt}\left(\operatorname{div}\vec{\omega}^{él}\right) - \operatorname{div}\left(\frac{d\vec{\omega}^{él}}{dt}\right)\right] \equiv 0 \tag{17.23}$$

so that the equation of continuity for the charges of rotation can be written

$$\frac{d\lambda}{dt} = -\operatorname{div}\vec{J} \tag{17.24}$$

Finally, it is still possible to establish an *energetic balance equation* from the equations *(17.15)* and *(17.9)*

$$-\vec{m}\vec{J} = \vec{m}\frac{d\vec{\omega}^{él}}{dt} - \vec{\phi}^{rot}\frac{d(n\vec{p}^{rot})}{dt} - \operatorname{div}\left(\frac{1}{2}\vec{\phi}^{rot} \wedge \vec{m}\right) \tag{17.25}$$

The relations thus obtained for the cosmic lattice in the local coordinates $Ox_1x_2x_3$ of ***GO***, in translation $\vec{\phi}_O(t)$ and in rotation $\dot{\vec{\omega}}_O(t)$ in the absolute referential, are reported in table 17.1, where they are compared with the *Maxwell's equations of electromagnetism in an electrically charged environment which is conductive, magnetic and dielectric.*

There is a *very strong analogy* between these two sets of equations, except that the evolution equations involve the total (material) derivative, while Maxwell's equations involve the partial derivative with respect to time. However, it must be remembered that the total derivative *(2.20)* in the local frame can be replaced by the partial derivative with respect to time if the strains are small enough and / or slow enough close to the origin of the local frame, which we did in table 17.1!

17.3 – Analogy with the Maxwell's equations of Electro-Magnetism

The analogy between the cosmological equations of a lattice taken at almost constant and homogeneous volume expansion and Maxwell's equations of electromagnetism is entirely remarkable, because it is absolutely complete, as clearly shown in the equations given in table 12.1 and 17.1. In fact, our equations contain an additional density of "rotational" flexion charges in the second pair of equations, which has no counterpart in the Maxwell's equations. By then assuming a cosmological lattice in which $\vec{\lambda}^{rot}$ can be neglected

Hypothesis 5: $$\vec{\lambda}^{rot} \approx 0 \tag{17.26}$$

the analogy between the equations of the cosmic lattice and the equations of Maxwell becomes absolutely exact, and deserve further comments.

On the analogy of the charges of rotation and the electrical charges

The equations of table 17.1 show a complete analogy between the density λ of charges of rotation and the density ρ of electrical charges, as well as the vectorial flow $\vec{J}$ of charges of rotation and the density of electrical current $\vec{j}$.

Table 17.1 - "Maxwellian" formulation of the equations of evolution of a cosmic lattice in the local framework $Ox_1x_2x_3$ of GO

$$\begin{cases} -\dfrac{\partial(2\vec{\omega}^{\acute{e}l})}{\partial t}+\overrightarrow{\mathrm{rot}}\,\vec{\phi}^{rot}\cong(2\vec{J}) \\ \mathrm{div}\,(2\vec{\omega}^{\acute{e}l})=(2\lambda) \end{cases} \quad\Leftrightarrow\quad \begin{cases} -\dfrac{\partial\vec{D}}{\partial t}+\overrightarrow{\mathrm{rot}}\,\vec{H}=\vec{j} \\ \mathrm{div}\,\vec{D}=\rho \end{cases}$$

$$\begin{cases} \dfrac{\partial(n\vec{p}^{rot})}{\partial t}\cong-\overrightarrow{\mathrm{rot}}\left(\dfrac{\vec{m}}{2}\right)+2K_2\vec{\lambda}^{rot} \\ \mathrm{div}(n\vec{p}^{rot})=0 \end{cases} \quad\Leftrightarrow\quad \begin{cases} \dfrac{\partial\vec{B}}{\partial t}=-\overrightarrow{\mathrm{rot}}\,\vec{E} \\ \mathrm{div}\,\vec{B}=0 \end{cases}$$

$$\begin{cases} (2\vec{\omega}^{\acute{e}l})=\dfrac{1}{(K_2+K_3)}\left(\dfrac{\vec{m}}{2}\right)+(2\vec{\omega}^{an})+\left(2\vec{\omega}_0(t)\right) \\ (n\vec{p}^{rot})=(nm)\left[\vec{\phi}^{rot}+(C_t-C_L)\vec{\phi}^{rot}+\left(\dfrac{1}{n}\left(\vec{J}_t^{rot}-\vec{J}_L^{rot}\right)\right)\right] \end{cases} \quad\Leftrightarrow\quad \begin{cases} \vec{D}=\varepsilon_0\vec{E}+\vec{P}+\vec{P}_0(t) \\ \vec{B}=\mu_0\left[\vec{H}+\left(\chi^{para}+\chi^{dia}\right)\vec{H}+\vec{M}\right] \end{cases}$$

$$\left\{ \dfrac{\partial(2\lambda)}{\partial t}\cong-\mathrm{div}(2\vec{J}) \right. \quad\Leftrightarrow\quad \left\{ \dfrac{\partial\rho}{\partial t}=-\mathrm{div}\,\vec{j} \right.$$

$$\begin{cases} -\left(\dfrac{\vec{m}}{2}\right)(2\vec{J})\cong \\ \quad\vec{\phi}^{rot}\dfrac{\partial(n\vec{p}^{rot})}{\partial t}+\left(\dfrac{\vec{m}}{2}\right)\dfrac{\partial(2\vec{\omega}^{\acute{e}l})}{\partial t}-\mathrm{div}\left(\vec{\phi}^{rot}\wedge\left(\dfrac{\vec{m}}{2}\right)\right) \end{cases} \quad\Leftrightarrow\quad \begin{cases} -\vec{E}\vec{j}= \\ \quad\vec{H}\dfrac{\partial\vec{B}}{\partial t}+\vec{E}\dfrac{\partial\vec{D}}{\partial t}-\mathrm{div}\left(\vec{H}\wedge\vec{E}\right) \end{cases}$$

$$\left\{ c_t=\sqrt{\dfrac{K_2+K_3}{mn}} \right. \quad\Leftrightarrow\quad \left\{ c=\sqrt{\dfrac{1}{\varepsilon_0\mu_0}} \right.$$

On the analogy between the anelasticity of the lattice and the dielectric properties of matter

The phenomenon of anelasticity introduced here by the term $2\vec{\omega}^{an}$ becomes in comparison with Maxwell's equations of electromagnetism, analogous to the dielectric polarization in the relationship $\vec{D}=\varepsilon_0\vec{E}+\vec{P}+\vec{P}_0(t)$, giving the electric displacement $\vec{D}$ versus electric field $\vec{E}$ and polarization of matter $\vec{P}$.

This analogy between fields $2\vec{\omega}^{an}$ and $\vec{P}$ is very strong since the possible phenomenological behavior of these two quantities are entirely similar, as shown in the relaxation, resonant or hysteresis behaviors described in section 7.8 and figures 7.7 and 7.10. For example, in the case of a pure relaxation, it is possible to connect $\vec{\omega}$ and $\vec{m}$ by means of a complex modulus, as it is possible to connect $\vec{D}$ and $\vec{E}$ via a similar complex dielectric coefficient in electromagnetism (in fact, a deeper comparison would show that the behaviors associated with thermal activation

also present analogies).
As for the term of homogeneous dielectric polarization $\vec{P}_0(t)$ we introduced here, it is the analogue of a term of global rotation of the local coordinate $Ox_1x_2x_3$ in the absolute ***GO*** referential. This term therefore disappears in the case where the local coordinate system $Ox_1x_2x_3$ is only in translation $\vec{\phi}_O(t)$ relative to the absolute referential.

On the analogy between mass transport in the lattice and magnetism of matter

As $n\vec{p}^{rot}$ represents both the average quantity of movement per unit volume of the solid and the average mass flow within the solid, we deduce that the mass flow within the solid is due at the same time to a transport of mass $nm\vec{\phi}^{rot}$ with velocity $\vec{\phi}^{rot}$ corresponding to the movement of the lattice, second to a mass transport $nm(C_I - C_L)\vec{\phi}^{rot}$ at velocity $\vec{\phi}^{rot}$ by the driving movement of the point defects by the lattice and finally to a mass transport $m\left(\vec{J}_I^{rot} - \vec{J}_L^{rot}\right)$ due to the phenomenon of self-diffusion of vacancies and interstitials.
Each of these mass transports has an analog in Maxwell's equations of electromagnetism. The mass transport $nm\vec{\phi}^{rot}$ by the lattice is analogous to the term $\mu_0\vec{H}$ of the *magnetic induction* in a vacuum. The mass transport $nm(C_I - C_L)\vec{\phi}^{rot}$ by dragging along the point defects by the lattice perfectly corresponds to the term $\mu_0(\chi^{para} + \chi^{dia})\vec{H}$ of magnetism, wherein the *magnetic susceptibility* is composed of two parts: the *positive paramagnetic susceptibility* χ^{para}, which becomes the analog of the concentration C_I of interstitials, and the *negative diamagnetic susceptibility* χ^{dia}, which is therefore analogous to the concentration of vacancies C_L.
With regards to the phenomena of auto-diffusion by the holes and interstitials, we have in these equations the term $m\left(\vec{J}_I^{rot} - \vec{J}_L^{rot}\right)$ which links the last part of $n\vec{p}^{rot}$ to velocities $\Delta\vec{\varphi}_L^{rot}$ and $\Delta\vec{\varphi}_I^{rot}$ of auto-diffusion of point defects

$$n\vec{p}^{rot}_{auto-diffusion} = m\left(\vec{J}_I^{rot} - \vec{J}_L^{rot}\right) = mn\left(C_I\Delta\vec{\varphi}_I^{rot} - C_L\Delta\vec{\varphi}_L^{rot}\right) \tag{17.27}$$

As an example we can imagine a hypothetical lattice in which the vacancies are tightly anchored to the lattice ($B_L(\tau,T) \to \infty$), while the interstitials are free to move ($B_I(\tau,T) \to 0$). The equations of movement *(7.61)* then become, by taking into account hypothesis 2

$$\begin{cases} \Delta\vec{\varphi}_L^{rot} \cong \dfrac{m}{B_L}\dfrac{\partial\vec{\phi}^{rot}}{\partial t} \\ m\dfrac{\partial\Delta\vec{\varphi}_I^{rot}}{\partial t} \cong -m\dfrac{\partial\vec{\phi}^{rot}}{\partial t} \end{cases} \tag{17.28}$$

The solutions to these hypothetical equations are then simply written, by introducing a constant velocity vector $\vec{v}_I^{rot}$

$$\begin{cases} \Delta\vec{\varphi}_L^{rot} \to 0 \\ \Delta\vec{\varphi}_I^{rot} \cong \vec{\phi}^{rot} + \vec{v}_I^{rot} \end{cases} \tag{17.29}$$

As a consequence, the quantity of movement $n\vec{p}^{rot}$ within the lattice can be written

$$n\vec{p}^{rot} = nm\left[\vec{\phi}^{rot} + (C_I - C_L)\vec{\phi}^{rot} + C_I\Delta\vec{\varphi}_I^{rot}\right] = nm\left[\vec{\phi}^{rot} + (2C_I - C_L)\vec{\phi}^{rot} + C_I\vec{v}_I^{rot}\right] \tag{17.30}$$

Mass transport $n\vec{p}^{rot}$ now has a term $\left(2C_I - C_L\right)\vec{\phi}^{rot}$ associated with both vacancies and interstitials, whose coefficient $\left(2C_I - C_L\right)$ is analogous to the magnetic susceptibility χ in electromagnetism, and that can take a positive or negative value depending on concentrations C_I and C_L of point defects. It further contains the term $nmC_I\vec{v}_I^{rot}$ associated with mass transport by inertial conservative interstitial movement, which is perfectly analogous to the permanent magnetization $\vec{M}$ of the ferromagnetic and antiferromagnetic materials in electromagnetism.

The presence of the constant term $nmC_I\vec{v}_I^{rot}$ in $n\vec{p}^{rot}$ clearly corresponds to a non-Markovian type of process, since the value must depend on the history of this hypothetical solid lattice. One could imagine for instance that the movement of interstitials is controlled by a dry type of friction with the lattice, in which case there would be a critical force of depinning for interstitials, which would lead to the emergence of cycles of hysteresis of $\Delta\vec{\varphi}_I^{rot}(t)$ as a function of $\vec{\phi}^{rot}(t)$. This is absolutely similar to the cycles of hysteresis of magnetization $\vec{M}(t)$ as a function of the magnetic field $\vec{H}(t)$ observed in ferromagnetic or antiferromagnetic materials!

On the complete analogy with the electromagnetism theory

The complete analogy between the parameters of our theory and the Maxwell's theory of electromagnetism is reported in table 17.2

Table 17.2 - The complete analogy with the Maxwell's theory of electromagnetism

$\vec{\omega}^{él}$	$\Leftrightarrow$	$\vec{D}$ = *electric field of displacement*
$n\vec{p}^{rot}$		$\vec{B}$ = *magnetic induction field*
$\vec{m}/2$		$\vec{E}$ = *electric field*
$\vec{\phi}^{rot}$		$\vec{H}$ = *magnetic field*
$2\vec{J}$	$\Leftrightarrow$	$\vec{j}$ = *electric current*
2λ		ρ = *density of electric charges*
$\vec{\lambda}^{rot}$		? = *unknown*
$1/(K_2+K_3)$	$\Leftrightarrow$	ε_0 = *dielectric permittivity of vacuum*
nm		μ_0 = *magnetic permeability of vacuum*
$2\vec{\omega}^{an}$	$\Leftrightarrow$	$\vec{P}$ = *dielectric polarization of matter*
(C_I-C_L)		$(\chi^{para}+\chi^{dia})$ = *paramagnetic and diamagnetic susceptibility of matter*
$(\vec{J}_I^{rot}-\vec{J}_L^{rot})/n$		$\vec{M}$ = *magnetization of matter*
$\vec{\phi}^{rot}\wedge\vec{m}/2$	$\Leftrightarrow$	$\vec{H}\wedge\vec{E}$ = *vector of Poynting*
$c_t=\sqrt{(K_2+K_3)/mn}$		$c=\sqrt{1/(\varepsilon_0\mu_0)}$ = *speed of light*

On the effects of volume expansion of the lattice in the absolute frame of the GO

In this analogy, the existence of a uniform nonzero translation $\vec{\phi}_O(t)$ of the lattice, equivalent to a translation of the local coordinate system $Ox_1x_2x_3$ relative to the absolute referential $Q\xi_1\xi_2\xi_3$ of the **GO** would be analogous in the Maxwell equations to a homogeneous magnetic field

$\vec{H}_0(t)$ in space. This last remark implies that if a solid lattice was expanding in the absolute referential frame of ***GO***, there should appear a field $\vec{\phi}_O(t)$ in the local referential frame $Ox_1x_2x_3$. This field $\vec{\phi}_O(t)$ should be similar to *a locally homogeneous magnetic field $\vec{H}_0(t)$ in space if the universe was expanding*, and which should point in the direction of movement of the local coordinate of the observer relative to absolute space!

On the non-existence of magnetic monopoles in this analogy

The equation $\operatorname{div}(n\vec{p}^{rot})=0$ reflects the fact that we consider a solid with a homogeneous field of static volume expansion. The existence of a non-null and constant value of $\operatorname{div}(n\vec{p}^{rot})$ such that

$$\operatorname{div}(n\vec{p}^{rot})=\operatorname{div}\left[mn\left(1+C_I-C_L\right)\vec{\phi}^{rot}\right]+\operatorname{div}\left[m\left(\vec{J}_I^{rot}-\vec{J}_L^{rot}\right)\right]\neq 0 \tag{17.31}$$

would imply that there exists a constant and divergent field of velocity $\vec{\phi}^{rot}$ of the sites of the lattice, and thus, with hypothesis $\tau=cste$, a non-zero source of sites of lattice S_n

$$\operatorname{div}\vec{\phi}^{rot}\cong\underbrace{\partial\tau/\partial t}_{=0}+\frac{S_n}{n}=\frac{S_n}{n} \tag{17.32}$$

or that we have a constant and divergent flow of auto-diffusion $m\left(\vec{J}_I^{rot}-\vec{J}_L^{rot}\right)$, and as a consequence, localized and non null sources of point defects S_{I-L}, S_L^{pl} and/or S_I^{pl}, which would be written, by taking into account the hypothesis that $C_I=cste$ and $C_L=cste$, as

$$\begin{cases} n\dfrac{\partial C_L}{\partial t}\cong 0 \quad\Rightarrow\quad \operatorname{div}\vec{J}_L^{rot}=\left(S_{I-L}+S_L^{pl}\right)-C_L\left(S_L^{pl}-S_I^{pl}\right) \\[2ex] n\dfrac{\partial C_I}{\partial t}\cong 0 \quad\Rightarrow\quad \operatorname{div}\vec{J}_I^{rot}=\left(S_{I-L}+S_I^{pl}\right)-C_I\left(S_L^{pl}-S_I^{pl}\right)\end{cases} \tag{17.33}$$

As part of the analogy with electromagnetism, a relationship $\operatorname{div}(n\vec{p}^{rot})=cste\neq 0$ would be like a $\operatorname{div}\vec{B}=cste\neq 0$ relationship. Now this last relationship shows the *well-known concept of magnetic monopoles, particles of unipolar magnetic charges*, suggested by some theories, but never observed experimentally, and who would therefore be localized and continuous source of lattice sites or of point defects in the lattice!

In fact, the existence of similarity between two theories is always a very fruitful and successful thing in physics by the reciprocal contribution of one theory to the other. In our case, it is clear that this analogy with the electromagnetic field theory will enable us subsequently to use the whole arsenal of theoretical tools developed for a long time in field theory, such as for example, the Lorentz transformation or delayed potential theory. In the other direction, the theory developed here is actually a much more complex theory that classical electromagnetism, since it stems from a tensorial theory, which can be reduced to a vectorial theory by contraction of tensor indices. We can also choose more specific cases with less restrictive hypothesis in the solid lattice. Considering the tensorial aspect of solid lattice theory and by relaxing the more restrictive hypothesis, the analogy will become particularly interesting and fruitful, as we shall see later.

On the possible existence of "vectorial electrical charges" in this analogy

One can legitimately ask what could be the analogy of the density of flexion charges $\vec{\lambda}^{rot}$ in the Maxwell's equations. If there were a quantity $\vec{\lambda}^{rot}$ similar in the Maxwell equations, one could hypothetically call it a density $\vec{\rho}$ of "*vectorial electric charges*" by postulating the following analogy

$$\vec{\rho} \Leftrightarrow \vec{\lambda}^{rot} \tag{17.34}$$

The equations of Maxwell would then be written a little differently from the known equations, with an extra term of charge but not in the equation $\operatorname{div}\vec{B}=0$ as suggested in the theories of magnetic monopoles, but in the equation $\partial\vec{B}/\partial t=-\overrightarrow{\operatorname{rot}}\vec{E}$, in the following way

$$\begin{cases} -\dfrac{\partial \vec{D}}{\partial t}+\overrightarrow{\operatorname{rot}}\vec{H}=\vec{j} \\ \operatorname{div}\vec{D}=\rho \end{cases} \quad \text{and} \quad \begin{cases} \dfrac{\partial \vec{B}}{\partial t}=-\overrightarrow{\operatorname{rot}}\vec{E}+\kappa\vec{\rho} \\ \operatorname{div}\vec{B}=0 \end{cases} \tag{17.35}$$

in which κ is *a new electric coefficient*, analogous to the modulus $2K_2$

$$\kappa \Leftrightarrow 2K_2 \tag{17.36}$$

In the static case, if such a vectorial charge did in fact exist, the equation containing it would be written as

$$\frac{\partial \vec{B}}{\partial t}=-\overrightarrow{\operatorname{rot}}\vec{E}+\kappa\vec{\rho}=0 \quad \Rightarrow \quad \overrightarrow{\operatorname{rot}}\vec{E}=\kappa\vec{\rho} \quad \Rightarrow \quad \overrightarrow{\operatorname{rot}}\vec{D}=\varepsilon_0\kappa\vec{\rho} \tag{17.37}$$

so that the density $\vec{\rho}$ of «*vectorial electric charges*» would be the source of a rotational electric field $\vec{E}$ and a rotational electric field of displacement $\vec{D}$, just as the scalar density ρ of electrical charges is the source of a divergent electric field of displacement $\vec{D}$

$$\begin{cases} \operatorname{div}\vec{D}=\rho \\ \overrightarrow{\operatorname{rot}}\vec{D}=\varepsilon_0\kappa\vec{\rho} \end{cases} \tag{17.38}$$

If we now compare the coefficients of both theories we obtain the following analogies

$$\varepsilon_0 \Leftrightarrow \frac{1}{(K_2+K_3)} \quad \textit{et} \quad \kappa \Leftrightarrow 2K_2 \quad \Rightarrow \quad \varepsilon_0\kappa \Leftrightarrow \frac{2K_2}{K_2+K_3} \tag{17.39}$$

However the experimental observations have never shown the existence of such "*vectorial electric charges*". Indeed, two reasons can be invoked to explain this state of affairs: either the "*vectorial electric charges*" simply do NOT exist, or the coefficient $\varepsilon_0\kappa$ is so small that we do not observe the presence of these «*vectorial electric charges*», so

$$|\varepsilon_0\kappa| << 1 \quad \Leftrightarrow \quad \left|\frac{2K_2}{K_2+K_3}\right| << 1 \tag{17.40}$$

Starting from the saying that "*everything which is not prohibited must exist*", we can here deduce a new conjecture for our theory

Conjecture 4: *the modules K_2 and K_3 must satisfy the relation $|K_2| << |K_3|$,*
<u>or else</u> the module K_2 could be null $(K_2=0)$ (17.41)

We will revisit this conjecture later as it is going to play a major role.

Chapter 18

Resolution of the Newton's equation in the presence of topological singularities

In this chapter, we show that in the presence of a topological singularity *Newton's equation of cosmic lattice* can be separated into two partial equations: a *"first partial equation of Newton"* that solves the problem of the field and elastic energy of the distortions associated with the presence of the topological singularity, and a *"second partial differential equation of Newton"* that controls the expansion perturbation field due to the elastic and potential energies of the topological singularity. Then we briefly discuss the use that will be made of the partial differential equations thus obtained through this separability of Newton's equation.

18.1 – Separability of the Newton's equation in two equations describing the "elastic distortions" and the "perturbations of expansion" associated with a topological singularity of the lattice

Assume the existence of a localized singularity of dislocation charges, of spherical or tubular or membrane shape, containing charge densities $\vec{\lambda}_i$, $\vec{\lambda}$ and/or λ, and suppose that one can neglect the anelasticity and the self-diffusion in the lattice, by assuming that

Hypothesis: $$\frac{dC_I}{dt} = \frac{dC_L}{dt} = 0 \quad \text{and} \quad \vec{\alpha}_i^{an} = 0 \, , \vec{\omega}^{\,an} = 0 \qquad (18.1)$$

The equation of Newton *(13.14)* gives us a lattice equation which is written

$$\begin{cases} n\,d\vec{p}/dt = -2(K_2+K_3)\overrightarrow{\mathrm{rot}}\,\vec{\omega}^{\acute{e}l} + \overrightarrow{\mathrm{grad}}(4K_2/3+2K_1)\tau + \overrightarrow{\mathrm{grad}}\,F_{dist} + 2K_2\,\vec{\lambda} \\ F_{dist} = K_2\sum_i(\vec{\alpha}_i^{\,el})^2 + 2K_3(\vec{\omega}^{\,el})^2 + K_1\tau^2 - K_0\tau \end{cases} \qquad (18.2)$$

The presence of a localized singularity of dislocation charges can be introduced into this equation, considering that the fields existing in the lattice are of three different types: the elastic fields due to the charges associated with the singularity, which will be indexed *(ch)*, the fields independent of the singularity, which are due for example to the other singularities, which will be indexed *(ext)*, the background field τ_0 of the volume expansion of the lattice and finally a disturbance field of the *volume expansion due the distortion energy F_{dist} stored in the lattice by the elastic field* of the singularity considered

$$\begin{cases} \tau = \left(\tau_0 + \tau^{ch} + \tau^{ext} + \tau^{(p)}\right) \\ \vec{\alpha}_i^{\acute{e}l} = \left(\vec{\alpha}_i^{ch} + \vec{\alpha}_i^{ext}\right) \\ \vec{\omega}^{\acute{e}l} = \left(\vec{\omega}^{ch} + \vec{\omega}^{ext}\right) \end{cases} \qquad \begin{cases} \vec{\lambda} = \left(\vec{\lambda}^{ch} + \vec{\lambda}^{ext}\right) \\ \\ \vec{p} = m\left(\vec{\phi}^{ch} + \vec{\phi}^{ext} + \vec{\phi}^{(p)}\right) \end{cases} \qquad (18.3)$$

Let's introduce these fields in the equation of Newton

$$nm\left(\frac{d\vec{\phi}^{ch}}{dt}+\frac{d\vec{\phi}^{ext}}{dt}+\frac{d\vec{\phi}^{(p)}}{dt}\right)=-2\left(K_2+K_3\right)\overrightarrow{\mathrm{rot}}\left(\vec{\omega}^{ext}+\vec{\omega}^{ch}\right)+2K_2\left(\vec{\lambda}^{ext}+\vec{\lambda}^{ch}\right)$$
$$+\overrightarrow{\mathrm{grad}}\begin{bmatrix}\left(4K_2/3+2K_1\right)\left(\tau^{ext}+\tau^{ch}+\tau^{(p)}\right)-K_0\left(\tau^{ext}+\tau^{ch}+\tau^{(p)}\right)\\+K_2\sum_i\left(\vec{\alpha}_i^{ext}+\vec{\alpha}_i^{ch}\right)^2+2K_3\left(\vec{\omega}^{ext}+\vec{\omega}^{ch}\right)^2+K_1\left(\tau_0+\tau^{ext}+\tau^{ch}+\tau^{(p)}\right)^2\end{bmatrix} \quad (18.4)$$

We develop these terms by grouping them appropriately

$$nm\left(\frac{d\vec{\phi}^{ch}}{dt}+\frac{d\vec{\phi}^{ext}}{dt}+\frac{d\vec{\phi}^{(p)}}{dt}\right)=-2\left(K_2+K_3\right)\overrightarrow{\mathrm{rot}}\left(\vec{\omega}^{ext}\right)-2\left(K_2+K_3\right)\overrightarrow{\mathrm{rot}}\left(\vec{\omega}^{ch}\right)+2K_2\vec{\lambda}^{ext}+2K_2\vec{\lambda}^{ch}$$
$$+\overrightarrow{\mathrm{grad}}\begin{bmatrix}+\left(4K_2/3+2K_1(1+\tau_0)-K_0\right)\tau^{ext}+\left(K_2\sum_i\left(\vec{\alpha}_i^{ext}\right)^2+2K_3\left(\vec{\omega}^{ext}\right)^2+K_1\left(\tau^{ext}\right)^2\right)\\+\left(4K_2/3+2K_1(1+\tau_0)-K_0\right)\tau^{ch}+\left(K_2\sum_i\left(\vec{\alpha}_i^{ch}\right)^2+2K_3\left(\vec{\omega}^{ch}\right)^2+K_1\left(\tau^{ch}\right)^2\right)\\+\left(4K_2/3+2K_1(1+\tau_0+\tau^{ext}+\tau^{ch})-K_0\right)\tau^{(p)}+K_1\left(\tau^{(p)}\right)^2\\+\left(2K_2\sum_i\vec{\alpha}_i^{ch}\vec{\alpha}_i^{ext}+4K_3\vec{\omega}^{ch}\vec{\omega}^{ext}+2K_1\tau^{ch}\tau^{ext}\right)\end{bmatrix} \quad (18.5)$$

We note that this equation is composed of three coupled equations that govern the different fields in the lattice. We will describe them in the following paragraphs

The equation of Newton for the fields external to the singularity

The fields external to the singularity satisfy their own equation of Newton

$$nm\frac{d\vec{\phi}^{ext}}{dt}=-2\left(K_2+K_3\right)\overrightarrow{\mathrm{rot}}\left(\vec{\omega}^{ext}\right)+\overrightarrow{\mathrm{grad}}\left[\left(4K_2/3+2K_1(1+\tau_0)-K_0\right)\tau^{ext}\right]$$
$$+\overrightarrow{\mathrm{grad}}\left(K_2\sum_i\left(\vec{\alpha}_i^{ext}\right)^2+2K_3\left(\vec{\omega}^{ext}\right)^2+K_1\left(\tau^{ext}\right)^2\right)+2K_2\vec{\lambda}^{ext} \quad (18.6)$$

However, this equation is in fact not perfectly independent from the other fields, due to the presence of the density of sites $n=n_0\,\mathrm{e}^{-\left(\tau_0+\tau^{ch}+\tau^{ext}+\tau^{(p)}\right)}$ in the expression of the quantity of movement associated with $\vec{\phi}^{ext}$. We will assume, to simplify the problem of treating the fields belonging to the singularity, that the external field τ^{ext} can be considered constant, that is to say, $\vec{\phi}^{ext}=0$ and $\tau^{ext}=\tau^{ext}(\vec{r})$, in which case the equation *(18.6)* in its static form becomes completely independent of the fields τ^{ch} and $\tau^{(p)}$

On the "first partial equation of Newton" for the elastic fields of distortion associated with the topological singularity

The fields $\vec{\omega}^{ch}$, τ^{ch} and $\tau^{(p)}$ associated to the singularity satisfy two partial equations of Newton which are tightly coupled. The first one treats the *elastic distortion fields $\vec{\omega}^{ch}$ and τ^{ch} associated with the charges due to the singularity*, and it is written in all generality

$$nm\frac{d\vec{\phi}^{ch}}{dt}=-2\left(K_2+K_3\right)\overrightarrow{\mathrm{rot}}\left(\vec{\omega}^{ch}\right)+\left(4K_2/3+2K_1(1+\tau_0)-K_0\right)\overrightarrow{\mathrm{grad}}\,\tau^{ch}+2K_2\vec{\lambda}^{ch} \qquad (18.7)$$

This equation is coupled to the fields τ^{ext} and $\tau^{(p)}$ by the expression $n=n_0\,\mathrm{e}^{-\left(\tau_0+\tau^{ch}+\tau^{ext}+\tau^{(p)}\right)}$ appearing in the expression for the quantity of movement associated with $\vec{\phi}^{ch}$. In the static case, this coupling disappears, so that we can be deduce the static fields of elastic distortions $\vec{\omega}^{ch}$ and τ^{ch} generated by the topological singularity totally independently of the fields τ^{ext} and $\tau^{(p)}$.

It is noted that this partial Newton equation depends on the density of flexion charges $\vec{\lambda}^{ch}$ of the singularity. The divergence of this equation in its static form then provides a static equation dependent on the density of curvature charges θ^{ch} of the singularity since the divergence of the density of flexion charges is equal to the density of curvature charges of the singularity

$$\Delta\left(\tau^{ch}_{statique}\right)=-\frac{2K_2}{4K_2/3+2K_1(1+\tau_0)-K_0}\mathrm{div}\,\vec{\lambda}^{ch}=-\frac{2K_2}{4K_2/3+2K_1(1+\tau_0)-K_0}\theta^{ch} \qquad (18.8)$$

On the "second partial equation of Newton" for the perturbation fields of expansion associated with the topological singularity

The last partial equation of Newton we can extract from *(18.5)* deals with the perturbation $\tau^{(p)}$ of the expansion field and the resulting elastic energy of the singularity stored in the lattice. It is written

$$nm\frac{d\vec{\phi}^{(p)}}{dt}=\overrightarrow{\mathrm{grad}}\left[\begin{array}{l}\left(4K_2/3+2K_1(1+\tau_0+\tau^{ext}+\tau^{ch})-K_0\right)\tau^{(p)}+K_1\left(\tau^{(p)}\right)^2\\+\left(K_2\sum_i\left(\vec{\alpha}_i^{ch}\right)^2+2K_3\left(\vec{\omega}^{ch}\right)^2+K_1\left(\tau^{ch}\right)^2\right)\\+\left(2K_2\sum_i\vec{\alpha}_i^{ext}\vec{\alpha}_i^{ch}+4K_3\vec{\omega}^{ext}\vec{\omega}^{ch}+2K_1\tau^{ext}\tau^{ch}\right)\end{array}\right] \qquad (18.9)$$

It is clear that this equation is itself very strongly coupled to the fields $\vec{\omega}^{ext}$, τ^{ext}, $\vec{\omega}^{ch}$ and τ^{ch} deduced from the other two Newton's equations. First there is a dynamical coupling via the term $n=n_0\,\mathrm{e}^{-\left(\tau_0+\tau^{ch}+\tau^{ext}+\tau^{(p)}\right)}$ appearing in the expression of the quantity of movement associated with $\vec{\phi}^{(p)}$. There also appears a coupling term associated with the modulus K_1 in the form $2K_1(1+\tau_0+\tau^{ext}+\tau^{ch})$. But the main terms of couplings are those due to the elastic energy of the singularity and the coupling energy of the singularity with external fields, which appear in two particular contributions and have very specific meanings:

- *the density of elastic energy* stored in the lattice by the elastic fields due to the singularity, that is, the *distortion energy density of the singularity*

$$F^{ch}_{dist}=K_2\sum_i\left(\vec{\alpha}_i^{ch}\right)^2+2K_3\left(\vec{\omega}^{ch}\right)^2+K_1\left(\tau^{ch}\right)^2 \qquad (18.10)$$

- *the density of coupling energy* of the singularity with the external fields, that is to say the density of *potential energy of the singularity*

$$F^{ch}_{pot}=2K_2\sum_i\vec{\alpha}_i^{ext}\vec{\alpha}_i^{ch}+4K_3\vec{\omega}^{ext}\vec{\omega}^{ch}+2K_1\tau^{ext}\tau^{ch} \qquad (18.11)$$

By assuming we know the two terms $F_{dist}^{ch}(\vec{r},t)$ and $F_{pot}^{ch}(\vec{r},t)$, obtained by the resolution of equations *(18.6)* and *(18.7)*, the equation of Newton for the perturbations of expansion $\tau^{(p)}$ due to the singularity can be written symbolically under the form

$$nm\frac{d\vec{\phi}^{(p)}}{dt} = \overrightarrow{\text{grad}}\left[\left(4K_2/3+2K_1(1+\tau_0+\tau^{ext}+\tau^{ch})-K_0\right)\tau^{(p)}+K_1\left(\tau^{(p)}\right)^2+F_{dist}^{ch}(\vec{r},t)+F_{pot}^{ch}(\vec{r},t)\right] \quad (18.12)$$

In the static case, if we have solved equations *(18.6)* and *(18.7)* taken in the static case, meaning that we know the equilibrium values of fields $\vec{\omega}^{ext}(\vec{r})$, $\tau^{ext}(\vec{r})$, $\vec{\omega}^{ch}(\vec{r})$ and $\tau^{ch}(\vec{r})$, the equation of equilibrium for the static fields of perturbation can be written

$$\overrightarrow{\text{grad}}\left\{\left[4K_2/3+2K_1\left(1+\tau_0+\tau^{ext}(\vec{r})+\tau^{ch}(\vec{r})\right)-K_0\right]\tau^{(p)}+K_1\left(\tau^{(p)}\right)^2+F_{dist}^{ch}(\vec{r})+F_{pot}^{ch}(\vec{r})\right\}=0 \quad (18.13)$$

and the solution is an equation of second order in $\tau^{(p)}(\vec{r})$

$$K_1\left(\tau^{(p)}(\vec{r})\right)^2+\left[4K_2/3+2K_1\left(1+\tau_0+\tau^{ext}(\vec{r})+\tau^{ch}(\vec{r})\right)-K_0\right]\tau^{(p)}(\vec{r})+\left(F_{dist}^{ch}(\vec{r})+F_{pot}^{ch}(\vec{r})\right)=cste=0 \quad (18.14)$$

in which the densities of energy $F_{dist}^{ch}(\vec{r})$ and $F_{pot}^{ch}(\vec{r})$, which are given by relations *(18.10)* and *(18.11)*, are calculated using the external fields $\vec{\alpha}_i^{ext}, \vec{\omega}^{ext}, \tau^{ext}$ and the elastic fields $\vec{\alpha}_i^{ch}, \vec{\omega}^{ch}, \tau^{ch}$ due to the singularity.
The constant $cste$ was introduced when we dealt with the gradient. But since $\tau^{(p)}(\vec{r})$ must be identically null if the energy $F_{dist}^{ch}(\vec{r})+F_{pot}^{ch}(\vec{r})$ is null, this constant must be null.

18.2 – On the consequences of the separability of the Newton's equation in the presence of a topological singularity of the lattice

On the method to find the fields associated with a topological singularity

The decomposition of the equation of Newton in three partial equations which we just went through shows an equation *(18.6)* for the external fields, an equation *(18.7)* for the elastic distortion fields associated with the presence of a topological singularity and an equation *(18.9)* for the perturbations of the expansion field due to the energy of elastic distortions associated with the topological singularity.
The methodology for solving the problem of fields associated with a topological singularity is then as follows:
- in a first step, one must solve independently the partial differential equation of Newton *(18.7)*, or equation *(18.8)* in a static case, in order to find the fields of distortions $\vec{\omega}^{ch}$ and τ^{ch} generated by the singularity, regardless of the disturbances of the expansion due to the energies $F_{dist}^{ch}(\vec{r},t)$ and $F_{pot}^{ch}(\vec{r},t)$ of the singularity,
- then, from the elastic field $\vec{\omega}^{ch}$ and τ^{ch} previously obtained from the partial differential equation of Newton *(18.7)*, we calculate using equation *(18.12)*, or equation *(18.14)* in the static case, the additional perturbations of the field of expansion due to the elastic energy $F_{dist}^{ch}(\vec{r},t)$ andc $F_{pot}^{ch}(\vec{r},t)$ of the singularity.

This process seems quite complex at first sight, but it contains a huge potential regarding the description and interpretation of the behavior of topological singularities in the cosmic lattice.

Indeed, we will show this later in this book by dealing, in detail, with the following themes:

On the link between the "first partial equation of Newton" for the elastic distortions and the Einstein's Special Relativity

The partial equation of Newton *(18.7)* which allows us to find the elastic distortions of fields associated with topological singularities will allow us to calculate the fields and the energies associated with screw dislocations, edge dislocations, screw disclination loops, edge dislocation loops and mixte dislocation loops (chapter 19). We will show that these fields are subject to a *relativistic dynamics* (chapter 20), which allow us to discuss the *"role of aether"* that the cosmic lattice plays vis-a-vis the topological singularities, and the similarities and differences with the *Special Relativity of Einstein*.

On the link between the "second partial equation of Newton" for the perturbation fields of expansion and the General Relativity of Einstein and the Quantum Physics

The partial equation of Newton *(18.9)* allowing us to find the volume expansion perturbations is very important too.

Indeed, we will see in chapters 22-26 that it, in its static form when applied to macroscopic cluster of singularities, allows to find gravitational effects. We will discuss the similarities and differences of our theory with the *Newton's gravitation*, *Einstein's General Relativity* and *modern cosmology of the universe*.

Then we will see in chapters 27-29 that this partial equation, when applied in its dynamic form to microscopic singularities, recovers the Quantum Physics, and we will discuss the similarities and differences with the *Schrödinger equation*, the *concepts of fermions and bosons*, the *principle of uncertainty of Heisenberg*, the *principle of exclusion of Pauli*, and the *concepts of spin and magnetic moment of elementary particles*.

Chapter 19

Topological singularities in a cosmic lattice

In this chapter, we use the first part of the equation of Newton of the cosmic lattice in the presence of a topological singularity to calculate the distortion fields, the rest energy, the classical kinetic energy and the inertial mass of screw and edge dislocations. Then we show that it is possible to define a *perfect cosmic lattice* satisfying certain specific conditions, which allows to find the expression of Einstein $E_0 = M_0c^2$ without any appeal to the principle of relativity, both for screw dislocations and edge dislocations.

We then calculate rotational and flexion fields associated respectively with the rotational charges and curvature charges of a localized macroscopic topological singularity within the lattice.

Finally, we describe the various elementary topological singularities which can be formed with dislocation and disclination loops. We then use the overall charges of rotation and curvature of these elementary singularities to calculate the fields away from these singularities.

Finally, we discuss why these elementary singularities could be the building blocks for constructing the elementary particles of the Standard Model.

19.1 – Fields, energies and inertial mass of a screw dislocation

The elastic fields of shear and local rotation of a screw dislocation

In the cosmic lattice, the rotation field of a dislocation string is the same as the one found in the case of a perfect solid in section 12.3

$$\vec{\omega}_{ext}^{screw} = \frac{\Lambda}{2\pi} \frac{\vec{r}}{r^2} = \frac{\Lambda}{2\pi} \frac{x_1\vec{e}_1 + x_3\vec{e}_3}{x_1^2 + x_3^2} \quad (r > R) \tag{19.1}$$

$$\vec{\omega}_{int}^{screw} = \frac{\Lambda}{2\pi R^2}\vec{r} = \frac{\Lambda}{2\pi R^2}(x_1\vec{e}_1 + x_3\vec{e}_3) \quad (r < R) \tag{19.2}$$

and the behavior of the module of $\vec{\omega}^{screw}$ as a function of distance r from the center of the string is the one shown in figure 12.2. The same goes for the external field of displacement by rotation, which is then written in cartesian coordinates and in polar coordinates

$$\vec{u}_{ext}^{screw} = -\vec{e}_2\frac{\Lambda}{\pi}\arctan\frac{x_1}{x_3} = \frac{\Lambda}{\pi}\left(\varphi - \frac{\pi}{2}\right)\vec{e}_2 \tag{19.3}$$

By using relations *(2.48)* which are valid outside the dislocation string, we can easily deduce the shear field outside the string from this external field of displacement

$$\alpha_{1\,ext}^{screw} = \frac{\Lambda}{2\pi}\frac{x_3}{x_1^2 + x_3^2}\vec{e}_2 \tag{19.4a}$$

$$\alpha_{2\,ext}^{screw} = \frac{\Lambda}{2\pi}\frac{x_3}{x_1^2+x_3^2}\vec{e}_1 - \frac{\Lambda}{2\pi}\frac{x_1}{x_1^2+x_3^2}\vec{e}_3 \tag{19.4b}$$

$$\alpha_{3\,ext}^{screw} = -\frac{\Lambda}{2\pi}\frac{x_1}{x_1^2+x_3^2}\vec{e}_2 \tag{19.4c}$$

The elastic energy of distortion of an immobile screw

The total rest energy of the screw dislocation string is obtained from the elastic potential energy $F = K_2\sum(\vec{\alpha}_i^{el})^2 + 2K_3(\vec{\omega}^{el})^2 + K_1(\tau)^2 - K_0\tau$ stored by unit volume outside the string. We have two terms of energy associated with the screw dislocation: first the term of elastic distortion F_{dist}^{screw} of local rotation and shear due to the linear charge Λ of the screw dislocation

$$F_{dist}^{screw} = K_2\sum_i(\vec{\alpha}_i^{screw})^2 + 2K_3(\vec{\omega}^{screw})^2 = (K_2+K_3)\frac{\Lambda^2}{2\pi^2}\frac{1}{r^2} \tag{19.5}$$

and, second, the energy term $F_{\tau^{(p)}}^{screw}$ associated with the perturbation $\tau^{(p)}$ of the volume expansion generated by the energy F_{dist}^{screw}. The perturbation is calculated thanks to equation *(18.13)*

$$F_{\tau^{(p)}}^{screw} = \left[2K_1\tau_0 - K_0\right]\tau^{(p)} + K_1\left(\tau^{(p)}\right)^2 \tag{19.6}$$

The term of energy of the pure elastic distortion of the screw depends primarily on the external deformation field as we have seen in section 12.3

$$E_{dist}^{screw} = \int_a^{R_\infty} F_{dist}^{screw}\,2\pi r\,dr = \cong (K_2+K_3)\int_a^{R_\infty}\frac{\Lambda^2}{2\pi^2 r^2}2\pi r\,dr = \frac{(K_2+K_3)\Lambda^2}{\pi}\ln\frac{R_\infty}{a} = \frac{(K_2+K_3)\vec{B}_{screw}^2}{4\pi}\ln\frac{R_\infty}{a} \tag{19.7}$$

in which R_∞ is the *external dimension* of the cosmic lattice and a is the lattice unit cell length of the cosmic lattice. We have $R_\infty >> a$.

With regards to the perturbation of expansion $\tau^{(p)}$ generated by the energy F_{dist}^{screw} and the energy $F_{\tau^{(p)}}^{screw}$ stored by these perturbations we will come back to those later.

On the kinetic energy associated with the distortions generated by a screw dislocation moving at low velocity

In the case where a screw dislocation moves in the direction Ox_1, with velocity $\vec{\mathbf{v}}$ that is small compared to the velocity of transversal waves c_t, the velocity field associated with the field of elastic displacement $\vec{u}_{ext}^{screw}$ of the dislocation is given by relation *(12.46)*

$$\vec{\phi}_{ext}^{screw} = -\frac{d\vec{u}}{dt} = \vec{e}_2\frac{\Lambda}{\pi}\frac{-x_3}{x_3^2+(x_1-\mathbf{v}t)^2}\mathbf{v} \tag{19.8}$$

and the kinetic energy stored in the lattice by this velocity field, expressed in unit length of the screw dislocation in movement, is given by relation *(12.49)*

$$E_{kin}^{screw} = \frac{mn\Lambda^2}{2\pi}\left(\ln\frac{R_\infty}{a}\right)\mathbf{v}^2 = \frac{mn\vec{B}_{screw}^2}{8\pi}\left(\ln\frac{R_\infty}{a}\right)\mathbf{v}^2 \tag{19.9}$$

By comparing the kinetic energy E_{kin}^{screw} stored in the lattice by the movement of the string with the potential energy E_{dist}^{screw} stored in the lattice by the presence of the same string, we have the following relationship between the rest energy and the kinetic energy of the screw dislocation

$$E_{kin}^{screw} = \frac{1}{2}\frac{mn}{(K_2+K_3)}\frac{(K_2+K_3)\Lambda^2}{\pi}\left(\ln\frac{R_\infty}{a}\right)\mathbf{v}^2 = \frac{1}{2}\frac{E_{dist}^{screw}}{c_t^2}\mathbf{v}^2 \qquad (19.10)$$

This kinetic energy E_{kin}^{screw} is stored in the solid lattice by the dynamic deformation of the lattice imposed by the mobile screw dislocation.

The inertial mass per unit length of a screw dislocation moving at small speed

With the relation *(19.10)*, we obtain the famous *Einstein expression* linking the *inertial mass* to the rest energy via the speed of transversal waves for a screw dislocation in the cosmic lattice

$$E_{kin}^{screw} = \frac{1}{2}M_0^{screw}\mathbf{v}^2 \quad \Rightarrow \quad M_0^{screw} = \frac{1}{c_t^2}E_{dist}^{screw} \quad \Rightarrow \quad E_{dist}^{screw} = M_0^{screw}c_t^2 \qquad (19.11)$$

This relation was found here *without any recourse to a relativistic dynamic of the string,* since it is due to the fact that the rest energy E_{dist}^{screw} and the kinetic energy E_{kin}^{screw} are respectively the elastic potential distortion energy (of shear and local rotation) and the kinetic energy that is stored within the lattice by the dynamic deformations due to the elastic distortion fields (shear and local rotation) of the mobile screw!

19.2 – Fields, energies and inertial mass of an edge dislocation

The equation of Newton for an edge dislocation in the cosmic lattice

In the presence of a screw or edge dislocation with linear charge Λ or $\vec{\Lambda}$, *the fields of elastic distortion* $\vec{\omega}^{ch}$ *and* τ^{ch} *associated with these charges satisfy the partial equation of Newton (18.7),* namely

$$nm\frac{d\vec{\phi}^{ch}}{dt} = -2(K_2+K_3)\overrightarrow{\mathrm{rot}}(\vec{\omega}^{ch}) + (4K_2/3+2K_1(1+\tau_0)-K_0)\overrightarrow{\mathrm{grad}}\,\tau^{ch} + 2K_2\vec{\lambda}^{ch} \qquad (19.12)$$

As we have seen in the previous chapter, this equation is coupled to the external fields τ^{ext} and to the perturbation of expansion $\tau^{(p)}$ due to the energy density of the charges by the term $n = n_0\,\mathrm{e}^{-(\tau_0+\tau^{ch}+\tau^{ext}+\tau^{(p)})}$ which appears in $nm\,d\vec{\phi}^{ch}/dt$.
However in the static case, this coupling disappears, so that equation *(19.12)* allows us to deduce the static fields $\vec{\omega}^{ch}$ and τ^{ch} generated by the singularity of charges in a manner totally independent of the fields τ^{ext} and $\tau^{(p)}$ thanks to the static equation

$$-2(K_2+K_3)\overrightarrow{\mathrm{rot}}\,\vec{\omega}^{ch} + (4K_2/3+2K_1(1+\tau_0)-K_0)\overrightarrow{\mathrm{grad}}\,\tau^{ch} = -2K_2\vec{\lambda} \qquad (19.13)$$

The elastic fields of rotation and volume expansion of an edge dislocation

Let's consider a *string of edge dislocation type,* which has the shape of a linear cylinder with infinite length and radius R, as shown in figure 12.1, which contains a vectorial density $\vec{\lambda}$ of flexion charges which is perpendicular to the direction of the string and points in direction Ox_3.

To find the static fields τ, $\vec{\omega}^{edge}$ and $\vec{\alpha}_i^{edge}$ associated with this dislocation when it is immobile within the cosmic lattice, we can use the partial equation of Newton *(19.13)* in a slightly modified form, by supposing that there is no anelasticity, no vacancies and no interstitials , in the form

$$\overrightarrow{\mathrm{rot}}\,\vec{\omega}^{edge} - \mathbb{C}\,\overrightarrow{\mathrm{grad}}\,\tau^{edge} = \begin{cases} K_2\vec{\lambda}/(K_2+K_3) & (r<R) \\ 0 & (r>R) \end{cases} \tag{19.14}$$

in which we introduce a dimensionless module $\mathbb{C}$ which is worth

$$\mathbb{C} = \frac{4K_2/3 + 2K_1(1+\tau_0) - K_0}{2(K_2+K_3)} \tag{19.15}$$

Outside the string of edge type $(r > R)$, we can take the solutions *(12.55)* which we had obtained in section 12.4. In this case, the vector of rotation $\vec{\omega}^{edge}$ must possess only one component along the axis Ox_2 which must depend on $\cos\varphi$ in cylindrical coordinates, while the scalar of expansion τ^{edge} must depend on $\sin\varphi$. Furthermore, the quantities $\vec{\omega}^{edge}$ and τ^{edge} must decrease in $1/r$ with distance r from the center of the dislocation, in the same fashion as the screw dislocation we just saw. Thus, the solution for equation *(19.15)* for $r > R$ must be written in the following form

$$\begin{cases} \vec{\omega}^{edge} = A\dfrac{\cos\varphi}{r}\vec{e}_2 = A\dfrac{x_1}{x_1^2+x_3^2}\vec{e}_2 \\ \tau^{edge} = B\dfrac{\sin\varphi}{r} = B\dfrac{x_3}{x_1^2+x_3^2} \end{cases} \quad (r>R) \tag{19.16}$$

Where A and B are integration constants. We can then calculate $\overrightarrow{\mathrm{rot}}\,\vec{\omega}^{edge}$ and $\overrightarrow{\mathrm{grad}}\,\tau^{edge}$

$$\overrightarrow{\mathrm{rot}}\,\vec{\omega}^{edge} = A\left[\frac{2x_1x_3}{(x_1^2+x_3^2)^2}\vec{e}_1 + \frac{x_3^2-x_1^2}{(x_1^2+x_3^2)^2}\vec{e}_3\right] \tag{19.17}$$

$$\overrightarrow{\mathrm{grad}}\,\tau^{edge} = -B\left[\frac{2x_1x_3}{(x_1^2+x_3^2)^2}\vec{e}_1 + \frac{x_3^2-x_1^2}{(x_1^2+x_3^2)^2}\vec{e}_3\right] \tag{19.18}$$

which, when introduced in the equilibrium equation *(19.15)*, give us the relationship that exists between A and B

$$A + \mathbb{C}B = 0 \tag{19.19}$$

Inside the string, we can integrate the equilibrium equation on the cylinder with radius R containing charge density $\vec{\lambda}$, with radius R and of unit length

$$\iiint_V \overrightarrow{\mathrm{rot}}\,\vec{\omega}_{int}^{edge}\,dV - \mathbb{C}\iiint_V \overrightarrow{\mathrm{grad}}\,\tau_{int}^{edge}\,dV \cong \frac{K_2}{K_2+K_3}\iiint_V \vec{\lambda}\,dV \tag{19.20}$$

Since the integral of $\vec{\lambda}$ on the cylinder of unit length gives us the linear charge $\vec{\Lambda}$ of the string, the previous relation is easily transformed in

$$\oiint_{cylinder} d\vec{S}\wedge\vec{\omega}_{int}^{edge} - \mathbb{C}\oiint_{cylinder} \tau_{int}^{edge}\,d\vec{S} \cong \frac{K_2}{K_2+K_3}\vec{\Lambda} \tag{19.21}$$

By symmetry, the integrals on the two sections (lateral faces) of the cylinder cancel each other, so that, with $d\vec{S} = R(\vec{e}_1\cos\varphi + \vec{e}_3\sin\varphi)d\varphi$

$$\begin{cases} \iint\limits_S d\vec{S}\wedge\vec{\omega}_{int}^{edge} = \iint\limits_S A\dfrac{\cos\varphi}{R}\left(d\vec{S}\wedge\vec{e}_2\right) = \int\limits_0^{2\pi} A\dfrac{\cos\varphi}{R}R\left(\vec{e}_3\cos\varphi-\vec{e}_1\sin\varphi\right)d\varphi = A\pi\vec{e}_3 \\ \iint\limits_S \tau_{int}^{edge}\,d\vec{S} = \iint\limits_S B\dfrac{\sin\varphi}{R}d\vec{S} = \int\limits_0^{2\pi} B\dfrac{\sin\varphi}{R}R\left(\vec{e}_1\cos\varphi+\vec{e}_3\sin\varphi\right)d\varphi = B\pi\vec{e}_3 \end{cases} \tag{19.22}$$

We then have as a result of the equation of equilibrium inside the string a relationship perfectly independent of the radius R of the string

$$A\pi\vec{e}_3 - \mathbb{C}B\pi\vec{e}_3 \cong \frac{K_2}{K_2+K_3}\vec{\Lambda} \quad\Rightarrow\quad A-\mathbb{C}B \cong \frac{K_2}{K_2+K_3}\frac{\vec{\Lambda}\vec{e}_3}{\pi} \tag{19.23}$$

From relations *(19.19)* and *(19.23)*, we deduce the constant values of A and B

$$A = \frac{K_2}{K_2+K_3}\frac{\vec{\Lambda}\vec{e}_3}{2\pi} \quad\text{and}\quad B = -\frac{K_2}{K_2+K_3}\frac{1}{\mathbb{C}}\frac{\vec{\Lambda}\vec{e}_3}{2\pi} \tag{19.24}$$

Thanks to relations *(19.16)*, and by using the relations of cylindrical coordinates

$$\frac{\cos\varphi}{r} = \frac{x_1}{x_1^2+x_3^2} \Rightarrow x_1 = r\cos\varphi \quad\text{and}\quad \frac{\sin\varphi}{r} = \frac{x_3}{x_1^2+x_3^2} \Rightarrow x_3 = r\sin\varphi \tag{19.25}$$

and remembering that the value of the linear charge $\vec{\Lambda}$ is worth $\vec{\Lambda} = \vec{\boldsymbol{B}}_{edge}\wedge\vec{t}$, the fields $\vec{\omega}^{edge}$ and τ^{edge} outside the dislocation string are written

$$\begin{cases} \vec{\omega}^{edge} = -\dfrac{1}{2}\overrightarrow{\mathrm{rot}}\,\vec{u}^{edge} = \dfrac{K_2}{K_2+K_3}\dfrac{\vec{\Lambda}\vec{e}_3}{2\pi}\dfrac{x_1}{x_1^2+x_3^2}\vec{e}_2 = \dfrac{K_2}{K_2+K_3}\dfrac{\vec{\Lambda}\vec{e}_3}{2\pi}\dfrac{\cos\varphi}{r}\vec{e}_2 \\ \tau^{edge} = -\mathrm{div}\,\vec{u}^{edge} = -\dfrac{K_2}{K_2+K_3}\dfrac{1}{\mathbb{C}}\dfrac{\vec{\Lambda}\vec{e}_3}{2\pi}\dfrac{x_3}{x_1^2+x_3^2} = -\dfrac{K_2}{K_2+K_3}\dfrac{1}{\mathbb{C}}\dfrac{\vec{\Lambda}\vec{e}_3}{2\pi}\dfrac{\sin\varphi}{r} \end{cases} \quad (r>R) \tag{19.26}$$

The elastic displacement field of an edge dislocation

It is possible to find the field of displacement $\vec{u}^{edge}$ outside the edge dislocation string. Indeed as the field $\vec{u}^{edge}$ can only have components along axis $\boldsymbol{O}x_1$ and $\boldsymbol{O}x_3$, and $\vec{u}^{edge}$ must not depend on x_2, we have

$$\vec{u}^{edge} = u_1^{edge}(x_1,x_3)\vec{e}_1 + u_3^{edge}(x_1,x_3)\vec{e}_3 \tag{19.27}$$

so that we have the following differential equations

$$\begin{cases} \vec{\omega}^{edge} = -\dfrac{1}{2}\overrightarrow{\mathrm{rot}}\,\vec{u}^{edge} = \dfrac{1}{2}\left(\dfrac{\partial u_3^{edge}}{\partial x_1}-\dfrac{\partial u_1^{edge}}{\partial x_3}\right)\vec{e}_2 \Rightarrow \dfrac{\partial u_3^{edge}}{\partial x_1}-\dfrac{\partial u_1^{edge}}{\partial x_3} = \dfrac{\vec{\Lambda}\vec{e}_3}{\pi}\dfrac{x_1}{x_1^2+x_3^2} \\ \tau^{edge} = -\mathrm{div}\,\vec{u}^{edge} = -\left(\dfrac{\partial u_1^{edge}}{\partial x_1}+\dfrac{\partial u_3^{edge}}{\partial x_3}\right) \Rightarrow \dfrac{\partial u_1^{edge}}{\partial x_1}+\dfrac{\partial u_3^{edge}}{\partial x_3} = \dfrac{\vec{\Lambda}\vec{e}_3}{2\pi}\dfrac{1}{\mathbb{C}}\dfrac{x_3}{x_1^2+x_3^2} \end{cases} \tag{19.28}$$

We can then try the classic solution *(12.65)* for the field of displacement $\vec{u}^{edge}$

$$\begin{cases} u_1^{edge}(x_1,x_3) = \alpha\arctan\left(\dfrac{x_3}{x_1}\right) + \beta\dfrac{x_1x_3}{x_1^2+x_3^2} \\ u_3^{edge}(x_1,x_3) = \gamma\ln\left(x_1^2+x_3^2\right) + \delta\dfrac{x_1^2-x_3^2}{x_1^2+x_3^2} \end{cases} \tag{19.29}$$

where $\alpha,\beta,\delta,\gamma$ are constants to be determined. The fields $\vec{\omega}^{edge}$ and τ^{edge} are written

$$\begin{cases} \vec{\omega}^{edge} = -\dfrac{1}{2}\overrightarrow{\text{rot}}\,\vec{u}^{edge} = \left[-\dfrac{1}{2}(\alpha+\beta-2\gamma)\dfrac{x_1}{x_1^2+x_3^2} + (\beta+2\delta)\dfrac{x_1 x_3^2}{\left(x_1^2+x_3^2\right)^2} \right]\vec{e}_2 \\ \tau^{edge} = -\text{div}\,\vec{u}^{edge} = \left[(\alpha-\beta-2\gamma)\dfrac{x_3}{x_1^2+x_3^2} + 2(\beta+2\delta)\dfrac{x_1^2 x_3}{\left(x_1^2+x_3^2\right)^2} \right] \end{cases} \tag{19.30}$$

By comparing these relations to *(19.26)*, we deduce three relations

$$\alpha - 2\gamma + \beta = -\frac{K_2}{K_2+K_3}\frac{\vec{\Lambda}\vec{e}_3}{\pi} \quad ; \quad \alpha - 2\gamma - \beta = -\frac{K_2}{K_2+K_3}\frac{1}{\mathbb{C}}\frac{\vec{\Lambda}\vec{e}_3}{2\pi} \quad ; \quad \beta + 2\delta = 0 \tag{19.31}$$

whose solutions are

$$\begin{cases} \alpha - 2\gamma = -\dfrac{\vec{\Lambda}\vec{e}_3}{2\pi}\dfrac{K_2}{K_2+K_3}\dfrac{2\mathbb{C}+1}{2\mathbb{C}} \\ \beta = -\dfrac{\vec{\Lambda}\vec{e}_3}{2\pi}\dfrac{K_2}{K_2+K_3}\dfrac{2\mathbb{C}-1}{2\mathbb{C}} \\ \delta = \dfrac{\vec{\Lambda}\vec{e}_3}{2\pi}\dfrac{K_2}{K_2+K_3}\dfrac{2\mathbb{C}-1}{4\mathbb{C}} \end{cases} \tag{19.32}$$

With this system, the constants α and γ are still undetermined and there exist a infinity of possible solutions of equilibrium. Let's solve this without making a choice for α and γ, by introducing an *adjustable parameter* ς such that we can write

$$\begin{cases} \alpha = -\dfrac{\vec{\Lambda}\vec{e}_3}{2\pi}\dfrac{K_2}{K_2+K_3}\varsigma \\ \gamma = \dfrac{\vec{\Lambda}\vec{e}_3}{2\pi}\dfrac{K_2}{K_2+K_3}\left(\dfrac{2\mathbb{C}+1}{4\mathbb{C}} - \dfrac{\varsigma}{2}\right) \end{cases} \quad \text{and} \quad \begin{cases} \beta = -\dfrac{\vec{\Lambda}\vec{e}_3}{2\pi}\dfrac{K_2}{K_2+K_3}\dfrac{2\mathbb{C}-1}{2\mathbb{C}} \\ \delta = \dfrac{\vec{\Lambda}\vec{e}_3}{2\pi}\dfrac{K_2}{K_2+K_3}\dfrac{2\mathbb{C}-1}{4\mathbb{C}} \end{cases} \tag{19.33}$$

The field of displacement only contains ς, and is written

$$\begin{cases} u_1^{edge}(x_1,x_3) = -\dfrac{\vec{\Lambda}\vec{e}_3}{2\pi}\dfrac{K_2}{K_2+K_3}\varsigma\arctan\left(\dfrac{x_3}{x_1}\right) - \dfrac{\vec{\Lambda}\vec{e}_3}{2\pi}\dfrac{K_2}{K_2+K_3}\dfrac{2\mathbb{C}-1}{2\mathbb{C}}\dfrac{x_1 x_3}{x_1^2+x_3^2} \\ u_3^{edge}(x_1,x_3) = \dfrac{\vec{\Lambda}\vec{e}_3}{2\pi}\dfrac{K_2}{K_2+K_3}\left(\dfrac{2\mathbb{C}+1}{4\mathbb{C}} - \dfrac{\varsigma}{2}\right)\ln\left(x_1^2+x_3^2\right) + \dfrac{\vec{\Lambda}\vec{e}_3}{2\pi}\dfrac{K_2}{K_2+K_3}\dfrac{2\mathbb{C}-1}{4\mathbb{C}}\dfrac{x_1^2-x_3^2}{x_1^2+x_3^2} \end{cases} \tag{19.34}$$

The elastic shear field of an edge dislocation

By applying relation *(2.48)*, namely

$$\vec{\alpha}_i^{edge} = -\overrightarrow{\text{grad}}\ u_i^{edge} + \frac{1}{2}\vec{e}_i \wedge \overrightarrow{\text{rot}}\,\vec{u}^{edge} + \frac{1}{3}\vec{e}_i\,\text{div}\,\vec{u}^{edge} = -\overrightarrow{\text{grad}}\ u_i^{edge} - \vec{e}_i \wedge \vec{\omega}^{edge} - \frac{1}{3}\vec{e}_i\tau^{edge} \tag{19.35}$$

we can also deduce the elastic shear tensor $\vec{\alpha}_i^{\acute{e}l}$ outside the dislocation string by using relations *(19.30)* and *(19.34)*. By calculating the shear tensor and expressing it in polar coordinates, we obtain

$$\begin{cases} \vec{\alpha}_1^{\,edge} = \dfrac{\vec{\Lambda}\vec{e}_3}{4\pi}\dfrac{K_2}{K_2+K_3}\left[-\left(2\varsigma-\dfrac{2}{3\mathbb{C}}+\dfrac{2\mathbb{C}-1}{\mathbb{C}}\cos 2\varphi\right)\dfrac{\sin\varphi}{r}\vec{e}_1+\left(2\varsigma-2+\dfrac{2\mathbb{C}-1}{\mathbb{C}}\cos 2\varphi\right)\dfrac{\cos\varphi}{r}\vec{e}_3\right] \\ \vec{\alpha}_2^{\,edge} = \dfrac{\vec{\Lambda}\vec{e}_3}{4\pi}\dfrac{K_2}{K_2+K_3}\left(\dfrac{2}{3\mathbb{C}}\right)\dfrac{\sin\varphi}{r}\vec{e}_2 \\ \vec{\alpha}_3^{\,edge} = \dfrac{\vec{\Lambda}\vec{e}_3}{4\pi}\dfrac{K_2}{K_2+K_3}\left[+\left(2\varsigma-2+\dfrac{2\mathbb{C}-1}{\mathbb{C}}\cos 2\varphi\right)\dfrac{\cos\varphi}{r}\vec{e}_1+\left(2\varsigma-\dfrac{4}{3\mathbb{C}}+\dfrac{2\mathbb{C}-1}{\mathbb{C}}\cos 2\varphi\right)\dfrac{\sin\varphi}{r}\vec{e}_3\right] \end{cases} \quad (19.36)$$

The elastic rest energy of an edge dislocation

The rest energy of an edge dislocation string is obtained from the elastic potential energy $F = K_2\sum(\vec{\alpha}_i^{\,el})^2 + 2K_3(\vec{\omega}^{\,el})^2 + K_1\tau^2 - K_0\tau$ stored by unit volume outside the string, in which the volume expansion is given by $\tau = \tau_0 + \tau^{edge} + \tau^{(p)}$. We have two terms of energy associated with the edge dislocation: the pure elastic energy F_{dist}^{edge} and the energy term $F_{\tau^{(p)}}^{edge}$ associated with the perturbation $\tau^{(p)}$ of volume expansion

$$\begin{cases} F_{dist}^{edge} = K_2\sum_i(\vec{\alpha}_i^{\,edge})^2 + 2K_3(\vec{\omega}^{\,edge})^2 + K_1\left(\tau^{edge}\right)^2 + \left(2K_1\tau_0 - K_0\right)\tau^{edge} \\ F_{\tau^{(p)}}^{edge} = \left[2K_1\left(\tau_0+\tau^{edge}\right) - K_0\right]\tau^{(p)} + K_1\left(\tau^{(p)}\right)^2 \end{cases} \quad (19.37)$$

The pure rest elastic energy of the edge dislocation is obtained by the following integration

$$\begin{aligned} E_{dist}^{edge} &= \int_a^{R_\infty}\int_0^{2\pi} F_{dist}^{edge}(r,\varphi)\, r\, d\varphi\, dr \\ &= \int_a^{R_\infty}\int_0^{2\pi}\left[K_2\sum_i(\vec{\alpha}_i^{\,edge})^2 + 2K_3(\vec{\omega}^{\,edge})^2 + K_1(\tau^{edge})^2 + \left(2K_1\tau_0 - K_0\right)\tau^{edge}\right] r\, d\varphi\, dr \end{aligned} \quad (19.38)$$

The integration on the the terms containing τ^{edge} gives us, thanks to *(19.26)*

$$E_{dist}^{edge}\Big|_{\tau^{edge}} = \frac{\vec{\Lambda}\vec{e}_3}{2\pi}\frac{K_2(K_0-2K_1\tau_0)}{K_2+K_3}\frac{1}{\mathbb{C}}\int_a^{R_\infty}\underbrace{\left[\int_0^{2\pi}\sin\varphi\, d\varphi\right]}_{=0}dr + \frac{K_1\vec{\Lambda}^2}{4\pi^2\mathbb{C}^2}\left(\frac{K_2}{K_2+K_3}\right)^2\int_a^{R_\infty}\underbrace{\left[\int_0^{2\pi}\sin^2\varphi\, d\varphi\right]}_{=\pi}\frac{dr}{r} \quad (19.39)$$

so that

$$E_{dist}^{edge}\Big|_{\tau^{edge}} = \frac{K_1\vec{\Lambda}^2}{4\pi}\left(\frac{K_2}{K_2+K_3}\right)^2\frac{1}{\mathbb{C}^2}\ln\frac{R_\infty}{a} \quad (19.40)$$

The integration on the term containing $(\vec{\alpha}_i^{\,edge})^2$ is obtained thanks to *(19.36)*

$$E_{dist}^{edge}\Big|_{\vec{\alpha}_i^{\,edge}} = \frac{K_2\vec{\Lambda}^2}{4\pi}\left(\frac{K_2}{K_2+K_3}\right)^2\left[4\varsigma^2 - 8\varsigma + \frac{2}{3}\left(\frac{1-3\mathbb{C}+9\mathbb{C}^2}{\mathbb{C}^2}\right)\right]\ln\frac{R_\infty}{a} \quad (19.41)$$

And finally the integration on the term containing $(\vec{\omega}^{\,coin})^2$ is obtained thanks to *(19.26)*

$$E_{dist}^{edge}\Big|_{\vec{\omega}^{\,edge}} = 2K_3\int_a^{R_\infty}\int_0^{2\pi}(\vec{\omega}^{\,edge})^2\, r\, d\varphi\, dr = \frac{K_3\vec{\Lambda}^2}{2\pi}\left(\frac{K_2}{K_2+K_3}\right)^2\ln\frac{R_\infty}{a} \quad (19.42)$$

The rest energy of the edge dislocation is thus worth, without taking into account the energy of the field of perturbation of expansion

$$E_{dist}^{edge} = \left[K_1 \frac{1}{\mathbb{C}^2} + K_2 \left[4\varsigma^2 - 8\varsigma + \frac{2}{3}\left(\frac{1-3\mathbb{C}+9\mathbb{C}^2}{\mathbb{C}^2} \right) \right] + 2K_3 \right] \frac{\vec{\Lambda}^2}{4\pi} \left(\frac{K_2}{K_2+K_3} \right)^2 \ln \frac{R_\infty}{a} \qquad (19.43)$$

We note that this rest energy contains a term dependent on modulus K_2 associated with the deformation of shear of the media, a term dependent on modulus K_1 associated with the deformations by volume expansion of the media and a term dependent on modulus K_3 associated with the rotation of the media. This rest energy does not depend on modulus K_0.
We will revisit later the terms of perturbation of expansion $\tau^{(p)}$ generated by the energy F_{dist}^{edge} and the energy $F_{\tau^{(p)}}^{edge}$ stored by these perturbations

The kinetic energy of an edge dislocation with low velocity

In the case where the edge dislocation is moving in the direction Ox_1, with velocity $\vec{\mathbf{v}}$ which is small compared to the speed of transversal waves c_t, the components of the field of velocity are obtained by replacing x_1 by $x_1(t) = x_1 - \mathbf{v}t$ in the expressions *(19.34)* of the components of the field of displacement, and by computing the derivation $\vec{\phi}^{edge} = -d\vec{u}^{edge} / dt$

$$\begin{cases} \phi_1^{edge}(x_1,x_3,t) = +\dfrac{\vec{\Lambda}\vec{e}_3}{2\pi} \dfrac{K_2}{K_2+K_3} \dfrac{\partial}{\partial t} \left[\varsigma \arctan\left(\dfrac{x_3}{x_1 - \mathbf{v}t} \right) + \dfrac{2\mathbb{C}-1}{2\mathbb{C}} \dfrac{(x_1-\mathbf{v}t)x_3}{(x_1-\mathbf{v}t)^2 + x_3^2} \right] \\ \phi_1^{edge}(x_1,x_3,t) = -\dfrac{\vec{\Lambda}\vec{e}_3}{2\pi} \dfrac{K_2}{K_2+K_3} \dfrac{\partial}{\partial t} \left[\left(\dfrac{2\mathbb{C}+1}{4\mathbb{C}} - \dfrac{\varsigma}{2} \right) \ln\left[(x_1-\mathbf{v}t)^2 + x_3^2 \right] + \dfrac{2\mathbb{C}-1}{4\mathbb{C}} \dfrac{(x_1-\mathbf{v}t)^2 - x_3^2}{(x_1-\mathbf{v}t)^2 + x_3^2} \right] \end{cases} \qquad (19.44)$$

After derivation in the coordinate system $Ox_1x_2x_3$, it is very useful to transform in the coordinate system $Ox_1'x_2'x_3'$ which is moving with velocity $\vec{\mathbf{v}}$ with the edge dislocation, and in which $x_1 - \mathbf{v}t$ becomes x_1'. We then have for the velocity field and by using polar coordinates $x_1' = r'\cos\varphi'$ and $x_3' = r'\sin\varphi'$ expressed in the plane of the mobile cylinder

$$\begin{cases} \phi_1^{edge}(r',\varphi') = +\dfrac{\vec{\Lambda}\vec{e}_3}{2\pi} \dfrac{K_2}{K_2+K_3} \dfrac{1}{r'} \left[\left(\varsigma + 1 - \dfrac{1}{2\mathbb{C}} \right) \cos^2\varphi' \sin\varphi' + \left(\varsigma - 1 + \dfrac{1}{2\mathbb{C}} \right) \sin^3\varphi' \right] \mathbf{v} \\ \phi_3^{edge}(r',\varphi') = -\dfrac{\vec{\Lambda}\vec{e}_3}{2\pi} \dfrac{K_2}{K_2+K_3} \dfrac{1}{r'} \left[\left(\varsigma - 3 + \dfrac{1}{2\mathbb{C}} \right) \sin^2\varphi' \cos\varphi' + \left(\varsigma - 1 - \dfrac{1}{2\mathbb{C}} \right) \cos^3\varphi' \right] \mathbf{v} \end{cases} \qquad (19.45)$$

The kinetic energy of the edge dislocation is then obtained by computing the following integral

$$E_{kin}^{edge} = \frac{1}{2} m \int_a^{R_\infty} \int_0^{2\pi} n \left(\vec{\phi}^{coin} \right)^2 (r',\varphi')\, r' d\varphi'\, dr' \qquad (19.46)$$

It is a rather lengthy calculation which finally leads to the kinetic energy of a edge dislocation. This energy is given by unit length, and it depends on parameters ς and module $\mathbb{C}$

$$E_{kin}^{edge} = \frac{mn\vec{\Lambda}^2}{8\pi} \left(\frac{K_2}{K_2+K_3} \right)^2 \left(\ln \frac{R_\infty}{a} \right) \left(2\varsigma^2 - 4\varsigma + \frac{1+12\mathbb{C}^2}{4\mathbb{C}^2} \right) \mathbf{v}^2 = \frac{1}{2} M_0^{coin} \mathbf{v}^2 \qquad (19.47)$$

On the inertial mass per unit length of the edge dislocation moving at small speed

We thus have the following relations for the elastic and kinetic energies of an edge dislocation moving in a cosmic lattice

$$\begin{cases} E_{dist}^{edge} \cong \dfrac{\vec{\Lambda}^2}{4\pi}\left(\dfrac{K_2}{K_2+K_3}\right)^2\left(\ln\dfrac{R_\infty}{a}\right)\left\{K_1\dfrac{1}{\mathbb{C}^2}+K_2\left[4\varsigma^2-8\varsigma+\dfrac{2}{3}\left(\dfrac{1-3\mathbb{C}+9\mathbb{C}^2}{\mathbb{C}^2}\right)\right]+2K_3\right\} \\ E_{kin}^{edge} \cong \dfrac{mn\vec{\Lambda}^2}{8\pi}\left(\dfrac{K_2}{K_2+K_3}\right)^2\left(\ln\dfrac{R_\infty}{a}\right)\left\{2\varsigma^2-4\varsigma+\dfrac{1+12\mathbb{C}^2}{4\mathbb{C}^2}\right\}\mathbf{v}^2=\dfrac{1}{2}M_0^{edge}\mathbf{v}^2 \end{cases} \quad (19.48)$$

So that the inertial mass of the edge dislocations is written

$$\begin{cases} M_0^{edge} \cong \left\{\dfrac{(K_2+K_3)\left(2\varsigma^2-4\varsigma+\dfrac{1+12\mathbb{C}^2}{4\mathbb{C}^2}\right)}{K_1\dfrac{1}{\mathbb{C}^2}+K_2\left[4\varsigma^2-8\varsigma+\dfrac{2}{3}\left(\dfrac{1-3\mathbb{C}+9\mathbb{C}^2}{\mathbb{C}^2}\right)\right]+2K_3}\right\}\dfrac{E_{dist}^{edge}}{c_t^2} \\ \mathbb{C} \cong \dfrac{4K_2+6K_1(1+\tau_0)-3K_0}{6(K_2+K_3)} \end{cases} \quad (19.49)$$

The relationship between the energy of distortion and the inertial mass of an edge dislocation differs from the Einstein relationship by the term in brackets, which depends on the parameter ς and moduli K_0, K_1, K_2, K_3, via the module $\mathbb{C}$.

19.3 – Conditions for an edge dislocation to satisfy Einstein's relation

To insure a complete analogy between the topological singularities of our theory and the particles of the real universe, we need for the edge dislocations to satisfy precisely the relation of Einstein. This is true if the term between brackets in *(19.49)* is equal to 1.

Let's start from the fact that the lattice under consideration is finite and that the boundary conditions are 'free', so that the value of the parameter ς is the one that minimizes the distortion energy of the edge dislocation. In section 12.5, we have shown that this condition implies that

Hypothesis 1: $\varsigma = 1$ *(19.50)*

With this condition, we obtain

$$M_0^{edge} \cong \left\{\dfrac{(K_2+K_3)\left(\dfrac{1+4\mathbb{C}^2}{4\mathbb{C}^2}\right)}{K_1\dfrac{1}{\mathbb{C}^2}+K_2\left[\dfrac{2-6\mathbb{C}+6\mathbb{C}^2}{3\mathbb{C}^2}\right]+2K_3}\right\}\dfrac{E_{dist}^{edge}}{c_t^2} \quad (19.51)$$

Let's suppose now *à priori* that conjecture 4 is satisfied, and also that the norm of modulus K_1 is also a lot smaller than the norm of K_3

Conjecture 5: *modulus* K_1 *satisfies relation* $|K_1| << |K_3|$ *(19.52)*

With this additional conjecture, the previous relation becomes a lot simpler

$$M_0^{edge} \cong \frac{1+4\mathbb{C}^2}{8\mathbb{C}^2}\frac{E_{dist}^{edge}}{c_t^2} \tag{19.53}$$

in this case, for the Einstein relation to be satisfied, it suffices that

$$\mathbb{C} = \pm 1/2 \quad \Leftrightarrow \quad K_0 \cong \pm K_3 \tag{19.54}$$

By using conjecture 1 $(K_2 + K_3 > 0)$ and conjecture 4 $(|K_2| << |K_3|)$, we deduce that the existence condition for transversal waves reduces to the fact that modulus K_3 is positive

Hypothesis 2: $K_3 > 0$ (19.55)

Let's make *à priori* a new conjecture (which we will see later is true), that modulus K_0 is positive

Conjecture 6: $K_0 > 0$ (19.56)

The only solution to relation *(19.54)* is then

$$K_0 \cong K_3 > 0 \quad \Rightarrow \quad \mathbb{C} = -1/2 \tag{19.57}$$

Thus, if the elastic moduli of a cosmic lattice satisfy the set of relations implied by conjectures 1 to 6, namely

Conjectures 0 to 6: *the «perfect cosmic lattice»:* $\begin{cases} K_0 = K_3 > 0, \\ 0 < K_1 << K_0 = K_3 \\ 0 \le K_2 << K_3 = K_0 \end{cases}$ (19.58)

the edge and screw dislocations both satisfy 'true' Einstein relations which were deduced from purely classical consideration, without any appeal to a 'principle of special relativity'!

$$M_0^{screw} \cong E_{dist}^{screw} / c_t^2 \quad \text{and} \quad M_0^{edge} \cong E_{dist}^{edge} / c_t^2 \tag{19.59}$$

19.4 – The «perfect cosmic lattice»

We will call this lattice the ***«perfect cosmic lattice»***. A lattice satisfying conjectures 0 to 6 is such that any elastic distortion of any dislocation type will satisfy exactly the relations of Einstein.

On the universality of the relation of Einstein in the perfect cosmic lattice

In a *perfect cosmic lattice*, we can express from relations *(19.48)* the values of potential energy and non-relativistic kinetic energy of an edge dislocation

$$\begin{cases} E_{dist}^{edge} \cong \left(\dfrac{K_2}{K_3}\right)^2 \dfrac{K_3 \vec{\Lambda}^2}{2\pi} \ln\dfrac{R_\infty}{a} \cong \left(\dfrac{K_2}{K_3}\right)^2 \dfrac{K_3 \vec{\boldsymbol{B}}_{edge}^2}{2\pi} \ln\dfrac{R_\infty}{a} \\ E_{kin}^{edge} \cong \left(\dfrac{K_2}{K_3}\right)^2 \dfrac{mn\vec{\Lambda}^2}{4\pi}\left(\ln\dfrac{R_\infty}{a}\right)\boldsymbol{v}^2 \cong \left(\dfrac{K_2}{K_3}\right)^2 \dfrac{mn\,\vec{\boldsymbol{B}}_{edge}^2}{4\pi}\left(\ln\dfrac{R_\infty}{a}\right)\boldsymbol{v}^2 \end{cases} \tag{19.60}$$

Compared to the potential energies and kinetic energies of a non-relativistic screw dislocation

$$\begin{cases} E_{dist}^{screw} = \dfrac{(K_2+K_3)\Lambda^2}{\pi}\ln\dfrac{R_\infty}{a} \cong \dfrac{K_3\Lambda^2}{\pi}\ln\dfrac{R_\infty}{a} \cong \dfrac{K_3\vec{B}_{screw}^2}{4\pi}\ln\dfrac{R_\infty}{a} \\ E_{kin}^{screw} = \dfrac{mn\Lambda^2}{2\pi}\left(\ln\dfrac{R_\infty}{a}\right)\mathbf{v}^2 = \dfrac{mn\vec{B}_{screw}^2}{8\pi}\left(\ln\dfrac{R_\infty}{a}\right)\mathbf{v}^2 \end{cases} \qquad (19.61)$$

we note that the potential energy and the kinetic energy of a non-relativistic edge dislocation in the *perfect cosmic lattice* are both *much smaller* than the potential energy and kinetic energy of a non relativistic screw dislocation with the same Burgers vector $\left|\vec{B}_{edge}\right| = \left|\vec{B}_{screw}\right|$, as we have according to conjecture 4

$$\frac{K_2}{K_3} << 1 \quad \Rightarrow \quad \begin{cases} E_{dist}^{edge} \cong 2(K_2/K_3)^2 E_{dist}^{screw} <<< E_{dist}^{screw} \\ E_{kin}^{edge} \cong 2(K_2/K_3)^2 E_{kin}^{screw} <<< E_{kin}^{screw} \end{cases} \qquad (19.62)$$

We will see later which important role we will attribute, in our analogy with the physical theories of the universe, to the fact that the edge dislocations follow exactly the Einstein relations and the fact that they also present energies which are a lot smaller than the screw dislocations.

19.5 – Spherical singularities of given charge of rotation

Let's imagine that there exists within a perfect cosmic lattice a macroscopic cluster of topological singularities with the shape of a sphere with radius $R_{cluster}$ containing a uniform density λ of charges of rotation, and let's try to calculate the elastic rotation $\vec{\omega}^{el}$ associated with this charge, both inside and outside the singularity.

The field of rotation due to a localized singularity of rotation

Let's consider a spherical coordinate system and let's apply the compatibility relation $\operatorname{div}\vec{\omega}^{el} = \lambda$ in the form of an integration on the spherical volume of radius r

$$\iiint_V \operatorname{div}\vec{\omega}^{el}\, dV = \iiint_V \lambda\, dV \qquad (19.63)$$

The divergence theorem allows us to transform this volume integral into an integral on the boundary surface of the volume

$$\oint\!\!\!\oint_S \vec{\omega}^{el}\vec{n}\, dS = \vec{\omega}^{el}\vec{n}\, 4\pi r^2 = \iiint_V \lambda\, dV \qquad (19.64)$$

where $\vec{n}$ is the normal vector to the spherical surface.

Let's introduce the *global charge* Q_λ given by the integration of the density λ in the volume of the cluster, or given by the sum of the elementary charges $q_{\lambda(i)}$ within the cluster

$$\iiint_V \lambda\, dV = \frac{4}{3}\pi R_{cluster}^3 \lambda = Q_\lambda = \sum_i q_{\lambda(i)} \qquad (19.65)$$

Outside the cluster, meaning for $r > R_{cluster}$, this relation gives us a external field $\vec{\omega}_{ext}^{él}$ of rotation due to the cluster of charges, which is independent of the radius $R_{cluster}$ of the cluster

$$\vec{\omega}_{ext}^{el}\vec{n} = \frac{\lambda R_{cluster}^3}{3}\frac{1}{r^2} = \frac{1}{4\pi}\left(\sum_i q_{\lambda(i)}\right)\frac{1}{r^2} = \frac{Q_\lambda}{4\pi}\frac{1}{r^2} \quad\Rightarrow\quad \vec{\omega}_{ext}^{el} = \frac{Q_\lambda}{4\pi}\frac{\vec{r}}{r^3} \qquad (r > R_{cluster}) \tag{19.66}$$

The field $\vec{\omega}_{int}^{el}$ inside the cluster, meaning for $r < R_{cluster}$, depends on $R_{cluster}$

$$\vec{\omega}_{int}^{el}\vec{n} = \frac{\lambda}{3}r = \frac{Q_\lambda}{4\pi R_{cluster}^3}r = \frac{\sum_i q_{\lambda(i)}}{4\pi R_{cluster}^3}r \quad\Rightarrow\quad \vec{\omega}_{int}^{el} = \frac{Q_\lambda}{4\pi R_{cluster}^3}\vec{r} \qquad (r < R_{cluster}) \tag{19.67}$$

So the norm of $\vec{\omega}$ as a function of r, the distance from the center of the charge, is shown in figure 19.1.

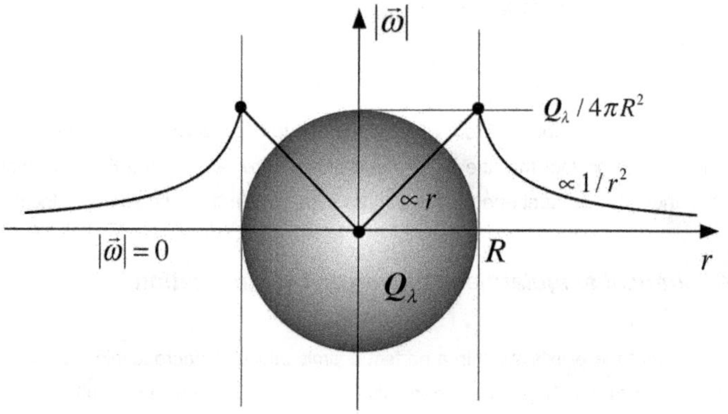

Figure 19.1 *- The norm of the field $\vec{\omega}$ both inside and outside a uniform charge of rotation $Q_\lambda > 0$*

The energy of the field of rotation of rotational singularity

To calculate the elastic energy stored in a lattice by the presence of a rotational field $\vec{\omega}^{el}$ of the singularity, namely the *elastic energy of elastic distortion* $E_{dist}^{(Q_\lambda)}$ *of the lattice due to a charge* Q_λ *of the cluster,* we should in principle calculate the energy associated with the field of rotation, and add the energy of the shear fields associated with said field of rotation. But in the case of a *perfect cosmic lattice,* we have relation $K_2 << K_3$ between the modules of rotation and shear, so that we can in principle neglect the energy associated with the shear and write approximatively, according to expression *(13.6)* of the density of elastic energy of distortion that

$$F^{el} \cong 2K_3(\vec{\omega}^{el})^2 \tag{19.68}$$

The elastic energy density stored by the presence of the cluster of singularities is thus worth $\approx 2K_3(\vec{\omega}^{el})^2$ by unit volume, so that the energy of distortion stored outside the singularity in a quasi infinite volume, meaning for a media such that $R_\infty >> R_{cluster}$, is written

$$E_{dist\,ext}^{(Q_\lambda)} \cong 2K_3\iiint_{V_{ext}}\frac{Q_\lambda^2}{16\pi^2 r^4}dV = 2K_3\int_{R_{cluster}}^{R_\infty}\frac{Q_\lambda^2}{16\pi^2 r^4}4\pi r^2\,dr \cong \frac{K_3 Q_\lambda^2}{2\pi R_{cluster}} \tag{19.69}$$

and the elastic energy of distortion stored inside the singularity

$$E_{dist\,int}^{(Q_\lambda)} \cong 2K_3 \iiint_{V_{int}} \frac{Q_\lambda{}^2 r^2}{16\pi^2 R_{cluster}^6} dV = 2K_3 \int_0^{R_{cluster}} \frac{Q_\lambda{}^2 r^2}{16\pi^2 R_{cluster}^6} 4\pi r^2\, dr = \frac{K_3 Q_\lambda{}^2}{10\pi R_{cluster}} \tag{19.70}$$

The elastic rest energy $E_{dist}^{(Q_\lambda)}$ of the spherical cluster of rotational charges Q_λ and of radius $R_{cluster}$ can be written, in the case of a perfect cosmic lattice, under the form

$$E_{dist}^{(Q_\lambda)} = E_{dist\,ext}^{(Q_\lambda)} + E_{dist\,int}^{(Q_\lambda)} \cong \frac{3K_3 Q_\lambda{}^2}{5\pi R_{cluster}} \tag{19.71}$$

We note that it is finite and essentially depends on the radius $R_{cluster}$ and the charge Q_λ of the cluster.

19.6 – Spherical singularities of given charge of curvature

A macroscopic singularity of radius $R_{cluster}$ can have, besides a global charge of rotation Q_λ, a global charge of curvature Q_θ. Indeed such a singularity can be composed of a cluster of elementary topological singularities of the lattice, such as prismatic loops of dislocations (fig. 9.36) which each possess an elementary curvature charge $q_{\theta(i)}$. If $Q_\theta > 0$, we speak of *a vacancy cluster* since we are missing lattice sites within the cluster, and if $Q_\theta < 0$, we speak of *an interstitial cluster*, since we have interstitials within the cluster.

The flexion field due to a localized singularity of curvature

A localized singularity of curvature is responsible for a non-null and divergent flexion field in it's vicinity. Indeed if we know the density $\theta^{ch}(\vec{r})$ of the charges of curvature within the singularity, we have according to *(8.39)*

$$\operatorname{div}\vec{\chi} = \theta^{ch}(\vec{r}) \tag{19.72}$$

The integration of this relation on a sphere of radius larger than R

$$\iiint_{\substack{sphere\\ r>R}} \operatorname{div}\vec{\chi}\, dV = \iiint_{\substack{sphere\\ r>R}} \theta^{ch}(\vec{r})\, dV = Q_\theta = \sum_i q_{\theta(i)} \tag{19.73}$$

allows us to write, thanks to the divergence theorem

$$\oiint_{\substack{sphere\\ r>R}} \vec{\chi}\, d\vec{S} = Q_\theta = \sum_i q_{\theta(i)} \tag{19.74}$$

and this integral implies the appearance of a field of flexion in the exterior of the singularity of curvature, linked to the spatial curvature of the lattice (see figure 3.3)

$$\vec{\chi}_{ext} = \frac{Q_\theta}{4\pi}\frac{\vec{r}}{r^3} = \frac{\sum_i q_{\theta(i)}}{4\pi}\frac{\vec{r}}{r^3} \tag{19.75}$$

The vectors of this flexion field converge towards the singularity if it is of interstitial nature (excess of lattice sites in the singularity), and is divergent from the singularity if it is of vacancy type (lack of sites within the singularity). Furthermore, we can also notice that the flexion field due to the cluster of charges of curvature does not depend on radius $R_{cluster}$ outside of the cluster.

19.7 – «Electrical» charge of rotation, energies and inertial mass of a twist disclination loop (Twist Loop - TL)

The simplest topological singularity which can have a localized charge of rotation q_λ, amongst all the topological singularities found in a solid lattice in chapter 9, is the twist disclination loop *(Twist Loop - TL)* described in figure 9.40, which could also be called a screw pseudo-dislocation loop *(9.77)*. We recall that such a loop is generated by a rotation $\vec{\Omega}_{TL}$ of the upper plane of a circular cut of the media with angle α_{TL} with respect to the inferior plane. The fact that we glue the two planes together, which have been displaced with respect to each other by a rotation, gives rise, on the plane of the loop, to a surface charge Π_{TL} of rotation. According to *(9.74)* and *(9.78)*, we have

$$\vec{\Omega}_{TL} = \alpha_{TL}\vec{n} = -\vec{n}\Pi_{TL} \quad \Rightarrow \quad q_{\lambda\, TL} = \pi R_{TL}^2 \Pi_{TL} = -\pi R_{TL}^2 \vec{\Omega}_{TL} \qquad (19.76)$$

This global charge $q_{\lambda\, TL}$ is in fact a global charge of rotation of the *TL* as seen at great distance from the loop. This means that such a loop can indeed behave like the source of a divergent field $\vec{\omega}$ of rotation within the solid media. We have also shown in chapter 9 that such a loop can be seen in a different way. Indeed the fact that we effect a rotation of the two planes with respect to each other creates a curvilinear displacement $R_{TL}\alpha_{TL}$ along the loop which is similar to that of a screw dislocation. The curvilinear Burgers vector $\vec{\underset{\frown}{B}}_{TL}$ and the linear charge Λ_{TL} of this loop of pseudo-dislocations is then worth

$$\vec{\underset{\frown}{B}}_{TL} = R_{TL}\alpha_{TL}\vec{t} \quad \Rightarrow \quad \Lambda_{TL} = -\vec{\underset{\frown}{B}}_{TL}\vec{t}\,/\,2 \quad \Rightarrow \quad q_{\lambda\, TL} = 2\pi R_{TL}\Lambda_{TL} = -\pi R_{TL}\vec{\underset{\frown}{B}}_{TL}\vec{t} \qquad (19.77)$$

We obtain the same value of the global charge tan that obtained by considering the surface charge Π_{TL}, which allows us to consider this topological singularity either as a twist disclination loop or a screw pseudo-dislocation loop!

The internal and external fields of rotation of a TL

A *TL* can be considered as a screw pseudo-dislocation with curvilinear Burgers vector $\vec{\underset{\frown}{B}}_{TL}$. Let's consider then the field of rotation within the torus encompassing a loop situated in the plane Ox_1x_3 (fig. 19.2), by introducing the distance ξ which separates, in a perpendicular section of the torus, a given point from the center (where the loops is). The norm of the rotational field near to this point ξ can be deduced from *(19.2)*

$$\left|\vec{\omega}_{int}^{torus}\right|(\xi) \cong \frac{|\Lambda|}{2\pi}\frac{1}{\xi} = \frac{\left|\vec{\underset{\frown}{B}}_{TL}\right|}{4\pi}\frac{1}{\xi} \qquad (19.78)$$

Let's now consider the rotational field away from the loop, which corresponds to the divergent external field of a charge $q_{\lambda\, TL}$ of rotation, and whose norm can be deduced from *(19.4)*

$$\left|\vec{\omega}_{ext}^{charge}\right| = \frac{\left|q_{\lambda\, TL}\right|}{4\pi}\frac{1}{r^2} = \frac{R_{TL}\left|\vec{\underset{\frown}{B}}_{TL}\right|}{4}\frac{1}{r^2} \qquad (19.79)$$

where r represents the distance that separates the point to center O of the disclination loop. In the case of the disclination loop, we go from a divergent near field of rotation with toroid symmetry to a divergent far field of rotation with spherical symmetry.

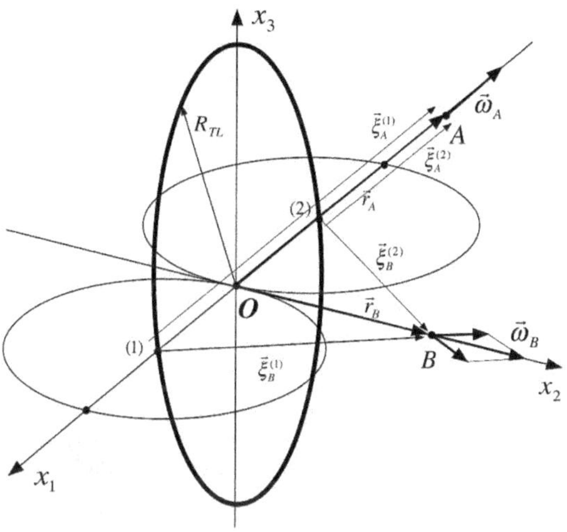

Figure 19.2 *- Twist disclination loop (TL) with radius R_{TL} in the plane Ox_1x_3 and the field of rotation $\vec{\omega}_A$ in the plane of the loop and $\vec{\omega}_B$ in the plane perpendicular to the loop*

To approximatively find the distance from which the field transition from toroidal symmetry to spherical symmetry, we must compare the expressions *(19.78) and (19.79)*, and suppose we go from one field to the other when the magnitudes of the fields becomes approximately equal, which means that

- in the plane of the loop, at point $\boldsymbol{A}$, we must have

$$\left|\vec{\omega}_{int\,A}^{torus}\right|(\xi_A^{(2)})-\left|\vec{\omega}_A^{torus}\right|(\xi_A^{(1)})\approx\left|\vec{\omega}_{ext\,A}^{charge}\right|(r_A) \qquad (19.80)$$

which is translated into relation

$$\frac{\left|\vec{\boldsymbol{B}}_{TL}\right|}{4\pi}\frac{1}{\xi_A^{(2)}}-\frac{\left|\vec{\boldsymbol{B}}_{TL}\right|}{4\pi}\frac{1}{\xi_A^{(2)}+2R_{TL}}\approx\frac{R_{TL}\left|\vec{\boldsymbol{B}}_{TL}\right|}{4}\frac{1}{(\xi_A^{(2)}+R_{TL})^2} \qquad (19.81)$$

The search for the values of $\xi_A^{(2)}$ and r_A for which the toroidal field is roughly equal to the spherical field in the plane of the loops gives the following values

$$\xi_A^{(2)}=\frac{(4-2\pi)\pm\sqrt{(2\pi-4)^2+8(\pi-2)}}{2(\pi-2)}R_{TL}\cong 0.66R_{TL}\quad\Rightarrow\quad r_A=R_{TL}+\xi_A^{(2)}\cong 1{,}66R_{TL} \qquad (19.82)$$

- in the plane perpendicular to the loop, at point $\boldsymbol{B}$, we have

$$\left|\vec{\omega}_{int\,B}^{torus}(\vec{\xi}_B^{(1)})+\vec{\omega}_{int\,B}^{torus}(\vec{\xi}_B^{(2)})\right|\approx\left|\vec{\omega}_{ext\,B}^{charge}\right|(r_B) \qquad (19.83)$$

which is translated into relation

$$\frac{\left|\vec{\boldsymbol{B}}_{TL}\right|}{2\pi}\frac{r_B}{R_{TL}^2+{r_B}^2}\approx\frac{R_{TL}\left|\vec{\boldsymbol{B}}_{TL}\right|}{4}\frac{1}{{r_B}^2} \qquad (19.84)$$

The search for the value r_B for which the toroidal field is roughly equal to the spherical field gives in that case the following value

$$2r_B^3 \approx \pi R_{TL}\left(R_{TL}^2 + r_B^2\right) \quad \Rightarrow \quad r_B \approx 2R_{TL} \tag{19.85}$$

We deduce that the transition between the internal field of toroidal symmetry and the external field of spherical symmetry is situated at a distance r from the center of the loop which is roughly $r_B \approx 2R_{TL}$. This realization will allow us to calculate the energies and inertial mass of the twist disclination loop.

The energies and the inertial mass of a twist disclination loop

The energy of distortion of a twist disclination loop is the energy stored by the rotation generated by the screw pseudo-dislocations with radius R_{TL} in a torus where the central fiber is the disclination loop and for which the radius of the section corresponds roughly to R_{TL}, to which we add the energy of the external field with spherical symmetry for distances larger than $r \geq 2R_{TL}$. The calculation of the exact value of this energy is rather complex, due in part to the fact that the field of rotation is exactly zero in the center of the loop. However, we can give an approximation of the distortion energy of the loop, by using the energy of a linear dislocation to calculate the energy of the curved dislocation. In the case where the radius R_{TL} of the loop is much greater than the core radius a of the screw pseudo-dislocation ($R_{TL} >> a$), this approximation must give the real value of the energy of distortion within the torus encompassing the loop and we can correct it by introducing a constant A_{TL} which corrects the value of the external radius of the torus to give us a better approximation of the actual energy. We will thus write the energy of distortion of a toroidal field from the energy of a screw dislocation by unit length *(19.7)*, as the energy contained in the torus encompassing the screw loop disclination and which is approximatively worth

$$E_{dist\ torus}^{TL} \cong 2\pi R_{TL}\left(\frac{(K_2+K_3)\Lambda_{TL}^2}{\pi}\ln\frac{A_{TL}R_{TL}}{a}\right) = \frac{1}{2}(K_2+K_3)R_{TL}\vec{\underline{B}}_{TL}^2\ln\frac{A_{TL}R_{TL}}{a} \tag{19.86}$$

where a is the core radius of the screw pseudo-dislocation, which is of the order of magnitude of the cosmic lattice step in the presence of a field of expansion τ, $A_{TL}R_{TL}$ is the reach of the toroidal field of the loop and A_{TL} is a constant which can only be obtained by the exact calculation of the energy of the loop, but which much be close to unity given our previous discussion. To simplify the rest of our expose, since the ratio $A_{TL}R_{TL}/a$ roughly does not depend on the background expansion, we will consider it constant in first approximation, and we will introduce a constant ζ_{TL} which is intrinsic to the twist disclination loop and which is worth

$$\zeta_{TL} = \ln\left(A_{TL}R_{TL}/a\right) \equiv cste \quad \text{with} \quad A_{TL} \approx 1 \tag{19.87}$$

So that we can write the energy of the toroidal field in the following simplified manner

$$E_{dist\ torus}^{TL} \cong 2(K_2+K_3)\zeta_{TL}R_{TL}\Lambda_{TL}^2 = \frac{1}{2}(K_2+K_3)\zeta_{TL}R_{TL}\vec{\underline{B}}_{TL}^2 \tag{19.88}$$

We can then compare this energy of the toroidal field of the loop to the energy associated with the spherical field of rotation in the far field of the loop, which is taken into account at distance greater than $2R_{TL}$ to the loop, and which is due to the rotational charge $\boldsymbol{q}_{\lambda TL}$, namely

$$\vec{\omega}_{ext}^{el} = \frac{q_{\lambda TL}}{4\pi}\frac{\vec{r}}{r^3} = \frac{1}{2}R_{TL}\Lambda_{TL}\frac{\vec{r}}{r^3} = -\frac{1}{4}R_{TL}\vec{B}_{TL}\vec{t}\frac{\vec{r}}{r^3} \qquad (r > 2R_{TL}) \tag{19.89}$$

The energy of the field is directly deduced from the value *(19.69)* and is consequently worth

$$E_{dist\,ext}^{TL} \cong 2K_3 \int_{2R_{TL}}^{R_\infty} \frac{q_{\lambda TL}^2}{16\pi^2 r^4} 4\pi r^2\,dr \cong \frac{K_3 q_{\lambda TL}^2}{4\pi R_{TL}} \cong \pi K_3 R_{TL}\Lambda_{TL}^2 \cong \frac{\pi}{4}K_3 R_{TL}\vec{B}_{TL}^2 \tag{19.90}$$

By comparing this value with the energy of the toroidal field *(19.88)*, we obtain the ratio

$$\frac{E_{dist\,ext}^{TL}}{E_{dist\,torus}^{TL}} \cong \frac{\pi K_3}{2(K_2+K_3)\zeta_{TL}} \cong \frac{\pi}{2\ln\left(A_{TL}R_{TL}/a\right)} \tag{19.91}$$

Let's admit the following new conjecture

Conjecture 7: *the radius of a twist disclination loop is much greater than the cosmic lattice step:*

$$\ln\left(A_{TL}R_{TL}/a\right) >> 1 \tag{19.92}$$

With this conjecture, the energy associated to the external field of rotation becomes perfectly negligeable vis-à-vis the toroidal energy of the loop. As a consequence the energy E_{dist}^{TL} of the twist disclination loop is essentially contained in the toroidal field of the loop

$$E_{dist}^{TL} \cong E_{dist\,torus}^{TL} \cong 2(K_2+K_3)\zeta_{TL}R_{TL}\Lambda_{TL}^2 = \frac{1}{2}(K_2+K_3)\zeta_{TL}R_{TL}\vec{B}_{TL}^2 \tag{19.93}$$

The relativistic kinetic energy of the *TL* is the energy which is stored by the movements of the lattice generated by the mobile screw pseudo-dislocation. By using the relation *(19.9)*, and admitting hypothesis *(19.92)*, the kinetic energy of the loop contained in the previous torus is roughly worth

$$E_{kin}^{TL} \cong 2\pi R_{TL}\left(\frac{mn\Lambda_{TL}^2\ln\left(A_{TL}R_{TL}/a\right)}{2\pi}\mathbf{v}^2\right) = mn\zeta_{TL}R_{TL}\Lambda_{TL}^2\mathbf{v}^2 = \frac{1}{4}mn\zeta_{TL}R_{TL}\vec{B}_{TL}^2\mathbf{v}^2 \tag{19.94}$$

Again, the external kinetic energy of rotation is negligeable in comparison with this kinetic energy, so that we can consider that the kinetic energy of the loop is essentially confined to the toroidal field of the loop. We thus deduce that the Einstein relation is perfectly applied to the non-relativistic kinetic energy of the twist disclination loop

$$E_{kin}^{TL} = \frac{1}{2}\frac{E_{dist}^{TL}}{c_t^2}\mathbf{v}^2 = \frac{1}{2}M_0^{TL}\mathbf{v}^2 \tag{19.95}$$

We deduce that the inertial mass of the twist disclination loop, expressed from the radius of the loop and it's Burgers pseudo-vector

$$M_0^{TL} = \frac{E_{dist}^{TL}}{c_t^2} \cong \frac{2}{c_t^2}(K_2+K_3)\zeta_{TL}R_{TL}\Lambda_{TL}^2 = \frac{1}{2c_t^2}(K_2+K_3)\zeta_{TL}R_{TL}\vec{B}_{TL}^2 \tag{19.96}$$

We know that the existence of elastic fields of distortion induces, via their energy, a field of perturbation of expansion. We will return in more detail later on this perturbation of the expansion associated with the twist disclination loop.

19.8 – «Electric» interaction between localized topological singularities

Supposed first that two twist disclination loops have charges $q_{\lambda TL(1)}$ and $q_{\lambda TL(2)}$ of rotation. There exists an interaction force between these two loops, of electrical type, and this interaction force can be deduced in a very generic fashion by using the force of Peach and Koehler. Indeed, the spherical field of external rotation generated by the charge $q_{\lambda TL(1)}$ situated in the center of the coordinate system is given by *(19.66)*

$$\vec{\omega}_{ext(1)}^{TL} = \frac{q_{\lambda TL(1)}}{4\pi}\frac{\vec{r}}{r^3} \quad (r > 2R_{TL}) \tag{19.97}$$

We deduce thanks to *(17.17)* the torsor of moments due to the charge $q_{\lambda TL(1)}$

$$\vec{m}_{ext(1)}^{TL} = 4\left(K_2 + K_3\right)\vec{\omega}_{ext(1)}^{TL} = \left(K_2 + K_3\right)\frac{q_{\lambda TL(1)}}{\pi r^3}\vec{r} \qquad (r > 2R_{TL}) \tag{19.98}$$

If a twist disclination loop with charge $q_{\lambda TL(2)}$ of rotation is then found in position $\vec{r} = \vec{d}$, the force acting on this charge due to the charge $q_{\lambda TL(1)}$ is the force of Peach and Koehler

$$\vec{F}_{PK(2)}^{TL} = q_{\lambda TL(2)}\vec{m}_{ext(1)}^{TL} = \frac{K_2 + K_3}{\pi}\frac{q_{\lambda TL(1)}q_{\lambda TL(2)}}{d^2}\frac{\vec{d}}{d} \tag{19.99}$$

Thus, the force between the two charges is repulsive if $q_{\lambda TL(1)}q_{\lambda TL(2)} > 0$ and attractive if $q_{\lambda TL(1)}q_{\lambda TL(2)} < 0$. This interaction force between the charges of rotation of the twist disclination loops is the perfect analog to the interaction force $\vec{F}^{electric} = q_{(1)}q_{(2)}\vec{d} / 4\pi\varepsilon_0 d^3$ between two electrical charges $q_{(1)}$ and $q_{(2)}$ in electromagnetism, and thus fits perfectly with the analogy we developed in section 17.3 with the Maxwell's equations.

Electric interaction between two macroscopic clusters of singularities

As relation *(19.99)* is perfectly independent of the size of the loops, it can be generalized without issue to two macroscopic clusters of topological singularities with macroscopic charges of rotation $Q_{\lambda(1)}$ and $Q_{\lambda(2)}$ separated by a distance d

$$\vec{F}_{PK(2)} = \frac{K_2 + K_3}{\pi}\frac{Q_{\lambda(1)}Q_{\lambda(2)}}{d^2}\frac{\vec{d}}{d} \tag{19.100}$$

In this case of two macroscopic clusters, the force acting between them does not depend on the respective radius $R_{amas(1)}$ and $R_{amas(2)}$ of the two clusters.

19.9 – «Gravitational» charge of curvature, energies and inertial mass of a prismatic edge dislocation loop (Edge Loop - EL)

The simplest microscopic structure of a localized charge of curvature q_θ, amongst all the topological singularities found in a solid lattice in chapter 9, is the prismatic edge dislocation loop *(EL)* described in figure 9.36b.

Such loops posses a Burgers vector which is perpendicular to the plan of the loop and exhibit a global scalar charge $q_{\theta EL}$ of curvature given by *(9.68)* to *(9.70)*. Since $\vec{B}_{EL} = \vec{t} \wedge \vec{\Lambda}_{EL}$,we have

$$q_{\theta EL} = -2\pi\vec{n}\vec{B}_{EL} = -2\pi\vec{n}\left(\vec{t}\wedge\vec{\Lambda}_{EL}\right) = 2\pi\vec{\Lambda}_{EL}\vec{m} \tag{19.101}$$

Given this last relation, we deduce that an *EL* of vacancy type has a product $\vec{\Lambda}_{EL}\vec{m}$ which is positive, and thus a charge $q_{\theta EL}$ which is positive, while an *EL* of interstitial type has a product $\vec{\Lambda}_{EL}\vec{m}$ which is negative, and thus a charge $q_{\theta EL}$ which is also negative.

The energies and the inertial mass of an EL

Let's consider an *EL* with radius R_{EL}, the distortions induced in the lattice are those of an edge dislocation. We can then calculate approximatively the elastic energy of distortion of this loop as the energy that is stored in the lattice by the distortions of this loop as the energy that is stored in the lattice by the elastic distortions generated by the edge dislocation in a torus centered on the loop. Using the same arguments we used for the *TL* we deduce that in a perfect cosmic lattice, the elastic energy of distortion of an *EL* is essentially contained in the toroidal fields surrounding the loop

$$E_{dist}^{EL} \cong E_{dist\ torus}^{EL} \cong 2\pi R_{EL}\left[\left(\frac{K_2}{K_3}\right)^2 \frac{K_3\vec{\Lambda}_{EL}^2}{2\pi}\ln\frac{A_{EL}R_{EL}}{a}\right] \cong \left(\frac{K_2}{K_3}\right)^2 K_3\zeta_{EL}R_{EL}\vec{B}_{EL}^2 \tag{19.102}$$

where A_{EL} is a constant close to 1, and which should be calculated exactly by the integration of energy on the fields within the torus, and where $\zeta_{EL} = \ln\left(A_{EL}R_{EL}/a\right)$ is a constant intrinsic to the *EL*.

Outside the *EL*, the fields due to the *EL* are reduced to *a field of flexion (19.75) with spherical symmetry*, which is then written

$$\vec{\chi}_{ext}^{EL} = \frac{q_{\theta EL}}{4\pi}\frac{\vec{r}}{r^3} = -\frac{1}{2}\vec{n}\vec{B}_{EL}\frac{\vec{r}}{r^3} = \frac{1}{2}\vec{\Lambda}_{EL}\vec{m}\frac{\vec{r}}{r^3} \qquad (r > 2R_{EL}) \tag{19.103}$$

It is clear that this field of flexion must be associated with a perturbation of the field of volume expansion which must posses a given energy. We will later revisit this problem, and we will show that the energy associated with this field of flexion is negligible vis-à-vis the energy of distortion *(19.102)*, so that the energy of the *EL* is essentially contained in the toroidal fields in the immediate vicinity of the loop.

The non-relativistic kinetic energy of the loop is essentially the kinetic energy stored in the lattice by the dynamic distortions generated by the *EL* in the torus centered on the loop. By using relations *(19.60)* we deduced for an *EL*, we deduce that in the cosmic lattice, we have approximately the following kinetic energy for an *EL*

$$E_{kin}^{EL} \cong \frac{1}{2}\left(\frac{K_2}{K_3}\right)^2 mnR_{EL}\vec{\Lambda}_{EL}^2\left(\ln\frac{A_{EL}R_{EL}}{a}\right)\mathbf{v}^2 = \frac{1}{2}\left(\frac{K_2}{K_3}\right)^2 mn\zeta_{EL}R_{EL}\vec{B}_{EL}^2\mathbf{v}^2 \tag{19.104}$$

Again, *this kinetic energy is essentially contained in the immediate vicinity of the loop,* and we notice that the *relation of Einstein* is perfectly in line with the non-relativistic kinetic energy of the dislocation loop of the *EL* since

$$E_{kin}^{EL} \cong \frac{1}{2}\frac{E_{dist}^{EL}}{c_t^2}\mathbf{v}^2 \tag{19.105}$$

We thus deduce the inertial mass of the *EL* in the cosmic lattice

$$M_0^{EL} = \frac{E_{dist}^{EL}}{c_t^2} \cong \left(\frac{K_2}{K_3}\right)^2 \frac{1}{c_t^2} K_3 \zeta_{EL} R_{EL} \vec{\Lambda}_{EL}^2 = \left(\frac{K_2}{K_3}\right)^2 \frac{1}{c_t^2} K_3 \zeta_{EL} R_{EL} \vec{B}_{EL}^2 \qquad (19.106)$$

We will revisit later the field of perturbation of expansion associated with the loop.

19.10 – «Electric» dipolar field of rotation, energies and inertial mass of a mixed sliding dislocation loop (Mixed Loop - ML)

The mixed dislocation loop *(ML)*, which is of vectorial nature, was obtained by sliding (a parallel translation in the plane of the loop) in the direction of the Burgers vector, so that the lattice does not show "extra-material" in this case (fig. 9.36a). On the other hand, the presence of a screw component in the regions where $\vec{B}_{ML} \parallel \vec{t}$ induces a dipolar field of rotation $\vec{\omega}_{dipolar}^{ML}(r,\theta,\varphi)$ in the vicinity of the sliding loop.

The energies and the inertial mass of a ML

if we consider a *ML* of radius R_{ML}, the distortions induced in the lattice at short distance are those of a screw dislocation for angles $\alpha = 0$ and $\alpha = \pi$, and those of an edge dislocation for angles $\alpha = \pi/2$ and $\alpha = 3\pi/2$. We can consider that we morph continuously as a function of angle α from a screw dislocation to an edge dislocation. The energy of distortion associated with the curved string is stored essentially in the torus centered on the loop. But as the two parts which are edge and the two parts with are loop are respectively of opposite charge, the field associated to the edge parts and the screw parts diminish very quickly at great distance from the loop. For example, the module of the field of rotation in the plane of the loop and on a diameter going through the screw parts behave as

$$\left|\vec{\omega}_{dist}^{ML}\right| \propto \frac{\left|\vec{B}_{ML}\right| R_{ML}}{4\pi r^2} \qquad (r >> R_{ML}) \qquad (19.107)$$

in the far field. If the radius R_{ML} of the loop is largely superior to the step a of the lattice, we can take into account this rapid decrease of the dipolar field by imagining that the field in the neighborhood of the string is that of a dislocation. We can thus calculate approximatively the rest energy of such a loop by integrating the energies by unit length of the string inside the torus for the screw and edge components as a function of angle α. We have

$$E_{dist}^{ML} \cong \int_0^{2\pi} \frac{(K_2+K_3)\left(\vec{B}_{ML}\cos\alpha\right)^2}{4\pi}\left(\ln\frac{A_{ML}R_{ML}}{a}\right) R_{ML}\,d\alpha + \int_0^{2\pi}\left(\frac{K_2}{K_3}\right)^2 \frac{K_3\left(\vec{B}_{ML}\sin\alpha\right)^2}{2\pi}\left(\ln\frac{A_{ML}R_{ML}}{a}\right) R_{ML}\,d\alpha \qquad (19.108)$$

A precise calculation of the energy will lead to the value of the constant A_{ML} characteristic to the geometry of the mixed loop and which must be close to 1, so that

$$E_{dist}^{ML} \cong \frac{(K_2+K_3)R_{ML}\vec{B}_{ML}^2}{4}\ln\frac{A_{ML}R_{ML}}{a} + \left(\frac{K_2}{K_3}\right)^2 \frac{K_3 R_{ML}\vec{B}_{ML}^2}{2}\ln\frac{A_{ML}R_{ML}}{a} \qquad (19.109)$$

Since $K_2 << K_3$ in the perfect cosmic lattice, we have approximatively

$$E_{dist}^{ML} = \frac{\zeta_{ML} R_{ML}\vec{B}_{ML}^2}{4}\left[K_2 + K_3\left(1+2\left(\frac{K_2}{K_3}\right)^2\right)\right] \cong \frac{1}{4}(K_2+K_3)\zeta_{ML}R_{ML}\vec{B}_{ML}^2 \qquad (19.110)$$

where $\zeta_{ML} = \ln\left(A_{ML}R_{ML} / a\right)$ is a constant parameter of the *ML*.
We should in theory take into account the energy of distortion due to the rotation field exterior to the loop. However it is smaller than the energy of distortion *(19.90)* associated with the field of external rotation of a twist disclination loop, so that we can neglect this energy in comparison to the energy of distortion E_{dist}^{ML} contained in the torus. This means, again, that the energy of the *ML* is essentially contained in the immediate vicinity of the dislocation loop.
The non-relativistic kinetic energy of the *ML* is approximatively computed like it's elastic energy of distortion

$$E_{kin}^{ML} = \int_0^{2\pi} \frac{mn\left(\vec{B}_{ML}\cos\alpha\right)^2}{8\pi}\left(\ln\frac{A_{ML}R_{ML}}{a}\right)\mathbf{v}^2 R_{ML}\,d\alpha + \int_0^{2\pi}\left(\frac{K_2}{K_3}\right)^2 \frac{mn\left(\vec{B}_{ML}\sin\alpha\right)^2}{4\pi}\left(\ln\frac{A_{ML}R_{ML}}{a}\right)\mathbf{v}^2 R_{ML}\,d\alpha \tag{19.111}$$

so that

$$E_{kin}^{ML} \cong \frac{mnR_{ML}\vec{B}_{ML}^2}{8}\zeta_{ML}\left[1+2\left(K_2/K_3\right)^2\right]\mathbf{v}^2 \cong \frac{1}{8}mn\zeta_{ML}R_{ML}\vec{B}_{ML}^2\mathbf{v}^2 \tag{19.112}$$

We notice that the energies E_{dist}^{ML} and E_{kin}^{ML} are in fact those given by the screw part of the sliding loop and that those are essentially contained in the immediate vicinity of the dislocation loop. The relation of Einstein is thus exact in the case of a *ML*, since

$$E_{kin}^{ML} \cong \frac{1}{2}\frac{E_{dist}^{ML}}{c_t^2}\mathbf{v}^2 \tag{19.113}$$

We deduce the inertial mass of this loop in a perfect cosmic lattice

$$M_0^{ML} = \frac{E_{dist}^{ML}}{c_t^2} \cong \frac{1}{4c_t^2}\left(K_2+K_3\right)\zeta_{ML}R_{ML}\vec{B}_{ML}^2 \tag{19.114}$$

With regards to the perturbations of the field of expansion associated with this loop, we will revisit them later in more detail in chapter 24, where we will see that the energy associated with the field is negligible vis-à-vis the energy of distortion E_{dist}^{ML} contained in the torus.

19.11 – Elementary topological building blocks for the world of fundamental particles

In table 19.1, we have shown the complete set of results we have obtained for the 3 types of elementary loops that one can find in a cosmic lattice *(TL, EL and ML)*. In our analogy with the real world, the three types of disclination and dislocation loops could constitute the topological building blocks of the cosmic lattice which could allow us to build more complex structures which would in turn be the analogs of the fundamental particles of the standard model.

The TL, the simplest topological singularity giving us an electrical charge

At a certain distance from the center of a *TL*, roughly at $2R_{TL}$, the external field of rotation of the *TL* behaves exactly like the external field of a spherical charge $q_{\lambda TL} = 2\pi R_{TL}\Lambda_{TL}$. We can then ask what should the radius R_{ch} of the spherical charge be so that it presents an elastic energy of distortion equal to the energy of distortion of the *TL*.

Table 19.1 - Energies and inertial masses of elementary loops of dislocation and disclination in a «perfect cosmic lattice»

The twist disclination loop (Twist Loop - TL)

$$\begin{cases} q_{\lambda TL} = 2\pi R_{TL}\Lambda_{TL} = -\pi R_{TL}\,\vec{\underset{\sim}{B}}_{TL}\vec{t} \quad \& \quad q_{\theta TL} = 0 \\ \vec{\omega}_{ext}^{TL} = \dfrac{q_{\lambda TL}}{4\pi}\dfrac{\vec{r}}{r^3} \end{cases}$$

$$\begin{cases} E_{dist}^{TL} \cong E_{dist\;torus}^{TL} \cong 2(K_2+K_3)\zeta_{TL}R_{TL}\Lambda_{TL}^2 = \dfrac{1}{2}(K_2+K_3)\zeta_{TL}R_{TL}\vec{\underset{\sim}{B}}_{TL}^2 \\ E_{kin}^{TL} \cong E_{kin\;torus}^{TL} \cong mn\zeta_{TL}R_{TL}\Lambda_{TL}^2\mathbf{v}^2 = \dfrac{1}{4}mn\zeta_{TL}R_{TL}\,\vec{\underset{\sim}{B}}_{TL}^2\mathbf{v}^2 \\ M_0^{TL} = \dfrac{E_{dist}^{TL}}{c_t^2} = \dfrac{2}{c_t^2}(K_2+K_3)\zeta_{TL}R_{TL}\Lambda_{TL}^2 = \dfrac{1}{2c_t^2}(K_2+K_3)\zeta_{TL}R_{TL}\vec{\underset{\sim}{B}}_{TL}^2 \\ \zeta_{TL} = \ln(A_{TL}R_{TL}/a) \end{cases}$$

The prismatic edge dislocation loop (Edge Loop - EL)

$$\begin{cases} q_{\lambda EL} = 0 \quad \& \quad q_{\theta EL} = -2\pi\vec{n}\left(\vec{t}\wedge\vec{\Lambda}_{EL}\right) = 2\pi\vec{\Lambda}_{EL}\vec{m} = -2\pi\vec{n}\vec{B}_{EL} \\ \vec{\chi}_{ext}^{EL} = \dfrac{q_{\theta EL}}{4\pi}\dfrac{\vec{r}}{r^3} \end{cases}$$

$$\begin{cases} E_{dist}^{EL} \cong E_{dist\,torus}^{EL} \cong \left(\dfrac{K_2}{K_3}\right)^2 K_3\zeta_{EL}R_{EL}\vec{\Lambda}_{EL}^2 \cong \left(\dfrac{K_2}{K_3}\right)^2 K_3\zeta_{EL}R_{EL}\,\vec{B}_{EL}^2 \\ E_{kin}^{EL} \cong E_{kin\,torus}^{EL} \cong \dfrac{1}{2}\left(\dfrac{K_2}{K_3}\right)^2 mn\zeta_{EL}R_{EL}\vec{\Lambda}_{EL}^2\mathbf{v}^2 = \dfrac{1}{2}\left(\dfrac{K_2}{K_3}\right)^2 mn\zeta_{EL}R_{EL}\,\vec{B}_{EL}^2\mathbf{v}^2 \\ M_0^{EL} = \dfrac{E_{dist}^{EL}}{c_t^2} = \left(\dfrac{K_2}{K_3}\right)^2\dfrac{1}{c_t^2}K_3\zeta_{EL}R_{EL}\vec{\Lambda}_{EL}^2 = \left(\dfrac{K_2}{K_3}\right)^2\dfrac{1}{c_t^2}K_3\zeta_{EL}R_{EL}\,\vec{B}_{EL}^2 \\ \zeta_{EL} \cong \ln(A_{EL}R_{EL}/a) \end{cases}$$

The sliding mixed dislocation loop (Mixed Loop - ML)

$$\begin{cases} q_{\lambda ML} = 0 \quad \& \quad q_{\theta ML} = 0 \\ \exists \text{ an external dipolar field of rotation } \vec{\omega}_{dipolar}^{ML}(r,\theta,\varphi) \end{cases}$$

$$\begin{cases} E_{dist}^{ML} \cong E_{dist\,tore}^{ML} \cong \dfrac{1}{4}(K_2+K_3)\zeta_{ML}R_{ML}\vec{B}_{ML}^2 \\ E_{kin}^{ML} \cong E_{kin\,torus}^{ML} \cong \dfrac{1}{8}mn\zeta_{ML}R_{ML}\vec{B}_{ML}^2\mathbf{v}^2 \\ M_0^{ML} = \dfrac{E_{dist}^{ML}}{c_t^2} \cong \dfrac{1}{4c_t^2}(K_2+K_3)\zeta_{ML}R_{ML}\vec{B}_{ML}^2 \\ \zeta_{ML} = \ln(A_{ML}R_{ML}/a) \end{cases}$$

With the charge value of $q_{\lambda TL}$ giving us a far field similar to the *TL*, the energy of the spherical charge of radius R_{ch} is worth, according to *(19.71)*

$$E_{dist}^{q_\lambda} \cong \frac{3K_3 q_{\lambda TL}^2}{5\pi R_{ch}} \cong \frac{12K_3\pi R_{TL}^2 \Lambda_{TL}^2}{5R_{ch}} \qquad (19.115)$$

For this global energy of the spherical charge to be equal to that of the *TL* with radius R_{TL} and with linear charge Λ_{TL}, the radius R_{ch} of the charge must satisfy the following relations, by remembering that $K_2 << K_3$ in the perfect cosmic lattice

$$E_{dist}^{q_\lambda} \cong E_{dist}^{TL} \quad \Leftrightarrow \quad \frac{12K_3\pi R_{TL}^2 \Lambda_{TL}^{2\;2}}{5R_{ch}} = 2(K_2+K_3)\zeta_{TL}R_{TL}\Lambda_{TL}^2 \cong 2K_3\zeta_{TL}R_{TL}\Lambda_{TL}^2 \qquad (19.116)$$

The radius R_{ch} of the spherical charge $q_{\lambda TL}$ which would have the same elastic energy as the *TL* is then

$$R_{ch} \cong \frac{6\pi R_{TL}}{5\zeta_{TL}} \cong R_{TL}\frac{6\pi}{5\ln\left(A_{TL}R_{TL}/a\right)} \qquad (19.117)$$

We note that the radius of a spherical charge which would have an energy of the field of rotation equivalent to the toroidal field of a *TL* should be *a lot smaller* than the radius of the *TL*. Since the *TL* is the simplest microscopic singularity of the lattice with a non-null charge $q_{\lambda TL}$ of rotation, the *TL* is the simplest analog of an electrical charge in our model.

The EL, the simplest topological singularity giving us a charge of spatial curvature

When we compare the elastic energy of distortion of an *EL* with the elastic energy of distorsion of a *TL* with the same Burgers vector, we note that, since $K_2 << K_3$, we have

$$M_0^{TL} \cong \frac{1}{2}\frac{K_3}{c_t^2}\zeta_{TL}R_{TL}\vec{B}_{TL}^2 \;>>\; \left(\frac{K_2}{K_3}\right)^2\frac{K_3}{c_t^2}\zeta_{EL}R_{EL}\vec{B}_{EL}^2 \cong M_0^{EL} \qquad (19.118)$$

Thus, the inertial mass of an *EL* is a lot smaller than the inertial mass of a *TL*.
Also since the *EL* has a non-null charge of curvature $q_{\theta EL} = -2\pi\vec{n}\vec{B}_{EL}$, which can be positive (for a vacancy loop) or negative (for an interstitial loop), it is necessarily associated with a flexion field $\vec{\chi}_{ext}^{EL}$ in the far field by curvature of the lattice which is given by

$$\vec{\chi}_{ext}^{EL} = \frac{q_{\theta EL}}{4\pi}\frac{\vec{r}}{r^3} = -\frac{1}{2}\vec{n}\vec{B}_{EL}\frac{\vec{r}}{r^3} = \frac{1}{2}\vec{\Lambda}_{EL}\vec{m}\frac{\vec{r}}{r^3} \qquad (r > R_{EL}) \qquad (19.119)$$

Thus, the *EL* is the simplest microscopic singularity which gives us a spatial curvature of the lattice by the divergent flexion field associated with it, while the *TL* is the simplest microscopic singularity of the lattice which is a source of spatial torsion of the lattice via the associated divergent field of rotation.

Just as we have in first approximation identified the *TL* with *an electron* in particle physics, the *EL*, which doesn't have a charge of rotation and whose rest mass is much weaker than the *TL* could very well be identified with *a neutrino,* which has no electrical charge and has no mass when compared to the electron.

If we admit in first approximation this analogy, the neutrino would in this case be a source of spatial curvature by flexion of the perfect cosmic lattice, corresponding to a curvature of space in general relativity, while the electron charge would be a source of spatial torsion by rotation of the perfect cosmic lattice, corresponding to the electric field of electromagnetism. This analogy with the two leptons of particle physics is very sketchy for now, and it could be that more complex combinations of these elementary loops in the form of dispiration loops with more complex structure will be needed to explain the different particles of particle physics, as we will see later.

The ML, the simplest topological singularity giving rise to an electric dipolar moment

Contrary to the *TL* and the *EL*, the *ML* does not have any far fields such as a divergent field of rotation or a divergent field of flexion. However this loop presents a dipolar moment of rotation $\vec{\omega}_{dipolar}^{ML}(r,\theta,\varphi)$ in it's vicinity, linked to the two opposed charges of rotation situated on each side of the loop. Thus the *ML* is the simplest singularity of the lattice which is at the source of a dipolar moment of rotation.
In our analogy with the 'real world' the *ML* could be the simplest structure giving us a dipolar electric moment for an elementary particle. Finding this dipole and measuring the dipolar electric moment in particle physics is actually an important topic of research in particle physics.

On the various physical properties carried by the loop singularities

From the previous discussion, it would seem that the *TL* would carry the electrical charge, the *EL* the curvature charge and the *ML* the electrical dipolar moment.
We can add to these three properties another property which could have a big role. In our analogy with the real world, it is difficult to imagine, in order to find an analog to the spin of a charge particle and the magnetic moment associated with it, that a singularity of spherical symmetry with a rotation charge as described in section 19.5 would rotate on itself. However if we consider that the analog to an electrical charge is a *TL*, as treated in section 19.7, the topology of this singularity allows us to naively imagine that this could turn about one of its diameters. In this case the distribution of charges of rotation, analogous to a distribution of the electrical charge in the form of a ring along the perimeter of the *TL*, would impose the emergence of a magnetic moment of the loop associate with the real movement of rotation. We will revisit this topic later.
There is a fifth fundamental property of particle physics which we can address with our analogy. It has to do with the fact that we can calculate the elastic energies of distortion E_{dist}^{loop} and the kinetic energies E_{kin}^{loop} of the various loops, and that we can deduce their inertial mass M_0^{loop}, and that they are essentially contained in the immediate vicinity of the loops! The important fact is that they all satisfy the Einstein relation

$$E_{dist}^{loop} = M_0^{loop} c_t^2 \tag{19.120}$$

which is a fundamental property of the loops which was derived without *any appeal to a relativistic principle.*
Furthermore, the inertial mass of the loops is a property linked to the inertial mass of the cosmic lattice in the absolute referential of the outside observer ***GO***. In an analogy with the 'real world' the inertial mass of the topological lattice would correspond to the famous *Higgs field* which had to be introduced to explain the mass of elementary particles, and the *Higgs particle* would then be the only real particle of the 'real world' since it would correspond to the fundamental massive

particle constituting the perfect cosmic lattice, while the other elementary particles of the Standard Model would correspond to topological singularities of the perfect cosmic lattice.

There is a lot of ground to be covered to find an analogy which would, via a combination of the different topological loops, give us the particles of the standard model and their physical properties. The main problem we will address in the following is to find analogies which could explain the *gravitational behavior* of the objects of the real world at a macroscopic scale (Newton gravitation, General Relativity), as well the *quantum behavior* of the world at a microscopic scale (Quantum Physics).

However, we will for now remember that many of the fundamental properties of elementary particles of the real world find a simple, and classical, explanation thanks to the analogies with the elementary loop singularities of a perfect cosmic lattice.

Chapter 20

Relativistic dynamics of topological singularities in the perfect cosmic lattice

In the two previous chapters, we calculated the kinetic energy associated with the movement of a dislocation or a loop of dislocation, or a loop of disclination in a perfect cosmic lattice, *implicitly* assuming that the distortion due to moving charge is transmitted within the lattice with a near infinite speed compared to the speed of the charge in the lattice. However, disturbances in a solid lattice are, in reality, transmitted with finite speeds c_t for transverse disturbances and c_l for longitudinal disturbances. To account for the effects of disturbances propagation with finite speeds in the solid lattice, when the speed of the charge becomes significant in comparison to the velocities of propagation of transverse and/or longitudinal waves, we will show that we have to introduce the *Lorentz transformation* as a mathematical tool allowing us to move from a stationary referential frame in the lattice to the mobile referential frame associated with the moving charge.

We apply here the Lorentz transformation to the singularities in motion in order to obtain, in the absolute frame of the lattice, the fields of dynamical distortions and velocities associated to screw and edge dislocations, localized rotation charges, twist loops and edge loops moving at relativistic speed. From these fields, their total energy will be calculated. The total energy is the sum of the potential energy stored by the dynamic distortions of the lattice created by the presence of the moving charge and the kinetic energy stored in the lattice by the movement of said charges. The total energy will be shown to satisfy a *relativistic dynamics.* Finally, we will show with the Lorentz transformation that a *relativistic term of force is* acting on the charges of rotation in movement, term that is perfectly analogous to the *Lorentz force* in electromagnetism.

20.1 – Mobile charges of rotation and Lorentz transformations

When topological singularities with charge $\vec{\lambda}_i$, $\vec{\lambda}$ or λ are moving in the referential frame $Ox_1x_2x_3$ attached to the solid lattice with speeds that are non negligible with respect to the speed of waves in the media (either longitudinal or transversal waves), we want to find the dynamical fields $\tau(\vec{r},t)$, $\vec{\omega}(\vec{r},t)$ and $\vec{\alpha}_i(\vec{r},t)$ that are generated by these singularities in the referential frame $Ox_1x_2x_3$. Solving the differential equations of the solid for moving singularities directly in the referential frame $Ox_1x_2x_3$ is not easy. But using a referential $O'x_1'x_2'x_3'$ that is co-moving with the singularities, in which the singularities appear immobile, must allow us to calculate more simply the statical fields in $O'x_1'x_2'x_3'$, and then to obtain the dynamical fields in $Ox_1x_2x_3$ *using some transformation laws which have to be defined.*

Example of the case of a moving screw dislocation

Let's consider an infinite screw dislocation along the axis Ox_2 and let's suppose that it is moving with velocity $\vec{\mathbf{v}}$ in the direction of axis Ox_1. In the reference frame $O'x_1'x_2'x_3'$ co-moving with the string, the field of displacement has to be the field of displacement for a statical screw dislocation, which we will write in $O'x_1'x_2'x_3'$ as

$$\vec{u}_{ext}^{screw\,\prime} = \vec{e}_2 \frac{\Lambda}{\pi} \arctan \frac{x_1'}{x_3'} \tag{20.1}$$

In order to transform this statical field in $O'x_1'x_2'x_3'$ in the dynamical field associated with the moving dislocation in $Ox_1x_2x_3$, we have to establish the laws of transformation which furnish the dynamical fields in $Ox_1x_2x_3$. And the dynamical fields thus obtained have to satisfy the spatiotemporal equations of evolution in $Ox_1x_2x_3$. As there is a translation of the frame $O'x_1'x_2'x_3'$ with regard to the frame $Ox_1x_2x_3$, the transformation law has to transform the coordinate x_1' of $O'x_1'x_2'x_3'$ to a coordinate which has naturally to depend on $(x_1 - \mathbf{v}t)$ in the frame $Ox_1x_2x_3$. We can emit *à priori* the following hypothesis for the transformation laws

Hypothesis 1:
$$\begin{cases} x_1' = \alpha(x_1 - \mathbf{v}t) \\ x_2' = \beta x_2 \\ x_3' = \beta x_3 \end{cases} \tag{20.2}$$

With this transformation law, the statical field of displacement $\vec{u}_{ext}^{screw\,\prime}$ becomes a dynamical one in $Ox_1x_2x_3$ which depends on the factor $(x_1 - \mathbf{v}t)$

$$\vec{u}_{ext}^{screw}(\vec{r},t) = \vec{e}_2 \frac{\Lambda}{\pi} \arctan \frac{x_1'}{x_3'} = \vec{e}_2 \frac{\Lambda}{\pi} \arctan \frac{\alpha(x_1 - \mathbf{v}t)}{\beta x_3} \tag{20.3}$$

It is possible to calculate the dynamical fields of rotation $\vec{\omega}_{ext}^{screw}(\vec{r},t)$ and lattice velocity $\vec{\phi}_{ext}^{screw}(\vec{r},t)$ associated with this moving screw dislocation in the frame $Ox_1x_2x_3$

$$\vec{\omega}_{ext}^{screw}(\vec{r},t) = \frac{1}{2}\overrightarrow{\mathrm{rot}}\,\vec{u}_{ext}^{screw}(\vec{r},t) = \frac{1}{2}\frac{\Lambda}{\pi}\frac{\alpha\beta(x_1 - \mathbf{v}t)\vec{e}_1 + \alpha\beta x_3 \vec{e}_3}{\alpha^2(x_1 - \mathbf{v}t)^2 + \beta^2 x_3^2} \tag{20.4}$$

$$\vec{\phi}_{ext}^{screw}(\vec{r},t) = \frac{\partial \vec{u}_{ext}^{screw}(\vec{r},t)}{\partial t} = -\vec{e}_2 \frac{\Lambda}{\pi}\frac{\alpha\beta \mathbf{v} x_3}{\alpha^2(x_1 - \mathbf{v}t)^2 + \beta^2 x_3^2} \tag{20.5}$$

But in the frame $Ox_1x_2x_3$, *these fields have to satisfy the following spatiotemporal equations of evolution*

$$\begin{cases} \dfrac{\partial(n\vec{p})}{\partial t} = -\dfrac{1}{2}\overrightarrow{\mathrm{rot}}\,\vec{m} \\ \mathrm{div}(n\vec{p}) = 0 \end{cases} \tag{20.6}$$

which can be written, using the fact that $\vec{m} = 4(K_2 + K_3)\vec{\omega}_{ext}^{screw}$ and $n\vec{p} = nm\vec{\phi}_{ext}^{screw}$, as

$$\begin{cases} \dfrac{\partial \vec{\phi}_{ext}^{screw}}{\partial t} = -2c_t^2\,\overrightarrow{\mathrm{rot}}\,\vec{\omega}_{ext}^{screw} \\ \mathrm{div}\,\vec{\phi}_{ext}^{screw} = 0 \end{cases} \tag{20.7}$$

It is easy to verify that the second relation is perfectly satisfied. For the first relation to be satisfied, by introducing $\vec{\phi}_{ext}^{screw}(\vec{r},t)$ and $\vec{\omega}_{ext}^{screw}(\vec{r},t)$ in it, the following relation has to be satisfied between the parameters α and β introduced in the transformation laws of hypothesis *(20.2)*

$$\alpha=\frac{\beta}{\sqrt{1-\mathbf{v}^2/c_t^2}}=\beta/\gamma_t \qquad \text{with} \qquad \gamma_t=\sqrt{1-\mathbf{v}^2/c_t^2} \tag{20.8}$$

in which it appears the well known factor $\gamma_t=\sqrt{1-\mathbf{v}^2/c_t^2}$ of the Lorentz transformations. Introducing this relation in *(20.3)*, *(20.4)* and *(20.5)*, we obtain the expressions of the fields

$$\begin{cases} \vec{u}_{ext}^{\,screw}(\vec{r},t)=\vec{e}_2\dfrac{\Lambda}{\pi}\dfrac{1}{\gamma_t}\arctan\dfrac{(x_1-\mathbf{v}t)}{x_3} \\ \vec{\omega}_{ext}^{\,screw}(\vec{r},t)=\dfrac{1}{2}\dfrac{\Lambda}{\pi}\dfrac{1}{\gamma_t}\dfrac{(x_1-\mathbf{v}t)\vec{e}_1+x_3\vec{e}_3}{(x_1-\mathbf{v}t)^2/\gamma_t^2+x_3^2} \\ \vec{\phi}_{ext}^{\,screw}(\vec{r},t)=-\vec{e}_2\dfrac{\Lambda}{\pi}\dfrac{1}{\gamma_t}\dfrac{\mathbf{v}x_3}{(x_1-\mathbf{v}t)^2/\gamma_t^2+x_3^2} \end{cases} \tag{20.9}$$

It is remarkable that these fields, which satisfy perfectly the spatiotemporal equations of evolution in frame $Ox_1x_2x_3$, do not depend on the parameter β, but only on the parameter γ_t, so that the parameter β can be freely chosen, and here we will admit the value of 1, so that the spatial transformation laws become simply

$$\begin{cases} x_1'=(x_1-\mathbf{v}t)/\gamma_t \\ x_2'=x_2 \\ x_3'=x_3 \end{cases} \tag{20.10}$$

The contraction of the dynamical fields in the direction of the screw dislocation motion

The expressions *(20.9)* for the dynamical fields $\vec{u}_{ext}^{\,screw}(\vec{r},t)$, $\vec{\omega}_{ext}^{\,screw}(\vec{r},t)$ and $\vec{\phi}_{ext}^{\,screw}(\vec{r},t)$ are effective solutions of the topological equations and the Newton equation for a screw dislocation moving in referential $Ox_1x_2x_3$. It is interesting to have a look on the behavior of these fields as a function of the velocity $\vec{\mathbf{v}}$ of the dislocation. We can take for example the projection ω_1 of the external vector of rotation in the direction of the dislocation movement, and report its value $\omega_1(t=0,x_3=0)$, taken at time $t=0$ and for coordinate $x_3=0$, as a function of x_1 for different values of the ratio $\mathbf{v}/c_t$, as illustrated in figure 20.1. We observe then that the horizontal component of the field of rotation seems contracted along the axis Ox_1. It is easy to calculate from *(20.9)* that the same value of $\omega_1(t=0,x_3=0)$ is observed at a distance Δx_1 of the origin, which corresponds to

$$\Delta x_1=\frac{\Lambda}{2\pi}\frac{\gamma_t}{\omega_1(t=0,x_3=0)}=\Delta x_1(\mathbf{v}=0)\sqrt{1-\mathbf{v}^2/c_t^2} \tag{20.11}$$

so that the field of rotation of the moving dislocation *is effectively contracted along the axis Ox_1 by a factor γ_t*.

The spatial contraction of a moving cluster of topological singularities of rotation

Imagine now a cluster of singularities of rotation which are bonded together through rotation fields (remember that the rotation field corresponds to the electrical field in our analogy with the real universe). If this cluster move along the axis Ox_1 in the reference frame $Ox_1x_2x_3$ of the

GO, the rotation fields associated with this cluster as to contract along the axis Ox_1 with a factor γ_t in order to satisfy the topological equations and the Newton equation of the lattice. The consequence is then that the cluster itself, which is bonded by these rotation fields, will have to contract along the axis Ox_1. If this cluster represents an "object" for the ***GO***, this "object" will contract along the axis Ox_1. But observed in its own frame $O'x_1'x_2'x_3'$, this "object" will remain exactly the same than at rest in the absolute frame $Ox_1x_2x_3$, and its shape does not change in the frame $O'x_1'x_2'x_3'$ whatever is the velocity $\vec{\boldsymbol{v}}$ of the "object" in $Ox_1x_2x_3$!

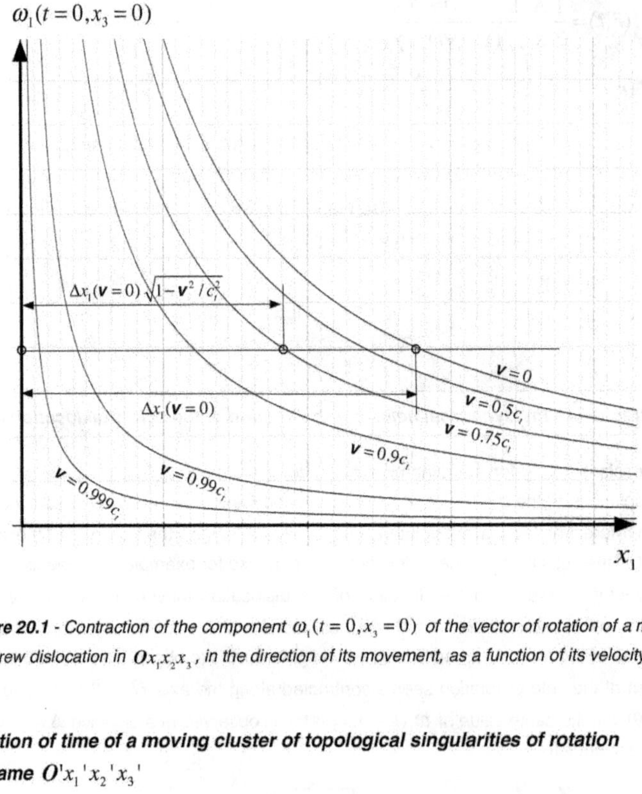

Figure 20.1 *- Contraction of the component $\omega_1(t=0, x_3=0)$ of the vector of rotation of a moving screw dislocation in $Ox_1x_2x_3$, in the direction of its movement, as a function of its velocity $\vec{\boldsymbol{v}}$*

The dilation of time of a moving cluster of topological singularities of rotation in the frame $O'x_1'x_2'x_3'$

Imagine now that the observer measure the time T_0 which is necessary for a transversal wave to travel a distance d_0 in the absolute frame $Ox_1x_2x_3$, to be reflected on a mirror, and to return to the point of emission. It is clear that he measures a time equal to

$$T_0 = 2d_0 / c_t \tag{20.12}$$

Such a device based on an "object" constituted by a cluster of singularities bonded by rotation fields can be used by the observer ***GO*** as a time base, *a clock giving the basic lapse time T_0*.
Imagine now that the same device, based on the same "object", but moving now at a speed $\vec{\boldsymbol{v}}$ along the axis Ox_1 in the reference frame $Ox_1x_2x_3$, is observed by the ***GO***. If the transversal wave is emitted in the moving frame $O'x_1'x_2'x_3'$ in the vertical direction inside this frame, the

emitted wave is seen by the **GO** as a non vertical wave in the frame $Ox_1x_2x_3$, as illustrated in figure 20.2. For the observer **GO,** the time T which is necessary for the wave to do the round-trip through the reflection on the mirror of the moving "object" is easily calculated using the tri-angle in the plane Ox_1x_3

$$c_t^2\frac{T^2}{4}=\mathbf{v}^2\frac{T^2}{4}+d_0^2 \quad\Rightarrow\quad T=\frac{2d_0/c_t}{\sqrt{1-\mathbf{v}^2/c_t^2}}=\frac{T_0}{\gamma_t} \tag{20.13}$$

This means that the basic time of the moving clock in frame $O'x_1'x_2'x_3'$, measured by the **GO** in its absolute reference frame $Ox_1x_2x_3$, seems to be dilated, expanded as a function of the velocity $\vec{\mathbf{v}}$ by a factor of $1/\gamma_t$. This means also that the clock of the moving "object" is slowing down in comparison with the absolute clock of the **GO.**

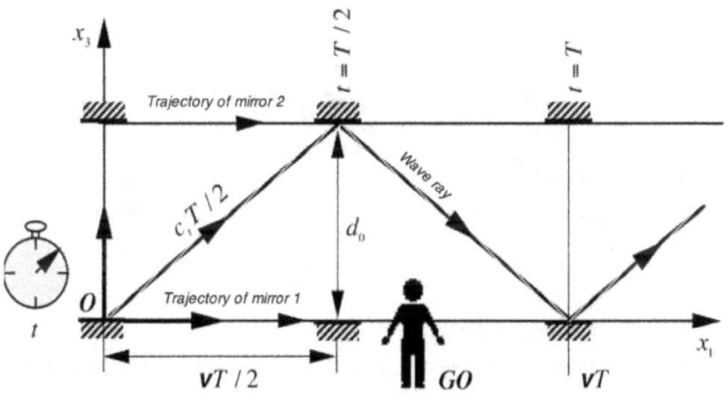

***Figure 20.2** - The trajectory of the transversal wave emitted vertically by the local clock of the moving "object" in the frame $O'x_1'x_2'x_3'$, as observed by the **GO** in its absolute frame $Ox_1x_2x_3$*

We can still ask if the time in the frame $O'x_1'x_2'x_3'$ of the "object" remains isotropic in this frame, in other words if a clock based on a horizontal trajectory of the transversal wave, give the same time than the vertical clock. If this horizontal clock is observed by the **GO** in its reference frame $Ox_1x_2x_3$, the trajectory of the wave can be illustrated as in figure 20.3.

In the trajectory diagram of figure 20.3, the trajectories of the moving mirrors are represented as two lines with a slope $\mathbf{v}$, separated by a distance d in the direction Ox_1. The trajectories of the transversal wave rays are represented by two lines with slopes $+c_t$ and $-c_t$ respectively, for the directions of the waves. In this diagram of the trajectories, we have

$$\begin{cases} x_1^{(1)}=c_tT_1=\mathbf{v}T_1+d=-c_tT_1+A \\ x_1^{(2)}=\mathbf{v}T\;=-c_tT\;+A \end{cases} \tag{20.14}$$

This system of equations as the following solution for the lapse time T which is necessary for the wave to do the roundtrip through the reflection on the mirror of the moving "object"

$$T=\frac{2d/c_t}{1-\mathbf{v}^2/c_t^2} \tag{20.15}$$

But for the **GO**, the distance d between the two mirrors associated to the moving "object" is contracted by the factor γ_t as we have seen in the previous section *(20.11)*, which furnishes the relation between the distance d and the distance d_0 at rest separating the two mirrors

$$d = \gamma_t d_0 = d_0\sqrt{1-\mathbf{v}^2 / c_t^2} \qquad (20.16)$$

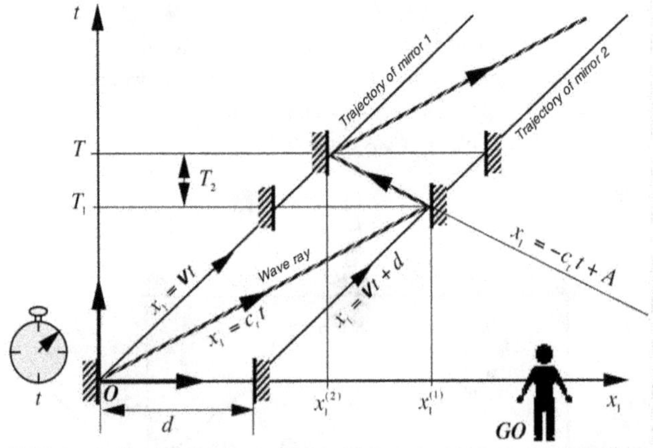

***Figure 20.3** - The trajectory of the transversal wave emitted horizontally by the local clock of the moving "object" in the frame $O'x_1'x_2'x_3'$, as observed by the **GO** in its absolute frame $Ox_1x_2x_3$*

Combining relations *(20.15)* and *(20.16)*, we deduce that

$$T = \frac{2d_0 / c_t}{\sqrt{1-\mathbf{v}^2 / c_t^2}} = \frac{T_0}{\gamma_t} \qquad (20.17)$$

which clearly shows, in comparison with relation *(20.13)*, that the two moving clocks, working respectively with a vertical wave propagation and a horizontal wave propagation in frame $O'x_1'x_2'x_3'$ furnish exactly the same local time, *meaning that a local time t' exists and that this local time t' remains isotropic in the mobile frame $O'x_1'x_2'x_3'$, independently of the direction of motion of the "object" inside the lattice.*

In the moving frame $O'x_1'x_2'x_3'$, the length that the wave has to travel along $O'x_1'$ or $O'x_3'$ inside the clock device is measured as the length d_0, and the local lapse of time to do the roundtrip through the reflection on the mirror is measured to be T_0 in the vertical as well as in the horizontal case by the local clock. This means that the wave velocity measured in the moving frame $O'x_1'x_2'x_3'$ has exactly the same value c_t than the one measured in the frame $Ox_1x_2x_3$, independently of the speed $\vec{\mathbf{v}}$ of the frame $O'x_1'x_2'x_3'$ inside the frame $Ox_1x_2x_3$.

Imagine then a transversal wave propagating along $O'x_1'$ in the frame $O'x_1'x_2'x_3'$

$$\vec{\omega}' = \omega_0 \vec{e}_3' \sin\left(k' x_1' - \omega' t'\right) = \omega_0 \vec{e}_3' \sin\left(\frac{\omega'}{c_t} x_1' - \omega' t'\right) \qquad (20.18)$$

In order to transform this wave in the frame $Ox_1x_2x_3$, we use the transformation *(20.10)* and the following relation in which ε and δ are parameters to be determined

$$t' = \varepsilon t + \delta x_1 \tag{20.19}$$

so that the expression of the wave *(20.18)* becomes, in the frame $Ox_1x_2x_3$

$$\vec{\omega} = \omega_0 \vec{e}_3 \sin\left[\omega'\left(\frac{1}{\gamma_t c_t} - \delta\right)x_1 - \omega'\left(\frac{\mathbf{v}}{\gamma_t c_t} + \varepsilon\right)t\right] = \omega_0 \vec{e}_3 \sin(kx_1 - \omega t) \tag{20.20}$$

This wave propagates also with the velocity c_t in $Ox_1x_2x_3$, so that

$$c_t = \frac{\omega}{k} = \quad \Rightarrow \quad c_t - \delta c_t^2 \gamma_t = \mathbf{v} + \varepsilon \gamma_t c_t \tag{20.21}$$

which can only be satisfied if the parameters ε and δ have the following values

$$\delta = -\frac{\mathbf{v}}{\gamma_t c_t^2} \quad \text{and} \quad \varepsilon = \frac{1}{\gamma_t} \tag{20.22}$$

The transformation law for t' becomes then

$$t' = \frac{t - \mathbf{v} x_1 / c_t^2}{\gamma_t} \tag{20.23}$$

The Lorentz transformation for a moving "object" bonded by fields of rotation

The fact that the fields of rotation, and as a consequence the moving "objects" bonded by rotation fields, are really contracted in the direction of motion by a factor $\gamma_t = \sqrt{1 - \mathbf{v}^2 / c_t^2}$, that the isotropic time measured by the clock of the moving "object" is really dilated by a factor $1/\gamma_t$ and that the velocities of transversal waves measured in $Ox_1x_2x_3$ and in $O'x_1'x_2'x_3'$ have exactly the same value c_t, means that the transformation laws *(20.10)* and *(20.23)* allowing to pass from one frame to the other is in fact the same than the well known *Lorentz transformation* of the electromagnetism

$$\begin{cases} x_1' = \dfrac{1}{\gamma_t}(x_1 - \mathbf{v}t) \\ x_2' = x_2 \\ x_3' = x_3 \\ t' = \dfrac{1}{\gamma_t}\left(t - \dfrac{\mathbf{v}x_1}{c_t^2}\right) \end{cases} \tag{20.24}$$

Note that this transformation has been initially used simply as *a mathematical tool* allowing one to calculate with the Maxwell's equations the electromagnetic fields generated by a moving electrical charge. Later, this transformation has been used in special relativity *by postulating* that the relation *(20.18)* is applicable *to any frames moving relatively to one another*, which corresponds in fact *to axiomatically admit the constance of the light velocity in any frame*!

Here, in the case of a solid lattice, the Lorentz transformation is obtained by a different approach based on the existence of a solid lattice in the absolute reference frame $Ox_1x_2x_3$ of the ***GO***, which is *the support (the aether) for the transversal wave propagation.* This approach allows us *to demonstrate the reality of the physical consequences of the Lorentz transformation,* as the spatial contraction and the time dilation in $Ox_1x_2x_3$ of moving "objects" constituted of topological singularities bonded by rotation fields. And this demonstration is based on the initial hypothesis *(20.2)* that the ***GO*** can introduce a relative frame $O'x_1'x_2'x_3'$ which is associated with the moving "object" in its absolute frame $Ox_1x_2x_3$. The use of the Lorentz transformation in the case of the cosmic lattice is then limited only to transform the fields between a moving relative frame $O'x_1'x_2'x_3'$ and the absolute frame $Ox_1x_2x_3$ of the ***GO***, which is fixed within the lattice. As a consequence, there is here absolutely no axiomatic hypothesis that the transformation of Lorentz is applicable to any frames moving relatively to one another!

20.2 – The two Lorentz transformations in the case of a cosmic lattice with an expansion background $\tau_0 > \tau_{0cr}$

Consider that the expansion background of the cosmic lattice satisfies the following hypothesis

Hypothesis 2: $\tau_0 > \tau_{0cr}$ *(20.25)*

In this particular case, both transversal waves as well as longitudinal waves can propagate inside the lattice.

If the observer ***GO*** has now to calculate the fields associated with a moving "object" constituted by topological singularities which are bonded at once by fields of rotation and by fields of expansion, the problem becomes much more complex than the previous one when only rotation fields are concerned. By supposing that the displacement of the bonded charges in the referential $Ox_1x_2x_3$ takes place with velocity $\vec{\mathbf{v}}$ in the direction Ox_1, we then define two mobile reference frameworks that are co-moving with the charges, $O'x_1'x_2'x_3'$ and $O''x_1''x_2''x_3''$, by attributing to each of these referential the transformation laws of Lorentz with velocities c_t and c_l respectively. We define

$$\gamma_t = \sqrt{1-\mathbf{v}^2/c_t^2} \quad \text{and} \quad \gamma_l = \sqrt{1-\mathbf{v}^2/c_l^2} \tag{20.26}$$

For $O'x_1'x_2'x_3'$, the Lorentz transformations and its inverse become respectively

$$\begin{cases} x_1' = \dfrac{1}{\gamma_t}(x_1 - \mathbf{v}t) \\ x_2' = x_2'' = x_2 \\ x_3' = x_3'' = x_3 \\ t' = \dfrac{1}{\gamma_t}\left(t - \mathbf{v}x_1/c_t^2\right) \end{cases} \quad \text{and} \quad \begin{cases} x_1 = \dfrac{1}{\gamma_t}(x_1' + \mathbf{v}t') \\ x_2 = x_2' = x_2'' \\ x_3 = x_3' = x_3'' \\ t = \dfrac{1}{\gamma_t}\left(t' + \mathbf{v}x_1'/c_t^2\right) \end{cases} \tag{20.27}$$

For $O''x_1''x_2''x_3''$, the Lorentz transformations are also easy to establish

$$\begin{cases} x_1'' = \dfrac{1}{\gamma_l}(x_1 - \mathbf{v}t) \\ x_2'' = x_2' = x_2 \\ x_3'' = x_3' = x_3 \\ t'' = \dfrac{1}{\gamma_l}\left(t - \mathbf{v}x_1/c_l^2\right) \end{cases} \quad \text{and} \quad \begin{cases} x_1 = \dfrac{1}{\gamma_l}(x_1'' + \mathbf{v}t'') \\ x_2 = x_2'' = x_2' \\ x_3 = x_3'' = x_3' \\ t = \dfrac{1}{\gamma_l}\left(t'' + \mathbf{v}x_1''/c_l^2\right) \end{cases} \tag{20.28}$$

Thanks to these transformations, we can also establish the relationships that exist between the expressions of the operators of time and space, to the first and second order, for $O'x_1'x_2'x_3'$

$$\begin{cases} \dfrac{\partial}{\partial x_1}=\dfrac{1}{\gamma_t}\dfrac{\partial}{\partial x_1'}-\dfrac{\mathbf{v}}{\gamma_t c_t^2}\dfrac{\partial}{\partial t'} \\ \dfrac{\partial}{\partial x_2}=\dfrac{\partial}{\partial x_2'} \\ \dfrac{\partial}{\partial x_3}=\dfrac{\partial}{\partial x_3'} \\ \dfrac{\partial}{\partial t}=\dfrac{1}{\gamma_t}\dfrac{\partial}{\partial t'}-\dfrac{\mathbf{v}}{\gamma_t}\dfrac{\partial}{\partial x_1'} \end{cases} \qquad \begin{cases} \dfrac{\partial^2}{\partial x_1^2}=\dfrac{1}{\gamma_t^2}\dfrac{\partial^2}{\partial x_1'^2}-\dfrac{2\mathbf{v}}{\gamma_t^2 c_t^2}\dfrac{\partial^2}{\partial x_1'\partial t'}+\dfrac{\mathbf{v}^2}{\gamma_t^2 c_t^4}\dfrac{\partial^2}{\partial t'^2} \\ \dfrac{\partial^2}{\partial x_2^2}=\dfrac{\partial^2}{\partial x_2'^2} \\ \dfrac{\partial^2}{\partial x_3^2}=\dfrac{\partial}{\partial x_3'^2} \\ \dfrac{\partial^2}{\partial t^2}=\dfrac{1}{\gamma_t^2}\dfrac{\partial^2}{\partial t'^2}-\dfrac{2\mathbf{v}}{\gamma_t^2}\dfrac{\partial^2}{\partial x_1'\partial t'}+\dfrac{\mathbf{v}^2}{\gamma_t^2}\dfrac{\partial^2}{\partial x_1'^2} \end{cases} \tag{20.29}$$

and for $O''x_1''x_2''x_3''$

$$\begin{cases} \dfrac{\partial}{\partial x_1}=\dfrac{1}{\gamma_t}\dfrac{\partial}{\partial x''_1}-\dfrac{\mathbf{v}}{\gamma_t c_t^2}\dfrac{\partial}{\partial t''} \\ \dfrac{\partial}{\partial x_2}=\dfrac{\partial}{\partial x''_2} \\ \dfrac{\partial}{\partial x_3}=\dfrac{\partial}{\partial x''_3} \\ \dfrac{\partial}{\partial t}=\dfrac{1}{\gamma_t}\dfrac{\partial}{\partial t''}-\dfrac{\mathbf{v}}{\gamma_t}\dfrac{\partial}{\partial x''_1} \end{cases} \qquad \begin{cases} \dfrac{\partial^2}{\partial x_1^2}=\dfrac{1}{\gamma_t^2}\dfrac{\partial^2}{\partial x''^2_1}-\dfrac{2\mathbf{v}}{\gamma_t^2 c_t^2}\dfrac{\partial^2}{\partial x''_1\partial t''}+\dfrac{\mathbf{v}^2}{\gamma_t^2 c_t^4}\dfrac{\partial^2}{\partial t''^2} \\ \dfrac{\partial^2}{\partial x_2^2}=\dfrac{\partial^2}{\partial x''^2_2} \\ \dfrac{\partial^2}{\partial x_3^2}=\dfrac{\partial}{\partial x''^2_3} \\ \dfrac{\partial^2}{\partial t^2}=\dfrac{1}{\gamma_t^2}\dfrac{\partial^2}{\partial t''^2}-\dfrac{2\mathbf{v}}{\gamma_t^2}\dfrac{\partial^2}{\partial x''_1\partial t'}+\dfrac{\mathbf{v}^2}{\gamma_t^2}\dfrac{\partial^2}{\partial x''^2_1} \end{cases} \tag{20.30}$$

An example of application of the two transformations of Lorentz in the frame $Ox_1x_2x_3$

Starting from the equations of evolution of a cosmic lattice in the presence of *weak perturbations $\tau^{(p)}$ of the volume expansion*, it is possible to find a simplified version of the equation of Newton *(13.14)* in $Ox_1x_2x_3$. In the very simple case where there are no charges, nor flows, nor auto-diffusion, namely if

$$\textbf{\textit{Hypothesis 3:}}\quad \begin{cases} \vec{J}=0 \;\;;\;\; \lambda=0 \;\;;\;\; \vec{\lambda}=0 \\ \vec{\omega}^{an}=0 \\ \vec{J}_I=\vec{J}_L=0 \end{cases} \tag{20.31}$$

the principal equations in the presence of weak perturbations can be summed up to the three following sets, by replacing d/dt by $\partial/\partial t$ in the vicinity of the origin of $Ox_1x_2x_3$

$$\begin{cases} \dfrac{\partial\vec{\omega}}{\partial t}\cong\left(\overrightarrow{\mathrm{rot}}\,\vec{\phi}\right)/2 \\ \operatorname{div}\vec{\omega}=0 \\ \dfrac{\partial\tau^{(p)}}{\partial t}\cong\operatorname{div}\vec{\phi} \end{cases} \tag{20.32a}$$

$$\left\{\; n\dfrac{\partial\vec{p}}{\partial t}t\cong-\left(\overrightarrow{\mathrm{rot}}\,\vec{m}\right)/2+\left[-K_0+4K_2/3+2K_1\left(1+\tau_0\right)\right]\overrightarrow{\mathrm{grad}}\,\tau^{(p)}\right.$$

$$\begin{cases} \vec{m} \cong 4(K_2+K_3)\vec{\omega} \\ \vec{p} \cong m(1+C_l-C_L)\vec{\phi} \\ n = n_0\,\mathrm{e}^{-\tau_0} \end{cases} \tag{20.32b}$$

We easily deduce a linearized wave equation for a velocity field $\vec{\phi}$

$$\frac{\partial^2\vec{\phi}}{\partial t^2} \cong -c_t^2\,\overrightarrow{\mathrm{rot}}\,\overrightarrow{\mathrm{rot}}\,\vec{\phi} + c_l^2\,\overrightarrow{\mathrm{grad}}\,\mathrm{div}\,\vec{\phi} \tag{20.33}$$

In which the velocity of transversal waves and longitudinal waves are respectively worth

$$c_t \cong \sqrt{(K_2+K_3)/mn(1+C_l-C_L)} \tag{20.34}$$

$$c_l \cong \sqrt{\left[-K_0+4K_2/3+2K_1(1+\tau_0)\right]/mn(1+C_l-C_L)} \tag{20.35}$$

By using a vector displacement approximatively given by relation $\vec{\phi} \cong -\partial\vec{u}/\partial t$, we can again write the wave equation under the form

$$\frac{\partial^2\vec{u}}{\partial t^2} \cong -c_t^2\,\overrightarrow{\mathrm{rot}}\,\overrightarrow{\mathrm{rot}}\,\vec{u} + c_l^2\,\overrightarrow{\mathrm{grad}}\,\mathrm{div}\,\vec{u} \tag{20.36}$$

Let's revisit equation *(20.27)* and calculate rotational and divergence by introducing the values $\vec{A}$ and B defined as

$$\vec{A} = \overrightarrow{\mathrm{rot}}\,\vec{u} \qquad \text{and} \qquad B = \mathrm{div}\,\vec{u} \tag{20.37}$$

We have then the following equations

$$\begin{cases} \dfrac{\partial^2\vec{A}}{\partial t^2} \cong -c_t^2\,\overrightarrow{\mathrm{rot}}\,\overrightarrow{\mathrm{rot}}\,\vec{A} = c_t^2\Delta\vec{A} \\ \dfrac{\partial^2 B}{\partial t^2} \cong c_l^2\,\mathrm{div}\,\overrightarrow{\mathrm{grad}}\,B = c_l^2\Delta B \end{cases} \tag{20.38}$$

Thus, in the referential $Ox_1x_2x_3$ linked to the solid lattice, the quantities $\vec{A}$ and B decouple from each other and each satisfies an independent wave equation, $\vec{A}$ governs the transversal displacement field and B the field of longitudinal displacement.

By applying the Lorentz transformations *(20.28) and (20.30)* to the wave equations *(20.38)*, we obtain the following relationships in frames $O'x_1'x_2'x_3'$ and $O''x_1''x_2''x_3''$ respectively

$$\begin{cases} \dfrac{\partial^2\vec{A}}{\partial t'^2} = c_t^2\Delta'\vec{A} = c_t^2\left(\dfrac{\partial^2\vec{A}}{\partial x_1'^2}+\dfrac{\partial^2\vec{A}}{\partial x_2'^2}+\dfrac{\partial^2\vec{A}}{\partial x_3'^2}\right) \\ \dfrac{\partial^2 B}{\partial t''^2} = c_l^2\Delta'' B = c_l^2\left(\dfrac{\partial^2 B}{\partial x_1''^2}+\dfrac{\partial^2 B}{\partial x_2''^2}+\dfrac{\partial^2 B}{\partial x_3''^2}\right) \end{cases} \tag{20.39}$$

We notice that the Lorentz transforms introduced previously insure that the fields $\vec{A}$ and B *satisfy the same waves equations* in both the co-moving referential and the immobile one with respect to the solid lattice. From which we can conclude that the transformations of Lorentz, leave *invariant* the physical laws in the mobile referential frames.

If a displacement field $\vec{u}(\vec{r},t)$ is generated in a lattice by localized charges in movement with velocity $\vec{\boldsymbol{v}}$ in the direction Ox_1, we apply the transformations defined above by using the frames $O'x_1'x_2'x_3'$ and $O''x_1''x_2''x_3''$ co-moving with the charges with velocity $\vec{v}$. The problem is then to resolve the static equations to describe fields $\vec{A}$ and B, which are due to the immobile charges in the frameworks $O'x_1'x_2'x_3'$ and $O''x_1''x_2''x_3''$ respectively. Knowing the static solutions for $\vec{A}$ and B in the mobile referential frames $O'x_1'x_2'x_3'$ and $O''x_1''x_2''x_3''$, it is in principle possible to find the dynamical solution for $\vec{u}(\vec{r},t)$ in the referential $Ox_1x_2x_3$ linked to the solid lattice.

20.3 – The only Lorentz transformation in the case of a cosmic lattice with an expansion background $\tau_0 < \tau_{0cr}$

The complete resolution of the previous type of problem for a density $\vec{\lambda}_i$ of mobile charges in $Ox_1x_2x_3$ when the expansion background satisfies $\tau_0 > \tau_{0cr}$ can be rather complex. Notably it can exist expansion fields which are non-homogenous within the lattice and which can propagate as longitudinal perturbations inside the lattice.

This is why we will treat for the remainder of this chapter only the particular case, *which is in fact the interesting case for our analogy with the universe,* of topological singularities that move in a perfect cosmic lattice presenting a constant and homogenous volume expansion which satisfies the following hypothesis

Hypothesis 4: *the volume expansion background of the cosmic lattice is constant and homogenous, and satisfies* $\tau_0 < \tau_{0cr}$ (20.40)

In this case, we know that *longitudinal waves cannot exist*, meaning that ***every perturbations of the distortion fields can only propagate with the transversal wave velocity***, and that the problem of determining the fields of moving singularities can be solved ***by applying the only transformation of Lorentz for the frame*** $O'x_1'x_2'x_3'$.

The problem of the fields of perturbation of the expansion linked to topological singularities will be treated later, in the chapters dealing with the *"gravitational field"* (the static perturbation of the expansion field due to topological singularities, chapter 22) on one hand and with the *"quantum field"* (the dynamical perturbations of the expansion field due to moving topological singularities when $\tau_0 < \tau_{0cr}$, chapter 27) on the other hand.

Application of the Lorentz transformation to scalar charges of rotation

Let's consider now mobile charges of rotation, with charge density λ, that moves within the lattice with velocity $\vec{v}$ along the axis Ox_1, the fields $\vec{\omega}$ generated by these charges will be dynamic fields which will evolve with the movement of the charges. As the transmission of information by the mobile charges in a given point of the lattice is made with velocity c_t of the transversal waves, we can use the Lorentz transformation of section 20.1 by associating a mobile referential $O'x_1'x_2'x_3'$ to the charges. It is interesting here to find the transformation relations concerning the couple of equations *(20.6)* that describe the dynamics inside the lattice but outside the charges, in the case where the volume expansion is homogeneous and constant ($n = cste$). This couple of equations can be written in components for p_i and m_i in the fixed referential $Ox_1x_2x_3$ and for p'_i and m'_i in the mobile referential $O'x_1'x_2'x_3'$ respectively in the following manner

$$\begin{cases} \dfrac{\partial(np_1)}{\partial t} = -\dfrac{1}{2}\left(\dfrac{\partial m_3}{\partial x_2} - \dfrac{\partial m_2}{\partial x_3}\right) \\ \dfrac{\partial(np_2)}{\partial t} = -\dfrac{1}{2}\left(\dfrac{\partial m_1}{\partial x_3} - \dfrac{\partial m_3}{\partial x_1}\right) \\ \dfrac{\partial(np_3)}{\partial t} = -\dfrac{1}{2}\left(\dfrac{\partial m_2}{\partial x_1} - \dfrac{\partial m_1}{\partial x_2}\right) \\ \dfrac{\partial(np_1)}{\partial x_1} + \dfrac{\partial(np_2)}{\partial x_2} + \dfrac{\partial(np_3)}{\partial x_3} = 0 \end{cases} \qquad \begin{cases} \dfrac{\partial(np_1)'}{\partial t'} = -\dfrac{1}{2}\left(\dfrac{\partial m_3'}{\partial x'_2} - \dfrac{\partial m_2'}{\partial x'_3}\right) \\ \dfrac{\partial(np_2)'}{\partial t'} = -\dfrac{1}{2}\left(\dfrac{\partial m_1'}{\partial x'_3} - \dfrac{\partial m_3'}{\partial x'_1}\right) \\ \dfrac{\partial(np_3)'}{\partial t'} = -\dfrac{1}{2}\left(\dfrac{\partial m_2'}{\partial x'_1} - \dfrac{\partial m_1'}{\partial x'_2}\right) \\ \dfrac{\partial(np_1)'}{\partial x'_1} + \dfrac{\partial(np_2)'}{\partial x'_2} + \dfrac{\partial(np_3)'}{\partial x'_3} = 0 \end{cases} \tag{20.41}$$

By applying in the referential $Ox_1x_2x_3$ the rules of transformation of Lorentz to these equations, we obtain, after some calculation, the set of equations in the $O'x_1'x_2'x_3'$ reference frame

$$\begin{cases} \dfrac{\partial(np_1)}{\partial t'} = -\dfrac{1}{2}\dfrac{\partial}{\partial x_2'}\left[\dfrac{1}{\gamma_t}(m_3 + 2\mathbf{v}np_2)\right] + \dfrac{1}{2}\dfrac{\partial}{\partial x_3'}\left[\dfrac{1}{\gamma_t}(m_2 - 2\mathbf{v}np_3)\right] \\ \dfrac{\partial}{\partial t'}\left[\dfrac{1}{\gamma_t}\left(np_2 + \dfrac{\mathbf{v}}{2c_t^2}m_3\right)\right] = -\dfrac{1}{2}\dfrac{\partial m_1}{\partial x_3'} + \dfrac{1}{2}\dfrac{\partial}{\partial x_1'}\left[\dfrac{1}{\gamma_t}(m_3 + 2\mathbf{v}np_2)\right] \\ \dfrac{\partial}{\partial t'}\left[\dfrac{1}{\gamma_t}\left(np_3 - \dfrac{\mathbf{v}}{2c_t^2}m_2\right)\right] = -\dfrac{1}{2}\dfrac{\partial}{\partial x_1'}\left[\dfrac{1}{\gamma_t}(m_2 - 2\mathbf{v}np_3)\right] + \dfrac{1}{2}\dfrac{\partial m_1}{\partial x_2'} \\ \dfrac{\partial(np_1)}{\partial x_1'} + \dfrac{\partial}{\partial x_2'}\left[\dfrac{1}{\gamma_t}\left(np_2 + \dfrac{\mathbf{v}}{2c_t^2}m_3\right)\right] + \dfrac{\partial}{\partial x_3'}\left[\dfrac{1}{\gamma_t}\left(np_3 - \dfrac{\mathbf{v}}{2c_t^2}m_2\right)\right] = 0 \end{cases} \tag{20.42}$$

which can be compared to the equations for the components of $(n\vec{p})'$ and $\vec{m}'$ in the reference frame $O'x_1'x_2'x_3'$. We deduce then the equations of transformation of the fields of quantity of movement $(n\vec{p})$ and of moment $\vec{m}$ in the reference frame $Ox_1x_2x_3$ and the fields $(n\vec{p})'$ and $\vec{m}'$ in the reference frame $O'x_1'x_2'x_3'$

$$\begin{cases} m_1' = m_1 \\ m_2' = (m_2 - 2\mathbf{v}np_3)/\gamma_t \\ m_3' = (m_3 + 2\mathbf{v}np_2)/\gamma_t \end{cases} \quad \text{and} \quad \begin{cases} np_1' = np_1 \\ np_2' = \left(np_2 + \mathbf{v}m_3/(2c_t^2)\right)/\gamma_t \\ np_3' = \left(np_3 - \mathbf{v}m_2/(2c_t^2)\right)/\gamma_t \end{cases} \tag{20.43}$$

Thanks to the transformation relations, we will be able to calculate the fields associated with the movement of different types of charges of rotation within the solid lattice, as well as their total energy, composed of their elastic potential energy and their kinetic energy.

20.4 – Relativistic dynamics of a screw or an edge dislocation line

Let's consider an infinite screw dislocation string and let's suppose that it is moving with velocity $\vec{\mathbf{v}}$ in the direction of axis Ox_1. In the reference frame $O'x_1'x_2'x_3'$ co-moving with the string, we have $p_1' = p_2' = p_3' = 0$, by definition, as well as $m_2' = 0$, so that from relations *(20.43)*, we deduce

$$\begin{cases} \omega_1 = \omega_1' \\ \omega_2 = 0 \\ \omega_3 = \omega_3'/\gamma_t \end{cases} \quad \text{and} \quad \begin{cases} \phi_1 = 0 \\ \phi_2 = -2\mathbf{v}\omega_3'/\gamma_t \\ \phi_3 = 0 \end{cases} \tag{20.44}$$

From expression *(19.1)* of the static field $\vec{\omega}_{ext}$ and the transformation relations of Lorentz *(20.27)*, we can deduce the dynamic fields $\vec{\omega}_{ext}$ and $\vec{\phi}$ expressed in referential $Ox_1x_2x_3$

$$\begin{aligned} \vec{\omega}_{ext}^{screw} &= \omega_1\vec{e}_1 + \omega_3\vec{e}_3 = \omega_1'\vec{e}_1 + \frac{1}{\gamma_t}\omega_3'\vec{e}_3 = \frac{\Lambda}{2\pi}\frac{x_1'}{x_1'^2 + x_3'^2}\vec{e}_1 + \frac{1}{\gamma_t}\frac{\Lambda}{2\pi}\frac{x_3'}{x_1'^2 + x_3'^2}\vec{e}_3 \\ &= \frac{\Lambda}{2\pi}\left[\frac{x_1'\vec{e}_1}{x_1'^2 + x_3'^2} + \frac{1}{\gamma_t}\frac{x_3'\vec{e}_3}{x_1'^2 + x_3'^2}\right] = \frac{\Lambda}{2\pi\gamma_t}\frac{(x_1 - \mathbf{v}t)\vec{e}_1 + x_3\vec{e}_3}{(x_1 - \mathbf{v}t)^2/\gamma_t^2 + x_3^2} \end{aligned} \tag{20.45a}$$

$$\vec{\phi}_{ext}^{screw} = \phi_2^{screw}\vec{e}_2 = -\frac{2\mathbf{v}\omega_3'\vec{e}_2}{\gamma_t} = -2\mathbf{v}\omega_3\vec{e}_2 = \frac{\Lambda}{\pi\gamma_t}\frac{-\mathbf{v}x_3\vec{e}_2}{(x_1 - \mathbf{v}t)^2/\gamma_t^2 + x_3^2} \tag{20.45b}$$

And we obtain exactly the solutions *(20.9)* which have been obtained in section 20.1 by using the method, which was proposed initially by Frank[1] in 1949, of expressing the field of displacement *(19.3)* in the referential $Ox_1x_2x_3$ by using the Lorentz transformations

$$\vec{u}_{ext}^{screw} = \vec{e}_2\frac{\Lambda}{\pi}\arctan\frac{x_1'}{x_3'} = \vec{e}_2\frac{\Lambda}{\pi}\arctan\frac{x_1 - \mathbf{v}t}{\gamma_t x_3} \tag{20.45c}$$

and of deducing directly the expressions *(20.45)* in the referential $Ox_1x_2x_3$ thanks to the two relationships $\vec{\omega}_{ext}^{screw} = (\overrightarrow{\text{rot}}\,\vec{u}_{ext}^{screw})/2$ and $\vec{\phi}_{ext}^{screw} = \partial\vec{u}_{ext}^{screw}/\partial t$.

The total relativistic energy of a moving screw dislocation

We deduce directly from relations *(20.45)* the density of elastic energy of distortion F_{dist}^{screw} and the density of kinetic energy F_{kin}^{screw} in the referential $Ox_1x_2x_3$

$$\begin{cases} F_{dist}^{screw} = 2(K_2 + K_3)\sum_i \omega_i^2 = 2(K_2 + K_3)\left(\omega_1'^2 + \omega_3'^2/\gamma_t^2\right) \\ F_{kin}^{screw} = mn\phi_2^2/2 = 2mn\mathbf{v}^2\omega_3'^2/\gamma_t^2 \end{cases} \tag{20.46}$$

and the total energy density $F_{tot}^{screw} = F_{dist}^{screw} + F_{kin}^{screw}$ by using relation $(K_2 + K_3)/mn = c_t^2$

$$F_{tot}^{screw} = F_{dist}^{screw} + F_{kin}^{screw} = 2(K_2 + K_3)\left[\omega_1'^2 + \omega_3'^2/\gamma_t^2 + \mathbf{v}^2\omega_3'^2/c_t^2\gamma_t^2\right] \tag{20.47}$$

The total energy $E_{\mathbf{v}}^{screw}$ by unit length of dislocation comes via the integration in $Ox_1x_2x_3$

$$E_{\mathbf{v}}^{screw} = \frac{2(K_2 + K_3)}{\gamma_t^2}\iiint_V\left[\gamma_t^2\omega_1'^2 + \omega_3'^2 + \mathbf{v}^2\omega_3'^2/c_t^2\right]dx_1dx_2dx_3 \tag{20.48}$$

But this integration can be carried out in a simpler fashion in the reference frame $O'x_1'x_2'x_3'$ since $dx_1 = \gamma_t dx_1'$, $dx_2 = dx_2'$ and $dx_3 = dx_3'$

$$E_{\mathbf{v}}^{screw} = \frac{2(K_2 + K_3)}{\gamma_t}\iiint_{V'}\left[\left(1 - \frac{\mathbf{v}^2}{c_t^2}\right)\omega_1'^2 + \omega_3'^2 + \frac{\mathbf{v}^2}{c_t^2}\omega_3'^2\right]dx_1'dx_2'dx_3' \tag{20.49}$$

In the reference frame $O'x_1'x_2'x_3'$, it is clear that $\omega_1'^2 = \omega_3'^2 = \vec{\omega}'^2/2$, so that

[1] *F. C. Frank, Proc. Phys. Soc., vol. A62, , p. 131, 1949*

$$E_{\mathbf{v}}^{screw} = \frac{1}{\gamma_t}\left[\left(1-\frac{\mathbf{v}^2}{2c_t^2}\right)+\frac{\mathbf{v}^2}{2c_t^2}\right]\left(2(K_2+K_3)\iiint_{V'}\vec{\omega}'^2\,dx_1'dx_2'dx_3'\right) \tag{20.50}$$

The value of the second parenthesis is the rest energy E_0^{screw} by unit length of the screw dislocation that we had obtained by the relation *(19.7)*, so that

$$E_{\mathbf{v}}^{screw} = \underbrace{\frac{1}{\gamma_t}\left(1-\frac{\mathbf{v}^2}{2c_t^2}\right)E_{dist}^{screw}}_{E_{\mathbf{v}}^{dist}} + \underbrace{\frac{1}{\gamma_t}\frac{\mathbf{v}^2}{2c_t^2}E_{dist}^{screw}}_{E_{\mathbf{v}}^{kin}} = \frac{E_{dist}^{screw}}{\gamma_t} = \frac{1}{\gamma_t}\frac{(K_2+K_3)\Lambda^2}{\pi}\ln\frac{R_\infty}{R} \tag{20.51a}$$

This expression of energy deserves a few comments:

- it is possible to transform somewhat relation *(20.51a)* to expressively show the inertial mass $M_0^{screw} \cong E_{dist}^{screw}/c_t^2$ at rest of the screw dislocation. We have

$$E_{\mathbf{v}}^{screw} = \underbrace{\frac{1}{\gamma_t}\left(1-\frac{\mathbf{v}^2}{2c_t^2}\right)E_{dist}^{screw}}_{E_{\mathbf{v}}^{dist}} + \underbrace{\frac{1}{\gamma_t}\frac{1}{2}M_0^{screw}\mathbf{v}^2}_{E_{\mathbf{v}}^{kin}} = \frac{E_{dist}^{screw}}{\gamma_t} = \frac{M_0^{screw}c_t^2}{\gamma_t} \tag{20.51b}$$

This remarkable expression allows us to understand the true physical origins of the relativistic terms $E_{\mathbf{v}}^{screw}$ and $E_{\mathbf{v}}^{screw}$ in our theory.
Indeed, under this form, the term $E_{\mathbf{v}}^{screw}$ corresponds to a relativistic correction of the energy of elastic distortion E_{dist}^{screw}, while the term $E_{\mathbf{v}}^{screw}$ corresponds to a relativistic correction of the kinetic energy $M_0^{screw}\mathbf{v}^2/2$.

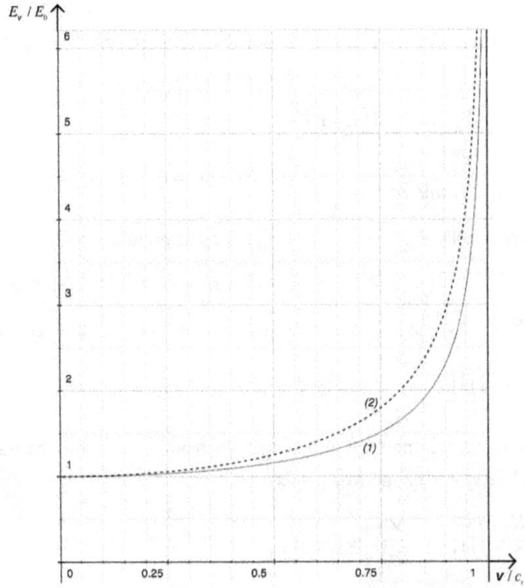

Figure 20.4 *- The total energy compared to the rest energy as a function of* $\mathbf{v}/c_t$*,*
in the case of a screw or edge dislocation (1) or a charge of rotation (2)

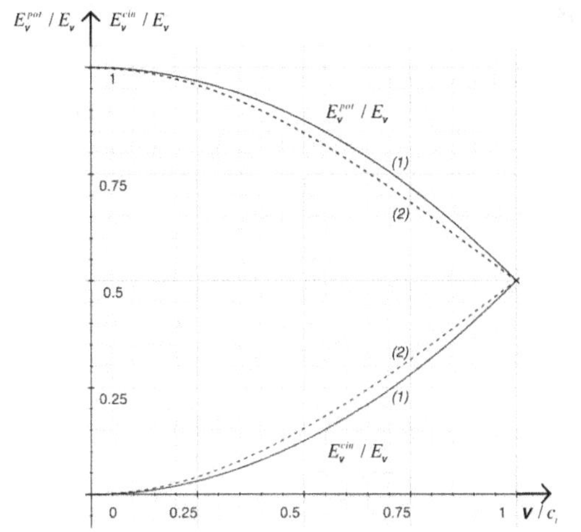

Figure 20.5 *- Fractions of total energy under a potential and kinetic form as a function of* $\mathbf{v} / c_t$ *, in the case of a screw or edge dislocation (1) or a spherical charge of rotation (2)*

- in the case of the screw dislocation, namely when the scalar charge density is found on a infinite linear string, the behavior of total energy $E_{\mathbf{v}}^{screw}$ is purely a *relativistic behavior*, which satisfies the famous relation of special relativity $E = E_0 / \gamma$,

- the total energy linked to the quantity of movement goes to an infinite value when velocity $\mathbf{v}$ tends towards the transversal speed c_t as shown in figure 20.4. This behavior is generated by the presence of the term $\gamma_t = \sqrt{1 - \mathbf{v}^2 / c_t^2}$ in the expression *(20.51)* of the energy, a term which is due to the relativistic contraction of the field of rotation in the direction of movement, according to the Lorentz transformation,

- the total energy $E_{\mathbf{v}}^{screw}$ associated with the charge in movement is not the energy stored in the singularity itself, but the movement of the singularity in the lattice which stored potential energy $E_{\mathbf{v}}^{dist}$ of elastic distortion of the lattice and of the newtonian kinetic energy $E_{\mathbf{v}}^{kin}$ of movement of the lattice in its vicinity,

- the fraction of total energy that is found under the form of an elastic potential energy of rotation and a kinetic energy of the lattice sites in movement is show in figure 20.5. We note among other things that the fraction of potential energy and kinetic energy are exactly equal when velocity $\mathbf{v}$ of a charge tends towards celerity c_t of transversal waves!

- the fact that we obtain very precisely a relativistic behavior is due to the particularity that the kinetic term $E_{\mathbf{v}}^{kin}$ is exactly compensated by an additional negative term in the potential energy $E_{\mathbf{v}}^{dist}$ in the case of the screw dislocation. We will see later on that this compensating effect is not systematic and that it depends essentially on the topology of the charge considered. We have as a consequence a paradoxal situation, namely that the relativistic dynamic of screw dislocations is *a consequence of the newtonian dynamic of the lattice in the absolute space* of the ***GO***, as it is the exact compensation of the newtonian kinetic energy $E_{\mathbf{v}}^{kin}$ of the lattice by the same negative term in the potential elastic energy $E_{\mathbf{v}}^{pot}$ which is responsible for it.

The relativistic energy of an edge dislocation in the perfect cosmic lattice

Let's revisit the case of an edge dislocation treated in section 19.2, by supposing the lattice a perfect cosmic lattice which satisfy conjectures *(19.58)* and hypothesis 4 *(20.40)*. In this case, we have seen via a purely classical calculation of the distortion energy and the kinetic energy of the edge dislocation that they satisfy a true Einstein relation *(19.59)*, just as the screw dislocation

$$E_{dist}^{edge} \cong M_0^{edge} c_t^2 \tag{20.52}$$

Thus, the conjectures *(19.58)*, namely that $K_0 = K_3 > 0$ and that $K_1 << K_0$ and $K_2 << K_3$, and the hypothesis 4 *(20.40)*, namely that $\tau_0 < \tau_{0cr}$ and that longitudinal waves do not exist, imply that the edge dislocations in a perfect cosmic lattice are subjected exactly to the same relativistic behaviors as the screw dislocation and thus we have

$$E_{\mathbf{v}}^{edge} = \frac{E_{dist}^{edge}}{\gamma_t} \tag{20.53}$$

which is easily verified since at low velocity we obtain a relation which agrees perfectly with the results obtained in section 19.2

$$E_{\mathbf{v}}^{edge} = \frac{E_{dist}^{edge}}{\sqrt{1-\mathbf{v}^2/c_t^2}} \underset{\mathbf{v}<<c_t}{\longrightarrow} E_{dist}^{edge}\left(1+\frac{1}{2}\mathbf{v}^2/c_t^2\right) = E_{dist}^{edge} + \frac{1}{2}\frac{E_{dist}^{edge}}{c_t^2}\mathbf{v}^2 = E_{dist}^{edge} + \frac{1}{2}M_0^{edge}\mathbf{v}^2 = E_{dist}^{edge} + E_{cin}^{edge} \tag{20.54}$$

Thus, the relation *(20.51b)* is also applicable to the edge dislocations in the perfect cosmic lattice when $\tau_0 < \tau_{0cr}$, so that

$$E_{\mathbf{v}}^{edge} = \underbrace{\frac{1}{\gamma_t}\left(1-\frac{\mathbf{v}^2}{2c_t^2}\right)E_{dist}^{edge}}_{E_{\mathbf{v}}^{dist}} + \underbrace{\frac{1}{\gamma_t}\frac{1}{2}M_0^{edge}\mathbf{v}^2}_{E_{\mathbf{v}}^{kin}} = \frac{E_{dist}^{edge}}{\gamma_t} = \frac{M_0^{edge}c_t^2}{\gamma_t} = \frac{1}{\gamma_t}\left(\frac{K_2}{K_3}\right)^2\frac{K_3\vec{\Lambda}^2}{2\pi}\ln\frac{R_\infty}{R} \tag{20.55}$$

The relativistic dynamic equation of a moving screw or edge dislocation

Let's suppose that a screw or an edge dislocation line, which is moving at velocity $\vec{\mathbf{v}}$ in a perfect cosmic lattice is submitted to a Peach and Koehler force $\vec{F}_{PK}$ per unit length. Due to the linear geometry of the dislocation, the vectors $\vec{\mathbf{v}}$ and $\vec{F}_{PK}$ can only be perpendicular to the dislocation line. The power transmitted to the dislocation by the force $\vec{F}_{PK}$ is written $\vec{F}_{PK}\vec{\mathbf{v}}$, and this power will increase the total energy $E_{\mathbf{v}}^{disloc}$ of the dislocation, so that the dynamic relativistic equation of the dislocation can be written

$$\frac{dE_{\mathbf{v}}^{disloc}}{dt} = \vec{F}_{PK}\vec{\mathbf{v}} \tag{20.56}$$

By supposing that vectors $\vec{\mathbf{v}}$ and $\vec{F}_{PK}$ are parallel, the equation can be written

$$\frac{dE_{\mathbf{v}}^{disloc}}{dt} = \frac{\partial E_{\mathbf{v}}^{disloc}}{\partial \mathbf{v}}\frac{d\mathbf{v}}{dt} = \frac{\partial}{\partial \mathbf{v}}\left(\frac{E_{dist}^{disloc}}{\gamma_t}\right)\mathbf{a} = E_{dist}^{disloc}\frac{\mathbf{v}}{\gamma_t^3 c_t^2}\mathbf{a} = F_{PK}\mathbf{v} \tag{20.57}$$

in which $\mathbf{a} = d\mathbf{v}/dt$ is the acceleration of the dislocation.

By using the relation $E_{dist}^{disloc} = M_0^{disloc}c_t^2$ *(19.59)*, the relativistic dynamic equation can be written in a vectorial form

$$\frac{E_{dist}^{disloc}}{\gamma_t^3 c_t^2}\vec{\boldsymbol{a}} = \frac{M_0^{disloc}}{\gamma_t^3}\vec{\boldsymbol{a}} = \vec{F}_{PK} \qquad (20.58)$$

By introducing the *relativistic quantity of movement* $\vec{\boldsymbol{P}}^{disloc}$ by unit of length of the screw or edge dislocation.

$$\vec{\boldsymbol{P}}^{disloc} = \frac{E_{dist}^{disloc}}{\gamma_t c_t^2}\vec{\boldsymbol{v}} = \frac{M_0^{disloc}}{\gamma_t}\vec{\boldsymbol{v}} \qquad (20.59)$$

we verify easily that the dynamic relativistic equation can be written

$$\frac{d\vec{\boldsymbol{P}}^{disloc}}{dt} = \frac{E_{dist}^{disloc}}{\gamma_t^3 c_t^2}\vec{\boldsymbol{a}} = \frac{M_0^{disloc}}{\gamma_t^3}\vec{\boldsymbol{a}} = \vec{F}_{PK} \qquad (20.60)$$

The expression *(20.57)* of the relativistic quantity of movement $\vec{\boldsymbol{P}}^{disloc}$ allows us to introduce a relativistic mass $M_{\boldsymbol{v}}^{disloc}$ of the dislocation in the quantity of movement from which we can deduce the total energy $E_{\boldsymbol{v}}^{disloc}$ and the quantity of movement $\vec{\boldsymbol{P}}^{disloc}$

$$M_{\boldsymbol{v}}^{disloc} = \frac{M_0^{disloc}}{\gamma_t} \quad \Rightarrow \quad \begin{cases} E_{\boldsymbol{v}}^{disloc} = \dfrac{M_0^{disloc} c_t^2}{\gamma_t} = M_{\boldsymbol{v}}^{disloc} c_t^2 \\ \vec{\boldsymbol{P}}^{disloc} = \dfrac{M_0^{disloc}}{\gamma_t}\vec{\boldsymbol{v}} = M_{\boldsymbol{v}}^{disloc}\vec{\boldsymbol{v}} \end{cases} \qquad (20.61)$$

Relations *(20.59)* through *(20.61)* are perfectly identical to the dynamic relations obtained in special relativity. We can also verify the classic relation in special relativity $E^2 = m^2c^4 + p^2c^2$

$$\left(E_{\boldsymbol{v}}^{disloc}\right)^2 = \left(M_0^{disloc}\right)^2 c_t^4 + \left(\vec{\boldsymbol{P}}^{disloc}\right)^2 c_t^2 \qquad (20.62)$$

An interesting remark can be done here: the total relativistic energy $E_{\boldsymbol{v}}^{disloc}$ associated with the dislocation is the sum of the potential energy $E_{\boldsymbol{v}}^{dist}$ of plastic deformations of the lattice and the newtonian kinetic energy $E_{\boldsymbol{v}}^{kin}$ of movement within the lattice. But by associating the total relativistic energy $E_{\boldsymbol{v}}^{disloc}$ to this moving string, and knowing that the rest energy of the string is given by E_{dist}^{disloc}, we could also consider that the energy of the moving string is equal to its rest energy E_{dist}^{disloc} and a *movement energy* E_{mvt}^{disloc} which corresponds to the additional energy generated by its displacement within the lattice, by writing

$$E_{mvt}^{disloc} = E_{\boldsymbol{v}}^{disloc} - E_{dist}^{disloc} = \left(M_{\boldsymbol{v}}^{disloc} - M_0^{disloc}\right)c_t^2 = M_0^{disloc}\left(\frac{1}{\gamma_t} - 1\right)c_t^2 \qquad (20.63)$$

In special relativity, this energy of movement E_{mvt}^{disloc} is often called the kinetic energy T *of the particle*. But in the case of the dislocation we consider here we know it is not really a kinetic energy since E_{mvt}^{disloc} is in fact the following combination of potential energy and kinetic energy of the particles of the lattice

$$E_{mvt}^{disloc} = \underbrace{\frac{1}{\gamma_t}\left(1 - \frac{\boldsymbol{v}^2}{2c_t^2}\right)E_{dist}^{disloc}}_{E_{\boldsymbol{v}}^{dist}} + \underbrace{\frac{1}{\gamma_t}\frac{\boldsymbol{v}^2}{2c_t^2}E_{dist}^{disloc}}_{E_{\boldsymbol{v}}^{kin}} - E_{dist}^{disloc} \qquad (20.64)$$

Finally, if we calculate the total energy for weak velocities ($\boldsymbol{v} << c_t$), we obtain

$$E_{\boldsymbol{v}}^{disloc} \cong E_{dist}^{disloc}\left(1 + \frac{1}{2}\frac{\boldsymbol{v}^2}{c_t^2} + \ldots\right) \cong E_{dist}^{disloc} + \frac{E_{dist}^{disloc}}{2c_t^2}\boldsymbol{v}^2 + \ldots = E_{dist}^{disloc} + \frac{1}{2}M_0^{disloc}\boldsymbol{v}^2 + \ldots \qquad (20.65)$$

and we find again the *inertial rest mass* M_0^{disloc} of the dislocation *(19.59)*, exactly as we had found it from classical means in chapter 19!

20.5 – Relativistic dynamics of loop singularities

We have seen in chapter 19 that the loop topological singularities in a *perfect cosmic lattice* all satisfy the Einstein relations

$$E_{dist}^{loop} = M_0^{loop} c_t^2 \tag{20.66}$$

which was obtained by a classical calculation of their elastic energy of distortion and their kinetic energy. This implies that the relativistic energy of loop singularities is deduced in the same way we deduce the relativistic energy of an edge dislocation. As a consequence ,we deduce the following relativistic energies for the loop singularities in a perfect cosmic lattice when $\tau_0 < \tau_{0cr}$, namely a twist disclination loop *(TL)*, an edge dislocation loop *(EL)* and a mixed dislocation loop *(ML)*

$$E_{\mathbf{v}}^{TL} = \underbrace{\frac{1}{\gamma_t}\left(1-\frac{\mathbf{v}^2}{2c_t^2}\right)E_{dist}^{TL}}_{E_{\mathbf{v}}^{dist}} + \underbrace{\frac{1}{\gamma_t}\frac{1}{2}M_0^{TL}\mathbf{v}^2}_{E_{\mathbf{v}}^{kin}} = \frac{E_{dist}^{TL}}{\gamma_t} = \frac{M_0^{TL}c_t^2}{\gamma_t} \tag{20.67}$$

$$E_{\mathbf{v}}^{EL} = \underbrace{\frac{1}{\gamma_t}\left(1-\frac{\mathbf{v}^2}{2c_t^2}\right)E_{dist}^{EL}}_{E_{\mathbf{v}}^{dist}} + \underbrace{\frac{1}{\gamma_t}\frac{1}{2}M_0^{EL}\mathbf{v}^2}_{E_{\mathbf{v}}^{kin}} = \frac{E_{dist}^{EL}}{\gamma_t} = \frac{M_0^{EL}c_t^2}{\gamma_t} \tag{20.68}$$

$$E_{\mathbf{v}}^{ML} = \underbrace{\frac{1}{\gamma_t}\left(1-\frac{\mathbf{v}^2}{2c_t^2}\right)E_{dist}^{ML}}_{E_{\mathbf{v}}^{dist}} + \underbrace{\frac{1}{\gamma_t}\frac{1}{2}M_0^{ML}\mathbf{v}^2}_{E_{\mathbf{v}}^{kin}} = \frac{E_{dist}^{ML}}{\gamma_t} = \frac{M_0^{ML}c_t^2}{\gamma_t} \tag{20.69}$$

We also deduce that in a perfect cosmic lattice, the relativistic dynamic equation of a loop singularity is identical to that of a screw or edge dislocation, namely

$$\frac{d\vec{\mathbf{P}}^{loop}}{dt} = \frac{E_{dist}^{loop}}{\gamma_t^3 c_t^2}\vec{\mathbf{a}} = \frac{M_0^{loop}}{\gamma_t^3}\vec{\mathbf{a}} = \vec{\mathbf{F}}^{loop} \tag{20.70}$$

in which $\vec{\mathbf{F}}^{loop}$ is the force acting globally on the loop and $\vec{\mathbf{P}}^{loop}$ is the relativistic quantity of movement of the loop, given by

$$\vec{\mathbf{P}}^{loop} = \frac{E_{dist}^{loop}}{\gamma_t c_t^2}\vec{\mathbf{v}} = \frac{M_0^{loop}}{\gamma_t}\vec{\mathbf{v}} \tag{20.71}$$

20.6 – Relativistic dynamic of a spherical charge of rotation

Let's consider now a spherical charge of rotation, like the one described in figure 19.1, which moves along the axis Ox_1 with velocity $\vec{v}$. In the referential $O'x_1'x_2'x_3'$ co-moving with the charge, we have since the charge is immobile that $p_1' = p_2' = p_3' = 0$. As a consequence, according to *(20.43)*, we have the relations

$$\begin{cases} m_1 = m_1' \\ m_2 = m_2'/\gamma_t \\ m_3 = m_3'/\gamma_t \end{cases} \quad \text{and} \quad \begin{cases} np_1 = 0 \\ np_2 = -\mathbf{v} m_3'/2c_t^2\gamma_t \\ np_3 = +\mathbf{v} m_2'/2c_t^2\gamma_t \end{cases} \tag{20.72}$$

which allow to deduce, since $m_i = 4(K_2+K_3)\omega_i$, $np_i = nm\phi_i$ and $(K_2+K_3)/mn = c_t^2$

$$\begin{cases} \omega_1 = \omega_1' \\ \omega_2 = \omega_2'/\gamma_t \\ \omega_3 = \omega_3'/\gamma_t \end{cases} \quad \text{and} \quad \begin{cases} \phi_1 = 0 \\ \phi_2 = -2\mathbf{v}\omega_3'/\gamma_t \\ \phi_3 = +2\mathbf{v}\omega_2'/\gamma_t \end{cases} \tag{20.73}$$

The total relativistic energy of a moving spherical charge of rotation

We deduce the potential density of energy $F_{dist}^{Q_\lambda}$ and of kinetic energy $F_{cin}^{Q_\lambda}$ in $Ox_1x_2x_3$

$$F_{dist}^{Q_\lambda} = 2(K_2+K_3)\sum_i \omega_i^2 = 2(K_2+K_3)\left(\omega_1'^2 + \frac{1}{\gamma_t^2}\omega_2'^2 + \frac{1}{\gamma_t^2}\omega_3'^2\right) \tag{20.74}$$

$$F_{kin}^{Q_\lambda} = \frac{1}{2}mn\left(\phi_2^2+\phi_3^2\right) = \frac{2mn\mathbf{v}^2}{\gamma_t^2}\left(\omega_2'^2+\omega_3'^2\right) \tag{20.75}$$

as well as the total density of energy $F_{tot}^{Q_\lambda} = F_{dist}^{Q_\lambda} + F_{kin}^{Q_\lambda}$ by using $(K_2+K_3)/mn = c_t^2$

$$F_{tot}^{Q_\lambda} = F_{dist}^{Q_\lambda} + F_{kin}^{Q_\lambda} = 2(K_2+K_3)\left[\omega_1'^2 + \frac{1}{\gamma_t^2}\left(\omega_2'^2+\omega_3'^2\right) + \frac{1}{\gamma_t^2}\frac{\mathbf{v}^2}{c_t^2}\left(\omega_2'^2+\omega_3'^2\right)\right] \tag{20.76}$$

The total energy $E_{\mathbf{v}}^{Q_\lambda}$ of the moving spherical charge of rotation is given by the integration on the volume of infinite solid in the referential $Ox_1x_2x_3$

$$E_{\mathbf{v}}^{Q_\lambda} = \frac{2(K_2+K_3)}{\gamma_t^2}\iiint_V \left[\gamma_t^2\omega_1'^2 + \left(\omega_2'^2+\omega_3'^2\right) + \frac{\mathbf{v}^2}{c_t^2}\left(\omega_2'^2+\omega_3'^2\right)\right] dx_1 dx_2 dx_3 \tag{20.77}$$

But this integration can also be done in a manner which is a lot simpler in the referential $O'x_1'x_2'x_3'$ since $dx_1 = \gamma_t dx_1'$, $dx_2 = dx_2'$ and $dx_3 = dx_3'$

$$E_{\mathbf{v}}^{Q_\lambda} = \frac{2(K_2+K_3)}{\gamma_t}\iiint_{V'} \left[\left(1-\frac{\mathbf{v}^2}{c_t^2}\right)\omega_1'^2 + \left(\omega_2'^2+\omega_3'^2\right) + \frac{\mathbf{v}^2}{c_t^2}\left(\omega_2'^2+\omega_3'^2\right)\right] dx_1' dx_2' dx_3' \tag{20.78}$$

In referential $O'x_1'x_2'x_3'$, it is clear that $\omega_1'^2 = \omega_2'^2 = \omega_3'^2 = \vec{\omega}'^2/3$, so that

$$E_{\mathbf{v}}^{Q_\lambda} = \frac{1}{\gamma_t}\left[\left(1-\frac{\mathbf{v}^2}{3c_t^2}\right) + \frac{2\mathbf{v}^2}{3c_t^2}\right]\left(2(K_2+K_3)\iiint_{V'} \vec{\omega}'^2 dx_1' dx_2' dx_3'\right) \tag{20.79}$$

The value of the second parenthesis is nothing else than the rest energy $E_{dist}^{Q_\lambda}$ of the charge which we had obtained via relation *(19.71)*. We have

$$E_{\mathbf{v}}^{Q_\lambda} = \underbrace{\frac{1}{\gamma_t}\left(1-\frac{\mathbf{v}^2}{3c_t^2}\right)E_{dist}^{Q_\lambda}}_{E_{pot}^{Q_\lambda}} + \underbrace{\frac{1}{\gamma_t}\frac{2\mathbf{v}^2}{3c_t^2}E_{dist}^{Q_\lambda}}_{E_{kin}^{Q_\lambda}} = \frac{1}{\gamma_t}\left(1+\frac{\mathbf{v}^2}{3c_t^2}\right)\underbrace{\frac{3(K_2+K_3)Q_\lambda^2}{5\pi R}}_{E_{dist}^{Q_\lambda}} \tag{20.80}$$

Given the term γ_t in the denominator, we find here a behavior of total energy $E_{\mathbf{v}}^{Q_\lambda}$ of the spherical charge which is similar to a relativistic behavior since it tends towards an infinite value when velocity $\mathbf{v}$ tends towards transversal waves velocity c_t as shown in figure 20.4 in which we show $E_{\mathbf{v}}^{Q_\lambda} / E_{dist}^{Q_\lambda}$ as a function of $\mathbf{v} / c_t$.

As is the case of an edge or screw dislocation, the total energy is found as an elastic potential energy of rotation of the lattice and as a kinetic energy of the nodes of the lattice, and the fraction which each represents depends on the value of $\mathbf{v} / c_t$ as shown in figure 20.5. We note that these fractions of energy under the potential form and kinetic form are perfectly equal when velocity $\mathbf{v}$ becomes equal to c_t !

However expression *(20.80)* of the total energy $E_{\mathbf{v}}^{Q_\lambda}$ is not the same as the classical relativistic behavior as $E_{\mathbf{v}}^{Q_\lambda} \neq E_{dist}^{Q_\lambda} / \gamma_t$. This difference is due to the fact that the additional negative term in the potential energy does not equally compensate the kinetic energy term (the kinetic energy term is twice superior to the absolute value of the additional term in the potential energy). If we calculate the total energy $E_{\mathbf{v}}^{Q_\lambda}$ for weak velocities ($\mathbf{v} << c_t$) by developing the term γ_t in the denominator, we obtain

$$E_{\mathbf{v}}^{Q_\lambda} \cong E_{dist}^{Q_\lambda}\left(1+\mathbf{v}^2/2c_t^2+...\right)\left(1+\mathbf{v}^2/3c_t^2\right) \cong E_{dist}^{Q_\lambda} + 5E_{dist}^{Q_\lambda}\mathbf{v}^2/6c_t^2+... \quad (20.81)$$

In this case, the energy of the moving charge is equal to its rest energy and the second term is proportional to the square of the velocity which can be assimilated to a term of *kinetic energy of the charge*. We can therefore assign a rest inertial mass M_0 to the spherical charge of rotation, given by

$$\frac{1}{2}M_0 \cong \frac{5}{6}\frac{E_{dist}^{Q_\lambda}}{c_t^2} \quad \Rightarrow \quad M_0 \cong \frac{5}{3}\frac{E_{dist}^{Q_\lambda}}{c_t^2} \quad \Rightarrow \quad E_{dist}^{Q_\lambda} \cong \frac{3}{5}M_0 c_t^2 \quad (20.82)$$

We note that the relation between the rest energy and the inertial mass for a spherical charge of rotation Q_λ differs from the famous Einstein equation of special relativity $E_0 = M_0 c^2$.

20.7 – On the paradox of the energy of electrons

We find in relation *(20.82)* our version of a famous paradox of classical electromagnetism. Indeed the same type of calculation done in classical electromagnetism, in order to find the energy stored by the electric field of an electron in movement gives us a very similar result, namely that $E_{electromagnetic}^{relativistic} \neq E_{electric\ field}^{rest} / \gamma$, and thus *the mass associated with the electromagnetic fields of the electron does not satisfy the principle of special relativity!* This famous result of electromagnetism has been widely discussed. Several models have been proposed to account for it, without much success. We can say here that it was never properly framed in classical electromagnetism or special relativity! A detailed discussion on this topic can be found in the famous lectures of R. P. Feynman [2].

This famous paradox of the electrical energy of the electron could find here a simple explanation, if we suppose that the electron has in fact a ring structure[3] similar to a twist disclination loop or a screw pseudo-dislocation loop and that the electrical field is analogous to the field of rotation. Indeed, the expression *(20.67)* of the relativistic energy of a loop of twist disclination

---

[2] *Richard P. Feynman, The Feynman Lectures on Physics, Addison-Wesly Publ. Company, 1970, chap. 28*

[3] *The idea of an electron with a ring shape has been initially proposed in 1915 by Parson (Smithsonian miscellaneous collections, nov. 1915), then developed by Webster (Amer. Acad., jan.. 1915) and Allen (Phil. Mag., 4, 1921, p. 113), and the proposition that an electron could be similar to a twist disclination loop has been proposed in 1996 by Unziker (arXiv:gr-qc/9612061v2).*

perfectly satisfies the expression of Einstein $E_v \cong E_0 / \gamma_t$, so that if the electron has the topological structure of a twist disclination loop in a cosmic lattice, we would have a localized charged q_λ of rotation which would present, in the far field, a divergent rotational field $\vec{\omega}^{el}$ just like the electron has a divergent electrical field $\vec{E}^{el}$, and which would satisfy, at low velocity ($\mathbf{v} << c_t$) the relation of special relativity since for a twist disclination loop we have

$$E_{dist}^{TL} = M_0^{TL} c_t^2 \tag{20.83}$$

20.8 – Peach and Koehler force and relativistic Lorentz force

In section 11.2, we have deduced the *force of Peach and Koehler as* $\vec{f}_{PK} = \lambda \vec{m} + \vec{\mathbf{v}} \wedge \vec{\mathbf{A}}$ which acts via the field $\vec{m}$ on the unit of volume of charges of rotation with a density λ. In this relation, the term $\lambda \vec{m}$ is analogous to the electric force $\vec{f} = \rho \vec{E}$ acting per unit volume on a density of electrical charges ρ in the equations of Maxwell of electromagnetism, while the term $\vec{\mathbf{v}} \wedge \vec{\mathbf{A}}$ was introduced to take into account the forces that gave no work. For a density λ of charges moving with velocity $\vec{\mathbf{v}}$ along the axis Ox_1 , the density of forces acting on the referential $O'x_1'x_2'x_3'$ linked to the charge is thus, since the field is immobile in this framework and that as a consequence $\vec{\mathbf{v}} = 0$ in that framework

$$\vec{f}'_{PK} = \lambda \vec{m}' \tag{20.84}$$

We can then find the force by unit volume acting on the same density of charges moving at velocity $\vec{\mathbf{v}}$ in the moving framework $Ox_1x_2x_3$, and using *(20.38)*

$$\begin{cases} f_1' = \lambda m_1' = \lambda m_1 = f_1 \\ f_2' = \lambda m_2' = \lambda \left(m_2 - 2\mathbf{v}_1 n p_3 \right) / \gamma_t = f_2 \\ f_3' = \lambda m_3' = \lambda \left(m_3 + 2\mathbf{v}_1 n p_2 \right) / \gamma_t = f_3 \end{cases} \tag{20.85}$$

We then have for the force $\vec{f}_{PK}$ in the framework $Ox_1x_2x_3$

$$\vec{f}_{PK} = \lambda m_1 \vec{e}_1 + \lambda \frac{1}{\gamma_t} \left(m_2 - 2\mathbf{v}_1 n p_3 \right) \vec{e}_2 + \lambda \frac{1}{\gamma_t} \left(m_3 + 2\mathbf{v}_1 n p_2 \right) \vec{e}_3 \tag{20.86}$$

we easily transform this, first by using the vectorial product $\vec{\mathbf{v}} \wedge n\vec{p}$

$$\vec{f}_{PK} = \lambda m_1 \vec{e}_1 + \frac{\lambda}{\gamma_t} \left(m_2 \vec{e}_2 + m_3 \vec{e}_3 \right) + \frac{2\lambda}{\gamma_t} \left(\vec{\mathbf{v}} \wedge n\vec{p} \right) \tag{20.87}$$

and by using the fact that $\vec{J}^{(\lambda)} = \lambda \vec{\mathbf{v}}$

$$\vec{f}_{PK} = \lambda m_1 \vec{e}_1 + \frac{\lambda}{\gamma_t} \left(m_2 \vec{e}_2 + m_3 \vec{e}_3 \right) + \frac{1}{\gamma_t} \left(\vec{J} \wedge 2n\vec{p} \right) \tag{20.88}$$

In the case where $|\vec{\mathbf{v}}| << c_t$, γ_t becomes close to unity and the force by unit of volume in the mobile framework $Ox_1x_2x_3$ becomes equal to

$$\vec{f}_{PK} = \lambda \vec{m} + \vec{J} \wedge 2n\vec{p} = 2\lambda \left(\frac{\vec{m}}{2} + \vec{\mathbf{v}} \wedge n\vec{p} \right) \tag{20.89}$$

which is the perfect analog to the *electromagnetic force of Lorentz*

$$\vec{f}_L = \rho \vec{E} + \vec{j} \wedge \vec{B} = \rho \left(\vec{E} + \vec{\mathbf{v}} \wedge \vec{B} \right) \tag{20.90}$$

The term $2\lambda(\vec{\mathbf{v}} \wedge n\vec{p})$ in the force $\vec{f}_{PK}$ is the $\vec{\mathbf{v}} \wedge \vec{\mathbf{A}}$ term which we had introduced in relation

(11.10) to take into account the forces that do no work, so that the vector $\vec{\boldsymbol{A}}$ now has a well known value

$$\vec{\boldsymbol{A}} = 2\lambda n\vec{p} \tag{20.91}$$

We can then apply relation *(20.84)* to the various topological singularities:

- in the case of *a linear dislocation,* the integration of *(20.84)* on the unit length of the dislocation gives us the following force acting per unit length of dislocation

$$\vec{\boldsymbol{F}}_{PK} = 2\Lambda\left(\frac{\vec{m}}{2} + \vec{\boldsymbol{v}} \wedge n\vec{p}\right) \tag{20.92}$$

However, if a linear screw dislocation is moving in a solid, its velocity $\vec{\boldsymbol{v}}$ is necessarily perpendicular to the dislocation line, and the force $\vec{\boldsymbol{F}}_{PK}$ will do work only if it is perpendicular to the line, so that only the component of $n\vec{p}$ along the string is capable of acting upon the dislocation through a force $\vec{\boldsymbol{F}}_{PK}$!

- in the case of *a spherical charge of rotation* $\boldsymbol{Q}_\lambda$, the relation *(20.89)* can be integrated on the volume of the charge, and we obtain the total force acting on the charge of rotation

$$\vec{\boldsymbol{F}}_{PK} = 2\boldsymbol{Q}_\lambda\left(\frac{\vec{m}}{2} + \vec{\boldsymbol{v}} \wedge n\vec{p}\right) \tag{20.93}$$

This relation corresponds directly to the expression of the electromagnetic force acting on an electrical charge q, namely $\vec{\boldsymbol{F}} = q\left(\vec{E} + \vec{\boldsymbol{v}} \wedge \vec{B}\right)$.

- in the case of *a twist disclination loop with charge* $\boldsymbol{q}_{\lambda TL} = 2\pi R\Lambda_{TL} = -\pi R\,\vec{\boldsymbol{B}}_{TL}\vec{\boldsymbol{t}}$, we can apply relation *(20.92)* or relation *(20.93)* to the loop, which allows us to write

$$\vec{\boldsymbol{F}}_{PK} = 2\boldsymbol{q}_{\lambda TL}\left(\frac{\vec{m}}{2} + \vec{\boldsymbol{v}} \wedge n\vec{p}\right) \tag{20.94}$$

Chapter 21

On the role of «aether» played by the cosmic lattice for a mobile cluster of singularities

In a perfect cosmic lattice satisfying $\tau_0 < \tau_{0cr}$, we have shown that all microscopical topological singularities like dislocation lines and dislocations/disclination loops satisfy Lorentz transformations based on the transversal wave velocity. As a consequence, a localized cluster of topological singularities which interact with each other via their rotation fields is also submitted globally to the Lorentz transformations.

On this base, we discuss the analogies which exist between our theory of the perfect cosmic lattice and the Special Relativity. We discuss among others the role of *«aether»* that the lattice plays vis-a-vis a cluster of singularities in movement interacting via their rotation fields. We show that this notion of *«aether»* gives us a completely new perspective on the theory of Special Relativity, as well as a very elegant explanation to the famous paradox of the twins in Special Relativity.

21.1 – The Lorentz transformation applied to a cluster of moving topological singularities that interact via their rotation fields

In chapter 20, we have seen that the displacement of a topological singularity in frame $Ox_1x_2x_3$ of a perfect cosmic lattice satisfying $\tau_0 < \tau_{0cr}$, with velocity $\vec{V}$ in the direction of axis Ox_1, can be described in a frame $O'x_1'x_2'x_3'$ co-moving with the singularity thanks to the Lorentz transformation based on the transversal wave velocity. At constant volume expansion, a cluster of singularities which are moving in the lattice, formed with localized singularities such as dislocation and disclination loops which interact via their fields of rotation, is also subject to the same *Lorentz transformation (20.24), with all its properties as time dilation and length contraction,* because the fields of rotation which give the interactions between the singularities satisfy this transformation.

On the strong mathematical analogy of the Lorentz transformations applied to the cosmic lattice and to the Special Relativity

There exists a strong *mathematical analogy* between the transformation of Lorentz used here for the transmission of information and interaction of the singularities via transversal waves within the cosmic lattice and the Lorentz transformation of the theory of Special Relativity to describe the relativistic dynamic of mobile objects in the universe in relation with the speed of light. But there exists also a serious *difference of physical interpretation* between these two theories, linked to the presence of an 'aether' for the topological singularities, which is the lattice itself and that confers a privileged status to the fixed singularities compared to those in movement, while in the theory of Special Relativity, all objects have the same status, hence the famous name of 'relativity'. This essential difference allows us to bring a new light on the phenomena of relativity. We will discuss those in the following.

On the primordial physical differences with Special Relativity:
role of "aether" played by the lattice and existence of an absolute reference frame

The dynamic of the singularities within a cosmic lattice is different from special relativity via the existence of an *absolute frame of reference* for the movement of singularities, and an *'aether'* for the propagation of transversal waves (longitudinal waves do not exist if $\tau_0 < \tau_{0cr}$).
Contrary to special relativity, the lattice can be described from the outside by an observer ***GO*** *(imaginary **Grand Observer**)* which has a universal clock and universal rulers in the absolute frame $Q\xi_1\xi_2\xi_3$. This external observer of the lattice is not subject to any constraint of speed of propagation of information, so that it is the only one who can observe qualitatively and measure quantitatively and precisely the notion *of instantaneity of events within the lattice.*

The local observers HS (Homo Sapiens)

We could imagine now very different observers, the local observers ***HS** (**Homo Sapiens**)*, which are embedded in the lattice and *made of the topological singularities of the lattice*. These particular observers then have a very different status from the observer ***GO*** since they are *integral parts of the lattice* and they are free to move about the lattice. But these observers are constrained by the fact that they transmit information from one point to another via the finite velocity of transversal waves or longitudinal waves. An ***HS*** observer has no access to an absolute definition of simultaneity of events such as that of the ***GO***, but only possesses a relativistic definition of the simultaneity, which depends on velocity $\vec{\boldsymbol{v}}$ of displacement vis-a-vis the lattice and the local value of volume expansion of the lattice.
For simplicity reasons, the ***GO*** can choose as universal rulers and universal clocks the rulers and clocks of any ***HS*** immobile with respect to the lattice, and which would be found at a point of the lattice which is immobile and with null expansion ($\tau = 0$).
Each ***HS*** is equipped with a local framework which has rulers and a clock which appear immutable for this ***HS***, while the length of it's rulers and the speed at which time is counted depend in reality, in the absolute referential of the ***GO***, on the volume expansion of the lattice at the point where ***HS*** is found and on it's velocity $\vec{\boldsymbol{v}}$ with respect to the lattice. As a consequence, the ***HS*** does not have direct access to the value of the volume expansion or to it's proper value of displacement velocity $\vec{\boldsymbol{v}}$ with respect to the lattice. Only the ***GO*** has access to this type of information!
The Lorentz transformations we have identified are actually ***GO*** tools, which can be used without problems in determining the rulers and local clocks of all ***HS*** attached to the lattice, or simply to calculate the various fields associated with topological singularities moving within the lattice. And the ***GO*** can apply these transformations anywhere on the lattice where it is possible to find a state of homogeneous and constant expansion, which may well be different from the zero expansion since the Lorentz transformations is based on the transmission velocity of transverse waves, which is perfectly determined regardless of the network expansion status

$\left(c_t(\tau)=c_{t0}\,\mathrm{e}^{\tau/2}\right)$. From this point of view, our interpretation of the Lorentz transformations is quite far from the interpretation of special relativity, for which these transformations are tools that can use any ***HS*** observer to switch to another Galilean framework in movement relative to the first, and for which the speed of light is an absolute constant! The main consequences of these essential differences will be analyzed in detail in the following sections.

21.2 – Contraction of length and dilation of time for an HS observer

On the real contraction of the length of an HS observer in movement inside the lattice

The transformations of Lorentz *(20.24)* imply that, for singularities moving at velocity $\vec{\mathbf{v}}$ in the direction Ox_1, the ruler in direction Ox_1 shortens by a factor γ_t. Indeed, let's consider a vector $\vec{d}=x_1\vec{e}_1$ in the direction Ox_1 at the instant $t=0$ in the framework $Ox_1x_2x_3$ immobile with respect to the lattice. This vector $\vec{d}$ can also be described in the mobile lattice $O'x_1'x_2'x_3'$ by writing

$$\vec{d}=x_1\vec{e}_1=x_1'\vec{e}_1' \tag{21.1}$$

By using the direct transformation laws of Lorentz *(20.27)*, taken at instant $t=0$, we obtain

$$\vec{d}=x_1\vec{e}_1=x_1'\vec{e}_1'=\frac{x_1}{\gamma_t}\vec{e}_1' \quad\Rightarrow\quad \vec{e}_1=\frac{1}{\gamma_t}\vec{e}_1' \tag{21.2}$$

We can also use the reverse Lorentz transformation *(20.27)*, taken at instant $t=0$, and we obviously obtain the same result

$$\left.\begin{array}{lll}\vec{d}=x_1\vec{e}_1=\dfrac{x_1'+\mathbf{v}t'}{\gamma_t}\vec{e}_1=x_1'\vec{e}_1' & \Rightarrow & \vec{e}_1=\dfrac{\gamma_t x_1'}{x_1'+\mathbf{v}t'}\vec{e}_1' \\ t=0=\dfrac{t'+\mathbf{v}x_1'/c_t^2}{\gamma_t} & \Rightarrow & t'=-\mathbf{v}x_1'/c_t^2\end{array}\right\}\Rightarrow\quad \vec{e}_1=\frac{1}{\gamma_t}\vec{e}_1' \tag{21.3}$$

These calculations show that, for the ***GO***, the ruler $\vec{e}_1'$ in the mobile framework $O'x_1'x_2'x_3'$ is effectively shortened by a factor γ_t compared to ruler $\vec{e}_1$ in the mobile framework $Ox_1x_2x_3$ in which the singularities move with velocity $\vec{\mathbf{v}}$

$$\vec{e}_1'=\gamma_t\vec{e}_1 \tag{21.4}$$

To interpret this shortening of rulers in the direction of movement, one has to imagine the architecture of the cluster as a set of topological singularities, linked by the interactions of their respective rotational fields (figure 21.1). These lattice singularities move with respect to the lattice with velocity $\vec{\mathbf{v}}$ in direction Ox_1, and the finite nature of velocity c_t and their interactions via the rotational field imposes that the complete architecture of the cluster of singularities contract in direction Ox_1. But this contraction does not affect the lattice, which conserves its state of original volume expansion, which we have represented in figure 21.1 for the case where two identical clusters move with velocities $\vec{\mathbf{v}}'$ and $\vec{\mathbf{v}}''$, measured with respect to the observer ***GO***.
Thus the relativistic effects on the rulers of observer ***HS***, associated to the collective movement of the singularities vis-à-vis the lattice, have nothing to do with the effects of volume expansion of the lattice, for which the modifications of the lengths of the rulers of ***HS*** will be associated with

real variation of the length of the unit cell of the cosmic lattice as we will see later in chapter 24! We should also note that these two effects are cumulative, namely that the rulers of an ***HS*** observer can be contracted or expanded by variation of the volume expansion and again contract by the movement of the singularity cluster with respect to the lattice! In this fashion, the contraction-expansion of the rulers and clocks of an ***HS*** observer depend both on the local expansion and the velocity $\vec{\mathbf{v}}$ of the ***HS*** with respect to the lattice. Furthermore, in the Lorentz transformation applied by the ***GO*** observer, the value of $\gamma_t = (1-\mathbf{v}^2/c_t^2)^{1/2}$ depends not only on the velocity $\vec{\mathbf{v}}$ of the ***HS*** with respect to the lattice, but also on the local velocity c_t of transversal waves, which depends on the volume expansion τ of the cosmic lattice since

$$c_t\Big|_{\tau\neq 0} = c_{t0}\Big|_{\tau=0}\, e^{\tau/2} \tag{21.5}$$

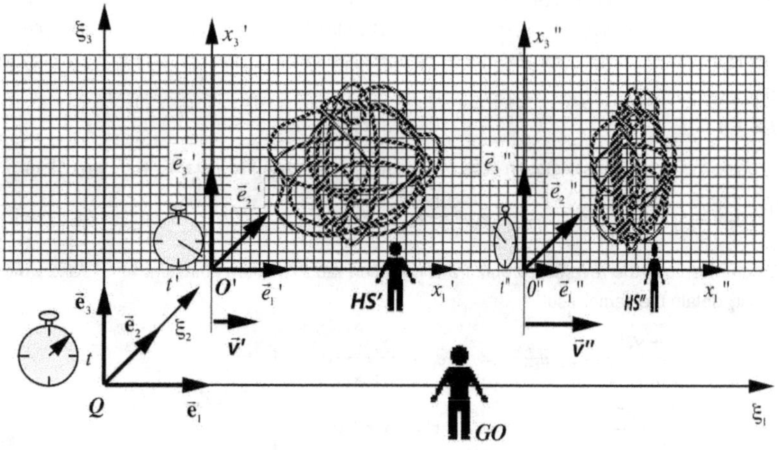

Figure 21.1 *- the mobile Lorentz frameworks of observers **HS'** and **HS''** in movement inside the lattice, as observed by the observer **GO***

On the real dilation of time of an HS observer in movement inside the lattice

The phenomenon of slowing down of the clock of the observer ***HS*** which is moving with respect to the lattice has already been explained in chapter 20 with the figures 20.2 and 20.3. Imagine that it is an observer ***HS*** who builds now its own clocks in his framework $O'x'_1x'_2x'_3$, by fixing two mirror face to face and at a distance d_0 one from the other, mirrors that have the property of reflecting transversal waves. By sending a transversal wave between the 2 mirrors, ***HS*** can perfectly use, as basis for time measurement, the time lapse $T_0 = 2d_0/c_t$ that flows between a back and forth of the wave between the two mirrors, because the distance d_0 and velocity c_t of the transversal waves are for him constants. If the observer ***HS*** is initially at rest with respect to the lattice, the ***GO*** can consider the time laps $T_0 = 2d_0/c_t$ as the basis for its proper time in $Ox_1x_2x_3$.

Let's imagine now that the ***HS*** observer is moving with respect to the lattice with velocity $\vec{\mathbf{v}}$ in the direction Ox_1, and that he places two clocks in 'quadrature', meaning that a first clock has 2 mirrors in one direction Ox_1' and the second clock has 2 mirrors along Ox_3' (or Ox_2'). In principle, in its framework $O'x'_1x'_2x'_3$, the time lapse $T_0 = 2d_0 / c_t$ measure by the ***HS*** with its two clocks is the same.

Let's take now the point of view of the ***GO***. In section 20.1, we have shown that the basic time of the moving clock of the ***HS*** observer in frame $O'x_1'x_2'x_3'$, measured by the ***GO*** in its absolute reference frame $Ox_1x_2x_3$, seems to be dilated, expanded as a function of the velocity $\vec{\boldsymbol{v}}$ by a factor of $1/\gamma_t$, identically for the two clocks in 'quadrature'

$$T = \frac{2d_0 / c_t}{\sqrt{1-\boldsymbol{v}^2 / c_t^2}} = \frac{T_0}{\gamma_t} \tag{21.8}$$

This means that a local time t' exists really for an ***HS*** observer, that this local time flows slower for an ***HS*** observer in movement with respect to the lattice, and that this local time t' remains *isotropic in the mobile frame* $O'x_1'x_2'x_3'$, independently of the direction of motion of the ***HS*** observer inside the lattice.

Concerning the dilation or contraction of time, there can also be a coupling between the relativistic effects and the effects of volume expansion. We will see later (chapter 24) that, in the case of a cosmic lattice, an observer ***HS'*** which would be placed in a zone with strong contraction ($\tau << 0$) would have a proper clock strongly slowed down with respect to the proper time of the ***GO***. Furthermore, if it moved with a velocity $\boldsymbol{v}$ close to c_t with respect to the lattice, it's proper time would also be strongly slowed down with respect to the proper time of the ***GO***, not only by the direct effect of volume contraction on the clock, but also by the effect of volume expansion on γ_t since

$$\gamma_t = \left(1-\frac{\boldsymbol{v}^2}{c_t^2}\right)^{1/2} = \left(1-\frac{\boldsymbol{v}^2}{c_{t0}^2}\mathrm{e}^{-\tau}\right)^{1/2} < \left(1-\frac{\boldsymbol{v}^2}{c_{t0}^2}\right)^{1/2} \quad if \quad \tau<0 \tag{21.9}$$

21.3 – The Michelson-Morley experiment and the Doppler-Fizeau effect in the cosmic solid lattice

It is clear that the lattice plays, vis-a-vis the singularities and the propagation of waves, the same role as the famous "aether" which was supposed to propagate the luminous waves and was discussed in the early 20th century. The experience of Michelson-Morley, which consisted on measuring, thanks to an interferometer, a difference in the velocity of propagation of luminous waves in the direction of displacement and transversely to the direction of said displacement, gave a negative result. It was concluded at the time that the aether did not exist. But in the two examples above, the calculation proposed in the solid lattice with two local clocks in quadrature shows that the result is identical to that obtained by the experiment of Michelson-Morley, namely that there is no difference in the time it takes for the signal to go through both perpendicular arms, which the ***HS*** interprets as the fact the velocity propagation does not depend on the direction in which it is measured. But in the case we have treated here, there exists an aether made of the cosmic lattice within which the singularities are moving and which are perfectly known by the ***GO***!

We deduce that, in the case of the solid cosmic lattice which acts as an aether, the singularities are moving with velocity $\vec{V}$ and have a proper clock which slows down as the **GO** measures a flight time T_0 with the **HS** clock immobile with respect to the lattice, but a time $T = T_0 / \gamma_t$ with an **HS** clock that would be moving with velocity $\vec{V}$ with respect to the lattice.
Furthermore, it is clear that if the **HS** measures the velocity c_t' of a transversal wave in its moving framework $O'x'_1 x'_2 x'_3$, with its own clocks and rulers, it will find exactly the same value as that measured by the **GO** in the lattice, since

$$c_t' = \frac{2d_0}{T_0} = \frac{2d / \gamma_t}{T / \gamma_t} = \frac{2d}{T} = c_t \quad \textit{in the direction } Ox_1' \tag{21.10}$$

The point of view of the HS observers in movement with respect to the lattice

To illustrate the point of view of the **HS** observers, and notably the fact that observers linked to the lattice do not have an absolute notion of simultaneity, like the **GO** does, we can imagine the following experiment.
In the first experience, we consider two simultaneous events observed by the **GO** in the referential $Ox_1x_2x_3$ at instant $t = 0$ and at coordinates $x_1^{(1)} = 0$ and $x_1^{(2)} = \Delta x_1$, so separated by a distance Δx_1. These two simultaneous events are then observed by an **HS** in its framework $O'x'_1 x'_2 x'_3$ moving with velocity $\vec{V}$ in direction Ox_1 with the following space-time coordinates, obtained from relations *(20.27)*

$$\begin{cases} x_1^{(1)\,\prime} = 0 & \text{and} \quad t_1' = 0 \\ x_1^{(2)\,\prime} = \dfrac{x_1^{(2)}}{\gamma_t} = \dfrac{\Delta x_1}{\gamma_t} & \text{and} \quad t_2' = -\dfrac{\mathbf{v} x_1^{(2)}}{c_t^2 \gamma_t} = -\dfrac{\mathbf{v} \Delta x_1}{c_t^2 \gamma_t} \end{cases} \tag{21.11}$$

We observe that the two events are not measured as simultaneous by the **HS**, but separated by a non-null time interval $\Delta t' = t_2'$, and the distance measured by the **HS** between the two events is equal to $\Delta x_1' = x_1^{(2)\,\prime} = \Delta x_1 / \gamma_t$ which is superior to the distance Δx_1 measured by the **GO**, and is a consequence of the contraction of ruler $\vec{e}_1'$ of the **HS** in the direction Ox_1.
In a second experiment, let's consider an event taking place at the origin of the referential $Ox_1x_2x_3$ of the **GO** and which lasts from $t_1 = 0$ and $t_2 = \Delta t$, so on a time lapse Δt. This event is then observed by an **HS** in its framework $O'x'_1 x'_2 x'_3$ moving with velocity $\vec{V}$ in the direction Ox_1 with the following spatiotemporal coordinates, obtained from relations *(20.27)*

$$\begin{cases} x_1^{(1)\,\prime} = 0 & \text{and} \quad t_1' = 0 \\ x_1^{(2)\,\prime} = -\dfrac{\mathbf{v} t_2}{\gamma_t} = -\dfrac{\mathbf{v} \Delta t}{\gamma_t} & \text{and} \quad t_2' = \dfrac{t_2}{\gamma_t} = \dfrac{\Delta t}{\gamma_t} \end{cases} \tag{21.12}$$

We notice that the event seems to be moving in the **HS** framework over a distance $\Delta x_1' = \left| x_1^{(2)\,\prime} \right| = \mathbf{v} \Delta t / \gamma_t$, longer than the absolute displacement $\mathbf{v} \Delta t$ of the framework $O'x'_1 x'_2 x'_3$ in the lattice, due to the ruler contraction $\vec{e}_1'$ used by the **HS**, and that the time lapse of the event for **HS** is worth $\Delta t' = t_2' = \Delta t / \gamma_t$, and thus seems longer for the **HS** than for the **GO**, which is at first rather strange since the **HS** clock moves slower than that of the **GO**! This phenomenon is due to the flight time of the transversal waves to reach the moving **HS** with respect to the lattice. This last experience shows that the time intervals measured by the **HS** are relative intervals since they depend on the finite propagation velocity of information within the lattice.

Relations between two HS observers in movement with respect to the lattice

In figure 21.1 we show two frameworks in translation along the axis Ox_1 at speeds $\vec{\mathbf{v}}'$ and $\vec{\mathbf{v}}''$ as measured by the ***GO*** observer. One wonders what form the relativity of speeds will take as measured by the ***HS***, including what is the relative speed $\vec{\mathbf{v}}_r$ which is measured by the observer ***HS'*** in its framework $O'x'_1x'_2x'_3$ for the movement of framework $O''x''_1x''_2x''_3$ of the observer ***HS"***. For ***GO***, the point O'' of the framework of ***HS"*** is moving in $Ox_1x_2x_3$ from $x_1^{(1)}$ to $x_1^{(2)}$ in a lapse of time which goes from t_1 to t_2, so that

$$\mathbf{v}'' = \frac{x_1^{(2)} - x_1^{(1)}}{t_2 - t_1} \tag{21.13}$$

If ***HS'*** observes the same displacement, it will find a relative velocity $\vec{\mathbf{v}}_r$ thanks to transformations *(20.22)* as

$$\mathbf{v}_r = \frac{x_1^{(2)}{'} - x_1^{(1)}{'}}{t_2{'} - t_1{'}} = \frac{\dfrac{x_1^{(2)} - \mathbf{v}' t_2}{\gamma_t{'}} - \dfrac{x_1^{(1)} - \mathbf{v}' t_1}{\gamma_t{'}}}{\dfrac{t_2 - \mathbf{v}' x_1^{(2)} / c_t^2}{\gamma_t{'}} - \dfrac{t_1 - \mathbf{v}' x_1^{(1)} / c_t^2}{\gamma_t{'}}} \tag{21.14}$$

Some transformations of this relation allow to write it under the form

$$\mathbf{v}_r = \frac{\dfrac{x_1^{(2)} - x_1^{(1)}}{t_2 - t_1} - \mathbf{v}'}{1 - \left(\dfrac{x_1^{(2)} - x_1^{(1)}}{t_2 - t_1}\right)\dfrac{\mathbf{v}'}{c_t^2}} = \frac{\mathbf{v}'' - \mathbf{v}'}{1 - \mathbf{v}''\mathbf{v}' / c_t^2} \tag{21.15}$$

The relative velocity of framework $O''x''_1x''_2x''_3$ measured by ***HS'*** corresponds to the classic relativistic composition of velocities. By symmetry, the relative velocity of framework $O'x'_1x'_2x'_3$ measured by ***HS"*** will be given by exactly the same expression with a changed sign!

Let's consider now two simultaneous events in the mobile framework $O''x''_1x''_2x''_3$, with coordinates $x_1^{(1)}{''} = 0$ and $x_1^{(2)}{''} = \Delta x_1{''}$ happening at instant $t'' = 0$. In the immobile referential $Ox_1x_2x_3$, the coordinates of these two events become two distinct events in time

$$\begin{cases} x_1^{(1)} = 0 & \text{and} \quad t_1 = 0 \\ x_1^{(2)} = \dfrac{x_1^{(2)}{''}}{\gamma_t{''}} = \dfrac{\Delta x_1{''}}{\gamma_t{''}} & \text{and} \quad t_2 = \dfrac{\mathbf{v}'' x_1^{(2)}{''}}{c_t^2 \gamma_t{''}} = \dfrac{\mathbf{v}'' \Delta x_1{''}}{c_t^2 \gamma_t{''}} \end{cases} \tag{21.16}$$

In the framework $O'x_1{'}x_2{'}x_3{'}$ of ***HS'***, the coordinates of these two events are written

$$\begin{cases} x_1^{(1)}{'} = 0 & \text{and} \quad t_1{'} = 0 \\ x_1^{(2)}{'} = \left(\dfrac{\Delta x_1{''}}{\gamma_t{''}} - \dfrac{\mathbf{v}'' \Delta x_1{''}}{c_t^2 \gamma_t{''}} \mathbf{v}'\right) / \gamma_t{'} & \text{and} \quad t_2{'} = \dfrac{\mathbf{v}'' \Delta x_1{''}}{c_t^2 \gamma_t{''}} - \dfrac{\mathbf{v}'}{c_t^2} \dfrac{\Delta x_1{''}}{\gamma_t{''}} \end{cases} \tag{21.17}$$

which we can write under the form of a spatial distance $\Delta x_1{'} = x_1^{(2)}{'}$ and a time interval

$\Delta t' = t_2'$ between these two events

$$\Delta x_1' = \left(1 - \frac{\mathbf{v'v''}}{c_t^2}\right)\frac{\Delta x_1''}{\gamma_t'\gamma_t''} \quad \text{and} \quad \Delta t' = (\mathbf{v''} - \mathbf{v'})\frac{\Delta x_1''}{c_t^2\gamma_t''\gamma_t'} \tag{21.18}$$

The two original simultaneous events separated by $\Delta x_1''$ in the framework $O''x''_1x''_2x''_3$ of ***HS"*** become two non-simultaneous events in the framework $O'x'_1x'_2x'_3$ of ***HS'***.

Let's consider now two successive events in the mobile framework $O''x''_1x''_2x''_3$, happening at the same place with coordinates $x_1^{(1)}''=0$ and happening at instants $t_1''=0$ and $t_2''=\Delta t''$. In the immobile referential $Ox_1x_2x_3$, the coordinates of the two events become two separate events in space

$$\begin{cases} x_1^{(1)} = 0 & \text{and} \quad t_1 = 0 \\ x_1^{(2)} = \dfrac{\mathbf{v''}t_2''}{\gamma_t''} = \dfrac{\mathbf{v''}\Delta t''}{\gamma_t''} & \text{and} \quad t_2 = \dfrac{t_2''}{\gamma_t''} = \dfrac{\Delta t''}{\gamma_t''} \end{cases} \tag{21.19}$$

In the framework $O'x_1'x_2'x_3'$ of ***HS'***, the coordinates of the two events are then written

$$\begin{cases} x_1^{(1)}{}' = 0 & \text{and} \quad t_1' = 0 \\ x_1^{(2)}{}' = \left(\dfrac{\mathbf{v''}\Delta t''}{\gamma_t''} - \dfrac{\mathbf{v'}\Delta t''}{\gamma_t''}\right) / \gamma_t' & \text{and} \quad t_2' = \left(\dfrac{\Delta t''}{\gamma_t''} - \dfrac{\mathbf{v'}}{c_t^2}\dfrac{\mathbf{v''}\Delta t''}{\gamma_t''}\right) / \gamma_t' \end{cases} \tag{21.20}$$

which we can write explicitly in the form of a spatial distance $\Delta x_1' = x_1^{(2)}{}'$ and a time interval $\Delta t' = t_2'$ between the two events

$$\Delta x_1' = (\mathbf{v''} - \mathbf{v'})\frac{\Delta t''}{\gamma_t'\gamma_t''} \quad \text{and} \quad \Delta t' = \left(1 - \frac{\mathbf{v'v''}}{c_t^2}\right)\frac{\Delta t''}{\gamma_t'\gamma_t''} \tag{21.21}$$

The two events happening at the origin of the framework $O''x''_1x''_2x''_3$ of ***HS"*** become then two separate events in the space of the framework $O'x'_1x'_2x'_3$ of ***HS'***.

The Doppler-Fizeau effects between singularities in movement

In figure 21.2, we show several experiments of exchange of signals at a given frequency between singularities in movement within the lattice via the transversal waves. By taking the point of view of the ***GO***, it is possible to easily describe these experiences that give rise to the Doppler-Fizeau effect. We suppose that all these experiences take place in a lattice which has a homogenous and constant value for the volume expansion, without which the description of the experiments would become a lot more complex.

First experiment: an observer ***HS'*** in the framework $O'x'_1x'_2x'_3$ in movement with velocity $\vec{\mathbf{v}}'$ in the direction Ox_1 with respect to the lattice emits a wave with frequency f_e', measured with its proper clock, towards an ***HS*** observer in a referential $Ox_1x_2x_3$ immobile with respect to the lattice (figure 21.2a). The transversal wave emitted in the framework $O'x'_1x'_2x'_3$ is written

$$\vec{\omega} = \vec{\omega}_0 \sin(\omega't' - k'x_1') \tag{21.22}$$

with $f_e' = \omega'/2\pi$ and $k' = \omega'/c_t$.

In the referential $Ox_1x_2x_3$, the same wave can be obtained by replacing the coordinates t' and x_1' of ***HS'*** by coordinates t and x_1 of ***HS***, by using the Lorentz transformations

$$\begin{aligned}\vec{\omega} &= \vec{\omega}_0 \sin\left(\omega'\frac{t - \mathbf{v}'x_1 / c_t^2}{\gamma_t'} - k'\frac{x_1 - \mathbf{v}'t}{\gamma_t'}\right) \\ &= \vec{\omega}_0 \sin\left[\left(\frac{\omega'}{\gamma_t'} + \frac{k'\mathbf{v}'}{\gamma_t'}\right)t - \left(\frac{k'}{\gamma_t'} + \frac{\omega'\mathbf{v}'}{c_t^2\gamma_t'}\right)x_1\right] = \vec{\omega}_0 \sin\left(\omega t - kx_1\right)\end{aligned} \tag{21.23}$$

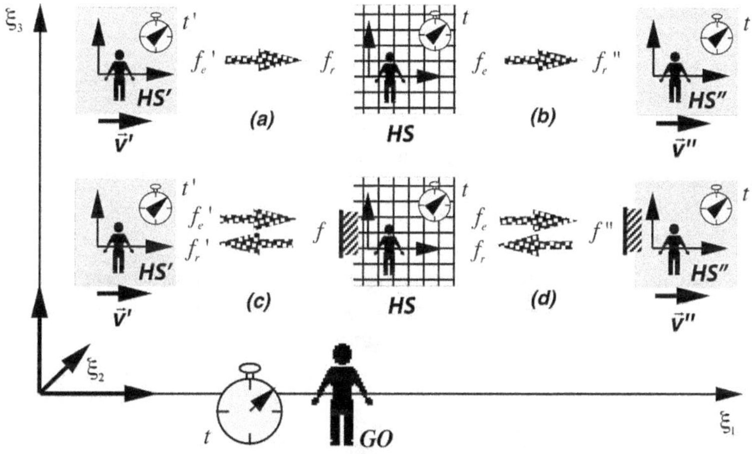

Figure 21.2 *- different configurations of measure of the Doppler-Fizeau effect*

We find as a consequence the relations giving ω and k from the values of ω' and k' in the framework $O'x'_1x'_2x'_3$

$$\begin{cases} \omega = \left(\omega' + k'\mathbf{v}'\right)/\gamma_t' \\ k = \left(k' + \omega'\mathbf{v}'/c_t^2\right)/\gamma_t' \end{cases} \tag{21.24}$$

As $k' = \omega'/c_t$ and $f_e' = \omega'/2\pi$, we deduce the relation existing between the frequency f_e' of the signal emitted by ***HS'*** and the frequency f_r measured by ***HS*** on the signal received with its proper clock

$$f_r = \frac{1 + \mathbf{v}'/c_t}{\gamma_t'} f_e' = f_e' \sqrt{\frac{1 + \mathbf{v}'/c_t}{1 - \mathbf{v}'/c_t}} \tag{21.25}$$

For $\mathbf{v}' > 0$, meaning when ***HS'*** gets closer to ***HS***, the frequency f_r of the signal received and measured by ***HS*** is higher than the frequency f_e' of the signal emitted and measured by ***HS'***. This is the *Doppler-Fizeau* effect and one usually talks about a *"blueshift of the signal"*. On the contrary, if ***HS'*** is moving away from ***HS*** ($\mathbf{v}' < 0$), the signal is received with a frequency f_r

which is inferior to the frequency $f_e{}'$ of the emitted signal, and one talks about a "redshift of the signal".

It is interesting to rewrite relation *(21.25)* under the following form

$$f_r = f_e{}'\sqrt{\frac{1+\mathbf{v}'/c_t}{1-\mathbf{v}'/c_t}} = \frac{1}{1-\mathbf{v}'/c_t}\gamma_t{}'f_e{}' \tag{21.26}$$

as, in this form, the relationship shows the term $(1-\mathbf{v}'/c_t)^{-1}$ of the purely classic Doppler effect, but which is applied to an emitted frequency $\gamma_t{}'f_e{}'$, which is nothing else than the frequency of the signal emitted by the ***HS'***, but measured by the ***HS*** with its proper clock, or by ***GO*** with a universal clock.

2nd experiment: an ***HS*** observer in the framework $Ox_1x_2x_3$ at rest with respect to the lattice sends a signal with frequency f_e, measured with its proper clock, towards an observer ***HS"*** which moves with velocity $\vec{\mathbf{v}}''$ in direction Ox_1 with respect to the lattice (figure 21.2b). With the same type of calculation that in the first case, it is easy to verify that the frequency $f_r{}''$ of the signal received by ***HS"*** and measured by him with its proper clock has the value

$$f_r{}'' = f_e\sqrt{\frac{1-\mathbf{v}''/c_t}{1+\mathbf{v}''/c_t}} = \left(1-\mathbf{v}''/c_t\right)\frac{f_e}{\gamma_t{}''} \tag{21.27}$$

For $\mathbf{v}''>0$, meaning when ***HS"*** is moving away from ***HS***, the frequency $f_r{}''$ of the signal received by ***HS"*** is lower than the frequency f_e of the signal emitted by ***HS***. It is again a *Doppler-Fizeau effect*. In the second form presented in *(21.27)*, the expression of $f_r{}''$ shows a term in $(1-\mathbf{v}''/c_t)$ due to the classic Doppler effect, but which is applied to a frequency $f_e/\gamma_t{}''$, which is nothing more than the frequency of the signal emitted by ***HS***, but such as it is measured by the clock of the ***HS"***.

3rd experiment: an ***HS'*** observer in the framework $O'x'_1x'_2x'_3$ moving with velocity $\vec{\mathbf{v}}'$ in the direction Ox_1 with respect to the lattice emits a wave at frequency $f_e{}'$, measured with its own clock, towards an observer ***HS"*** which is moving with velocity $\vec{\mathbf{v}}''$ in the direction Ox_1 with respect to the lattice (figure 21.2a-b). The frequency $f_r{}''$ of the signal received by the ***HS"*** and measured by him with his proper clock is easily obtained by combining relations *(21.25)* and *(21.27)*. We obtain

$$f_r{}'' = f_e{}'\sqrt{\frac{1+\mathbf{v}'/c_t}{1-\mathbf{v}'/c_t}}\sqrt{\frac{1-\mathbf{v}''/c_t}{1+\mathbf{v}''/c_t}} = \left(\frac{1-\mathbf{v}''/c_t}{1-\mathbf{v}'/c_t}\right)\frac{\gamma_t{}'}{\gamma_t{}''}f_e{}' \tag{21.28}$$

Under the second form presented in *(21.28)*, the expression of $f_r{}''$ explicitly shows in parenthesis the classic Doppler effect due to the movement of two observers with respect to the lattice as well as the frequency $\gamma_t{}'f_e{}'/\gamma_t{}''$ which is nothing more than the frequency of the signal emitted by ***HS'***, but measured by the clock of the ***HS"***.

4th experiment: an observer ***HS'*** in the framework $O'x'_1x'_2x'_3$ moving with velocity $\vec{\mathbf{v}}'$ in the direction Ox_1 with respect to the lattice emits a wave with a frequency $f_e{}'$ measured by its proper clock, which is reflected by a mirror associated to the framework $Ox_1x_2x_3$ immobile with respect to the lattice, and receives the echo of the wave of which he measures the frequency $f_r{}'$, always with its proper clock (figure 21.2c). It is easy to find the value of $f_r{}'$ by using relations *(21.25)* and *(21.27)* in which we introduce the frequency f received and re-emitted by the mirror in the framework of ***HS,*** namely

$$f = f_e'\sqrt{\frac{1+\mathbf{v}'/c_t}{1-\mathbf{v}'/c_t}} = \frac{1}{1-\mathbf{v}'/c_t}\gamma_t' f_e' \quad \text{and} \quad f_r' = f\sqrt{\frac{1+\mathbf{v}'/c_t}{1-\mathbf{v}'/c_t}} = \left(1+\mathbf{v}'/c_t\right)\frac{f}{\gamma_t'} \qquad (21.29)$$

The combination of these two relations shows us that, in this case, the effect measured by ***HS'*** is purely a *classic Doppler effect*, which is logical since we use the proper clock of ***HS'*** to measure f_e' and f_r'

$$f_r' = \frac{1+\mathbf{v}'/c_t}{1-\mathbf{v}'/c_t} f_e' \qquad (21.30)$$

5th experiment: an ***HS*** observer in the framework $Ox_1x_2x_3$ immobile with respect to the lattice emits a wave at frequency f_e, measured with its proper clock, which bounces on a mirror associated with a framework $O''x''_1x''_2x''_3$ moving with velocity $\vec{\mathbf{v}}''$ in the direction Ox_1 with respect to the static lattice, and receives the echo of these waves with frequency f_r, measured with its proper clock (figure 21.2d). It is easy to find the value of f_r by using relations *(21.25)* and *(21.27)* in which we introduce the frequency f'' received and re-emitted by the mirror in the framework of the ***HS"***. The combination of these two relations shows us that, in this case, the effect measured by the ***HS*** is also a purely *classic Doppler effect*, since ***HS*** uses its own clock to measure f_e and f_r

$$f_r = \frac{1-\mathbf{v}''/c_t}{1+\mathbf{v}''/c_t} f_e \qquad (21.31)$$

6th experiment: an observer ***HS'*** in the framework $O'x'_1x'_2x'_3$ moving with velocity $\vec{\mathbf{v}}'$ in the direction Ox_1 with respect to the lattice emits a wave with frequency f_e', measured with its own clock, which bounces of a mirror associated with framework $O''x''_1x''_2x''_3$ moving in direction $\vec{\mathbf{v}}''$ in the direction Ox_1 with respect to the lattice, and receives the echo of said wave for which it measures frequency f_r', always with its own clock (figure 21.2c-d). It is easy to find the value of f_r' by using twice the relation *(21.28)*. We again find that in this case, the effect measured by the ***HS'*** is purely a classic Doppler effect, since it uses its proper clock to measure f_e' and f_r'

$$f_r' = \frac{\left(1+\mathbf{v}'/c_t\right)\left(1-\mathbf{v}''/c_t\right)}{\left(1-\mathbf{v}'/c_t\right)\left(1+\mathbf{v}''/c_t\right)} f_e' \qquad (21.32)$$

21.4 – On the explanation of the famous twins paradox of special relativity

The existence of a lattice, and thus of an *«aether»*, allows us to give a very simple and elegant explanation of the famous paradox of the twins in special relativity.

On the impossibility for an HS observer to measure its own velocity with respect to the lattice

We have already seen that a local observer ***HS"*** in its framework $O''x''_1x''_2x''_3$ mobile with

velocity $\vec{\mathbf{v}}''$ with respect to the lattice in the direction Ox_1 is in principle not capable of measuring that velocity $\vec{\mathbf{v}}''$ since its clock and its proper rulers do not change, which has as a consequence that experiments of the Michelson-Morley type do not bring any useful information. We can however ask ourselves if experiments of the type Doppler-Fizeau with another observer ***HS'*** mobile with velocity $\vec{\mathbf{v}}'$ with respect to the lattice in direction Ox_1 could bring more information. In relation with observer ***HS'***, the observer ***HS"*** can perform three types of measurements:
- it can measure the relative velocity $\mathbf{v}_r$ of ***HS'*** with respect to it, given by *(21.15)*

$$\mathbf{v}_r = \frac{\mathbf{v}' - \mathbf{v}''}{1 - \mathbf{v}''\mathbf{v}' / c_t^2} \tag{21.33}$$

- it can measure the ratio of frequencies f_r''/f_e' of a given known event which happens in his framework and in the framework of the ***HS'***, given by *(21.28)*

$$\frac{f_r''}{f_e'} = \sqrt{\frac{1+\mathbf{v}'/c_t}{1-\mathbf{v}'/c_t}}\sqrt{\frac{1-\mathbf{v}''/c_t}{1+\mathbf{v}''/c_t}} \tag{21.34}$$

- it can measure the ratio of frequencies f_r''/f_e'' of a given known signal which he sent itself and which is reflected by a mirror in the framework of ***HS'***, given by *(21.32)*

$$\frac{f_r''}{f_e''} = \frac{(1+\mathbf{v}'/c_t)(1-\mathbf{v}''/c_t)}{(1-\mathbf{v}'/c_t)(1+\mathbf{v}''/c_t)} \tag{21.35}$$

We can show that these three experimental measurements do not allow to the ***HS"*** to determine univocally $\vec{\mathbf{v}}'$ and $\vec{\mathbf{v}}''$. Indeed, the two relations *(21.34)* and *(21.35)* are absolutely equivalent and thus do not allow us to solve the problem. As for relations *(21.33) and (21.34)*, it is easy to show that

$$\frac{f_r''}{f_e'} = \sqrt{\frac{1+\mathbf{v}'/c_t}{1-\mathbf{v}'/c_t}}\sqrt{\frac{1-\mathbf{v}''/c_t}{1+\mathbf{v}''/c_t}} = \sqrt{\frac{1+\mathbf{v}_r/c_t}{1-\mathbf{v}_r/c_t}} \tag{21.36}$$

and thus this system is also not determined, so that the ***HS"*** has no way of finding its relative velocity $\vec{\mathbf{v}}''$ with respect to the lattice by using experiments of the type Doppler-Fizeau.

On the paradox of the twins which is only a paradox in the mind of the observers HS!

The relation *(21.36)* is very interesting, as it shows that ***HS"*** can deduce the relative velocity $\mathbf{v}_r$ of ***HS'*** with respect to itself by measuring the frequency ratio f_r''/f_e' of a given known event in its framework and in the framework of ***HS'***. For itself, in its framework $O''x''_1x''_2x''_3$, this ratio of frequencies is of relativistic type. But the observer ***HS'***, in its framework $O'x'_1x'_2x'_3$, could in theory perform the same measurement, and it would obtain exactly the same result! Thus, for an ***HS*** observer which does not have access to absolute velocities with respect to the lattice (and thus to the aether), their relativistic principles are exactly the same principles as those of *"Special Relativity"*. Notably, by applying the Lorentz transformations, ***HS"*** will have the impression that ***HS'*** ages less than it, while ***HS'*** will also have the impression that ***HS"*** is aging slower than it. This strange situation is called the *twin paradox* in special relativity.
But this paradoxical conclusion of the twins ***HS'*** and ***HS"*** is only a paradox in the mind of the observers ***HS'*** and ***HS"***. Indeed, for the ***GO*** which can read the relativistic velocities of the ***HS***

with respect to the lattice, it is perfectly clear that it is the ***HS*** which is moving with respect to the lattice which is aging slower than the ***HS*** which remains immobile within the lattice. Thus, if a couple of ***HS*** twins perform the famous experiment of the Langevin twins, namely that one of the twins leaves earth on a rocket with subliminal speeds and comes back towards its twin who stayed at the point of origin, the ***GO*** will be able to say without error that it was the ***HS***, who travelled with respect to the lattice with a great speed, who will be the youngest when they meet after the trip. And the ***GO*** knows perfectly that this effect *is an effect that happened all along the trip, even during the periods where the velocity of the traveling twin will have been constant with respect to the lattice!*

This new interpretation of the twin paradox based on the existence of the cosmic lattice (the "aether") gives a very logical and elegant answer to numerous questions and interpretations of the twin paradox suggested by Special Relativity and General Relativity [1].

---

[1] *See for example:*
http://fr.wikipedia.org/wiki/Paradoxe_des_jumeaux
http://en.wikipedia.org/wiki/Twin_paradox

PARTIE II

C

Gravitation and Cosmology

Gravitational fields
of localized topological singularities

Gravitational mass

Gravitational interactions
between topological singularities

Newton's law of universal gravitation
and Einstein's general relativity

«Weak interaction force»
binding dispiration loops

Cosmological evolution of the universe
and «dark matter»

Hubble's law, redshift
and cosmic microwave background

Chapter 22

"Gravitational" perturbations of expansion by localized topological singularities

This chapter uses the second partial equation of Newton of section 18.1 to address the problem of external perturbation of the field of expansion due to a spherical topological singularity at rest, containing an energy $E_{dist}^{cluster} + V_{pot}^{cluster}$ associated with the complex internal structure of a spherical singularity (for example, dislocation and/or disclination loops), or containing an overall rotation charge Q_λ and/or an overall curvature charge Q_θ.

It will be shown that the perturbations of the expansion fields associated with a localized topological singularity are in fact the expression of the existence of a "gravitational field" at long distance outside of this singularity.

We then show that the collapse of clusters of lacunar or interstitial-type singularities leads to rather singular macroscopic topological singularities: a sort of macroscopic hole in the network, in the case of vacancies, or a piece of additional lattice, a kind of macroscopic interstitial within the lattice, in the case of the collapse of interstitials. The description of the "gravitational" fields of these two complementary types of singularities shows that the macroscopic singularity of vacancy type can behave like a real black hole, while the macroscopic singularity of interstitials does not have this property! By analogy, we can compare the macroscopic vacancies to the black holes of General Relativity, while the macroscopic interstitials would be comparable to pulsars (neutron stars).

22.1 – Localized singularity with a given energy of distortion

Let's consider a localized singularity at rest, with volume $V_{cluster}$, made up of a loop or a cluster of loops of dislocation and disclination, and assume that we know, because we can calculate them with relations *(18.10)* and *(18.11)*, the densities of energy of distortion $F_{dist}^{cluster}(\vec{r})$ and the potential energy $F_{pot}^{cluster}(\vec{r})$ within that singularity. The equilibrium of the field of perturbation $\tau_{int}^{(E)}(\vec{r})$ of expansion within this singularity is given by the second degree equation *(18.14)*, for which the solution is, inside the singularity

$$\tau_{int}^{(E)}(\vec{r}) = \frac{4K_2/3 + 2K_1(1+\tau_0+\tau^{ext}(\vec{r})+\tau^{cluster}(\vec{r})) - K_0}{2K_1}\left[-1+\sqrt{1-\frac{4K_1\left(F_{dist}^{cluster}(\vec{r})+F_{pot}^{cluster}(\vec{r})\right)}{\left(4K_2/3+2K_1(1+\tau_0+\tau^{ext}(\vec{r})+\tau^{cluster}(\vec{r}))-K_0\right)^2}}\right] \tag{22.1}$$

We cannot make here an exact calculation of $\tau_{int}^{(E)}(\vec{r})$ since that would necessitate to treat a concrete case of a singularity to know exactly the distribution of density of energy of distortion

$F_{dist}^{cluster}(\vec{r})$ and potential energy $F_{pot}^{cluster}(\vec{r})$ within the singularity. However, we can treat this problem in an approximated fashion with average values for the various fields under consideration. This is what we will do in the rest of the chapter.

In the case of a singularity with a given energy density, we can calculate the *global rest energy* $E_{dist}^{cluster}+V_{pot}^{cluster}$ of it, with the relation

$$\left(E_{dist}^{cluster}+V_{pot}^{cluster}\right)=\iiint\limits_{V_{cluster}}\left(F_{dist}^{cluster}(\vec{r})+F_{pot}^{cluster}(\vec{r})\right)dV \tag{22.2}$$

which allows us to introduce average values of densities within them

$$\overline{F}_{dist}^{cluster}+\overline{F}_{pot}^{cluster}=\frac{1}{V_{cluster}}\left(E_{dist}^{cluster}+V_{pot}^{cluster}\right) \tag{22.3}$$

By supposing that we can neglect $\tau^{ext}(\vec{r})+\tau^{cluster}(\vec{r})$ vis-à-vis $1+\tau_0$, we can deduce the average value $\overline{\tau}_{int}^{(E)}$ of the internal field of perturbation of expansion as

$$\overline{\tau}_{int}^{(E)}\cong\frac{4K_2/3+2K_1(1+\tau_0)-K_0}{2K_1}\left[-1+\sqrt{1-\frac{4K_1\left(E_{dist}^{cluster}+V_{pot}^{cluster}\right)}{\left(4K_2/3+2K_1(1+\tau_0))-K_0\right)^2V_{cluster}}}\right] \tag{22.4}$$

It is clear that the average field is purely virtual, in other words, it does not really exist, but it represents a form of the average value of all the accidents of the field of perturbation $\tau_{int}^{(E)}(\vec{r})$ within the singularity, accidents which are pronounced in the event that we are dealing with a cluster of several topological singularities.

The condition of existence of an average virtual field of static perturbation of expansion

For such a virtual static solution $\overline{\tau}_{int}^{(E)}$ to exist, one must have a positive quantity under the root in equation *(22.4)*, this implies that a condition of existence of the average density within the singularity be

Hypothesis 1: $$\frac{E_{dist}^{cluster}+V_{pot}^{cluster}}{V_{cluster}}\leq\frac{1}{4K_1}\left(4K_2/3+2K_1(1+\tau_0)-K_0\right)^2 \tag{22.5}$$

This condition depends expressly on the volume expansion of the background τ_0. It can never be satisfied in a domain of volume expansion τ_0 centered on $\tau_0=\tau_{0cr}$, and covering the following domain

$$\tau_{0cr}-\sqrt{\frac{1}{K_1}\frac{E_{dist}^{cluster}+V_{pot}^{cluster}}{V_{cluster}}}\leq\tau_0\leq\tau_{0cr}+\sqrt{\frac{1}{K_1}\frac{E_{dist}^{cluster}+V_{pot}^{cluster}}{V_{cluster}}} \tag{22.6}$$

We calculate the critical value of expansion τ_{0cr}, by using the relations *(19.58)* from the conjectures 0-6 of the perfect cosmic lattice and the relation *(15.4)*

$$\tau_{0cr}=\frac{K_0}{2K_1}-\frac{2K_2}{3K_1}-1>1 \tag{22.7}$$

In this domain, centered on $\tau_0=\tau_{0cr}$, we can only have a *dynamical solution of the equation of Newton (18.12) for the perturbation expansion*. This dynamic solution allows us to pass from the static solution of domain $\tau_0<\tau_{0cr}$ to the static solution in domain $\tau_0>\tau_{0cr}$ and vice-versa.

The condition for a field of perturbation with a null average value

In areas where there is a static solution of non-zero virtual value $\bar{\tau}_{int}^{(E)}$ of the disturbance of expansion field within the singularity, the volume expansion outside the singularity must also be affected by static disturbance field $\tau_{ext}^{(E)}(r)$ which is spherical symmetric and which must compensate the expansion or contraction of the local network due to the singularity. In fact, the value of the average field of the entire network must be equal to τ_0, so that the average expansion of disturbances due to the presence of singularity must be zero. In the absence of an overall bending charge of the singularity $Q_\theta = 0$, the external adjustment field $\tau_{ext}^{(E)}(r)$ must satisfy the balance equation *(18.8)* with $\theta^{ch} = 0$, so

$$\Delta\left(\tau_{ext}^{(E)}(r)\right) = 0 \qquad (22.8)$$

With the spherical symmetry of the problem, this Laplacian has only *one solution* which tends towards a null value for $r \to R_\infty$

$$\tau_{ext}^{(E)}(r) = \frac{A}{r} \qquad (22.9)$$

To determine the constant A, one must make sure that the introduction of the singularity in the lattice did no modify the average value τ_0 of expansion of the lattice. This condition is written

$$\tau_0 = \frac{3}{4\pi R_\infty^3}\left[\iiint_{V_{cluster}}\left(\tau_0 + \bar{\tau}_{int}^{(E)}\right)dV + \int_{R_{cluster}}^{R_\infty}\left(\tau_0 + \frac{A}{r}\right)4\pi r^2\,dr\right] \;\Rightarrow\; \bar{\tau}_{int}^{(E)}V_{cluster} + 4\pi A\int_{R_{cluster}}^{R_\infty} r\,dr = 0 \qquad (22.10)$$

The calculation of this expression gives us the following value for the A *constant*

$$-\bar{\tau}_{int}^{(E)}V_{cluster} = 4\pi A\left(\frac{R_\infty^2}{2} - \frac{R_{cluster}^2}{2}\right) \;\Rightarrow\; A \cong -\bar{\tau}_{int}^{(E)}\frac{V_{cluster}}{2\pi R_\infty^2} \qquad (22.11)$$

We directly deduce the perturbation field $\tau_{ext}^{(E)}(r)$ of expansion generated by the singularity and which depends directly on the virtual average field $\bar{\tau}_{int}^{(E)}$

$$\tau_{ext}^{(E)}(r) = -\frac{\bar{\tau}_{int}^{(E)}V_{cluster}}{2\pi R_\infty^2}\frac{1}{r} \qquad (22.12)$$

The simplest cases of static fields of perturbation of expansion

If the condition of existence *(22.5)* is largely satisfied, we can issue a new hypothesis concerning the value of the internal energy density of the singularity, which is stronger than the condition of existence *(22.5)*

Hypothesis 2: $$\frac{E_{dist}^{cluster} + V_{pot}^{cluster}}{V_{cluster}} << \frac{\left(4K_2/3 + 2K_1(1+\tau_0) - K_0\right)^2}{4K_1} \qquad (22.13)$$

This allows us to develop in first order the square root of expression *(22.4)*, so that the static solution for the internal average field $\bar{\tau}_{int}^{(E)}$ can be approximatively written as

$$\bar{\tau}_{int}^{(E)} \cong -\frac{1}{4K_2/3 + 2K_1(1+\tau_0)) - K_0}\frac{E_{dist}^{cluster} + V_{pot}^{cluster}}{V_{cluster}} \qquad (22.14)$$

By combining the expression *(22.13)* with that value of $\bar{\tau}_{int}^{(E)}$, we obtain that $\left|\bar{\tau}_{int}^{E}\right|$ must, with hypothesis 2, satisfy the following relation

$$\left|\overline{\tau}_{int}^{(E)}\right| << \frac{\left|4K_2/3+2K_1(1+\tau_0))-K_0\right|}{4K_1} \tag{22.15}$$

The condition associated with hypothesis 2 depends on the volume expansion of the background τ_0 of the lattice, so that it can only be satisfied if the background expansion satisfies one of the following conditions

$$\tau_0 >> \tau_{0cr} + \sqrt{\frac{1}{K_1}\frac{E_{dist}^{cluster}+V_{pot}^{cluster}}{V_{cluster}}} \quad \text{or} \quad \tau_0 << \tau_{0cr} - \sqrt{\frac{1}{K_1}\frac{E_{dist}^{cluster}+V_{pot}^{cluster}}{V_{cluster}}} \tag{22.16}$$

The expressions *(22.12)* and *(22.14)* allow us to deduce an approximate average value of the static field $\tau_{ext}^{(E)}(r)$ of perturbations of external expansions of the singularity

$$\tau_{ext}^{(E)}(r) \cong -\frac{1}{2\pi\left(K_0-2K_1(1+\tau_0)-4K_2/3\right)R_\infty^2}\frac{E_{dist}^{cluster}+V_{pot}^{cluster}}{r} \tag{22.17}$$

It is remarkable that this field of perturbation depends *only* on the total energy $E_{dist}^{cluster}+V_{pot}^{cluster}$ of the cluster and further more that it *does not depend on the volume or the radius of the said cluster.*

The energy of the field of perturbations of the expansion

The total energy of the field of expansion can be calculated

$$E_{dist}^{(\tau)} = \iiint_V \left(K_1\tau^2 - K_0\tau\right)dV = \left[\begin{array}{l} K_1 \iiint\limits_{V_{cluster}} \left(\tau_0+\overline{\tau}_{int}^{(E)}\right)^2 dV + K_1 \int\limits_{R_{cluster}}^{R_\infty} \left(\tau_0+\dfrac{A}{r}\right)^2 4\pi r^2\, dr \\ -K_0 \iiint\limits_{V_{cluster}} \left(\tau_0+\overline{\tau}_{int}^{(E)}\right)^2 dV - K_0 \int\limits_{R_{cluster}}^{R_\infty} \left(\tau_0+\dfrac{A}{r}\right) 4\pi r^2\, dr \end{array}\right] \tag{22.18}$$

By using relation *(22.10)*, we immediately deduce that the terms in K_0 have a contribution equal to $-VK_0\tau_0$ to the energy, and thus do not depend on the perturbations of the singularity. As $K_0 >> K_1$, the relation *(22.10)* corresponds in fact *to minimize the energy* $E_{dist}^{(\tau)}$, which justifies *à posteriori* our choice of the condition of a perturbation field with a null average value used to determine the value of the constant A. It is thus the *terms in* K_1 *which give a contribution to the energy of the field of perturbation due to the singularity.* If we subtract the energy terms which only depend on τ_0, we obtain the increase in energy $E_{grav}^{(E)}$ associated exclusively with the «gravitational» nature of the singularity and which is worth

$$E_{grav}^{E} \cong K_1\left(\overline{\tau}_{int}^{(E)}\right)^2 V_{cluster} \cong \frac{K_1}{\left(4K_2/3+2K_1(1+\tau_0)-K_0\right)^2}\frac{\left(E_{dist}^{cluster}\right)^2}{V_{cluster}} \tag{22.19}$$

We notice that this energy is positive, that it is proportional to the product of the volume of the singularity by the square of the internal field of perturbation of the singularity, and that it only depends on the module of elasticity K_1 which, as we remind ourselves here, is much smaller than K_0 and K_3 in the perfect cosmic lattice.

It is interesting to compare the gravitational energy of the singularity with its elastic energy of distortion by calculating the ratio of the two energies. One obtains

$$\frac{E_{grav}^{E}}{E_{dist}^{cluster}} \cong \frac{K_1\left(\overline{\tau}_{int}^{(E)}\right)^2 V_{cluster}}{E_{dist}^{cluster}} \cong \frac{K_1}{\left(4K_2/3+2K_1(1+\tau_0)-K_0\right)^2} \frac{E_{dist}^{cluster}}{V_{cluster}} \cong \frac{K_1}{K_0^2}\frac{E_{dist}^{cluster}}{V_{cluster}} \qquad \left(\tau_0 << \tau_{0cr}\right) \qquad (22.20)$$

The ratio K_1 / K_0^2 is extremely small, so that the gravitational energy of the singularity is surely much smaller than its elastic energy of distortion in the area $\tau_0 << \tau_{0cr}$. This point will be treated in the following.

As the gravitational energy depends on the square of $\overline{\tau}_{int}^{(E)}$ and of the volume of the singularity, we deduce that the calculation of the real gravitational energy in a cluster of N singularities, must be done from the sum of the energies of gravitation of each of the singularities in the cluster, and, if there exists an average static field of expansion within the singularities, a sum of the products of the volumes and the squares of the real expansion of each singularity

$$E_{grav\ cluster}^{(E)} \cong \sum_{i=1}^{N} E_{grav\ sing\ i}^{(E)} \cong K_1 \sum_{i=1}^{N} V_{sing\ i}\left(\overline{\tau}_{int\ sing\ i}^{(E)}\right)^2 \qquad (22.21)$$

in which $V_{sing\ i}$ is the volume occupied by the singularity i and $\overline{\tau}_{int\ sing\ i}^{(E)}$ is the internal volume expansion of the i singularity due to its energy.

The effects of expansion due to a cluster of topological singularities of given energy

We notice that the external field of perturbation taken on the surface of the singularity, has an absolute value which is $\left|\tau_{ext}^{(E)}(R)\right| = 2R_{cluster}^2 \left|\overline{\tau}_{int}^{(E)}\right| / 3R_\infty^2$ which is *largely greater* to that of the average internal field $\left|\overline{\tau}_{int}^{(E)}\right|$ since $\left|\tau_{ext}^{(E)}(R)\right|$ contains R_∞^2 in the denominator. This implies that the average virtual densities of mass of the lattice are very different inside and outside the singularities

$$mn_{ext}(R_{cluster}) \cong mn_0\, e^{-\tau_0}\, e^{-\tau_{ext}^{(E)}(R_{cluster})} \quad \text{and} \quad m\overline{n}_{int} \cong mn_0\, e^{-\tau_0}\, e^{-\overline{\tau}_{int}^{(E)}} \qquad (22.22)$$

With the conjectures 0 to 6 of the perfect cosmological lattice *(19.58)*, if you are in the area of expansion in which there are no longitudinal waves $\tau_0 < \tau_{0cr}$, the average virtual internal field of the singularity is positive. As for the actual external field of the singularity $\tau_{ext}^{(E)}$, it is negative and thus fully satisfies conjecture 2 deducted from the curvature of the wave rays near the singularity (fig. 22.1a).

However, if you are in the area of expansion where $\tau_0 > \tau_{0cr}$ there are longitudinal waves, the fields given in figure 22.1b are reversed with respect to the fields of figure 22.1a : the average virtual internal field of the singularity becomes negative and the real external field becomes positive, so that it no longer meets the conjecture 2 deduced from the curvature of the wave ray in the vicinity of the singularity !

The domains of solutions of the perturbations of expansion of a cluster of singularities with a given energy as a function of the background lattice expansion

We have seen in this section that a given cluster of energy singularities can have static or dynamic solutions for the disturbance of expansion fields associated with it, and that the domains that involve these solutions are given by the conditions *(22.6)*.

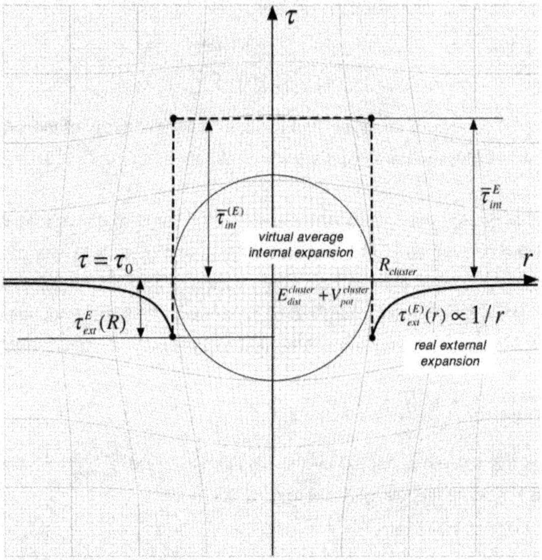

Figure 22.1a *- The field of expansion inside and outside the singularity with a given density of energy in the domain* $\tau_0 < \tau_{0cr}$

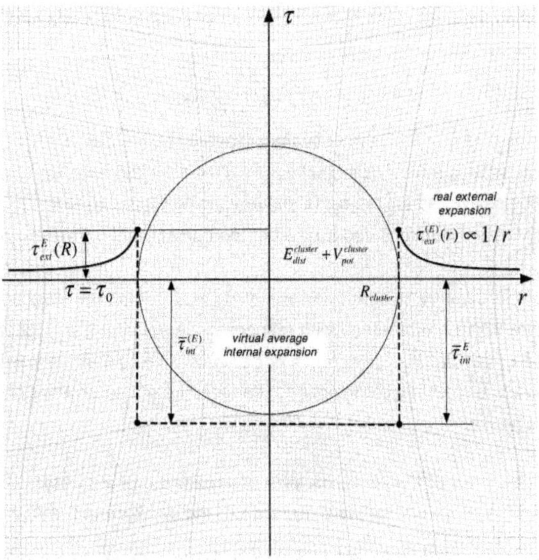

Figure 22.1b *- The field of expansion inside and outside the singularity with a given energy density in the domain* $\tau_0 > \tau_{0cr}$

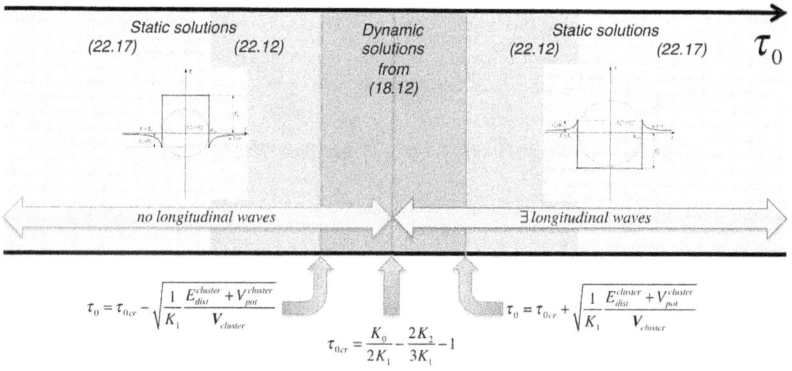

***Figure 22.2** - The domains of the solutions of perturbation of expansion of a singularity of a given density of energy given as a function of the background expansion τ_0*

We also looked for simple solutions *(22.17)* for static perturbation fields, where τ_0, the expansion of the background, obeys conditions *(22.16)*. The domains of the solutions of perturbation of expansion associated with a cluster of singularities of energy are reported in a synoptic fashion in figure 22.2.

One can also ask whether a given energy singularity could be a black hole. For this to happen, it is necessary that disruption of the external expansion field be "negative", which can only occur in the field of satisfactory expansion $\tau_0 < \tau_{0cr}$, and furthermore we need conditions *(15.3)* to be satisfied.

By then assuming a cluster with a strong value of $\overline{\tau}_{int}^{(E)}$, the expression *(22.17)* can be used for the outfield $\tau_{ext}^{(E)}(r)$. The first condition (15.3) involves a critical radius of the cluster such that

$$\left.\frac{\partial \tau_{ext}^{(E)}(r)}{\partial r}\right|_{r_{cr}} = \frac{2}{r_{cr}} \quad \Rightarrow \quad r_{cr} = \frac{E_{dist}^{cluster} + V_{pot}^{cluster}}{4\pi\left(K_0 - 2K_1(1+\tau_0) - 4K_2/3\right)R_\infty^2} \qquad (\tau_0 < \tau_{0cr}) \qquad (22.23)$$

and the second condition *(15.3)* implies that

$$r_{cr} > R_{cluster} \quad \Rightarrow \quad \frac{E_{dist}^{cluster} + V_{pot}^{cluster}}{R_{cluster}} > 4\pi\left(K_0 - 2K_1(1+\tau_0) - 4K_2/3\right)R_\infty^2 \qquad (\tau_0 < \tau_{0cr}) \qquad (22.24)$$

But it is also necessary that this condition be satisfied below the domain of dynamic solutions *(22.6)*. We can then show that these only appear if the energy of the singularity satisfies the following, stronger, inequality

$$\left(E_{dist}^{cluster} + V_{pot}^{cluster}\right)R_{cluster} > 48\pi K_1 R_\infty^4 \qquad (\tau_0 < \tau_{0cr}) \qquad (22.25)$$

Final discussion

We just calculated the fields of perturbations of the expansion due to the overall energy $E_{dist}^{cluster} + V_{pot}^{cluster}$ of a topological singularity, or a cluster of topological singularities, by basing ourselves on a strong approximation of assuming a constant and homogenous average

$\overline{F}_{dist}^{cluster} + \overline{F}_{pot}^{cluster}$ energy densities inside the singularity. This procedure does actually provide *a virtual value* $\overline{\tau}_{int}^{(E)}$ *of an average field of internal disturbances*. However, it provides a *correct value of the external disturbance field* $\tau_{ext}^{(E)}(r)$ *(22.17), which does not depend on the size of the singularity*, even if the energy distribution was not homogeneous within the singularity.

22.2 – Localized singularity with a given curvature charge

A localized singularity with a radius $R_{cluster}$ possesses a given rest energy. It can also have an overall charge of curvature Q_θ. Indeed, such a singularity may consist of a mass of discrete topological singularities, such as prismatic dislocation loops (fig. 9.36) which each have an elementary curvature charge $q_{\theta BC(i)}$. If $Q_\theta > 0$, we speak of *a singularity of vacancy nature* as it represents holes of the lattice within the cluster, and if $Q_\theta < 0$ we speak of *a singularity of interstitial nature* because it represents extra lattice sites in the cluster.
A curvature singularity is responsible for a nonzero bending field and divergent in its neighborhood as we have shown in section 19.2.

The expansion fields due to a curvature singularity

The equilibrium condition *(18.8) of the field of expansion of such a singularity* requires for us to know the density $\theta^{cluster}(\vec{r})$ of charges of curvature inside the singularity. The equilibrium equation *(18.8)* is equivalent to rewriting the Laplacian of $\tau_{int}^{(Q_\theta)}$ inside the singularity

$$\Delta\left(\tau_{int}^{(Q_\theta)}\right) = -\frac{2K_2\theta^{cluster}(\vec{r})}{4K_2/3+2K_1(1+\tau_0)-K_0} \tag{22.26}$$

Without knowing the density $\theta^{cluster}(\vec{r})$ which is proper to a given singularity, we will use the approximation of an average homogenous density of curvature within the singularity, so that

$$\overline{\theta}^{cluster} = Q_\theta / \left(4\pi R_{cluster}^3 / 3\right) \tag{22.27}$$

We then obtain the following simplified condition of equilibrium for the virtual field

$$\Delta\left(\tau_{int}^{(Q_\theta)}\right) \cong -\frac{3K_2}{2\pi\left(4K_2/3+2K_1(1+\tau_0)-K_0\right)}\frac{Q_\theta}{R_{cluster}^3} \tag{22.28}$$

And we verify that the solution to this equation has a spherical symmetry, and is written

$$\tau_{int}^{(Q_\theta)}(r) \cong -\frac{K_2 Q_\theta}{4\pi\left(4K_2/3+2K_1(1+\tau_0)-K_0\right)}\frac{r^2}{R_{cluster}^3} \tag{22.29}$$

Furthermore, outside the singularity, we have the equilibrium condition

$$\Delta\left(\tau_{ext}^{(Q_\theta)}(r)\right) = 0 \tag{22.30}$$

With the spherical symmetry of the problem, the Laplacian possesses only one solution which tends towards a null value for $r \to R_\infty$, namely

$$\tau_{ext}^{(Q_\theta)}(r) = A/r \tag{22.31}$$

The condition for a null average field of perturbation

As before, the constant of integration A must be such that the value of the average field inside the lattice be equal to the background expansion τ_0 of the lattice so that

$$\tau_0 = \frac{3}{4\pi R_\infty^3}\left[\int_0^{R_{cluster}} \left(\tau_0 + \tau_{int}^{(Q_\theta)}(r)\right) 4\pi r^2\, dr + \int_{R_{cluster}}^{R_\infty} \left(\tau_0 + \tau_{ext}^{(Q_\theta)}(r)\right) 4\pi r^2\, dr\right]$$

$$\Rightarrow \quad \int_0^{R_{cluster}} \tau_{int}^{(Q_\theta)}(r) r^2\, dr + \int_{R_{cluster}}^{R_\infty} \tau_{ext}^{(Q_\theta)}(r) r^2\, dr = 0 \tag{22.32}$$

The solution to this equation gives us

$$A \cong \frac{K_2}{10\pi\left(4K_2/3 + 2K_1(1+\tau_0) - K_0\right) R_\infty^2} R_{cluster}^2 Q_\theta \tag{22.33}$$

And the fields of perturbation of expansion due to the charges of curvature are thus

$$\begin{cases} \tau_{int}^{(Q_\theta)}(r) \cong \dfrac{K_2}{3\left(K_0 - 2K_1(1+\tau_0) - 4K_2/3\right)} \left(\dfrac{Q_\theta}{4\pi R_{cluster}^3/3}\right) r^2 \\ \tau_{ext}^{(Q_\theta)}(r) \cong -\dfrac{K_2}{10\pi\left(K_0 - 2K_1(1+\tau_0) - 4K_2/3\right) R_\infty^2} \dfrac{R_{cluster}^2 Q_\theta}{r} \end{cases} \tag{22.32a}$$

The particular case of the edge dislocation loop

We try now to find the external field of expansion perturbations in the case of an edge dislocation loop, by introducing the value *(19.101)* of $q_{\theta EL} = -2\pi \vec{B}_{EL}\vec{n}$ and the real volume of the loop, meaning the volume $V_{tore} = 2\pi R_{EL}^3$ of the torus containing the essential part of the elastic energy of the loop. The external field of perturbations of the expansion due to the edge loop becomes

$$\tau_{ext}^{(q_{\theta EL})}(r) = \frac{K_2}{15\pi\left(K_0 - 2K_1(1+\tau_0) - 4K_2/3\right) R_\infty^2} \pi R_{EL}^2 \left(\vec{B}_{EL}\vec{n}\right) \frac{1}{r} \tag{22.35}$$

We find that the term $\pi R_{EL}^2\left(\vec{B}_{EL}\vec{n}\right)$ is *the effective volume of the edge loop*, which is directly related to the number N_{sites} of lattice sites to be removed or added to the lattice in order to form an edge dislocation edge loop, as

$$N_{sites} = \pi R_{EL}^2 \left|\vec{B}_{EL}\vec{n}\right| / a^3 \quad \Rightarrow \quad \tau_{ext}^{(q_{\theta EL})}(r) = \frac{K_2}{15\pi\left(K_0 - 2K_1(1+\tau_0) - 4K_2/3\right) R_\infty^2} N_{sites} a^3 \frac{1}{r} \tag{22.36}$$

The energy of the field of perturbations of the expansion

The total energy of the field of expansion can be calculated

$$E_{dist}^{(\tau)} = \iiint_V \left(K_1\tau^2 - K_0\tau\right) dV = \left[\begin{array}{l} K_1 \displaystyle\int_0^{R_{cluster}} \left(\tau_0 + \tau_{int}^{(Q_\theta)}(r)\right)^2 4\pi r^2\, dr + K_1 \int_{R_{cluster}}^{R_\infty} \left(\tau_0 + \tau_{ext}^{(Q_\theta)}(r)\right)^2 4\pi r^2\, dr \\ -K_0 \displaystyle\int_0^{R_{cluster}} \left(\tau_0 + \tau_{int}^{(Q_\theta)}(r)\right) 4\pi r^2\, dr - K_0 \int_{R_{cluster}}^{R_\infty} \left(\tau_0 + \tau_{ext}^{(Q_\theta)}(r)\right) 4\pi r^2\, dr \end{array}\right] \tag{22.37}$$

In this one, because of condition *(22.33)*, the terms in K_0 give the value $-VK_0\tau_0$ and do not depend on the singularity. As $K_0 >> K_1$, the relation *(22.32)* corresponds in fact to minimize the energy $E_{dist}^{(\tau)}$, which justifies *à posteriori* our choice of the condition of a perturbation field with a null average value used to determine the value of the constant A. It is therefore the terms in K_1 which give us a contribution to the energy of the field of perturbation of expansion of the singularity. If we subtract the terms of energy which only depend on τ_0, we obtain the increase in energy $E_{grav}^{(Q_\theta)}$ associated with the gravitational field of the singularity of curvature

$$E_{grav}^{(Q_\theta)} = \frac{K_1 K_2^{\,2}}{\pi\left(K_0 - 2K_1(1+\tau_0) - 4K_2/3\right)^2}\left(\frac{1}{28} + \frac{R_{cluster}^3}{25R_\infty^3}\right) R_{cluster} Q_\theta^{\,2} \tag{22.38}$$

which can be simplified if the charge is spherical with a radius $R_{cluster}$

$$E_{grav}^{(Q_\theta)} \cong \frac{K_1}{28\pi\left(K_0 - 2K_1(1+\tau_0) - 4K_2/3\right)^2} K_2^2 R_{cluster} Q_\theta^{\,2} \tag{22.39}$$

The effects of expansion due to a singularity with a given curvature charge

The fields $\tau_{ext}^{(Q_\theta)}(R_{cluster})$ and $\tau_{int}^{(Q_\theta)}(R_{cluster})$ at the surface of the singularity have very different values, which is due to the term in $R_{cluster}^2 / R_\infty^2$ in the expression of the external field

$$\begin{cases} \tau_{int}^{(Q_\theta)}(R_{cluster}) \cong \dfrac{K_2}{4\pi\left(K_0 - 2K_1(1+\tau_0) - 4K_2/3\right)}\left(\dfrac{Q_\theta}{R_{cluster}}\right) \\[2ex] \tau_{ext}^{(Q_\theta)}(R_{cluster}) \cong -\dfrac{K_2}{10\pi\left(K_0 - 2K_1(1+\tau_0) - 4K_2/3\right)} \dfrac{R_{cluster}^2}{R_\infty^2}\left(\dfrac{Q_\theta}{R_{cluster}}\right) \end{cases} \tag{22.40}$$

The fields of expansion do *not match at the interface of the singularity* (fig. 22.3, a and b), this implies that the mass densities of the lattice are different inside and outside the cluster

$$mn_{ext}(R_{cluster}) \cong mn_0\, e^{-\tau_0}\, e^{-\tau_{ext}^{(Q_\theta)}(R_{cluster})} \quad \text{and} \quad mn_{int}(R_{cluster}) \cong mn_0\, e^{-\tau_0}\, e^{-\tau_{int}^{(Q_\theta)}(R_{cluster})} \tag{22.41}$$

These expressions also imply that, if the fields exist, meaning if $K_2 \neq 0$, these have a singularity which is infinite when the background expansion τ_0 attains the critical value τ_{0cr} which renders the denominator null. The signs of the term $K_0 - 2K_1(1+\tau_0) - 4K_2/3$ and of module K_2 play an important role here. By using the conjectures 0 to 6 of the *perfect cosmological lattice, we have that:*

- if $K_2 / \left[K_0 - 2K_1(1+\tau_0) - 4K_2/3\right] > 0$, meaning if $\tau_0 < \tau_{0cr}$, *only the vacancy singularities,* with $Q_\theta > 0$, satisfy conjecture 2 concerning the curvature by attraction of the wave rays in the vicinity of the singularity, while the *interstitial singularities,* with $Q_\theta < 0$ push the wave rays away,
- if $K_2 / \left[K_0 - 2K_1(1+\tau_0) - 4K_2/3\right] < 0$, meaning if $\tau_0 > \tau_{0cr}$, only the singularities of interstitial nature, with $Q_\theta < 0$, satisfy conjecture 2 concerning the curvature by attraction of wave rays in the vicinity of the singularity, while the singularities of vacancy type, with $Q_\theta > 0$, push the wave rays away.

The dependency of a curvature singularity on the background expansion of the lattice

The domains of solutions of perturbation of expansion associated with a curvature singularity are reported synoptically in figure 22.4.

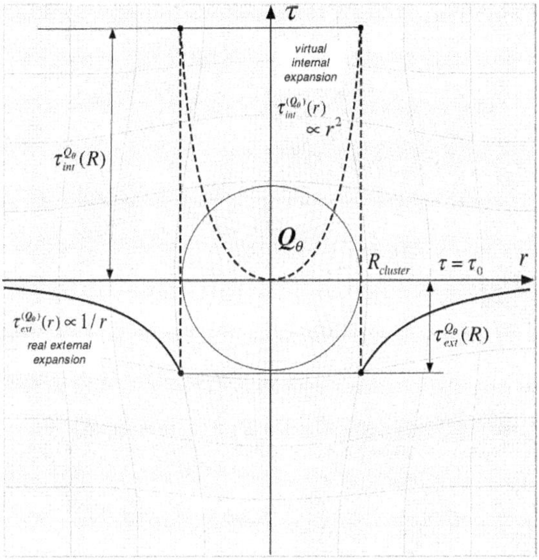

Figure 22.3a *- The field of expansion inside and outside of a charge of curvature* $Q_\theta < 0$ *(of interstitial type) if* $\tau_0 > \tau_{0cr}$ *or a curvature charge* $Q_\theta > 0$ *(of vacancy type) if* $\tau_0 < \tau_{0cr}$

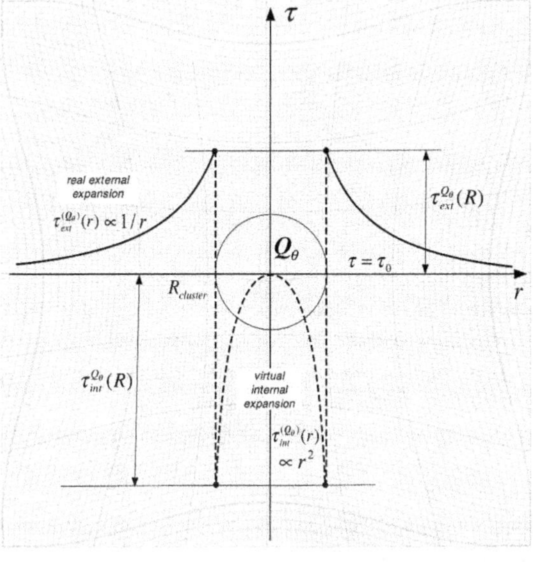

Figure 22.3b *- The field of expansion inside and outside the curvature charge* $Q_\theta > 0$ *(of vacancy type) if* $\tau_0 > \tau_{0cr}$ *or a curvature charge* $Q_\theta < 0$ *(interstitial type) if* $\tau_0 < \tau_{0cr}$

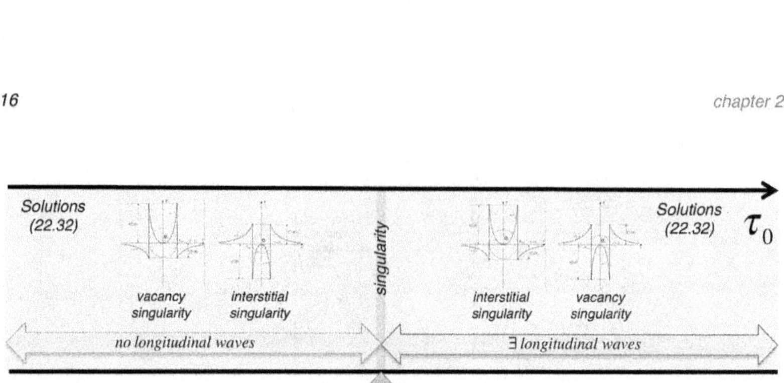

$$\tau_{0cr} = \frac{K_0}{2K_1} - \frac{2K_2}{3K_1} - 1 > 1$$

Figure 22.4 *- The domains of solutions of perturbation of expansion of a curvature singularity as a function of* τ_0

We ask whether a pure singularity of curvature can behave as a black hole. For this, it is needed that the first condition *(15.3)* be satisfied, and thus, by using relation *(22.33)*, that

$$\left.\frac{\partial \tau_{ext}^{(Q_\theta)}(r)}{\partial r}\right|_{r_{cr}} = \frac{2}{r_{cr}} \Rightarrow r_{cr} = \frac{K_2}{30\left(K_0 - 2K_1(1+\tau_0) - 4K_2/3\right)R_\infty^2}\frac{Q_\theta}{V_{cluster}}R_{cluster}^5 \Leftrightarrow \begin{cases} Q_\theta < 0\ (\tau_0 > \tau_{0cr}) \\ Q_\theta > 0\ (\tau_0 < \tau_{0cr}) \end{cases} \quad (22.42)$$

With regards to the second condition *(15.3)*, it implies that

$$r_{cr} > R_{cluster} \Rightarrow \frac{|Q_\theta|}{V_{cluster}}R_{cluster}^4 > \frac{30\left|K_0 - 2K_1(1+\tau_0) - 4K_2/3\right|}{K_2}R_\infty^2 \Leftrightarrow \begin{cases} Q_\theta < 0\ (\tau_0 > \tau_{0cr}) \\ Q_\theta > 0\ (\tau_0 < \tau_{0cr}) \end{cases} \quad (22.43)$$

And thus implies that the module of the charge of curvature be sufficiently large. However the numerator of the right part of the inequality goes through zero in the vicinity of the singularity $\tau_0 = \tau_{0cr}$, so that a curvature singularity goes through a 'black hole' stage for $\tau_0 \le \tau_{0cr}$ in the case of vacancy singularities and $\tau_0 \ge \tau_{0cr}$ in the case of interstitial singularities.

Final discussion

We have calculated the fields of disturbance of expansion due to an overall charge of curvature Q_θ of a topological singularity, or a cluster of topological singularities. We based our calculation on a strong approximation of expressing an average density of charges of curvature $\overline{\theta}^{amas}$ homogeneous in medium within the singularity. This approximation provides an approximate value of virtual expansion $\tau_{int}^{(Q_\theta)}(r)$ inside the singularity, that would be correct if the curvature of charge density is effectively homogeneous within the singularity. However, it does give a correct value of the external disturbance field $\tau_{ext}^{(Q_\theta)}(r)$ *(22.34)*, a singularity with global charge Q_θ, even if the charge distribution $\theta^{amas}(\vec{r})$ was not homogeneous within that singularity. This is proved by the value of the external field obtained for an edge dislocation loop, which depends exactly of the effective volume $\pi R_{EL}^2\left(\vec{B}_{EL}\vec{n}\right)$ of the edge dislocation loop.

22.3 – Localized singularity with a given rotation charge

Let's imagine now that there exists within the perfect lattice a localized singularity of volume $V_{cluster}$ and of global charge of rotation Q_λ , composed from a cluster of elementary charges of rotation $q_{\lambda TL(i)}$ or containing a density $\lambda(\vec{r})$ of charges of rotation.
Such a singularity has an external field of perturbations of the expansion related to the internal elastic energy of rotation of the singularity, which can be expressed from *(22.17)*

$$\tau_{ext}^{(E_{dist(rot\,int)})}(r) \cong -\frac{1}{2\pi\left(K_0-2K_1(1+\tau_0)-4K_2/3\right)R_\infty^2}\frac{E_{dist(rot\,int)}^{cluster}}{r} \tag{22.44}$$

And this field does not depend on the volume $V_{cluster}$ or the radius $R_{cluster}$ of the singularity of rotation.
But an other field has to be added due to the distorsion energy associated to the external rotation field $\vec{\omega}_{ext}^{el}$ generated by the singularity. The external rotation field generated by the singularity is given by relation *(19.66)*, and it does not depend on the radius or the volume of the singularity

$$\vec{\omega}_{ext}^{el} = \frac{Q_\lambda}{4\pi}\frac{\vec{r}}{r^3} \tag{22.45}$$

This external perturbation $\tau_{ext}^{(\vec{\omega}_{ext}^{el})}(r)$ of the expansion can be deducted from equation *(18.14)*, in which the elastic energy density is due to the rotaion field of the singularity, and in which τ^{ext} and τ^{ch} are null

$$K_1\left(\tau_{ext}^{(\vec{\omega}_{ext}^{el})}(r)\right)^2+\left(4K_2/3+2K_1(1+\tau_0)-K_0\right)\tau_{ext}^{(\vec{\omega}_{ext}^{el})}(r)+2\left(K_2+K_3\right)\left(\vec{\omega}_{ext}^{el}(r)\right)^2=0 \tag{22.46}$$

The condition of existence of a static field of perturbation of expansion

The solution $\tau_{ext}^{(Q_\lambda)}(r)$ to this second degree equation is the following

$$\tau_{ext}^{(\vec{\omega}_{ext}^{el})}(r)=\frac{\left(4K_2/3+2K_1(1+\tau_0)-K_0\right)}{2K_1}\left(-1\pm\sqrt{1-\frac{K_1\left(K_2+K_3\right)}{2\pi^2\left(4K_2/3+2K_1(1+\tau_0)-K_0\right)^2}\frac{Q_\lambda^2}{r^4}}\right) \tag{22.47}$$

As a matter of fact this static solution only exists if the argument under the square root is positive, which gives us a condition of existence on the value of charge Q_λ , if we take $1/r^4$ at its maximum, meaning in $r = R_{amas}$

$$\textbf{Hypothesis 1:}\quad \left(\frac{Q_\lambda}{R_{cluster}^2}\right)^2 \leq \frac{2\pi^2\left(4K_2/3+2K_1(1+\tau_0)-K_0\right)^2}{K_1\left(K_2+K_3\right)} \tag{22.48}$$

With conjectures 0 to 6 of the *perfect cosmological lattice*, this condition can never be satisfied in a domain of volume expansion τ_0 centered on $\tau_0=\tau_{0cr}$, and covering the following domain

$$\tau_{0cr}-\sqrt{\frac{K_2+K_3}{8\pi^2K_1}}\frac{|Q_\lambda|}{R_{cluster}^2}\leq\tau_0\leq\tau_{0cr}+\sqrt{\frac{K_2+K_3}{8\pi^2K_1}}\frac{|Q_\lambda|}{R_{cluster}^2} \tag{22.49}$$

In this domain, *there can only be dynamical solutions to the equation of Newton (18.12) for the perturbation of expansion associated with the singularity of rotation*, which allows us to pass from the static solution in domain $\tau_0<\tau_{0cr}$ to the other static solution in the *domain* $\tau_0>\tau_{0cr}$

and vice-versa.

A simpler case of fields of static perturbations of the expansion

Rather than look for an exact solution of the field of perturbation deduced from equation *(22.43)* in the case where the condition of existence *(22.44)* is satisfied, we can emit a stronger condition than hypothesis 1, namely

$$\textbf{Hypothesis 2:}\quad \tau_0 << \tau_{0cr} - \sqrt{\frac{K_2+K_3}{8\pi^2 K_1}\frac{|Q_\lambda|}{R_{cluster}^2}} \quad \text{or} \quad \tau_0 >> \tau_{0cr} + \sqrt{\frac{K_2+K_3}{8\pi^2 K_1}\frac{|Q_\lambda|}{R_{cluster}^2}} \tag{22.50}$$

and develop to first order the square root in expression *(22.47)*, so that the static field satisfies approximately the following relation

$$\tau_{ext}^{(\vec{\omega}_{ext}^{el})}(r) \cong -\frac{2(K_2+K_3)\left(\vec{\omega}_{ext}^{el}(r)\right)^2}{4K_2/3+2K_1(1+\tau_0)-K_0} \tag{22.51}$$

By introducing the value *(22.45)* of the elastic field of rotation outside the singularity we obtain the field $\tau_{ext}^{(\vec{\omega}_{ext}^{el})}(r)$ of expansion perturbations due to the external rotation field of the singularity

$$\tau_{ext}^{(\vec{\omega}_{ext}^{el})}(r) \cong -\frac{(K_2+K_3)}{8\pi^2\left(4K_2/3+2K_1(1+\tau_0)-K_0\right)}\frac{Q_\lambda^{\ 2}}{r^4} \tag{22.52}$$

Condition for a nul average field of the perturbations of expansion due to the external field of rotation of the singularity

However, as the average value of disturbance field must be zero for the background field to be equal to τ_0, the external field $\tau_{ext}^{(\vec{\omega}_{ext}^{el})}(r)$ should also include an additional field correction $\tau_{ext}^{cor}(r)$ ensuring that conservation of the number of lattice sites, which must satisfy the Laplacian $\Delta\left(\tau_{ext}^{cor}(r)\right)=0$, so a spherically symmetric field tending towards zero value as $r \to R_\infty$, and therefore written $\tau_{ext}^{cor}(r)=A/r$. The disturbance field of volumetric expansion due to the presence of the rotational load can thus be written

$$\tau_{ext}^{(E_{dist(rot\,ext)})}(r) \cong \tau_{ext}^{(\vec{\omega}_{ext}^{el})}(r)+\frac{A}{r} = -\frac{(K_2+K_3)}{8\pi^2\left(4K_2/3+2K_1(1+\tau_0)-K_0\right)}\frac{Q_\lambda^{\ 2}}{r^4}+\frac{A}{r} \tag{22.53}$$

The constant of integration A is determined by the same condition as before

$$\tau_0 = \frac{3}{4\pi R_\infty^3}\left[\int_{R_{amas}}^{R_\infty}\left(\tau_0+\tau_{ext}^{(E_{dist(rot\,ext)})}(r)\right)4\pi r^2\,dr\right] \Rightarrow \int_{R_{amas}}^{R_\infty}\tau_{ext}^{(E_{dist(rot\,ext)})}(r)r^2\,dr = 0 \tag{22.54}$$

Which leads to the following condition for constant A

$$A = \frac{2(K_2+K_3)Q_\lambda^{\ 2}}{8\pi^2\left(4K_2/3+2K_1(1+\tau_0)-K_0\right)R_\infty^2}\frac{1}{R_{cluster}} \tag{22.55}$$

The external field of expansion perturbations, due to the external field of rotation of the charge Q_λ is thus written

$$\tau_{ext}^{(E_{dist(rot\,ext)})}(r) \cong \frac{(K_2+K_3)Q_\lambda^{\ 2}}{8\pi^2\left(K_0-4K_2/3-2K_1(1+\tau_0)\right)}\left[\frac{1}{r^4}-\frac{2}{R_\infty^2 R_{cluster}}\frac{1}{r}\right] \tag{22.56}$$

The global external field of perturbations of expansion of a singularity of rotation

The global external field of expansion perturbations due to a singularity of rotation Q_λ can be written, under the condition *(22.50)*

$$\tau_{ext}^{(Q_\lambda)}(r) = \tau_{ext}^{(E_{dist(rot\,int)})}(r) + \tau_{ext}^{(E_{dist(rot\,ext)})}(r)$$
$$\cong \frac{1}{2\pi\left(K_0 - 4K_2/3 - 2K_1(1+\tau_0)\right)}\left[\frac{(K_2+K_3)Q_\lambda^{\,2}}{4\pi r^4} - \frac{1}{R_\infty^2}\left(E_{dist(rot\,int)}^{cluster} + \frac{2(K_2+K_3)Q_\lambda^{\,2}}{4\pi R_{cluster}}\right)\frac{1}{r}\right] \tag{22.57}$$

From this expression, the global external field $\tau_{ext}^{(Q_\lambda)}(r)$ has to change its sign at a definite critical distance r_{cr} of the singularity

$$r_{cr} = \sqrt[3]{\frac{R_\infty^2 R_{cluster}}{2\left(1+\dfrac{2\pi E_{dist(rot\,int)}^{cluster} R_{cluster}}{(K_2+K_3)Q_\lambda^{\,2}}\right)}} > R_{cluster} \tag{22.58}$$

which cannot be higher than the radius $R_{cluster}$ of the singularity.

One finds that the global external field of expansion perturbations of a rotation singularity presents a long range component depending on $1/r$, which is associated with the internal elastic rotation energy of the singularity, and a short range component depending on $1/r^4$, which is due to the elastic energy of the external rotation field of the singularity. The superposition of these two fields is schematically reported in figures 22.6a and 22.6b, in the csases $\tau_0 < \tau_{0cr}$ and $\tau_0 > \tau_{0cr}$ respectively.

The particular case of a spherical singularity with a rotation charge Q_λ

In the case of a spherical rotation singularity, as described in section 19.5, the energy $E_{dist(rot\,int)}^{amas}$ can be deduced from *(19.70)*, and one obtains

$$\tau_{ext}^{(Q_\lambda)}(r) \cong \frac{1}{2\pi\left(K_0 - 4K_2/3 - 2K_1(1+\tau_0)\right)}\frac{3(K_2+K_3)Q_\lambda^{\,2}}{5\pi R_{cluster}}\left[\frac{5R_{cluster}}{12r^4} - \frac{1}{R_\infty^2}\frac{1}{r}\right] \tag{22.59}$$

The global external field of expansion perturbations depends in fact on the total distortion energy $E_{dist}^{(Q_\lambda)}$ *(19.71)* of the rotation cluster, so that

$$\tau_{ext}^{(Q_\lambda)}(r) \cong \frac{1}{2\pi\left(K_0 - 4K_2/3 - 2K_1(1+\tau_0)\right)} E_{dist}^{(Q_\lambda)}\left[\frac{5R_{cluster}}{12}\frac{1}{r^4} - \frac{1}{R_\infty^2}\frac{1}{r}\right] \tag{22.60}$$

and that the critical radius where this field is canceling is given by

$$r_{cr} = \sqrt[3]{\frac{5}{12}R_{cluster}R_\infty^2} > R_{cluster} \tag{22.61}$$

The particular case of the twist disclination loop

In the case of the twist disclination loop described at section 19.7, the energy $E_{dist(rot\,int)}^{amas}$ becomes the energy $E_{dist\,torus}^{TL}$ found by the relation *(19.88)* and, on the other hand, the rotation charge becomes $Q_\lambda = q_{\lambda TL} = -\pi R_{TL}\vec{B}_{TL}\vec{t}$, found by the relation *(19.77)*. One obtains

$$\tau_{ext}^{(q_{\lambda TL})}(r) \cong \frac{1}{2\pi\left(K_0 - 4K_2/3 - 2K_1(1+\tau_0)\right)} \frac{1}{2}\left(K_2 + K_3\right)\zeta_{TL} R_{TL} \vec{B}_{TL}^2 \left[\frac{R_{TL}}{2\pi\zeta_{TL} r^4} - \frac{2\pi + 1/\zeta_{TL}}{2\pi R_\infty^2} \frac{1}{r}\right] \tag{22.62}$$

The global external field of expansion perturbations depends in fact on the distortion energy $E_{dist\,torus}^{TL}$ situated in the torus around the loop *(19.71)*, so that

$$\tau_{ext}^{(q_{\lambda BV}\,global)}(r) \cong \frac{1}{2\pi\left(K_0 - 4K_2/3 - 2K_1(1+\tau_0)\right)} E_{dist\,torus}^{TL} \left[\frac{R_{TL}}{2\pi\zeta_{TL} r^4} - \frac{1}{R_\infty^2} \frac{1}{r}\right] \tag{22.63}$$

and that the critical radius where this field is canceling is given by

$$r_{cr} = \sqrt[3]{\frac{1}{2\pi\zeta_{TL}} R_{TL} R_\infty^2} > R_{TL} \tag{22.64}$$

In the two cases (spherical singularity of rotation or twist disclination loop), one finds that this critical distance is clearly higher than the radius of the singularity, so that *there is always a critical radius where the field of expansion perturbations changes its sign.*

As a consequence, the perturbations of the expansion field at the interface of the singularity are positive in the expansion $\tau_0 < \tau_{0cr}$ (fig. 22.6a) and negative in the expansion area $\tau_0 > \tau_{0cr}$ (fig. 22.6b).

On the other hand, as the distortion energies $E_{dist}^{(Q_\lambda)}$ and $E_{dist\,torus}^{BV}$ are always positive, there is no dissymmetry of the expansion field between charges and anti-charges of rotation. But it appears an inversion of these fields between the area $\tau_0 < \tau_{0cr}$ and $\tau_0 > \tau_{0cr}$.

The internal field of expansion perturbations

Inside the singularity, there is an average internal virtual field of expansion $\overline{\tau}_{int}^{(Q_\lambda)}$ given by the relations *(22.4)* or *(22.14)*

$$\overline{\tau}_{int}^{(Q_\lambda)} \cong \frac{1}{K_0 - 4K_2/3 - 2K_1(1+\tau_0)} \frac{E_{dist\,(rot\,int)}^{cluster}}{V_{cluster}} \tag{22.65}$$

which is due to the elastic energy $E_{dist\,(rot\,int)}^{cluster}$ of internal rotation of the singularity, and which is compensated by the external field $\tau_{ext}^{(E_{dist\,(rot\,int)})}(r)$ of expansion perturbations given by *(22.44)*.

Concerning the external field $\tau_{ext}^{(\vec{\omega}_{ext}^{el})}(r)$ depending on $1/r^4$ *(22.48)*, it is compensated by an external field depending on $1/r$, which is included in the field $\tau_{ext}^{(E_{dist\,(rot\,ext)})}(r)$ given by *(22.56)*.

The particular case of a homogenous spherical rotation charge

In the particular case of a homogenous spherical rotation charge, the condition *(22.50)* allows us to deduce the exact statical internal field $\tau_{int}^{(Q_\lambda)}(r)$ by the following relation

$$\tau_{int}^{(Q_\lambda)}(r) \cong -\frac{2\left(K_2 + K_3\right)\left(\vec{\omega}_{int}^{el}(r)\right)^2}{4K_2/3 + 2K_1(1+\tau_0) - K_0} \tag{22.66}$$

Using expression *(19.67)* of the internal rotation field

$$\vec{\omega}_{int}^{el}(r) = \frac{Q_\lambda}{4\pi R_{cluster}^3} \vec{r} \tag{22.67}$$

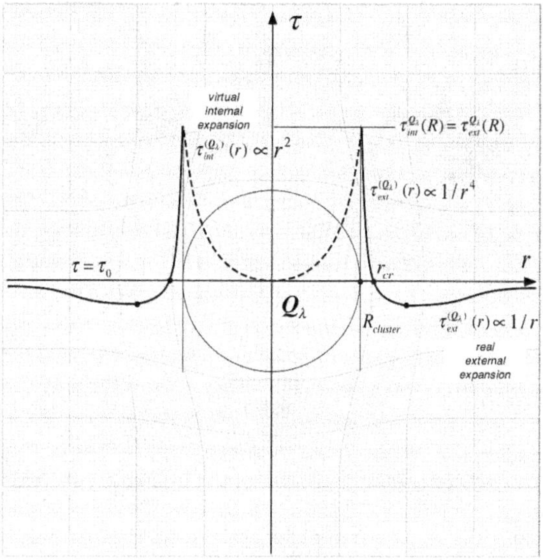

Figure 22.6a *- the field of expansion inside and outside a charge of rotation* Q_λ *in the case where* $\tau_0 < \tau_{0cr}$

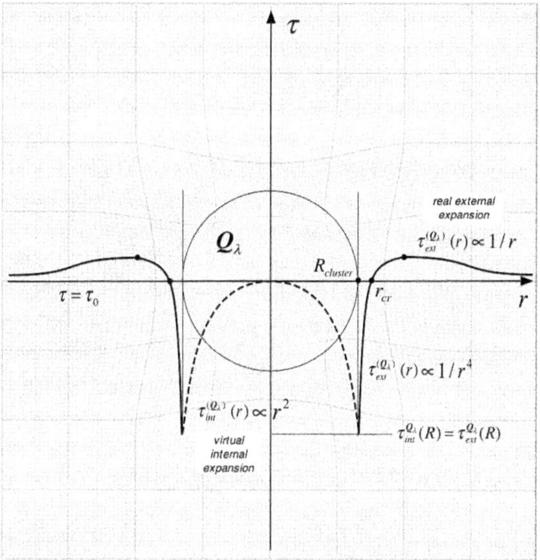

Figure 22.6b *- The field of expansion inside and outside the charge of rotation* Q_λ *in the case where* $\tau_0 > \tau_{0cr}$

one obtains the following internal field of expansion perturbations

$$\tau_{int}^{(Q_\lambda)}(r) \cong -\frac{(K_2+K_3)}{8\pi^2\left(4K_2/3+2K_1(1+\tau_0)-K_0\right)}\frac{Q_\lambda^2}{R_{cluster}^6}r^2 \tag{22.68}$$

It is this internal field of expansion perturbations inside a spherical rotation charge which is reported in figures 22.6a and 22.6b. At the interface of the singularity, this internal field $\tau_{int}^{(E_{dist(rot\,int)}^{cluster})}(r)$ and the global external field $\tau_{ext}^{(Q_\lambda)}(r)$ have almost the same value since

$$\begin{cases} \tau_{int}^{(Q_\lambda)}(R_{cluster}) \cong \dfrac{(K_2+K_3)}{8\pi^2\left(K_0-4K_2/3-2K_1(1+\tau_0)\right)}\dfrac{Q_\lambda^2}{R_{cluster}^4} \\ \tau_{ext}^{(Q_\lambda)}(R_{cluster}) \cong \dfrac{1}{2\pi\left(K_0-4K_2/3-2K_1(1+\tau_0)\right)}\dfrac{3(K_2+K_3)Q_\lambda{}^2}{5\pi R_{cluster}^4}\left[\dfrac{5}{12}-\dfrac{R_{cluster}^2}{R_\infty^2}\right] \\ \qquad\qquad \cong \dfrac{(K_2+K_3)}{8\pi^2\left(K_0-4K_2/3-2K_1(1+\tau_0)\right)}\dfrac{Q_\lambda{}^2}{R_{cluster}^4} \end{cases} \tag{22.69}$$

The energy of the field of expansion perturbations of a singularity of rotation

The energy of the field of expansion perturbations associated to the internal elastic rotation energy of a singularity is deduced from *(22.14)* and *(22.19)*

$$E_{grav}^{(E_{dist(rot\,int)})} \cong \frac{K_1}{\left(K_0-2K_1(1+\tau_0)-4K_2/3\right)^2}\frac{\left(E_{dist(rot\,int)}^{cluster}\right)^2}{V_{cluster}} \tag{22.70}$$

The energy of the field of expansion perturbations associated to the external elastic rotation energy of a singularity can be obtained in the following way

$$E_{grav}^{(E_{dist(rot\,ext)})} \cong \iiint\limits_{V_{external}} K_1\tau^2\,dV = K_1\int_{R_{cluster}}^{R_\infty}\left(\tau_{ext}^{(\vec{\omega}_{ext}^{el})}(r)\right)^2 4\pi r^2\,dr \tag{22.71}$$

Calculation of this expression furnishes

$$E_{grav}^{(E_{dist(rot\,ext)})} \cong \frac{K_1}{\left(K_0-4K_2/3-2K_1(1+\tau_0)\right)^2}\frac{(K_2+K_3)^2}{80\pi^3}\frac{Q_\lambda{}^4}{R_{cluster}^5} \tag{22.72}$$

The global gravitational energy associated to a rotation singularity becomes then

$$E_{grav}^{(Q_\lambda)} = E_{grav}^{(E_{dist(rot\,int)})} + E_{grav}^{(E_{dist(rot\,ext)})} \cong \frac{K_1}{\left(K_0-2K_1(1+\tau_0)-4K_2/3\right)^2}\left[\frac{\left(E_{dist(rot\,int)}^{cluster}\right)^2}{V_{cluster}}+\frac{(K_2+K_3)^2}{80\pi^3}\frac{Q_\lambda{}^4}{R_{cluster}^5}\right] \tag{22.73}$$

The particular case of the gravitational energy of a spherical singularity of rotation

In the particular case of a spherical singularity of rotation, the energy $E_{dist(rot\,int)}^{amas}$ is given by *(19.70)*, so that its global gravitational energy can be written

$$E_{grav}^{(Q_\lambda)} \cong \frac{K_1}{\left(K_0-2K_1(1+\tau_0)-4K_2/3\right)^2}\frac{(K_2+K_3)^2 Q_\lambda{}^4}{50\pi^3 R_{cluster}^5} \tag{22.74}$$

One can compare this gravitational energy with the global elastic energy of distortion of the spherical singularity, given by*(19.71)*, and one obtains

$$\frac{E_{grav}^{(Q_\lambda)}}{E_{dist}^{(Q_\lambda)}} \cong \frac{K_1}{\left(K_0-2K_1(1+\tau_0)-4K_2/3\right)^2}\frac{2}{9}\frac{E_{dist}^{(Q_\lambda)}}{V_{cluster}} \cong \frac{K_1}{\left(K_0\right)^2}\frac{2}{9}\frac{E_{dist}^{(Q_\lambda)}}{V_{cluster}} \qquad \left(\tau_0 \ll \tau_{0cr}\right) \qquad (22.75)$$

As the ratio K_1 / K_0^2 is extremely small, the gravitational energy of the rotation singularity is surely much smaller than the elastic energy of distortion of the singularity in the area $\tau_0 \ll \tau_{0cr}$.

The dependency of a singularity of rotation on the background expansion of the lattice

We have seen that the expansion field $\tau_{ext}^{(Q_\lambda)}(r) \propto 1/r^4$ due to the external rotation field of the singularity can only be dynamic in a small area *(22.49)* centered around τ_{0cr}, and that it is possible to find simple static solutions in the area given by *(22.50)*. We can roughly see these behaviors in the expansion of a rotation singularity as was done in figure 22.7.
Concerning the expansion perturbations $\tau_{ext}^{(E_{rotation})}(r) \propto 1/r$ due to the internal energy of rotation of the singularity, it can only be dynamic in a small area already represented in figure 22.2.

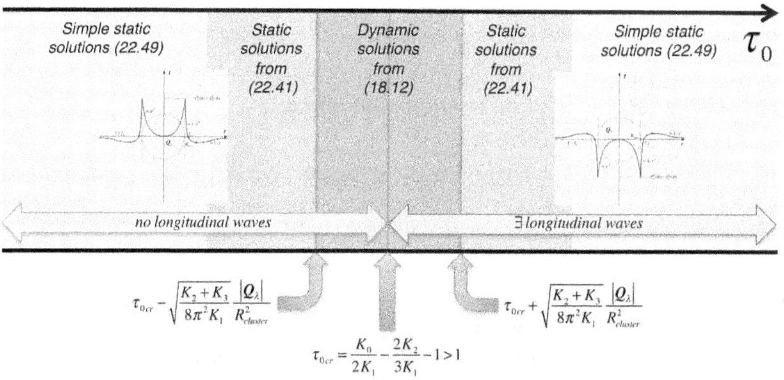

***Figure 22.7** - the domains of solutions of perturbation of expansion of a rotational singularity as a function of τ_0*

As the expansion field associated with the immediate vicinity of such a singularity can be negative only if $\tau_0 > \tau_{0cr}$, the existence of a black hole effect can only appear in this area. To determine the conditions for the appearance of such a black hole, apply the conditions *(15.3)* in the field $\tau_{ext}^{(Q_\lambda)}(r) \propto 1/r^4$ given by *(22.53)*. The first condition applies mainly on the $1/r^4$ component of the field

$$\left.\frac{\partial \tau_{ext}^{(Q_\lambda)}(r)}{\partial r}\right|_{r_{cr}} = \frac{2}{r_{cr}} \Rightarrow r_{cr}^4 = \frac{\left(K_2+K_3\right)Q_\lambda^2}{4\pi^2\left(4K_2/3+2K_1(1+\tau_0)-K_0\right)} \qquad (22.76)$$

and it allows us to obtain the following critical radius for the black hole

$$\left(\frac{r_{cr}}{R_{cluster}}\right)^4 \cong \frac{(K_2+K_3)}{4\pi^2\left(4K_2/3+2K_1(1+\tau_0)-K_0\right)}\left(\frac{Q_\lambda}{R_{cluster}^2}\right)^2 \qquad (\tau_0 > \tau_{0cr}) \tag{22.77}$$

The second condition implies that

$$r_{cr} > R_{cluster} \quad \Rightarrow \quad \frac{Q_\lambda^{\;2}}{R_{cluster}^4} > \frac{4\pi^2\left(4K_2/3+2K_1(1+\tau_0)-K_0\right)}{(K_2+K_3)} \qquad (\tau_0 > \tau_{0cr}) \tag{22.78}$$

But it is necessary that this condition be satisfied beyond the domain of the dynamic solutions, which is true only if the following condition is satisfied

$$\frac{Q_\lambda^{\;2}}{R_{cluster}^4} > \frac{8\pi^2 K_1}{(K_2+K_3)} \qquad (\tau_0 > \tau_{0cr}) \tag{22.79}$$

Final discussion

We have calculated the expansion of disturbance field due to an overall rotation charge Q_λ of a topological singularity, or a cluster of topological singularities, or a twist disclination loop. One finds that there exists two external fields of expansion perturbations *(22.53)* due to such a charge, the field $\tau_{ext}^{(E_{rotation})}(r)$ associated to the internal elastic energy of rotation of the charge, and the field $\tau_{ext}^{(Q_\lambda)}(r)$ due to the external rotation field of the charge. And these two fields are absolutely independent of the volume or the radius of the topological singularity of charge Q_λ.

22.4 – Macroscopic vacancies within the lattice

Imagine a cluster of vacancies, that is to say, singularities carrying positive bending charges, such as prismatic dislocation loops of vacancy type for example, that collapses on itself (under the "gravitational attractive forces " to be described later). If the initial cluster is neutral vis-à-vis the rotational charges, the individual peculiarities of the cluster are summed, losing their own identity (dislocation or disclination loops) to form a single macroscopic hole within the network, as a kind of macroscopic gap formed by individual vacancies (fig. 22.8, a, b and c). This means that there appears a macroscopic vacancy of N_L lattice sites. The radius of this macroscopic gap, assumed to be spherical, is then worth, in the imaginary case where the lattice showed a homogeneous expansion with $\tau = 0$

$$R_{L0} = \sqrt[3]{3N_L/(4\pi n_0)} \tag{22.80}$$

The condition of equilibrium for the expansion field of a macroscopic vacancy

In the actual lattice, the presence of this gap will generate a macroscopic field of spherical volume expansion $\tau_{ext}^{(L)}(r)$ to be determined. On the surface of the singularity , the expansion field is equal to the sum of the field generated by the singularity $\tau_{ext}^{(L)}(R_L)$, and the background expansion of τ_0 of the lattice and an external field of expansion due to other singularities located in the vicinity of the macroscopic gap. On the surface of the singularity, the total field has to be adjusted so that the pressure at the interface of the hole is zero (fig. 22.8, a, b and c). The zero *pressure condition at the interface* is written then using the equation of state *(13.8)*

$$p(r=R_L)=K_0-2K_1\left(\tau_0+\tau_{ext}^{(L)}(R_L)+\tau^{external}(R_L)\right)+K_0\left(\tau_0+\tau_{ext}^{(L)}(R_L)+\tau^{external}(R_L)\right)$$
$$-K_1\left(\tau_0+\tau_{ext}^{(L)}(R_L)+\tau^{external}(R_L)\right)^2=0 \quad (22.81)$$

This second degree equation has two roots for $\left(\tau_0+\tau_{ext}^{(L)}(R_L)+\tau^{external}(R_L)\right)$

$$\left(\tau_0+\tau_{ext}^{(L)}(R_L)+\tau^{external}(R_L)\right)=\frac{\left(K_0-2K_1\right)\pm\sqrt{K_0^2+4K_1^2}}{2K_1} \quad (22.82)$$

with conjecture 5 of the *perfect cosmic lattice,* namely that $K_0 >> K_1$, we then obtain

$$\left(\tau_0+\tau_{ext}^{(L)}(R_L)+\tau^{external}(R_L)\right)\cong\frac{\left(K_0-2K_1\right)\pm K_0}{2K_1}=\begin{cases}-1\\ \dfrac{K_0-K_1}{K_1}\cong\dfrac{K_0}{K_1}>>1\end{cases} \quad (22.83)$$

The second solution, namely $K_0/K_1>>1$, does not make physical sense since it represents a gigantic hole in the lattice. However, the first solutions makes a lot of sense as it can be recovered directly from expression *(22.81)* and admitting the conjecture 5. Indeed

$$p(r=R_L)\cong K_0+K_0\left(\tau_0+\tau_{ext}^{(L)}(R_L)+\tau^{external}(R_L)\right)=0 \quad\Rightarrow\quad \left(\tau_0+\tau_{ext}^{(L)}(R_L)+\tau^{external}(R_L)\right)=-1 \quad (22.84)$$

The field of expansion $\tau_{ext}^{(L)}(r)$ of macroscopic vacancy must satisfy $\Delta\tau_{ext}^{(L)}(r)\cong 0$, so

$$\tau_{ext}^{(L)}(r)=A/r \quad (22.85)$$

With the equilibrium condition of the surface pressure at the interface, we obtain the external field

$$\tau_{ext}^{(L)}(r)\cong-\left(1+\tau_0+\tau^{external}(R_L)\right)\frac{R_L}{r} \quad (22.86)$$

As for the real radius R_L of the macroscopic vacancy, it is then worth

$$R_L=R_{L0}\,\mathrm{e}^{\frac{\tau_0+\tau_{ext}^{(L)}(R_L)+\tau^{external}(R_L)}{3}}=\sqrt[3]{\frac{3N_L}{4\pi n_0}}\,\mathrm{e}^{\frac{-1}{3}}=\sqrt[3]{\frac{3N_L}{4\pi n_0\,\mathrm{e}}} \quad (22.87)$$

and we note that the real radius R_L of the macroscopic vacancy is a number which must only depend on the number N_L of sites of lattice vacancies.

We can then calculate the average field of expansion in the presence of a macroscopic singularity

$$\bar{\tau}=\frac{1}{\frac{4}{3}\pi R_\infty^3}\int_{R_L}^{R_\infty}\left(\tau_0+\tau_{ext}^{(L)}(r)\right)4\pi r^2\,dr=\tau_0-\left(1+\tau_0+\tau^{external}(R_L)\right)\frac{3R_L}{R_\infty^3}\left(\frac{R_\infty^2}{2}-\frac{R_L^2}{2}\right)\cong\tau_0 \quad (22.88)$$

and we do find an average field equal to the expansion of the background τ_0 of the lattice!

The energy of a macroscopic vacancy

The total energy of the field of expansion is written

$$E_{dist}^{(\tau)}=K_1\int_{R_L}^{R_\infty}\left(\tau_0+\tau_{ext}^{(L)}(r)\right)^2 4\pi r^2\,dr-K_0\int_{R_L}^{R_\infty}\left(\tau_0+\tau_{ext}^{(L)}(r)\right)4\pi r^2\,dr \quad (22.89)$$

If we subtract the energy due exclusively to the field τ_0, we obtain the gravitational energy of the perturbation of the field of expansion associated with the presence of the macroscopic vacancy

$$E_{grav}^{(L)} = 4\pi K_1 \int_{R_L}^{R_\infty} \left(\tau_{ext}^{(L)}(r)\right)^2 r^2\, dr + 8\pi K_1 \tau_0 \int_{R_L}^{R_\infty} \tau_{ext}^{(L)}(r) r^2\, dr - 4\pi K_0 \int_{R_L}^{R_\infty} \tau_{ext}^{(L)}(r) r^2\, dr \qquad (22.90)$$

By carrying the integrations with value *(22.86)* of $\tau_{ext}(r)$, we obtain

$$E_{grav}^{(L)} = \begin{bmatrix} -2\pi\left(K_0 - 2K_1\tau_0\right)\left(1+\tau_0+\tau^{external}(R_L)\right)R^3 - 4\pi K_1\left(1+\tau_0+\tau^{external}(R_L)\right)^2 R^3{}_L \\ +4\pi K_1\left(1+\tau_0+\tau^{external}(R_L)\right)^2 R_\infty R^2{}_L + 2\pi\left(K_0 - 2K_1\tau_0\right)\left(1+\tau_0+\tau^{external}(R_L)\right)R_\infty^2 R_L \end{bmatrix} \qquad (22.91)$$

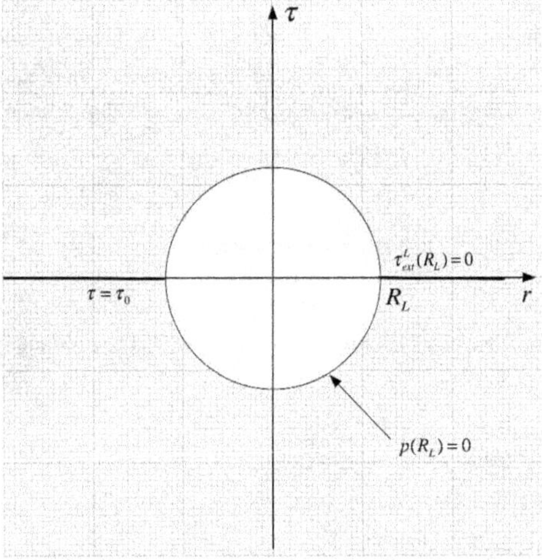

Figure 22.8a *- The null field of expansion $\tau_{ext}^{(L)}(r)$ of a macroscopic vacancy*

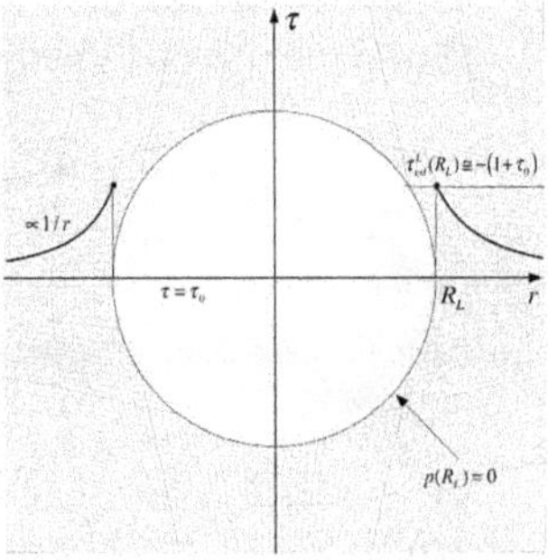

Figure 22.8b *- The field of expansion of the same macroscopic vacancy with 10'000 sites of lattice when $\tau_0 < -\left(1+\tau^{external}(R_L)\right)$, represented on the same scale as figure 22.8a*

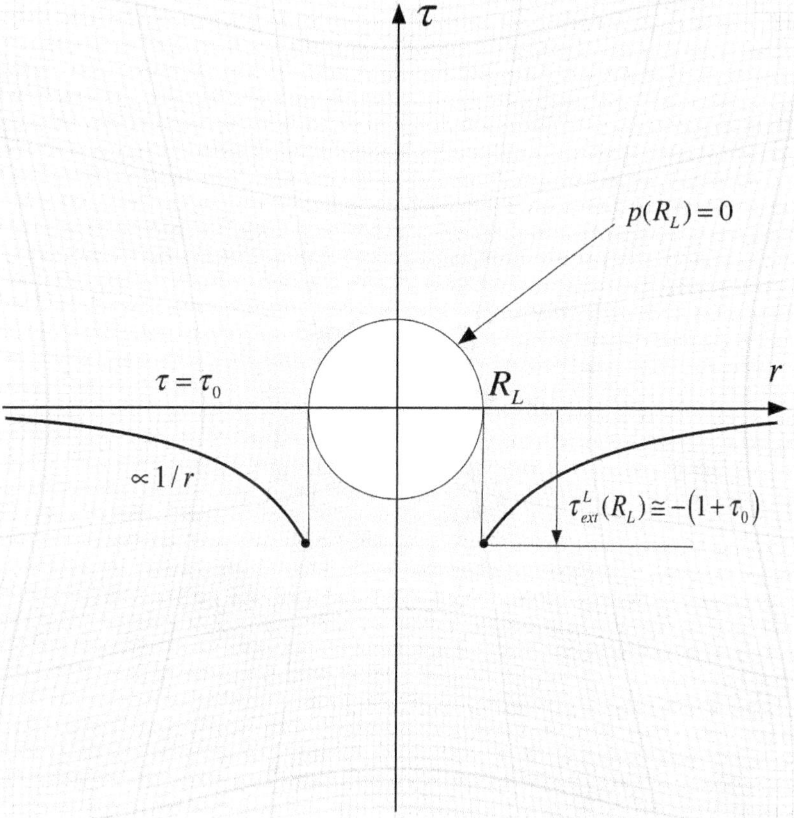

***Figure 22.8c** - The field of expansion of the same macroscopic vacancy with 10'000 sites of lattice when $\tau_0 > -\left(1+\tau^{external}(R_L)\right)$, represented on the same scale as figure 22.8a*

which becomes, by keeping the dominant term

$$E_{grav}^{(L)} \cong 2\pi\left(K_0-2K_1\tau_0\right)\left(1+\tau_0+\tau^{external}(R_L)\right)R_\infty^2 R_L \cong \sqrt[3]{\frac{6\pi^2 N_L}{n_0\,\mathrm{e}}}\left(K_0-2K_1\tau_0\right)\left(1+\tau_0+\tau^{external}(R_L)\right)R_\infty^2 \qquad (22.92)$$

We notice that the energy of formation of a microscopic vacancy does not depend on the background expansion τ_0 of the lattice and is null for two values namely

$$E_{grav}^{(L)} \to 0 \quad \Leftrightarrow \quad \tau_0 \to -\left(1+\tau^{external}(R_L)\right) \quad \text{and} \quad \tau_0 \to \frac{K_0}{2K_1} \qquad (22.93)$$

Between these two values, namely in the interval $-\left(1+\tau^{external}(R_L)\right)<\tau_0<K_0/2K_1$, the energy of formation of a macroscopic vacancy is positive, while it becomes negative outside that interval.

Furthermore, inside the interval $-\left(1+\tau^{external}(R_L)\right)<\tau_0<K_0/2K_1$, the energy of formation of two macroscopic vacancies of N_L sites is higher than the formation of a single macroscopic

vacancy of $2N_L$ sites since

$$E_{grav}^{(L)}\Big|_{2N_L} \cong \frac{1}{\sqrt[3]{4}} 2E_{grav}^{(L)}\Big|_{N_L} < 2E_{grav}^{(L)}\Big|_{N_L} \quad \Leftrightarrow \quad -\left(1+\tau^{external}(R_L)\right) < \tau_0 < K_0 / 2K_1 \qquad (22.94)$$

Two macroscopic vacancies have an energetic advantage to fuse together when the field of expansion is within the interval $-\left(1+\tau^{external}(R_L)\right) < \tau_0 < K_0 / 2K_1$.

When a macroscopic lattice becomes a real black hole!

In the presence of a macroscopic lattice, the first condition *(15.3)* for the appearance of a black hole implies that

$$\left.\frac{\partial \tau_{ext}^{(L)}(r)}{\partial r}\right|_{r_{cr}} = \frac{2}{r_{cr}} \quad \Rightarrow \quad r_{cr} = \frac{1}{2}\left(1+\tau_0+\tau^{external}(R_L)\right)R_L \qquad (22.95)$$

and a second condition *(15.3)* that

$$r_{cr} > R_L \quad \Rightarrow \quad \tau_0 > 1-\tau^{external}(R_L) \qquad (22.96)$$

From which we deduce that a macroscopic vacancy, whatever it's size and energy, *necessarily becomes a 'black hole' as soon as the volume of background expansion* τ_0 *exceeds the value* $1-\tau^{external}(R_L)$. This conclusion is very interesting since this stable macroscopic vacancy *is the real analogous of a black hole in general relativity* (fig. 22.8c). This topological singularity of vacancy type behaves like a *white hole* which pushes away the wave rays when condition $\tau_0 < -\left(1+\tau^{external}(R_L)\right)$ is satisfied (fig. 22.8b).

22.5 – Macroscopic interstitials within the lattice

Now imagine a mass of interstitial singularities, that is to say, singularities with negative curvature charges, such as prismatic dislocation loops of the interstitial type for example, which collapses onto itself under the effect of attractive "gravitational" forces as described in the previous chapter. If the initial cluster is neutral vis-à-vis the rotational charges, the individual peculiarities of the cluster are combined, losing their own identity (dislocation or disclination loops) to form a single piece of macroscopic local lattice, and formed of N_I sites (fig. 22.9).

This means there is locally an excess of N_I lattice sites forming a macroscopic interstitial. The radius of this macroscopic embedding, supposed spherical, is then, in the imaginary case where $\tau = 0$

$$R_{I0} = \sqrt[3]{3N_I / (4\pi n_0)} \qquad (22.97)$$

we can obviously consider that this macroscopic interstitial with N_I sites of lattice corresponds to the *anti-singularity of the macroscopic vacancy of* $N_L = N_I$ sites, in the sense that the combination of the two leaves the lattice unchanged globally since the N_L sites are matched by the N_I sites!

The condition of equilibrium of the field of expansion of a macroscopic interstitial

In the presence of this cluster within the lattice there is no coherence of the two lattices and the condition of equilibrium means that the pressure at the interface must be equal on both sides so

$$p_{ext}(R_I) = p_{int} \quad \Rightarrow \quad \tau_{int}^{(I)} = \tau_{ext}^{(I)}(R_I) \tag{22.98}$$

The external field $\tau_{ext}^{(I)}(r)$ satisfies the equation $\Delta\tau_{ext}^{(I)}(r) \cong 0$, and is thus written

$$\tau_{ext}^{(I)}(r) = A / r \tag{22.99}$$

to find the constant value of A , one must assure that the number of lattice meshes is equal to that after the introduction of the singularity, so that

$$n_0 \int_{R_I}^{R_\infty} e^{-\tau_0}\, e^{-\tau_{ext}^{(I)}(r)}\, 4\pi r^2\, dr = n_0\, e^{-\tau_0} \left(\frac{4}{3} \pi R_\infty^3 \right) \tag{22.100}$$

If we assume a priori that $\tau_{ext}^{(I)}(r) < 1$, the solution to this condition with field *(22.99) roughly* gives us

$$\left(\frac{R_\infty^3}{3} - \frac{R_I^3}{3} \right) - A\left(\frac{R_\infty^2}{2} - \frac{R_I^2}{2} \right) = \frac{1}{3} R_\infty^3 \quad \Rightarrow \quad A = -\frac{2R_I^3}{3\left(R_\infty^2 - R_I^2\right)} \cong -\frac{2R_I^3}{3R_\infty^2} \tag{22.101}$$

which does not contradict *a fortiori* the hypothesis $\tau_{ext}^{(I)}(r) < 1$. We then have for external field

$$\tau_{ext}^{(I)}(r) = -\frac{2R_I^2}{3R_\infty^2} \frac{R_I}{r} = -\frac{2R_I^3}{3R_\infty^2} \frac{1}{r} \tag{22.102}$$

and for the internal field, homogenous within the macroscopic interstitial of the lattice

$$\tau_{int}^{(I)} = \tau_{ext}^{(I)}(R_I) = -\frac{2R_I^2}{3R_\infty^2} << 1 \tag{22.103}$$

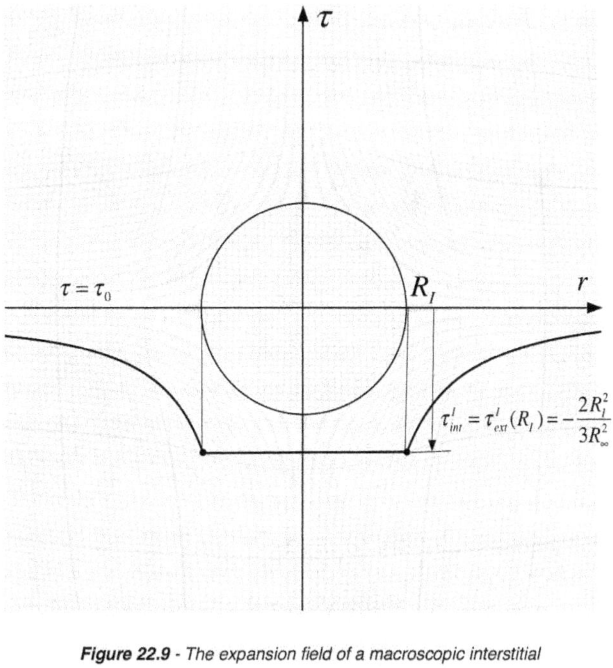

***Figure 22.9** - The expansion field of a macroscopic interstitial*

The two fields of expansion, associated with the macroscopic singularity are presented in figure 22.9. By supposing that there exists in the vicinity of the macroscopic interstitial a field of expansion $\tau^{external}(R_I)$ due to the other singularities situated in its vicinity, the real radius of the macroscopic interstitial will depend on the background expansion τ_0, of it's own internal expansion field $\tau_{int}^{(I)}$ and of the external field $\tau^{external}(R_I)$ via the relation

$$R_I = R_{I0}\,\mathrm{e}^{\frac{\tau_0+\tau_{int}^{(I)}+\tau^{external}(R_I)}{3}} = \sqrt[3]{\frac{3N_I}{4\pi n_0}}\,\mathrm{e}^{\frac{\tau_0+\tau^{external}(R_I)}{3}}\,\mathrm{e}^{-\frac{2R_I^2}{9R_\infty^2}} \cong \sqrt[3]{\frac{3N_I}{4\pi n_0}}\,\mathrm{e}^{\frac{\tau_0+\tau^{external}(R_I)}{3}} \tag{22.104}$$

This expression of R_I can be used to express differently the field of external perturbations *(22.102)*, under the form

$$\tau_{ext}^{(I)}(r) = -\frac{2R_I^3}{3R_\infty^2}\frac{1}{r} = -\frac{1}{R_\infty^2}\frac{N_I}{2\pi n}\frac{1}{r} = -\frac{1}{R_\infty^2}\frac{N_I}{2\pi\frac{1}{a^3}}\frac{1}{r} = -\frac{a^3}{2\pi R_\infty^2}N_I\frac{1}{r} \tag{22.105}$$

which shows that the external field of expansion perturbations is simply proportional to the number of additional lattice sites associated to the macroscopic interstitial.

The energy of a macroscopic interstitial

The energy of the total field of expansion is then written, taking into account both internal and external fields

$$E_{dist}^{(\tau)} = \left[\begin{array}{l} K_1\int_{R_I}^{R_\infty}\left(\tau_0+\tau_{ext}^{(I)}(r)\right)^2 4\pi r^2\,dr - K_0\int_{R_I}^{R_\infty}\left(\tau_0+\tau_{ext}^{(I)}(r)\right)4\pi r^2\,dr \\ +K_1\int_0^{R_I}\left(\tau_0+\tau_{int}^{(I)}\right)^2 4\pi r^2\,dr - K_0\int_0^{R_I}\left(\tau_0+\tau_{int}^{(I)}\right)4\pi r^2\,dr \end{array}\right] \tag{22.106}$$

The 'gravitational' energy of the field of expansion associated with this macroscopic interstitial then becomes, if we subtract the energy due exclusively to the background field τ_0, a value which depends on the energy density as shown in brackets in this equation

$$E_{grav}^{(I)} = \frac{4}{3}\pi R_I^3\left[\left(K_0-2K_1\tau_0\right)+\left(6K_1\tau_0-K_0\right)\frac{R_I^2}{3R_\infty^2}+K_1\frac{4R_I^3}{3R_\infty^3}-K_1\frac{8R_I^4}{9R_\infty^4}\right] \tag{22.107}$$

and which simplifies, using *(22.104)*, under the form

$${}^{I)}_{rav} \cong \frac{4}{3}\pi R_I^3\left(K_0-2K_1\tau_0\right) \cong \left(K_0-2K_1\tau_0\right)\frac{N_I}{n_0}\mathrm{e}^{\tau_0+\tau^{external}(} \tag{22.108}$$

Whereas the *macroscopic interstitial is the anti-singularity of the vacancy one,* it's formation energy is always positive while $\tau_0 < K_0/2K_1$, and infinitely smaller than the energy of formation of a macroscopic hole, which introduces a *huge dissymmetry* between the two singularities!

Can a macroscopic interstitial be a black hole?

In the presence of a macroscopic interstitial of the lattice, the first condition *(15.3)* for the appearance of a black hole implies that

$$\left.\frac{\partial \tau_{ext}^{(I)}(r)}{\partial r}\right|_{r_{cr}} = \frac{2}{r_{cr}} \quad \Rightarrow \quad r_{cr} = \frac{R_I^3}{3R_\infty^2} \tag{22.109}$$

and the second condition *(15.3)* that

$$r_{cr} > R_I \quad \Rightarrow \quad R_I > \sqrt{3}R_\infty \tag{22.110}$$

From which we deduce that a macroscopic interstitial, whatever its size and its energy, can <u>never behave</u> as a black hole. This conclusion is very interesting, as we have here a topological object which presents a great dissymmetry from its anti-singularity, the macroscopic vacancy, which becomes a black hole as son as $\tau_0 > 1$.

Table 22.1 - The long range fields of a topological singularity

"Electric" field of rotation due to a rotation charge Q_λ

$$\vec{\omega}_{ext}^{\acute{e}l}(\vec{r}) = \frac{Q_\lambda}{4\pi}\frac{\vec{r}}{r^3}$$

"Gravitational" field of flexion due to a curvature charge Q_θ

$$\vec{\chi}_{ext}(\vec{r}) = \frac{Q_\theta}{4\pi}\frac{\vec{r}}{r^3}$$

"Gravitational" field of expansion due to the energies E_{dist}^{ch} and V_{pot}^{ch}

$$\tau_{ext}^{(E)}(r) \cong -\frac{1}{2\pi\left(K_0 - 2K_1(1+\tau_0) - 4K_2/3\right)R_\infty^2}\left(E_{dist}^{cluster} + V_{pot}^{cluster}\right)\frac{1}{r}$$

"Gravitational" field of expansion due to the curvature charge Q_θ

$$\tau_{ext}^{(Q_\theta)}(r) \cong -\frac{1}{2\pi\left(K_0 - 2K_1(1+\tau_0) - 4K_2/3\right)R_\infty^2}\frac{K_2 R_{cluster}^2 Q_\theta}{5}\frac{1}{r}$$

"Gravitational" field of expansion due to the charge of rotation Q_λ

$$\tau_{ext}^{(Q_\lambda)}(r) \cong \frac{1}{2\pi\left(K_0 - 4K_2/3 - 2K_1(1+\tau_0)\right)R_\infty^2}\left[\begin{array}{l}\dfrac{(K_2+K_3)R_\infty^2 Q_\lambda^2}{4\pi}\dfrac{1}{r^4} \\ -\left(E_{dist\,(rot\,int)}^{cluster} + \dfrac{2(K_2+K_3)Q_\lambda^2}{4\pi R_{cluster}}\right)\dfrac{1}{r}\end{array}\right]$$

22.6 – Analogies with the 'electric' and 'gravitational' fields

In the analogy between our theory and the great theories of physics, the field of rotational expansion is typically a field of 'electric' nature, while the fields of flexion and of volume expansion are of "gravitational" nature.

In table 22.1, we have reported the "electric" fields and the "gravitational" fields generated *at long distance* by a topological singularity of global energy $E_{dist}^{ch}+V_{pot}^{ch}$, of global rotational charge $\boldsymbol{Q}_{\lambda}$ and/or global curvature charge $\boldsymbol{Q}_{\theta}$.

In table 22.2, we have shown the «gravitational» long-range fields and the energies of the vacancies and interstitials with macroscopic radius R_L and R_I, corresponding respectively to N_L holes of lattice sites or N_I interstitials sites.

Table 22.2 - The long-range fields and energies of vacancy and interstitial singularities

«Gravitational» field of expansion of a macroscopic vacancy singularity

$$\begin{cases} \tau_{ext}^{(L)}(r) \cong -\left(1+\tau_0+\tau^{external}(R_L)\right)\dfrac{R_L}{r} \\ R_L = \sqrt[3]{\dfrac{3N_L}{4\pi n_0\,\mathrm{e}}} \end{cases}$$

Energy of a macroscopic vacancy

$$E_{grav}^{(L)} \cong 2\pi\left(K_0-2K_1\tau_0\right)\left(1+\tau_0+\tau^{external}(R_L)\right)R_\infty^2 R_L \cong \sqrt[3]{\frac{6\pi^2 N_L}{n_0\,\mathrm{e}}}\left(K_0-2K_1\tau_0\right)\left(1+\tau_0+\tau^{external}(R_L)\right)$$

«Gravitational» field of expansion of a macroscopic interstitial singularity

$$\begin{cases} \tau_{int}^{(I)} = \tau_{ext}^{(I)}(R_I) = -\dfrac{2R_I^2}{3R_\infty^2} \ll 1 \\ \tau_{ext}^{(I)}(r) = -\dfrac{2R_I^3}{3R_\infty^2}\dfrac{1}{r} \ll 1 \\ R_I \cong \sqrt[3]{\dfrac{3N_I}{4\pi n_0}}\,\mathrm{e}^{\frac{\tau_0+\tau^{external}(R_I)}{3}} \end{cases}$$

Energy of a macroscopic interstitial singularity

$$E_{grav}^{(I)} \cong \frac{4\pi R_I^3}{3}\left(K_0-2K_1\tau_0\right) \cong \left(K_0-2K_1\tau_0\right)\frac{N_I}{n_0}\mathrm{e}^{\tau_0+\tau^{external}(R_I}$$

On the existence of many fields of 'gravitational' nature

Regarding the nature of the "gravitational" field, it is very encouraging to see that there is here a first expansion field directly associated with the energy $E_{dist}^{cluster}+V_{pot}^{cluster}$ of the cluster of topological singularities similar to the gravitational field of General Relativity of Einstein. It is also due to the *energy-momentum tensor of matter.*

But there are in our theory two other "gravitational" fields of expansion associated respectively to the overall charge of curvature Q_θ and to the overall rotation charge Q_λ of the cluster of topological singularities. *These fields do not have equivalents in the General Relativity of Einstein.*

The existence of the second "gravitational" field of expansion, due to the *curvature charge* Q_θ, is subject to the condition that the shear modulus K_2 of the perfect lattice is not zero. So there is still an opportunity to discuss the existence or not of this field according to the value to be assigned to the module K_2 in our analogy with the real world, knowing that this module should anyway be very small vis-à-vis module K_3, as was already specified in conjecture 4 of the perfect cosmic lattice.

The third "gravitational" field of expansion is associated with the rotational charge Q_λ of the cluster of topological singularities considered. As part of our analogy, this third field of expansion must exist if the cluster has a non-zero Q_λ charge, since the K_3 module must exist to meet the analogy with Maxwell's equations! But this field has no direct analog in the theory of General Relativity of Einstein and Particle Physics .

Note also that the three "gravitational" fields have nonzero gravitational energy given by the relations *(22.19), (22.39)* and *(22.73)* respectively

$$\left\{\begin{array}{l} E_{grav}^{E} \cong \dfrac{K_1}{\left(4K_2/3+2K_1(1+\tau_0)-K_0\right)^2}\dfrac{\left(E_{dist}^{cluster}\right)^2}{V_{amas}} \\ E_{grav}^{(Q_\theta)} \cong \dfrac{K_1}{\left(K_0-2K_1(1+\tau_0)-4K_2/3\right)^2}\dfrac{K_2^2 R_{cluster} Q_\theta^{\ 2}}{28\pi} \\ E_{grav}^{(Q_\lambda)} \cong \dfrac{K_1}{\left(K_0-2K_1(1+\tau_0)-4K_2/3\right)^2}\left[\dfrac{\left(E_{dist\,(rot\,int)}^{cluster}\right)^2}{V_{cluster}}+\dfrac{\left(K_2+K_3\right)^2}{80\pi^3}\dfrac{Q_\lambda^{\ 4}}{R_{cluster}^5}\right] \end{array}\right. \tag{22.111}$$

As these energies depend on the coefficient $K_1/K_0^2 <<< 1$, which must be very small in the perfect cosmic lattice, the gravitational energy of the singularities is surely negligible vis-à-vis the elastic energy of distortion of the topological singularities!

The fact that it appears two " gravity " fields of volume expansion which apparently have no analogues in the theory of General Relativity of Einstein and Particle Physics is very interesting to test our theory. We will return in the following chapters on possible roles of the three fields of volume expansion associated respectively with the energy $E_{dist}^{cluster}+V_{pot}^{cluster}$, with the charge of curvature Q_θ and with the charge of rotation Q_λ of a singularity or of a localized cluster of singularities.

On the dynamic field of perturbations of expansion due to singularities with a large density of energy or a large charge of rotation in the vicinity of the critical expansion

If the background field of expansion of the lattice increases or decreases so as to go through the critical value τ_{0cr}, there appears a range where we cannot have static solutions to the equation of Newton *(18.12)*, which means that it must necessarily appear a dynamic solution that transforms the static field from one range $\tau_0 < \tau_{0cr}$ to the other $\tau_0 > \tau_{0cr}$, and vice versa. The non-existence of a static solution to the differential equation *(18.12)* when the energy density and/or rotational charges of singularity become too important is mathematically quite similar to the fact that we have already seen in section 12.9 in the case of the *Frank-Read sources*, when there were no more static solutions to the deformation of a dislocation string anchored when the stress exceeded a certain critical limit, and we will see in the following parts that the appearance of pure dynamic solutions for the expansion of disturbance field actually have a very close relationship with quantum physics!

On the possibility for a localized singularity to behave like a statical black hole in the vicinity of the critical expansion

A singularity possessing a large value of the energy, of the curvature charge or of the rotational charge, can behave like a statical black hole in the vicinity of the critical value of expansion τ_{0cr}, and the existence of such a behavior is subjected to the following conditions:

- in the case where a singularity or a cluster of given energies $E_{dist}^{cluster} + V_{pot}^{cluster}$

$$\left(E_{dist}^{cluster} + V_{pot}^{cluster}\right) R_{cluster} > 48\pi K_1 R_\infty^4 \quad \Leftrightarrow \quad \tau_0 < \tau_{0cr} \tag{22.112}$$

- in the case of a singularity or a cluster of charges of curvature Q_θ

$$\left|Q_\theta\right| R_{cluster} > \frac{20\pi\left|K_0 - 2K_1(1+\tau_0) - 4K_2/3\right| R_\infty^2}{K_2} \quad \Leftrightarrow \quad \begin{cases} Q_\theta < 0 \ \& \ \tau_0 > \tau_{0cr} \\ Q_\theta > 0 \ \& \ \tau_0 < \tau_{0cr} \end{cases} \tag{22.113}$$

- in the case of a singularity or cluster of charges of rotation Q_θ

$$\frac{Q_\lambda^{\ 2}}{R_{cluster}^4} > \frac{8\pi^2 K_1}{\left(K_2 + K_3\right)} \quad \Leftrightarrow \quad \tau_0 > \tau_{0cr} \tag{22.114}$$

On the possible analogy between the vacancy singularities and the black holes and between the interstitial singularities and the neutron stars

The macroscopic interstitial is the perfect anti-singularity of the macroscopic gap if $N_I = N_L$ since the combination of the two singularities completely restores the perfect lattice. But there is a huge difference between these two singularities, since their formation energies are very different and the macroscopic gap becomes a black hole as soon as $\tau_0 > 1$ in the cosmological system, whereas the macroscopic interstitial can never become a black hole !

Considering that these two topological objects may form by gravitational collapse of clusters of singularities, we find, by analogy, the formation of black holes and neutron stars by gravitational collapse! But if we have sufficient initial mass of the cluster, a condition to achieve a gravitational collapse, in our theory it would not be the initial mass of the cluster which determines the evolution of a black hole or a neutron star but *the very nature of the initial cluster* !

If we admit the following conjecture:

Conjecture 8: *the singularities of vacancy type correspond by analogy to anti-matter and the singularities of interstitial type to matter* (22.115) black holes would then be residues of collapsed clusters of antimatter while neutron stars of collapsed clusters of matter.

In this analogy, the black holes by their constitution of "vacancy holes", can retain no memory of the initial mass of singularities from which they come, apart from the number of gaps, ie the number of missing lattice sites in the initial singularities, that remains invariant. On the other hand, neutron stars, by their very constitution of "interstitials" of non-coherent pieces of lattice could retain, not only the amount of interstitials, that is to say the number of excess lattice sites in the initial interstitial singularities, but also the memory of the rotational angular momentum of the initial cluster of interstitial singularities from which they come, in the form of a very fast rotation of the cluster, which correspond well with huge rotation speeds observed in the case of neutron stars, also called pulsars because they emit electromagnetic pulses at a fixed frequency due to their fast rotation!

Chapter 23

"Gravitational" properties of topological loops in the cosmic lattice

In this chapter , we use the results obtained in previous chapters to address in detail the "gravitational" properties of elementary loop singularities: the loops of twist disclination *(TL)*, the prismatic edge dislocation loops *(EL)* and the mixed dislocation loops by sliding *(ML)*. We will search for external disturbances of the expansion field and deduce several extremely interesting consequences, including the existence, in the case of the edge dislocation loop, of an equivalent gravitational curvature mass that is higher than the inertial mass, which may even be negative in the case of interstitial loops, a result which is very surprising and promising!

23.1 – The twist disclination loop (TL)

The twist disclination loop *(TL)* has already been described in detail in section 19.7. Regarding its "gravitational" properties, ie its long-range external field of perturbation of expansion, one can directly deduce them from *(22.63)*

$$\begin{cases} \tau^{TL}_{ext\ LR}(r) = \tau^{q_{\lambda TL}}_{ext\ LR}(r) \cong -\dfrac{1}{2\pi\left(K_0 - 2K_1(1+\tau_0) - 4K_2/3\right)R_\infty^2} E^{TL}_{dist} \dfrac{1}{r} \\ \\ \tau^{TL}_{ext\ SR}(r) = \tau^{q_{\lambda TL}}_{ext\ SR}(r) \cong \dfrac{1}{2\pi\left(K_0 - 2K_1(1+\tau_0) - 4K_2/3\right)} \dfrac{R_{TL} E^{TL}_{dist}}{2\pi\zeta_{TL}} \dfrac{1}{r^4} \end{cases} \tag{23.1}$$

As these fields are perturbations of the volume expansion, they correspond in our analogy to the real world to *gravitational fields, acting at short range (SR) and at range (LR)*! As a matter of fact, it is very interesting here to replace in these expressions the energy of distortion E^{TL}_{dist} by the mass of inertia M^{TL}_0 of the loop, by using the expression $E^{TL}_{dist} = M^{TL}_0 c_t^2$ of the *relation of Einstein (19.96)*

$$\begin{cases} \tau^{TL}_{ext\ LR}(r) \cong -\dfrac{c_t^2}{2\pi\left(K_0 - 2K_1(1+\tau_0) - 4K_2/3\right)R_\infty^2} \dfrac{M^{TL}_0}{r} \\ \\ \tau^{TL}_{ext\ SR}(r) \cong \dfrac{c_t^2}{2\pi\left(K_0 - 2K_1(1+\tau_0) - 4K_2/3\right)} \dfrac{R_{TL} M^{TL}_0}{2\pi\zeta_{TL}} \dfrac{1}{r^4} \end{cases} \tag{23.2}$$

In this form, the field of long-range expansion disturbances *(LR)* only depends on the inertial mass of the loop, and does not depend on the size of the loop, which further strengthens our analogy between this field of perturbation of expansion and a natural gravitational field!

On the "gravitational" energy due to the perturbations of expansion of the twist loop

The fields of perturbation of expansion that we just found possess an energy which we should compare to the elastic energy E_{dist}^{TL} of the loop. The global energy of the field of perturbation of expansion due to the elastic energy of the loop is deduced thanks to relations *(22.73)*

$$E_{grav}^{(q_{\lambda TL})} = \frac{K_1}{\left(K_0 - 2K_1(1+\tau_0) - 4K_2/3\right)^2}\left[\frac{\left(E_{dist\,torus}^{TL}\right)^2}{V_{TL}} + \frac{\left(K_2+K_3\right)^2}{80\pi^3}\frac{q_{\lambda TL}^4}{R_{TL}^5}\right] \tag{23.3}$$

By introducing here the values of $E_{dist\,torus}^{TL}$ and of $q_{\lambda TL}$ obtained respectively from *(19.88)* and *(19.77)*, one has

$$E_{grav}^{(q_{\lambda TL})} \cong \frac{K_1\left(K_2+K_3\right)^2}{\left(K_0 - 2K_1(1+\tau_0) - 4K_2/3\right)^2}\left[\frac{3\zeta_{TL}^2}{16\pi} + \frac{\pi}{80}\right]\frac{\vec{B}_{TL}^4}{R_{TL}} \tag{23.4}$$

As $\zeta_{TL} = \ln\left(A_{TL}R_{TL}/a\right) > 1$, one obtains the approximation

$$E_{grav}^{(q_{\lambda TL})} \cong \frac{K_1\left(K_2+K_3\right)^2}{\left(K_0 - 2K_1(1+\tau_0) - 4K_2/3\right)^2}\frac{3\zeta_{TL}^2\vec{B}_{TL}^4}{16\pi R_{TL}} \tag{23.5}$$

We compare this energy with the elastic energy, by calculating the ratio and introducing the value $E_{dist\,torus}^{TL}$ taken from *(19.88)*

$$\frac{E_{grav}^{(q_{\lambda TL})}}{E_{dist}^{TL}} \cong \frac{K_1\left(K_2+K_3\right)}{\left(K_0 - 2K_1(1+\tau_0) - 4K_2/3\right)^2}\frac{3\zeta_{TL}\vec{B}_{TL}^2}{8\pi R_{TL}^2} \tag{23.6}$$

By applying the conditions *(19.58)* of the perfect cosmic lattice, one obtains approximatively

$$\frac{E_{grav}^{(q_{\lambda TL})}}{E_{dist}^{TL}} \cong \frac{3}{8\pi}\frac{K_1}{K_0}\zeta_{TL}\left(\frac{\left|\vec{B}_{TL}\right|}{R_{TL}}\right)^2 << 1 \tag{23.7}$$

because $K_1 << K_0$.

We deduce consequently that the "gravitational" energy $E_{grav}^{(q_{\lambda TL})}$ of the twist disclination loop due to its charge of rotation $q_{\lambda TL}$ is perfectly negligible vis-à-vis it's energy in the perfect cosmic lattice.

The essential properties of the twist disclination loop

In table 23.1, we have presented the important properties that we have deduced so far for a twist disclination loop in a perfect cosmic lattice, namely it's charge, it's inertial mass, it's fields of rotation and perturbation of expansion at short distance, it's kinetic and elastic energies and finally its relativistic behavior. We note that the inertial mass of the loop M_0^{TL} not only controls the dynamic properties of the loop, such as its kinetic energy E_{kin}^{TL}, but that it is also responsible for the gravitational fields $\tau_{ext\,LR}^{TL}(r)$ and $\tau_{ext\,SR}^{TL}(r)$ of external perturbations of expansion.

The 'gravitational' energy $E_{grav}^{(q_{\lambda TL})}$ of the fields of perturbations of expansion due to the elastic energies associated to the rotational charge $q_{\lambda BV}$ of the *TL* is perfectly negligible vis-à-vis the elastic energy E_{dist}^{BV} of the loop. This "gravitational" energy does not figure in the table of essential properties.

With regards to the field of perturbation of expansion within the torus encompassing the loop, and the energy which is associated with this internal field, we will revisit it later, when we look at the problem of spin and magnetic moment of the twist disclination loop *(TL)*.

Table 23.1 - Essential properties of a twist disclination loop (TL)

$$\begin{cases} q_{\lambda TL} = -\pi R_{TL}^2 \boldsymbol{\Omega}_{TL} = 2\pi R_{TL} \Lambda_{TL} = -\pi R_{TL} \vec{\underline{B}}_{TL} \vec{t} \\ M_0^{TL} = \dfrac{2}{c_t^2}(K_2+K_3)\zeta_{TL} R_{TL} \Lambda_{TL}^2 = \dfrac{1}{2c_t^2}(K_2+K_3)\zeta_{TL} R_{TL}^3 \boldsymbol{\Omega}_{TL}^2 \end{cases}$$

$$\begin{cases} \vec{\omega}_{ext}^{TL}(\vec{r}) = \dfrac{q_{\lambda TL}}{4\pi} \dfrac{\vec{r}}{r^3} \\ \tau_{ext\,LR}^{TL}(\vec{r}) \cong -\dfrac{c_t^2}{2\pi\left(K_0 - 2K_1(1+\tau_0) - 4K_2/3\right) R_\infty^2} \dfrac{M_0^{TL}}{r} \\ \tau_{ext\,SR}^{TL}(r) \cong \dfrac{c_t^2}{2\pi\left(K_0 - 2K_1(1+\tau_0) - 4K_2/3\right)} \dfrac{R_{TL} M_0^{TL}}{2\pi\zeta_{TL}} \dfrac{1}{r^4} \end{cases}$$

$$\begin{cases} E_{\mathbf{v}}^{TL} = \underbrace{\dfrac{1}{\gamma_t}\left(1 - \dfrac{\mathbf{v}^2}{2c_t^2}\right) E_{dist}^{TL}}_{E_{\mathbf{v}}^{dist}} + \underbrace{\dfrac{1}{\gamma_t}\dfrac{1}{2} M_0^{TL} \mathbf{v}^2}_{E_{\mathbf{v}}^{kin}} = \dfrac{E_{dist}^{TL}}{\gamma_t} = \dfrac{M_0^{TL} c_t^2}{\gamma_t} \\ E_{dist}^{TL} \cong M_0^{TL} c_t^2 \quad if \quad \mathbf{v} = 0 \\ E_{cin}^{TL} \cong \dfrac{1}{2} M_0^{TL} \mathbf{v}^2 \quad if \quad \mathbf{v} \ll c_t \end{cases}$$

23.2 – The prismatic edge dislocation loop (EL)

The prismatic edge dislocation loop *(EL)* which we have described in section 19.9 possesses a curvature charge $q_{\theta EL}$ given by relation *(19.101)*, which makes it a fundamental brick of charge of curvature of the lattice in our analogy with the real world. This charge is responsible for an external divergent flexion field, analogous to a field of geometrical curvature, described by relation *(19.103)*

Knowing the elastic energy *(19.102)* of the *EL*, we can use relation *(22.17)* to calculate the external field $\tau_{ext}^{(E)}(r)$ of perturbations of expansion associated with the elastic energy E_{dist}^{EL} of the edge dislocation, if we neglect here its potential energy V_{pot}^{EL}

$$\tau_{ext}^{(E)}(r) \cong -\frac{K_3}{2\pi\left(K_0 - 2K_1(1+\tau_0) - 4K_2/3\right)R_\infty^2}\left(\frac{K_2}{K_3}\right)^2 \frac{\zeta_{EL} R_{EL} \vec{B}_{EL}^2}{r} \tag{23.8}$$

We can also use relation *(22.34)* to calculate the external field $\tau_{ext}^{(q_{\theta EL})}(r)$ of perturbations of expansion associated with the rotational charge $q_{\theta EL}$ of the loop

$$\tau_{ext}^{(q_{\theta EL})}(r) \cong \frac{K_2}{5\left(K_0 - 2K_1(1+\tau_0) - 4K_2/3\right)R_\infty^2} \frac{R_{EL}^2 \vec{n}\vec{B}_{EL}}{r} \tag{23.9}$$

The total field of perturbations of expansion can then be written as

$$\tau_{ext}^{EL}(r) \cong \frac{K_2}{5\left(K_0 - 2K_1(1+\tau_0) - 4K_2/3\right)R_\infty^2}\left[-\frac{5}{2\pi}\left(\frac{K_2}{K_3}\right)\zeta_{EL}\vec{n}\vec{B}_{EL} + R_{EL}\right]\frac{R_{EL}\vec{n}\vec{B}_{EL}}{r} \tag{23.10}$$

By using conjecture 4 *(17.41)*, we can emit here the plausible hypothesis that

Hypothesis 1: $K_2 \ll K_3$ and $\zeta_{EL}\left|\vec{B}_{EL}\right| = \left|\vec{B}_{EL}\right| \ln\left(\frac{A_{EL} R_{EL}}{a}\right) \leq R_{EL}$ (23.11)

which allows us to say that the first term within the brackets in *(23.10)* is most likely largely smaller than the second term. Thus, in the case of the *EL*, the field of perturbations of expansion due to the charge of curvature is larger than the field of perturbation of expansion due to the elastic energy of distortion of the loop, contrary to the *EL*!

By introducing the *inertial mass* M_0^{EL} *of the edge loop (19.106)*

$$M_0^{EL} \cong \frac{E_{dist}^{EL}}{c_t^2} \cong \left(\frac{K_2}{K_3}\right)^2 \frac{K_3}{c_t^2} \zeta_{EL} R_{EL} \vec{B}_{EL}^2 \tag{23.12}$$

as well as an equivalent curvature mass $M_{curvature}^{EL}$ which is worth

$$M_{curvature}^{EL} = -\frac{2\pi K_2}{5c_t^2} R_{EL}^2 \vec{n}\vec{B}_{EL} \tag{23.13}$$

and which can be positive or negative

$$\begin{cases} M_{curvature}^{EL} > 0 \text{ if vacancy loop } (\vec{n}\vec{B}_{EL} < 0) \\ M_{curvature}^{EL} < 0 \text{ if interstitial loop } (\vec{n}\vec{B}_{EL} > 0) \end{cases} \tag{23.14}$$

We can write the field of gravitation as

$$\tau_{ext\,LR}^{EL}(r) \cong -\frac{c_t^2}{2\pi\left(K_0 - 4K_2/3 - 2K_1(1+\tau_0)\right)R_\infty^2} \frac{M_{curvature}^{EL} + M_0^{EL}}{r} \tag{23.15}$$

According to hypothesis *(23.11)*, the equivalent mass of curvature in this expression of the field of gravitation obeys the following relation

$$\left|M_{curvature}^{EL}\right| \gg M_0^{EL} \tag{23.16}$$

Furthermore the equivalent mass of curvature can be positive or negative. These two facts will have very remarkable implications as we will see!

On the gravitational energy due to the perturbation of expansion of the edge loop

The field of perturbation of expansion that we just found have an energy which we will want to compare to the elastic energy E_{dist}^{EL} of the *EL*. The energy of the field of perturbation of expansion due to the elastic energy of the loop is deduced thanks to relations *(22.19) and (22.14)*

$$E_{grav}^{(E)} \cong K_1 \frac{3}{4\pi} \frac{1}{\left(4K_2/3+2K_1(1+\tau_0)\right)-K_0\right)^2} \frac{\left(E_{dist}^{EL}\right)^2}{R_{EL}^3} \tag{23.17}$$

Comparing this energy with the elastic energy and taking the ratio and introducing the value E_{dist}^{EL} taken from *(19.102)* and conditions *(19.58)* of the perfect lattice

$$\frac{E_{grav}^{(E)}}{E_{dist}^{EL}} \cong \frac{3}{4\pi} \frac{K_1 K_3\left(\dfrac{K_2}{K_3}\right)^2}{\left(4K_2/3+2K_1(1+\tau_0)\right)-K_0\right)^2} \zeta_{EL}\left(\frac{\vec{B}_{EL}}{R_{EL}}\right)^2 \cong \frac{3}{4\pi}\frac{K_1 K_2^2}{K_0^3}\zeta_{EL}\left(\frac{\vec{B}_{EL}}{R_{EL}}\right)^2 <<< 1 \tag{23.18}$$

as $K_1 K_2^2 <<< K_0^3$.

We deduce that the gravitational energy $E_{grav}^{(E)}$ of the *EL* due to an elastic energy E_{dist}^{EL} is perfectly negligible vis-a-vis the elastic energy of the defect in the perfect lattice.

Let's calculate the energy associated to the field of perturbation of expansion due to the charge of curvature of the *EL*. By using relation *(22.39)*, we obtain the energy of the field of gravitation associated with the curvature charge

$$E_{grav}^{(q_{\theta EL})} \cong K_1 \frac{1}{28\pi} \frac{K_2^{\,2}}{\left(K_0-2K_1(1+\tau_0)-4K_2/3\right)^2} R_{EL} q_{\theta EL}^2 \tag{23.19}$$

Let's compare this energy with elastic energy, by taking the ratio and introducing the value of E_{dist}^{EL} taken from *(19.102)* and the value of $q_{\theta EL}$ taken from *(19.101)*

$$\frac{E_{grav}^{(q_{\theta EL})}}{E_{dist}^{EL}} \cong \frac{K_1 \dfrac{1}{28\pi} \dfrac{K_2^{\,2}}{\left(K_0-2K_1(1+\tau_0)-4K_2/3\right)^2} R_{EL}\left(2\pi\vec{n}\vec{B}_{EL}\right)^2}{\left(\dfrac{K_2}{K_3}\right)^2 \dfrac{K_3}{c_t^2}\zeta_{EL} R_{EL}\vec{B}_{EL}^2} \cong \frac{\pi}{7}\frac{K_1}{K_0}\frac{1}{\zeta_{EL}} << 1 \tag{23.20}$$

since $K_1 << K_0$. We deduce again that the gravitational energy $E_{grav}^{(q_{\theta EL})}$ of the edge dislocation loop due to the curvature charge $q_{\theta EL}$ is negligible compared to the elastic energy E_{dist}^{EL} in the perfect lattice.

The essential properties of the edge dislocation loop

In table 23.2 , we show all the important properties we deduced to date for an *EL* in a perfect cosmic lattice, namely its bending charge, its mass-inertia, its gravitational mass equivalent, its bending fields and disruptions to long-range expansion, its elastic and kinetic energy, and ultimately his relativistic behavior. It shows among other things that it is its elastic energy E_{dist}^{EL} and inertial mass M_0^{EL} that controls its dynamic properties, such as its kinetic energy E_{kin}^{EL}, but that it is its equivalent gravitational mass $M_{curvature}^{EL}$ of curvature which essentially controls its external gravitational field disturbances in the far field.

As the gravitational energies $E_{grav}^{(E)}$ and $E_{grav}^{(q_{\theta EL})}$ of the fields of perturbation of expansion due to the elastic energies E_{dist}^{EL} and the curvature charge $q_{\theta EL}$ of the *EL* are negligible compared to the elastic energy E_{dist}^{EL} of the loop, these gravitational energies do not feature in the table of

essential properties.
Also, the equivalent gravitational mass of curvature $M^{EL}_{curvature}$ is not only much larger than the inertial mass M^{EL}_0, but it can even be negative in the case where the edge loops are interstitials! This is a striking result as it corresponds to the existence of a possible negative gravitational field, and is a new result compared to the General Relativity of Einstein! The possible consequences of this striking result will be explored in more detail in the rest of the book!

Table 23.2 - Essential properties of the prismatic edge dislocation loop (EL)

$$\left\{\begin{array}{l} q_{\theta EL} = -2\pi\vec{n}\left(\vec{t}\wedge\vec{\Lambda}_{EL}\right) = 2\pi\vec{\Lambda}_{EL}\vec{m} = -2\pi\vec{n}\vec{B}_{EL} \begin{cases} >0 \text{ if vacancy loop } (\vec{n}\vec{B}_{EL}<0) \\ <0 \text{ if interstitial loop } (\vec{n}\vec{B}_{EL}>0) \end{cases} \\ M_0^{EL} = \dfrac{E^{EL}_{dist}}{c_t^2} \cong \left(\dfrac{K_2}{K_3}\right)^2 \dfrac{K_3}{c_t^2}\zeta_{EL}R_{EL}\vec{\Lambda}^2_{EL} = \left(\dfrac{K_2}{K_3}\right)^2 \dfrac{K_3}{c_t^2}\zeta_{EL}R_{EL}\vec{B}^2_{EL} \\ M^{EL}_{curvature} = -\dfrac{2\pi K_2}{5c_t^2}R^2_{EL}\vec{n}\vec{B}_{EL} = \dfrac{K_2}{5c_t^2}R^2_{EL}q_{\theta EL} \begin{cases} >0 \text{ if vacancy loop } (q_{\theta EL}>0) \\ <0 \text{ if interstitial loop } (q_{\theta EL}<0) \end{cases} \\ \left|M^{EL}_{curvature}\right| >> M_0^{EL} \end{array}\right.$$

$$\left\{\begin{array}{l} \vec{\chi}^{EL}_{ext}(\vec{r}) = \dfrac{q_{\theta EL}}{4\pi}\dfrac{\vec{r}}{r^3} \\ \tau^{EL}_{ext}(r) \cong -\dfrac{c_t^2}{2\pi\left(K_0-4K_2/3-2K_1(1+\tau_0)\right)R_\infty^2}\dfrac{M^{EL}_{curvature}+M_0^{EL}}{r} \end{array}\right.$$

$$\left\{\begin{array}{l} E_{\mathbf{v}}^{EL} = \underbrace{\dfrac{1}{\gamma_t}\left(1-\dfrac{\mathbf{v}^2}{2c_t^2}\right)E^{EL}_{dist}}_{E_{\mathbf{v}}^{dist}} + \underbrace{\dfrac{1}{\gamma_t}\dfrac{1}{2}M_0^{EL}\mathbf{v}^2}_{E_{\mathbf{v}}^{kin}} = \dfrac{E^{EL}_{dist}}{\gamma_t} = \dfrac{M_0^{EL}c_t^2}{\gamma_t} \\ E^{EL}_{dist} \cong M_0^{EL}c_t^2 \quad if \quad \mathbf{v}=0 \\ E^{EL}_{kin} \cong \dfrac{1}{2}M_0^{EL}\mathbf{v}^2 \quad if \quad \mathbf{v} << c_t \end{array}\right.$$

23.3 – The mixed sliding dislocation loop (ML)

The mixed dislocation loop *(ML)* which we described in section 19.10 does not possess a charge of rotation nor a charge of curvature but instead a dipolar moment $\vec{\omega}^{ML}_{dipolar}(r,\theta,\varphi)$ of the field of rotation, analogous to a dipolar electric moment. Knowing the elastic energy of distortion *(19.110)* of the *ML* in the perfect lattice, we can use relation *(22.17)* to calculate the external field $\tau_{ext}^{E}(r)$ of perturbations associated with the elastic energy E_{dist}^{ML}, by neglecting here the potential energy V_{pot}^{ML}. Also, it becomes interesting to replace in the expression of $\tau_{ext}^{E}(r)$ the energy of distortion E_{dist}^{ML} by the inertial mass M_0^{ML} of the loop by using the relation $E_{dist}^{ML} = M_0^{ML} c_t^2$ *(19.114)*

$$\begin{aligned} \tau_{ext}^{E}(r) &\cong -\frac{1}{2\pi\left(K_0 - 4K_2/3 - 2K_1\left(1+\tau_0\right)\right)R_\infty^2}\frac{E_{dist}^{ML}}{r} \\ &\cong -\frac{c_t^2}{2\pi\left(K_0 - 4K_2/3 - 2K_1\left(1+\tau_0\right)\right)R_\infty^2}\frac{M_0^{ML}}{r} \end{aligned} \tag{23.21}$$

Table 23.3 - Essential properties of the mixed dislocation loop (ML)

$$\left\{ M_0^{ML} \cong \frac{K_2 + K_3}{4c_t^2}\zeta_{ML} R_{ML}\, \vec{B}_{ML}^2 \right.$$

$$\left\{ \begin{array}{l} \exists \text{ an external dipolar field of rotation } \vec{\omega}_{dipolar}^{ML}(r,\theta,\varphi) \\ \tau_{ext}^{ML}(r) \cong -\dfrac{c_t^2}{2\pi\left(K_0 - 4K_2/3 - 2K_1\left(1+\tau_0\right)\right)R_\infty^2}\dfrac{M_0^{ML}}{r} \end{array} \right.$$

$$\left\{ \begin{array}{l} E_{\mathbf{v}}^{ML} = \underbrace{\dfrac{1}{\gamma_t}\left(1 - \dfrac{\mathbf{v}^2}{2c_t^2}\right)E_{dist}^{ML}}_{E_{\mathbf{v}}^{dist}} + \underbrace{\dfrac{1}{\gamma_t}\dfrac{1}{2}M_0^{ML}\mathbf{v}^2}_{E_{\mathbf{v}}^{kin}} = \dfrac{E_{dist}^{ML}}{\gamma_t} = \dfrac{M_0^{ML}c_t^2}{\gamma_t} \\ E_{dist}^{ML} \cong M_0^{ML} c_t^2 \quad \textit{if} \quad \mathbf{v} = 0 \\ E_{cin}^{ML} \cong \dfrac{1}{2} M_0^{ML}\mathbf{v}^2 \quad \textit{ifi} \quad \mathbf{v} << c_t \end{array} \right.$$

On the gravitational energy due to the perturbations of expansion of the mixed loop

The fields of perturbation of expansion that we just found possess an energy which we want to compare to the elastic energy E_{dist}^{ML} of the loop. The energy of the field of perturbation of expansion due to the elastic energy of the loop is deduced thanks to relations *(22.19) and (22.14)*

$$E_{grav}^{(E)} \cong K_1 \frac{3}{4\pi}\frac{1}{\left(4K_2/3 + 2K_1(1+\tau_0)) - K_0\right)^2}\frac{\left(E_{dist}^{ML}\right)^2}{R_{ML}^3} \tag{23.22}$$

Let's compare this energy with the elastic energy, by taking the ratio and introducing the value of E_{dist}^{ML} taken from *(19.110)*

$$\frac{E_{grav}^{(E)}}{E_{dist}^{ML}} \cong \frac{3}{16\pi} \frac{K_1(K_2+K_3)}{\left(4K_2/3+2K_1(1+\tau_0)\right)-K_0\right)^2} \zeta_{ML}\left(\frac{\vec{B}_{ML}}{R_{ML}}\right)^2 \cong \frac{3}{16\pi}\frac{K_1}{K_0}\zeta_{ML}\left(\frac{\vec{B}_{ML}}{R_{ML}}\right)^2 << 1 \qquad (23.23)$$

since $K_1 << K_0$.
We thus deduce that the gravitational energy $E_{grav}^{(E)}$ of the *ML* due to its elastic energy E_{dist}^{ML} is perfectly negligible compared to the elastic energy of the perfect lattice.

The essential properties of the mixed dislocation loop

In table 23.3, we show the properties of the *ML* in the perfect lattice. While this loop does not possess neither charge of rotation nor charge of curvature generating an external field of rotation or of flexion in the far field, it is equipped with a dipolar external field $\vec{\omega}_{dipolar}^{ML}(r,\theta,\varphi)$ in the near field, which is analogous to a dipolar electric field. The external field of perturbation of expansion is due to the elastic energy of distortion of the loop and depends on the inertial mass M_0^{ML}.
Also as the gravitational energy $E_{grav}^{(E)}$ of the perturbation field of expansion due to the elastic energy E_{dist}^{ML} of the *ML* is negligible compared to the elastic energy E_{dist}^{ML} of the loop, this gravitational energy does not figure in the table of the essential properties.

23.4 – The various properties of the elementary topological loops

On the negligible energy of the field of perturbation of expansion linked to the loops

We have shown in this chapter the bulk of the essential properties of three elementary loops that we can find in the perfect lattice and we have established the expressions for the field of external expansion which corresponds to gravitational fields and will play an important role in the rest of the book!
We have also shown that $E_{grav}^{(E)}, E_{grav}^{(q_{\lambda TL})}, E_{grav}^{(q_{\theta EL})}$ of the gravitational fields associated with the elastic energies, the charge of rotation and the charge of curvature of elementary loops are negligible compared to the elastic energies associated with the loops, and thus we can ignore them in our calculations.
We thus have addressed the questions we asked in chapter 19 concerning the field of perturbation of expansion of the elementary loops.

Can the elementary topological loops be black holes?

We can ask ourselves whether the elementary topological loops can become black holes. For that, we apply conditions *(15.3)* to the fields of expansion $\tau_{ext\,LR}^{TL}(r)$, $\tau_{ext}^{EL}(r)$ and $\tau_{ext}^{ML}(r)$ respectively.
By applying the conditions to the twist disclination loop in the case $\tau_0 << \tau_{0cr}$, we obtain the following condition for this loop to become a black hole

$$\left|\vec{\underset{\sim}{B}}_{TL}\right| \geq 2R_\infty\sqrt{\frac{2\pi}{\zeta_{TL}}} \qquad \Leftrightarrow \qquad \tau_0 << \tau_{0cr} \qquad (23.24)$$

which would imply that the pseudo-vector of Burgers of the loop be on the order of the radius of the lattice, which of course, is non-sensical.
The same holds for the mixed dislocation loop since the condition for it to be a black hole becomes

$$\left|\vec{B}_{ML}\right| > 4R_\infty\sqrt{\frac{\pi}{\zeta_{ML}}} \qquad \Leftrightarrow \qquad \tau_0 << \tau_{0cr} \tag{23.30}$$

In the case of the edge dislocation loop, the condition is expressed differently since the gravitational mass of gravitation becomes $M_{curvature}^{EL} + M_0^{EL}$. Since $\left|M_{curvature}^{EL}\right| >> M_0^{EL}$, we obtain

$$R_{EL}\left|\vec{B}_{EL}\right| \geq 10\frac{K_0}{K_2}R_\infty^2 \qquad \Leftrightarrow \qquad \tau_0 << \tau_{0cr} \tag{23.31}$$

But since the module K_2 must be smaller than K_0 according to conjecture 4, this condition can never be satisfied.
As a consequence, it is clear that the three elementary loops *cannot be black holes* in the domain $\tau_0 << \tau_{0cr}$.

On the remarkable properties of the elementary topological loops

In the analogy between our theory and the great theories of physics, the three types of elementary we have discussed in the previous chapter have remarkable properties we will enumerate here:

- they are respectively the elementary building block of the *electrical charge*, of the *curvature charge* and the *electrical dipolar moment*, from which it should be possible to form dispirations, by combining several of these loops in order to find topological singularities which are analogous to the elementary particles of the 'real world',

- their rest energy and their kinetic energy are essentially confined in the *toroidal field around the loops*,

- as the energies associated with the gravitational fields of perturbation of expansion are perfectly negligible, they satisfy the *relation of Einstein* $E_{dist}^{loop} \cong M_0^{loop}c_t^2$, which is obtained in our theory as a purely classical property of the topological singularities within a lattice, without any recourse to a 'relativistic principle',

- they perfectly satisfy *special relativity*, with an original explanation of relativistic energy $E_{\mathbf{v}}^{loop} = E_{\mathbf{v}}^{dist} + E_{\mathbf{v}}^{kin} = E_{dist}^{loop} / \gamma_t = M_0^{loop} c_t^2 / \gamma_t$ as the sum of a relativistic energy of elastic distortion and a term of relativistic kinetic energy,

- they satisfy a *relativistic dynamic equation* given by relations *(20.50)* and *(20.51)*,

- the twist disclination loop *(TL)*, carries a rotational charge equivalent to the electrical charge, which satisfy the *Maxwell's equations* (table 17.1) and the *Lorentz force (20.74)*,

- the three types of loops exhibit a field of disturbance of volume expansion in the far field, which is analogous to a *gravitational field* which decreases in $1/r$, *and* which *only depends on the gravitational mass* $M_0^{loop} + M_{curvature}^{loop}$ of the loops. The gravitational mass is then composed on the inertial mass and the equivalent curvature mass of the loops, without depending directly on the size of the loops R_{loop},

- the *gravitational masses* of the loops of twist disclination and mixed dislocation are strictly equal to their inertial masses while the gravitational mass of the prismatic edge is composed of the inertial mass and the curvature mass of the loop, with the curvature mass a lot larger than the inertial mass, and which can even be negative in the case of interstitial loops,

- the gravitational mass of the edge dislocation loop *(EL)* contains two terms. The first term which dominates is the curvature mass $M^{EL}_{curvature}$, and is positive or negative depending on whether the loop is of vacancy or interstitial nature. The second term, the mass of inertia M_0^{EL} is always positive. This means that the global gravitational mass $M^{EL}_{curvature} + M_0^{EL}$ is not symmetric between a loop of vacancy type and a loop of interstitial type. There is here a *weak asymmetry on the absolute value of the gravitation mass between the loop of interstitial type and its anti-loop of vacancy type,* which is expressed by the fact that

$$M^{EL(I)}_{curvature} + M_0^{EL(I)} > 0 \quad ; \quad M^{EL(i)}_{curvature} + M_0^{EL(i)} < 0 \quad ; \quad \left| M^{EL(i)}_{curvature} + M_0^{EL(i)} \right| \tilde{<} M^{EL(I)}_{curvature} + M_0^{EL(I)} \qquad (23.27)$$

- all these properties are perfectly analogous to the fundamental properties of elementary particles of the real world, except for the *gravitational mass* $M^{EL}_{curvature} + M_0^{EL}$ of the prismatic edge dislocation loops, which have a strong analogy with the neutrinos. This very special property of the loops of prismatic edge dislocation will be discussed in the following chapters, in which we will look at the gravitational interaction of the loops.

Chapter 24

"Gravitational" interactions of singularities composed of twist disclination loops

In this chapter, we study in detail the gravitational interactions of twist disclination loops (TL), which will yield a strong analogy with Newtonian gravitation in the far-field but that will exhibit differences in the near-field. We will also exhibit a dependence of the constant of gravitation on the volume expansion of the lattice.
Next, we focus on the *Maxwell* formulation of the equations of evolution presented in chapter 17, which corresponded to the expression of the local laws of physics, such as electromagnetism, as seen by an imaginary ***Grand Observer (GO)***. In this chapter we focus on a hypothetical local observer we call the ***Homo Sapiens observer (HS)*** which would be linked to a local framework, and himself composed of clustered singularities of the lattice. This ***HS*** observer only knows of local measures with local rods and local clocks constituting his local reference frame. It will be shown that this makes for him the Maxwell equations to become invariant with respect to volume expansion. There then appears a relativistic notion of time for the local ***HS*** observers, which will present a strong analogy with the time in the theory of *General relativity of Einstein*. We will discuss in detail the analogies, and the differences and advantages when compared to General Relativity.

24.1 – Long range «gravitational» interaction of a cluster of twist disclination loops (TL)

On the dependance of inertial mass and curvature charge on local expansion

In the previous chapter, we have shown that the inertial mass of twist loops is much greater than the inertial mass of edge loop and mixed loops, so that the perturbations of the field of expansion will be mostly due to the twist disclination loops. We will then first analyze the field of perturbation of expansion at long range due to those.
For an TL, the inertial mass depends on the square of the charge $q_{\lambda TL}$ of rotation

$$\begin{cases} q_{\lambda BV} = -\pi R_{BV}^2 \boldsymbol{\Omega}_{BV} = 2\pi R_{BV} \Lambda_{BV} = -\pi R_{BV} \underset{\frown}{\vec{\boldsymbol{B}}}_{BV} \vec{t} \\ M_0^{BV} = \dfrac{1}{2\pi^2 c_t^2}(K_2+K_3)\zeta_{BV} \dfrac{q_{\lambda BV}^2}{4\pi^2 R_{BV}} = \dfrac{1}{2c_t^2}(K_2+K_3)\zeta_{BV} R_{BV}^3 \boldsymbol{\Omega}_{BV}^2 \end{cases} \tag{24.1}$$

If such a loop is in a field of expansion τ which is variable, then the quantities depend on it. We can assume *a priori* that the radius of the loop is linked to the lattice step, so that $R_{TL} = R_{TL0}\,\mathrm{e}^{\tau/3}$, and that the angle of rotation $\boldsymbol{\Omega}_{TL}$ must correspond to an angle satisfying the *symmetry of the lattice*, which means a multiple of $\pi/2$ for a cubic lattice, or a multiple of $\pi/3$

for a hexagonal lattice. It follows that Ω_{TL} should not depend on background expansion. However, we do not know the exact nature of the cosmic lattice, so that, in all generality, we will need to suppose dependencies of R_{TL} and Ω_{TL} as if the loop could have an extension of its radius and a torsion depending on the volume expansion of the lattice

$$\textbf{\textit{Hypothesis 1:}} \quad \begin{cases} R_{TL} = R_{TL0}\,\mathrm{e}^{\alpha_{TL}\tau} \\ \Omega_{TL} = \Omega_{TL0}\,\mathrm{e}^{\beta_{TL}\tau} \end{cases} \tag{24.2}$$

with this hypothesis, the charge and the inertial mass of the TL read

$$\begin{cases} q_{\lambda TL} = -\pi R_{TL0}^2\,\mathrm{e}^{2\alpha_{TL}\tau}\,\Omega_{TL0}\,\mathrm{e}^{\beta_{TL}\tau} = -\pi R_{TL0}^2 \Omega_{TL0}\,\mathrm{e}^{(2\alpha_{TL}+\beta_{TL})\tau} = q_{\lambda TL0}\,\mathrm{e}^{(2\alpha_{TL}+\beta_{TL})\tau} \\ M_0^{TL} = \dfrac{1}{2c_{t0}^2\,\mathrm{e}^{\tau}}\left(K_2+K_3\right)\zeta_{TL} R_{TL0}^3\,\mathrm{e}^{3\alpha_{TL}\tau}\,\Omega_{TL0}^2\,\mathrm{e}^{2\beta_{TL}\tau} = M_{00}^{TL}\,\mathrm{e}^{(3\alpha_{TL}+2\beta_{TL}-1)\tau} \end{cases} \tag{24.3}$$

The far-field perturbations of a cluster of twist loops

The long range field of expansion of the loop is *(23.2)*

$$\tau_{ext\,LR}^{TL}(r) \cong -\frac{c_t^2}{2\pi\left(K_0-4K_2/3-2K_1(1+\tau_0)\right)R_\infty^2}\frac{M_0^{TL}}{r} \tag{24.4}$$

As this field does not depend on the radius of the loop or the charge of rotation of the loop, we can generalize the mass and charge of rotation of the cluster

$$\begin{cases} Q_\lambda^{cluster} = \sum\limits_i q_{\lambda TL(i)} = q_{\lambda TL0(i)}\,\mathrm{e}^{(2\alpha_{TL}+\beta_{TL})\tau} = Q_{\lambda 0}^{cluster}\,\mathrm{e}^{(2\alpha_{TL}+\beta_{TL})\tau} \\ M_0^{cluster} = \sum\limits_i M_{0(i)}^{TL} = \sum\limits_i M_{00(i)}^{TL}\,\mathrm{e}^{(3\alpha_{TL}+2\beta_{TL}-1)\tau} = M_{00}^{cluster}\,\mathrm{e}^{(3\alpha_{TL}+2\beta_{TL}-1)\tau} \end{cases} \tag{24.5}$$

so that

$$\tau_{ext\,LR}^{cluster}(r) \cong -\frac{c_t^2}{2\pi\left(K_0-4K_2/3-2K_1(1+\tau_0)\right)R_\infty^2}\frac{M_0^{cluster}}{r} \tag{24.6}$$

The gravitational force of interaction between two clusters of twist loops

Consider two clusters of TL situated at a distance d one from the other which will interact via their long-range expansion fields. We can calculate the energy of clusters (1) and (2) from their inertial mass

$$E_{(i)}^{cluster}(\tau) = M_{0(i)}^{cluster}c_t^2 = M_{00(i)}^{cluster}c_{t0}^2\,\mathrm{e}^{(3\alpha_{TL}+2\beta_{TL})\tau} \tag{24.7}$$

The two clusters which are separated with distance d interact via their perturbation fields. As we know, the elastic energy of a loop is essentially stored in the vicinity of the loop, and we know that the elastic energy of the cluster is essentially in the heart of the cluster, so that their respective energy is influenced by the presence of the other cluster in the following fashion

$$\begin{cases} E_{(1)}^{cluster} = E_{(1)}^{cluster}(\tau+\tau_{ext\,LR(2)}^{cluster}(d)) = M_{00(1)}^{cluster}c_{t0}^2\,\mathrm{e}^{(3\alpha_{TL}+2\beta_{TL})\left(\tau+\tau_{ext\,LR(2)}^{cluster}(d)\right)} \\ E_{(2)}^{cluster} = E_{(2)}^{cluster}(\tau+\tau_{ext\,LR(1)}^{cluster}(d)) = M_{00(2)}^{cluster}c_{t0}^2\,\mathrm{e}^{(3\alpha_{TL}+2\beta_{TL})\left(\tau+\tau_{ext\,LR(1)}^{cluster}(d)\right)} \end{cases} \tag{24.8}$$

There appears as a result an increase ΔE_{grav} of the energy of two interacting clusters, which is written

$$\Delta E_{grav}(d)=\left[\begin{array}{l}\left(M_{00(1)}^{cluster}c_{t0}^{2}\,\mathrm{e}^{(3\alpha_{TL}+2\beta_{TL})\left(\tau+\tau_{ext\,LR(2)}^{cluster}(d)\right)}-M_{00(1)}^{cluster}c_{t0}^{2}\,\mathrm{e}^{(3\alpha_{TL}+2\beta_{TL})\tau}\right)\\+\left(M_{00(2)}^{cluster}c_{t0}^{2}\,\mathrm{e}^{(3\alpha_{TL}+2\beta_{TL})\left(\tau+\tau_{ext\,LR(1)}^{cluster}(d)\right)}-M_{00(2)}^{cluster}c_{t0}^{2}\,\mathrm{e}^{(3\alpha_{TL}+2\beta_{TL})\tau}\right)\end{array}\right]$$
$$=M_{0(1)}^{cluster}c_{t}^{2}\left(\mathrm{e}^{(3\alpha_{TL}+2\beta_{TL})\tau_{ext\,LR(2)}^{cluster}(d)}-1\right)+M_{0(2)}^{cluster}c_{t}^{2}\left(\mathrm{e}^{(3\alpha_{TL}+2\beta_{TL})\tau_{ext\,LR(1)}^{cluster}(d)}-1\right) \quad (24.9)$$

The total force of interaction between the two clusters is given by the derivative with respect to d of the variation of energy ΔE_{grav} of two clusters, namely

$$\boldsymbol{F}_{grav}(d)=\frac{\partial\Delta E_{grav}(d)}{\partial d}=M_{0(1)}^{cluster}c_{t}^{2}\frac{\partial\mathrm{e}^{(3\alpha_{TL}+2\beta_{TL})\tau_{ext\,LR(2)}^{cluster}(d)}}{\partial d}+M_{0(2)}^{cluster}c_{t}^{2}\frac{\partial\mathrm{e}^{(3\alpha_{TL}+2\beta_{TL})\tau_{ext\,LR(1)}^{cluster}(d)}}{\partial d}$$
$$=M_{0(1)}^{cluster}c_{t}^{2}\,\mathrm{e}^{(3\alpha_{TL}+2\beta_{TL})\tau_{ext\,LR(2)}^{cluster}(d)}\frac{\partial(3\alpha_{TL}+2\beta_{TL})\tau_{ext\,LR(2)}^{cluster}(d)}{\partial d}+M_{0(2)}^{cluster}c_{t}^{2}\,\mathrm{e}^{(3\alpha_{TL}+2\beta_{TL})\tau_{ext\,LR(1)}^{cluster}(d)}\frac{\partial(3\alpha_{TL}+2\beta_{TL})\tau_{ext\,LR(1)}^{cluster}(d)}{\partial d} \quad (24.10)$$

The derivatives in both these expressions are deduced from *(24.6)*, so that

$$\frac{\partial(3\alpha_{TL}+2\beta_{TL})\tau_{ext\,LR(i)}^{cluster}(d)}{\partial d}=\frac{(3\alpha_{TL}+2\beta_{TL})c_{t}^{2}}{2\pi\left(K_{0}-4K_{2}/3-2K_{1}(1+\tau_{0})\right)R_{\infty}^{2}}\frac{M_{0(i)}^{cluster}}{d^{2}} \quad (24.11)$$

and so, as a consequence

$$\boldsymbol{F}_{grav}(d)=\frac{(3\alpha_{TL}+2\beta_{TL})c_{t}^{4}}{\pi\left(K_{0}-4K_{2}/3-2K_{1}(1+\tau_{0})\right)R_{\infty}^{2}}\frac{M_{0(1)}^{cluster}M_{0(2)}^{cluster}}{d^{2}}\frac{1}{2}\left(\mathrm{e}^{(3\alpha_{TL}+2\beta_{TL})\tau_{ext\,LR(2)}^{cluster}(d)}+\mathrm{e}^{(3\alpha_{TL}+2\beta_{TL})\tau_{ext\,LR(1)}^{cluster}(d)}\right) \quad (24.12)$$

24.2 – Analogies and differences from newtonian gravity

The expression *(24.14)* depends on the gravitational masses divided by the distance separating the two clusters, which looks exactly like newtonian gravity. We thus introduce a "gravitation constant" $\boldsymbol{G}_{grav}$ such that

$$\boldsymbol{G}_{grav}=\frac{(3\alpha_{TL}+2\beta_{TL})c_{t}^{4}}{\pi\left(K_{0}-4K_{2}/3-2K_{1}(1+\tau_{0})\right)R_{\infty}^{2}} \quad (24.13)$$

The interaction force reads

$$\boldsymbol{F}_{grav}(d)=\boldsymbol{G}_{grav}\frac{M_{0(1)}^{cluster}M_{0(2)}^{cluster}}{d^{2}}\left(\frac{1}{2}\mathrm{e}^{(3\alpha_{TL}+2\beta_{TL})\tau_{ext\,LR(2)}^{cluster}(d)}+\frac{1}{2}\mathrm{e}^{(3\alpha_{TL}+2\beta_{TL})\tau_{ext\,LR(1)}^{cluster}(d)}\right) \quad (24.14)$$

which we recognize if we develop the exponential terms in first order

$$\frac{1}{2}\mathrm{e}^{(3\alpha_{TL}+2\beta_{TL})\tau_{ext\,LR(2)}^{cluster}(d)}+\frac{1}{2}\mathrm{e}^{(3\alpha_{TL}+2\beta_{TL})\tau_{ext\,LR(1)}^{cluster}(d)}\cong 1-\boldsymbol{G}_{grav}\frac{M_{0(1)}^{cluster}+M_{0(2)}^{cluster}}{4c_{t}^{2}d}+\ldots \quad (24.15)$$

since we now see the *newtonian gravitational force* between the two loops

$$\boldsymbol{F}_{grav}(d)=\boldsymbol{G}_{grav}\frac{M_{0(1)}^{cluster}M_{0(2)}^{cluster}}{d^{2}}\left(1-\boldsymbol{G}_{grav}\frac{M_{0(1)}^{cluster}+M_{0(2)}^{cluster}}{4c_{t}^{2}d}+\ldots\right) \quad (24.16)$$

For large distances d between the loops, the second order term in parenthesis can be neglected so that we will find a perfect analogy with the *Newton gravity*, with a 'gravitational constant' G_{grav} *which is very weak* since we have R_∞^2 in the denominator

$$F_{grav}(d) \cong G_{grav} \frac{M_{0(1)}^{cluster} M_{0(2)}^{cluster}}{d^2} \quad (24.17)$$

With the 'gravitational constant» G_{grav}, we can also rewrite the gravitational field *(24.8)* of a cluster under the form

$$\tau_{ext\,LR}^{cluster}(r) \cong -\frac{G_{grav} M_0^{cluster}}{2(3\alpha_{TL} + 2\beta_{TL}) c_t^2 r} \quad (24.18)$$

On the variable nature of the 'constant' of gravitation

What is clear in looking at the constant of gravitation is that it is not a 'constant' since it depends on the average expansion of the lattice τ_0 via the values of c_t^4 and R_∞^2, as well as the factor in front of module K_1, which renders G_{grav} positive if $\tau_0 < \tau_{0cr}$ and negative if $\tau_0 > \tau_{0cr}$

$$G_{grav}(\tau_0) = \frac{(3\alpha_{TL} + 2\beta_{TL}) c_{t0}^4}{\pi(K_0 - 4K_2/3 - 2K_1(1+\tau_0)) R_{\infty 0}^2} e^{4\tau_0/3} \begin{cases} >0 & if \quad \tau_0 < \tau_{0cr} \\ <0 & if \quad \tau_0 > \tau_{0cr} \end{cases} \quad (24.19)$$

This strong dependence of G_{grav} in the expansion of the background lattice should play a primordial role in the evolution of the universe during cosmological expansion. We will later revisit this point.

On the differences with Newton gravity at short distances

At short distances d, we have a *corrective term to Newton's law* which takes the form of a multiplier containing exponentials as shown in relation *(24.14)*.
This second order term involves the gravitational mass of two clusters and will only be noticeable when the clusters are close to each other.
The second order term in expression *(24.16)* of the law of gravity will modify the interactions between the two clusters if they are close. However, contrary to the results obtained in General Relativity thanks to the Schwarzschild metric, which predicts a small increase in the attractive force in $1/d^4$ at very short distances d, the corrective term in second order of our theory leads to a $1/d^3$ dependency of the correction of the attractive force at short distance d!
For example, in the case of Mercury which is rather close to the Sun, the corrective term in *(24.16)* has the following value

$$\frac{G_{grav}}{4c_t^2} \frac{M_{Sun}}{d_{Mercury}} \cong 5{,}6 \cdot 10^{-7}$$

with

$$M_{Sun} \cong 2 \cdot 10^{30}[kg], d_{Mercury} \cong 5{,}8 \cdot 10^{10}[m], G_{grav} \cong 6{,}6 \cdot 10^{-11}[m^3/kg \cdot s^2], c_t^2 \cong 10^{17}[m^2/s^2]$$

which gives us for the revolution of Mercury an increase of 2,128 seconds on the period of 88 days ($7{,}6 \cdot 10^6$ seconds) as calculated with the Newton Gravity.

24.3 – On the local rods and clocks of an HS observer

Let's consider a local framework $Ox_1x_2x_3$ defined by the ***GO*** observer *(the **Grand Observer**)* with respect to his absolute referential $Q\xi_1\xi_2\xi_3$. This local framework $Ox_1x_2x_3$ is in fact a convenience used by the ***GO*** to solve the problems of local evolution of the solid lattice, notably in the regions of the solid presenting a non-null volume expansion, but which can be considered *constant and homogenous* in the vicinity of the origin of the framework $Ox_1x_2x_3$, for example by using the *maxwellian formulation* described in chapter 17.

But let's imagine now that there exists another category of local observer we will call the ***HS*** *observer (**Homo Sapiens**)* which is truly in the local framework $Ox_1x_2x_3$ as it is composed of singularities of the lattice, most notably twist disclination loops, that interact via their field of rotation generated by their rotation charge. In his local framework, the ***HS*** does not have access to the 'global' view of the lattice in the absolute referential, as it only knows about the local frame. The ***HS*** observer will only be able to define its own rods $\vec{e}_{yi}$ in his frame $Oy_1y_2y_3$, by defining them *from the dimension of objects contained in the lattice in which it lives.* This is illustrated in figure 24.1 for two observers ***HS*** and ***HS'*** living in tow different parts of the lattice, where the respective expansions of the background τ and τ' are different.

The rods and clocks of an HS observer

If the lattice has a certain volume expansion τ, the rods of ***HS*** should satisfy the following

Hypothesis 2: $\vec{e}_{yi} = e^{a\tau}\,\vec{e}_i$ *(24.20)*

where the a constant is unknown and should be determined.

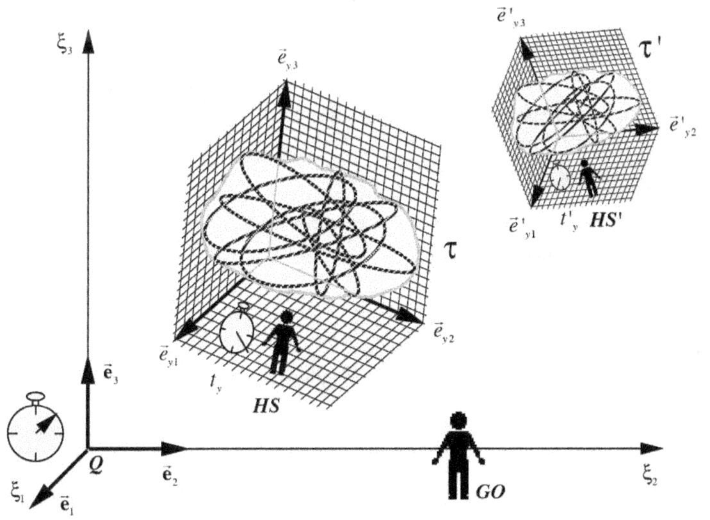

Figure 24.1** - the local rods and clocks in the frame of the local observers **HS** and **HS'

This implies that the rods of ***HS*** will be of different length than that of the ***GO*** if the lattice is locally in contraction ($\tau < 0$) or expansion ($\tau > 0$). If a given point in space is marked by vector $\vec{r}$, is observed simultaneously by the ***GO*** and by ***HS***, the vector will read in $Ox_1x_2x_3$ and in $Oy_1y_2y_3$ under the form

$$\vec{r} = \sum_i x_i \vec{e}_i = \sum_i x_i \, \mathrm{e}^{-a\tau} \, \vec{e}_{yi} = \sum_i y_i \vec{e}_{yi} \tag{24.21}$$

so that the coordinates of a point in space transform as

$$y_i = \mathrm{e}^{-a\tau} \, x_i \tag{24.22}$$

The time measured in $Ox_1x_2x_3$ with $\tau \neq 0$ must also be different from the time measured when $\tau = 0$, so the proper clock of ***HS*** in it's frame $Oy_1y_2y_3$ will indicate a time t_y different from the absolute time of the ***GO***, but linked to it by a relation

Hypothesis 3: $t_y = \mathrm{e}^{b\tau} \, t$ *(24.23)*

On the clock of the observer HS based on the speed of transversal waves

The clock of the ***HS*** has to be locally built as he doesn't have access to the absolute time of the ***GO***. He can build a simplistic clock by using local rods and the speed of transversal waves which he can measure in his own frame. Consider a rod of length d_0 measured by ***GO*** in a lattice with null volume expansion. The length of the same rods placed in the ***HS*** frame becomes, as measured by the ***GO***

$$d = d_0 \, \mathrm{e}^{a\tau} \tag{24.24}$$

To cover this distance, the transversal waves uses a lapse of time Δt measured by the ***GO***, and given by

$$\Delta t = \Delta t_0 \, \mathrm{e}^{-b\tau} \tag{24.25}$$

In the framework $Ox_1x_2x_3$, which may have a non-null background expansion τ, the initial length d_0 of the rods becomes $d_0 \, \mathrm{e}^{a\tau}$ and the time taken to cover it by transversal waves is $\Delta t_0 \, \mathrm{e}^{-b\tau}$, so that the velocity of transversal waves in the presence of expansion τ is

$$c_t = \frac{d}{\Delta t} = \frac{d_0 \, \mathrm{e}^{a\tau}}{\Delta t_0 \, \mathrm{e}^{-b\tau}} = \frac{d_0}{\Delta t_0} \mathrm{e}^{(a+b)\tau} = c_{t0} \, \mathrm{e}^{(a+b)\tau} = c_{t0} \, \mathrm{e}^{\tau/2} \tag{24.26}$$

This implies that time flows differently for the ***HS*** in $Oy_1y_2y_3$ depending on whether the lattice is in contraction or expansion. We then have the following relation between the values of a and b due to how the ***HS*** clock works

$$a + b = 1/2 \tag{24.27}$$

Invariance of the Maxwell equations for the HS observer in his local frame

From relations *(24.20) and (24.23)*, we deduce the expression linking the partial derivatives in both frames

$$\frac{\partial}{\partial x_i} = \frac{\partial}{\partial y_i}\frac{\partial y_i}{\partial x_i} = \mathrm{e}^{-a\tau}\frac{\partial}{\partial y_i} \quad \Rightarrow \quad \begin{cases} \overrightarrow{\mathrm{grad}}\, A = \mathrm{e}^{-a\tau}\, \overrightarrow{\mathrm{grad}}_y\, A \\ \overrightarrow{\mathrm{rot}}\, \vec{A} = \mathrm{e}^{-a\tau}\, \overrightarrow{\mathrm{rot}}_y\, \vec{A} \\ \mathrm{div}\, \vec{A} = \mathrm{e}^{-a\tau}\, \mathrm{div}_y\, \vec{A} \end{cases} \tag{24.28}$$

and the temporal derivative in both frames

$$\frac{d}{dt}=\frac{d}{dt_y}\frac{\partial t_y}{\partial t}=\mathrm{e}^{b\tau}\frac{d}{dt_y} \tag{24.29}$$

By applying these relations to Maxwell's equation in table 17.1, and by replacing n by $n_0\,\mathrm{e}^{-\tau}$, we obtain the following equations

$$\begin{cases} -\dfrac{d(2\vec{\omega}^{el})}{dt_y}+\overrightarrow{\mathrm{rot}}_y\left(\mathrm{e}^{-(a+b)\tau}\,\vec{\phi}^{rot}\right)=(2\,\mathrm{e}^{-b\tau}\,\vec{J}) \\ \mathrm{div}_y\,(2\vec{\omega}^{el})=(2\,\mathrm{e}^{a\tau}\,\lambda) \end{cases}$$

$$\begin{cases} \dfrac{d\left(n_0\,\mathrm{e}^{(a+b-1)\tau}\,\vec{p}^{rot}\right)}{dt_y}=-\overrightarrow{\mathrm{rot}}_y\left(\dfrac{\vec{m}}{2}\right)+2K_2(\mathrm{e}^{a\tau}\,\vec{\lambda}^{rot}) \\ \mathrm{div}_y\left(n_0\,\mathrm{e}^{(a+b-1)\tau}\,\vec{p}^{rot}\right)=0 \end{cases} \tag{24.30}$$

$$\begin{cases} (2\vec{\omega}^{el})=\dfrac{1}{(K_2+K_3)}\left(\dfrac{\vec{m}}{2}\right)+(2\vec{\omega}^{an}) \\ \left(n_0\,\mathrm{e}^{(a+b-1)\tau}\,\vec{p}^{rot}\right)=(n_0 m)\begin{bmatrix} \mathrm{e}^{(a+b-1)\tau}\,\vec{\phi}^{rot}+(C_t-C_L)\mathrm{e}^{(a+b-1)\tau}\,\vec{\phi}^{rot} \\ +\left(\dfrac{1}{n_0}\left(\mathrm{e}^{(a+b)\tau}\,\vec{J}_I^{rot}-\mathrm{e}^{(a+b)\tau}\,\vec{J}_L^{rot}\right)\right) \end{bmatrix} \end{cases}$$

and

$$\begin{cases} \dfrac{d(2\,\mathrm{e}^{a\tau}\,\lambda)}{dt_y}=-\,\mathrm{div}_y(2\,\mathrm{e}^{-b\tau}\,\vec{J}) \end{cases}$$

$$\begin{cases} -\left(\dfrac{\vec{m}}{2}\right)(2\,\mathrm{e}^{-b\tau}\,\vec{J})= \\ \left(\mathrm{e}^{-(a+b)\tau}\,\vec{\phi}^{rot}\right)\dfrac{d(n_0\,\mathrm{e}^{(a+b-1)\tau}\,\vec{p}^{rot})}{dt_y}+\left(\dfrac{\vec{m}}{2}\right)\dfrac{d(2\vec{\omega}^{el})}{dt_y}-\mathrm{div}_y\left(\left(\mathrm{e}^{-(a+b)\tau}\,\vec{\phi}^{rot}\right)\wedge\left(\dfrac{\vec{m}}{2}\right)\right) \end{cases} \tag{24.31}$$

$$\begin{cases} c_{t0}=\sqrt{\dfrac{K_2+K_3}{mn_0}} \end{cases}$$

To insure the coherence and invariance of these equations, it is necessary that the coefficients of $\vec{\phi}^{rot}$ be the same everywhere, and thus $a+b=1/2$, and we find again the relation *(24.27)* linked to the clock of the observer, which allows to write the Maxwell equations in the following form

$$\begin{cases} -\dfrac{d(2\vec{\omega}^{el})}{dt_y}+\overrightarrow{\mathrm{rot}}_y\left(\mathrm{e}^{-\tau/2}\,\vec{\phi}^{rot}\right)=(2\,\mathrm{e}^{-b\tau}\,\vec{J}) \\ \mathrm{div}_y\,(2\vec{\omega}^{el})=(2\,\mathrm{e}^{a\tau}\,\lambda) \end{cases} \tag{24.32a}$$

$$\begin{cases} \dfrac{d\,(n_0\,\mathrm{e}^{-\tau/2}\,\vec{p}^{rot})}{dt_y}=-\overrightarrow{\mathrm{rot}}_y\left(\dfrac{\vec{m}}{2}\right)+2K_2(\mathrm{e}^{a\tau}\,\vec{\lambda}^{rot}) \\ \mathrm{div}_y\,(n_0\,\mathrm{e}^{-\tau/2}\,\vec{p}^{rot})=0 \end{cases}$$

$$\begin{cases} (2\vec{\omega}^{el}) = \dfrac{1}{(K_2+K_3)}\left(\dfrac{\vec{m}}{2}\right) + (2\vec{\omega}^{an}) \\ (n_0\,e^{-\tau/2}\,\vec{p}^{rot}) = (n_0 m)\left[\begin{array}{l} e^{-\tau/2}\,\vec{\phi}^{rot} + (C_I - C_L)\,e^{-\tau/2}\,\vec{\phi}^{rot} \\ +\left(\dfrac{1}{n_0}\left(e^{\tau/2}\,\vec{J}_I^{rot} - e^{\tau/2}\,\vec{J}_L^{rot}\right)\right) \end{array}\right] \end{cases}$$

$$\left\{ \dfrac{d(2\,e^{a\tau}\,\lambda)}{dt_y} = -\operatorname{div}_y(2\,e^{-b\tau}\,\vec{J}) \right. \tag{24.32b}$$

$$\begin{cases} -\left(\dfrac{\vec{m}}{2}\right)(2\,e^{-b\tau}\,\vec{J}) = \\ \left(e^{-\tau/2}\,\vec{\phi}^{rot}\right)\dfrac{d(n_0\,e^{-\tau/2}\,\vec{p}^{rot})}{dt_y} + \left(\dfrac{\vec{m}}{2}\right)\dfrac{d(2\vec{\omega}^{el})}{dt_y} - \operatorname{div}_y\left(\left(e^{-\tau/2}\,\vec{\phi}^{rot}\right)\wedge\left(\dfrac{\vec{m}}{2}\right)\right) \end{cases}$$

$$\left\{ c_{t0} = \sqrt{\dfrac{K_2+K_3}{m n_0}} \right.$$

Let's introduce the following local values in frame $Oy_1y_2y_3t_y$ of the ***HS:***

- the fields of rotation, of torque, of velocity and of linear momentum

$$\begin{cases} \vec{\omega}_{(y)}^{el} = \vec{\omega}^{el} \\ \vec{m}_{(y)} = \vec{m} \end{cases} \qquad \begin{cases} \vec{\phi}_{(y)}^{rot} = e^{-\tau/2}\,\vec{\phi}^{rot} \\ \vec{p}_{(y)}^{rot} = e^{-\tau/2}\,\vec{p}^{rot} \end{cases} \tag{24.33}$$

- the values associated with densities and flows of charges of rotation

$$\begin{cases} \lambda_{(y)} = e^{a\tau}\,\lambda = e^{\alpha_{BV}\tau}\,\lambda \\ \vec{\lambda}_{(y)}^{rot} = e^{a\tau}\,\vec{\lambda}^{rot} = e^{\alpha_{BV}\tau}\,\vec{\lambda}^{rot} \\ \vec{J}_{(y)} = \lambda_{(y)}\vec{\boldsymbol{v}}_y = \lambda\,e^{a\tau}\,\vec{\boldsymbol{v}}\,e^{-\tau/2} = e^{-b\tau}\,\vec{J} = e^{-(1/2-\alpha_{BV})\tau}\,\vec{J} \end{cases} \tag{24.34}$$

- the quantities associated with the concentration of vacancies and interstitials and the flows associated with those defects

$$\begin{cases} C_{I(y)} = C_I \\ C_{L(y)} = C_L \\ \vec{J}_{I(y)}^{rot} = C_{I(y)} n_0 \vec{\phi}_{I(y)}^{rot} = C_I n\,e^{\tau}\,\vec{\phi}_I^{rot}\,e^{-\tau/2} = e^{\tau/2}\,\vec{J}_{I(y)}^{rot} \\ \vec{J}_{L(y)}^{rot} = C_{L(y)} n_0 \vec{\phi}_{L(y)}^{rot} = C_L n\,e^{\tau}\,\vec{\phi}_L^{rot}\,e^{-\tau/2} = e^{\tau/2}\,\vec{J}_{L(y)}^{rot} \end{cases} \tag{24.35}$$

- and finally the following quantities

$$\begin{cases} n_y = n_0 = n\,e^{\tau} \\ c_{ty} = c_{t0} = c_t\,e^{-\tau/2} \\ \vec{\boldsymbol{v}}_y = \vec{\boldsymbol{v}}\,e^{-\tau/2} \end{cases} \tag{24.36}$$

In these expressions, we have logically asserted $\vec{\omega}_{(y)}^{el} = \vec{\omega}^{el}$ and $\vec{m}_{(y)} = \vec{m}$, since these quantities are associated with angles of rotation. Also we have asserted that $\vec{\phi}_{(y)}^{rot} = e^{-\tau/2}\,\vec{\phi}^{rot}$ and $\vec{p}_{(y)}^{rot} = e^{-\tau/2}\,\vec{p}^{rot}$, since they are associated with velocities. By this change of variable, we obtain a set of *Maxwell equations which are invariant in frame* $Oy_1y_2y_3t_y$ *of the* ***HS*** *observer* (table

Table 24.1 - <u>invariance</u> of the Maxwellian equations of evolution of a cosmic lattice in the local frame $Oy_1y_2y_3t_y$ of the HS

$$\begin{cases} -\dfrac{\partial(2\vec{\omega}^{el}_{(y)})}{\partial t_y}+\overrightarrow{\mathrm{rot}}_y\,\vec{\phi}^{rot}_{(y)}\cong(2\vec{J}_{(y)}) \\ \mathrm{div}_y\,(2\vec{\omega}^{el}_{(y)})=(2\lambda_{(y)}) \end{cases} \quad\Leftrightarrow\quad \begin{cases} -\dfrac{\partial\vec{D}}{\partial t}+\overrightarrow{\mathrm{rot}}\,\vec{H}=\vec{j} \\ \mathrm{div}\,\vec{D}=\rho \end{cases}$$

$$\begin{cases} \dfrac{\partial\left(n_0\vec{p}^{rot}_{(y)}\right)}{\partial t_y}\cong-\overrightarrow{\mathrm{rot}}_y\left(\dfrac{\vec{m}_{(y)}}{2}\right)+2K_2\vec{\lambda}^{rot}_{(y)} \\ \mathrm{div}_y\left(n_0\vec{p}^{rot}_{(y)}\right)=0 \end{cases} \quad\Leftrightarrow\quad \begin{cases} \dfrac{\partial\vec{B}}{\partial t}=-\overrightarrow{\mathrm{rot}}\,\vec{E} \\ \mathrm{div}\,\vec{B}=0 \end{cases}$$

$$\begin{cases} (2\vec{\omega}^{el}_{(y)})=\dfrac{1}{(K_2+K_3)}\left(\dfrac{\vec{m}_{(y)}}{2}\right)+(2\vec{\omega}^{an}_{(y)}) \\ \left(n_0\vec{p}^{rot}_{(y)}\right)=(n_0m)\left[\vec{\phi}^{rot}_{(y)}+\left(C_{I(y)}-C_{L(y)}\right)\vec{\phi}^{rot}_{(y)}+\left(\dfrac{1}{n_0}\left(\vec{J}^{rot}_{I(y)}-\vec{J}^{rot}_{L(y)}\right)\right)\right] \end{cases} \quad\Leftrightarrow\quad \begin{cases} \vec{D}=\varepsilon_0\,\vec{E}+\vec{P} \\ \vec{B}=\mu_0\left[\vec{H}+\left(\chi^{para}+\chi^{dia}\right)\vec{H}+\vec{M}\right] \end{cases}$$

$$\left\{ \dfrac{\partial(2\lambda_{(y)})}{\partial t_y}\cong-\mathrm{div}_y(2\vec{J}_{(y)}) \right. \quad\Leftrightarrow\quad \left\{ \dfrac{\partial\rho}{\partial t}=-\mathrm{div}\,\vec{j} \right.$$

$$\begin{cases} -\left(\dfrac{\vec{m}_{(y)}}{2}\right)(2\vec{J}_{(y)})\cong \\ \vec{\phi}^{rot}_{(y)}\dfrac{\partial(n_0\vec{p}^{rot}_{(y)})}{\partial t_y}+\left(\dfrac{\vec{m}_{(y)}}{2}\right)\dfrac{\partial(2\vec{\omega}^{el}_{(y)})}{\partial t_y}-\mathrm{div}_y\left(\vec{\phi}^{rot}_{(y)}\wedge\left(\dfrac{\vec{m}_{(y)}}{2}\right)\right) \end{cases} \quad\Leftrightarrow\quad \begin{cases} -\vec{E}\vec{j}= \\ \vec{H}\dfrac{\partial\vec{B}}{\partial t}+\vec{E}\dfrac{\partial\vec{D}}{\partial t}-\mathrm{div}\left(\vec{H}\wedge\vec{E}\right) \end{cases}$$

$$\left\{ c_{t0}=\sqrt{\dfrac{K_2+K_3}{mn_0}} \right. \quad\Leftrightarrow\quad \left\{ c=\sqrt{\dfrac{1}{\varepsilon_0\mu_0}} \right.$$

24.1), by which we mean that they do not depend on the value of the local expansion τ. We notice also that the speed of transversal waves becomes an invariant constant for the observer ***HS***, no matter what the state of expansion of the lattice in which it lives is.

We also notice that only the quantities associated with the densities and flows of charges of rotation transform depending on the value of α, while all the other quantities transform in a logical and predictable way.

The fact that Maxwell equations for the ***HS*** observers are invariant (do not depend on the local expansion τ) implies that the local ***HS*** observers cannot measure the local state of expansion of the lattice in which it lives based solely on electromagnetic (EM) measurements, which themselves are described by Maxwell equations. Notably, the measure of speed of transversal waves by the ***HS*** observers always gives us an invariant quantity, whatever the state of expansion of the lattice! Thus, the local observers ***HS*** are submitted essentially to the laws of physics described by EM, and are only aware of the gravitational effects associated with the field of expansions and only via indirect observations of their effects such as the movement of planets or the slowing down of clocks in the field of gravitation. This is why the ***HS*** will seek to explain phenomenas linked to gravitation with ad-hoc theories (Newton Gravitation, Einstein General Relativity) which at first glance seem independent of the laws of electromagnetism, but which he will seek to unify!

24.4 – HS observer in the gravitational field of a cluster

Figure 24.1 illustrated the existence of a strong analogy between our theory and the Einstein theory of General Relativity. Indeed, the rods and clocks of an ***HS*** living in a given point of the lattice depend on the local volume expansion τ, in an analog fashion to what is stipulated in General Relativity for the rods and clocks of an observer embedded in a given gravitation field. In the case of a lattice as illustrated in figure 24.1, we see that the lattice plays the role of an 'aether' which imposes the size of rods of the ***HS*** *observer*, while it is the speed of transversal waves within the lattice (which represents information transport) which imposes how the clocks of the ***HS*** behave.
Furthermore, the existence of three degrees of freedom in the parameters a, b, α_{TL} and β_{TL} is quite surprising, as it implies that there exist another possible choice at that level, which cannot be determined on the basis of our current understanding of the cosmic lattice. As a matter of fact, an arbitrary choice of a, α_{TL} and β_{TL} should not entail any incoherence in the system and the cosmological lattice thus obtained could be entirely viable. As a consequence, we must consider that parameters a, α_{TL} and β_{TL} are *truly intrinsic properties of the cosmic lattice*, just as the elastic modules K_i or the inertial mass m by site of the lattice. The determination of these constants involves experimentation and the measure of said properties of the cosmic lattice!

On the laws of transformation in the gravitational field of a cluster

By combining relations *(24.18), (24.20) and (24.22)*, we can write the transformations to go from one local frame $Oy_1y_2y_3t_y$ of an ***HS*** situated at distance r in the field of gravity of a cluster of mass $M_0^{cluster}$, as a function of the unknown parameters a, α_{TL} and β_{TL}

$$\tau_{ext\,LR}^{cluster}(r) \cong -\frac{G_{grav} M_0^{cluster}}{(6\alpha_{TL}+4\beta_{TL})c_t^2 r} \quad \Rightarrow \quad \begin{cases} \vec{e}_{yi} = e^{a\tau}\,\vec{e}_i = e^{-\frac{a}{6\alpha_{TL}+4\beta_{TL}}\frac{G_{grav}M_0^{cluster}}{c_t^2 r}}\,\vec{e}_i \\ y_i = e^{-a\tau}\,x_i = e^{+\frac{a}{6\alpha_{TL}+4\beta_{TL}}\frac{G_{grav}M_0^{cluster}}{c_t^2 r}}\,x_i \\ t_y = e^{b\tau}\,t = e^{-\frac{b}{6\alpha_{TL}+4\beta_{TL}}\frac{G_{grav}M_0^{cluster}}{c_t^2 r}}\,t \end{cases} \qquad (24.37)$$

The equations of transformation *(24.40)* can, at a certain distance of the cluster of mass M_0^{amas}, be described approximatively by developing the exponentials

$$\begin{cases} \vec{e}_{yi} \cong \left(1 - \dfrac{a}{6\alpha_{TL} + 4\beta_{TL}} \dfrac{G_{grav} M_0^{amas}}{c_t^2 r} + ...\right) \vec{e}_i \\ y_i \cong \left(1 + \dfrac{a}{6\alpha_{TL} + 4\beta_{TL}} \dfrac{G_{grav} M_0^{amas}}{c_t^2 r} + ...\right) x_i \\ t_y \cong \left(1 - \dfrac{b}{6\alpha_{TL} + 4\beta_{TL}} \dfrac{G_{grav} M_0^{amas}}{c_t^2 r} + ...\right) t \end{cases} \tag{24.38}$$

In General relativity, the dependance of the radial rods and the clock of an observer submitted to a field of gravitation from a mass $M_0^{cluster}$ are deduced from the *Schwarzschild metric.* This metric is obtained in the case of a massive object with spherical symmetry by postulating an invariant metric vis-a-vis the rotations, which is written

$$ds^2 = c_t^2 \left(1 - \frac{2G_{grav}}{c_t^2} \frac{M_0^{cluster}}{r}\right) dt^2 - \left(1 - \frac{2G_{grav}}{c_t^2} \frac{M_0^{cluster}}{r}\right)^{-1} dr^2 + r^2 d\theta^2 + r^2 \sin^2\theta \, d\varphi^2 \tag{24.39}$$

We observe that the rods and clocks of an observer depend symmetrically on distance r with expressions similar to *(24.41)*, but with a coefficient 1 in front of $G_{grav} M_0^{cluster} / c_t^2 r$

$$\begin{cases} r_y \cong \left(1 - \dfrac{G}{c^2} \dfrac{M_0}{r}\right)^{-1} r \cong \left(1 + \dfrac{G}{c^2} \dfrac{M_0}{r} + ...\right) r \\ t_y \cong \left(1 - \dfrac{G}{c^2} \dfrac{M_0}{r}\right) t \end{cases} \tag{24.40}$$

The time dilatation in a gravity field, represented by the second relation in *(24.40)*, was verified with great experimental precision[1], on distance differences as small as 1 meter of the surface of the earth, and we take into account this effect in the navigational systems like *GPS*. This effect, which is experimentally verified can be used to determine parameters a , α_{TL} and β_{TL} , by requiring that the last relation of *(24.38)* correspond to the second relation *(24.40)*, and thus that

$$\frac{b}{6\alpha_{TL} + 4\beta_{TL}} = 1 \quad \Rightarrow \quad b = 6\alpha_{TL} + 4\beta_{TL} \tag{24.41}$$

On the curvature of rays of transversal waves near massive cluster

We can also test our analogies with the theory of General Relativity of Einstein, by example by calculating the curvature of rays of transversal waves next to a massive cluster (fig. 24.2), since the measure of the effect at the beginning of the 20th century was the first experimental verification of the General Relativity of Einstein.

In the vicinity of a massive cluster, the speed of the transversal waves $c_t(r)$ depends on the distance to the center of the cluster. This implies that the direction associated with two parallel rays impinging on the center of the cluster at distances r et $r + dr$ of the cluster will present an infinitesimal angle $d\alpha$ such that

[1] see for example: *C. W. Chou, D. B. Hume, T. Rosenband, D. J. Wineland : "Optical Clocks and Relativity", Science, vol. 329, 5999, pp. 1630-1633*

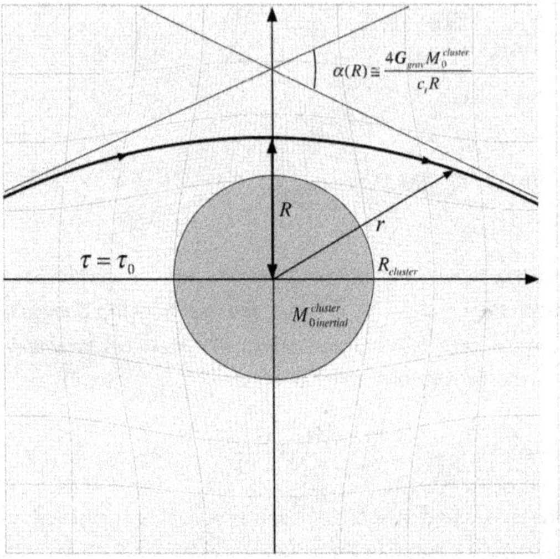

Figure 24.2 *- curvature of transversal rays in the vicinity of a cluster.*

$$\Delta\alpha(r) = \frac{c_t(r+dr)\Delta t - c_t(r)\Delta t}{dr} = \frac{\partial c_t}{\partial r}\Delta t = \frac{\partial c_t}{\partial r}\frac{\Delta t}{\Delta r}\Delta r = \frac{1}{c_t}\frac{\partial c_t}{\partial r}\Delta r \tag{24.42}$$

We use the dependency of speed $c_t(r)$ in the gravitational field $\tau_{ext\,LD}^{amas}(r)$, by writing

$$\Delta\alpha(r) = \frac{1}{c_t}\frac{\partial c_t}{\partial r}\Delta r = \frac{1}{c_t}\frac{\partial c_t\left(\tau_{ext\,LR}^{cluster}(r)\right)}{\partial \tau_{ext\,LR}^{cluster}(r)}\frac{\partial \tau_{ext\,LR}^{cluster}(r)}{\partial r}\Delta r = \frac{1}{2}\frac{\partial \tau_{ext\,LR}^{cluster}(r)}{\partial r}\Delta r \tag{24.43}$$

Since $c_t(\tau) = c_{t0}\,e^{\tau/2}$, by using expression *(24.40)* of $\tau_{ext\,LR}^{cluster}(r)$, we have

$$\Delta\alpha(r) = \frac{G_{grav}M_0^{cluster}}{4(3\alpha_{TL} + 2\beta_{TL})c_t^2 r^2}\Delta r \tag{24.44}$$

As the wave will travel the path represented in figure 24.2, the tangents at infinity of the incident wave and the waves which is deviated have a total angle $\alpha(R)$ which depends on the minimum distance R of the wave ray to the center of the cluster. Half of the total angle $\alpha(R)$ can then be deduced approximatively by integrating $\Delta\alpha(r)$ for distances r to the center of the cluster going from R to R_∞, so that

$$\alpha(R) \cong 2\int_R^{R_\infty} \frac{G_{grav}M_0^{cluster}}{4(3\alpha_{TL} + 2\beta_{TL})c_t^2 r^2}dr = \frac{G_{grav}M_0^{cluster}}{2(3\alpha_{TL} + 2\beta_{TL})c_t^2}\int_R^{R_\infty}\frac{dr}{r^2} = \frac{G_{grav}M_0^{cluster}}{2(3\alpha_{TL} + 2\beta_{TL})c_t^2 R} \tag{24.45}$$

General Relativity predicts a curvature worth $\alpha(R) \cong 4G_{grav}M_0^{cluster} / c_t^2 R$, and the calculation of this value for a luminous ray grazing the Sun gives us a deviation of 1,75" arc degrees.

The experimental values of the deviation of light by the Sun, measured by *Eddington* at the beginning of the 20th century (may 1919) during a solar eclipse, have given us 1,98 ± 0,12" (at Sobral in Brazil) and 1,61 ± 0,31" of arc degrees (at Sao Tomé-et-Principe in the Golf of Guinea), which corresponds quite nicely to the value calculated theoretically by General Relativity, in spite of the numerous difficulties in achieving this experimental measure!

Starting from this experimental result, our calculation must match that of General Relativity which gives us a relation on parameters α_{TL} and β_{TL}

$$\frac{1}{6\alpha_{TL}+4\beta_{TL}}=4 \quad \Rightarrow \quad 6\alpha_{TL}+4\beta_{TL}=1/4 \tag{24.46}$$

Finally, combining relations *(24.27)*, *(24.41)* and *(24.46)* allows to define the following relations for the parameters a, b, α_{TL} and β_{TL} if we are to satisfy the experimental observations

$$\begin{cases} a+b=1/2 \\ b=6\alpha_{TL}+4\beta_{TL} \\ 6\alpha_{TL}+4\beta_{TL}=1/4 \end{cases} \quad \Rightarrow \quad \begin{cases} a=1/4 \\ b=1/4 \\ 6\alpha_{TL}+4\beta_{TL}=1/4 \end{cases} \tag{24.47}$$

On the symmetry of the transformations of space-time

During the Lorentz transformation described in chapter 20, the rules of transformation are symmetrical with respect to time and space, just like the transformation in General Relativity in the case of the Schwarzschild metric *(24.40)*. Amidst all of the possible results for parameters a, b, α_{TL} and β_{TL}, on the results *(24.47)* obtained from the dilation of time and curvature of the wave rays in a field of gravity conveniently give us symmetric laws of transformation as $a=b$, and also that

$$\tau_{ext\,LR}^{cluster}(r)\cong-\frac{4G_{grav}M_0^{cluster}}{c_t^2 r} \quad \Rightarrow \quad \begin{cases} \vec{e}_{yi}=e^{\tau/4}\,\vec{e}_i=e^{-\frac{G_{grav}M_0^{cluster}}{c_t^2 r}}\,\vec{e}_i\cong\left(1-\dfrac{G_{grav}M_0^{cluster}}{c_t^2 r}\right)\vec{e}_i \\ y_i=e^{-\tau/4}\,x_i=e^{+\frac{G_{grav}M_0^{cluster}}{c_t^2 r}}\,x_i\cong\left(1+\dfrac{G_{grav}M_0^{cluster}}{c_t^2 r}\right)x_i \\ t_y=e^{\tau/4}\,t=e^{-\frac{G_{grav}M_0^{cluster}}{c_t^2 r}}\,t\cong\left(1-\dfrac{G_{grav}M_0^{cluster}}{c_t^2 r}\right)t \end{cases} \tag{24.48}$$

We can consequently emit the following conjecture which contains at the same time the effects of time dilatation and of wave rays curvature in weak gravity:

Conjecture 9: *the metric of our theory in a weak gravitational field must be the same than the Schwarzschild metric in General Relativity* (24.49)

On the choice of the values α_{TL} *and* β_{TL}

It remains one degree of liberty for the choice of the values of the parameters α_{TL} and β_{TL}, which have to satisfy the expression $3\alpha_{TL}+2\beta_{TL}=1/8$. However, knowing exactly the values of α_{TL} and β_{TL} is not important concerning the gravitational properties of the twist loop, as the relation $3\alpha_{TL}+2\beta_{TL}=1/8$ involves that the inertial mass M_0^{BV} and the distortion energy E^{BV} of a loop, as well as the values $\boldsymbol{G}_{grav}$, $\tau_{ext\,LR}^{cluster}(r)$ and $\boldsymbol{F}_{grav}(d)$, do not depend explicitly on α_{TL} and β_{TL}, as shown by the relations of table 24.2.

The exact choice of the parameters α_{TL} and β_{TL} depend obviously on the deep nature of the cosmic lattice. But the simplest solution we can imagine is that β_{TL} be null, because in this

Table 24.2 - The laws of transformation linked to the expansion and the gravitational behavior of a cluster of twist disclination loops

Dependency of the twist disclination loops on expansion

$$\begin{cases} R_{TL} = R_{TL0}\, e^{\alpha_{TL}\tau} \\ \Omega_{TL} = \Omega_{TL0}\, e^{\beta_{TL}\tau} \end{cases} \Rightarrow \begin{cases} q_{\lambda TL} = q_{\lambda TL0}\, e^{(2\alpha_{TL}+\beta_{TL})\tau} \\ Q_{\lambda}^{cluster} = \sum_i q_{\lambda TL(i)} = Q_{\lambda 0}^{cluster}\, e^{(2\alpha_{TL}+\beta_{TL})\tau} \end{cases}$$

with $3\alpha_{TL} + 2\beta_{TL} = 1/8$

Dependency of the mass and the energy of the twist loops on expansion

$$\begin{cases} M_0^{TL} = M_{00}^{TL}\, e^{(3\alpha_{TL}+2\beta_{TL}-1)\tau} = M_{00}^{TL}\, e^{-7\tau/8} \\ M_0^{cluster} = \sum_i M_{0(i)}^{TL} = M_{00}^{cluster}\, e^{-7\tau/8} \end{cases}$$

$$\begin{cases} E^{TL} = M_0^{TL} c_t^2 = M_{00}^{TL} c_{t0}^2\, e^{(3\alpha_{TL}+2\beta_{TL})\tau} = M_{00}^{TL} c_{t0}^2\, e^{\tau/8} \\ E^{cluster}(\tau) = \sum_i M_{0(i)}^{TL} c_t^2 = M_{00}^{cluster} c_{t0}^2\, e^{\tau/8} \end{cases}$$

Gravitational behavior of the twist loops

$$\begin{cases} G_{grav} = \dfrac{c_t^4}{8\pi\big(K_0 - 4K_2/3 - 2K_1(1+\tau_0)\big) R_\infty^2} \\ \tau_{ext\,LR}^{cluster}(r) \cong -\dfrac{4 G_{grav} M_0^{cluster}}{c_t^2 r} \\ F_{grav}(d) = G_{grav} \dfrac{M_{0(1)}^{cluster} M_{0(2)}^{cluster}}{d^2}\left(1 - G_{grav}\dfrac{M_{0(1)}^{cluster} + M_{0(2)}^{cluster}}{4c_t^2 d} + \ldots\right) \end{cases}$$

Transformation laws between frames $Ox_1x_2x_3$ *et* $Oy_1y_2y_3$

$$\begin{cases} \vec{e}_{yi} = e^{\tau/4}\, \vec{e}_i = e^{-\frac{G_{grav} M_0^{cluster}}{c_t^2 r}}\, \vec{e}_i \cong \left(1 - \dfrac{G_{grav} M_0^{cluster}}{c_t^2 r} + \ldots\right) \vec{e}_i \\ y_i = e^{-\tau/4}\, x_i = e^{\frac{G_{grav} M_0^{cluster}}{c_t^2 r}}\, x_i \cong \left(1 + \dfrac{G_{grav} M_0^{cluster}}{c_t^2 r} + \ldots\right) x_i \\ t_y = e^{\tau/4}\, t = e^{-\frac{G_{grav} M_0^{cluster}}{c_t^2 r}}\, t \cong \left(1 - \dfrac{G_{grav} M_0^{cluster}}{c_t^2 r} + \ldots\right) t \end{cases}$$

case the torsion Ω_{TL} of the twist disclination becomes a constant independent of the lattice expansion, for example a multiple of $\pi / 2$ for a cubic lattice, or a multiple of $\pi / 3$ for a hexagonal lattice, which seems evidently the most logical solution.
But a null value of β_{TL} implies also that α_{TL} becomes equal to 1/24, and than that the loop radius $R_{TL} = R_{TL0}\,\mathrm{e}^{\tau/24}$ depends much less of the lattice expansion than the lattice parameter $a = a_0\,\mathrm{e}^{\tau/3}$ or the length $\vec{e}_{yi} = \mathrm{e}^{\tau/4}\,\vec{e}_i$ of the local rods of the observer ***HS***.
As a choice of explicit values for the parameters α_{TL} and β_{TL} cannot be simply obtained, we leave open this problem, reminiscing that this degree of freedom on the values α_{TL} and β_{TL} exists, with the constraint that $3\alpha_{TL} + 2\beta_{TL} = 1/8$.

24.5 – Analogies and differences with the General Relativity of Einstein

On the Schwarzschild sphere of a black hole

Assume a twist disclination loop with inertial mass M_0^{TL} located at one end of the lattice and a cluster of singularities with inertial mass $M_0^{cluster} >> M_0^{TL}$ in the center of the lattice. The gravitational force acting on the loop is approximatively written

$$\boldsymbol{F}_{grav}^{TL}(r) \cong \boldsymbol{G}_{grav}\frac{M_0^{cluster} M_0^{TL}}{r^2} \qquad (24.50)$$

With this force, the twist loop will undergo an acceleration in the direction of the cluster of singularities, and the gravitational potential energy of the loop will transform progressively into a kinetic energy of the loop which will be worth as a function of its distance d to the cluster

$$\boldsymbol{E}_{kinetic}^{TL}(r) = \left|\int_{R_\infty}^{R} \boldsymbol{F}_{grav}^{TL}(r)dr\right| = \boldsymbol{G}_{grav} M_0^{cluster} M_0^{BV}\left|\int_{R_\infty}^{R}\frac{dr}{r^2}\right| \cong \boldsymbol{G}_{grav}\frac{M_0^{cluster} M_0^{TL}}{R} \qquad (R << R_\infty) \qquad (24.51)$$

However the kinetic energy of the loop is also written in the non-relativistic case as

$$\boldsymbol{E}_{kinetic}^{TL} = M_0^{TL}\boldsymbol{v}^2 / 2 \qquad (24.52)$$

so that the velocity of the loop as a function of distance d that separates it from the cluster depends on the *gravitational constant* and *the inertial mass of the cluster*

$$\boldsymbol{v}^2(R) = 2\boldsymbol{G}_{grav} M_0^{cluster} / R \qquad (24.53)$$

As the *inertial mass of the loop does not figure in this relation*, this relation remains valid for relativistic velocities of the loop. However, we know that the relativistic energy of the loop $E_{\boldsymbol{v}}^{TL}(\boldsymbol{v})$ tends towards infinity when $\boldsymbol{v}$ tends towards the velocity of transversal waves c_t (see figure 20.1), so that the following condition holds before we reach the limit velocity

$$\boldsymbol{v}^2(R) = 2\boldsymbol{G}_{grav} M_0^{cluster} / R \leq c_t^2 \qquad (24.54)$$

This condition implies the existence of a *critical distance* R_{cr} for which the energy of the loop becomes infinite

$$R_{cr} = 2\boldsymbol{G}_{grav} M_0^{cluster} / c_t^2 \qquad (24.55)$$

This critical distance R_{cr} only depends on the mass $M_0^{cluster}$ of the cluster, and only exists if the radius of the cluster is smaller than d_{cr}. It is called the *Schwarzschild radius* of the cluster and corresponds to the limit beyond which the loop cannot escape the cluster as it would need an infinite energy to do so. Thus, the cluster of mass $M_0^{cluster}$ for which the radius $R_{cluster}$ satisfies

$$R_{cluster} < R_{cr} = R_{Schwarzschild} = 2G_{grav}M_0^{cluster} / c_t^2 \tag{24.56}$$

would actually be a *black hole* which would absorb all singularities which would come close to it within a distance $R \leq R_{Schwarzschild}$.

This *Schwarzschild radius* [2] $R_{Schwarzschild}$ is obtained by the same consideration in Relativity so that it is identical in our theory and General Relativity.

On the sphere of perturbations of a black hole

In our theory, we have already touched upon the notion of a black hole by defining conditions *(15.3)* so that a singularity of the gravitational field behaves like a black hole *vis-à-vis transversal waves,* by defining a radius r_{cr} of the *sphere of perturbations* around the singularity, namely the sphere beyond which any transversal wave are captured by the singularity. By applying conditions *(15.3)* to the cluster of mass $M_0^{cluster}$ generating the gravitational field *(24.43)*

$$\tau_{ext\,LR}^{cluster}(r) \cong -\frac{G_{grav}M_0^{cluster}}{2(3\alpha_{TL} + 2\beta_{TL})c_t^2 r} \tag{24.57}$$

we obtain the following expression of radius r_{sphere} of the *sphere of perturbations* of the cluster

$$\left.\frac{\partial \tau_{ext\,LR}^{cluster}(r)}{\partial t}\right|_{r_{sphere}} = \frac{2}{r_{sphere}} \quad \Rightarrow \quad r_{sphere} = \frac{G_{grav}M_0^{cluster}}{4(3\alpha_{TL} + 2\beta_{TL})c_t^2} = \frac{2G_{grav}M_0^{cluster}}{c_t^2} \tag{24.58}$$

With values *(24.50)* of parameters a, b, α_{TL} and β_{TL}, we then obtain the radius of the sphere of perturbation $R_{perturbations\,sphere} = 2GM / c^2$, which is to say the same value as the Schwarzschild sphere. In General Relativity, one defines the *sphere of photons*, meaning the limit in the vicinity of the black hole from which no photon can escape and whose radius is $R_{photons\,sphere} = 3GM / c^2$, or 3/2 of the Schwarzschild radius. We will come back on this troubling difference.

On the limit radius for infinite dilation of time of an observer falling in a black hole

In our theory, the proper time of an ***HS*** is exactly written as an exponential with a term proportional to $-1/r$

$$\frac{t_y}{t} = e^{-\frac{b}{6\alpha_{TL}+4\beta_{TL}}\frac{G_{grav}M_0^{cluster}}{c_t^2 r}} = e^{-\frac{G_{grav}M_0^{cluster}}{c_t^2 r}} \tag{24.59}$$

With this new exponential expression, we note that the proper time of the ***HS*** observer which gets close to a cluster *dilates infinitely when r tends to zero.*

In General Relativity, we say that the proper time of the ***HS*** seems to dilate infinitely when the ***HS*** gets closer than a critical limit r_{cr} calculated on the basis of the Schwarzschild metric, thanks to relation *(24.40)*

---

[2] *Schwarzschild, K. (1916). On the gravitational field of a point mass according to Einstein's theory. Sitzber. Preuss. Akad. Wiss., Physik-Math. Kl., Vol. 189, pp.189-196. (translated by Helga and Roger Stuewer).*

$$t_y \rightarrow 0 \quad \Rightarrow \quad \left(1-\frac{G_{grav}}{c_t^2}\frac{M_0^{amas}}{r}\right) \rightarrow 0 \quad \Rightarrow \quad r_{cr}=\frac{G_{grav}M_0^{amas}}{c_t^2} \tag{24.60}$$

This limit distance is smaller than the Schwarzschild sphere, the point of no-return of a black hole. However it seems rather hard to imagine a time that freezes when the ***HS*** reaches a critical distance, and it is surprising that this critical distance be the half of the radius of Schwarzschild, and not the Schwarzschild radius itself, or the null radius as in our theory!

On the differences of the characteristic radii of blacks holes

In General relativity, the simplest black holes are characterized by 3 critical radii: *the photon sphere* which is worth $R_{photons\ sphere}=3GM/c^2$,the radius of the horizon or the point of no-return, called the *Schwarzschild radius*, which is worth $R_{Schwrzschild}=2GM/c^2$ and the radius for which the time dilation of the observer goes to infinity which is approximately $R_{dilation\ time\rightarrow\infty}\cong GM/c^2$. The fact that there exist 3 different radii for black holes in General Relativity is rather intriguing, as is the existence of infinite time dilation at a given, non-null, radius. It is also mainly for this reason that we cannot describe the dynamics of things falling into black holes using General Relativity beyond the Schwarzschild radius.
In our theory, the radius of the sphere of perturbations and the point of no-return are both similar to the Schwarzschild radius ($2GM/c^2$), what is very satisfying for the spirit as there is only one limit representing the *horizon of a black hole.* Also, there isn't anywhere in our theory a non-null radius where time dilation would be infinite, so that our theory is not limited for the description of a black hole beyond the Schwarzschild field. However our theory is equivalent to general relativity as long as the field of gravitation is weak and satisfies the condition $\tau_{ext\ LR}^{cluster}(r)<1$. The reason is that it is the two experimentally verified facts, namely time dilatation and curvature of light rays in weak fields, which we have chosen to render our theory identical to that of General Relativity in weak fields. However, our theory is different for the strong gravitational fields, as is obvious in expressions *(24.37)* compared to *(24.40)*, which explains the differences that we just described as far as characteristic radii of black holes are concerned.

On the formal analogy of spatial 3D flexion of the cosmic lattice and the 4D Einstein field equations in General Relativity

The spatial curvature of the local cosmic lattice, such as measured by the observer ***GO***, is characterized by a *flexion vector*, given by relation *(8.35)*, namely

$$\vec{\chi}=-\sum_k \vec{e}_k \wedge \overrightarrow{\mathrm{rot}}\,\vec{\alpha}_k-\frac{2}{3}\overrightarrow{\mathrm{grad}}\,\tau=-\overrightarrow{\mathrm{rot}}\,\vec{\omega}-\sum_k \vec{e}_k\wedge\vec{\lambda}_k=-\overrightarrow{\mathrm{rot}}\,\vec{\omega}+\vec{\lambda} \tag{24.61}$$

It is remarkable to notice that this field of flexion can be directly obtained via the *Newton equation (13.14)* of the cosmic lattice. We thus obtain the value of a field of flexion in the cosmic lattice under the form

$$\vec{\chi}=\underbrace{-\frac{4K_2/3+2K_1}{2(K_2+K_3)}\overrightarrow{\mathrm{grad}}\,\tau}_{\text{"gravitational" field}}+\underbrace{\frac{1}{2(K_2+K_3)}\left(n\frac{d\vec{p}}{dt}-\overrightarrow{\mathrm{grad}}\,F^{def}\right)}_{\text{energy–impulsion}}+\underbrace{\frac{K_3}{K_2+K_3}\vec{\lambda}}_{\text{flexion charge}} \tag{24.62}$$

We conclude that the existence of a local topological curvature of the lattice seen by the ***GO*** via the vector of flexion depends on 3 terms:

- the gradient of expansion of the volume, which is nothing more than the gradient of the *field of gravitation* τ *within the lattice,*

- the temporal variations of the local quantity of movement of the lattice and the gradient of density of elastic energy F^{el} stored in the lattice, term we could call the *«vector of "energy-momentum"* due to the singularities present within the lattice,

- the density $\vec{\lambda}$ of charges of flexion within the lattice, which reflects *the presence of topological singularities within the lattice*, such as dislocations and/or disclinations.

On the other hand, for a local observer ***HS***, both its reference rods and its clock depend on *local volume expansion*, so that an equation similar to equation *(24.62)* should necessarily become a 4-dimensional equation of curvature of space-time, which we will not strive to establish here.

The equation *(8.39)*, namely

$$\operatorname{div}\vec{\chi} = \operatorname{div}\vec{\lambda} = \theta \tag{24.63}$$

permits rewriting *(24.62)* under the following form

$$\operatorname{div}\vec{\chi} = \operatorname{div}\left[\frac{1}{2(K_2+K_3)}\left(n\frac{d\vec{p}}{dt} - \overrightarrow{\operatorname{grad}} F^{déf}\right) - \frac{4K_2/3+2K_1}{2(K_2+K_3)}\overrightarrow{\operatorname{grad}}\,\tau + \frac{K_3}{K_2+K_3}\vec{\theta}\right] = \theta \tag{24.64}$$

This relation shows that the *divergence of the vector of flexion is equal to the density of charges of curvature* due to topological singularities in the lattice, which is null if there are no charges of curvature. Furthermore, in the case where there are no charges of curvature, the divergence of the flexion vector *<u>is</u> the equation of Newton of the lattice*!

Relation *(24.62)* which gives us the spatial curvature from the Newton equation of the lattice is the *3D analog of the 4D equation of the Einstein Field in General Relativity*[3], which is written

$$\boldsymbol{G} = 8\pi\boldsymbol{T} \tag{24.65}$$

in which $\boldsymbol{G}$ is the Einstein curvature tensor *(Einstein tensor)*, which is expressed in terms of the *Ricci curvature tensor, corresponding to a certain part of the tensor of Riemann* which describes the curvatures of space-time

$$G_{\mu\nu} = R_{\mu\nu} - \frac{1}{2}g_{\mu\nu}R \tag{24.66}$$

As far as tensor $\boldsymbol{T}$ is concerned, it is a *«geometrical objet»* called the tensor of energy-momentum *(stress-energy tensor)* which characterizes the matter contained in the space.

This equation of the field of Einstein shows how the tensor of energy-momentum of matter generates an average curvature of space-time in its vicinity. It allows us to calculate the static curvature of a massive object, or the dynamic generation of gravitational waves by a massive object. Also it contains the equations of movement («Newton's equations») for the matter which generates the curvature of space-time.

In the case of the equations of field of Einstein, we should also note that the tensor of energy-momentum is a tensor with null divergence

$$\vec{\nabla}\bullet\boldsymbol{T} = 0 \tag{24.67}$$

---

[3] *A very good report of the General Relativity of Einstein can be found in:*
Ch. W. Misner, K. S. Thorne, J. A. Wheeler, Gravitation, W.H. Freeman and Co, San Francisco 1973

which translates the conservation of energy and momentum. This equation *(24.67)* represents the equation of movement in General Relativity.

There is a strong analogy between the field of Einstein *(24.65)* and the equation of the field of flexion *(24.62)* in the case of the *cosmic lattice*, because they also link a "geometrical vector" of spatial curvature to a sort of energy-momentum vector within the solid, which contains both the temporal variations of the local quantity of movement, the gradient of volume expansion, and the gradient of elastic energy F^{el} stored in the lattice, quantities which are all influenced by the presence of torsion charges or of curvature charges within the lattice. Furthermore, this equation *(24.62)* derives directly from the equation of Newton of the lattice. However, contrary to the field equations of Einstein, this equation is deduced by the ***GO*** who is lucky to posses a global clock, so that there is no 'curvature of time' for him, and that, as a consequence, it's curvature equation is purely spatial!

Additionally, in the case of the field equations of Einstein, equation *(24.67)* of the divergence of the energy-momentum tensor, represents the *equation of movement of matter* in General Relativity, just as the equation of divergence of curvature *(24.66)* represents the divergence of the equation of movement of Newton of the lattice in the absence of charges of curvature.

On the other hand, in the equations of field of Einstein, the notion of charges of curvature associated with topological singularities does not exist, which is explicit in the fact that the divergence of the energy-momentum tensor is always null, so that there is no equivalent to *(24.64)* in General Relativity!

It is tempting at this stage to say that the *notion of charges of curvature*, and the pure geometric equation $\operatorname{div}\vec{\chi} = \operatorname{div}\vec{\lambda} = \theta$ responds to the question that Einstein asked when he was referring to the left term in field equation *(24.65)*, the one with the Ricci curvature tensor, as made of fine marble while that on the right, the "energy-momentum tensor", as made of bad quality wood. What he meant is that the right term is only a phenomenological description of the matter parachuted in the field equations and that it is not derived from first principles, while the term on the left (curvature) was. In our theory the right term of *(24.62)* is also 'derived' and of "fine marble", since it derives from the first principle of the Newton's equation of the cosmic lattice!

Chapter 25

Long-range "gravitational" interactions between topological singularities of the lattice

Clusters of twist disclination loops (TL) satisfy Newton's gravity, and most of the results of General Relativity. We will now study how the other types of loops, such as dislocation loops, mixed loops, and macroscopic interstitials/holes, interact with gravity. We will deduce here the full set of long-range interactions of the different kinds of singularity and their behavior.

25.1 – On the dependency of topological singularities on the background expansion

The twist disclination loop (TL) as the source of electrical charge

In the case of a twist disclination loop (TL), we have seen in the previous chapter (table 24.2) that the dependencies on the volume expansion of the radius of the loop, of its torsion angle, of its charge of rotation and of its inertial mass are written

$$\begin{cases} R_{TL} = R_{TL0}\,\mathrm{e}^{\alpha_{TL}\tau} \\ \boldsymbol{\Omega}_{TL} = \boldsymbol{\Omega}_{TL0}\,\mathrm{e}^{\beta_{TL}\tau} \end{cases} \quad \text{and} \quad \begin{cases} q_{\lambda TL} = q_{\lambda TL0}\,\mathrm{e}^{(2\alpha_{TL}+\beta_{TL})\tau} \\ M_0^{TL} = M_{00}^{TL}\,\mathrm{e}^{-7\tau/8} \end{cases} \tag{25.1}$$

This permits us to deduce the energy of distortion of the loop on the expansion

$$E_{dist}^{TL} \cong \left[\frac{1}{2}(K_2+K_3)\zeta_{TL}R_{TL0}^3\boldsymbol{\Omega}_{TL0}^2\right]\mathrm{e}^{\tau/8} \tag{25.2}$$

In the presence of an external field $\tau^{external}$ generated by other singularities, and a background expansion τ_0 of the cosmic lattice, we can then write

$$E_{dist}^{TL} \cong \left[\frac{1}{2}(K_2+K_3)\zeta_{TL}R_{TL0}^3\boldsymbol{\Omega}_{TL0}^2\right]\mathrm{e}^{(\tau_0+\tau^{external})/8} = E_{dist}^{TL}(\tau_0)\,\mathrm{e}^{\tau^{external}/8} \cong M_0^{TL}c_t^2\,\mathrm{e}^{\tau^{external}/8} \tag{25.3}$$

We deduce the 3 important relations in the presence of an external field $\tau^{external}$

$$\begin{cases} \tau_{ext\ LR}^{TL}(r) \cong -\dfrac{4G_{grav}}{c_t^2}\dfrac{M_0^{TL}}{r} \\ E_{dist}^{TL}(\tau^{external}) \cong M_0^{TL}c_t^2\,\mathrm{e}^{\tau^{external}/8} \\ M_0^{TL} \cong \dfrac{1}{2c_t^2}(K_2+K_3)\zeta_{TL}R_{TL}^3\boldsymbol{\Omega}_{TL}^2 \cong \left[\dfrac{1}{2c_{t0}^2}(K_2+K_3)\zeta_{TL}R_{TL0}^3\boldsymbol{\Omega}_{TL0}^2\right]\mathrm{e}^{-7\tau_0/8} \end{cases} \tag{25.4}$$

The Prismatic Edge Dislocation Loop (EL) as the source of curvature charge

In the case of an edge dislocation loop (EL), we do not know a-priori the dependencies of the various quantities, radius of the loop, Burgers vector and curvature charge. We thus introduce the *expansion constants* α_{EL} and β_{EL} such that

$$\begin{cases} R_{EL} = R_{EL0}\, \mathrm{e}^{\alpha_{EL}\tau} \\ \vec{\boldsymbol{B}}_{EL} = \vec{\boldsymbol{B}}_{EL0}\, \mathrm{e}^{\beta_{EL}\tau} \\ q_{\theta EL} = q_{\theta EL0}\, \mathrm{e}^{\beta_{EL}\tau} \end{cases} \tag{25.5}$$

These constants of expansion α_{EL} and β_{EL} are unknown. The dependency of the radius of the loop should be similar to that of the TL, so $\alpha_{EL} = 1/4$, but it could also be equal to the dependency of the lattice step, meaning $\alpha_{EL} = 1/3$. As regards the dependency of the Burgers vector, which must be a vector in the lattice, it should in principle take the value $\beta_{EL} = 1/3$. But for now, we will not fix these values and we will keep α_{EL} and β_{EL} , as the exact values of these parameters are not called to play a crucial role in the rest of the theory. The expressions *(25.5)* allow us to write

$$E_{dist}^{EL} \cong \left[\left(\frac{K_2}{K_3} \right)^2 K_3 \zeta_{EL} R_{EL0}\, \vec{\boldsymbol{B}}_{EL0}^2 \right] \mathrm{e}^{(\alpha_{EL}+2\beta_{EL})\tau} \tag{25.6}$$

In the presence of an external field $\tau^{external}$ generated by other singularities, and a background expansion τ_0 of the cosmic lattice, we can write

$$E_{dist}^{EL} \cong \left[\left(\frac{K_2}{K_3} \right)^2 K_3 \zeta_{EL} R_{EL0}\, \vec{\boldsymbol{B}}_{EL0}^2 \right] \mathrm{e}^{(\alpha_{EL}+2\beta_{EL})(\tau_0+\tau^{external})} = E_{dist}^{EL}(\tau_0)\mathrm{e}^{(\alpha_{EL}+2\beta_{EL})\tau^{external}} \cong M_0^{EL} c_t^2\, \mathrm{e}^{(\alpha_{EL}+2\beta_{EL})\tau^{external}} \tag{25.7}$$

From which we deduce 4 important relations for an edge dislocation loop (EL) in the presence of an external field $\tau^{external}$

$$\begin{cases} \tau_{ext\ LR}^{EL}(r) \cong -\dfrac{4G_{grav}}{c_t^2} \dfrac{M_{curvature}^{EL} + M_0^{EL}}{r} \\ E_{dist}^{EL}(\tau^{external}) \cong M_0^{EL} c_t^2\, \mathrm{e}^{(\alpha_{EL}+2\beta_{EL})\tau^{external}} \\ M_0^{EL} \cong \left(\dfrac{K_2}{K_3} \right)^2 \dfrac{K_3 \zeta_{EL} R_{EL}\, \vec{\boldsymbol{B}}_{EL}^2}{c_t^2} \cong \left[\left(\dfrac{K_2}{K_3} \right)^2 \dfrac{K_3 \zeta_{EL} R_{EL0}\, \vec{\boldsymbol{B}}_{EL0}^2}{c_{t0}^2} \right] \mathrm{e}^{(\alpha_{EL}+2\beta_{EL}-1)\tau_0} \\ M_{curvature}^{EL} = -\dfrac{2\pi K_2}{5c_t^2} R_{EL}^2 \vec{n}\vec{\boldsymbol{B}}_{EL} = \left[-\dfrac{2\pi K_2}{5c_{t0}^2} R_{EL0}^2 \vec{n}\vec{\boldsymbol{B}}_{EL0} \right] \mathrm{e}^{(2\alpha_{EL}+\beta_{EL}-1)\tau_0} \end{cases} \tag{25.8}$$

recalling that the curvature charge and the gravitational mass associated with the curvature charge can be either negative or positive depending on the interstitial nature of the charge

$$M_{curvature}^{EL} \ \text{ and } \ q_{\theta EL} \begin{cases} >0 & \textit{if vacancy loop} \\ <0 & \textit{if interstitial loop} \end{cases} \tag{25.9}$$

The Mixed Dislocation Loop (ML) as the source of dipolar electric moment

In the case of a mixed dislocation loop (ML), we also do not know a priori the dependencies on volume expansion of the radius of the loop and its Burgers vector. We introduce the following *constants of expansion* α_{ML} and β_{ML} so as to write

$$\begin{cases} R_{ML} = R_{ML0}\,\mathrm{e}^{\alpha_{ML}\tau} \\ \vec{B}_{ML} = \vec{B}_{ML0}\,\mathrm{e}^{\beta_{ML}\tau} \end{cases} \tag{25.10}$$

They allow us to deduce the dependency of the energy of distortion of the loop of expansion

$$E_{dist}^{ML} \cong \left[\frac{K_2+K_3}{4}\zeta_{ML}R_{ML0}\,\vec{B}_{ML0}^2\right]\mathrm{e}^{(\alpha_{ML}+2\beta_{ML})\tau} \tag{25.11}$$

In the presence of an external field $\tau^{external}$ generated by other singularities, and a background expansion τ_0 of the cosmic lattice, we can write

$$\begin{aligned} E_{dist}^{ML} &\cong \left[\frac{K_2+K_3}{4}\zeta_{ML}R_{ML0}\,\vec{B}_{ML0}^2\right]\mathrm{e}^{(\alpha_{BM}+2\beta_{BM})(\tau_0+\tau^{external})} \\ &\cong \left[\frac{K_2+K_3}{4}\zeta_{ML}R_{ML0}\,\vec{B}_{ML0}^2\right]\mathrm{e}^{(\alpha_{BM}+2\beta_{BM})\tau_0}\,\mathrm{e}^{(\alpha_{BM}+2\beta_{BM})\tau^{external}} \cong \left(M_0^{ML}\,\mathrm{e}^{(\alpha_{BM}+2\beta_{BM})\tau^{external}}\right)c_t^2 \end{aligned} \tag{25.12}$$

We deduce the four following relations for a ML in the presence of an external field τ^{ext}

$$\begin{cases} \tau_{ext\ LR}^{ML}(r) \cong -\dfrac{4G_{grav}}{c_t^2}\dfrac{M_0^{ML}}{r} \\ E_{dist}^{ML}(\tau^{external}) \cong M_0^{ML}c_t^2\,\mathrm{e}^{(\alpha_{ML}+2\beta_{ML})\tau^{external}} \\ M_0^{ML} \cong \dfrac{K_2+K_3}{4c_t^2}\zeta_{ML}R_{ML}\vec{B}_{ML}^2 \cong \left[\dfrac{K_2+K_3}{4c_{t0}^2}\zeta_{ML}R_{ML0}\,\vec{B}_{ML0}^2\right]\mathrm{e}^{(\alpha_{ML}+2\beta_{ML}-1)\tau_0} \end{cases} \tag{25.13}$$

The macroscopic vacancy (macroscopic lattice hole), the analog of a black hole

The expansion field *(22.86)* associated with a macroscopic lattice can be written by using the gravitational "constant' in the following form

$$\tau_{ext}^{(L)}(r) \cong -\left(1+\tau_0+\tau^{external}\right)\frac{R_L}{r} = -\frac{4G_{grav}}{c_t^2}\frac{\left(1+\tau_0+\tau^{external}\right)c_t^2R_L}{4G_{grav}}\frac{1}{r} \tag{25.14}$$

where $\tau^{external}$ represents the expansion field due to the other singularities that are in the vicinity of the macroscopic vacancy. If this field is weak ($\left|\tau^{external}\right| \ll 1$), we can introduce a *gravitational mass* $M_{grav}^{(L)}$ *of the macroscopic vacancy* by writing

$$\tau_{ext}^{(L)}(r) \cong -\frac{4G_{grav}}{c_t^2}\left(\frac{\left(1+\tau_0\right)c_t^2R_L}{4G_{grav}}\right)\frac{1}{r} \cong -\frac{4G_{grav}}{c_t^2}\frac{M_{grav}^{(L)}}{r} \tag{25.15}$$

in which

$$M_{grav}^{(L)} \cong \frac{c_t^2}{4G_{grav}}R_L\left(1+\tau_0\right) \cong \frac{1}{c_t^2}\sqrt[3]{\frac{6\pi^2N_L}{n_0\,\mathrm{e}}}\left(K_0-4K_2/3-2K_1(1+\tau_0)\right)\left(1+\tau_0\right)R_\infty^2 \tag{25.16}$$

And recalling that $K_2 \ll K_0$ and $K_1 \ll K_0$ in the perfect cosmological lattice, by deducing the following relation

$$\sqrt[3]{\frac{6\pi^2N_L}{n_0\,\mathrm{e}}}\left(K_0-2K_1\tau_0\right)\left(1+\tau_0\right)R_\infty^2 \cong M_{grav}^{(L)}c_t^2 \tag{25.17}$$

which allows us to write the energy *(22.92)* of the macroscopic lattice under the form

$$\begin{aligned} E_{grav}^{(L)} &\cong \sqrt[3]{\frac{6\pi^2 N_L}{n_0 \mathrm{e}}}\left(K_0-2K_1\tau_0\right)\left(1+\tau_0+\tau^{external}\right)R_\infty^2 \\ &\cong \sqrt[3]{\frac{6\pi^2 N_L}{n_0 \mathrm{e}}}\left(K_0-2K_1\tau_0\right)\left(1+\tau_0\right)R_\infty^2+\sqrt[3]{\frac{6\pi^2 N_L}{n_0 \mathrm{e}}}\left(K_0-2K_1\tau_0\right)\left(1+\tau_0\right)R_\infty^2\frac{\tau^{external}}{\left(1+\tau_0\right)} \\ &\cong \sqrt[3]{\frac{6\pi^2 N_L}{n_0 \mathrm{e}}}\left(K_0-2K_1\tau_0\right)\left(1+\tau_0\right)R_\infty^2\left(1+\frac{\tau^{external}}{\left(1+\tau_0\right)}\right) \cong M_{grav}^{(L)}c_t^2\left(1+\frac{\tau^{external}}{\left(1+\tau_0\right)}\right) \end{aligned} \tag{25.18}$$

By using the fact that $c_{t0}^2=\left(K_2+K_3\right)/mn_0$, we obtain the following relations for a macroscopic hole placed in a external field $\tau^{external}$, written with the gravitational mass of the hole M_{grav}^{L}

$$\begin{cases} R_L=\sqrt[3]{\dfrac{3N_L}{4\pi n_0 \mathrm{e}}} \\ \tau_{ext}^{(L)}(r)\cong-\left(1+\tau_0+\tau^{external}\right)\dfrac{R_L}{r}\cong-\dfrac{4G_{grav}}{c_t^2}\dfrac{M_{grav}^{(L)}}{r} \\ E_{grav}^{(L)}\cong M_{grav}^{(L)}c_t^2\left(1+\dfrac{\tau^{external}}{\left(1+\tau_0\right)}\right) \\ M_{grav}^{(L)}\cong\dfrac{R_\infty^2}{c_t^2}\sqrt[3]{\dfrac{6\pi^2 N_L}{n_0 \mathrm{e}}}\left(K_0-4K_2/3-2K_1(1+\tau_0)\right)\left(1+\tau_0\right) \\ \qquad\cong\left[mR_{\infty 0}^2\sqrt[3]{6\pi^2\mathrm{e}^{-1}n_0^2 N_L}\right]\dfrac{\left(K_0-4K_2/3-2K_1(1+\tau_0)\right)\left(1+\tau_0\right)}{K_2+K_3}\mathrm{e}^{-\frac{\tau_0}{3}} \end{cases} \tag{25.19}$$

We note that the gravitational mass of the hole has the property of changing signs for two different values of the volume expansion of the lattice

$$M_{grav}^{(L)}(\tau_0)\begin{cases} <0 & \Leftrightarrow \quad \tau_0<-1 \\ >0 & \Leftrightarrow \quad -1<\tau_0<\tau_{0cr} \\ <0 & \Leftrightarrow \quad \tau_0>\tau_{0cr} \end{cases} \tag{25.20}$$

Let's recall that the macroscopic vacancy is the only singularity which will necessarily become a *black hole* when the volume expansion of the lattice satisfies relation *(22.100)*, namely

$$\tau_0>1-\tau^{external} \tag{25.21}$$

The macroscopic interstitial, the analog of a neutron star

The expansion field *(22.105)* associated with the macroscopic interstitial can be written by using the gravitational constant under the following form

$$\tau_{ext}^{(I)}(r)=-\frac{2R_I^3}{3R_\infty^2}\frac{1}{r}=-\frac{4G_{grav}}{c_t^2}\left(\frac{c_t^2R_I^3}{6G_{grav}R_\infty^2}\right)\frac{1}{r}=-\frac{4G_{grav}}{c_t^2}\frac{M_{grav}^{(I)}}{r} \tag{25.22}$$

in which we introduce a *gravitational mass* $M_{grav}^{(I)}$ *of the interstitial*

$$M_{grav}^{(I)}=\frac{c_t^2R_I^3}{6G_{grav}R_\infty^2}=\frac{c_t^2}{6G_{grav}R_\infty^2}\frac{3N_I}{4\pi n}=\frac{1}{c_t^2}\left(K_0-4K_2/3-2K_1(1+\tau_0)\right)\frac{N_I}{n} \tag{25.23}$$

Recalling that $K_2 << K_0$ and $K_1 << K_0$ in the perfect lattice, we deduce the following relation

$$(K_0 - 2K_1\tau_0)\frac{N_I}{n} \cong M_{grav}^{(I)} c_t^2 \tag{25.24}$$

which allows us to write the energy *(22.108)* of the macroscopic interstitial under the form

$$E_{grav}^{(I)} \cong \frac{4\pi}{3}(K_0 - 2K_1\tau_0)R_I^3 = (K_0 - 2K_1\tau_0)\frac{N_I}{n}e^{\tau^{external}} \cong \left(M_{grav}^{(I)} e^{\tau^{external}}\right)c_t^2 \tag{25.25}$$

By using the fact that $c_{t0}^2 = (K_2 + K_3)/mn_0$, we obtain the following relations for a macroscopic interstitial placed in an external field $\tau^{external}$, expressed by using the gravitational mass M_{grav}^I of the interstitial

$$\begin{cases} R_I \cong \sqrt[3]{\dfrac{3N_I}{4\pi n}} \cong \sqrt[3]{\dfrac{3N_I}{4\pi n_0}}\, e^{\frac{\tau_0}{3}} \\ \tau_{ext}^{(I)}(r) \cong -\dfrac{4G_{grav}}{c_t^2}\dfrac{M_{grav}^{(I)}}{r} \\ E_{grav}^{(I)} \cong M_{grav}^{(I)} c_t^2\, e^{\tau^{external}} \\ M_{grav}^{(I)} \cong \dfrac{N_I}{nc_t^2}\left(K_0 - 4K_2/3 - 2K_1(1+\tau_0)\right) \cong mN_I\, \dfrac{K_0 - 4K_2/3 - 2K_1(1+\tau_0)}{K_2 + K_3} \end{cases} \tag{25.26}$$

We notice that the gravitational mass of the interstitial has the property of changing sign as a function of the volume expansion of the background

$$M_{grav}^{(I)}(\tau_0) \begin{cases} > 0 & \Leftrightarrow \quad \tau_0 < \tau_{0cr} \\ < 0 & \Leftrightarrow \quad \tau_0 > \tau_{0cr} \end{cases} \tag{25.27}$$

25.2 – Gravitational interactions between the different topological singularities of the cosmic lattice

On the interactions between two twist loops (TL-TL)

If a twist disclination loop *(1)* interact with another twist disclination loop *(2)*, the elastic energy of distortion of the loop *(1)* is written

$$E_{dist(1)}^{BV} \cong \left(M_{0(1)}^{BV} e^{\tau_{ext\,LD(2)}^{BV}(d)/8}\right)c_t^2 \cong \left(M_{0(1)}^{BV} e^{-\frac{G_{grav}}{2c_t^2}\frac{M_{0(2)}^{BV}}{d}}\right)c_t^2 \cong M_{0(1)}^{BV} c_t^2\left(1 - \frac{G_{grav}}{2c_t^2}\frac{M_{0(2)}^{BV}}{d}\right) \tag{25.28}$$

The increase in energy of the loop *(1)* due to the presence of the loop *(2)* is worth

$$\Delta E_{dist(1)}^{TL} \cong -\frac{G_{grav}}{2}\frac{M_{0(1)}^{TL} M_{0(2)}^{TL}}{d} \tag{25.29}$$

The increase in energy of the two loops corresponds to the interaction energy of the loops, namely gravitational energy

$$\Delta E_{grav}^{TL-TL} \cong -G_{grav}\frac{M_{0(1)}^{TL} M_{0(2)}^{TL}}{d} \tag{25.30}$$

and the derivative of this energy with respect to distance corresponds to the force of gravitation between the two loops

$$F_{grav}^{TL-TL} \cong \boldsymbol{G}_{grav} \frac{M_{0(1)}^{TL} M_{0(2)}^{TL}}{d^2} \tag{25.31}$$

On the interaction between two edge loops (EL-EL)

In the case of two edge dislocation loops, the energy of loop *(1)* is written

$$\begin{aligned} E_{dist(1)}^{EL} &\cong \left(M_{0(1)}^{EL} \, \mathrm{e}^{(\alpha_{EL}+2\beta_{EL})\tau_{ext\,LR(2)}^{EL}(d)} \right) c_t^2 \cong M_{0(1)}^{EL} c_t^2 \, \mathrm{e}^{-\frac{4(\alpha_{EL}+2\beta_{EL})\boldsymbol{G}_{grav}}{c_t^2} \frac{M_{curvature(2)}^{EL}+M_{0(2)}^{EL}}{d}} \\ &\cong M_{0(1)}^{EL} c_t^2 \left(1 - \frac{4(\alpha_{EL}+2\beta_{EL})\boldsymbol{G}_{grav}}{c_t^2} \frac{M_{curvature(2)}^{EL}+M_{0(2)}^{EL}}{d} \right) \end{aligned} \tag{25.32}$$

and the increase in energy of loop *(1)* due to loop *(2)* is worth

$$\Delta E_{dist(1)}^{EL} \cong -(\alpha_{EL}+2\beta_{EL})\boldsymbol{G}_{grav} \frac{M_{0(1)}^{EL}\left(M_{curvature(2)}^{EL} + M_{0(2)}^{EL} \right)}{d} \tag{25.33}$$

The gravitational energy of the two loops becomes

$$\Delta E_{grav}^{EL-EL} \cong -(\alpha_{EL}+2\beta_{EL})\boldsymbol{G}_{grav} \frac{M_{0(1)}^{EL} M_{curvature(2)}^{EL} + M_{curvature(1)}^{EL} M_{0(2)}^{EL} + 2 M_{0(1)}^{EL} M_{0(2)}^{EL}}{d} \tag{25.34}$$

which leads us to a gravitational force

$$F_{grav}^{EL-EL} \cong (\alpha_{EL}+2\beta_{EL})\boldsymbol{G}_{grav} \frac{M_{curvature(1)}^{EL} M_{0(2)}^{EL} + M_{curvature(2)}^{EL} M_{0(1)}^{EL}}{d^2} + 2(\alpha_{EL}+2\beta_{EL})\boldsymbol{G}_{grav} \frac{M_{0(1)}^{EL} M_{0(2)}^{EL}}{d^2} \tag{25.35}$$

On the interaction between two mixed loops (ML-ML)

In the case of two mixed dislocation loops interacting we have the following energies

$$\begin{aligned} E_{dist(1)}^{ML} &\cong \left(M_{0(1)}^{ML} \, \mathrm{e}^{(\alpha_{ML}+2\beta_{ML})\tau_{ext\,LR(2)}^{ML}(d)} \right) c_t^2 \cong M_{0(1)}^{ML} c_t^2 \, \mathrm{e}^{-(\alpha_{ML}+2\beta_{ML})\frac{4\boldsymbol{G}_{grav}}{c_t^2} \frac{M_{0(2)}^{ML}}{d}} \\ &\cong M_{0(1)}^{ML} c_t^2 \left(1 - (\alpha_{ML}+2\beta_{ML}) \frac{4\boldsymbol{G}_{grav}}{c_t^2} \frac{M_{0(2)}^{ML}}{d} \right) \\ \Rightarrow \quad \Delta E_{dist(1)}^{ML} &\cong -\,4(\alpha_{ML}+2\beta_{ML})\boldsymbol{G}_{grav} \frac{M_{0(1)}^{ML} M_{0(2)}^{ML}}{d} \end{aligned} \tag{25.36}$$

which lead to the following gravitational interaction

$$\Delta E_{grav}^{ML-ML} \cong -2(\alpha_{ML}+2\beta_{ML})\boldsymbol{G}_{grav} \frac{M_{0(1)}^{ML} M_{0(2)}^{ML}}{d} \tag{25.37}$$

we deduce the gravitational interaction force between the two loops

$$F_{grav}^{ML-ML} \cong 2(\alpha_{ML}+2\beta_{ML})\boldsymbol{G}_{grav} \frac{M_{0(1)}^{ML} M_{0(2)}^{ML}}{d^2} \tag{25.38}$$

On the interaction between a twist loop and an edge loop (TL-EL)

In the case of the interaction between a twist loop and an edge loop (TL-EL), the energy of the twist disclination loop (TL) becomes

$$E_{dist}^{TL} \cong \left(M_0^{TL}\, \mathrm{e}^{\tau_{ext\,LR}^{EL}(d)/8}\right)c_t^2 \cong M_0^{TL} c_t^2\, \mathrm{e}^{-\frac{G_{grav}}{2c_t^2}\frac{M_{courbure}^{EL}+M_0^{EL}}{d}}$$
$$\cong M_0^{TL} c_t^2\left(1-\frac{G_{grav}}{2c_t^2}\frac{M_{courbure}^{EL}+M_0^{EL}}{d}\right) \qquad (25.39a)$$
$$\Rightarrow \quad \Delta E_{dist}^{TL} \cong -\frac{G_{grav}}{2}\frac{M_0^{TL}\left(M_{courbure}^{EL}+M_0^{EL}\right)}{d}$$

And that of the edge dislocation loop (EL)

$$E_{dist}^{EL} \cong \left(M_0^{EL}\, \mathrm{e}^{(\alpha_{EL}+2\beta_{EL})\tau_{ext\,LD}^{TL}(d)}\right)c_t^2 \cong M_0^{EL} c_t^2\, \mathrm{e}^{-\frac{4(\alpha_{EL}+2\beta_{EL})G_{grav}}{c_t^2}\frac{M_0^{TL}}{d}}$$
$$\cong M_0^{EL} c_t^2\left(1-\frac{4(\alpha_{EL}+2\beta_{EL})G_{grav}}{c_t^2}\frac{M_0^{TL}}{d}\right) \qquad (25.39b)$$
$$\Rightarrow \quad \Delta E_{dist}^{EL} \cong -4(\alpha_{EL}+2\beta_{EL})G_{grav}\frac{M_0^{TL}M_0^{EL}}{d}$$

which gives us a total energy

$$\Delta E_{grav}^{TL-EL} \cong -\frac{G_{grav}}{2}\frac{M_0^{TL}M_{courbure}^{EL}}{d}-\left(\frac{1}{2}+4(\alpha_{EL}+2\beta_{EL})\right)G_{grav}\frac{M_0^{TL}M_0^{EL}}{d} \qquad (25.40)$$

and hence the following gravitational force between the two loops

$$F_{grav}^{TL-EL} \cong \frac{1}{2}G_{grav}\frac{M_{courbure}^{EL}M_0^{TL}}{d^2}+\left(\frac{1}{2}+4(\alpha_{EL}+2\beta_{EL})\right)G_{grav}\frac{M_0^{TL}M_0^{EL}}{d^2} \qquad (25.41)$$

On the interaction between a twist loop and a mixed loop (TL-ML)

In the case of a twist loop-mixed loop (TL-ML) interaction, the energy of the TL reads

$$E_{dist}^{TL} \cong \left(M_0^{TL}\, \mathrm{e}^{\tau_{ext\,LR}^{ML}(d)/8}\right)c_t^2 \cong M_0^{TL} c_t^2\, \mathrm{e}^{-\frac{G_{grav}}{2c_t^2}\frac{M_0^{ML}}{d}} \cong M_0^{TL} c_t^2\left(1-\frac{G_{grav}}{2c_t^2}\frac{M_0^{ML}}{d}\right)$$
$$\Rightarrow \quad \Delta E_{dist}^{TL} \cong -\frac{G_{grav}}{2}\frac{M_0^{TL}M_0^{ML}}{d} \qquad (25.42a)$$

That of the ML

$$E_{dist}^{ML} \cong \left(M_0^{ML}\, \mathrm{e}^{(\alpha_{ML}+2\beta_{ML})\tau_{ext\,LD}^{TL}(d)}\right)c_t^2 \cong M_0^{ML} c_t^2\, \mathrm{e}^{-\frac{4(\alpha_{ML}+2\beta_{ML})G_{grav}}{c_t^2}\frac{M_0^{TL}}{d}}$$
$$\cong M_0^{ML} c_t^2\left(1-\frac{4(\alpha_{ML}+2\beta_{ML})G_{grav}}{c_t^2}\frac{M_0^{TL}}{d}\right) \qquad (25.42b)$$
$$\Rightarrow \quad \Delta E_{dist}^{ML} \cong -4(\alpha_{ML}+2\beta_{ML})G_{grav}\frac{M_0^{TL}M_0^{ML}}{d}$$

for a total energy

$$\Delta E_{grav}^{TL-ML} \cong -\left(\frac{1}{2}+4\left(\alpha_{ML}+2\beta_{ML}\right)\right)\mathbf{G}_{grav}\frac{M_0^{TL}M_0^{ML}}{d} \tag{25.43}$$

which gives us the force

$$F_{grav}^{TL-ML} \cong \left(\frac{1}{2}+4\left(\alpha_{ML}+2\beta_{ML}\right)\right)\mathbf{G}_{grav}\frac{M_0^{TL}M_0^{ML}}{d^2} \tag{25.44}$$

On the interaction between an edge loop and a mixed loop (EL-ML)

In the case of an edge loop-mixed loop (EL-ML) interaction, the energy of the EL reads

$$\begin{aligned} E_{dist}^{EL} &\cong \left(M_0^{EL}\, e^{(\alpha_{EL}+2\beta_{EL})\tau_{ext\,LR}^{ML}(d)}\right)c_t^2 \cong M_0^{EL}c_t^2\, e^{-\frac{4(\alpha_{EL}+2\beta_{EL})\mathbf{G}_{grav}}{c_t^2}\frac{M_0^{ML}}{d}} \\ &\cong M_0^{EL}c_t^2\left(1-\frac{4\left(\alpha_{EL}+2\beta_{EL}\right)\mathbf{G}_{grav}}{c_t^2}\frac{M_0^{ML}}{d}\right) \end{aligned} \tag{25.45a}$$

$$\Rightarrow \quad \Delta E_{dist}^{EL} \cong -4\left(\alpha_{EL}+2\beta_{EL}\right)\mathbf{G}_{grav}\frac{M_0^{EL}M_0^{ML}}{d}$$

that of the ML

$$\begin{aligned} E_{dist}^{ML} &\cong \left(M_0^{ML}\, e^{(\alpha_{ML}+2\beta_{ML})\tau_{ext\,LR}^{EL}(d)}\right)c_t^2 \cong M_0^{ML}c_t^2\, e^{-\frac{4(\alpha_{ML}+2\beta_{ML})\mathbf{G}_{grav}}{c_t^2}\frac{M_{curvature}^{EL}+M_0^{EL}}{d}} \\ &\cong M_0^{ML}c_t^2\left(1-\frac{4\left(\alpha_{ML}+2\beta_{ML}\right)\mathbf{G}_{grav}}{c_t^2}\frac{M_{curvature}^{EL}+M_0^{EL}}{d}\right) \end{aligned} \tag{25.45b}$$

$$\Rightarrow \quad \Delta E_{dist}^{ML} \cong -4\left(\alpha_{ML}+2\beta_{ML}\right)\mathbf{G}_{grav}\frac{M_0^{ML}\left(M_{curvature}^{EL}+M_0^{EL}\right)}{d}$$

which gives us the gravitational energy

$$\Delta E_{grav}^{EL-ML} \cong -4\left(\alpha_{ML}+2\beta_{ML}\right)\mathbf{G}_{grav}\frac{M_{curvature}^{EL}M_0^{ML}}{d}-4\left(\alpha_{EL}+2\beta_{EL}+\alpha_{ML}+2\beta_{ML}\right)\mathbf{G}_{grav}\frac{M_0^{EL}M_0^{ML}}{d} \tag{25.46}$$

from which we compute the force

$$F_{grav}^{EL-ML} \cong 4\left(\alpha_{ML}+2\beta_{ML}\right)\mathbf{G}_{grav}\frac{M_{curvature}^{EL}M_0^{ML}}{d^2}+4\left(\alpha_{EL}+2\beta_{EL}+\alpha_{ML}+2\beta_{ML}\right)\mathbf{G}_{grav}\frac{M_0^{EL}M_0^{ML}}{d^2} \tag{25.47}$$

On the interaction between a twist loop and a macroscopic vacancy (TL-V)

In the case of an interaction (TL-V) between a twist loop (TL) and a macroscopic vacancy (L=lacuna), the energy of the TL reads

$$\begin{aligned} E_{dist}^{TL} &\cong \left(M_0^{TL}\, e^{\tau_{ext}^{(L)}(d)/8}\right)c_t^2 \cong M_0^{TL}c_t^2\, e^{\tau_{ext}^{(L)}(d)/8} \cong M_0^{TL}c_t^2\, e^{-\frac{\mathbf{G}_{grav}}{2c_t^2}\frac{M_{grav}^{(L)}}{d}} \\ &\cong M_0^{TL}c_t^2\left(1-\frac{\mathbf{G}_{grav}}{2c_t^2}\frac{M_{grav}^{(L)}}{d}\right) \end{aligned} \tag{25.48a1}$$

$$\Rightarrow \quad \Delta E_{dist}^{TL} \cong -\frac{\boldsymbol{G}_{grav}}{2}\frac{M_0^{TL}M_{grav}^{(L)}}{d} \tag{25.48a2}$$

and the energy of the vacancy reads

$$E_{grav}^{(L)} \cong M_{grav}^{(L)}c_t^2\left(1+\frac{\tau_{ext\,LR}^{TL}(d)}{(1+\tau_0)}\right) \cong M_{grav}^{(L)}c_t^2\left(1-\frac{4\boldsymbol{G}_{grav}}{c_t^2(1+\tau_0)}\frac{M_0^{TL}}{d}\right)$$

$$\Rightarrow \quad \Delta E_{grav}^{(L)} \cong -\frac{4\boldsymbol{G}_{grav}}{(1+\tau_0)}\frac{M_0^{TL}M_{grav}^{(L)}}{d} \tag{25.48b}$$

for a total gravitational energy

$$\Delta E_{grav}^{TL-L} \cong -\boldsymbol{G}_{grav}\left[\frac{1}{2}+\frac{4}{(1+\tau_0)}\right]\frac{M_0^{TL}M_{grav}^{(L)}}{d} \cong -\boldsymbol{G}_{grav}\frac{9+\tau_0}{2(1+\tau_0)}\frac{M_0^{TL}M_{grav}^{(L)}}{d} \tag{25.49}$$

from which we deduce the force

$$F_{grav}^{TL-L} \cong \frac{1}{2}\boldsymbol{G}_{grav}\frac{9+\tau_0}{1+\tau_0}\frac{M_0^{TL}M_{grav}^{(L)}}{d^2} \tag{25.50}$$

which is presented under a simpler form by using values $\boldsymbol{G}_{grav}$ and $M_{grav}^{(L)}$

$$F_{grav}^{TL-L} \cong \frac{c_t^2}{8}(9+\tau_0)\frac{M_0^{TL}R_L}{d^2} \tag{25.51}$$

On the interaction between an edge loop and a macroscopic vacancy (EL-L)

In the case of an interaction between an edge loop (EL) and a macroscopic vacancy (L), the energy of EL and the energy of L read

$$E_{dist}^{EL} \cong \left(M_0^{EL}\,\mathrm{e}^{(\alpha_{EL}+2\beta_{EL})\tau_{ext}^{(L)}(d)}\right)c_t^2 \cong M_0^{EL}c_t^2\,\mathrm{e}^{-\frac{4(\alpha_{EL}+2\beta_{EL})\boldsymbol{G}_{grav}}{c_t^2}\frac{M_{grav}^{(L)}}{d}} \cong M_0^{EL}c_t^2\left(1-\frac{4(\alpha_{EL}+2\beta_{EL})\boldsymbol{G}_{grav}}{c_t^2}\frac{M_{grav}^{(L)}}{d}\right) \tag{25.52a}$$

$$\Rightarrow \quad \Delta E_{dist}^{EL} \cong -4(\alpha_{EL}+2\beta_{EL})\boldsymbol{G}_{grav}\frac{M_0^{EL}M_{grav}^{(L)}}{d}$$

$$E_{grav}^{(L)} \cong M_{grav}^{(L)}c_t^2\left(1+\frac{\tau_{ext\,LR}^{EL}(d)}{(1+\tau_0)}\right) \cong M_{grav}^{(L)}c_t^2\left(1-\frac{4\boldsymbol{G}_{grav}}{(1+\tau_0)c_t^2}\frac{M_{curvature}^{EL}+M_0^{EL}}{d}\right)$$

$$\Rightarrow \quad \Delta E_{grav}^{(L)} \cong -\frac{4\boldsymbol{G}_{grav}}{(1+\tau_0)}\frac{\left(M_{curvature}^{EL}+M_0^{EL}\right)M_{grav}^{(L)}}{d} \tag{25.52b}$$

which leads to the following gravitational energy

$$\Delta E_{grav}^{EL-L} \cong -\frac{4\boldsymbol{G}_{grav}}{(1+\tau_0)}\frac{M_{curvature}^{EL}M_{grav}^{(L)}}{d} - 4\boldsymbol{G}_{grav}\left[\frac{1+(\alpha_{EL}+2\beta_{EL})(1+\tau_0)}{1+\tau_0}\right]\frac{M_0^{EL}M_{grav}^{(L)}}{d} \tag{25.53}$$

and gravitational force

$$F_{grav}^{EL-L} \cong 4\boldsymbol{G}_{grav}\frac{1}{1+\tau_0}\frac{M_{curvature}^{EL}M_{grav}^{(L)}}{d^2} + 4\boldsymbol{G}_{grav}\frac{1+(\alpha_{EL}+2\beta_{EL})(1+\tau_0)}{1+\tau_0}\frac{M_0^{EL}M_{grav}^{(L)}}{d^2} \tag{25.54}$$

which we write, by using values of $\boldsymbol{G}_{grav}$ and $M_{grav}^{(L)}$, under the form

$$F_{grav}^{EL-L} \cong c_t^2\frac{M_{curvature}^{EL}R_L}{d^2} + c_t^2\left[1+(\alpha_{EL}+2\beta_{EL})(1+\tau_0)\right]\frac{M_0^{EL}R_L}{d^2} \tag{25.55}$$

On the interaction between a mixed loop and a macroscopic vacancy (ML-L)

In the case of an interaction between a mixed loop and a macroscopic vacancy (ML-L), the energy of the ML reads

$$E_{dist}^{ML} \cong \left(M_0^{ML} \, e^{(\alpha_{ML}+2\beta_{ML})\tau_{ext}^{(L)}(d)} \right) c_t^2 \cong M_0^{ML} c_t^2 \, e^{\frac{4(\alpha_{ML}+2\beta_{ML})G_{grav}}{c_t^2}\frac{M_{grav}^{(L)}}{d}}$$

$$\cong M_0^{ML} c_t^2 \left(1 - \frac{4(\alpha_{ML}+2\beta_{ML})G_{grav}}{c_t^2} \frac{M_{grav}^{(L)}}{d} \right) \qquad (25.56a)$$

$$\Rightarrow \quad \Delta E_{dist}^{ML} \cong -4(\alpha_{ML}+2\beta_{ML}) G_{grav} \frac{M_0^{ML} M_{grav}^{(L)}}{d}$$

and for the vacancy (L)

$$E_{grav}^{(L)} \cong M_{grav}^{(L)} c_t^2 \left(1 + \frac{\tau_{ext\,LR}^{ML}(d)}{(1+\tau_0)} \right) \cong M_{grav}^{(L)} c_t^2 \left(1 - \frac{4G_{grav}}{(1+\tau_0)c_t^2} \frac{M_0^{ML}}{d} \right)$$

$$\Rightarrow \quad \Delta E_{grav}^{(L)} \cong -\frac{4G_{grav}}{(1+\tau_0)} \frac{M_0^{ML} M_{grav}^{(L)}}{d} \qquad (25.56b)$$

One obtains for the total energy

$$\Delta E_{grav}^{ML-L} \cong -4G_{grav} \frac{1+(\alpha_{ML}+2\beta_{ML})(1+\tau_0)}{(1+\tau_0)} \frac{M_0^{ML} M_{grav}^{(L)}}{d} \qquad (25.57)$$

and for the gravitational force

$$F_{grav}^{ML-L} \cong 4G_{grav} \frac{1+(\alpha_{ML}+2\beta_{ML})(1+\tau_0)}{1+\tau_0} \frac{M_0^{ML} M_{grav}^{(L)}}{d^2} \qquad (25.58)$$

which we rewrite by introducing the values of G_{grav} and $M_{grav}^{(L)}$

$$F_{grav}^{ML-L} \cong c_t^2 \left[1+(\alpha_{ML}+2\beta_{ML})(1+\tau_0) \right] \frac{M_0^{ML} R_L}{d^2} \qquad (25.59)$$

On the interaction between a twist loop and a macroscopic interstitial (TL-I)

In the case of an interaction between a twist loop and a macroscopic interstitial (TL-I), the energies become

$$E_{dist}^{TL} \cong \left(M_0^{TL} \, e^{\tau_{ext}^{(I)}(d)/8} \right) c_t^2 \cong M_0^{TL} c_t^2 \, e^{-\frac{G_{grav}}{2c_t^2}\frac{M_{grav}^{(I)}}{d}} \cong M_0^{TL} c_t^2 \left(1 - \frac{G_{grav}}{2c_t^2} \frac{M_{grav}^{(I)}}{d} \right)$$

$$\Rightarrow \quad \Delta E_{dist}^{TL} \cong -\frac{G_{grav}}{2} \frac{M_0^{TL} M_{grav}^{(I)}}{d} \qquad (25.60a)$$

$$E_{grav}^{(I)} \cong \left(M_{grav}^{(I)} \, e^{\tau_{ext\,LR}^{TL}(d)} \right) c_t^2 \cong M_{grav}^{(I)} c_t^2 \, e^{-\frac{4G_{grav}}{c_t^2}\frac{M_0^{TL}}{d}} \cong M_{grav}^{(I)} c_t^2 \left(1 - \frac{4G_{grav}}{c_t^2} \frac{M_0^{TL}}{d} \right)$$

$$\Rightarrow \quad \Delta E_{grav}^{(I)} \cong -4G_{grav} \frac{M_0^{TL} M_{grav}^{(I)}}{d} \qquad (25.60b)$$

For the total energy, one has

$$\Delta E_{grav}^{TL-I} \cong -4\boldsymbol{G}_{grav}\frac{M_0^{TL}M_{grav}^{(I)}}{d} - \frac{\boldsymbol{G}_{grav}}{2}\frac{M_0^{TL}M_{grav}^{(I)}}{d} \cong -\frac{9}{2}\boldsymbol{G}_{grav}\frac{M_0^{TL}M_{grav}^{(I)}}{d} \tag{25.61}$$

and for the gravitational force

$$F_{grav}^{TL-I} \cong \frac{9}{2}\boldsymbol{G}_{grav}\frac{M_0^{TL}M_{grav}^{(I)}}{d^2} \tag{25.62}$$

which we write by using the values of $\boldsymbol{G}_{grav}$ and $M_{grav}^{(I)}$, as

$$F_{grav}^{TL-I} \cong \frac{3c_t^2}{4R_\infty^2}\frac{M_0^{TL}R_I^3}{d^2} \tag{25.63}$$

On the interaction between an edge loop and a macroscopic interstitial (EL-I)

The energy of the edge loop (EL) interacting with a macroscopic interstitial (I) reads

$$\begin{aligned} E_{dist}^{EL} &\cong \left(M_0^{EL}\,\mathrm{e}^{(\alpha_{EL}+2\beta_{EL})\tau_{ext}^{(I)}(d)}\right)c_t^2 \cong M_0^{EL}c_t^2\,\mathrm{e}^{-\frac{4(\alpha_{EL}+2\beta_{EL})\boldsymbol{G}_{grav}}{c_t^2}\frac{M_{grav}^{(I)}}{d}} \\ &\cong M_0^{EL}c_t^2\left(1-\frac{4(\alpha_{EL}+2\beta_{EL})\boldsymbol{G}_{grav}}{c_t^2}\frac{M_{grav}^{(I)}}{d}\right) \\ \Rightarrow\quad \Delta E_{dist}^{EL} &\cong -4(\alpha_{EL}+2\beta_{EL})\boldsymbol{G}_{grav}\frac{M_0^{EL}M_{grav}^{(I)}}{d} \end{aligned} \tag{25.64a}$$

and the energy of the interstitial

$$\begin{aligned} E_{grav}^{(I)} &\cong \left(M_{grav}^{(I)}\,\mathrm{e}^{\tau_{ext\,LR}^{EL}(d)}\right)c_t^2 \cong M_{grav}^{(I)}c_t^2\,\mathrm{e}^{-\frac{4\boldsymbol{G}_{grav}}{c_t^2}\frac{M_{curvature}^{EL}+M_0^{EL}}{d}} \\ &\cong M_{grav}^{(I)}c_t^2\left(1-\frac{4\boldsymbol{G}_{grav}}{c_t^2}\frac{M_{curvature}^{EL}+M_0^{EL}}{d}\right) \\ \Rightarrow\quad \Delta E_{grav}^{(I)} &\cong -4\boldsymbol{G}_{grav}\frac{M_{grav}^{(I)}\left(M_{curvature}^{EL}+M_0^{EL}\right)}{d} \end{aligned} \tag{25.64b}$$

which gives us the following gravitational energy

$$\Delta E_{grav}^{EL-I} \cong -4\boldsymbol{G}_{grav}\frac{M_{curvature}^{EL}M_{grav}^{(I)}}{d} - 4\boldsymbol{G}_{grav}\left(1+\alpha_{EL}+2\beta_{EL}\right)\frac{M_0^{EL}M_{grav}^{(I)}}{d} \tag{25.65}$$

from which we deduce the force of gravitational interaction between the PDL and MI

$$F_{grav}^{EL-I} \cong 4\boldsymbol{G}_{grav}\frac{M_{curvature}^{EL}M_{grav}^{(I)}}{d^2} + 4\boldsymbol{G}_{grav}\left(1+\alpha_{EL}+2\beta_{EL}\right)\frac{M_0^{EL}M_{grav}^{(I)}}{d^2} \tag{25.66}$$

which can also be writen, by introducing the values of $\boldsymbol{G}_{grav}$ and $M_{grav}^{(I)}$, under the form

$$F_{grav}^{EL-I} \cong \frac{2c_t^2}{3R_\infty^2}\frac{M_{curvature}^{EL}R_I^3}{d^2} + \frac{2c_t^2}{3R_\infty^2}\left(1+\alpha_{EL}+2\beta_{EL}\right)\frac{M_0^{EL}R_I^3}{d^2} \tag{25.67}$$

On the interaction between a mixed loop and a macroscopic interstitial (ML-I)

For an interaction between a mixed loop and a macroscopic interstitial (ML-I), we have

$$E_{dist}^{ML} \cong \left(M_0^{ML} \, \mathrm{e}^{(\alpha_{ML}+2\beta_{ML})\tau_{ext}^{(I)}(d)} \right) c_t^2 \cong M_0^{ML} c_t^2 \, \mathrm{e}^{-\frac{4(\alpha_{ML}+2\beta_{ML})G_{grav}}{c_t^2}\frac{M_{grav}^{(I)}}{d}}$$

$$\cong M_0^{ML} c_t^2 \left(1 - \frac{4(\alpha_{ML}+2\beta_{ML})G_{grav}}{c_t^2} \frac{M_{grav}^{(I)}}{d} \right) \tag{25.68a}$$

$$\Rightarrow \quad \Delta E_{dist}^{ML} \cong -4(\alpha_{ML}+2\beta_{ML}) G_{grav} \frac{M_0^{ML} M_{grav}^{(I)}}{d}$$

$$E_{grav}^{(I)} \cong \left(M_{grav}^{(I)} \, \mathrm{e}^{\tau_{ext\,LD}^{BM}(d)} \right) c_t^2 \cong M_{grav}^{(I)} c_t^2 \, \mathrm{e}^{-\frac{4G_{grav}}{c_t^2}\frac{M_0^{ML}}{d}} \cong M_{grav}^{(I)} c_t^2 \left(1 - \frac{4G_{grav}}{c_t^2} \frac{M_0^{ML}}{d} \right)$$

$$\Rightarrow \quad \Delta E_{grav}^{(I)} \cong -4G_{grav} \frac{M_0^{ML} M_{grav}^{(I)}}{d} \tag{25.68b}$$

which gives us a gravitational energy

$$\Delta E_{grav}^{ML-I} \cong -4G_{grav} \left[1+\alpha_{ML}+2\beta_{ML} \right] \frac{M_0^{ML} M_{grav}^{(I)}}{d} \tag{25.69}$$

and a gravitational force

$$F_{grav}^{ML-I} \cong 4G_{grav} \left[1+\alpha_{ML}+2\beta_{ML} \right] \frac{M_0^{ML} M_{grav}^{(I)}}{d^2} \tag{25.70}$$

Which we rewrite with the values of G_{grav} and $M_{grav}^{(I)}$

$$F_{grav}^{ML-I} \cong \frac{2c_t^2}{3R_\infty^2} \left[1+\alpha_{ML}+2\beta_{ML} \right] \frac{M_0^{ML} R_I^3}{d^2} \tag{25.71}$$

On the interaction between two macroscopic vacancies (L-L)

In the case of an interaction between two macroscopic vacancies (L-L), the energies become

$$E_{grav(1)}^{(L)} \cong M_{grav(1)}^{(L)} c_t^2 \left(1 + \frac{\tau_{ext(2)}^{(L)}(d)}{(1+\tau_0)} \right) \cong M_{grav(1)}^{(L)} c_t^2 \left(1 - \frac{4G_{grav}}{c_t^2 (1+\tau_0)} \frac{M_{grav(2)}^{(L)}}{d} \right) \tag{25.72a}$$

$$\Rightarrow \quad \Delta E_{grav(1)}^{(L)} \cong -\frac{4G_{grav}}{(1+\tau_0)} \frac{M_{grav(1)}^{(L)} M_{grav(2)}^{(L)}}{d} \tag{25.72b}$$

which leads to the following gravitational energy

$$\Delta E_{grav}^{L-L} \cong -\frac{8G_{grav}}{(1+\tau_0)} \frac{M_{grav(1)}^{(L)} M_{grav(2)}^{(L)}}{d} \tag{25.73}$$

and the interaction force

$$F_{grav}^{L-L} \cong \frac{8G_{grav}}{(1+\tau_0)} \frac{M_{grav(1)}^{(L)} M_{grav(2)}^{(L)}}{d^2} \tag{25.74}$$

which can also be written, with the values of G_{grav} and $M_{grav(i)}^{(L)}$, under the form

$$F_{grav}^{L-L} \cong \frac{c_t^4 (1+\tau_0)}{2G_{grav}} \frac{R_{L(1)} R_{L(2)}}{d^2} \tag{25.75}$$

On the interaction between to macroscopic interstitials (I-I)

In the case of an interaction between to macroscopic interstitials (I-I), the energies are

$$E_{grav(1)}^{(I)} \cong \left(M_{grav(1)}^{(I)}\, e^{\tau_{ext(2)}^{(I)}(d)}\right)c_t^2 \cong M_{grav(1)}^{(I)}c_t^2\, e^{-\frac{4G_{grav}}{c_t^2}\frac{M_{grav(2)}^{(I)}}{d}} \cong M_{grav(1)}^{(I)}c_t^2\left(1-\frac{4G_{grav}}{c_t^2}\frac{M_{grav(2)}^{(I)}}{d}\right) \quad (25.76)$$

$$\Rightarrow \quad \Delta E_{grav(1)}^{(I)} \cong -4G_{grav}\frac{M_{grav(1)}^{(I)}M_{grav(2)}^{(I)}}{d}$$

The gravitational energy becomes

$$\Delta E_{grav}^{I-I} \cong -2G_{grav}\frac{M_{grav(1)}^{(I)}M_{grav(2)}^{(I)}}{d} \quad (25.77)$$

and the gravitational force

$$F_{grav}^{I-I} \cong 2G_{grav}\frac{M_{grav(1)}^{(I)}M_{grav(2)}^{(I)}}{d^2} \quad (25.78)$$

which we rewrite, with the values of G_{grav} and $M_{grav(i)}^{(I)}$, under the form

$$F_{grav}^{I-I} \cong \frac{c_t^4}{18G_{grav}R_\infty^4}\frac{R_{I(1)}^3R_{I(2)}^3}{d^2} \quad (25.79)$$

On the interaction between a macroscopic vacancy and a macroscopic interstitial (L-I)

In the case of an L-I interaction, the energies read

$$E_{grav}^{(L)} \cong M_{grav}^{(L)}c_t^2\left(1+\frac{\tau_{ext}^{(I)}(d)}{(1+\tau_0)}\right) \cong M_{grav}^{(L)}c_t^2\left(1-\frac{4G_{grav}}{(1+\tau_0)c_t^2}\frac{M_{grav}^{(I)}}{d}\right)$$

$$\Rightarrow \quad \Delta E_{grav}^{(L)} \cong -\frac{4G_{grav}}{(1+\tau_0)}\frac{M_{grav}^{(L)}M_{grav}^{(I)}}{d} \quad (25.80a)$$

$$E_{grav}^{(I)} \cong \left(M_{grav}^{(I)}\, e^{\tau_{ext}^{(L)}(d)}\right)c_t^2 \cong M_{grav}^{(I)}c_t^2\, e^{-\frac{4G_{grav}}{c_t^2}\frac{M_{grav}^{(L)}}{d}} \cong M_{grav}^{I}c_t^2\left(1-\frac{4G_{grav}}{c_t^2}\frac{M_{grav}^{(L)}}{d}\right)$$

$$\Rightarrow \quad \Delta E_{grav}^{(I)} \cong -4G_{grav}\frac{M_{grav}^{(L)}M_{grav}^{(I)}}{d} \quad (25.80b)$$

For the gravitational energy, one obtains

$$\Delta E_{grav}^{L-I} \cong -4G_{grav}\frac{M_{grav}^{(L)}M_{grav}^{(I)}}{d} - \frac{4G_{grav}}{(1+\tau_0)}\frac{M_{grav}^{(L)}M_{grav}^{(I)}}{d} \cong -4G_{grav}\frac{2+\tau_0}{1+\tau_0}\frac{M_{grav}^{(L)}M_{grav}^{(I)}}{d} \quad (25.81)$$

and for the gravitational force

$$F_{grav}^{L-I} \cong 4G_{grav}\frac{2+\tau_0}{1+\tau_0}\frac{M_{grav}^{(L)}M_{grav}^{(I)}}{d^2} \quad (25.82)$$

which we rewrite, with the values of G_{grav}, $M_{grav}^{(L)}$ and $M_{grav}^{(I)}$, as

$$F_{grav}^{L-I} \cong \frac{c_t^4}{6R_\infty^2}\frac{2+\tau_0}{G_{grav}}\frac{R_LR_I^3}{d^2} \quad (25.83)$$

On the diverse gravitational forces between the different topological singularities

We can now draw the complete table for the expressions of gravitational interactions between the various topological singularities within the perfect cosmic lattice (table 25.1).

Table 25.1 - The gravitational forces between the different topological singularities

$$\begin{cases} F_{grav}^{TL-TL} \cong G_{grav} \dfrac{M_{0(1)}^{TL} M_{0(2)}^{TL}}{d^2} \\ F_{grav}^{EL-EL} \cong \left(\alpha_{EL}+2\beta_{EL}\right) G_{grav} \dfrac{M_{curvature(1)}^{EL} M_{0(2)}^{EL} + M_{curvature(2)}^{EL} M_{0(1)}^{EL}}{d^2} + 2\left(\alpha_{EL}+2\beta_{EL}\right) G_{grav} \dfrac{M_{0(1)}^{EL} M_{0(2)}^{EL}}{d^2} \\ F_{grav}^{ML-ML} \cong 2\left(\alpha_{ML}+2\beta_{ML}\right) G_{grav} \dfrac{M_{0(1)}^{ML} M_{0(2)}^{ML}}{d^2} \end{cases}$$

$$\begin{cases} F_{grav}^{TL-EL} \cong \dfrac{1}{2} G_{grav} \dfrac{M_{curvature}^{EL} M_0^{TL}}{d^2} + \left(\dfrac{1}{2} + 4(\alpha_{EL}+2\beta_{EL})\right) G_{grav} \dfrac{M_0^{TL} M_0^{EL}}{d^2} \\ F_{grav}^{TL-ML} \cong \left(\dfrac{1}{2} + 4(\alpha_{ML}+2\beta_{ML})\right) G_{grav} \dfrac{M_0^{TL} M_0^{ML}}{d^2} \\ F_{grav}^{EL-ML} \cong 4\left(\alpha_{ML}+2\beta_{ML}\right) G_{grav} \dfrac{M_{curvature}^{EL} M_0^{ML}}{d^2} + 4\left(\alpha_{EL}+2\beta_{EL}+\alpha_{ML}+2\beta_{ML}\right) G_{grav} \dfrac{M_0^{EL} M_0^{ML}}{d^2} \end{cases}$$

$$\begin{cases} F_{grav}^{TL-L} \cong \dfrac{1}{2} G_{grav} \dfrac{9+\tau_0}{1+\tau_0} \dfrac{M_0^{TL} M_{grav}^{(L)}}{d^2} \cong \dfrac{c_t^2}{8}\left(9+\tau_0\right) \dfrac{M_0^{TL} R_L}{d^2} \\ F_{grav}^{EL-L} \cong 4G_{grav} \dfrac{1}{1+\tau_0} \dfrac{M_{curvature}^{EL} M_{grav}^{(L)}}{d^2} + 4G_{grav} \dfrac{1+\left(\alpha_{EL}+2\beta_{EL}\right)\left(1+\tau_0\right)}{1+\tau_0} \dfrac{M_0^{EL} M_{grav}^{(L)}}{d^2} \\ \qquad \cong c_t^2 \dfrac{M_{curvature}^{EL} R_L}{d^2} + c_t^2\left[1+\left(\alpha_{EL}+2\beta_{EL}\right)\left(1+\tau_0\right)\right] \dfrac{M_0^{EL} R_L}{d^2} \\ F_{grav}^{ML-L} \cong 4G_{grav} \dfrac{1+\left(\alpha_{ML}+2\beta_{ML}\right)\left(1+\tau_0\right)}{1+\tau_0} \dfrac{M_0^{ML} M_{grav}^{(L)}}{d^2} \cong c_t^2\left[1+\left(\alpha_{ML}+2\beta_{ML}\right)\left(1+\tau_0\right)\right] \dfrac{M_0^{ML} R_L}{d^2} \end{cases}$$

$$\begin{cases} F_{grav}^{TL-I} \cong \dfrac{9}{2} G_{grav} \dfrac{M_0^{TL} M_{grav}^{(I)}}{d^2} \cong \dfrac{3c_t^2}{4R_\infty^2} \dfrac{M_0^{TL} R_I^3}{d^2} \\ F_{grav}^{EL-I} \cong 4G_{grav} \dfrac{M_{curvature}^{EL} M_{grav}^{(I)}}{d^2} + 4G_{grav}\left(1+\alpha_{EL}+2\beta_{EL}\right) \dfrac{M_0^{EL} M_{grav}^{(I)}}{d^2} \\ \qquad \cong \dfrac{2c_t^2}{3R_\infty^2} \dfrac{M_{curvature}^{EL} R_I^3}{d^2} + \dfrac{2c_t^2}{3R_\infty^2}\left(1+\alpha_{EL}+2\beta_{EL}\right) \dfrac{M_0^{EL} R_I^3}{d^2} \\ F_{grav}^{ML-I} \cong 4G_{grav}\left(1+\alpha_{ML}+2\beta_{ML}\right) \dfrac{M_0^{ML} M_{grav}^{(I)}}{d^2} \cong \dfrac{2c_t^2}{3R_\infty^2}\left(1+\alpha_{ML}+2\beta_{ML}\right) \dfrac{M_0^{ML} R_I^3}{d^2} \end{cases}$$

$$\begin{cases} F_{grav}^{L-L} \cong \dfrac{8G_{grav}}{\left(1+\tau_0\right)} \dfrac{M_{grav(1)}^{(L)} M_{grav(2)}^{(L)}}{d^2} \cong \dfrac{c_t^4\left(1+\tau_0\right)}{2G_{grav}} \dfrac{R_{L(1)} R_{L(2)}}{d^2} \\ F_{grav}^{I-I} \cong 2G_{grav} \dfrac{M_{grav(1)}^{(I)} M_{grav(2)}^{(I)}}{d^2} \cong \dfrac{c_t^4}{18G_{grav} R_\infty^4} \dfrac{R_{I(1)}^3 R_{I(2)}^3}{d^2} \\ F_{grav}^{L-I} \cong 4G_{grav} \dfrac{2+\tau_0}{1+\tau_0} \dfrac{M_{grav}^{(L)} M_{grav}^{(I)}}{d^2} \cong \dfrac{c_t^4}{6R_\infty^2} \dfrac{2+\tau_0}{G_{grav}} \dfrac{R_L R_I^3}{d^2} \end{cases}$$

In that table, we notice two important things:

- only the interaction between the twist disclination loops which carry an electrical charge of rotation, satisfy exactly the Newton law of gravitation. It should be noted that these loops have a much higher energy than the other types of loops and that they dominate the interactions between loops,

- all the other interactions have a slightly modified form of gravitational interaction. Amongst the various loops, all interactions depends on the 'constant of gravitation' G_{grav}, but with different numerical factors, which can contain the unknown parameters $\alpha_{EL}, \beta_{EL}, \alpha_{ML}, \beta_{ML}$,

- in the case where an edge dislocation loop intervenes, there are always two terms of interaction, one depending on the curvature mass $M^{EL}_{curvature}$ of the edge loop, and the other depends on the inertial mass M^{EL}_{0} of the edge loop. As the mass of curvature of the edge loop has a greater value than it's inertial mass, the term containing the curvature mass largely dominates in the expression of the force. Also this term can be attractive or repulsive since the curvature mass of the loop is positive if the loop is of vacancy type and negative if the loop is interstitial,

- in the case where a macroscopic vacancy or a macroscopic interstitial are involved, there are two possible formulations of the gravitational force of interaction: the formulation which features the gravitational masses $M^{(L)}_{grav}$ et $M^{(I)}_{grav}$ and which resemble the formulation of Newton's law, but which presents the disadvantage that the masses $M^{(L)}_{grav}$ and $M^{(I)}_{grav}$ depend strongly on the background expansion of the lattice, to the point of changing signs in certain domains of expansion. This is why we will use the second formulation which features radius R_L and R_I of the macroscopic singularities, and that has the advantage of being a lot simpler to analyze when it comes to the sign of the interaction (attractive or negative),

- on the basis of these expressions the 'gravitational' forces of interaction between singularities, we can deduce the attractive or repulsive nature of the interactions in table 25.1 as a function of background expansion τ_0 (fig. 25.1),

In figure 25.1, we did not respect the scale on the axis of expansion τ_0, specifically τ_{0cr} is much larger ($\tau_{0cr} >> 1$) since $K_0 >> K_1$ in the case of a cosmic lattice.

In this figure, we note that the interactions evolved under the effect of the background expansion of the lattice. Gravitational forces go from an attractive to a repulsive mode, or vice-versa for certain values of expansion. The changes in sign correspond either to a through-zero for the force of interaction, or to the emergence of a singularity of the attractive force.

It is evident in figure 25.1 that all of this has important implications for the cosmological behavior of singularities within the lattice, meaning on the evolution of singularities during the cosmological expansion of the lattice. We will revisit this important topic in the next chapter.

In figure 25.1, we also have reported certain phenomenas associated with the expansion of the background:

- first off, for $\tau_0 = \tau_{0cr}$, the transition from the domain without transversal waves, which is dominated by localized longitudinal modes (which we will talk about later), to the domain of expansion where there is a real propagation of longitudinal waves and *where quantum mechanics disappears as we will see later,*

- domains of expansion where macroscopic vacancies are or are not black holes, with a transition for the expansion value $\tau_0 = 1$,

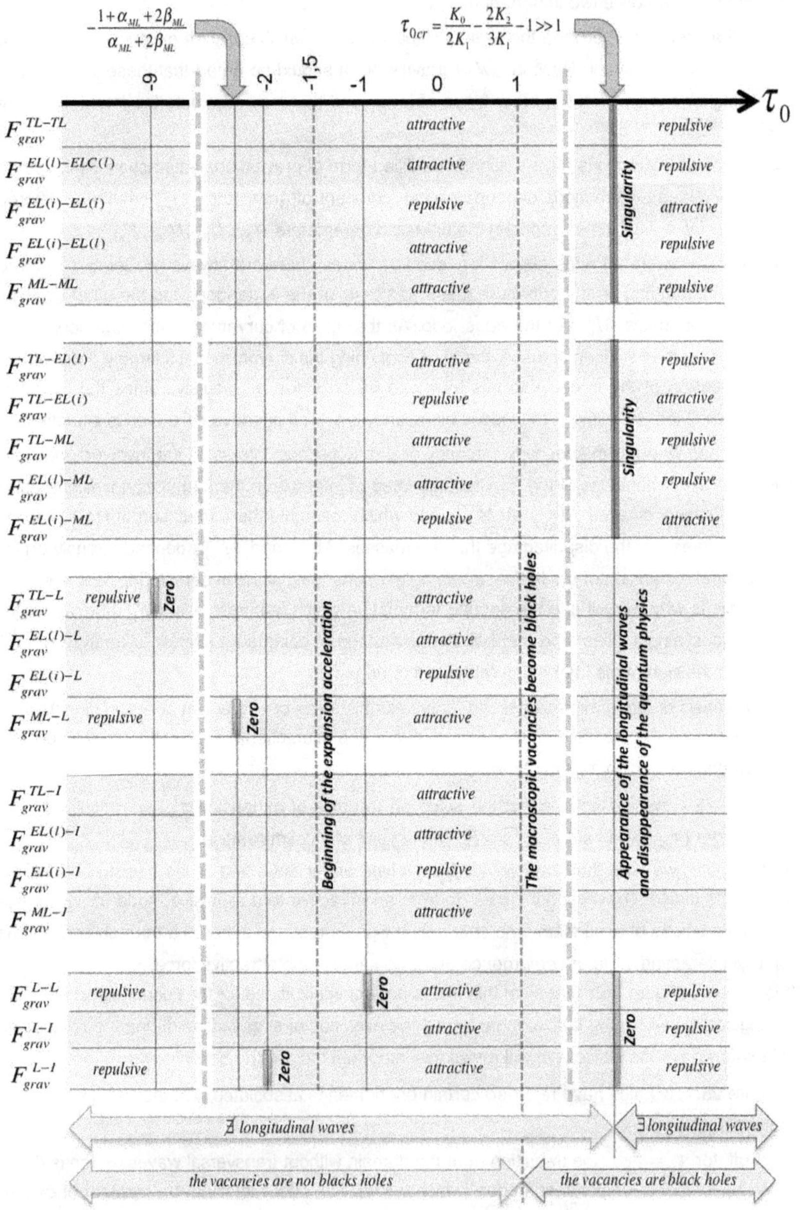

***Figure 25.1** - Attractive or repulsive behaviors of the gravitational interactions forces between singularities as the function of the background expansion τ_0 of the cosmic lattice*

- the expansion value $\tau_0 = -3/2$ for which the evolution of the cosmic lattice (figure 16.8) presents a transition from the domain of end of inflation with reduction of expansion velocity to the domain of expansion with accelerating velocity of expansion,
- one has not shown in figure 25.1 the steps of cosmic evolution that follow $\tau_0 = \tau_{0cr}$, namely the end of acceleration of expansion, the transition through a null velocity of expansion and finally a return to the contraction of the cosmic lattice (see figure 16.8).

On the consequences of values $\alpha_{EL}, \beta_{EL}, \alpha_{ML}, \beta_{ML}$

We have deduced in chapter 24, starting from the dilation of time and the curvature of rays in a weak gravitational field, that the values of the parameters α_{TL} and β_{TL} associated with the twist disclination can take any values, but satisfying the condition $3\alpha_{TL} + 2\beta_{TL} = 1/8$. For the edge and mixed dislocation loops, there are no well known experimental facts that would allow us to deduce the values for parameters $\alpha_{EL}, \beta_{EL}, \alpha_{ML}, \beta_{ML}$. However, nothing prevents us from trying to guess what these values are and what the consequences would be. As EL and ML are topological defects due to translations in the lattice, we can for example assume that these parameters directly follow expansions of the lattice and we therefore have the values

Example I:
$$\begin{cases} \alpha_{EL} = \alpha_{ML} = 1/3 \\ \beta_{EL} = \beta_{ML} = 1/3 \end{cases} \tag{25.84}$$

This case is simple, and the principal consequences would be that the diverse properties of these loops follow the following rules of expansion in the cosmic lattice

$$\begin{cases} M_0^{EL} = M_{00}^{EL}\, \mathrm{e}^{(\alpha_{EL}+2\beta_{EL}-1)\tau_0} = M_{00}^{EL} \neq M_0^{EL}(\tau) \\ M_{courbure}^{EL} = M_{courbure0}^{EL}\, \mathrm{e}^{(2\alpha_{EL}+\beta_{EL}-1)\tau_0} = M_{courbure0}^{EL} \neq M_{courbure}^{EL}(\tau) \\ q_{\theta EL} \propto q_{\theta EL0}\, \mathrm{e}^{\beta_{EL}\tau} \propto q_{\theta EL0}\, \mathrm{e}^{\tau/3} \end{cases} \tag{25.85a}$$

$$\begin{cases} M_0^{ML} = M_{00}^{ML}\, \mathrm{e}^{(\alpha_{ML}+2\beta_{ML}-1)\tau_0} = M_{00}^{ML} \neq M_0^{ML}(\tau) \\ \vec{\omega}_{dipolaire}^{ML} \propto \vec{\omega}_{dipolaire0}^{ML}\, \mathrm{e}^{(\alpha_{ML}+\beta_{ML})\tau_0} \propto \vec{\omega}_{dipolaire0}^{ML}\, \mathrm{e}^{2\tau/3} \end{cases} \tag{25.85b}$$

and thus that the inertial mass and the curvature of these loops would be invariant in the expansion and that the charge of curvature of the EL and the dipolar moment of the ML would both depend on the lattice expansion.

However we can also consider a second possibility based on the fact that the dispiration which is analogous to an electron is in fact composed of the association of a TL and an EL. In this case, in order to maintain the analogy between the dispiration TL-EL and an electron, we need that the radius of the EL presents the same dependency than that of the TL, meaning that $\alpha_{ELC} = \alpha_{TL}$. And by extension, also that the mixed loop satisfied $\alpha_{ML} = \alpha_{TL}$. Furthermore, as the Burgers vectors of the edge loop and the mixed loop have to be translation vectors of the lattice, we will assume a $\mathrm{e}^{\tau/3}$ dependency, and thus that $\beta_{EL} = \beta_{ML} = 1/3$, so that

Example II:
$$\begin{cases} \alpha_{EL} = \alpha_{ML} = \alpha_{TL} \\ \beta_{EL} = \beta_{ML} = 1/3 \end{cases} \tag{25.86}$$

with the following physical consequences

$$\begin{cases} M_0^{EL} = M_{00}^{EL}\, e^{(\alpha_{EL}+2\beta_{EL}-1)\tau_0} = M_{00}^{EL}\, e^{(\alpha_{TL}-1/3)\tau_0} \\ M_{courbure}^{EL} = M_{courbure0}^{EL}\, e^{(2\alpha_{EL}+\beta_{EL}-1)\tau_0} = M_{courbure0}^{EL}\, e^{2(\alpha_{TL}-1/3)\tau_0} \\ q_{\theta EL} \propto q_{\theta EL0}\, e^{\beta_{EL}\tau} \propto q_{\theta EL0}\, e^{\tau/3} \end{cases} \tag{25.87a}$$

$$\begin{cases} M_0^{BM} = M_{00}^{BM}\, e^{(\alpha_{BM}+2\beta_{BM}-1)\tau_0} = M_{00}^{BM}\, e^{(\alpha_{BV}-1/3)\tau_0} \\ \vec{\omega}_{dipolaire}^{BM} \propto \vec{\omega}_{dipolaire0}^{BM}\, e^{(\alpha_{BM}+\beta_{BM})\tau_0} \propto \vec{\omega}_{dipolaire0}^{BM}\, e^{(\alpha_{BV}+1/3)\tau_0} \end{cases} \tag{25.87b}$$

In the case where the torsion of the twist loop is zero, when $\beta_{TL}=0$, the value of the parameters would be then $\alpha_{EL}=\alpha_{ML}=\alpha_{TL}=1/24$.
The values of the parameters $\alpha_{TL},\beta_{TL},\alpha_{EL},\beta_{EL},\alpha_{ML},\beta_{ML}$ are not known, but have to satisfy $3\alpha_{TL}+2\beta_{TL}=1/8$. Their real values must have physical consequences which should be within experimental reach, albeit difficult to reach!

On the possibility of existence of a kind of magnetic monopoles in the case of a variation of the background expansion of the cosmic lattice

There is another disconcerting physical consequence! Indeed if the radius of the edge dislocation loop has a value different of 1/3 ($\alpha_{EL}\neq 1/3$) which assures us of the existence of dispirations analogous to electrons, that would mean that the number of holes or interstitials of the EL should vary if the local expansion changes. However the only possibility for this number to vary is that the *edge loop behaves like a sink or source of vacancies or interstitials in the presence of a variation of the expansion* τ. This effect would have very striking implications for Maxwell equations, as in this case there exists divergent flux of vacancies or interstitials, meaning that the equation of Maxwell $\operatorname{div}(n\vec{p}^{rot})=0$ should be replaced by $\operatorname{div}(n\vec{p})\neq 0$, with as the analog $\operatorname{div}\vec{B}\neq 0$. In other words, in the case where $\alpha_{EL}\neq 1/3$, *the edge dislocation loops behave like magnetic monopoles in the presence of variation of the expansion of the cosmic lattice!*
This ***effect should be measurable by the HS observers***, which would measure a very small monopole magnetic component linked to particles containing edge loops and caused by a local expansion *(which* ***HS*** *should not in principle be able to measure!)*. And this monopole component of the particles containing edge loops must exist due to the effect of cosmological expansion of the universe, which could open a very exciting chapter of this theory!

Chapter 26

Short range «gravitational» interaction and the weak force of cohesion of a dispiration

The previous chapter dealt with the long-distance gravitational interactions between the diverse singularities of the cosmic lattice. This chapter will look at the very short distance gravitational interaction between a twist disclination loop (TL) and an edge dislocation loop (EL) due to the charge of curvature of the EL and the charge of rotation of the TL and their respective perturbations to the field of expansion.

We show that this *interaction between charges of rotation and curvature corresponds to a repulsive force at very short distance that scales in* $1/d^5$ *when the loops are separated, but that it is an attractive force between the two loops when they form a dispiration.*

This gravitational interaction between a charge of rotation and a charge of curvature presents numerous analogies with the famous *'weak force' of particle physics.*

26.1 – Long and short distance interactions between a twist disclination loop (TL) and an edge dislocation loop (EL)

If a twist loop with radius R_{TL} gets close enough to an edge loop with radius R_{EL}, the energy of interaction will feature an energy associated with long range gravitational fields, but also an interaction energy due to *the short range perturbations of the expansion field (23.2)*. We express the short range (SR) field of the twist loop TL in the following way

$$\tau_{ext\,SR}^{TL}(r) \cong \frac{4G_{grav}R_\infty^2}{c_t^2}\frac{R_{TL}M_0^{TL}}{2\pi\zeta_{TL}}\frac{1}{r^4} \tag{26.1}$$

The energy of distortion *(25.7)* of the edge loop EL is

$$E_{dist}^{EL} \cong M_0^{EL}c_t^2\,\mathrm{e}^{(\alpha_{EL}+2\beta_{EL})\tau_{ext\,SR}^{TL}(d)} \cong M_0^{EL}c_t^2\,\mathrm{e}^{(\alpha_{EL}+2\beta_{EL})\frac{2G_{grav}R_\infty^2}{c_t^2}\frac{R_{TL}M_0^{TL}}{\pi\zeta_{TL}}\frac{1}{d^4}} \tag{26.2}$$

The total gravitational energy between the two loops is written, by using *(25.39a)* and *(25.39b)*

$$E_{dist}^{TL-EL} \cong M_0^{TL}c_t^2\,\mathrm{e}^{-\frac{G_{grav}}{2c_t^2}\frac{M_{curvature}^{EL}+M_0^{EL}}{d}} + M_0^{EL}c_t^2\,\mathrm{e}^{-\frac{4(\alpha_{EL}+2\beta_{EL})G_{grav}}{c_t^2}\frac{M_0^{TL}}{d}} + M_0^{EL}c_t^2\,\mathrm{e}^{(\alpha_{EL}+2\beta_{EL})G_{grav}\frac{2R_\infty^2}{\pi c_t^2\zeta_{TL}}\frac{R_{TL}M_0^{TL}}{d^4}} \tag{26.3}$$

Taking into account that $\left|M_{curvature}^{EL}\right| >> M_0^{EL}$ *(23.16)*, the total energy of the two loops can be written

$$E_{dist}^{TL-EL} \cong M_0^{TL}c_t^2\,\mathrm{e}^{-\frac{G_{grav}}{2c_t^2}\frac{M_{curvature}^{EL}}{d}} + M_0^{EL}c_t^2\,\mathrm{e}^{-\frac{4(\alpha_{EL}+2\beta_{EL})G_{grav}}{c_t^2}\frac{M_0^{TL}}{d}+(\alpha_{EL}+2\beta_{EL})G_{grav}\frac{2R_\infty^2}{\pi c_t^2\zeta_{TL}}\frac{R_{TL}M_0^{TL}}{d^4}} \tag{26.4}$$

We can then find the increase in energy due to the interaction between the two loops by developing the exponentials and taking into account that $\left|M_{curvature}^{EL}\right| >> M_0^{EL}$, under the form

$$\Delta E_{inter}^{TL-EL} \cong G_{grav}\left(\frac{2(\alpha_{EL}+2\beta_{EL})R_\infty^2 R_{TL}}{\pi\zeta_{TL}}\frac{M_0^{TL}M_0^{EL}}{d^4}-\frac{M_{courbure}^{EL}M_0^{TL}}{2d}\right) \tag{26.5}$$

At long distance, it is the term in $1/d$ which dominates, so that the energy of interaction is negative if $M_{curvature}^{EL}>0$, meaning if the edge loop (EL) is of vacancy type and positive if $M_{curvature}^{EL}<0$, meaning the EL is interstitial (fig. 26.1).
At short distance, it is the term in $1/d^4$ which dominates, so that the energy of interaction becomes positive (fig. 26.1). In the case of a vacancy edge loop (EL), we have $M_{curvature}^{EL}>0$, so that the energy of interaction goes through zero for $d=d_0$ which is worth

$$d_0=\sqrt[3]{\frac{4}{\pi\zeta_{TL}}(\alpha_{EL}+2\beta_{EL})R_\infty^2 R_{TL}\frac{M_0^{EL}}{M_{courbure}^{EL}}} \tag{26.6}$$

From the energy increase ΔE_{inter}^{TL-EL}, we can deduce the gravitational force between the loops

$$F_{grav}^{TL-EL}(d)=\frac{\partial E_{inter}^{TL-EL}}{\partial d}\cong G_{grav}\left(\frac{M_{courbure}^{EL}M_0^{TL}}{2d^2}-\frac{8(\alpha_{EL}+2\beta_{EL})R_\infty^2 R_{TL}}{\pi\zeta_{TL}}\frac{M_0^{TL}M_0^{EL}}{d^5}\right) \tag{26.7}$$

At long distance it is the term in $1/d^2$ which dominates so that the interaction force is negative and repulsive if $M_{curvature}^{EL}<0$, meaning if the edge loop (EL) is interstitial, and it is positive and attractive if $M_{curvature}^{EL}>0$, meaning if the EL is of vacancy type.
At short distance, it is the term in $1/d^5$ which dominates, so that the interaction force becomes negative and repulsive. In the case of a vacancy edge loop for which $M_{curvature}^{EL}>0$, the force of interaction goes through zero for $d=d_1$ which is worth

$$d_1=\sqrt[3]{\frac{16}{\pi\zeta_{TL}}(\alpha_{EL}+2\beta_{EL})R_\infty^2 R_{TL}\frac{M_0^{EL}}{M_{courbure}^{EL}}}=d_0\sqrt[3]{4} \tag{26.8}$$

and it presents a maximum for $d=d_2$ which is worth

$$d_3=\sqrt[3]{\frac{40}{\pi\zeta_{TL}}(\alpha_{EL}+2\beta_{EL})R_\infty^2 R_{TL}\frac{M_0^{EL}}{M_{courbure}^{EL}}}=d_1\sqrt[3]{10} \tag{26.9}$$

26.2 – The coupling energy of a Twist-Edge dispiration Loop (TEL) formed of a twist loop (TL) and an edge loop (EL)

If a twist disclination loop (TL) with radius R_{TL} is coupled with an edge dislocation loop (EL) with radius $R_{EL}=R_{TL}$, we obtain a *Twist-Edge dispiration Loop (TEL)*. The elastic energy and the kinetic energy of this dispiration loop is due to the field of rotation of the screw loop, to the field of expansion and of shear of the edge loop and to the velocity fields of the two loops. As the various fields of the two loops are all orthogonal and contained in the torus around the dispiration loop, the relativistic energy is the sum of the relativistic energies *(20.47)* and *(20.48)*

$$E_{\mathbf{v}}^{TEL}=\underbrace{\frac{1}{\gamma_t}\left(1-\frac{\mathbf{v}^2}{2c_t^2}\right)\left(E_{dist}^{TL}+E_{dist}^{RL}\right)}_{E_{\mathbf{v}}^{dist}}+\underbrace{\frac{1}{\gamma_t}\frac{1}{2}\left(M_0^{TL}+M_0^{EL}\right)\mathbf{v}^2}_{E_{\mathbf{v}}^{kin}}=\frac{\left(E_{dist}^{TL}+E_{dist}^{EL}\right)}{\gamma_t}=\frac{\left(M_0^{TL}+M_0^{EL}\right)c_t^2}{\gamma_t} \tag{26.10}$$

However if we consider the external field of perturbation of expansion associated with this dispiration loop, we will superpose the fields *(23.2)* of the twist loop with the fields of the edge loop, both long and short distance, so that

$$\tau_{ext}^{TEL}(r) \cong \underbrace{-\frac{4G_{grav}}{c_t^2}\frac{M_0^{TL}}{r}}_{\tau_{ext\,LD}^{TL}(r)} \underbrace{+\frac{4G_{grav}}{c_t^2}\frac{R_\infty^2 R_{TL} M_0^{TL}}{2\pi\zeta_{TL} r^4}}_{\tau_{ext\,CD}^{TL}(r)} \underbrace{-\frac{4G_{grav}}{c_t^2}\frac{M_0^{EL}+M_{curvature}^{EL}}{r}}_{\tau_{ext\,LR}^{EL}(r)} \tag{26.11}$$

The gravitational energy of this field of perturbation is calculated in the following way

$$\begin{aligned} E_{grav}^{TEL} &\cong \iiint_V dV\left[K_1\left(\tau_{ext}^{TEL}(r)\right)^2 - \left(K_0 - 2K_1\tau_0\right)\tau_{ext}^{TEL}(r)\right] \\ &\cong \iiint_V \begin{bmatrix} K_1\left(\tau_{ext\,LR}^{TL}(r)+\tau_{ext\,SR}^{TL}(r)+\tau_{ext\,LR}^{EL}(r)\right)^2 \\ -\left(K_0-2K_1\tau_0\right)\left(\tau_{ext\,LR}^{TL}(r)+\tau_{ext\,SR}^{TL}(r)+\tau_{ext\,LR}^{EL}(r)\right) \end{bmatrix} dV \end{aligned} \tag{26.12}$$

We find the individual energies of the field of perturbation of each of the loops, but we add a new coupling term between the two loops, due to the interaction between TL and EL which intervene in the square term $\left(\tau_{ext\,LR}^{TL}(r)+\tau_{ext\,SR}^{TL}(r)+\tau_{ext\,LR}^{EL}(r)\right)^2$, and which gives us the following increase in energy

$$\begin{aligned} \Delta E_{coupling}^{TEL} &\cong 2K_1 \iiint_V \left(\tau_{ext\,LR}^{EL}(r)\tau_{ext\,LR}^{TL}(r)+\tau_{ext\,LR}^{EL}(r)\tau_{ext\,SR}^{TL}(r)\right)dV \\ &\cong \frac{128\pi K_1 G_{grav}^2}{c_t^4}\left(M_0^{EL}+M_{curvature}^{EL}\right)M_0^{TL}\int_R^{R_\infty}\left(1-\frac{R_\infty^2 R_{TL}}{2\pi\zeta_{TL}}\frac{1}{r^3}\right)dr \end{aligned} \tag{26.13}$$

By carrying the integration and recalling that $M_0^{EL} << \left|M_{curvature}^{EL}\right|$ and by conserving the most important term of this integration, we obtain the coupling energy between the external gravitational fields of the two loops comprising the dispiration

$$\Delta E_{coupling}^{TEL} \cong -\frac{32K_1 G_{grav}^2}{\zeta_{TL} c_t^4}\frac{R_\infty^2}{R_{TL}} M_{courbure}^{EL} M_0^{TL} \tag{26.14}$$

In fact, this energy is due to the coupling of the rotational charge of the twist loop and the curvature charge of the edge loopL. It is then an interaction between charges of rotation and curvature of the two loops of the TEL! We note that this coupling energy is negative if $M_{curvature}^{EL} > 0$, meaning if the edge loop is of vacancy type and positive if $M_{curvature}^{EL} < 0$, meaning if the EL is interstitial

On the binding energy of a TEL dispiration

Let's compare now the gravitational energy at short distance $\Delta E_{inter\,SR}^{TL-EL}(d)$ between two loops separated by distance d with coupling energy $\Delta E_{coupling}^{TEL}$ within the dispiration loop. We have

$$\left|\frac{\Delta E_{couplage}^{DCV}}{\Delta E_{inter\,CD}^{BV-BC}(d)}\right| = \frac{2K_1}{K_0 - 4K_2/3 - 2K_1(1+\tau_0)} \frac{1}{\left(\alpha_{BC}+2\beta_{BC}\right)} \frac{\left|M_{courbure}^{BC}\right|}{M_0^{BC}} \frac{d^4}{R_\infty^2 R_{BV}^2} \tag{26.15}$$

For a distance on the order of $d \approx 2R_{BV}$, we have, if $\tau_0 << \tau_{0cr}$ and recalling that $K_3 = K_0$ in the perfect cosmic lattice

$$\left|\frac{\Delta E_{coupling}^{TEL}}{\Delta E_{inter\,SR}^{TL-EL}(d)}\right|_{d\cong 2R} \cong \frac{32}{(\alpha_{EL}+2\beta_{EL})}\frac{K_1}{K_0}\frac{R_{TL}^2}{R_\infty^2}\frac{\left|M_{curvature}^{EL}\right|}{M_0^{EL}} \cong \frac{64\pi}{5(\alpha_{EL}+2\beta_{EL})\zeta_{EL}}\frac{K_1}{K_2}\frac{R_{TL}^2 R_{EL}}{R_\infty^2\left|\vec{B}_{EL}\right|} < 1 \qquad (26.16)$$

In this expression, there appears the ratio $R_{TL}^2 R_{EL} / R_\infty^2 \left|\vec{B}_{EL}\right|$ which most likely is very small compared to 1, so that the absolute value $\left|\Delta E_{coupling}^{TEL}\right|$ of the coupling energy of the dispiration is assuredly smaller than the interaction energy ΔE_{inter}^{TL-EL} between the two loops when they are separated by a distance $d \approx 2R_{TL}$.

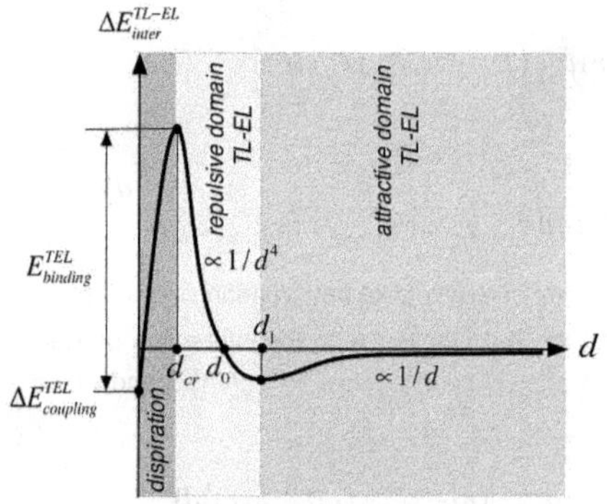

vacancy edge dislocation loop - twist disclination loop

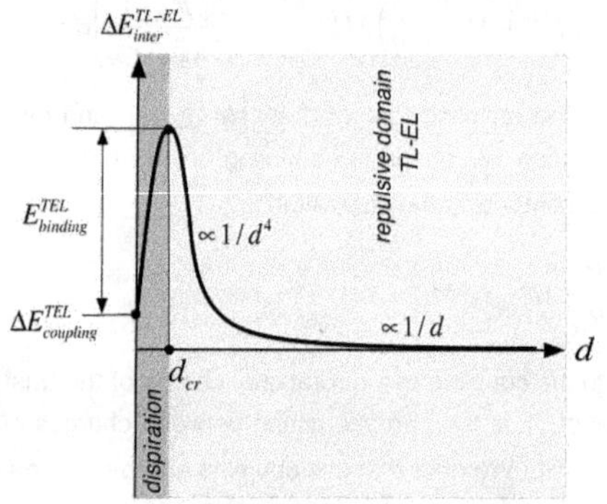

interstitial edge dislocation loop - twist disclination loop

Figure 26.1 *- Interaction potentials between an edge dislocation loop (EL) (of vacancy or interstitial type) and a twist disclination loop (TL) to form a twist-edge dispiration loop (TEL)*

From which we deduce that the potential energy of interaction ΔE_{inter}^{TL-EL} between the two loops, as a function of distance d between the two loops in fact behaves like a *capture potential* such as that represented in figure 26.1 in the domain "dispiration". This potential holds the two loops together within the dispiration, with a binding energy $E_{binding}^{TEL}$ corresponding to the difference of energy between the maximum of the potential energy of interaction ΔE_{inter}^{TL-EL} of the TL and EL and the coupling energy $\Delta E_{coupling}^{TEL}$ of the dispiration (fig. 26.1).

The potential energy of interaction ΔE_{inter}^{TL-EL} has a repulsive force that scales as $1/d^5$ as soon as the loops are separated by a critical value on the order of $d_{cr} \approx 2R_{TL}$. To separate the two dispiration loops, it suffices:
- either a fluctuation in energy equal or superior to the binding energy $E_{binding}^{TEL}$ of the dispiration loop, so that the individual loops are at a distance $d > d_{cr}$ and repulse each other indefinitely,
- either the edge loop crosses the barrier $d > d_{cr}$ by a quantic tunnel effect and the loops start repulsing each other indefinitely.

26.3 – On the analogy with the weak interaction of the standard model of particle physics

The short range interaction that we just described between the charges of rotation and curvature, meaning between a twist disclination loop (TL) and an edge dislocation loop (EL) respectively, present a remarkable analogy with t*he weak interaction of the Standard Model of particle physics.*

The weak interaction of the standard Model is one of the four fundamental forces of nature. It is responsible for the radio-active decay of fundamental particles and it applies to all fermions, meaning electrons, neutrinos and quarks. In the standard model, the weak interaction is linked to the exchange of massive bosons $\overline{W}^+$, W^- and Z^0, and it allows us to explain the weak interactions for leptons, semi-leptons and hadrons. It is because this interaction is of very short range and is very weak, compared to the strong force, that it is called the weak interaction. Also this interaction breaks the parity symmetry ***P*** and the symmetry ***CP***. It is also directly linked to the electrical charge since the electromagnetic interactions and the weak interaction have been unified as two different aspects of electro-weak interaction.

The analogy between the short range interaction between TL and EL and the weak interaction of the standard model is rather self-evident if you consider:
- both are the foundation for a weak binding at the heart of particles or loops,
- both are very short distance interactions,
- both allow the decay of a particle or a loop in other particles or loops,
- the decay of a particle or a dispiration loop can be obtained by a local fluctuation of energy, or a quantum tunnel effect, which is random, just like radio-active decay, which is a statistical phenomena,
- the weak interaction participates in the breaking of the ***P*** and ***CP***, symmetry, which coincides with the fact that there is an asymmetry of interaction between a TL and an EL of vacancy or interstitial nature (fig. 26.1).

As an example, let's consider the weak decay of a muon μ^- into an electron e^- as shown in figure 26.2. This weak interaction of leptonic type shows an initial decay of a muon μ^- in a muon neutrino ν_μ and a massive boson W^-, and the decomposition of a massive boson W^- and an electron e^- and an electronic anti-neutrino $\overline{\nu}_e$.

Let's consider conjecture 8 which stated that *"singularities of a vacancy nature correspond by analogy to anti-matter and interstitial singularities to matter»*. On this basis, let's assume for example that the combination of a twist loop TL^- with an interstitial edge loop $EL_{(1)}^{(i)}$ in a twist-edge dispiration loop (TEL) is the analog of an electron e^-, and let's imagine then that the combination of a twist loop TL^- with an interstitial edge loop $EL^{(i)}_{(2)}$ *with slightly different topologies* (see chapter 30) makes a dispiration loop which is the analogous of a muon μ^-. The weak decay changing the muon μ^- in an electron e^- represented in figure 26.2(a) would posses then a decay analogous to the decay of loops represented in figure 26.2(b). The initial dispiration corresponding to muon μ^-, and composed of the couple $TL^- + EL^{(i)}_{(2)}$ linked by the weak force, decomposes in a twist loop TL^- carrying the charge of rotation, and is analogous to the massive boson W^- *carrying the electrical charge*, and an interstitial *(i)* edge loop $EL^{(i)}_{(2)}$ analogous to the muonic neutrino ν_μ. Then the twist loop BV^- is combined with an interstitial *(i)* edge loop $EL^{(i)}_{(1)}$ to form a dispiration $TL^- + EL^{(i)}_{(1)}$, analogous to the electron e^-, by emitting a vacancy *(l)* edge loop $EL^{(l)}_{(1)}$, anti-loop of the interstitial *(i)* edge loop $EL^{(i)}_{(1)}$, and analogous to the electronic anti-neutrino $\bar{\nu}_e$.

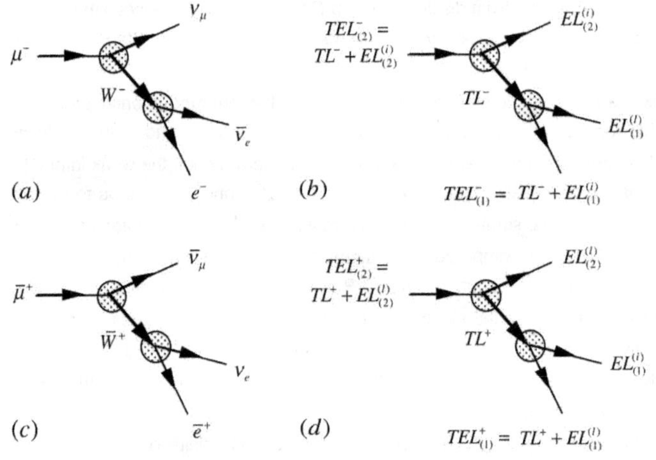

Figure 26.2 *- Analogy between weak leptonic interactions and the decay and formation of dispiration loops*

The weak interaction corresponds to the transformation of the anti-muon $\bar{\mu}^+$ and a positron $\bar{e}^+$ represented by figure 26.2(c) and also has a perfect analog with the loops presented in figure 26.2(d). But this time, the TL^- is replaced by the anti-loop TL^+ with a charge of rotation opposed and the vacancy edge loop $EL^{(l)}_{(1)}$ is replaced by an interstitial edge loop $EL^{(i)}_{(1)}$ and vice-versa. If we consider now the interaction between loops of figure 26.2(b) and 26.2(d), we can imagine immediately that there exist an asymmetry between these two reactions due to the potentials of interaction that are slightly different in the case of vacancy or interstitial edge loops (fig. 26.1). This asymmetry becomes then analogous to the violations of ***P*** parity and ***CP*** symmetry in the case of weak interactions. We will return in detail in chapter 31 on the topological structure of twist loops and edge loops which could be associated with weak interactions.

Chapter 27

A plausible scenario of cosmological evolution towards our current Universe

As we study the analogies of our theory with the great theories of physics, we start this chapter, which is *essentially qualitative and speculative*, by emitting some hypothesis as regards the composition of matter and anti-matter and by supposing that particles and anti-particles are made up of clusters of topological loop singularities, such as screw disclination, edge dislocations and mixed dislocations, of a perfect cosmic lattice. The weak asymmetry that exist between matter and anti-matter is introduced by assuming that matter is based on the edge dislocation loops of interstitial nature and that anti-matter is based on the edge dislocation loops of vacancy type. By using the results of the previous chapter (table 25.1), we will see that the interactions of gravitational nature between particles and anti-particles of the Universe are almost all attractive, while showing a weak scaling of the intensity of interaction depending on the type of particles that are interacting. Only the forces which feature at least one particle based exclusively on a cluster of interstitial nature (which we will interpret as neutrinos) present a repulsive nature!

On the basis of these considerations, it is possible to imagine a scenario of cosmological evolution of the topological singularities which form after the big-bang. This scenario explains the formation of galaxies, the phenomenon of *dark matter* of the astro-physicists, the disappearance of anti-matter from the Universe, the formation of massive black holes in the center of galaxies of matter, the formation of stars and the formation of neutron stars during the gravitational collapse of matter.

Then we will seek to interpret the Hubble constant, the redshift of galaxies and the background cosmic radiation in the frame of our theory.

27.1 – Matter, anti-matter and their gravitational interactions

The existence of 15 gravitational forces depending on the nature of the singularities involved (tableau 25.1), as well as the behavior of the forces as a function of the background expansion of the lattice (figure 25.1), allow us to elaborate a simple scenario for the cosmic evolution of our Universe.

On matter and anti-matter

Starting from conjecture 8, which stipulated that the singularities of vacancy nature correspond by analogy to anti-matter and the singularities of interstitial nature to matter, we introduce the following hypothesis:

- the particles of matter (électron e^-, neutrino ν^0, neutron n^0, proton p^+, etc.) of the Uni-

verse are made of twist disclination loops, which confer them an electrical charge, of mixed dislocation loops, which give them a dipolar electric charge, and of edge dislocation loops of interstitial nature, which gives them a *negative curvature charge*,
- the particles of anti-matter (positron $\overline{e}^{+}$, anti-neutrino $\overline{\nu}^{0}$, anti-neutron $\overline{n}^{0}$, anti-proton $\overline{p}^{-}$, etc.) of the Universe are made of twist disclination loops, which give them an electrical charge, of mixed dislocation loops, which give them a dipolar moment, and of edge dislocation loops of vacancy type which gives them a *positive curvature charge.*

On the asymmetry between particles and anti-particles

If we accept this distinction between particles and anti-particles, the existence of a positive or a negative charge of curvature, due to whether the edge loop is of interstitial or vacancy type, does not appear in General Relativity, nor in the Standard Model. This introduces a weak asymmetry between particles and anti-particles which only exists in our theory.
This asymmetry is similar to the asymmetry observed experimentally between particles and anti-particles in particle physics, but which has no explanation. This asymmetry is seen in certain properties of fundamental particles (like the violation of ***CP*** symmetry, an action combined of a charge conjugation ***C*** and a reflexion symmetry ***P***), but not the rest mass of these particles (linked to the non-violation of the ***CPT*** symmetry, an action combined of a charge conjugation ***C***, a reflexion symmetry ***P*** and a time inversion ***T***)! However, in modern physics, be it the Standard Model or General Relativity, the notion of charge of curvature doesn't exist as it only appears due to the presence of topological defects of the lattice as developed in this work! This property of curvature charge which is linked to the topological singularities could therefore be a great candidate to explain the experimentally observed asymmetry between particles and anti-particles.
To simplify let's call:
- *particle* X a matter particle like an electron e^{-}, a muon μ^{-}, a tau τ^{-}, a neutrino n^{0}, a proton p^{+} (or any other elementary particle composed of quarks) which involves twist loops, with electrical charges, eventually mixed loops in the case of a dipolar electrical field and a majority of edge loops of interstitial nature, and thus *a negative curvature charge*,
- *anti-particle* $\overline{X}$ an anti-matter particle like a positron $\overline{e}^{+}$, an anti-muon $\overline{\mu}^{+}$, an anti-tau $\overline{\tau}^{+}$, an anti-neutron $\overline{n}^{0}$, an anti-proton $\overline{p}^{-}$ (or any other particle made of quarks) which involves twist loops and their electrical charges, eventually mixed loops in the case of a dipolar electric field and a majority of edge loops of vacancy nature and thus *a positive curvature charge*,
- *neutrino* ν^{0} a particle of matter corresponding to the electronic neutrino ν_{e}, to the muonic neutrino ν_{μ} or to the tau neutrino ν_{τ}, which contains no twist loops and no mixed loops, but only edge loops of interstitial nature and therefore a *negative curvature charge*,
- *anti-neutrino* $\overline{\nu}^{0}$ a particle of anti-matter corresponding to the electronic anti-neutrino $\overline{\nu}_{e}$, the muon anti-neutrino $\overline{\nu}_{\mu}$ or the tau anti-neutrino $\overline{\nu}_{\tau}$, which contains no twist loops, no mixed loops but only edge loops of vacancy type, and thus a *positive curvature charge.*
To these 4 types of particles or anti-particles, we can, thanks to the previous chapters, assign inertial masses and equivalent masses of curvature, which satisfy the following relations in the case of particles and anti-particles and in the case of neutrinos and anti-neutrinos

$$\begin{cases} M_0^X = M_0^{\bar{X}} > 0 \\ M_{curvature}^{\bar{X}} > 0 \quad ; \quad M_{curvature}^X < 0 \\ \left| M_{curvature}^X \right| = M_{curvature}^{\bar{X}} << M_0^X = M_0^{\bar{X}} \end{cases} \tag{27.1}$$

$$\begin{cases} M_0^{\nu^0} = M_0^{\bar{\nu}^0} > 0 \\ M_{curvature}^{\bar{\nu}^0} > 0 \quad ; \quad M_{curvature}^{\nu^0} < 0 \\ \left| M_{curvature}^{\nu^0} \right| = M_{curvature}^{\bar{\nu}^0} >> M_0^{\nu^0} = M_0^{\bar{\nu}^0} \end{cases} \tag{27.2}$$

On the effect of the asymmetry between particles and anti-particles on the gravitational interactions

Thus, without knowing a-priori the exact composition of the various particles in terms of singular loops, we can deduce thanks to these relationships between mass of inertia and mass of curvature, relevant informations with regard to the behavior of the forces of gravitational interaction between the diverse particles. Indeed, in the case of the interactions between particles X and $\bar{X}$, we have, thanks to table 25.1, that

$$\begin{cases} \boldsymbol{F}_{grav}^{X-X} \cong \boldsymbol{G}_{grav} \dfrac{\left(M_0^X\right)^2}{d^2} + \boldsymbol{G}_{grav} \dfrac{M_{curvature}^X M_0^X}{d^2} \cong \boldsymbol{G}_{grav} \dfrac{\left(M_0^X\right)^2}{d^2} - \boldsymbol{G}_{grav} \dfrac{M_{curvature}^{\bar{X}} M_0^X}{d^2} \\ \boldsymbol{F}_{grav}^{X-\bar{X}} \cong \boldsymbol{G}_{grav} \dfrac{M_0^X M_0^{\bar{X}}}{d^2} + \dfrac{1}{2}\boldsymbol{G}_{grav} \dfrac{M_{curvature}^X M_0^{\bar{X}}}{d^2} + \dfrac{1}{2}\boldsymbol{G}_{grav} \dfrac{M_{curvature}^{\bar{X}} M_0^X}{d^2} \cong \boldsymbol{G}_{grav} \dfrac{\left(M_0^X\right)^2}{d^2} \\ \boldsymbol{F}_{grav}^{\bar{X}-\bar{X}} \cong \boldsymbol{G}_{grav} \dfrac{\left(M_0^{\bar{X}}\right)^2}{d^2} + \boldsymbol{G}_{grav} \dfrac{M_{curvature}^{\bar{X}} M_0^{\bar{X}}}{d^2} \cong \boldsymbol{G}_{grav} \dfrac{\left(M_0^X\right)^2}{d^2} + \boldsymbol{G}_{grav} \dfrac{M_{curvature}^{\bar{X}} M_0^{\bar{X}}}{d^2} \end{cases} \tag{27.3}$$

From which we deduce the following inequality relationships between the forces of gravitational interaction

$$\boldsymbol{F}_{grav}^{X-X} < \boldsymbol{F}_{grav}^{X-\bar{X}} < \boldsymbol{F}_{grav}^{\bar{X}-\bar{X}} \tag{27.4}$$

As $M_{curvature}^{\bar{X}} << M_0^{\bar{X}}$, the difference between the forces of interaction remain weak, but it assures us nonetheless that there is an asymmetry between particles and anti-particles (gravitationally the particles attract each other a little less than the anti-particles) which will play an important role in the evolution of the Universe as we will see in the next section!
In the case of the interaction between particles X and Y, we have

$$\begin{cases} \boldsymbol{F}_{grav}^{X-Y} \cong \boldsymbol{G}_{grav} \dfrac{M_0^X M_0^Y}{d^2} - \dfrac{1}{2}\boldsymbol{G}_{grav} \dfrac{M_{curvature}^{\bar{X}} M_0^Y}{d^2} - \dfrac{1}{2}\boldsymbol{G}_{grav} \dfrac{M_{curvature}^{\bar{Y}} M_0^X}{d^2} \\ \boldsymbol{F}_{grav}^{\bar{X}-Y} \cong \boldsymbol{G}_{grav} \dfrac{M_0^X M_0^Y}{d^2} + \dfrac{1}{2}\boldsymbol{G}_{grav} \dfrac{M_{curvature}^{\bar{X}} M_0^Y}{d^2} - \dfrac{1}{2}\boldsymbol{G}_{grav} \dfrac{M_{curvature}^{\bar{Y}} M_0^X}{d^2} \\ \boldsymbol{F}_{grav}^{X-\bar{Y}} \cong \boldsymbol{G}_{grav} \dfrac{M_0^X M_0^Y}{d^2} - \dfrac{1}{2}\boldsymbol{G}_{grav} \dfrac{M_{curvature}^{\bar{X}} M_0^Y}{d^2} + \dfrac{1}{2}\boldsymbol{G}_{grav} \dfrac{M_{curvature}^{\bar{Y}} M_0^X}{d^2} \\ \boldsymbol{F}_{grav}^{\bar{X}-\bar{Y}} \cong \boldsymbol{G}_{grav} \dfrac{M_0^X M_0^Y}{d^2} + \dfrac{1}{2}\boldsymbol{G}_{grav} \dfrac{M_{curvature}^{\bar{X}} M_0^Y}{d^2} + \dfrac{1}{2}\boldsymbol{G}_{grav} \dfrac{M_{curvature}^{\bar{Y}} M_0^X}{d^2} \end{cases} \tag{27.5}$$

From which we deduce the following inequality relations between the gravitational forces

$$\boldsymbol{F}_{grav}^{X-Y} < \boldsymbol{F}_{grav}^{\bar{X}-Y} \cong \boldsymbol{F}_{grav}^{X-\bar{Y}} < \boldsymbol{F}_{grav}^{\bar{X}-\bar{Y}} \tag{27.6}$$

These relations show again that particles attract each other a little less than anti-particles!
As regards neutrinos

$$\begin{cases} \boldsymbol{F}_{grav}^{\nu^0-\nu^0} \cong 2(\alpha_{BC}+2\beta_{BC})\boldsymbol{G}_{grav}\dfrac{M_{curvature}^{\nu^0}M_0^{\nu^0}}{d^2} \cong -2(\alpha_{BC}+2\beta_{BC})\boldsymbol{G}_{grav}\dfrac{M_{curvature}^{\bar{\nu}^0}M_0^{\nu^0}}{d^2} < 0 \\ \boldsymbol{F}_{grav}^{\nu^0-\bar{\nu}^0} \cong (\alpha_{BC}+2\beta_{BC})\boldsymbol{G}_{grav}\dfrac{M_{curvature}^{\nu^0}M_0^{\bar{\nu}^0}}{d^2} + (\alpha_{BC}+2\beta_{BC})\boldsymbol{G}_{grav}\dfrac{M_{curvature}^{\bar{\nu}^0}M_0^{\nu^0}}{d^2} + 2(\alpha_{BC}+2\beta_{BC})\boldsymbol{G}_{grav}\dfrac{M_0^{\bar{\nu}^0}M_0^{\nu^0}}{d^2} \\ \qquad\quad \cong 2(\alpha_{BC}+2\beta_{BC})\boldsymbol{G}_{grav}\dfrac{\left(M_0^{\nu^0}\right)^2}{d^2} \\ \boldsymbol{F}_{grav}^{\bar{\nu}^0-\bar{\nu}^0} \cong 2(\alpha_{BC}+2\beta_{BC})\boldsymbol{G}_{grav}\dfrac{M_{curvature}^{\bar{\nu}^0}M_0^{\bar{\nu}^0}}{d^2} \cong 2(\alpha_{BC}+2\beta_{BC})\boldsymbol{G}_{grav}\dfrac{M_{curvature}^{\bar{\nu}^0}M_0^{\nu^0}}{d^2} > 0 \end{cases} \tag{27.7}$$

From which we deduce the following relations

$$\boldsymbol{F}_{grav}^{\nu^0-\nu^0} < 0 \quad ; \quad \boldsymbol{F}_{grav}^{\bar{\nu}^0-\bar{\nu}^0} > 0 \quad ; \quad \boldsymbol{F}_{grav}^{\nu^0-\bar{\nu}^0} \cong 0 \quad ; \quad \boldsymbol{F}_{grav}^{\nu^0-\nu^0} = -\boldsymbol{F}_{grav}^{\bar{\nu}^0-\bar{\nu}^0} \tag{27.8}$$

In other words, neutrinos repulse each other, with the same amplitude as the force with which anti-neutrinos attract each other. With regard to the interaction between neutrinos and anti-neutrinos, it is very weak as it involves $(M_0^{\nu^0})^2$!
Finally, with regards the interactions between particles and neutrinos, we have

$$\begin{cases} \boldsymbol{F}_{grav}^{X-\nu^0} \cong \boldsymbol{G}_{grav}\dfrac{M_0^X M_0^{\nu^0}}{d^2} - \dfrac{1}{2}\boldsymbol{G}_{grav}\dfrac{M_0^{\nu^0}M_{curvature}^{\bar{X}}}{d^2} - \dfrac{1}{2}\boldsymbol{G}_{grav}\dfrac{M_0^X M_{curvature}^{\bar{\nu}^0}}{d^2} \cong -\dfrac{1}{2}\boldsymbol{G}_{grav}\dfrac{M_0^{\nu^0}M_{curvature}^{\bar{X}}}{d^2} - \dfrac{1}{2}\boldsymbol{G}_{grav}\dfrac{M_0^X M_{curvature}^{\bar{\nu}^0}}{d^2} \\ \boldsymbol{F}_{grav}^{X-\bar{\nu}^0} \cong \boldsymbol{G}_{grav}\dfrac{M_0^X M_0^{\nu^0}}{d^2} - \dfrac{1}{2}\boldsymbol{G}_{grav}\dfrac{M_0^{\nu^0}M_{curvature}^{\bar{X}}}{d^2} + \dfrac{1}{2}\boldsymbol{G}_{grav}\dfrac{M_0^X M_{curvature}^{\bar{\nu}^0}}{d^2} \cong -\dfrac{1}{2}\boldsymbol{G}_{grav}\dfrac{M_0^{\nu^0}M_{curvature}^{\bar{X}}}{d^2} + \dfrac{1}{2}\boldsymbol{G}_{grav}\dfrac{M_0^X M_{curvature}^{\bar{\nu}^0}}{d^2} \\ \boldsymbol{F}_{grav}^{\bar{X}-\nu^0} \cong \boldsymbol{G}_{grav}\dfrac{M_0^X M_0^{\nu^0}}{d^2} + \dfrac{1}{2}\boldsymbol{G}_{grav}\dfrac{M_0^{\nu^0}M_{curvature}^{\bar{X}}}{d^2} - \dfrac{1}{2}\boldsymbol{G}_{grav}\dfrac{M_0^X M_{curvature}^{\bar{\nu}^0}}{d^2} \cong +\dfrac{1}{2}\boldsymbol{G}_{grav}\dfrac{M_0^{\nu^0}M_{curvature}^{\bar{X}}}{d^2} - \dfrac{1}{2}\boldsymbol{G}_{grav}\dfrac{M_0^X M_{curvature}^{\bar{\nu}^0}}{d^2} \\ \boldsymbol{F}_{grav}^{\bar{X}-\bar{\nu}^0} \cong \boldsymbol{G}_{grav}\dfrac{M_0^X M_0^{\nu^0}}{d^2} + \dfrac{1}{2}\boldsymbol{G}_{grav}\dfrac{M_0^{\nu^0}M_{curvature}^{\bar{X}}}{d^2} + \dfrac{1}{2}\boldsymbol{G}_{grav}\dfrac{M_0^X M_{curvature}^{\bar{\nu}^0}}{d^2} \cong +\dfrac{1}{2}\boldsymbol{G}_{grav}\dfrac{M_0^{\nu^0}M_{curvature}^{\bar{X}}}{d^2} + \dfrac{1}{2}\boldsymbol{G}_{grav}\dfrac{M_0^X M_{curvature}^{\bar{\nu}^0}}{d^2} \end{cases} \tag{27.9}$$

which implies that

$$\boldsymbol{F}_{grav}^{X-\nu^0} < 0 \quad ; \quad \boldsymbol{F}_{grav}^{X-\bar{\nu}^0} \cong 0 \quad ; \quad \boldsymbol{F}_{grav}^{\bar{X}-\nu^0} \cong 0 \quad ; \quad \boldsymbol{F}_{grav}^{\bar{X}-\bar{\nu}^0} > 0 \tag{27.10}$$

Thus, the interaction between a neutrino and a particle is repulsive. Between an anti-neutrino and an anti-particle, it is attractive. And between a neutrino and an anti-particle, or an anti-neutrino and a particle, the interaction can be slightly positive or negative, but with less amplitude than in the first two cases.

27.2 – A plausible scenario for the cosmological evolution of topological singularities in a perfect cosmic lattice

The effects of cosmological expansion of the lattice on the gravitational interactions

With the inequality relations between the gravitational interactions that we just developed we can return to the cosmic evolution of the perfect cosmic lattice (fig. 16.8 and 16.11g), and include the behavior of gravitational forces between particles. We obtain figure 27.1.

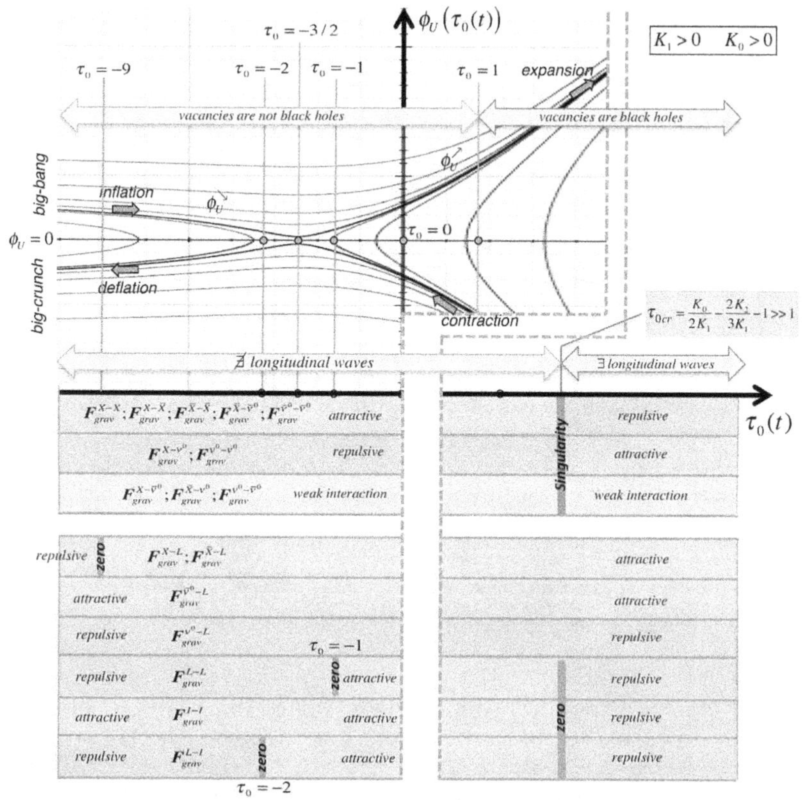

Figure 27.1 *- Behavior of gravitational forces as a function of cosmic evolution of background expansion*

As seen in this figure there appears a series of characteristic values of expansion $\tau_0(t)$ for which there are sudden modifications, either of the behavior of cosmologic expansion, either of the behavior of the diverse forces of gravitational interaction. In this figure, we also have reported the gravitational interaction concerning the macroscopic vacancies (black holes as soon as $\tau_0(t) \geq 1$) and the macroscopic interstitials (neutron stars). Amid the characteristic values of cosmic expansion that are important, we have

- $\tau_0(t \to t_{init}) \to -\infty$, which represents the *'big-bang'* of the lattice at time t_{init},
- $\tau_0(t) = -9$, which represents the value of expansion for which the interaction force between a macro vacancy and a particle or anti-particle goes from repulsive to attractive,
- $\tau_0(t) = -2$, which represents the value of expansion for which the interaction force between a macro vacancy and a macro interstitial goes from repulsive to attractive,
- $\tau_0(t) = -3/2$ *(16.15)*, which corresponds to the transition from the *inflation phase*, during which the velocity of expansion $\phi_U(\tau_0(t))$ decreases to the *expansion phase* during which the

expansion velocity $\phi_U\left(\tau_0(t)\right)$ grows again,

• $\tau_0(t)=-1$, which corresponds to the transition from the repulsion stage between macroscopic vacancies to the attractive stage between macroscopic vacancies,

• $\tau_0(t)=1$, which corresponds to the value of expansion from which macroscopic vacancies become black holes,

• $\tau_0(t)=\tau_{0cr}$, which represents the critical expansion from which we have longitudinal waves within the lattice, but no more local proper mode of vibration of expansion, and which also corresponds to the critical expansion for which many gravitational forces change sign, either by going through a singularity, or a null value.

Beyond τ_{0cr}, the cosmic evolution of the lattice goes from the *expansion phase*, during which the velocity of expansion $\phi_U\left(\tau_0(t)\right)$ is positive, towards a *contraction phase* during which the velocity of expansion $\phi_U\left(\tau_0(t)\right)$ becomes negative, and which finishes, after phases of *contraction and deflation*, in a *«big-crunch»* followed by a new *«big-bang»* of the lattice, and thus a *«big-bounce»* of the lattice due to the kinetic energy stored, as shown in figures 16.8 and 16.11g.

On the basis of figure 27.1, we will develop a plausible scenario for the evolution of topological singularities in a cosmic lattice, which implies several steps.

On a hypothetical liquefaction and solidification of the lattice during the «big-bounce» and on the formation of an initial «hot primordial soup» of singularity loops

In the scenario of a *«big-bounce»* universe as represented in figures 16.8 and 16.11g, the intense contraction of the lattice in the end of *«big-crunch»* must heat the lattice since its kinetic energy becomes enormous, which could lead to a *«liquefaction»*. It is evident that such a phenomenon, based on our knowledge of matter, is not easy to imagine, and that the nodes of the lattice are associated with 'strange particles', which would be responsible for the mass of the lattice and could correspond to the famous Higgs particles of the standard model! For the lattice to present a transition phenomenon of *«liquefaction»*, it would be necessary for its complete state function to not only contain the free energy terms of deformation *(13.6)*, but also thermic terms leading to its phase transition!

By assuming thus that the *«big-bang»* following the *«big-crunch»* happens from a very hot liquid of 'strange particles' which have mass, the inflation phase of the cosmologic evolution should lead to a cooling of the liquid (a reducing of its thermal agitation) and a sudden "liquid-solid" phase transition leading to the cosmic lattice which we introduced in chapter 13! During this phase transition, there could appear structural defects of the lattice as dislocations, disclinations, loops, vacancies and interstitials, and even grain boundaries, in a way that is very similar to that observed during a rapid liquid-solid transition of a metal.

We could talk about 'primordial hot soup' of the loop singularities, the term 'soup' alluding to the facts that we have an initial homogenous distribution of diverse types of loops of singularities and a great mobility of these loops as in a liquid, while the term 'hot' representing a lattice that is very hot, meaning containing a large quantity of transversal wave modes (photons) and localized modes of longitudinal vibrations (gravitons), implying *a strong thermal agitation of the initial loops.*

On the inflation and condensation of singularity loops in particles and anti-particles

During inflation, and thus the cooling of the cosmic lattice, and as soon as temperature will be low enough, the various loops will congregate within the "hot soup" to form complex localized topological dispirations, forming loops of dislocations and disclinations linked by the *weak interaction force* (section 25.3), which will correspond to the various particles of matter (electron e^-, neutrino ν^0, neutron n^0, proton p^+, etc.) and anti-matter (positron $\overline{e}^+$, anti-neutrino $\overline{\nu}^0$, anti-neutron $\overline{n}^0$, anti-proton $\overline{p}^-$, etc.) of our universe. The existence of such combinations of loops in a local form, which could correspond to the various particles of our Universe, will be discussed later in this book.

On the precipitation of matter and anti-matter within the sea of neutrinos and on the formation of galaxies

Within the hot soup, a homogeneous mixture of particles and anti-particles, there are particles and anti-particles whose interaction is attractive (electron e^-, neutron n^0, proton p^+, positron $\overline{e}^+$, anti-neutrino $\overline{\nu}^0$, anti-neutron $\overline{n}^0$, anti-proton $\overline{p}^-$, etc.), but there are also the various neutrinos ν^0 for which gravitational interaction with the other particles (such as electron e^-, neutron n^0, proton p^+, positron $\overline{e}^+$, anti-neutron $\overline{n}^0$, anti-proton $\overline{p}^-$, etc.) is repulsive or non existent (with the anti-neutrinos $\overline{\nu}^0$), and there is also a sea of highly energetic photons interacting strongly with the charged particles and anti-particles by the Compton diffusion mechanism. This situation linked to the component of edge loops and their curvature charge is unique to our theory, and will necessarily lead to a known phenomenon, which is really hard to explain at the moment by the other theories, which is the *initial formation of galaxies*.

Indeed, we can build a very simplified model of the initial, homogeneous hot soup of particles and anti-particles to describe the formation of galaxies. Let's consider that the initial soup forms a sort of liquid composed of attractive X on one hand (electron e^-, neutron n^0, proton p^+, positron $\overline{e}^+$, anti-neutrino $\overline{\nu}^0$, anti-neutron $\overline{n}^0$, anti-proton $\overline{p}^-$, etc.) and of neutrinos ν^0 which are repulsive on the other hand (electronic neutrino ν_e, muon neutrino ν_μ and tau neutrino ν_τ,), and let's try to express the free energy of interaction $f^{interaction}$ per particles within this liquid mixture[1]. By introducing the concentrations C_{ν^0} and $C_X = 1 - C_{\nu^0}$ of repulsive neutrinos ν^0 and attractive particles X within the mixture, the free energy of interaction can be written like a sum of a term of free energy of interaction and a term of entropy

$$f^{interaction} = \frac{z}{2} \sum_{x,y \in (\nu^0, X)} e_{x-y}^{interaction} C_x C_y - kT_{lattice} \sum_{x,y \in (\nu^0, X)} C_x \ln C_y \qquad (27.11)$$

where z is the average coordination number, which represents the average neighboring number with which a particle can form an interaction of pairs and where the 1/2 factor is introduced to not count twice the same interaction.

By introducing now an average value for the inertial mass of attractive particles $\overline{M}_0^X > 0$ and of neutrinos $\overline{M}_0^{\nu^0} > 0$, as well as the average curvature $\overline{M}_{curvature}^{\nu^0} < 0$ of matter neutrinos,

[1] *see section 7.6 in «Théorie eulérienne des milieux déformables: charges de dislocation et de désinclinaison dans les solides», G. Gremaud, Presses Polytechniques et Universitaires Romandes, Lausanne 2013, ISBN 978-2-88074-964-4 (751 pages).*

and by supposing an average distance $\bar{d}(\tau_0)$ between the particles in the initial hot soup, we can express very approximatively the free energy of interaction per particle under the form

$$f^{interaction} \cong \frac{z}{2}\left[-G_{grav}\frac{(\bar{M}_0^X)^2}{\bar{d}(\tau_0)}C_X + \frac{1}{2}G_{grav}\frac{\bar{M}^{v^0}_{curvature}\bar{M}_0^X}{\bar{d}(\tau_0)}C_X C_{v^0} + 2(\alpha_{EL}+2\beta_{EL})G_{grav}\frac{\bar{M}^{v^0}_{curvature}\bar{M}_0^{v^0}}{\bar{d}(\tau_0)}C_{v^0}\right] - kT_{lattice}\left[C_{v^0}\ln C_{v^0} + C_X \ln C_X\right] \quad (27.12)$$

In this expression, the term containing the factor $(\alpha_{EL}+2\beta_{EL})$ is negligeable with respect to the two other terms so that we write with $C_X = 1 - C_{v^0}$

$$f^{interaction} \cong \frac{z}{2}\left[-G_{grav}\frac{(\bar{M}_0^X)^2}{\bar{d}(\tau_0)}\left(1-C_{v^0}\right) + \frac{1}{2}G_{grav}\frac{\bar{M}^{v^0}_{curvature}\bar{M}_0^X}{\bar{d}(\tau_0)}C_{v^0}\left(1-C_{v^0}\right)\right] - kT_{lattice}\left[C_{v^0}\ln C_{v^0} + \left(1-C_{v^0}\right)\ln\left(1-C_{v^0}\right)\right] \quad (27.13)$$

If we represent this free energy as a function of concentration C_{v^0} of neutrinos for different temperature of the lattice (figure 27.2), we notice that at high temperature the minimum free energy is obtained with a homogeneous mixture of attractive particles X and of repulsive neutrinos v^0. But if the temperature of the lattice drops sufficiently, there appears two minima of free energy as a function of concentration C_{v^0} : a minimum corresponding to a weak concentration of neutrinos and a minimum corresponding to a strong concentration of neutrinos. In fact there is a *phase transition by precipitation* which tends to separate the attractive particles X and the repulsive neutrinos v^0. There will be *precipitates*, clusters of attractive particles X, within a sea of repulsive neutrinos v^0. At low temperature, the energy minima correspond to concentrations $C_{v^0} = 0$ and $C_{v^0} = 1$, which corresponds to a complete separation of attractive particles and repulsive neutrinos!

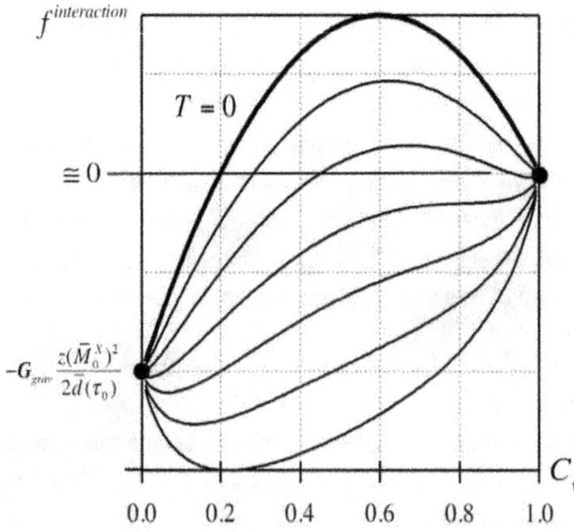

Figure 27.2 *- The free energy of interaction per particle within the initial hot soup of particles as a function of the concentration of repulsive neutrinos for different temperatures of the lattice*

The *phase transition by precipitation* of attractive particles and anti-particles in the form of localized clusters corresponds to the *formation of galaxies in our Universe.* In this model, it is the existence of repulsive neutrinos that becomes the driver for the formation of galaxies. And it is very interesting to note that the repulsive nature of neutrinos of matter is due exclusively to the charge of curvature of neutrinos, which is a concept that exists neither in General Relativity nor in the Standard Model. Also, we already know that the curvature charge is also at the origin of a small asymmetry between matter and anti-matter, which confirms the strong link existing between this experimentally observed asymmetry and the initial formation of galaxies and of the structures of our actual Universe.

On the «dark matter» of astrophysicists

With regards to the formation of a '*sea of repulsive neutrinos*' in which our galaxies bathe, it explains perfectly the phenomenon of 'dark matter' of astrophysics. Indeed when one observes a galaxy and we measure the velocities of stars that compose it as a function of their distance to the center of the galaxy, we notice that the velocities of the stars situated in the periphery of our galaxy are too high compared to the velocities obtained by applying the Newton law of gravitation with the mass of stars (which we can calculate experimentally based on their brilliance). Everything happens as if there was a halo of matter, invisible to our eyes, in the periphery of the galaxy, which, through its gravitational effect, forces the stars to rotate faster to compensate for this attractive effect. This halo of invisible matter was called *dark matter* by astrophysicists, and the quest for the nature of this dark matter is actually one of the great topics in fundamental physics. In our theory, the concept of dark matter is no longer necessary, as it is replaced by the concept of *«sea of repulsive neutrinos»* in which all the galaxies are bathed, the globular clusters, and the other structures of the visible Universe. Indeed, consider a galaxy submitted to the repulsive force of the neutrino sea. This repulsive force corresponds to a compression force which applies to the stars of the galaxy suburbs. To resist to this compression force, the peripheral stars have to turn more quickly than the velocity calculated by Newton's gravity on the basis of the visible mass, in order to equilibrate the compression force of the repulsive neutrino sea by an additional centripetal force of rotation.

On the formation and separation of matter and anti-matter within galaxies

Let's look at what happens inside birthing galaxies, during this phenomenon of precipitation of attractive particles and anti-particles. Within the liquid phase that is precipitating, the attractive gravitational interactions present slight differences, depending on whether we are dealing with particles or anti-particles. Let's consider for example a family of particles X and anti-particles $\overline{X}$. The gravitational forces of interaction between these particles can be written, from *(27.3)*

$$\begin{cases} \boldsymbol{F}_{grav}^{X-X} \cong \boldsymbol{G}_{grav} \dfrac{\left(M_0^X\right)^2}{d^2} - \boldsymbol{G}_{grav} \dfrac{M_{curvature}^{\overline{X}} M_0^X}{d^2} \\ \boldsymbol{F}_{grav}^{X-\overline{X}} \cong \boldsymbol{G}_{grav} \dfrac{\left(M_0^X\right)^2}{d^2} \\ \boldsymbol{F}_{grav}^{\overline{X}-\overline{X}} \cong \boldsymbol{G}_{grav} \dfrac{\left(M_0^X\right)^2}{d^2} + \boldsymbol{G}_{grav} \dfrac{M_{curvature}^{\overline{X}} M_0^{\overline{X}}}{d^2} \end{cases} \qquad (27.14)$$

With the mass of inertia M_0^X of these particles, we deduce, thanks to the classic Newton equation, the acceleration that these particles undergo during their interaction

$$\begin{cases} \boldsymbol{a}_{grav}^{X-X} \cong \dfrac{\boldsymbol{G}_{grav}}{d^2}\left(M_0^X - M_{curvature}^{\overline{X}}\right) \\ \boldsymbol{a}_{grav}^{X-\overline{X}} \cong \dfrac{\boldsymbol{G}_{grav}}{d^2} M_0^X \\ \boldsymbol{a}_{grav}^{\overline{X}-\overline{X}} \cong \dfrac{\boldsymbol{G}_{grav}}{d^2}\left(M_0^X + M_{curvature}^{\overline{X}}\right) \end{cases} \tag{27.15}$$

We deduce that the anti-particles $\overline{X}$ attract each other more strongly than the particles X, and we must therefore see a progressive segregation phenomenon of anti-particles and particles, during which the anti-particles will have a tendency to regroup in the center of the nascent galaxy, leaving the particles in its periphery!

It is clear that this segregation phenomenon must be accompanied by an intense activity of annihilation between particles and anti-particles, in a zone situated around the center of the galaxy, and that would necessarily be *a source of gamma radiation.* But there also must be a combining activity between particles and anti-particles to form matter and anti-matter (initially of hydrogen and anti-hydrogen atoms and helium and anti-helium atoms). These processes of annihilation and recombination must follow through until there is an effective separation between the heart of the galaxy, composed essentially of anti-matter and the periphery of the galaxy composed essentially of matter. We again notice that this separation of matter and anti-matter is *due to the existence of the curvature charge of the edge dislocation loops,* since these charges are responsible for the equivalent mass of curvature $M_{curvature}^{\overline{X}}$, which is itself responsible for the small difference of gravitational interaction between matter and anti-matter!

On the formation of a cosmological radiation background

Initially, all the particles and anti-particles are in thermal equilibrium with a sea of photons, via interactions by Compton diffusion, while their temperature has not dropped enough to form atoms. But as soon as the temperature drops below 3'000 K, there is formation of helium, anti-helium, hydrogen and anti-hydrogen, assuring us of the electric neutrality of matter and anti-matter. At that instant there is also a decoupling of photons and neutral matter and anti-matter. The universe becomes
transparent to photons, which fill the whole space in the form of a *cosmic radiation background.* This cosmic radiation background is known as the cosmological microwave background and has been studied and observed experimentally. It is almost isotropic and present the spectrum of a perfect black body radiation, meaning a *Planck distribution of density of energy $U(\nu)$ of photons,* centered on a temperature T which is measured currently at a value of 2,7 K

$$U(\nu) = \frac{8\pi h\nu^3}{c^3}\frac{1}{e^{\frac{h\nu}{kT}}-1}d\nu \tag{27.16}$$

with c being the speed of light, h the Planck's constant, k the Boltzmann constant, T the temperature of the black body and ν the frequency of photons.

We will revisit, in the next section, the process by which the background radiation 'cools'.

On the gravitational collapse and the disappearance of anti-matter by the formation of gigantic black holes in the center of galaxies

The formation by precipitation of galaxies composed of particles and anti-particles that attract within the sea of repulsive neutrinos will lead to large pressures at the heart of galaxies as they evolve. The emergence of twist in galaxies allows to partially balance the attractive gravitational forces within the galaxies and the compression force of the sea of neutrinos. But at the center of the galaxies, the compression forces could reach values large enough to lead to a gravitational collapse at the heart of the galaxies. If such a collapse happens, as the heart of the galaxies is formed essentially of anti-matter, it will be responsible for the appearance of macroscopic lattice vacancies, as during the collapse the disclination twist loops will annihilate each other (if anti-matter is electrically neutral), while the vacancy edge dislocation loops, which characterize anti-matter, combine to form the macroscopic lattice vacancies in the center of the galaxies.
The macroscopic vacancy created in the center of a galaxy by the gravitational collapse of anti-matter *is a giant topological singularity* which becomes a *large black hole* as soon as the expansion of the background exceeds unity ($\tau_0 \geq 1$). *This phenomena of gravitational collapse of anti-matter at the heart of galaxies would explain simply, and at once both, the experimental observation that there exist gigantic black holes at the center of galaxies and that anti-matter seems to have disappeared from our present Universe.*

On the coalescence of matter in galaxies and on the formation of stars

The matter which composes galaxies after the collapse of the heart of anti-matter in a black hole will coalesce bit by bit under the effect of gravitational attraction to form hydrogen gas and helium gas, various types of stars and planetary systems, such as observed in our actual universe.

On the gravitational collapse of stars and on the formation of neutron stars

As the galaxies are essentially made of matter, based on interstitial edge dislocation loops, all gravitational collapse of a large star under the effect of its own gravity will lead to a localized topological singularity of macroscopic interstitial type and not of the vacancy type. As a consequence, there cannot be the emergence of a vacancy black hole after the gravitational collapse of a massive star of matter!
Experimentally, we sometimes suddenly observe this gravitational collapse of massive stars of matter under the form of *'supernovas'* and in the form of the residual gas after the initial explosion of the star, which extends at great velocity, with, in the center of the supernova a rather small and massive object, which should correspond to a residual interstitial singularity, which we usually call a *pulsar* (due to its emission properties of electromagnetic pulses corresponding to a frequency of rapid rotation of the object on itself) or *neutron stars* (due to the high mass density of the object).
The rest of the story is well known and described by astrophysicists with the formation of atoms of increasing size by nuclear fusion of hydrogen and light elements inside the stars, and by the dispersion of these elements by supernovas, which leads finally to the apparition of all elements of the table of Mendeleïev and the formation of stars which are more and more complex, plane-

tary systems, etc

On the future of our universe

In the scenario of the universe *«big-bounce»* which was depicted in figures 16.8 and 16.11g, which corresponds actually to our own universe, the expansion phase with constant velocity, in the condition where there are no longitudinal waves, is found between values $\tau_0 = -3/2$ and $\tau_0 = \tau_{0cr} >> 1$. Our actual universe must then be in this range of values of the background expansion since recent observations have shown that the expansion of the universe is probably done at increasing speed. We can even say that the expansion of the background should be in the domain $1 < \tau_0 < \tau_{0cr}$, since massive black holes seem to have been observed in the center of most galaxies, notably in the center of our galaxy, the milky way!
As this range $(1 < \tau_0 < \tau_{0cr} >> 1)$ is very large, it is hard to know where we are actually and how much time we will need to reach critical expansion τ_{0cr}. What we can already say however is that when it gets close to τ_{0cr}, there will appear titanic transformation of celestial bodies, of matter, of black holes and of the sea of repulsive neutrinos since the following will be:
- the gravitational constant G_{grav} will become negative by going through a singularity at τ_{0cr},
- the localized vibration models will disappear to the profit of longitudinal models (which should correspond to the disappearance of quantum physics as we will see later in this treatment!).

These two phenomenas should be cataclysmic. But we can go further by considering the phenomena appearing during the re-contraction phase of the cosmic lattice, specifically during the transition through τ_{0cr} in the inverse direction, where the gravitational constant becomes positive again and where there appears again proper modes of localized vibrations in lieu of longitudinal waves. These predictions are possibly in the domain of the possible with our theory, but surely very hard and approximative. Actually, we are now in science fiction land.
It remains so that our theory goes much further in the explanations and predictions than General Relativity and that many exotic phenomenas such as instantaneous displacement in space time via wormholes as described in general relativity, and which delight theoretical physicist and science fiction writers alike, are pure delirium (aka bullshit) in our theory.

27.3 – Hubble constant, redshift of galaxies and 'cooling' of the cosmic background radiation

Experimentally, we notice that the light from galaxies shows a 'redshift' in the spectral emission of atoms. This redshift was attributed by Hubble to the velocity of galaxies moving away from us with apparent velocity $\boldsymbol{V}$ of galaxies as a function of their distance d due to the expansion of the Universe. The experimental relationship between velocity of recession and distance d was measured by Hubble, who found $\boldsymbol{V} = H_0 d$, where H_0 is the *Hubble constant* which is approximately worth 70 (km/s)/Mpc (70 kilometers per second per megaparsec). The interpretation of this initial observation as a Doppler-Fizeau effect due to the velocity of escape of distant galaxies leads us to the conclusion that galaxies situated further than 4'000 megaparsecs away, would be moving with velocities superior to that of the speed of light, which is non-sense according to special relativity. The solution to this problem, can be deduced in General Relativity, for which the expansion of the Universe must not be interpreted by a movement of galaxies in space but by that of the fabric of space itself, which implies a progressive distance of object contained there-in. We have the same solution in our theory, where the perfect cosmic lattice is expanding.

On the Hubble constant in our theory

Let's take a look at the Hubble constant in our theory, which is different than General Relativity, if anything because of the existence of a scalar of volume expansion τ of the cosmic background. Assume thus a cosmic evolution as that described in figures 16.8 and 16.11g. During the evolution, suppose that the ***GO*** observes a certain region in graph 16.11g, in which the cosmic lattice is in expansion, and indeed in expansion with growing velocity. This situation can be simulated with a development at second order of the expansion as a function of time, under the following form

$$\tau(t)=\tau_i+\left.\frac{\partial\tau}{\partial t}\right|_i t+\left.\frac{\partial^2\tau}{\partial t^2}\right|_i t^2 \quad \text{with} \quad \left.\frac{\partial\tau}{\partial t}\right|_i>0 \quad \text{and} \quad \left.\frac{\partial^2\tau}{\partial t^2}\right|_i>0 \tag{27.16}$$

Suppose then that the ***GO*** observes two galaxies which are originally distant by d_0 at moment $t=0$. If these galaxies <u>*do not move*</u> with respect to the cosmic lattice in expansion, the initial distance d_0 will evolve during time, and the ***GO*** will observe that the distance between the two galaxies will grow as

$$d(t)=d_0\,\mathrm{e}^{(\tau(t)-\tau_i)/3}=d_0\,\mathrm{e}^{\left(\left.\frac{\partial\tau}{\partial t}\right|_i t+\left.\frac{\partial^2\tau}{\partial t^2}\right|_i t^2\right)/3} \tag{27.17}$$

so that two galaxies will move away from each other with relative velocity

$$\boldsymbol{V}(t)=\frac{\partial d(t)}{\partial t}=\left(\frac{1}{3}\left.\frac{\partial\tau}{\partial t}\right|_i+2\left.\frac{\partial^2\tau}{\partial t^2}\right|_i t\right)d_0\,\mathrm{e}^{\left(\left.\frac{\partial\tau}{\partial t}\right|_i t+\left.\frac{\partial^2\tau}{\partial t^2}\right|_i t^2\right)/3}=\left(\frac{1}{3}\left.\frac{\partial\tau}{\partial t}\right|_i+2\left.\frac{\partial^2\tau}{\partial t^2}\right|_i t\right)d(t) \tag{27.18}$$

At instant $t=0$, the relationship existing between velocity $\boldsymbol{V}$ and distance d_0 is written

$$\boldsymbol{V}=\frac{1}{3}\left.\frac{\partial\tau}{\partial t}\right|_i d_0=H_0\,d_0 \quad\Rightarrow\quad H_0=\frac{1}{3}\left.\frac{\partial\tau}{\partial t}\right|_i \tag{27.19}$$

This relationship is indeed equivalent to Hubble. But if the ***GO*** observes the relationship at an instant $t\neq 0$, he will find that

$$\boldsymbol{V}=\left(\frac{1}{3}\left.\frac{\partial\tau}{\partial t}\right|_i+2\left.\frac{\partial^2\tau}{\partial t^2}\right|_i t\right)d=H_0(t)d \quad\Rightarrow\quad H_0(t)=\frac{1}{3}\left.\frac{\partial\tau}{\partial t}\right|_i+2\left.\frac{\partial^2\tau}{\partial t^2}\right|_i t \tag{27.20}$$

and that the Hubble 'constant' becomes dependent on time. Notably it increases in a domain where the velocity of expansion is increasing.

On the redshift of galaxies in our theory

Let's consider now two galaxies separated by distance d_0 at instant $t=0$ for the ***GO***. If, at this instant $t=0$, the galaxy 1 emits a signal towards galaxy 2, the signal will carry a distance dx

during a time dt such that $dx = c_t dt$. But as this lattice is in expansion according to relation *(27.16)*, and by neglecting the acceleration of expansion, we have that

$$dx = c_t dt = c_{t0}\,\mathrm{e}^{\tau/2}\,dt = c_{t0}\,\mathrm{e}^{\left(\tau_i + \left.\frac{\partial \tau}{\partial t}\right|_i t\right)/2}\,dt = c_{t(i)}\,\mathrm{e}^{\left(\left.\frac{\partial \tau}{\partial t}\right|_i t\right)/2}\,dt \qquad (27.21)$$

where $c_{t(i)}$ is the velocity of the transversal waves at instant $t = 0$. The distance $d(T)$ between the two galaxies will thus be covered in a time T such that

$$x(T) \cong \int_0^T c_{t(i)}\,\mathrm{e}^{\left(\left.\frac{\partial \tau}{\partial t}\right|_i t\right)/2}\,dt = \frac{2c_{t(i)}}{\left.\frac{\partial \tau}{\partial t}\right|_i}\left[\mathrm{e}^{\left(\left.\frac{\partial \tau}{\partial t}\right|_i T\right)/2} - 1\right] = d(T) \qquad (27.22)$$

But $d(T)$ is given by *(27.17)*, so that

$$\mathrm{e}^{\left(\left.\frac{\partial \tau}{\partial t}\right|_i T\right)/2} - 1 \cong \left.\frac{\partial \tau}{\partial t}\right|_i \frac{d_0}{2c_{t(i)}}\,\mathrm{e}^{\left(\left.\frac{\partial \tau}{\partial t}\right|_i T\right)/3} \qquad (27.23)$$

By supposing then the case where $\left(\partial \tau / \partial t|\right)_i T >> 1$, we obtain the time lapse T measured by ***GO*** between the emission of the signal by galaxy 1 and the reception of the signal by galaxy 2

$$\mathrm{e}^{\left(\left.\frac{\partial \tau}{\partial t}\right|_i T\right)/6} \cong \left.\frac{\partial \tau}{\partial t}\right|_i \frac{d_0}{2c_{t(i)}} \quad \Rightarrow \quad T \cong \frac{6}{\left.\partial \tau / \partial t\right|_i} \ln\left(\frac{d_0}{2c_{t(i)}} \left.\frac{\partial \tau}{\partial t}\right|_i\right) \qquad (27.24)$$

During the reception of the signal by galaxy 2, the expansion of the universe will reach a critical value of expansion τ_f worth according to *(27.16)*

$$\tau_f \cong \tau_i + \left.\frac{\partial \tau}{\partial t}\right|_i T \cong \tau_i + 6\ln\left(\frac{d_0}{2c_{t(i)}} \left.\frac{\partial \tau}{\partial t}\right|_i\right) \qquad (27.25)$$

Suppose that the signal emitted by galaxy 1 at instant $t = 0$ is measured by a local observer ***HS(1)*** as having frequency $\nu_{emitted}^{HS^{(1)}}$ for a given atomic spectral emission. The frequency of the received signal $\nu_{received}^{HS^{(2)}}$ is measured by an observer ***HS(2)*** in galaxy 2 at instant $t = T$ will be different due to the increase in expansion $\Delta\tau = \tau_f - \tau_i$ of the Universe which appeared during propagation. Thus, an ***HS(2)*** observer situated in galaxy 2 will be able to compare the frequency $\nu_{received}^{HS^{(2)}}$ of this signal received with frequency $\nu_{emitted}^{HS^{(2)}}$ of the same spectral ray of the atom emitted in it's own laboratory, and we will call *redshift of galaxy 1* the ratio between the two frequencies

$$\textit{redshift} = \frac{\nu_{received}^{HS^{(2)}}}{\nu_{emitted}^{HS^{(2)}}} \qquad (27.26)$$

To calculate this redshift, we must represent schematically how we link the physical measurements as measured by observer ***HS(1)*** in galaxy 1 at instant $t = 0$ with the same physical values measured by the ***GO*** at instant $t = 0$ and at instant $t = T$ and by the observer ***HS(2)*** in galaxy 2 at instant $t = T$.

For that, we recall that the elapsed time Δt_y measured by an ***HS*** in its proper referential $Oy_1y_2y_3$ are perceived by the ***GO*** in its local referential $Ox_1x_2x_3$ as lapses of time Δt linked by expression *(24.23)*

$$\Delta t_y^{HS} = \Delta t^{GO}\, \mathrm{e}^{\tau/4} \tag{27.27}$$

in which τ is the expansion of the lattice at the place where the ***HS*** is. This expression allows us to link the measures of the frequencies done by an ***HS*** and by the ***GO***

$$\nu_y^{HS} = \nu^{GO}\, \mathrm{e}^{-\tau/4} \tag{27.28}$$

With this relation, we can schematically represent how the measures of frequency and length behave during the transmission experience between galaxies 1 and 2 (fig. 27.3).

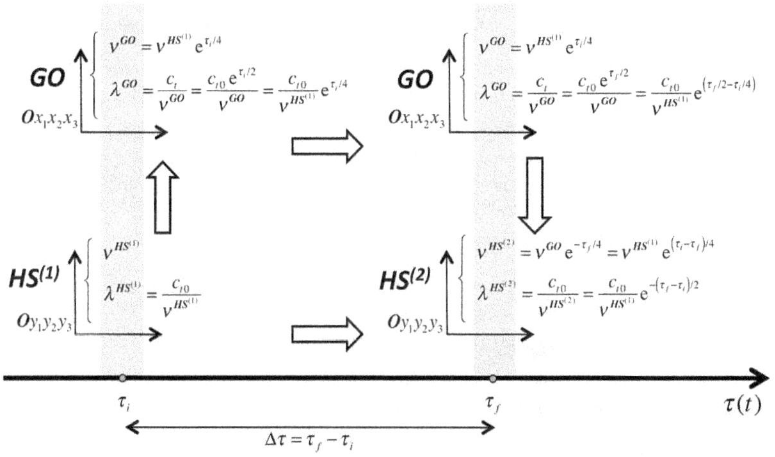

Figure 27.3 *- Schematic representation of the physical values measured by the observers **HS(1)**, **GO** and **HS(2)** during the transmission of a signal between galaxies 1 and 2.*

We notice then that the «redshift» measured by observer ***HS(2)*** is worth

$$\text{"redshift"} = \frac{\nu_{received}^{HS^{(2)}}}{\nu_{emitted}^{HS^{(2)}}} = \frac{\nu_{emitted}^{HS^{(1)}}\, \mathrm{e}^{(\tau_i-\tau_f)/4}}{\nu_{emitted}^{HS^{(2)}}} = \mathrm{e}^{-\Delta\tau/4} \tag{27.29}$$

By using the value $\Delta\tau = \tau_f - \tau_i$ obtained in *(27.25)*, we obtain for the redshift measured by the observer ***HS(2)***

$$\text{"redshift"} = \mathrm{e}^{-\Delta\tau/4} = \mathrm{e}^{-\frac{3}{2}\ln\left(\frac{d_0}{2c_{t(i)}}\left.\frac{\partial\tau}{\partial t}\right|_i\right)} = \left(\frac{d_0}{2c_{t(i)}}\left.\frac{\partial\tau}{\partial t}\right|_i\right)^{-\frac{3}{2}} = \frac{\left(2c_{t(i)}\right)^{3/2}}{\left(\partial\tau/\partial t|_i\right)^{3/2}}\frac{1}{d_0^{3/2}} \tag{27.30}$$

The redshift depends simultaneously on the instantaneous velocity of expansion $\partial\tau/\partial t|_i$ and the distance d_0 between the two galaxies. But we can also link the redshift to the "instantaneous constant" of Hubble H_0 with the relative velocity $\boldsymbol{V}$ *of recession of the galaxies* by using relation *(27.19)*

$$\text{"redshift"} = \frac{\left(2c_{t(i)}\right)^{3/2}}{\left(\partial\tau/\partial t|_i\right)^{3/2}}\frac{1}{d_0^{3/2}} \cong \left(\frac{2}{3}c_{t(i)}\right)^{3/2}\frac{1}{\left(H_0 d_0\right)^{3/2}} \cong \left(\frac{2}{3}c_{t(i)}\right)^{3/2}\frac{1}{\boldsymbol{V}^{3/2}} \tag{27.31}$$

The redshift observed by observer ***HS(2)*** will thus be proportional to $1/d_0^{3/2}$ which means that the ratio between frequencies is smaller as the initial distance d_0 between the galaxies is great, and thus that the "shift towards the red" of the spectral ray augments with the increase in initial distance d_0. The redshift is also proportional to $1/(H_0)^{3/2}$, which means that it gets smaller if the Hubble constant gets bigger. Finally it is proportional to $1/\boldsymbol{V}^{3/2}$. But the velocity at which two galaxies separate from each other is not limited by the velocity of transversal waves in our theory, since this velocity is associated with the absolute velocity of the lattice in the absolute space of the ***GO,*** which satisfies a purely newtonian dynamic! Thus, the redshift measured can tend towards 0 if the velocity of recession ***V*** due to expansion tends towards infinity!

It is noteworthy that the calculations were made by making the following two restrictive suppositions:

- we find ourselves in a limited region of the graph 16.11g in which the cosmic lattice is in expansion, and which can be approximated by a development to second order in expansion $\tau(t)$ in time *(27.16)*. For the phenomenas which would spread over larger time periods, the calculations get singularly more complicated, since we should then know the exact function $\tau(t)$ of the cosmic expansion of the lattice!

- that galaxies do not move with respect to the cosmic lattice in expansion. If that was not the case, for example due to gravitational interactions between galaxies, we should add to the redshift due to the expansion of the lattice a Doppler-Fizeau effect due to the displacement of the galaxies with respect to the lattice, such as described in section 21.3!

On the mechanism of "cooling" of the cosmic background radiation in our theory

The diffuse cosmic background radiation is actually observed as the *spectrum of a perfect black body radiation,* following precisely the *distribution (27.16) of Planck of the density of energy* $U(v)$ *for photons*, centered on a temperature T worth 2,7 K. We suppose that this radiation carries over from the big-bang and was formed during the decoupling of the photons and the particles during the formation of neutral helium and hydrogen atomes, and that, as a consequence, it was first emitted at a temperature of 3'000 K. We can then ask ourselves what is the cooling mechanism of this light radiation in our theory. For that it suffices to look at figure 27.3. If we suppose that the decoupling of photons and matter appeared when the Universe had an expansion τ_i and that the actual expansion of the universe is τ_f, the frequency of emission of the diffuse background is given by $v_{émise}^{HS^{(1)}}$ and the observed frequency actually observed by a ***HS(2)*** is worth

$$v_{measured}^{HS^{(2)}} = v_{emitted}^{HS^{(1)}} e^{(\tau_i - \tau_f)/4} = v_{emitted}^{HS^{(1)}} e^{-\Delta\tau/4} \qquad (27.32)$$

According to the Planck distribution, there exists a ratio between the frequency of the black body and it's temperature of color, measure at expansions τ_f and τ_i

$$\frac{hv_{measured}^{HS^{(2)}}}{kT_{measured}^{HS^{(2)}}} = \frac{hv_{emitted}^{HS^{(1)}}}{kT_{emitted}^{HS^{(1)}}} \Rightarrow T_{measured}^{HS^{(2)}} = T_{emitted}^{HS^{(1)}} \frac{v_{measured}^{HS^{(2)}}}{v_{emitted}^{HS^{(1)}}} = T_{emitted}^{HS^{(1)}} e^{-\Delta\tau/4} \qquad (27.33)$$

We deduce a numerical value of the variation of expansion of the lattice between the moment of decoupling of matter and photon and the actual time

$$e^{-\Delta\tau/4} = \frac{T^{HS^{(2)}}_{mesurée}}{T^{HS^{(1)}}_{émise}} \Rightarrow \Delta\tau = 4\ln\frac{T^{HS^{(1)}}_{émise}}{T^{HS^{(2)}}_{mesurée}} \cong 4\ln(3'000) \cong 32 \tag{27.34}$$

This augmentation of expansion corresponds in fact to an augmentation of volume of the basic mesh of the lattice, which is approximatively

$$\frac{V^{HS^{(2)}}}{V^{HS^{(1)}}} = e^{\Delta\tau} \cong e^{32} \cong 8 \cdot 10^{13} \tag{27.35}$$

which must take place during the inflation phase of the cosmic lattice (fig. 16.11g).
We notice that the apparent 'cooling' of the background radiation is a direct effect of the expansion of the lattice, which modifies strongly the behavior of local clocks and frames $Oy_1y_2y_3$.
It is interesting to notice that, for the ***GO***, the frequency $\nu^{GO}_{measured}$ of the cosmic background will not change during the expansion since it will always be the same value during it's emission

$$\nu^{GO}_{measured} = \nu^{HS^{(1)}}_{emitted}\, e^{\tau_i/4} \tag{27.36}$$

But on the contrary, for the ***GO***, it is the wavelength $\lambda^{GO}_{measured}$ which will evolve with the expansion, since it will then be equal to

$$\lambda^{GO}_{measured} = \frac{c_t}{\nu^{GO}_{measured}} = \frac{c_{t0}\, e^{\tau(t)/2}}{\nu^{GO}_{measured}} = \frac{c_{t0}}{\nu^{HS^{(1)}}_{emitted}}\, e^{\tau(t)/2-\tau_i/4} \tag{27.37}$$

The points of view of the local observers ***HS***$^{(i)}$ and the external observer ***GO*** are thus very different, and that is due to the fact that the velocity of the transversal waves of rotation (speed of light) is a universal constant c_{t0} for the ***HS***$^{(i)}$ observers, independently of the state of expansion of the lattice in which they are placed, while the velocity of the transversal waves of rotation $c_t = c_{t0}\, e^{\tau(t)/2}$ varies greatly as a function of instantaneous expansion $\tau(t)$ of the lattice for the external observer ***GO.***

PART II

D

Quantum physics and Standard Model of particles

Gravitational fluctuations associated to the topological singularities: quantum physics, Schrödinger's equation, uncertainty principle, bosons, fermions and exclusion principle

Spin et intrinsic magnetic momentum of the singularities

Quantified transversal fluctuations: photons, wave-particle duality, entanglement and decoherence

Ingredients of the Standard Model of elementary particles: strong force and families of leptons and quarks

Chapter 28

Gravitational fluctuations associated with mobile singularities: the Schrödinger's equation

Intuitively, we can see that Quantum Mechanics (QM) could be linked to the existence of dynamical solutions of the second partial equation of Newton *(18.9)*, under the form of temporal fluctuations of the field of expansion, associated with the topological singularities of the cosmic lattice when it is without longitudinal waves in the domain $\tau_0 < \tau_{0cr}$. We will show in this chapter a wave function directly deduced from the second partial derivative equation of Newton for the perturbations of expansion of the lattice which is intimately linked to the moving topological singularities of the lattice, whether they are clusters of singularities or isolated single loops. We will thus give a rather simple '*wave interpretation*' of quantum mechanics: the quantum wave function represents the amplitude and phase of gravitational fluctuations coupled to topological singularities. This interpretation implies that the square of the amplitude of the normalized wave function is indeed linked to the probability of presence of the topological singularity which is associated with it!
At the same time, we will recover the Heisenberg incertitude principle, the QM notions of bosons, fermions, and non-discernibility, the Pauli exclusion principle, as well as a physical comprehension of intriguing phenomenas such as entanglement and quantum decoherence.

28.1 – Dynamic "gravitational" fluctuations of the field of expansion within the lattice

Wave equation of the dynamical fluctuations of the gravitational field of expansion

Suppose the existence of longitudinal fluctuations in the cosmic lattice. These fluctuations must satisfy the dynamical version of the second partial derivative equation of Newton *(18.9)*. We obtain in first order in $\tau^{(p)}$, by taking into account the geometro-kinetic equation for $\tau^{(p)}$, the following equations

$$\begin{cases} \dfrac{\partial \vec{\phi}^{(p)}}{\partial t} \cong -\dfrac{\alpha}{mn}\overrightarrow{\text{grad}}\,\tau^{(p)} \\[2ex] \dfrac{\partial \tau^{(p)}}{\partial t} \cong \text{div}\,\vec{\phi}^{(p)} \end{cases} \tag{28.1}$$

with

$$\frac{\alpha}{mn} \cong \frac{K_0 - 4K_2/3 - 2K_1(1+\tau_0)}{mn} \tag{28.2}$$

which can be boiled down to one equation on gravitational fluctuations $\tau^{(p)}$

$$\frac{\partial^2 \tau^{(p)}}{\partial t^2} \cong -\frac{\alpha}{nm} \Delta \tau^{(p)} \tag{28.3}$$

Suppose then that we find ourselves in a perfect cosmic lattice in a domain rather far from the critical value τ_{0cr}, and thus with the following hypothesis

Hypothesis 1: $\tau_0 << \tau_{0cr} \quad ; \quad K_2 << K_0 \quad ; \quad 2K_1(1+\tau_0) << K_0$ (28.4)

The wave equation for gravitational fluctuations can be written

$$\frac{\partial^2 \tau^{(p)}}{\partial t^2} \cong -\frac{K_0}{mn} \Delta \tau^{(p)} = -c_t^2 \Delta \tau^{(p)} \tag{28.5}$$

We can then propose a solution of gravitational perturbations with frequency ω_f, which we will write out of convenience under the complex form

$$\underline{\tau}^{(p)} \cong \underline{\psi}\, \mathrm{e}^{(\pm)i\omega_f t} \tag{28.6}$$

where $\mathrm{e}^{(\pm)i\omega_f t}$ corresponds to an *oscillation of the fluctuation* and the *complex wave function* $\underline{\psi}(\vec{r},t)$ represents *the phase and amplitude of said oscillation.*
By introducing this solution *(28.6)* in the Newton equation *(28.5)* of the fluctuations of the lattice, we obtain a *wave equation for the complex function* $\underline{\psi}(\vec{r},t)$ in the following form

$$\frac{\partial^2 \underline{\psi}}{\partial t^2}(\pm)2i\omega_f \frac{\partial \underline{\psi}}{\partial t} - \omega_f^2 \underline{\psi} \cong -c_t^2 \Delta \underline{\psi} \tag{28.7}$$

Isolated gravitational fluctuations within the cosmic lattice

First let's propose a simple solution, local and independent of time, by supposing that frequency ω_f is a constant independent of time and space, in the following form

$$\underline{\psi}(\vec{r}) = \underline{\psi}_0\, \mathrm{e}^{-\frac{|x_1|}{\delta_1}} \mathrm{e}^{-\frac{|x_2|}{\delta_2}} \mathrm{e}^{-\frac{|x_3|}{\delta_3}} \tag{28.8}$$

By introducing this solution in *(28.7)*, we obtain

$$\frac{1}{\delta_1^2} + \frac{1}{\delta_2^2} + \frac{1}{\delta_3^2} = \frac{\omega_f^2}{c_t^2} \tag{28.9}$$

So that there appears here a *localized* gravitational fluctuation

$$\underline{\tau}^{(p)} \cong \underline{\psi}(\vec{r})\mathrm{e}^{(\pm)i\omega_f t} \cong \underline{\psi}_0\, \mathrm{e}^{-\frac{|x_1|}{\delta_1}} \mathrm{e}^{-\frac{|x_2|}{\delta_2}} \mathrm{e}^{-\frac{|x_3|}{\delta_3}} \mathrm{e}^{(\pm)i\omega_f t} \tag{28.10}$$

This fluctuation corresponds to a local vibrational regime, non damped, with frequency ω_f and which decays symmetrically and exponentially in the vicinity of the origin, with spatial characteristic length worth δ_1, δ_2 and δ_3, which are correlated to each other and which decrease with the frequency via relationship *(28.9)*.
We deduce that, in a perfect cosmic lattice which satisfies hypothesis *(28.4)*, there very well may be stable localized fluctuations of volume expansion. We will revisit this in chapter 32.

Gravitational fluctuations associated with a mobile topological singularity

Let's imagine that to the topological singularities that are moving within the lattice, we also associate longitudinal dynamical fluctuations. They must satisfy the dynamical version of the second partial equation of Newton *(28.5)* outside the topological singularity.

We know that, in the cosmic lattice, the second partial equation of Newton *(18.9)* introduces gravitational perturbations of the field of expansion with the moving singularities themselves, perturbations that depend directly on the elastic distortion energy of said singularities. In section 25.1, we showed that this field of perturbations of expansion is directly responsible for an external static gravitational field which also depends on the elastic energy of distortion of the singularity via parameter $4G_{grav}/c_t^2$

$$\tau_{ext\ LD}^{boucle}(r) \cong -\frac{4G_{grav}}{c_t^2}\frac{M_0^{boucle}}{r} \quad \text{where} \quad \tau_{ext\ LD}^{amas}(r) \cong -\frac{4G_{grav}}{c_t^2}\frac{M_0^{amas}}{r} \qquad (28.11)$$

It is therefore likely that the dynamical field of gravitational perturbations $\tau^{(p)}$ of expansion *(28.6)* outside the singularity also depends on the elastic energy of the singularity, but also the energy associated with the movement of the singularity. However, we know that a topological singularity which is in movement in the lattice is characterized fully by a *total relativistic energy* E_v and a *total relativistic translational momentum* $\vec{\boldsymbol{P}}_v$. As a consequence, the frequency $\underline{\omega}_f$ of fluctuations of expansion which are associated with it must depend also on relativistic energy $E_v(\vec{r},t)$ of the singularity as well as it's relativistic momentum $\vec{\boldsymbol{P}}_v(\vec{r},t)$. We can then emit the hypothesis that there should be the following relationship for $\underline{\omega}_f$

Hypothesis 2: $\underline{\omega}_f \cong \underline{\omega}_f\left(E_v(\vec{r},t),\vec{\boldsymbol{P}}_v(\vec{r},t)\right) = \underline{\omega}_f(\vec{r},t)$ (28.12)

The frequency $\underline{\omega}_f$ of fluctuations of expansion should then depend on time and space around the singularity, via the temporal and spatial dependency of energy $E_v(\vec{r},t)$ and the momentum $\vec{\boldsymbol{P}}_v(\vec{r},t)$. We can then propose the following solution to gravitational perturbations with frequency $\underline{\omega}_f(\vec{r},t)$

$$\underline{\tau}^{(p)}(\vec{r},t) \cong \underline{\psi}(\vec{r},t)\,e^{(\pm)i\underline{\omega}_f\left(E_v(\vec{r},t),\vec{\boldsymbol{P}}_v(\vec{r},t)\right)t} \cong \underline{\psi}(\vec{r},t)\,e^{(\pm)i\underline{\omega}_f(\vec{r},t)t} \qquad (28.13)$$

where $e^{(\pm)i\underline{\omega}_f(\vec{r},t)t}$ corresponds to *the oscillation of fluctuation* and the *complex wave function* $\underline{\psi}(\vec{r},t)$ represents the phase and amplitude of this oscillation.

By introducing this solution *(28.13) in the equation of* Newton *(28.5)* of the fluctuations of the lattice, and by *supposing that we can neglect the spatial and temporal derivatives of* $\underline{\omega}_f(\vec{r},t)$ *vis-à-vis the temporal and spatial derivatives of* $\underline{\psi}(\vec{r},t)$, we obtain a wave equation for the complex function $\underline{\psi}(\vec{r},t)$ in the following form

$$\frac{\partial^2\underline{\psi}}{\partial t^2}(\pm)2i\underline{\omega}_f(\vec{r},t)\frac{\partial\underline{\psi}}{\partial t} - \underline{\omega}_f^2(\vec{r},t)\underline{\psi} \cong -c_t^2\Delta\underline{\psi} \qquad (28.14)$$

On the conjecture of existence of energy and momentum operators

In this equation *(28.14)*, there appears explicitly the frequency $\underline{\omega}_f(\vec{r},t)$ of gravitational fluctuations, which we suppose depends on energy $E_v(\vec{r},t)$ and on momentum $\vec{\boldsymbol{P}}_v(\vec{r},t)$ of the singularity via relation *(28.12)*, and the *complex wave function* $\underline{\psi}(\vec{r},t)$ which represents the *phase and amplitude of gravitational fluctuations.*

If the frequency $\underline{\omega}_f(\vec{r},t)$ of gravitational fluctuations depends effectively on $E_v(\vec{r},t)$ and $\vec{\boldsymbol{P}}_v(\vec{r},t)$, these two quantities are implicitly contained in the spatial and temporal behavior of the spatial complex function $\underline{\psi}(\vec{r},t)$ associated with the mobile singularity.

This wave function $\underline{\psi}(\vec{r},t)$ associated with the wave equation *(28.14)* reminds us of the wave function that appears in quantum physics, and quantum physics says it is possible to define operators measuring E_v and $\vec{\boldsymbol{P}}_v$ from the quantum wave function $\underline{\psi}(\vec{r},t)$. When they are applied locally to the wave function, these quantum operators correspond to the partial derivatives with respect to time and space of the quantum wave function, *giving us the total energy and the total momentum that the singularity has at this point, multiplied by the wave function* $\underline{\psi}(\vec{r},t)$.

By analogy with quantum physics, we can *conjecture a priori an operator with similar properties*

Conjecture 10: *There exists two operators which, when applied to the wave function $\underline{\psi}(\vec{r},t)$ of a mobile singularity in the lattice, allows us to measure its relativistic energy and its relativistic momentum, namely*

$$\begin{cases} i\hbar\dfrac{\partial}{\partial t}\underline{\psi} \to E_v\underline{\psi} \\ -i\hbar\dfrac{\partial}{\partial x_i}\underline{\psi} \to \boldsymbol{P}_{vi}\,\underline{\psi} \end{cases} \tag{28.15}$$

In these relations, we have introduced, as in quantum mechanics, the constant $\hbar$ allowing us to normalize the partial derivatives of the wave function to energy terms (and which is in the case of quantum mechanics the *Planck constant).*

It is rather simple to verify that these operators, applied twice to the same wave function give us the square of total energy and the square of the total momentum

$$\begin{cases} -\hbar^2\dfrac{\partial^2}{\partial t^2}\underline{\psi} \to E_v^2\underline{\psi} \\ -\hbar^2\Delta\underline{\psi} \to \vec{\boldsymbol{P}}_v^2\underline{\psi} \end{cases} \tag{28.16}$$

We can normalize equation *(28.14)* under the form of an equation composed of terms corresponding to squares of energy by multiplying by $\hbar^2$

$$\hbar^2\frac{\partial^2\underline{\psi}}{\partial t^2}(\pm)2\left(\hbar\underline{\omega}_f(\vec{r},t)\right)i\hbar\frac{\partial\underline{\psi}}{\partial t}-\left(\hbar\underline{\omega}_f(\vec{r},t)\right)^2\underline{\psi} \cong -c_t^2\hbar^2\Delta\underline{\psi} \tag{28.17}$$

From which we deduce that each term in the equation must represent the product of the square of an energy by the non-dimensional wave function $\underline{\psi}(\vec{r},t)$. We deduce that the terms $\hbar\omega_f$ has *the dimension of an energy.*

Furthermore, the second derivative operator with respect to time gives us *the square of the total relativistic energy* E_v of the singularity at the spot where we apply the operator. The laplacian operator gives us the square of the *total relativistic momentum* $\vec{\boldsymbol{P}}_v$ of the singularity at the spot where we apply the operator. From this definition of operators of second derivative and of laplacian, we deduce the operators giving directly the total energy of the singularity and the components of the quantity of movement at a given spot.

Armed with these hypothesized operators modeled after quantum mechanics, we can try to apply them to relation *(28.17)* which we deduced from the second partial derivative equation of

Netwon. We have

$$E_v^2(\vec{r},t)(\mp)2\hbar\underline{\omega}_f(\vec{r},t)E_v(\vec{r},t)+\left(\hbar\underline{\omega}_f(\vec{r},t)\right)^2=\left(E_v(\vec{r},t)(\mp)\hbar\omega_f(\vec{r},t)\right)^2\cong-c_t^2\vec{\boldsymbol{P}}_v^2(\vec{r},t) \tag{28.18}$$

We thus obtain the relation that should exist between the *relativistic energy* E_v of the singularity, the *energy of motion* $\sqrt{c_t^2\vec{\boldsymbol{P}}_v^2}$ of the singularity and the *frequency* $\underline{\omega}_f(\vec{r},t)$ *of gravitational fluctuations associated with the singularity and which becomes a complex number*

$$E_v(\vec{r},t)(\mp)\hbar\underline{\omega}_f(\vec{r},t)\cong[\pm]i\sqrt{c_t^2\vec{\boldsymbol{P}}_v^2(\vec{r},t)} \tag{28.19}$$

Wave equation of gravitational fluctuations of a mobile singularity

In the case where a singularity moves with relativistic velocity, it must satisfy the relativistic relations *(20.41)* and *(20.42)* of chapter 20, namely

$$E_v^2=M_0^2c_t^4+\vec{\boldsymbol{P}}_v^2c_t^2 \tag{28.20}$$

In which E_v is the *total relativistic energy of the singularity*, M_0 is *the rest mass of the singularity*, which is directly linked to its *rest energy* E_0^{dist} and its *potential energy* $V(\vec{r},t)$, and $\vec{\boldsymbol{P}}_v$ is the *relativistic momentum*, given by

$$\begin{cases} E_v=\dfrac{M_0c_t^2}{\gamma}=\dfrac{E_0^{dist}+V(\vec{r},t)}{\gamma} \\ M_0c_t^2=E_0^{dist}+V(\vec{r},t) \qquad \text{with} \quad \gamma=\sqrt{1-\dfrac{\vec{\boldsymbol{v}}^2}{c_t^2}} \\ \vec{\boldsymbol{P}}_v=\dfrac{M_0}{\gamma}\vec{\boldsymbol{v}}=\dfrac{E_0^{dist}+V(\vec{r},t)}{\gamma c_t^2}\vec{\boldsymbol{v}} \end{cases} \tag{28.20}$$

These expressions lead us to the following relation giving the complex frequency $\underline{\omega}_f(\vec{r},t)$ of gravitational fluctuations associated with a topological singularity (an elementary loop or a cluster of elementary loops linked between themselves) and moving with velocity $\boldsymbol{v}$

$$\hbar\underline{\omega}_f=(\pm)E_v[\pm]i\sqrt{c_t^2\vec{\boldsymbol{P}}_v^2}=(\pm)E_v[\pm]i\sqrt{E_v^2-M_0^2c_t^4}=(\pm)E_v\left(1[\pm]i\frac{\boldsymbol{v}}{c_t}\right)=(\pm)\frac{E_0^{dist}+V(\vec{r},t)}{\gamma}\left(1[\pm]i\frac{\boldsymbol{v}}{c_t}\right) \tag{28.22}$$

The complex relativistic pulsation of $\underline{\omega}_f(\vec{r},t)$ presents two conjugated solutions

$$\begin{cases} \underline{\omega}_f(\vec{r},t)=(\pm)\dfrac{E_v}{\hbar}\left(1[+]i\dfrac{\boldsymbol{v}}{c_t}\right)=\pm\dfrac{1}{\hbar}\dfrac{E_0^{dist}+V(\vec{r},t)}{\gamma}\left(1[+]i\dfrac{\boldsymbol{v}}{c_t}\right) \\ \underline{\omega}_f^*(\vec{r},t)=(\pm)\dfrac{E_v}{\hbar}\left(1[-]i\dfrac{\boldsymbol{v}}{c_t}\right)=(\pm)\dfrac{1}{\hbar}\dfrac{E_0^{dist}+V(\vec{r},t)}{\gamma}\left(1[-]i\dfrac{\boldsymbol{v}}{c_t}\right) \end{cases} \tag{28.23}$$

It is interesting to notice that we do find an expression that doesn't contradict hypothesis 2 by doing the product of the conjugated values of the pulsation, and we notice that the norm of the complex pulsation $\underline{\omega}_f(\vec{r},t)$ is a simple function of E_v and $\vec{\boldsymbol{P}}_v$

$$\left|\underline{\omega}_f\right|(\vec{r},t)=\sqrt{\underline{\omega}_f\underline{\omega}_f^*}=\frac{1}{\hbar}E_v\sqrt{1+\frac{\vec{\boldsymbol{v}}^2}{c_t^2}}=\frac{1}{\hbar}\sqrt{E_v^2+c_t^2\vec{\boldsymbol{P}}_v^2} \tag{28.24}$$

The wave equation *(28.17)* for $\underline{\psi}(\vec{r},t)$ possesses two relativistic versions due to the sign $[\pm]$ which appears in $\left(1[\pm]i\boldsymbol{v}/c_t\right)$, while the first sign $(\pm)$ in *(28.22)* does not appear anymore

$$\hbar^2 \frac{\partial^2 \underline{\psi}}{\partial t^2} + 2E_{\mathbf{v}}\left(1[\pm]i\frac{\mathbf{v}}{c_t}\right) i\hbar \frac{\partial \underline{\psi}}{\partial t} - E_{\mathbf{v}}^2\left(1[\pm]i\frac{\mathbf{v}}{c_t}\right)^2 \underline{\psi} \cong -c_t^2 \hbar^2 \Delta \underline{\psi} \qquad (28.24)$$

Gravitational perturbations associated with a mass singularity moving with relativistic speeds

Let's go back to the relativistic wave equation *(28.24)* and let's try to find a solution for a mass singularity which would move 'almost-free' with relativistic velocity $\mathbf{v}$ *more or less constant in the direction* Ox_2, which implies that the following hypothesis be satisfied.

Hypothesis: $E_{\mathbf{v}}(\vec{r},t)$ *varies slowly in space and time* (28.25)

Under this hypothesis the total relativistic energy of the singularity varies slowly in space and time, so that we can admit that the wave function $\underline{\psi}(\vec{r},t)$, which represents the amplitude and phase of the oscillation at frequency $\omega_f(\vec{r},t)$, is in fact a function of position x_2 along axis Ox_2. Lets then propose a wave solution to the relativistic wave equation along the axis Ox_2, which does not depend explicitly on time

$$\underline{\psi} \cong \psi_0 \, \mathrm{e}^{i\underline{k}(\vec{r},t)x_2} \qquad (28.26)$$

where the complex number $\underline{k}(\vec{r},t)$ also varies very slowly in space and time. By injecting this solution into *(28.24)*, we obtain the following expression for the value of the complex wave number $\underline{k}$

$$\underline{k}(\vec{r},t) \cong \{\pm\} i \frac{E_{\mathbf{v}}}{\hbar c_t}\left(1[\pm]i\frac{\mathbf{v}}{c_t}\right) \qquad (28.27)$$

We notice that $\underline{k}(\vec{r},t)$ depends in fact on $\vec{r}$ and t uniquely through the dependency of the relativistic energy $E_{\mathbf{v}}(\vec{r},t)$ in $\vec{r}$ and t. The wave function $\underline{\psi}(\vec{r},t)$ associated with the singularity is written as a consequence

$$\underline{\psi} = \psi_0 \, \mathrm{e}^{\{\mp\}\frac{E_{\mathbf{v}}}{\hbar c_t}x_2} \, \mathrm{e}^{\{\mp\}[\mp]i\frac{E_{\mathbf{v}}}{\hbar c_t}\frac{\mathbf{v}}{c_t}x_2} \qquad (28.28)$$

which allows to express the fluctuations of expansion *(28.3)* associated with the relativistic singularity

$$\underline{\tau}^{(p)}(\vec{r},t) \cong \underline{\psi}(\vec{r},t)\mathrm{e}^{(\pm)i\omega_f(\vec{r},t)t} \cong \psi_0 \, \mathrm{e}^{\{\mp\}\frac{E_{\mathbf{v}}}{\hbar c_t}x_2} \, \mathrm{e}^{\{\pm\}[\mp]i\frac{E_{\mathbf{v}}}{\hbar c_t}\frac{\mathbf{v}}{c_t}x_2} \, \mathrm{e}^{(\pm)i\omega_f(\vec{r},t)t} \qquad (28.29)$$

By introducing the value *(28.12)* of $\omega_f(\vec{r},t)$, we obtain

$$\tau^{(p)}(\vec{r},t) \cong \psi_0 \, \mathrm{e}^{\{\mp\})\frac{E_{\mathbf{v}}}{\hbar c_t}x_2} \, \mathrm{e}^{\{\pm\}[\mp]i\frac{E_{\mathbf{v}}}{\hbar c_t}\frac{\mathbf{v}}{c_t}x_2} \, \mathrm{e}^{i\frac{E_{\mathbf{v}}}{\hbar}t} \, \mathrm{e}^{[\mp]\frac{E_{\mathbf{v}}}{\hbar}\frac{\mathbf{v}}{c_t}t} \qquad (28.30)$$

Only two solutions among the four really have a physical significance, so that, finally, the solution can be written as

$$\tau^{(p)}(\vec{r},t) \cong \psi_0 \, \mathrm{e}^{\{-\}\frac{E_{\mathbf{v}}}{\hbar c_t}\left|x_2[\mp]\mathbf{v}t\right|} \, \mathrm{e}^{i\frac{E_{\mathbf{v}}}{\hbar}\left(t[\pm]\frac{\mathbf{v}}{c_t^2}x_2\right)} \qquad (28.31)$$

which allows us to explicitly write the real fluctuations of expansion of a relativistic singularity, by taking the real part, in the following form

$$\tau_{réel}^{(p)}(\vec{r},t) \cong \psi_0 \, e^{(-)\frac{E_v}{\hbar c_t}|x_2[\mp]\mathbf{v}t|} \cos\left[\frac{E_v}{\hbar}\left(t[\pm]\frac{\mathbf{v}\, x_2}{c_t^2}\right)\right] \tag{28.32}$$

To discuss this expression, we decompose the cosine by writing

$$\tau_{réel}^{(p)}(\vec{r},t) \cong \psi_0 \, e^{(-)\frac{E_v}{\hbar c_t}|x_2[\mp]\mathbf{v}t|} \left[\cos\left(\frac{E_v}{\hbar}t\right)\cos\left(\frac{E_v}{\hbar}\frac{\mathbf{v}}{c_t^2}x_2\right)[\mp]\sin\left(\frac{E_v}{\hbar}t\right)\sin\left(\frac{E_v}{\hbar}\frac{\mathbf{v}}{c_t^2}x_2\right)\right] \tag{28.33}$$

We notice that this function represents the product of oscillations in time and oscillations in space. The oscillations in time have a frequency given by

$$f = \frac{E_v}{2\pi\hbar} = \frac{E_0^{dist} + V(\vec{r},t)}{2\pi\hbar\gamma} = \frac{E_0^{dist} + V(\vec{r},t)}{2\pi\hbar\sqrt{1-\frac{\vec{\mathbf{v}}^2}{c_t^2}}} \tag{28.34}$$

And the oscillations in space a wavelength given by

$$\lambda = \frac{2\pi\hbar c_t^2}{E_v \mathbf{v}} = \frac{2\pi\hbar c_t^2 \gamma}{\left(E_0^{dist} + V(\vec{r},t)\right)\mathbf{v}} = \frac{2\pi\hbar c_t^2}{\left(E_0^{dist} + V(\vec{r},t)\right)\mathbf{v}}\sqrt{1-\frac{\vec{\mathbf{v}}^2}{c_t^2}} \tag{28.35}$$

The frequency f of temporal oscillations is thus an increasing function of velocity $\mathbf{v}$ and relativistic energy E_v of the singularity, and it tends towards an infinite value for $\mathbf{v} \to c_t$. It depends directly on the position $\vec{r}$ of the singularity and on the time t via the dependencies in potential $V(\vec{r},t)$.

With regards to the wavelength λ of these spatial oscillations, it diminishes as a function of velocity $\mathbf{v}$ and of relativistic energy E_v of the singularity, and it tends towards 0 for $\mathbf{v} \to c_t$. It is also modulated in space and time via the potential function $V(\vec{r},t)$.

The amplitude of these temporal and spatial oscillations is modulated by an envelope which is exponentially decreasing on both sides of the average position $x_2(t) = [\pm]\mathbf{v}t$ of the singularity (which thus moves in direction of the axis Ox_2 or in opposite direction depending on the sign + or -). How the envelope decreases $e^{(-)|x_2[\mp]\mathbf{v}t|/\delta}$ is linked to the 'range' δ *of the envelope of oscillations* which is worth

$$\delta = \frac{\hbar c_t}{E_v} = \frac{\hbar c_t \gamma}{E_0^{dist} + V(\vec{r},t)} = \frac{\hbar c_t}{E_0^{dist} + V(\vec{r},t)}\sqrt{1-\frac{\vec{\mathbf{v}}^2}{c_t^2}} \tag{28.36}$$

The range δ of the envelope of oscillation decreases when the velocity $\mathbf{v}$ of the singularity increases and when it's relativistic energy E_v increases. It tends towards 0 for $\mathbf{v} \to c_t$. It is then modulated in space and time via the potential $V(\vec{r},t)$ to which the singularity is submitted. All of this implies that the gravitational fluctuations associated with the total relativistic energy of the singularity are of very short range and become negligible for heavy singularities, such as a cluster of loops. For example, for an electron with non-relativistic velocity, the reach of gravitational perturbations associated with its rest energy E_0^{dist} is already small, on the order of $2 \cdot 10^{-12}\, m$.

Finally, we notice that the dynamic fluctuations of expansion associated with the relativistic singularity are contracted along the axis Ox_2 of movement of the singularity as a function of the singularity velocity $\mathbf{v}$, as we can see on the wavelength λ of spatial oscillations and on the range δ of the envelope of oscillations. These effects *correspond exactly to the relativistic effect of contraction of rods of a cluster of singularities as described in section* 21.2 !

28.2 – Schrödinger's equation of the gravitational fluctuations of expansion for a non-relativistic singularity

The "reduced" wave equation of a singularity moving at non-relativistic speeds

The treatment of the previous section applies to gravitational perturbations associated with relativistic energy of a singularity which is massive enough not to be influenced by the random gravitational perturbations that would exist within the perfect cosmic lattice (see chapter 32). But for a singularity small enough to be subjected to the effects of gravitational fluctuations, there must exist a wave equation which takes these effects into account. Let's consider thus the case of a microscopic singularity (a twist disclination loop or an edge dislocation loop) in non-relativistic regime $|\mathbf{v}| << c_t$, and let's rewrite the wave equation *(28.24)*

$$\hbar^2 \frac{\partial^2 \underline{\psi}}{\partial t^2} + 2E_{\mathbf{v}}\, i\hbar \frac{\partial \underline{\psi}}{\partial t} - E_{\mathbf{v}}^2 \underline{\psi} \cong -c_t^2 \hbar^2 \Delta \underline{\psi} \tag{28.38}$$

By using the first relation *(28.16)*, namely

$$-\hbar^2 \frac{\partial^2 \underline{\psi}}{\partial t^2} \to E_{\mathbf{v}}^2 \underline{\psi} \tag{28.39}$$

the wave equation can transform in a 'reduced' form which only contains the first time derivative

$$i\hbar \frac{\partial \underline{\psi}}{\partial t} \cong -\frac{c_t^2 \hbar^2}{2E_{\mathbf{v}}} \Delta \underline{\psi} + E_{\mathbf{v}} \underline{\psi} \tag{28.40}$$

But the total energy of the non-relativistic energy can be approximatively written as

$$E_{\mathbf{v}} = \frac{M_0 c_t^2}{\gamma} \underset{\gamma \to 1}{\cong} M_0 c_t^2 = E_0^{dist} + V(\vec{r},t) \tag{28.41}$$

So that we can express the reduced wave equation by doing the following replacements

$$i\hbar \frac{\partial \underline{\psi}}{\partial t} \cong -\frac{\hbar^2}{2M_0} \Delta \underline{\psi} + \left(E_0^{dist} + V(\vec{r},t)\right) \underline{\psi} \tag{28.42}$$

We then recognize a wave equation which starts looking a lot like the *Schrödinger wave equation of quantum mechanics*, except for the fact that we have the term $\left(E_0^{dist} + V(\vec{r},t)\right)\underline{\psi}$ in lieu of $V(\vec{r},t)\underline{\psi}$. Using definitions *(28.15)* and *(28.16)* of the operators to interpret this wave equation, we have

$$E_{\mathbf{v}} \cong \frac{\vec{\mathbf{P}}_{\mathbf{v}}^2}{2M_0} + \left(E_0^{dist} + V(\vec{r},t)\right) = E_0^{dist} + T + V(\vec{r},t) \tag{28.43}$$

The wave equation *(28.42)* expresses thus the fact that total energy $E_{\mathbf{v}}$ of a singularity is equal to the sum of its rest energy E_0^{dist} , its kinetic energy T and its potential energy $V(\vec{r},t)$.
The solution for dynamic gravitational perturbations associated with the singularity is written as a consequence as the product of the wave function $\underline{\psi}(\vec{r},t)$ deduced from the wave equation *(28.42)* with the oscillatory term with frequency $\omega_{E_{\mathbf{v}}} = E_{\mathbf{v}} / \hbar$

$$\underline{\tau}^{(p)}(\vec{r},t) \cong \underline{\psi}(\vec{r},t) e^{(\pm) i \omega_{E_{\mathbf{v}}} t} \cong \underline{\psi}(\vec{r},t) e^{(\pm) i \frac{E_{\mathbf{v}}}{\hbar} t} \cong \underline{\psi}(\vec{r},t) e^{(\pm) i \frac{E_0^{dist} + T + V(\vec{r},t)}{\hbar} t} \tag{28.44}$$

The Schrödinger wave equation of a non-relativistic singularity

We start again with solution *(28.44)* for the dynamical gravitational perturbations, and rewrite it under the following modified form

$$\underline{\tau}^{(p)}(\vec{r},t) \cong \underline{\psi}(\vec{r},t)\mathrm{e}^{(\pm)i\frac{E_0^{dist}}{\hbar}t}\mathrm{e}^{(\pm)i\frac{T+V(\vec{r},t)}{\hbar}t} = \underline{\psi}_H(\vec{r},t)\mathrm{e}^{(\pm)i\frac{T+V(\vec{r},t)}{\hbar}t} = \underline{\psi}_H(\vec{r},t)\mathrm{e}^{(\pm)i\frac{H}{\hbar}t} \qquad (28.45)$$

which introduces an oscillatory term of frequency $\omega_H = H/\hbar$ corresponding to the *hamiltonian* H of the singularity, meaning the sum of its kinetic energy and its potential energy, by introducing the proper rest frequency $\omega_0 = E_0^{dist}/\hbar$ of the singularity in the wave equation $\underline{\psi}_H(\vec{r},t)$ associated with that hamiltonian. It is easy to show that the wave function $\underline{\psi}_H(\vec{r},t)$ derives then from the following wave equation

$$i\hbar\frac{\partial\underline{\psi}_H}{\partial t} \cong -\frac{\hbar^2}{2M_0}\Delta\underline{\psi}_H + V(\vec{r},t)\underline{\psi}_H \qquad (28.46)$$

which is exactly the ***Schrödinger wave equation of quantum mechanics***. Indeed, we can verify thanks to the interpretation of this wave equation with the operators *(28.15)* and *(28.16)*

$$E_H \cong \frac{\vec{\boldsymbol{P}}_{\mathbf{v}}^2}{2M_0} + V(\vec{r},t) = T + V(\vec{r},t) = H \qquad (28.47)$$

that this wave equation gives us a wave function $\underline{\psi}_H(\vec{r},t)$ linked to the hamiltonian of the singularity, namely the sum of its kinetic and its potential energy.

On the interpretation of the Schrödinger wave equation for a non-relativistic singularity

The wave equation *(28.46)* corresponds very precisely to the *equation of Schrödinger of quantum mechanics for a non-relativistic particle*, if we admit that the universal constant $\hbar$ which we have introduced is effectively the *Planck constant* $\hbar$ *of quantum mechanics.*

This perfect similarity is not gratuitous and allows us <u>for the first time</u> to give a comprehensible interpretation to quantum mechanics by saying that:

> *"The Schrödinger equation is an equation deduced from the second partial equation of Newton of a perfect cosmic lattice in the domain $\tau_0 << \tau_{0cr}$, which allows us to* ***calculate the wave function $\underline{\psi}_H(\vec{r},t)$ of a topological singularity, representing the amplitude and phase of dynamic gravitational fluctuations with frequency $\omega_H = H/\hbar$ associated with its hamiltonian****".*

On the static wave equation of a singularity placed in a static potential

If the potential in which we place the topological singularity is a static potential $V(\vec{r})$, the left term of equation *(28.40)* is an operator giving total energy $E_{\mathbf{v}}$ of the singularity, which must evidently be a constant since the moving singularity is in a static potential. We then have

$$i\hbar\frac{\partial\underline{\psi}}{\partial t} \cong -\frac{\hbar^2}{2M_0}\Delta\underline{\psi} + \left(E_0^{dist} + V(\vec{r})\right)\underline{\psi} = E_{\mathbf{v}}\underline{\psi} \qquad (28.49)$$

which we rewrite under the following form

$$-\frac{\hbar^2}{2M_0}\Delta\underline{\psi}+V(\vec{r})\underline{\psi}=\left(E_v-E_0^{dist}\right)\underline{\psi}=\left(E^{kin}+V\right)\underline{\psi}=H\underline{\psi} \qquad (28.50)$$

We find here the expression of the ***static Schrödinger equation of quantum mechanics***, which is as we know from quantum mechanics a problem with eigenvalues, which means that the hamiltonian $H=E^{kin}+V=E_v-E_0^{dist}$ is a constant which can take *various eigenvalues* H_n depending on potential $V(\vec{r})$, so that the wave function presents *eigenstates* $\underline{\psi}_n$ satisfying the following equation

$$-\frac{\hbar^2}{2M_0}\Delta\underline{\psi}_n+V(\vec{r})\underline{\psi}_n=H_n\underline{\psi}_n \qquad (28.51)$$

On the basis of solution $\underline{\psi}_n(\vec{r})$ of the wave equation, we deduce the *real physical value, namely the static perturbations of expansion* associated with the singularity via potential $V(\vec{r})$, given by *(28.45)*

$$\underline{\tau}_n^{(p)}(\vec{r},t)\cong\underline{\psi}_n(\vec{r})\,\mathrm{e}^{\pm\frac{H_n}{\hbar}t} \qquad (28.52)$$

and this real part of $\underline{\tau}_n^{(p)}(\vec{r},t)$ will represent the real perturbations of expansion.

28.3 – On the consequences of the Schrödinger wave equations of gravitational fluctuations of a non-relativistic singularity

On operator commutators and the Heisenberg uncertainty principle

We know the power of the dynamic Schrödinger equation *(28.42)* and the static Schrödinger equation *(28.51)* in quantum mechanics. Many consequences linked to the Schrödinger equation are described in Appendix B, and these consequences are applicable in our theory, such as:

- the *commutators of operators*: relations *(B.5) to (B.8)*,
- the *Heisenberg uncertainty principle:* relations *(B.9)*,
- the static eigenstates of a particle in different kinds of potentials (harmonic oscillator, anharmonic oscillator, particle in a box, rotation of two linked particles, particle in a central potential): relations *(B.12) to (B.23)*,
- the density of states in phase space: relations *(B.24) to (B.27)*.

Due to this perfect correspondence between our theory of gravitational perturbations associated with topological singularities and the Schrödinger wave equation of quantum physics, which is experimentally verified, we deduce that our conjecture 10 is *à posteriori* justified. As a consequence we have a "classical" interpretation of quantum mechanics, namely that quantum mechanics is a consequence of the second partial equation of Newton applied to describe the expansion perturbations of the lattice.

On the probabilistic interpretation of the square of the wave function

Whereas the complex wave functions $\underline{\psi}(\vec{r},t)$ do not give us any indication as to the position and trajectory of the singularity, we can find a very physical interpretation for them. As these wave functions correspond to a complex representation of amplitude and phase of the gravitational fluctuations with frequency $\omega_H = H / \hbar$ *associated to the singularity,* it is logical and probable that, if there are locally no gravitational fluctuations, meaning if the wave function $\underline{\psi}(\vec{r},t)$ is very small in certain parts of space, there will be no chance of finding the topological singularity there, while if the fluctuations become maximal in other parts of space then you will probably find the topological singularity there!

We thus arrive at an interesting interpretation of the complex wave function $\underline{\psi}(\vec{r},t)$: it must be associated with the probability of presence of the topological singularity to which the gravitational fluctuations $e^{i\omega_H t}$ are associated! The function $\underline{\psi}(\vec{r},t)$ is in fact a complex mathematical object representing the amplitude and phase of gravitational fluctuations $e^{i\omega_H t}$, while a probability of presence is a mathematical object, a positive scalar, whose sum over the whole space must be equal to one. As a consequence, a possibility of extracting a quantitative value of probability of presence of the topological singularity from function $\underline{\psi}(\vec{r},t)$ is to use the fact that the square of an oscillating function gives us a positive scalar. In the case of a complex quantity such as $\underline{\psi}(\vec{r},t)$, it is the product $\underline{\psi}(\vec{r},t)\bullet\underline{\psi}^*(\vec{r},t)$ of the complex function $\underline{\psi}(\vec{r},t)$ by its complex conjugate $\underline{\psi}^*(\vec{r},t)$ which represents the square of amplitude of the function. It suffices thus to normalize the product $\underline{\psi}(\vec{r},t)\bullet\underline{\psi}^*(\vec{r},t)$ taken on a portion of space ΔV by this product taken over all space V which could contain the singularity, to obtain a probability P of finding the singularity in a portion of space ΔV

$$P = \iiint_{\Delta V} \underline{\psi}(\vec{r},t)\bullet\underline{\psi}^*(\vec{r},t)dV \Big/ \iiint_{V} \underline{\psi}(\vec{r},t)\bullet\underline{\psi}^*(\vec{r},t)dV \tag{28.53}$$

We find the usual interpretation of the wave function of quantum mechanics, while giving it a conceptual grounding!

The fact that the complex function $\underline{\psi}(\vec{r},t)$ allows us to deduce, not the position of singularities at a given moment, but their probability of presence in a given place at a given time, signifies also that the wave equations *(28.42)* and *(28.46)* which allow us to calculate $\underline{\psi}(\vec{r},t)$ (or $\underline{\psi}_n(\vec{r})$ in the stationary case), and which are nothing else than results of the *Newton equation of the lattice* applied to the gravitational fluctuations, are at the same time a *new form of equations for the dynamics of topological singularities inside the lattice.*

On the "stochastic walk" of topological singularities

As the exact microscopic movements of the topological singularities are not computable via their complex wave function $\underline{\psi}(\vec{r},t)$, but only their probability of presence, when submitted to a potential $V(\vec{r},t)$ or $V(\vec{r})$, can be obtained, the real movements of the singularities within the lattice must be stochastic and chaotic movements.

We can imagine for example that random gravitational fluctuations (see chapter 32), with different frequencies ω_H, that appear and disappear in the vicinity of the singularity, could move it about by giving it random accelerations. These accelerations would contribute then to the stochastic movements of the singularities. But as the stochastic march of the singularity must also be coupled to its gravitational fluctuations with proper frequency ω_H, this stochastic march will have to present a statistical distribution of presence which will manifest via the probability presence *(28.53)* deduced from the wave function.

There exists two physical phenomena that are observable, which present strong analogies with such a stochastic march of topological singularities:
- in solids, the dislocations can present a microscopic march under the effect of random thermal fluctuations (due to phonons) that can push them. There appears then a stochastic movement of the dislocations, called "*brownian motion*", such as described for example in the article *«overview on dislocation-point defect interaction: the brownian picture of dislocation motion»*[1] !
- recent macroscopic experiences, realized in labs with droplets bouncing on a liquid surface vibrating at a given frequency, present rather striking results. The drops move randomly on the surface of the liquid, which is the reason they have been called "walkers" [2]. This "walk" is attributed to a resonant interaction of the drop with its own wave field[3]. The measure of the probability of distribution of the droplet on a given surface can then present regularities that look similar to the probability of presence of quantum particle confined in a potential well[4].

28.4 – Superposition of topological singularities, bosons, fermions and the exclusion principle

We can legitimately ask ourselves what becomes of the field of gravitational fluctuations when two topological singularities are next to each other.

On the stationary state of superposition of two identical topological singularities

Let's imagine two singularities *(a)* and *(b)* which evolve in the same space and the same potential. We search for the standing wave of 'superposition', meaning the way we can write the perturbations of volume expansion due to the two singularities at the same time. By supposing that the Schrödinger equations for standing waves are valid for both singularities, we have

$$\begin{cases} -\dfrac{\hbar^2}{2M_0}\Delta\underline{\psi}_n(\vec{r}_a)+V(\vec{r}_a)\underline{\psi}_n(\vec{r}_a)=H_n\underline{\psi}_n(\vec{r}_a) \\ -\dfrac{\hbar^2}{2M_0}\Delta\underline{\psi}_m(\vec{r}_b)+V(\vec{r}_b)\underline{\psi}_m(\vec{r}_b)=H_m\underline{\psi}_m(\vec{r}_b) \end{cases} \quad (28.54)$$

We try to combine these two relations, by multiplying the first one by $\underline{\psi}_m(\vec{r}_b)$ and the second one by $\underline{\psi}_n(\vec{r}_a)$, and by summing it all. We have

$$-\frac{\hbar^2}{2M_0}\Delta\left[\underline{\psi}_n(\vec{r}_a)\underline{\psi}_m(\vec{r}_b)\right]+\left[V(\vec{r}_a)+V(\vec{r}_b)\right]\underline{\psi}_n(\vec{r}_a)\underline{\psi}_m(\vec{r}_b)=\left(H_n+H_m\right)\underline{\psi}_n(\vec{r}_a)\underline{\psi}_m(\vec{r}_b) \quad (28.55)$$

which is the Schrödinger equation for the superposition wave function $\underline{\psi}_n(\vec{r}_a)\underline{\psi}_m(\vec{r}_b)$. We de-

[1] *G. Gremaud, Materials Science and Engineering A 370 (2004) 191-198*

[2] *Stéphane Perrard: "Une mémoire ondulatoire: états prores, chaos et probabilités", thèse de doctorat, 2014, Université Paris Diderot (https://tel.archives-ouvertes.fr/tel-01158368)*

[3] *D. M. Harris, J. Moukhtar, E. Fort, Y. Couder, J. W. M. Bush: "Wavelike statistics from pilot-wave dynamics in a circular corral" Physical Review E, 88, 011001(R), 2013*

[4] *R. Brady, R. Anderson: "Why bouncing droplets are a pretty good model of quantume mechanics", University of Cambridge Computer Laboratory, 2014 (arXiv:1401.4356v1)*

duce that the oscillatory perturbations of volume expansion due to the superposition of two singularities is written

$$\underline{\tau}(\vec{r}_a,\vec{r}_b,t)=\underline{\psi}_n(\vec{r}_a)\mathrm{e}^{(\pm)i\frac{H_n}{\hbar}t}\,\underline{\psi}_m(\vec{r}_b)\mathrm{e}^{[\pm]i\frac{H_m}{\hbar}t} \qquad (28.56)$$

We note that there are two types of possible superpositions, following the signs of the exponents, which present global oscillation frequencies that are different

$$\begin{cases} \underline{\tau}_{boson}(\vec{r}_a,\vec{r}_b,t)=\underline{\psi}_n(\vec{r}_a)\underline{\psi}_m(\vec{r}_b)\mathrm{e}^{(\pm)i\frac{1}{\hbar}(H_n+H_m)t} \\ \underline{\tau}_{fermion}(\vec{r}_a,\vec{r}_b,t)=\underline{\psi}_n(\vec{r}_a)\underline{\psi}_m(\vec{r}_b)\mathrm{e}^{(\pm)i\frac{1}{\hbar}(H_n-H_m)t} \end{cases} \qquad (28.57)$$

By analogy with quantum physics, we will call *"bosons"* the singularities corresponding to the first solution of superposition, whose frequency is $(H_n+H_m)/\hbar$, because the two singularities can occupy the same level of energy without disappearance of the oscillatory disturbances of expansion. With respect to the singularities which correspond to the second frequency $|H_n-H_m|/\hbar$, we call them *"fermions"* because they cannot be superimposed in the same energy level as in this case the gravitational perturbations disappear!

This state of fact on how singularities will superpose then directly shows the famous *Pauli exclusion principle:* the singularities which combine according to the second possibility *(28.57)*, namely fermions, cannot be found in the same state (in $\vec{r}_a=\vec{r}_b$).

In usual quantum physics, where we only talk about the wave functions and where we ignore the physical significance of these wave functions in terms of amplitude and phase of the oscillatory gravitational perturbations, we can manifest a difference between bosons and fermions directly in the superposition $\underline{\psi}_n(\vec{r}_a)\underline{\psi}_m(\vec{r}_b)$ of the wave functions. For that we note that for a given energy E of the system, there are two possible solutions to equation *(28.55)* for the wave function $\underline{\Psi}$ of superposition, which correspond simply to exchanging two identical singularities

$$\underline{\Psi}=\underline{\psi}_n(\vec{r}_a)\underline{\psi}_m(\vec{r}_b) \quad \text{and} \quad \underline{\Psi}=\underline{\psi}_n(\vec{r}_b)\underline{\psi}_m(\vec{r}_a) \qquad (28.58)$$

However, one of the fundamental properties of homogeneous linear differential equations is that any linear combination of solutions is also a solution, so that the most generic form of solution for the Schrödinger equation *(28.55)* can be written as the following superposition

$$\underline{\Psi}=\alpha\underline{\psi}_n(\vec{r}_a)\underline{\psi}_m(\vec{r}_b)+\beta\underline{\psi}_n(\vec{r}_b)\underline{\psi}_m(\vec{r}_a) \qquad (28.59)$$

This expression would indicate that there exist a large number of stationary states for a system with two singularities. Nonetheless, we must now take into account the fact that due to the uncertainty principle linked to the commutation of operators, the identical singularities lose their individuality. We say the singularities are indiscernible which simply means that it is not possible to follow the trajectory of a given singularity over time. If we consider the wave function $\underline{\Psi}$ *(28.59)* of the system, we know that $\underline{\Psi}^2$ determines the possibility to find the two singularities in a portion of space. If we exchange the two singularities, it is clear that $\underline{\Psi}^2$ must remain unchanged. However, the phase of $\underline{\Psi}$ can be modified by this exchange, so that $\underline{\Psi}\to\underline{\Psi}e^{i\eta}$. If we proceed to a second exchange of singularities, we have evidently $\underline{\Psi}\to\underline{\Psi}e^{2i\eta}$ and we find ourselves in the initial state $\underline{\Psi}$, so that $e^{2i\eta}=1$. For that to be true it suffices that $e^{2i\eta}=+1$ or $e^{2i\eta}=-1$.

In the case where the wave function $\underline{\Psi}$ transforms as $\underline{\Psi}\to+\underline{\Psi}$ during the exchange of two singularities, the wave function is said to be *symmetric*, and the singularities are called *bosons*. The wave function $\underline{\Psi}$ is written with a normalization factor α

$$\underline{\Psi}=\alpha\left(\underline{\psi}_n(\vec{r}_a)\underline{\psi}_m(\vec{r}_b)+\underline{\psi}_n(\vec{r}_b)\underline{\psi}_m(\vec{r}_a)\right) \tag{28.60}$$

If the wave function $\underline{\Psi}$ transforms as $\underline{\Psi}\rightarrow-\underline{\Psi}$, the wave function is said to be *antisymmetric*, and the singularities are called *fermions*. The wave function $\underline{\Psi}$ is written with a normalization factor β

$$\underline{\Psi}=\beta\left(\underline{\psi}_n(\vec{r}_a)\underline{\psi}_m(\vec{r}_b)-\underline{\psi}_n(\vec{r}_b)\underline{\psi}_m(\vec{r}_a)\right) \tag{28.61}$$

The indistinguishability of the two singularities is clearly shown with the previous two expressions of the wave function $\underline{\Psi}$. We notice also that, for an anti-symmetric wave function, it is not possible that both singularities be in the same state as $\underline{\Psi}$ would then be null: this is the mathematical expression at the level of the wave function itself, of the *exclusion principle* associated to expressions *(28.57)*, which states that two fermions cannot occupy the same state simultaneously as the gravitational perturbations disappear in that case!

In the case of a system with N identical singularities, the previous concepts are easily generalized. In the case of bosons, the symmetric wave function $\underline{\Psi}_{sym}$ of the system can be written

$$\underline{\Psi}_{sym}=\alpha\sum_{a,b,\ldots,N}\psi_n(\vec{r}_a)\psi_m(\vec{r}_b)\ldots\psi_k(\vec{r}_N) \tag{28.62}$$

where the sum refers to all the possible permutations of all the different states of the system. If the system possesses n_1 singularities with energies n, n_2 singularities with energy m, n_3 singularities with energy k, etc., the number of terms comprising the wave function $\underline{\Psi}_{sym}$ is

$$P=\frac{N!}{n_1!n_2!n_3!\ldots} \tag{28.63}$$

In the case of fermions, the anti-symmetric wave function $\underline{\Psi}_{anti}$ of the system can be written in the form of a determinant

$$\underline{\Psi}_{anti}=\beta\begin{vmatrix}\psi_n(\vec{r}_a) & \psi_n(\vec{r}_b) & \ldots\\ \psi_m(\vec{r}_a) & \psi_m(\vec{r}_b) & \ldots\\ \ldots & \ldots & \ldots\end{vmatrix} \tag{28.64}$$

In effect, the permutations of two columns of a determinant changes the sign of the determinant, which assures us of the anti-symmetry of wave functions $\underline{\Psi}_{anti}$ under the exchange of two singularities. Also, we know that a determinant is null if two lines are the same, which corresponds to the expression of the *Pauli exclusion principle*, namely that a given state cannot be occupied by more than a fermion.

28.5 – On the analogy with quantum physics

It is entirely remarkable that the wave function associated with gravitational perturbations of volume expansion is perfectly similar to the quantum wave function of a particle, and that it satisfies a wave equation identical to Schrödinger's equation. This deserves a more in depth discussion!

On the strong analogy with the quantum wave equation and the Schrödinger equation

The relativistic wave equation *(28.24)* deduced from the second partial equation of Newton *(28.3)* allowed us to describe gravitational fluctuations of expansion associated with a massive singularity moving with relativistic speeds within the lattice. With regards to the non-relativistic wave equation of a singularity linked to a potential, it is absolutely identical to the *Schrödinger equation* of quantum physics since their respective interpretations in terms of probability of presence of a particle are *identical !* The key passages used to reach the Schrödinger equation *(28.46)* of a singularity from the second partial equation of Newton *(28.3)* for the gravitational perturbations of expansion are, first *conjecture 10* which postulates the physical significance of operators of space and time applied to the wave function, and secondly the reduction of the wave equation allowing one to go from the wave equation *(28.38)* of second degree in the spatial derivatives to the wave equation *(28.40)* which is of first degree in the spatial derivatives, by using here again *conjecture 10*. It is these two key passages that allow us to establish a physical theory completely similar to the quantum mechanics to describe the microscopic behavior of topological singularities within the cosmic lattice which does not present longitudinal waves.
But we are still missing in our theory a physical explanation of these two passages and their reason for being. Notably, we can legitimately ask why the Planck constant exists, where does its value come from, and whether it is really a universal constant, or whether one can deduce it from other constants in our theory. A response to these questions would afford us a deeper comprehension about quantum mechanics.

On the demystification of quantum mechanics

In our theory, the complex wave function $\underline{\psi}(\vec{r},t)$ and the Schrödinger wave equation are physically demystified, as they become the mathematical expressions of the envelope and phase of the vibrational fluctuations of expansion, and thus of gravitational fluctuations correlated with topological singularities.
From this innovative interpretation of quantum mechanics, we have the possibility to have singularities of the type *«bosons»* and of the type *«fermions»*, we have the fact that we cannot discern between topological singularities when they contribute to the same field of gravitational fluctuations, and the fact that singularities of the type «fermions» must satisfy *an exclusion principle* similar to the Pauli exclusion principle. This is probably the most remarkable and striking result of our calculations, as it demystifies a side of QM that was always rather obscure.
Finally, it is just as remarkable to note that all these properties, such as the *property of superposition* (the symmetry of the wave function $\underline{\Psi}_{sym}$ of singularities of type "bosons" and the antisymmetry of the wave function $\underline{\Psi}_{antisym}$ of singularities of type «fermions», and the indistinguishability of topological singularities and the exclusion principle) are direct consequences of the fact that the gravitational fluctuations *associated with one of more singularities must satisfy the second partial equation of Newton of the cosmic lattice!*
In fact, the image of a field with gravitational fluctuations correlated with a topological singularity

has great potential to explain simply the quantum phenomena we have observed and calculated, but remain mysterious within the framework of usual quantum mechanics. Let's think about the following examples:

- the concept of wave-particle duality of quantum mechanics finds here an immediate and simple explanations since the particle is the topological singularity and the wave is the gravitational fluctuations associated with it,
- the quantum interference experiences obtained by the passage of particles through two slits, but with one particle at a time, can be explained by the fact that the singularity can only go through one slit while the gravitational field perturbations go through both slits, so that there is a possibility of interference of these perturbations, which leads to a coupling with the singularity and a modification of its trajectory, and finally there will be a statistical distribution of impact points on a screen placed after the two slits,
- the Heisenberg uncertainty principle, which is evidently satisfied in our theory since it admits *à fortiori* the same interpretation of operators acting on the wave function of quantum mechanics, and satisfies thus all the relations of *Appendix B.* The uncertainty relations are then directly linked to the existence of perturbations of the gravity field linked to the singularity,
- the very mysterious experiences of entanglement and quantum decoherence. We can imagine that entanglement is embodied in the fact that two or more singularities can possess a common field of gravitational fluctuations, in which case, acting on one singularity is going to modify the common gravitational fluctuations, which will act on the other singularities, namely the decoupling of the topological singularities via a decoupling of the gravitational fluctuating fields.

«God does not play dice»

Einstein used to say *«God does not play dice»* referring to QM. What he meant was that he considered that the theories of his age were incomplete theories. In his opinion, there had to be a rational and pragmatic explanation to account for the *probabilistic* nature of QM. This opinion of Einstein has been widely discussed, even scorned! It was proven that there could not be local hidden variable to explain quantum mechanics, however there could be non-local hidden variables, and that is precisely the case with gravitational fluctuations linked to topological singularities! We must concede here that Einstein was right, and that there exists a rational explanation to quantum mechanics!

There is a highly ironic tone in the famous sentence of Einstein, since QM would be explained by the gravitational fluctuations of the volume field of expansion and by a stochastic movement of the topological singularities interacting with said gravitational fluctuations. These are ingredients with which God WOULD be playing dice. Ironically it was Einstein himself who proposed both General Gravity <u>and</u> brownian motion, which got him a Nobel prize!

Thus, our explanation for quantum mechanics goes with Einstein and shows that it is the expression of gravitational fluctuations on a very small scale in a cosmic lattice without longitudinal propagation. As a consequence, all the modern attempts at quantifying gravity are bound to fail since QM is precisely the expression of the gravitational fluctuations at a microscopic scale!

Chapter 29

Gravitational fluctuations within topological singularities: spin and magnetic moment

In this chapter, we will find a solution to the second partial equation of Newton with the torus around a SDL. We will show that there are no static solutions to this equation and that, as a consequence, we will have to search for a dynamic solution for the gravitational perturbations of expansion in the immediate vicinity of the loop. This dynamic solution will turn out to be a *quantized movement of rotation of the loop on itself.* This solution satisfies the second partial equation of Newton, which becomes in this case the Schrödinger equation as we have seen in the previous chapter!

This movement of rotation of the loop about itself is nothing else than the *«spin»* of the loop, and we can show that a *magnetic moment* is associated with it, which corresponds exactly to the magnetic moment of particle physics! Furthermore we will show that, within our theory, *this is a real movement of rotation*, and that it does not infringe on special relativity, contrary to what the early pioneers of quantum mechanics thought of spin!

29.1 – Internal field of "gravitational" perturbations of the expansion of a twist disclination loop (TL)

We have already calculated the static external fields *(23.2)* of gravitational perturbations of expansion of a twist disclination loop (TL) and we have seen that these fields are responsible for long distance gravitational attraction via the *gravitational force* (section 24.2), but they are also responsible for short distance coupling effects with other loops via the *weak force* (chapter 26).

The existence condition of an internal static field of expansion

We have not yet considered the case of fields of perturbations of expansion in the immediate vicinity of the twist disclination loop (TL). We must calculate this field within the torus encompassing the TL. By using a simplified static version of the second partial equation of Newton *(18.14)*, we have, within the torus

$$\tau_{int}^{TL}(\vec{\xi})=\frac{4K_2/3+2K_1(1+\tau_0)-K_0}{2K_1}\left[-1+\sqrt{1-\frac{4K_1F_{dist}^{TL}(\vec{\xi})}{\left(4K_2/3+2K_1(1+\tau_0+)-K_0\right)^2}}\right] \qquad (29.1)$$

in which $\vec{\xi}$ represents the position vector giving us a point of the torus in relation to the center of the section (fig. 19.2). In this second degree equation, the energy of distortion $F_{dist}^{TL}(\vec{\xi})$ is associated to the fields of rotation and shear of the TL and is worth, according to *(19.5)*

$$F_{dist}^{TL}(\vec{\xi}) \cong (K_2+K_3)\frac{\Lambda^2}{2\pi^2}\frac{1}{\xi^2} \cong (K_2+K_3)\frac{\vec{B}_{TL}^2}{8\pi^2}\frac{1}{\xi^2} \tag{29.2}$$

A static field of perturbations would then be deduced from equation

$$\tau_{int}^{TL}(\vec{\xi}) = \frac{4K_2/3+2K_1(1+\tau_0)-K_0}{2K_1}\left[-1+\sqrt{1-\frac{K_1(K_2+K_3)}{(4K_2/3+2K_1(1+\tau_0+)-K_0)^2}\frac{\vec{B}_{TL}^2}{2\pi^2}\frac{1}{\xi^2}}\right] \tag{29.3}$$

For this equation to have a real solution, it would be necessary for the argument under the root to be positive, and thus that the distance $\vec{\xi}$ to the center of the loop satisfy relation

$$\xi^2 > \frac{K_1(K_2+K_3)}{(4K_2/3+2K_1(1+\tau_0+)-K_0)^2}\frac{\vec{B}_{TL}^2}{2\pi^2} \tag{29.4}$$

Knowing that the rotational charge of a TL satisfies the following relation

$$q_{\lambda TL}^2 = \pi^2 R_{TL}^2 \vec{B}_{TL}^2 \quad \Rightarrow \quad \vec{B}_{TL}^2 = \frac{q_{\lambda TL}^2}{\pi^2 R_{TL}^2} \tag{29.5}$$

we deduce the condition for the existence of a static solution to equation *(29.3)*, notably in the domain of expansion of the background $\tau_0 << \tau_{0cr}$

$$\xi^2 > \frac{K_1(K_2+K_3)}{2\pi^4(4K_2/3+2K_1(1+\tau_0+)-K_0)^2}\frac{q_{\lambda TL}^2}{R_{TL}^2} \cong \frac{K_1}{2\pi^4 K_0}\frac{q_{\lambda TL}^2}{R_{TL}^2} \tag{29.6}$$

Which is equivalent to the condition that

$$\xi > \frac{1}{\pi^2}\sqrt{\frac{K_1}{2K_0}}\frac{|q_{\lambda TL}|}{R_{TL}} \tag{29.7}$$

We will try to express the condition that must exist on module K_1 for this condition to be equivalent to distance $\vec{\xi}$ being superior to radius R_{TL}. We quite easily find that this condition is written

$$K_1 > K_{1cr} = K_0\frac{2\pi^4 R_{TL}^2}{q_{\lambda TL}^2} \quad \Rightarrow \quad \xi > R_{TL} \tag{29.8}$$

What this means is that if module K_1 is bigger than critical value K_{1cr}, there are no static solution to equation *(29.3)* within the whole volume of the torus around the TL!

It is now interesting to consider numerical values inspired by "the real world", by using for example an analogy to electrons, namely that they have an electrical charge worth $q_{\lambda TL} = q_{électron} = 1.6\cdot 10^{-19}[C]$, and a radius estimated to be on the order of $10^{-18}[m]$ and that the elastic modules $K_3 = K_0$ are actually the analogs to the dielectric constant of vacuum, so that we can use numerical values: $K_3 = K_0 = 1/\varepsilon_0^{dielectric} = 1.1\cdot 10^{11}[m/F]$. With these numerical values, the condition *(29.8)* would then imply that $K_{1cr} = K_0 2\pi^4 R_{TL}^2 / q_{\lambda TL}^2 \cong 10^{-21}$.

On the existence and nature of an internal dynamic field of expansion

The condition that module K_1 be superior to some critical value $K_{1cr} \cong 10^{-21}$ is likely to be true in the presence of twist loops according to the numerical values obtained with an electron from the 'real world'. Let's make the following conjecture

Conjecture 11: K_1 *satisfies the following equation in the cosmic lattice*

$$K_1 \geq K_{1cr} = K_0\, 2\pi^4\, R_{TL}^2 \,/\, q_{\lambda TL}^2 \tag{29.9}$$

If we admit this conjecture then the gravitational field within the torus around the TL cannot be a static solution *(29.3)*, and must consequently become a *dynamical* gravitational perturbation field. But the Newton equation for the dynamic perturbations is nothing else than the Schrödinger equation *(28.46)*. We must then find a movement of the loop which is not a translation but a movement confined to the same volume. The only possible movement is *a rotation about itself!*

29.2 – Angular momentum, spin and magnetic moment of a twist disclination loop (TL)

Let us consider a TL with radius R_{TL} as represented in figure 19.2, and let's imagine that it can turn about an axis of direction $\vec{e}_{axis}$ contained in the plane of the loop with *an angular velocity* ω_{TL}, which is not impossible since the loop corresponds to a *pseudo-screw dislocation*. If we first treat this problem in a classical way, we can use polar coordinates to define the *angular momentum* $\vec{L}_{TL}$ of the loop about it's axis of rotation, *by supposing that the mass of the loop is uniformly distributed at the surface of the loop*

$$\vec{L}_{TL} = \oint \vec{r} \wedge \vec{v}\, dm = \vec{e}_{axis} \int_0^{R_{TL}} 4 \int_0^{\pi/2} \underbrace{r\cos\theta}_{|\vec{r}|}\, \underbrace{r\cos\theta\, \omega_{TL}}_{|\vec{v}|}\, \underbrace{\frac{M_0^{TL}}{\pi R_{TL}^2} r d\theta dr}_{dm} = \frac{M_0^{TL} R_{TL}^2}{4} \omega_{TL} \vec{e}_{axis} \tag{29.10}$$

As we only roughly know the mass distribution in the vicinity of the loop, we will introduce a numerical factor δ_1 of correction such that

$$\vec{L}_{TL} \cong \delta_1 \frac{M_0^{TL} R_{TL}^2}{4} \omega_{TL} \vec{e}_{axis} \quad \text{with} \quad \delta_1 \cong 1 \tag{29.11}$$

We can then introduce the *moment of inertia* of the loop about the axis of rotation

$$\vec{L}_{TL} \cong I_{TL} \omega_{TL} \vec{e}_{axis} \quad \Rightarrow \quad I_{TL} = \delta_1 \frac{M_0^{TL} R_{TL}^2}{4} \quad \text{with} \quad \delta_1 \cong 1 \tag{29.12}$$

An twist loop will also possess a charge of rotation $q_{\lambda TL}$ (analogous to the electric charge), so it also possesses a *«magnetic moment»* $\vec{\mu}_{TL}$ in the direction of the axis of rotation, which we can calculate by supposing that the charge is located on the contour of the loop, as

$$\vec{\mu}_{TL} = \frac{1}{2} \oint \vec{r} \wedge \vec{v}\, dq = \vec{e}_{axis} 2 \int_0^{\pi/2} \underbrace{R_{TL}\cos\theta}_{|\vec{r}|}\, \underbrace{R_{TL}\cos\theta\, \omega_{TL}}_{|\vec{v}|}\, \underbrace{\frac{q_{\lambda TL}}{2\pi} d\theta}_{dq} = \frac{R_{TL}^2 q_{\lambda TL}}{4} \omega_{TL} \vec{e}_{axis} \tag{29.13}$$

As we only approximatively know the charge distribution in the vicinity of the loop, we will introduce a numerical factor δ_2 of correction such that

$$\vec{\mu}_{TL} \cong \delta_2 \frac{R_{TL}^2 q_{\lambda TL}}{4} \omega_{TL} \vec{e}_{axis} \quad \text{with} \quad \delta_2 \cong 1 \tag{29.14}$$

We find now a direct relation between the *«magnetic moment»* $\vec{\mu}_{TL}$ and the *angular momentum* $\vec{L}_{BV}$ of the loop, which is called the *gyro-magnetic ratio*

$$\vec{\mu}_{TL} \cong 2\frac{\delta_2}{\delta_1}\frac{q_{\lambda TL}}{2M_0^{TL}}\vec{L}_{TL} = g_{TL}\frac{q_{\lambda TL}}{2M_0^{TL}}\vec{L}_{TL} \quad \text{with} \quad g_{BV} = 2\frac{\delta_2}{\delta_1} \cong 2 \tag{29.15}$$

We can furthermore calculate the *kinetic energy* $E^{kin}_{rotation\,TL}$ associated with this movement of rotation, under the form

$$E^{kin}_{rotation\,TL} = \frac{1}{2}\oint \vec{v}^2\, dm = \int_0^{R_{TL}} 2\int_0^{\pi/2} \underbrace{r^2\cos^2\theta\,\omega^2}_{\vec{v}^2}\, \underbrace{\frac{M_0^{TL}}{\pi R_{TL}^2} r d\theta dr}_{dm} = \frac{M_0^{TL} R_{TL}^2}{8}\omega_{TL}^2 \tag{29.16}$$

With the corrective factor δ_1 on the mass distribution within the loop, we obtain

$$E^{kin}_{rotation\,TL} \cong \delta_1 \frac{M_0^{TL} R_{TL}^2}{8}\omega_{TL}^2 \quad \text{with} \quad \delta_1 \cong 1 \tag{29.17}$$

We have the following links between *kinetic energy, angular momentum* and *moment of inertia*

$$E^{kin}_{rotation\,TL} \cong \frac{2\vec{L}_{TL}^2}{\delta_1 M_0^{TL} R_{TL}^2} \cong \frac{\vec{L}_{TL}^2}{2I_{TL}} \tag{29.18}$$

On the quantification of the angular momentum of the twist loop

If we admit that the twist loop (TL) does indeed turn about itself, this microscopic movement of rotation must be calculated with the static version of the Schrödinger equation *(28.46)*. We sum up in *appendix B* the treatment of the movement of rotation of a microscopic object about it's axis, obeying Schrödinger's stationary equation. The energy is quantized by

$$\varepsilon_j = \frac{\hbar^2}{2I} j(j+1) \tag{29.19}$$

For each value of the energy ε_j corresponding to a given angular velocity, there are $2j+1$ eigenstates corresponding classically to different orientations of the axis of rotation. We say that the energy state ε_j is $2j+1$ times degenerated.
With regards to the *quantum magnetic number* m_z , which characterizes the projection of the angular momentum along a certain axis z, it can take the following values

$$m_z = j, j-1, ..., 1-j, -j \tag{29.20}$$

so that the projection L_z on the Oz axis takes the values

$$L_z = \hbar m_z \tag{29.21}$$

The kinetic energy and the angular momentum of the loop are thus worth

$$\begin{cases} E^{kin}_{rotation\,TL} = \dfrac{\hbar^2}{2I_{TL}} j(j+1) \\ \vec{L}_{TL}^2 = 2I_{TL}E^{kin}_{rotation\,TL} = \hbar^2 j(j+1) \;\Rightarrow\; \left|\vec{L}_{TL}\right| = \hbar\sqrt{j(j+1)} \end{cases} \tag{29.22}$$

The *magnetic moment of the loop* along Oz is then written

$$\vec{\mu}_{TLz} \cong g_{TL}\frac{q_{\lambda TL}}{2M_0^{TL}} L_z\vec{e}_z = g_{TL}\frac{\hbar q_{\lambda TL}}{2M_0^{TL}} m_z\,\vec{e}_z \quad \text{with} \quad g_{TL} = 2\frac{\delta_2}{\delta_1} \cong 2 \tag{29.23}$$

where g_{TL} is the *Landé g-factor* of the TL, roughly equal to 2, but which would depend on the distribution of mass and charge in the case of other topological singularities.

We note that, in the expression *(29.23)*, we find the famous value of the *Bohr Magneton*, namely $\hbar q_{\lambda TL} / 2M_0^{TL}$.

On the classic interpretation of the spin of a particle

In quantum mechanics, the spin of a charged particle like the electron was initially attributed to a self rotation of the particle. However, at the time, the electron was considered to be a spherical particle, very small in size. This made people doubt the classical interpretation of spin. The strongest argument against the classical interpretation of spin was the fact that the calculated equatorial velocities of the electron would be much larger than the speed of light, thereby violating special relativity!

But it is completely different in our theory. Indeed, if we try to calculate the equatorial velocity in the case of a TL, which we obtain from its radius R_{TL} and its angular velocity ω_{TL}

$$\mathbf{v}_{equatorial} = R_{TL}\omega_{TL} \qquad (29.24)$$

To determine the angular velocity ω_{TL}, we identify the *kinetic rotation energy* of the loop *(29.17)* with its *kinetic energy (29.22)* which we determined from the Schrödinger equation. We have

$$E_{rotation\,TL}^{kin} \cong \delta_1 \frac{M_0^{TL} R_{TL}^2}{8}\omega_{TL}^2 = \frac{\hbar^2}{2I_{BV}} j(j+1) \;\Rightarrow\; \omega_{TL} = \pm\frac{4\hbar}{\delta_1 M_0^{TL} R_{TL}^2}\sqrt{j(j+1)} \qquad (29.25)$$

By introducing this value of ω_{TL} in the expression of the equatorial velocity $\mathbf{v}_{equatorial}$, we obtain

$$\mathbf{v}_{equatorial} = R_{TL}\omega_{TL} = \pm\frac{4\hbar}{\delta_1 M_0^{TL} R_{TL}}\sqrt{j(j+1)} \qquad (29.26)$$

Numerically, let's use the known numerical values of the electron, namely the mass $M_0^{TL} \equiv M_0^{électron} = 9{,}1\bullet 10^{-31}\,[kg]$, its radius on the order of $10^{-18}\,[m]$, the value δ_1 close to 1, the value of the Planck constant $\hbar \equiv 6{,}6\bullet 10^{-34}\,[m^2 kg/s]$, and its known spin of $j = 1/2$. We then have the velocity

$$\mathbf{v}_{equatorial} \cong 2{,}5\bullet 10^{15}\,[m/s] \qquad (29.27)$$

We find also that this equatorial velocity is largely superior to the speed of light in the lattice, namely $c_{t0} \equiv c = 3{,}3\bullet 10^8\,[m/s]$, as was found by the pioneers of QM.

However in our theory, there is a new fact, which is that the static volume expansion in the immediate vicinity of the singularity is very large. We can express the static volume expansion at the limits of the torus where the perturbations of expansion are static. In this limit, the local volume expansion is maximal and is given by the unique solution to equation *(29.3)* when the term under the root is null, so that

$$\tau_{static\,max}^{TL} = \frac{K_0 - 4K_2/3 - 2K_1(1+\tau_0)}{2K_1} \cong \frac{K_0}{2K_1} \qquad (29.28)$$

From which we deduce that the real velocity of transversal waves in the immediate vicinity of the loop is actually worth

$$c_t\big|_{limit} \cong c_{t0}\,e^{\tau_{static\,max}^{TL}/2} \cong c_{t0}\,e^{\frac{K_0}{4K_1}} \qquad (29.29)$$

As a consequence, the speed $c_t\big|_{limit}$ is assuredly much larger than $c_{t0} \equiv c = 3{,}3\bullet 10^8\,[m/s]$

since $K_0 / K_1 >> 1$. As a matter of fact, it suffices that

$$c_t\big|_{limit} \cong c_{t0}\, e^{\frac{K_0}{4K_1}} > \mathbf{v}_{equatorial} \cong 2{,}5 \bullet 10^{15}\left[m/s\right] \qquad (29.30)$$

for the equatorial velocity of the loop to be possible.

We can try to determine what is the limit value of module K_1 for the rotation of the loop to be possible. We obtain

$$K_1 / K_0 < 1{,}6 \bullet 10^{-2} \qquad (29.31)$$

And this condition is always satisfied since $K_1 / K_0 << 1$. In our theory, we are thus assured that the movement of rotation of the loop on itself is not only possible but may very well be necessary as it is the only possible solution to the second partial equation of Newton!!!

We conclude that there exists a very classical explanation for the spin of a particle, as a real quantified movement of rotation of the loop about an axis and which does not violate special relativity. This explanation takes away the magical nature of the notion of spin in QM. It also explains perfectly the existence of a quantized magnetic moment of the electron, directly associated to the real rotation of the charged loop!

29.3 – On the problem of the value of the spin of a topological loop

If the existence of a proper rotation of loops is a necessity in our theory to satisfy the second partial equation of Newton in the immediate vicinity of the loop, we have a new question: what is the value of the spin we should attribute to the loop?

Formulated otherwise, this question is equivalent to looking for a value to attribute to the azimuthal quantum number j which characterizes the quantification of the energy of rotation and of the angular momentum as in *(29.22)* for the loop, as well as it's magnetic moment *(29.23)*.

Experimentally, we know that the spin of the electron is worth $j = 1/2$ and that the spin of the boson W^- is $j = 1$. But the deeper reason for which these particles possess these particular values remain very mysterious! It is the same case in our theory: besides the fact that spin $j = 1/2$ and $j = 1$ are the weakest, and thus correspond to the lowest possible kinetic energies, no reasonable argument allows us for now to make a choice regarding the value of j to be chosen for a twist loop! Let's take a look at the consequence of a spin $j = 1/2$ or a spin $j = 1$ on the TL.

Twist disclination loop with spin 1/2

Consider a TL with spin $j = 1/2$. No matter what the direction of the axis of rotation is, there can only be 2 eigenvalues for the loop, corresponding to a left or right handed rotation about the axis, as the degeneracy of energy is in this case $2j + 1 = 2$.

The kinetic energy of the loop is then worth

$$\begin{cases} E^{kin}_{rotation\,TL} = \dfrac{\hbar^2}{2I_{TL}}\, j\,(j+1) = \dfrac{3\hbar^2}{8I_{TL}} \\ \vec{L}^2_{TL} = \hbar^2\, j\,(j+1) = \dfrac{3}{4}\hbar^2 \;\Rightarrow\; \left|\vec{L}_{TL}\right| = \dfrac{\sqrt{3}}{2}\hbar \end{cases} \qquad (29.32)$$

With regards to the magnetic quantum number m_z, it can take the following values

$$m_z = j, j-1, ..., 1-j, -j = \pm\frac{1}{2} \tag{29.33}$$

so that the projection L_z of the angular momentum along an axis Oz takes the values $\pm\hbar/2$

$$L_z = \hbar m_z = \pm\frac{\hbar}{2} \tag{29.34}$$

We find here exactly the notion of a particle of spin *1/2*.

The magnetic moment of the loop along Oz is then written

$$\vec{\mu}_{TLz} \cong g_{TL}\frac{\hbar q_{\lambda TL}}{2M_0^{TL}} m_z\vec{e}_z = \pm g_{TL}\frac{\hbar q_{\lambda TL}}{4M_0^{TL}}\vec{e}_z \quad \text{with} \quad g_{TL} \cong 2 \tag{29.35}$$

Twist disclination loop with spin 1

Consider a TL with spin $j=1$. Whatever the direction of the rotation may be, there can only be 3 proper states for the loop, a left handed rotation, a right handed rotation and no rotation at all, as the degeneracy of the energy is worth $2j+1=3$.

The kinetic energy and the angular momentum are written

$$\begin{cases} E^{kin}_{rotation\,TL} = \dfrac{\hbar^2}{2I_{TL}}\, j(j+1) = \dfrac{\hbar^2}{I_{TL}} \\ \vec{L}^2_{TL} = \hbar^2\, j(j+1) = 2\hbar^2 \quad \Rightarrow \quad \left|\vec{L}_{TL}\right| = \hbar\sqrt{2} \end{cases} \tag{29.36}$$

The magnetic quantum number m_z can take the following values

$$m_z = j, j-1, ..., 1-j, -j = \begin{cases} 0 \\ \pm 1 \end{cases} \tag{29.37}$$

so that the projection L_z of the angular momentum along axis Oz takes values $\pm\hbar/2$

$$L_z = \hbar m_z = \begin{cases} 0 \\ \pm\hbar \end{cases} \tag{29.38}$$

We recover here exactly the notion of particle with spin *1*.

The magnetic moment of the loop along Oz is written

$$\vec{\mu}_{TLz} \cong g_{TL}\frac{\hbar q_{\lambda TL}}{2M_0^{TL}} m_z\vec{e}_z = \begin{cases} 0 \\ \pm g_{TL}\dfrac{\hbar q_{\lambda TL}}{2M_0^{TL}}\vec{e}_z \end{cases} \quad \text{with} \quad g_{TL} \cong 2 \tag{29.39}$$

On the existing link between bosons, fermions and spin

The question of knowing if a loop singularity behaves like a fermion or a boson in the case of superposition of various loops (see section 29.4) and the question of the value of spin for a loop singularity are linked. Indeed, we know from QM that fermions have a spin 1/2 and that bosons have a spin 1. From QM, we also know that the spin component of the wave function Ψ of two particles is symmetric when the spin of the two particles are parallel, and anti-symmetric if the spins are anti-parallel, thus we have the following possibilities for the wave function Ψ of two particles:

• *Fermions:* anti-symmetric wave function $\Longrightarrow$ parallel spins and anti-symmetric spatial component, or anti-parallel spins and symmetric spatial component.

• *Bosons:* symmetric wave function $\Longrightarrow$ parallel spins and symmetric spatial component, or anti-parallel spins and anti-symmetric spatial component.

It would be very interesting to look deeper into this problem, and to see what topological interpretation to give it in our theory of topological loops. But this problem is, for now, beyond this book, and will be left as 'open problem' in the model of the cosmic lattice!

Chapter 30

Quantified transversal fluctuations: photons

In chapter 14, we have demonstrated that the propagation of a linearly polarized transversal "electromagnetic" wave of rotation is accompanied with a 'gravitational' wavelet. In this chapter, we will focus on what happens in the case of a localized wave packet. We will show that these wave packets can only appear with a *non-null helicity* so that their total energy does not depend on time. By supposing that these wave packets are emitted when a topological singularity changes state suddenly, it becomes understandable that they present a 'quantification of energy'. These wave packets behave as energetic quasi-particles of 'electromagnetic' fluctuations that we could call *«photons»* and which have properties very close to that of photons, as helicity, momentum, wave-particle duality, non-locality and entanglement, etc.

30.1 – Localized transversal electromagnetic fluctuations

We have seen in chapter 14 that the propagation of a polarized transversal wave within the cosmic lattice is constrained by a perturbation correlated to the expansion of the lattice, and that only the circularly polarized waves are transversally 'pure' without wavelets of expansion associated with them. We can reasonably ask ourselves if this striking property couldn't be at the origin of quantized "electromagnetic" fluctuations, who would then look like the famous photons of QM?

The localized wave packets of rotation, real 'electromagnetic' fluctuations

Let's consider transversal waves propagating along Ox_2, with a polarization of the field of rotation along axis Ox_1 and of velocity along Ox_3. The linearized field of the fluctuations are deduced from *(14.13)*, *(14.14)* et *(14.18)* and are written, by supposing a perfect cosmic lattice in the domain $\tau_0 < \tau_{0cr}$

$$\frac{\partial^2 \omega_1}{\partial t^2} \cong c_t^2 \frac{\partial^2 \omega_1}{\partial x_2^2} \qquad \begin{cases} \dfrac{\partial \tau^{(p)}}{\partial t} \cong \dfrac{\partial \phi_2}{\partial x_2} \\ \dfrac{\partial \phi_2}{\partial t} \cong -c_t^2 \dfrac{\partial \tau^{(p)}}{\partial x_2} + c_t^2 \dfrac{\partial}{\partial x_2} (\omega_1)^2 \end{cases} \qquad \begin{cases} \dfrac{\partial \tau^{(p)}}{\partial t} \cong \dfrac{\partial \phi_2}{\partial x_2} \\ \dfrac{\partial \phi_2}{\partial t} \cong -c_t^2 \dfrac{\partial \tau^{(p)}}{\partial x_2} + c_t^2 \dfrac{\partial}{\partial x_2} (\omega_1)^2 \end{cases} \tag{30.1}$$

We try to form a transversal packet of waves, with frequency ω, and an exponential envelope with ranges δ_2 and $\delta_1 = \delta_3 = \delta$, expressed in complex form with a rotational field given by

$$\underline{\omega}_1(x_2,t) = \omega_{10}\, e^{-\frac{|x_1|}{\delta}}\, e^{-\frac{|x_3|}{\delta}}\, e^{-\frac{|x_2 - c_t t|}{\delta_2}}\, e^{i(kx_2 - \omega t)} \tag{30.2}$$

By using the wave equations *(30.1)*, we can show that we obtain the following fields

$$\begin{cases} \underline{\omega}_1(x_2,t) = \omega_{10}\, e^{-\frac{|x_1|}{\delta}}\, e^{-\frac{|x_3|}{\delta}}\, e^{-\frac{|x_2-c_t t|}{\delta_2}}\, e^{i\frac{\omega}{c_t}(x_2-c_t t)} \\ \underline{\phi}_3(x_2,t) = -2c_t\omega_{10}\, e^{-\frac{|x_1|}{\delta}}\, e^{-\frac{|x_3|}{\delta}}\, e^{-\frac{|x_2-c_t t|}{\delta_2}}\, e^{i\frac{\omega}{c_t}(x_2-c_t t)} \\ \underline{\tau}^{(p)}(x_2,t) = \frac{1}{2}\omega_{10}^2\, e^{-\frac{|x_1|}{\delta}}\, e^{-\frac{|x_3|}{\delta}}\, e^{-2\frac{|x_2-c_t t|}{\delta_2}}\, e^{i\frac{2\omega}{c_t}(x_2-c_t t)} \\ \underline{\phi}_2(x_2,t) = -\frac{1}{2}c_t\omega_{10}^2\, e^{-\frac{|x_1|}{\delta}}\, e^{-\frac{|x_3|}{\delta}}\, e^{-2\frac{|x_2-c_t t|}{\delta_2}}\, e^{i\frac{2\omega}{c_t}(x_2-c_t t)} \end{cases} \tag{30.3}$$

By writing the real fields , the real part of complex fields, we then have

$$\begin{cases} \omega_1(x_2,t) = \omega_{10}\, e^{-\frac{|x_1|}{\delta}}\, e^{-\frac{|x_3|}{\delta}}\, e^{-\frac{|x_2-c_t t|}{\delta_2}} \cos\left[\frac{\omega}{c_t}(x_2-c_t t)\right] \\ \phi_3(x_2,t) = -2c_t\omega_{10}\, e^{-\frac{|x_1|}{\delta}}\, e^{-\frac{|x_3|}{\delta}}\, e^{-\frac{|x_2-c_t t|}{\delta_2}} \cos\left[\frac{\omega}{c_t}(x_2-c_t t)\right] \\ \tau^{(p)}(x_2,t) = \frac{1}{2}\omega_{10}^2\, e^{-\frac{|x_1|}{\delta}}\, e^{-\frac{|x_3|}{\delta}}\, e^{-2\frac{|x_2-c_t t|}{\delta_2}} \cos\left[2\frac{\omega}{c_t}(x_2-c_t t)\right] \\ \phi_2(x_2,t) = -\frac{1}{2}c_t\omega_{10}^2\, e^{-\frac{|x_1|}{\delta}}\, e^{-\frac{|x_3|}{\delta}}\, e^{-2\frac{|x_2-c_t t|}{\delta_2}} \cos\left[2\frac{\omega}{c_t}(x_2-c_t t)\right] \end{cases} \tag{30.4}$$

The energy per unit of volume of this fluctuation in a cosmic lattice where the expansion satisfies $\tau_0 < \tau_{0cr}$ is written in this case

$$e^{fluctuation} = \underbrace{2K_3\omega_1^2}_{e_{\vec{\omega}}^{dist}} + \underbrace{K_1\left(\tau^{(p)}\right)^2 - 2\left(K_0 - 2K_1\tau_0\right)\tau^{(p)}}_{e_{\tau^{(p)}}^{dist}} + \underbrace{\frac{1}{2}mn\vec{\phi}^2}_{e^{kin}} \tag{30.5}$$

so that, as $c_t^2 = K_3 / mn$, we have the following density of energy

$$e^{fluctuation} = \left[\begin{aligned} &4K_3\omega_{10}^2\, e^{-2\frac{|x_1|}{\delta}}\, e^{-2\frac{|x_3|}{\delta}}\, e^{-2\frac{|x_2-c_t t|}{\delta_2}} \cos^2\left[\frac{\omega}{c_t}(x_2-c_t t)\right] \\ &+\frac{1}{8}\left(K_3+2K_1\right)\omega_{10}^4\, e^{-2\frac{|x_1|}{\delta}}\, e^{-2\frac{|x_3|}{\delta}}\, e^{-4\frac{|x_2-c_t t|}{\delta_2}} \cos^2\left[2\frac{\omega}{c_t}(x_2-c_t t)\right] \\ &-\left(K_0-2K_1\tau_0\right)\omega_{10}^2\, e^{-\frac{|x_1|}{\delta}}\, e^{-\frac{|x_3|}{\delta}}\, e^{-2\frac{|x_2-c_t t|}{\delta_2}} \cos\left[2\frac{\omega}{c_t}(x_2-c_t t)\right] \end{aligned}\right] \tag{30.6}$$

On the necessity to introduce a helicity of the rotational wave packet

The energy of this fluctuation by oscillation of the field of rotation only contains terms in $\cos^2 \omega t$ and $\cos\omega t$, so that it is not independent of time while it should in principle be! To get rid of the terms in $\cos^2 \omega t$, we can add a rotation oscillation $\omega_1(x_2,t)$ in $\cos\omega t$ along the axis Ox_2 and an oscillation of rotation $\omega_3(x_2,t)$ in $\sin\omega t$ along the Ox_3 axis. These oscillations satisfy the following relations

$$\begin{cases} \dfrac{\partial \omega_1}{\partial t} \cong \dfrac{1}{2}\dfrac{\partial \phi_3}{\partial x_2} \\ \dfrac{\partial \phi_3}{\partial t} \cong 2c_t^2 \dfrac{\partial \omega_1}{\partial x_2} \end{cases}$$
$$\begin{cases} \dfrac{\partial \omega_3}{\partial t} \cong -\dfrac{1}{2}\dfrac{\partial \phi_1}{\partial x_2} \\ \dfrac{\partial \phi_1}{\partial t} \cong -2c_t^2 \dfrac{\partial \omega_3}{\partial x_2} \end{cases} \tag{30.7}$$
$$\begin{cases} \dfrac{\partial \tau^{(p)}}{\partial t} \cong \dfrac{\partial \phi_2}{\partial x_2} \\ \dfrac{\partial \phi_2}{\partial t} \cong -c_t^2 \dfrac{\partial \tau^{(p)}}{\partial x_2} + c_t^2 \dfrac{\partial}{\partial x_2}(\omega_1)^2 + c_t^2 \dfrac{\partial}{\partial x_2}(\omega_3)^2 \end{cases}$$

By written the wave packets of rotation $\underline{\omega}_1(x_2,t)$ and $\underline{\omega}_3(x_2,t)$ under the complex form

$$\begin{cases} \underline{\omega}_1(x_2,t) = \omega_{10}\, e^{-\frac{|x_1|}{\delta}} e^{-\frac{|x_3|}{\delta}} e^{-\frac{|x_2-c_t t|}{\delta_2}} e^{i\frac{\omega}{c_t}(x_2-c_t t)} \\ \underline{\omega}_3(x_2,t) = (\pm) i\omega_{10}\, e^{-\frac{|x_1|}{\delta}} e^{-\frac{|x_3|}{\delta}} e^{-\frac{|x_2-c_t t|}{\delta_2}} e^{i\frac{\omega}{c_t}(x_2-c_t t)} \end{cases} \tag{30.8}$$

we show thanks to relations *(30.7)* that the wave packets associated to the lattice velocities $\underline{\phi}_1(x_2,t)$ and $\underline{\phi}_3(x_2,t)$ are then written

$$\begin{cases} \underline{\phi}_3(x_2,t) = -2c_t\omega_{10}\, e^{-\frac{|x_1|}{\delta}} e^{-\frac{|x_3|}{\delta}} e^{-\frac{|x_2-c_t t|}{\delta_2}} e^{i\frac{\omega}{c_t}(x_2-c_t t)} \\ \underline{\phi}_1(x_2,t) = \{\pm\} i2c_t\omega_{10}\, e^{-\frac{|x_1|}{\delta}} e^{-\frac{|x_3|}{\delta}} e^{-\frac{|x_2-c_t t|}{\delta_2}} e^{i\frac{\omega}{c_t}(x_2-c_t t)} \end{cases} \tag{30.9}$$

Thanks to the last relation *(30.7)*, we deduce also that the fields $\underline{\phi}_2(x_2,t)$ and $\tau^{(p)}(x_2,t)$ due to $\underline{\omega}_1(x_2,t)$ and $\underline{\omega}_3(x_2,t)$ cancel simultaneously, as a matter of fact simply because

$$\underline{\omega}_1^2 + \underline{\omega}_3^2 = 0 \quad \Rightarrow \quad \frac{\partial \underline{\omega}_1^2}{\partial x_2} + \frac{\partial \underline{\omega}_3^2}{\partial x_2} = 0 \tag{30.10}$$

This implies that, if there exists a field $\tau^{(p)}(x_2,t)$ associated with the rotational wave packet, it is independent of fields $\underline{\omega}_1(x_2,t)$ and $\underline{\omega}_3(x_2,t)$, and it must satisfy the following wave equation, identical to relation *(28.5)*

$$\frac{\partial^2 \tau^{(p)}}{\partial t^2} \cong -c_t^2 \frac{\partial^2 \tau^{(p)}}{\partial x_2^2} \tag{30.11}$$

The energy of the rotational wave packet

The wave packed defined by *(30.8)* and *(30.9)* represents an electromagnetic wave packet, which ***MUST*** posses a right or left helicity for it's energy to be independent of time, and more importantly so that it not be associated with a gravitational perturbation. In the real representation we obtain in this case

$$\begin{cases} \omega_1(x_2,t)=\omega_{10}\,e^{-\frac{|x_1|}{\delta}}\,e^{-\frac{|x_3|}{\delta}}\,e^{-\frac{|x_2-c_tt|}{\delta_2}}\cos\left[\frac{\omega}{c_t}(x_2-c_tt)\right] \\ \omega_3(x_2,t)=(\mp)\omega_{10}\,e^{-\frac{|x_1|}{\delta}}\,e^{-\frac{|x_3|}{\delta}}\,e^{-\frac{|x_2-c_tt|}{\delta_2}}\sin\left[\frac{\omega}{c_t}(x_2-c_tt)\right] \\ \phi_3(x_2,t)=-2c_t\omega_{10}\,e^{-\frac{|x_1|}{\delta}}\,e^{-\frac{|x_3|}{\delta}}\,e^{-\frac{|x_2-c_tt|}{\delta_2}}\cos\left[\frac{\omega}{c_t}(x_2-c_tt)\right] \\ \underline{\phi}_1(x_2,t)=\{\mp\}2c_t\omega_{10}\,e^{-\frac{|x_1|}{\delta}}\,e^{-\frac{|x_3|}{\delta}}\,e^{-\frac{|x_2-c_tt|}{\delta_2}}\sin\left[\frac{\omega}{c_t}(x_2-c_tt)\right] \end{cases} \tag{30.12}$$

Knowing that $mnc_t^2 = K_3$ in the perfect cosmic lattice for $\tau_0 < \tau_{0cr}$, the volume density of energy of this wave packet is given by

$$e^{fluctuation} = \underbrace{2K_3\left(\omega_1^2+\omega_3^2\right)}_{e_\omega^{dist}} + \underbrace{\frac{1}{2}mn\left(\phi_1^2+\phi_3^2\right)}_{e^{cin}} = 4K_3\omega_{10}^2\,e^{-2\frac{|x_1|}{\delta}}\,e^{-2\frac{|x_3|}{\delta}}\,e^{-2\frac{|x_2-c_tt|}{\delta_2}} \tag{30.13}$$

By effecting the variable change $x_2 - c_tt = y$, we can quite easily calculate the total energy of this fluctuation of the field of rotation, which is then worth

$$E^{fluctuation} = 8\int_0^\infty dx_1\int_0^\infty dx_3\int_0^\infty dy\left(4K_3\omega_{10}^2\,e^{-2\frac{x_1}{\delta}}\,e^{-2\frac{x_3}{\delta}}\,e^{-2\frac{y}{\delta_2}}\right) = 4K_3\omega_{10}^2\delta^2\delta_2 \tag{30.14}$$

30.2 – Quantification of the energy of 'electromagnetic' fluctuations and analogy with 'photons' quasi-particles

If we consider that the perfect cosmic lattice is a real representation of our Universe, then rotational wave packets that we just described must correspond to photons. By supposing that these wave packets are emitted when a topological singularity suddenly changes state (for example during the transition of an electron in an atom), it is then very simple to explain that they have a quantification. Let's assume that a singularity goes from a higher energetic state *(a)* to a lower energetic state *(b)*. According to relation *(28.22)* expressed in the non-relativistic case, we then have the following energetic transition

$$\left(E_0^{dist}+V\right)^{(a)} = E_0^{dist}+V^{(a)} = \hbar\underline{\omega}_f^{(a)} \quad\rightarrow\quad \left(E_0^{dist}+V\right)^{(b)} = E_0^{dist}+V^{(b)} = \hbar\underline{\omega}_f^{(b)} \tag{30.15}$$

with

$$V^{(a)} > V^{(b)} \quad\Rightarrow\quad \underline{\omega}_f^{(a)} > \underline{\omega}_f^{(b)} \tag{30.16}$$

During the transition, the singularity loses the following energy

$$\Delta E_{lost} = E_0^{dist}+V^{(a)} - \left(E_0^{dist}+V^{(b)}\right) = V^{(a)}-V^{(b)} = \hbar\underline{\omega}_f^{(a)} - \hbar\underline{\omega}_f^{(b)} = \hbar\left(\underline{\omega}_f^{(a)}-\underline{\omega}_f^{(b)}\right) \tag{30.17}$$

and this energy is dissipated in the form of a photon, and thus in the form of a transversal wave carrying this lost energy by the singularity, so that

$$\boxed{E^{fluctuation} = 4K_3\omega_{10}^2\delta^2\delta_2 = \hbar\left(\underline{\omega}_f^{(a)}-\underline{\omega}_f^{(b)}\right) = \hbar\underline{\omega}_{fluctuation}} \tag{30.18}$$

This relationship is then quite remarkable, as it shows that the *energy of the transversal fluctuation is quantified with the value* $\hbar\left(\underline{\omega}_f^{(a)}-\underline{\omega}_f^{(b)}\right)$, and that *the frequency* $\underline{\omega}_{fluctuation}$ *of the transversal wave is nothing else than the difference of frequencies of the gravitational perturbations of the singularity in states (a) and (b).* We find again as a consequence the experimental observation that the energy of the photons is quantified, as was first proposed by Einstein, and that the energy of a photon does indeed possess a fixed value proportional to its frequency $\underline{\omega}_{fluctuation}$ via the Planck constant!

On the non-locality of the rotational wave packet

The wave packet thus formed possesses a *"volume"* $\delta^2\delta_2$, an amplitude ω_{10} and an energy $E^{fluctuation}=\hbar\omega$. As its energy is bound to remain constant, it implies that neither the amplitude ω_{10}, neither the «volume» $\delta^2\delta_2$ are predetermined, but they are simply linked by the following relation

$$\omega_{10}^2=\frac{E^{fluctuation}}{4K_3}\frac{1}{\delta^2\delta_2}=\frac{\hbar\omega}{4K_3}\frac{1}{\delta^2\delta_2} \tag{30.19}$$

The wave packet presents a sort of 'malleability" or "plasticity": it can for example extend or contract in the axis of propagation Ox_2, or expand or contract alongs the axis Ox_1 and Ox_3 perpendicularly to the direction of propagation, or expand and contract in an isotropic fashion, as long as the product $\omega_{10}^2\delta^2\delta_2$ remains a constant equal to $E^{fluctuation}/4K_3=\hbar\omega/4K_3$.
If the wave packet is 'scrunched up' meaning if its volume $\delta^2\delta_2$ is very small and its amplitude ω_{10} is large, it will behave as if it was a *localized quasi-particule with energy* $E^{fluctuation}$. But during its propagation, it can also expand and occupy a 'volume' $\delta^2\delta_2$ which would be very large with a small ω_{10} amplitude and behave in this case more like a wave, which is then capable of interference and diffraction just like a wave does. We find here the property of 'non-locality' of the particle during its propagation, in the QM sense.

On the momentum of the 'photon' quasi-particle

In the guise of a quasi-particle, meaning when the wave packet is contracted and has a small volume, the wave packet does not contain any inertial mass, but it has a *non-null momentum.* We deduce this particularity from the fact that the wave packet moves with velocity c_t and that it must then satisfy the *relativistic energy equation (20.42)*, with an inertial mass $M_0=0$, namely

$$\left(E_{\mathbf{v}}^{photon}\right)^2=\left(\vec{\mathbf{P}}^{photon}\right)^2c_t^2 \tag{30.20}$$

Which implies a non-null momentum (quantity of movement) in the direction of propagation Ox_2

$$\vec{\mathbf{P}}^{photon}=\frac{E_{\mathbf{v}}^{photon}}{c_t}\vec{e}_2=\frac{\hbar\omega}{c_t}\vec{e}_2 \tag{30.21}$$

On the wave-particle duality of the rotation waves packet

The waves packet exhibits a wave-particle duality similar to that of QM. The only restriction imposed to this waves packet, due to the fact that it propagates in a medium satisfying $\tau_0<\tau_{0cr}$,

is that it must remain *a single entity*, with a given energy and helicity, so that its energy $\hbar\omega$ remains constant and there are no expansion perturbations!
This implies that such a waves packet, if it is very extended and if it must go through a slit for example must necessarily contract sufficiently to go through the slit as a single entity. But nothing prevents the wave nature of the entity to interact with the slit during its flight, and therefore for the trajectory of the quasi-particle to be modified!
In the same fashion, if this extended waves packet finds a double slit, it can go through the two slits by contracting locally and recombine after the fact, assuming that its structural integrity was not touched during said movement. But the sum of the wave entities after passage through both slits creates waves interference, so that the probability of finding a quasi-particle in the space after the slits has similar fringes to the interference fringes to that of a plane wave going through both slits!
This implies that if the waves packet, which extends during its propagation, starts to be absorbed by an obstacle, the condition that its energy remains constant during time forces it to contract so that the absorption of energy is a very local phenomenon. We could reasonably talk about 'dematerialization' of the wave packet in the form of a quasi-particle. It must then behave as a very local quasi-0particle during its creation and annihilation.
It should be noted that what we call the *"measurement problem"* in QM corresponds exactly to this type of phenomena. Any attempt at touching the waves packet is going to force it to modify so that its energy remains constant in time. Thus a measure on this waves packet is necessarily an action which will change this waves packet and modify its characteristics.

On the creation of pairs of "photon" quasi-particles

As the 'photon' quasi-particle possesses a momentum due to it's relativistic behavior, the creation of a unique photon would violate the conservation of quantity of movement. This implies that the photons can only be created as 'pairs of photons' with the same frequencies, that propagate in two opposite directions so that their global momentum is null.

On the phenomenon of entanglement of two virtual quasi-particle 'photons'

Initially, during the creation of a pair of photons, there could exist only a single packet created locally, in which case it must extend at velocity c_t on both side of the axis of propagation to ensure that the global momentum be null! We could say in this case that the unique waves packet of energy $2\hbar\omega$ represents the two quasi-particles with momentum $\vec{\boldsymbol{P}}^{photon} = \pm(\hbar\omega / c_t)\vec{e}_2$ corresponding then to an *entangled state of the two quasi-particles.* However if one of the extremities of this waves packet is suddenly 'materialized' in the form of a quasi-particle (photon 1), transferring an energy $E_{deposited} = \hbar\omega$ to an "object' interacting with it, the second end of the waves packet will regroup and will posses an energy $E^{fluctuation} = \hbar\omega$ and the quantity of movement $\vec{\boldsymbol{P}}^{photon} = \pm(\hbar\omega / c_t)\vec{e}_2$. It will transform then necessarily in a wave packet representing photon 2, which will be materialized in the form of a quasi-particle. It should be noted that the initial wave packet possessed, at the time of 'materialization" of the first quasi-particle, a polarization and a helicity which were measured and that this polarization and helicity that are measured become the property of the residual wave packet. It is exactly what QM predicts when it talks about *entanglement of two photons*! And there is therefore no 'instantaneous' transmission of information between a quasi-particle (photon 1) to the other quasi-particle (photon 2) since it is during 'materialization' of the first quasi-particle (photon 1) that the wave packet associated with the second quasi-particle (photon 2) forms and acquires the complementary characteristics to the first quasi-particle (photon 1), characteristics which will be observed during the 'materialization' of the second quasi-particle (photon 2)!

On the phenomenon of decoherence

As we just saw, a waves packet with energy $E^{fluctuation} = 2\hbar\omega$ represents the two photons initially created which can elongate along a single axis over great distances. But this elongation implies that the amplitude ω_{10} of the wave packet diminishes. As the wave packet 'extends' it will become more and more sensitive to its environment, meaning to the fluctuations of the field that it encounters, until it find a fluctuation strong enough to 'break' the initial waves packet and divide it in two independent waves packets, which will no longer be entangled. At this point, the two waves packets become independent. We can then talk about the *phenomena of decoherence,* in the sense that the 'materialization' of two waves packets in the form of two individual photons will not behave as we have just described in the previous section. This phenomenon is similar to the phenomena of 'decoherence' of QM to explain the passage from microscopic to the classic macroscopic world!

On the analogy with the photons, the quantum fluctuations of the vacuum, the multiverses and the gravitons

The results obtained in this chapter are very interesting, as they signify that the cosmic lattice, which does not present longitudinal waves for $\tau_0 < \tau_{0cr}$, can contain local perturbations of pure transversal waves of circular polarization which have all the characteristics of photons (quantification, duality wave-particle, entanglement, etc.). In chapter 31, we will show that there could exist a superposition of local longitudinal fluctuations of expansion whose energy is null, resembling what we call *the quantum fluctuations of the vacuum* in QM. We will mention also hypothetical correlated fluctuations of expansion, whose energy would essentially be kinetic, and which could represent, at the macroscopic scale, *multi-verses in expansion and contraction*, and, at the microscopic scale, quantified perturbations which could be identified to *'gravitons'* but who would be very different from the 'gravitons' postulated in General Relativity.

Chapter 31

Ingredients of an analogy with the standard model of particle physics

We have shown previously that the perfect lattice presents strong analogies with the great theories of modern physics, namely the equations of electro-magnetism, general relativity, special relativity, black holes, cosmology, dark energy and quantum mechanics, and that we can have 3 types of basic topological loop singularities which possess respectively the analogue of an electrical charge, an electric dipole moment or a curvature charge by flexion. It should be noted that the curvature charge is unique to our theory, and explain rather simply several mysterious phenomena, such as the weak coupling force of two topological loops, the dark matter, the galactic black holes, and the disappearance of anti-matter.

In this chapter, we will strive to find and describe the ingredients which could explain, on the basis of topological singularities, the existence of the standard model of particle physics. In other words, we will strive to find mechanisms to generate the fundamental particles such as leptons and quarks, and what could cause three generations of these fundamental particles, and from whence could come the strong force which binds quarks to form baryons and mesons.

This chapter does not pretend to give an elaborate theory or a final, quantitative solution to explain the standard model of particle physics, but rather to show with a few specific arguments that it could be the choice of a microscopic structure of the lattice that could answer the questions of the standard model. Here we will show that there appears a full 'zoology' of loops in a well choosen structure of the lattice, and that it resembles the elementary particles of the standard model. We will show the presence of an asymptotic strong force which could bind the topological loops!

31.1 – On the problems of the standard model of fundamental particles

Currently, particle physics explains the fine structure of matter with the help of a model called *"the Standard Model of Particle Physics"* (see annex C). In this model there appears *fermions,* particles of matter which present different families, the family of *leptons* and the family of *quarks,* as well as 3 types of interactions which can appear between fermions: the electromagnetic interaction, the weak interaction and the strong interaction.

The interactions between fermions happen via the exchange of particles called *gauge bosons,* corresponding to the quantas of the quantum fields of interaction. The electromagnetic interaction involves *photon* γ , the weak interaction the three gauge bosons Z^0 , W^+ and W^- , and the strong interaction involves 8 gauge bosons called *"gluons"*.

The mass of particles is introduced in the standard model via a new interaction associated with the quantum Higgs fields, and the mediating particle is called the *'Higgs boson'.*

On the problems related to the Standard Model that already have a solution in our theory

In spite of its undeniable success, the standard model has many open questions. In this chapter we will try to see if an approach to the standard model via our theory of the cosmic lattice can bring an answer to these various questions. We are not trying here to give a final answer to these problems, but to show, very qualitatively, how the cosmic lattice would go about answering these questions. Some of the open problems of the standard model already have some answers as shown in the previous chapters.
Let's then take a look at those problems that have a partial answer in our theory and how the lattice gives some answers to these various problems:

- *On the absence of gravitational interaction in the standard model:* gravitational interaction is intrinsically woven to the theory of the cosmic lattice, as the static solution to the Newton equations of motion, and it is this equation, under its dynamical form that allowed us to introduce a simple explanation to quantum mechanics and the notion of spin of loop singularities.

- *On the necessity of the Higgs boson and the impossibility to calculate the masses of the different fermions and bosons in the standard model:* in the basic standard mode, fermions have no mass, and theoretical physicists had to introduce an ad-hoc mechanism, the interaction with the Higgs field via the Higgs boson, which confers inertial mass to elementary particles. However, in the standard model, it is not possible to obtain quantitative values for the inertial mass of the particles, which are experimentally derived values. The cosmic lattice theory contains in fact a mechanism analogous to the Higgs field: it is the field of inertial masses of 'objects' of the lattice (which are then analogous to Higgs boson with spin 0) as well as the elastic energy of distortion of the lattice which are together responsible for the inertial relativistic properties of topological singularities, without having to set values to experiments.

- *On the physical nature of electromagnetic interaction in the standard model:* electromagnetic interaction, as well as its boson carrier, the photon, with it's diverse quantum properties, are part of the cosmic lattice theory, and have a very simple physical explanation on the basis of the field of rotation within the cosmic lattice.

- *On the physical nature of the weak interaction of the standard model:* a weak interaction presenting an analogy with the weak interaction of the standard model was obtained in the cosmic lattice theory (chapter 26), under the form of a force with very short range, linking «topological fermions» together (twist disclination loop TL to edge dislocation loop EL), via a coupling of their respective charges of rotation and curvature.

- *On the violation of **CP** invariance (charge/parity) in the standard model:* in the real universe, we observe a violation of the **CP** invariance (charge/parity) which theoreticians estimate to be the probable cause for the asymmetry of matter and anti-matter. In the cosmic lattice, this weak asymmetry between matter and anti-matter exists, and can be explained by the existence of a curvature charge due to edge loops. This charge has no equivalent in the standard model. This phenomenon also explains the dark matter of astrophysics and the disappearance of anti-matter during the cosmic evolution of the universe.

- *On the absence of explanation of dark energy and dark matter in the standard model:* these two concepts invented by theoretical physicists to explain the acceleration of expansion and the gravitational behavior of galaxies have explanations in the cosmic lattice: the energy of elastic *distortion for dark energy and the gravitational repulsion of neutrinos as regards dark matter.*

The problems of standard model which are not yet explained by the cosmic lattice theory

Amongst the problems of the standard model of particle physics, there are some which do not yet find a plausible explanation in the cosmic lattice. We have:

- *the existence of fermions under the form of three generations of leptons and quarks:* if fermions correspond to topological singularities in the theory of the cosmic lattice, the existence of fermions under the form of leptons and quarks, as well as the existence of three generations of fermions, should find an explanation with a particular choice of the cosmic lattice structure and the building of elementary particles as topological singularities in the form of dispiration loops.

- *the existence of three massive gauge bosons in the weak interaction:* as the weak interaction already appeared in our theory as a force linking the twist loops to the edge loops, we must find what are the massive gauge bosons, carriers of this interaction, in the cosmic lattice theory.

- *the existence of a strong interaction linking quarks by a confinement mechanism:* the strong interaction, with its confinement mechanism and its carrier bosons, the gluons, are the only interaction which has not appeared in our theory. But we already have encountered mechanisms which could be potential candidates to explain this force and its asymptotic behavior, as the mechanisms that generate a stacking fault energy within the lattice, such as the dissociation of a dislocation for example.

- *the existence of quantified electrical charges with relative values,1, 1/3 and 2/3:* the electrical charge of fermions possess relative values 1, 1/3 and 2/3 between the charge of electrons and the charge of quarks. These quantified values do not have any explanation in the standard model, but one can conjecture that a judicious choice of a particular structure for the cosmic lattice could explain this problem.

In the rest of this chapter, we will try to find answers to these interrogations about the standard model, by exclusively focusing on the structure of the cosmic lattice and on the properties of the topological singularities contained therein.

In the standard model, 26 different parameters are required (in the case where neutrinos have a mass) to obtain a functional theory, such that the mass of particles and the intensity of the various forces, and these parameters become 'set' on the values observed and obtained experimentally. We can conjecture that the cosmic lattice model will reduce the number of parameters one can adjust, simply because it can explain phenomena that the standard model cannot directly explain!

31.2 – A 'colored' cubic lattice to explain the first family of quarks and leptons of the standard model

In the perfect cosmic lattice, we have seen that the topological singularity that explains the electrical charge is the twist disclination loop (TL). As we have seen in chapter 24, for the gravitational interaction of TL to map to observed behaviors (time dilatation, curvature of wave rays), it suffices that coefficients α_{TL} and β_{TL}, in the expressions $R_{TL} = R_{TL0}\,e^{\alpha_{TL}\tau}$ and $\Omega_{TL} = \Omega_{TL0}\,e^{\beta_{TL}\tau}$ giving the dependency of the radius and angle of torsion of the TL as a function of the background expansion, have to satisfy the relation $3\alpha_{TL} + 2\beta_{TL} = 1/8$. This implies that torsion Ω_{TL} could

(i) be a constant independent of expansion, in which case, $\alpha_{TL} = 1/24$ and $\beta_{TL} = cste$, which allows for a topological explanation of the discrete, and independent of expansion, values for angle Ω_{TL},

(ii) either depend on volume expansion, in which case Ω_{TL} cannot take a discrete value which would be directly linked to the structure of the lattice, since this angle would disappear in a continuous fashion during volume expansion.

Thus if quantized values appear, such as the charge of the electron, but also fractional charges 1/3 and 2/3 as it is the case for the quarks in the standard model, we would need that the rotation angle of the two internal planes intrinsic to the twist loop corresponds to symmetries of the lattice itself, for example $\pi/2$, π, $3\pi/2$, ... in the case of a cubic lattice, or $\pi/3$, $2\pi/3$, π, ... in the case of a hexagonal lattice.

We will choose as a consequence the following hypothesis concerning the cosmic lattice

Conjecture 12: The angle Ω_{TL} takes discrete values linked to the symmetry of the lattice and independent in volume expansion $(\beta_{TL} = cste)$ *(31.1)*

A 'colored' cubic lattice with peculiar rules of stacking and rotation

Let us imagine *à priori* a cubic lattice which is rather peculiar (fig. 31.1), with a lattice step of a, which would be constituted of *«colored planes of particles»* indexed with 3 fundamental colors ***R, G, B*** (these artificial colors are just a convenient representation and do not, at least as of yet, have any relationship with the 'colors' used by the standard model to explain the color charge of quarks and gluons). Even if we do not know the physical reason for the existence of "colored planes", we will suppose that this alternation of colors of the planes of particles is a necessary condition of existence of the cosmic lattice in the absence of topological singularities, and that if the alternation ***R, G, B*** of the planes of the lattice is broken by the presence of a topological singularity, there could appear fault energies between the planes of particles. Let's also postulate *à priori* stacking and rotation rules of the colored plane in this very peculiar lattice as such:

Conjecture 13: The stacking of planes ***R, G, B*** follows three elementary rules: *(31.2)*

Rule 1: the *alternation of planes* ***R, G, B*** cannot be broken (either by impossibility or by a very large energy associated with a surface stacking fault energy γ_1),

Rule 2: in a given direction of space, there may appear a stacking fault corresponding to a shift in the alternation of planes ***R, G, B***, which possesses a surface stacking fault energy γ_0 which is not null (fig. 31.1, image 4).

Rule 3: if a plane with a given color undergoes a rotation by an angle $\pm\pi/2$, $\pm\pi$ or $\pm 3\pi/2$, it changes color according to table 31.1, which corresponds to the existence of a given axial property of the lattice.

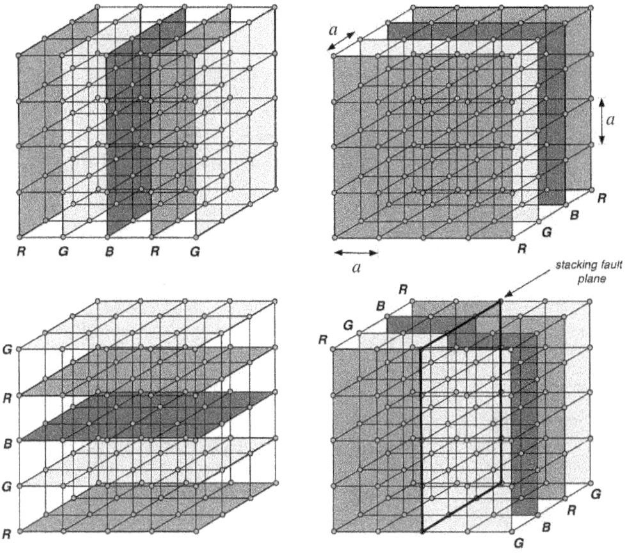

***Figure 31.1** - a cosmic lattice, cubic and isotropic, with planes **RGB** presenting a regular stacking in the 3 directions of space, as well as a stacking fault of the colored planes.*

rotation angle Ω_{TL}	color change	colors R, G, B and complementary colors $\bar{R}$, $\bar{G}$, $\bar{B}$
$\begin{cases} +3\pi/2 \\ 0 \\ -3\pi/2 \end{cases}$	$\begin{cases} \mathbf{R} \to \mathbf{R} \\ \mathbf{G} \to \mathbf{G} \\ \mathbf{B} \to \mathbf{B} \end{cases}$	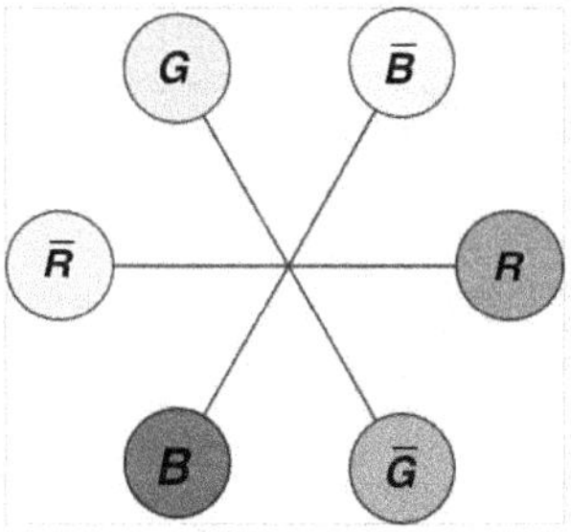
$\begin{cases} +\pi/2 \\ -\pi \end{cases}$	$\begin{cases} \mathbf{R} \to \mathbf{G} \\ \mathbf{G} \to \mathbf{B} \\ \mathbf{B} \to \mathbf{R} \end{cases}$	
$\begin{cases} -\pi/2 \\ +\pi \end{cases}$	$\begin{cases} \mathbf{R} \to \mathbf{B} \\ \mathbf{G} \to \mathbf{R} \\ \mathbf{B} \to \mathbf{G} \end{cases}$	

***Tableau 31.1** - the color changes of a plane by rotation, and the complementary colors to red, green and blue, namely cyan, magenta and yellow.*

On the necessity to combine a twist disclination loop and an edge dislocation loop in such a lattice and the existence of quarks

We can introduce a twist loop (TL) in our lattice (represented symbolically in figure 31.2,h with an angle of rotation of the inferior plane by $\pm\pi/2$, or $\pm\pi$, or $\pm 3\pi/2$. But according to rule 3, a rotation of $\pm\pi/2$ or $\pm\pi$ induced on the inferior planes of the loop will change their color, with, according to rule 2, the genesis of a cylinder of stacking faults with a surface fault energy γ_0, as schematically shown in figures 31.2, a through d.

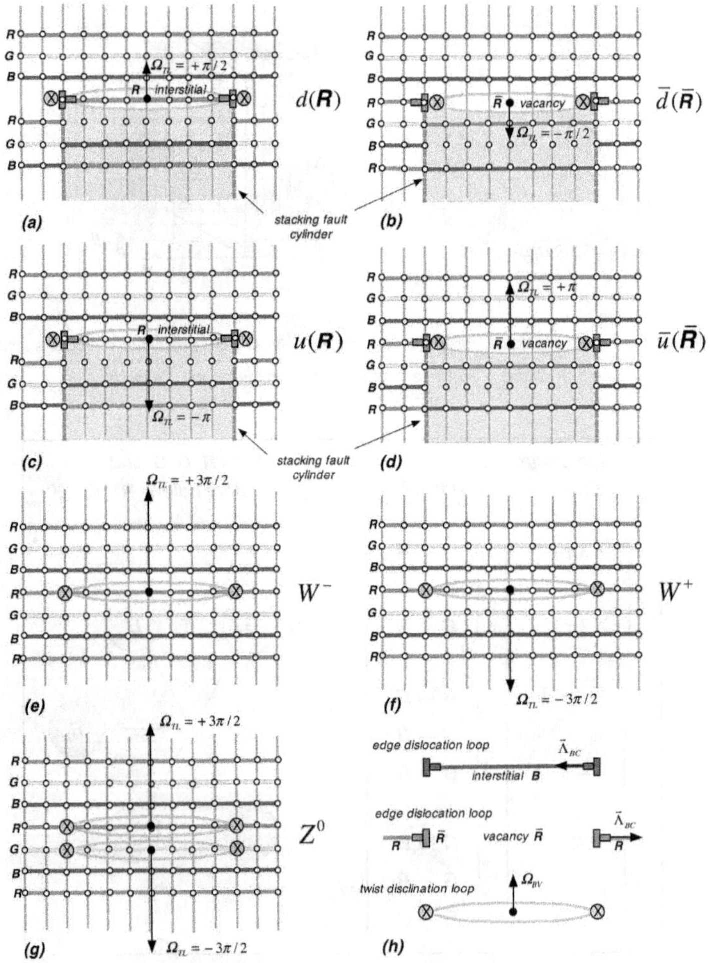

***Figure 31.2** - the combinations of twist disclination loops with angles $\pm\pi/2$, $\pm\pi$ or $\pm 3\pi/2$ with edge dislocation loops that assure the continuity of alternation of planes **RGB**.*

The change of color of the lower planes in the case of rotations $\pm\pi/2$ or $\pm\pi$ implies that if rule 1 is violated at the level of the twist loop, and in order to satisfy the stacking of colored planes, we must associate an edge loop to the twist loop (represented symbolically in figure 31.2, h), of interstitial type if the angle of rotation is $+\pi/2$ or $-\pi$ (fig. 31.2, a and c), or of vacancy type if the angle of rotation is worth $-\pi/2$ or $+\pi$ (fig. 31.2, b and d).
The interstitial plane in the case of an interstitial edge loop will have one of 3 colors ***R*, *G*, *B*,** while the missing plane in the case of a vacancy edge loop will have the anti-color of the plane which was interrupted, namely colors $\bar{\boldsymbol{R}}$, $\bar{\boldsymbol{G}}$ or $\bar{\boldsymbol{B}}$ (we use in the figures the complementary colors to ***R*, *G*, *B*,** which are *cyan, magenta and yellow,* see table 31.1). In the four cases (fig. 31.2, a to d), the twist loop is evidently linked to the edge loop by the weak force described in section 25.3, but also by the necessity to introduce the edge loop to assure the alternance of colors of planes at the level of the twist loop!
The dispirations thus formed, possess a 'color', which corresponds to the color of the interstitial plane or the anti-color of the vacancy plane (the anti-color or complementary color to the color of the plane in which the vacancy loop appears).
With respect to the disclination loops with angle $\pm 3\pi/2$ (fig. 31.2, e to g), they do not need to be combined with edge loops since these rotations do not entail any color change in the inferior planes, and thus as a consequence no cylinder of stacking faults under it.

name	$\boldsymbol{\Omega}_{TL}$	$q_{\lambda TL}$	***edge loop***	$q_{\theta EL}$	***color***
d	$+\pi/2$	$-\pi^2 R_{TL}^2/2$	*interstitial*	$-2\pi a$	***R*, *G*** or ***B***
u	$-\pi$	$+\pi^2 R_{TL}^2$	*interstitial*	$-2\pi a$	***R*, *G*** or ***B***
$\bar{d}$	$-\pi/2$	$+\pi^2 R_{TL}^2/2$	*vacancy*	$+2\pi a$	$\bar{\boldsymbol{R}}$, $\bar{\boldsymbol{G}}$ or $\bar{\boldsymbol{B}}$
$\bar{u}$	$+\pi$	$-\pi^2 R_{TL}^2$	*vacancy*	$+2\pi a$	$\bar{\boldsymbol{R}}$, $\bar{\boldsymbol{G}}$ or $\bar{\boldsymbol{B}}$
W^-	$+3\pi/2$	$-3\pi^2 R_{TL}^2$	-	0	-
W^+	$-3\pi/2$	$+3\pi^2 R_{TL}^2$	-	0	-
Z^0	$(+3\pi/2)+(-3\pi/2)$	0	-	0	-

Table 31.2 *- The seven singularities composed of a twist loop and combined (or not) with an edge loop*

On the existence of intermediary gauge bosons

In table 31.2, we have shown the different properties of topological singularities thus formed, by

giving them as in figure 31.2, a name chosen "randomly", and by using the fact that the two dispirations on the right (a and c) in figure 31.2 are clearly anti-loops of the left loops (b and d).
In table 31.2, we notice that the charges of rotation $q_{\lambda TL}$, analogous to the electrical charge, present 3 different values, corresponding respectively to 1/3x, 2/3x and 1x the charge of loops W^- or W^+. Also, only the dispirations d, u, $\overline{d}$ and $\overline{u}$ present a charge $q_{\theta EL}$ of curvature via a non-null flexion and the sign of these charges, positive in the case of a vacancy edge loop and negative in the case of an interstitial edge loop, imply as we have already postulated in conjecture 8, that the particles d and u correspond by analogy to matter and that their anti-particles $\overline{d}$ and $\overline{u}$ correspond to anti-matter. With respect to particles W^-, W^+ and Z^0 which do not posses a curvature charge $q_{\theta EL}$, they must have an important mass since they are twist disclination loops with a high angle of rotation Ω_{TL}.

On the weak interaction of quarks via intermediary bosons

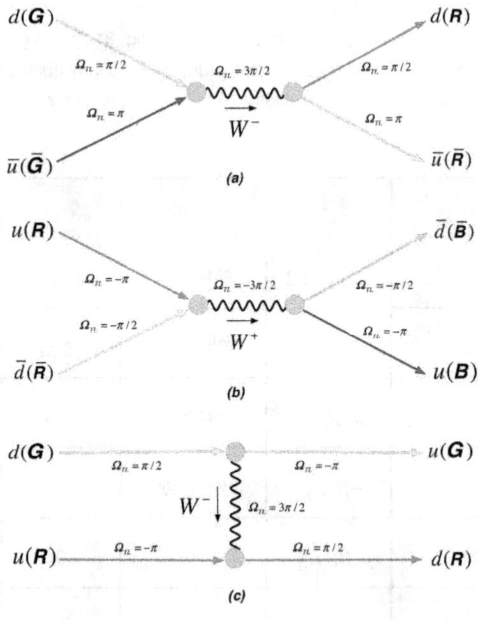

Figure 31.3 *- Feynman diagrams of the combination and exchange between quarks of table 31.2*

It is interesting to note here that the combination of two disparitions d and $\overline{u}$, or $\overline{d}$ and u, contribute to create a pure twist loop W^- or W^+, which can again transform in a pair d and $\overline{u}$, or $\overline{d}$ and u. We can also imagine an exchange of a loop W^- or W^+ between two dispirations d and $\overline{u}$, or $\overline{d}$ and u, which will change their nature, or to speak more poetically their "flavor".
These combinations and exchanges are illustrated in figure 31.3 in the form of *Feynman dia-*

grams. They are characterized by the fact that the total rotation Ω_{TL} is conserved, which assures us at the same time the conservation of rotation charge $q_{\lambda TL}$. We also note that the total charge $q_{\theta EL}$ is conserved in the reactions. It is then undeniable that these reactions have a strange similarity with the weak interactions of the standard model as shown in table C.3!

On the existence of «baryons» and localized «mesons», formed from 3 and 2 dispirations

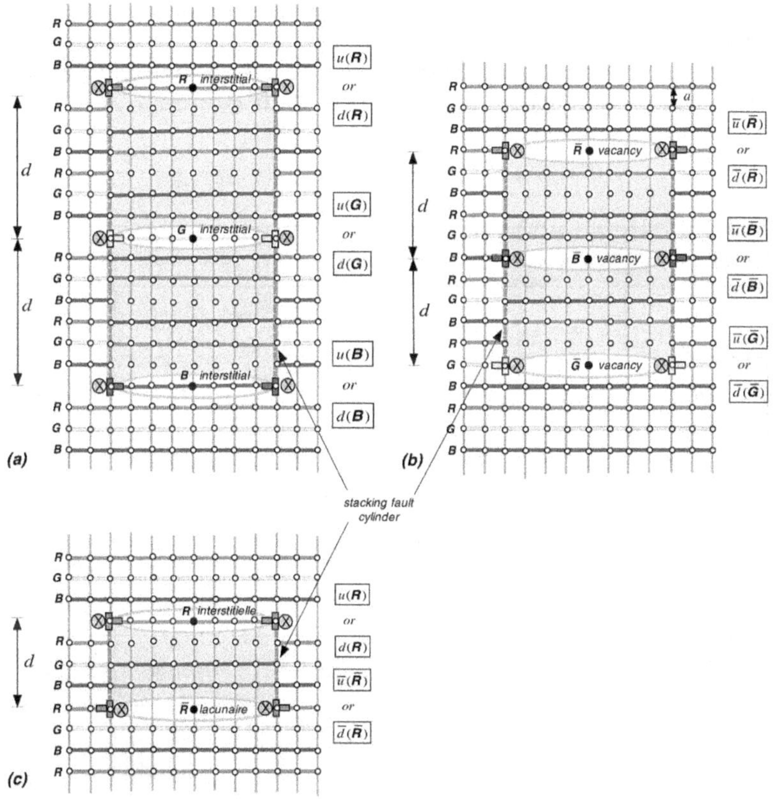

Figure 31.4 *- the 3 possibilities of combination of 2 or 3 dispirations allowing one to form topological singularities which are local*

Each of the dispirations (a to d) of table 31.2 gives us a cylinder of stacking faults which possesses an energy proportional to the lateral surface of the cylinder (fig. 31.2). As a consequence, it is impossible for these disparitions to disappear in an isolated fashion, as the cylinder would then have a very large length $\sim R_{\infty}$, and thus a great energy! We can then wonder how we can generate singularities composed of such dispirations, and that they be of a reasonable energy!

In fact, there exists 3 ways of combining the 4 dispirations of table 31.2 so that the topological singularity thus formed be localized, meaning a tube of stacking faults with finite length:
- *the combination of 3 singularities* u *or* d (fig.31.4,a),
- *the combination of 3 anti-singularities* $\overline{u}$ *or* $\overline{d}$ (fig.31.4,b),
- *the combination of a singularity* u *or* d *with an anti-singularity* $\overline{u}$ *or* $\overline{d}$ (fig.31.4,c).

For the 3 rules we formulated earlier to be satisfied in these various combinations, it is necessary that:
- *the sum of the angles* Ω_{TL} *of rotation of all the dispirations be null or a multiple of* $3\pi/2$, which allows the tube to be of finite length,
- *that the color of the assembly thus formed be "white",* so that the assembly has the 3 colors ***R***, ***G***, ***B*** (fig.31.3,a), or be the sum of the 3 anti-colors $\overline{\boldsymbol{R}}$, $\overline{\boldsymbol{G}}$, $\overline{\boldsymbol{B}}$ (fig.31.3,b), or the sum of a color ***R***, ***V***, ***B*** with its respective anti-color $\overline{\boldsymbol{R}}$, $\overline{\boldsymbol{G}}$, $\overline{\boldsymbol{B}}$ (fig. 31.3,c).

combination	***symbol***	Ω_{TL}	$q_{\lambda TL}$	***edge loop***	$q_{\theta EL}$
ddd	Δ^-	$+3\pi/2$	$-3\pi^2 R_{TL}^2/2$	*interstitial*	$-6\pi a$
dud	n,Δ^0	0	0	*interstitial*	$-6\pi a$
udu	p,Δ^+	$-3\pi/2$	$+3\pi^2 R_{TL}^2/2$	*interstitial*	$-6\pi a$
uuu	Δ^{++}	-3π	$+3\pi^2 R_{TL}^2$	*interstitial*	$-6\pi a$
$\overline{d}\,\overline{d}\,\overline{d}$	$\overline{\Delta}^+$	$-3\pi/2$	$+3\pi^2 R_{TL}^2/2$	*vacancy*	$6\pi a$
$\overline{d}\,\overline{u}\,\overline{d}$	$\overline{n},\overline{\Delta}^0$	0	0	*vacancy*	$6\pi a$
$\overline{u}\,\overline{d}\,\overline{u}$	$\overline{p},\overline{\Delta}^-$	$+3\pi/2$	$-3\pi^2 R_{TL}^2/2$	*vacancy*	$6\pi a$
$\overline{u}\,\overline{u}\,\overline{u}$	$\overline{\Delta}^{--}$	$+3\pi$	$-3\pi^2 R_{TL}^2$	*vacancy*	$6\pi a$

Table 31.3 - *White baryons formed of 3 dispirations*

In table 31.3, we have reported the 8 different combinations that are possible with the 3 dispirations of table 31.2, with their properties, by giving them a symbol and calling them *baryons*, by analogy with standard model.

In this table, the analogy with baryons of the standard model, composed with quark triplets of u and d or triplets of anti-quarks $\overline{u}$ and $\overline{d}$, is obvious and perfect! Not only do we see particles composed of *quarks* with fractional charges of rotations $q_{\lambda TL}$ corresponding to the electrical charges of the standard model, but it is here added to the charge of curvature by flexion $q_{\theta EL}$ which does not have an equivalent in the standard model, and which corresponds perfectly with our conjecture 8 (22.91), namely that the *singularities of vacancy nature correspond by analogy to the anti-matter and the singularities of interstitial nature to matter!* The fact that the particles of the standard model appear with two different symbols in this table for combination dud, udu, $\bar{d}\bar{u}\bar{d}$ and $\bar{u}\bar{d}\bar{u}$ is explained by the notion of spin of the twist loops developed in chapter 29. Indeed, if each quark possesses a spin $\pm 1/2$, then the composition of the spins can create a global spin $\pm 1/2$ in the case of particles n *(neutron) and* p *(proton)* and the anti-particles $\bar{n}$ *(anti-neutron)* and $\bar{p}$ *(anti-proton)*, or a spin $\pm 3/2$ in the case of particles Δ^0 and Δ^+ and of anti-particules $\bar{\Delta}^0$ and $\bar{\Delta}^-$. In the case of combinations ddd, uuu, $\bar{d}\bar{d}\bar{d}$ and $\bar{u}\bar{u}\bar{u}$, the spins of the 3 quarks are necessarily aligned (for a reason that remains to be explained, but which is strongly linked to the principle of exclusion!) and the composition of the spins can then only yield a global spin of $\pm 3/2$ in the case of particles Δ^- and Δ^{++} and of anti-particules $\bar{\Delta}^+$ and $\bar{\Delta}^{--}$.

In table 31.4, we have reported the different possible combinations of 2 dispirations of table 31.2, with their properties, by giving them a symbol and by calling them 'mesons' by analogy with the standard model.

combination	symbol	Ω_{TL}	$q_{\lambda TL}$	edge loop	$q_{\theta EL}$
$d\bar{d}$	π^0,ρ^0	0	0	-	0
$d\bar{u}$	π^-,ρ^-	$+3\pi/2$	$-3\pi^2 R_{TL}^2/2$	-	0
$\bar{d}u$	π^+,ρ^+	$-3\pi/2$	$+3\pi^2 R_{TL}^2/2$	-	0
$u\bar{u}$	η^0,ω^0	0	0	-	0

Table 31.4 *- The white mesons formed by 2 dispirations*

In this table, the analogy with the mesons of the standard model, composed of doublets of *quarks* u or d with anti-quarks $\bar{u}$ or $\bar{d}$, is obvious and perfect! We see particles composed of quarks (with fractional charges of rotation $q_{\lambda TL}$) which correspond to the mesons of the standard model, but with a null charge of curvature by flexion $q_{\theta EL}$, which implies that the topological singularities cannot be tagged as anti-matter (singularities of vacancy type) or of matter (singularity of interstitial nature)!

The fact that the particles of the standard model appear with two different symbols in the table is explained also by the notion of spin of the twist loop developed in chapter 29. Indeed, if each quark possesses a spin $\pm 1/2$, then the composition of spins can create global null spin in the case of particles π^0, π^-, π^+ *and* η^0, or a spin ± 1 in the case of particles ρ^0, ρ^-, ρ^+ and ω^0.

On the strong force and its asymptotic behavior

The quarks composing the particles of table 31.3 and 31.4 are linked by the cylinder of stacking faults, so that the energy of the topological singularity grows as $E_\gamma \sim \gamma_0 2\pi R_{TL} d$ if the distance d separating two dispirations grows. The force linking the dispirations is thus of "asymptotic nature": it is a strong force in the *sense that the linking force grows if we try to separate dispirations*. It is a phenomenon similar to the case of the energy of stacking faults between two partial dislocations in an *FCC* lattice (see fig. 9.9), or in the case of the energy of fault of mapping between 3 dislocations in an axial cubic lattice (see fig. 9.33). The equilibrium distance d between the dispirations is thus controlled by a competitive mechanism similar to those described in figures 9.9 et 9.33.

On the strong interaction between quarks via gauge bosons, the gluons

In the standard model, the quantum treatment of 'colors' of the quarks is done in Quantum Chromo Dynamics (QCD). In that theory, there exist 8 bosons with color gauges, vectors of the strong force, called *gluons*.

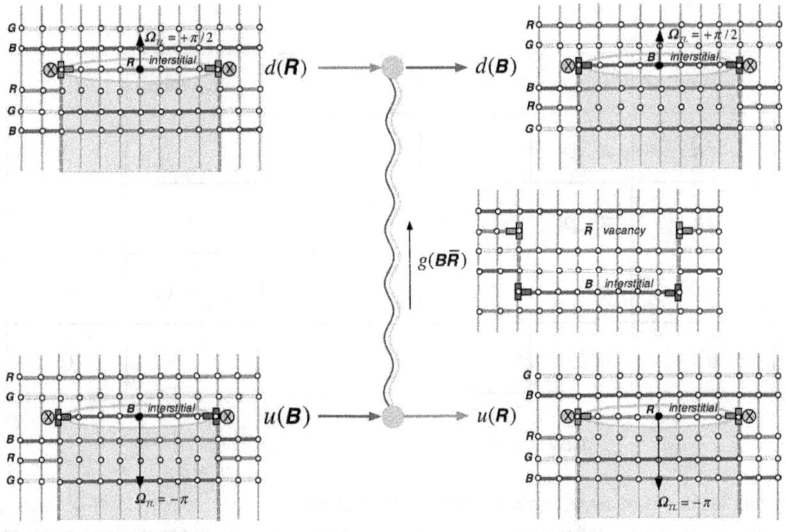

Figure 31.5 *- Feynman diagram of the exchange of colors of two quarks by the exchange of a bicolor gluon*

It is the exchange of a colored gluon between two quarks that makes these quarks change color, by an interaction that can be represented by a Feynman diagram (fig. 31.5) illustrated by the configuration of the topological singularities concerned.

The colored gluons correspond then to two edge dislocation loops, one of interstitial nature and one of vacancy nature and their charges of rotation $q_{\lambda EL}$ is null. The edge loops are linked to each other by the existence of a cylinder of stacking faults and are therefore submitted to the strong force. With respect to their curvature charge $q_{\theta EL}$, it is null since we have $q_{\theta EL} = (+2\pi a) + (-2\pi a)$, so that the *energy associated with the distortions of this pair of loops must be really weak*, and that, as a consequence, *the mass of gluons must be almost null, while it possesses a non null energy coming from the cylinder of stacking faults!* From this standpoint, the gluons are like photons.

In QCD, we think that it is this mechanism of exchange of gluons between neutrons and protons of the atomic core which explains the coherence of atoms. We are looking at a secondary effect of the strong force since the exchange of colored gluons perturb distances d between the dispirations composing neutrons and protons, with the consequence that they perturb the energies of protons and neutrons.

On the constitution of leptons and intermediary bosons of the standard model

In the standard model, there also exists a first family of quasi-punctual particles, which we call leptons and which are represented by the electron e^- , the anti-electron or positron e^+ , the electronic neutrino ν_e and the electronic anti-neutrino $\overline{\nu}_e$.

In the cosmic lattice, we already had postulated the existence of a neutrino in the form of an edge loop of interstitial nature, while the anti-neutrino would be an edge loop of vacancy type. It is actually what allowed us to deduce the gravitational properties for the neutrino, which are quite remarkable. These are due to a curvature charge of flexion which dominates the attractive gravitational effects due to its inertial mass. In the case of a 'colored cosmic lattice" as in figure 31.1, to respect the 3 rules that the lattice must abide by, the neutrino can only correspond to the insertion of 3 consecutive planes of color ***R***, ***G***, ***B***, and the anti-neutrino to the subtraction of 3 consecutive planes $\overline{\boldsymbol{R}}$, $\overline{\boldsymbol{G}}$, $\overline{\boldsymbol{B}}$ (fig. 31.6, a and b), so as to form an interstitial or vacancy which has no color (white). Under this form, the neutrinos and anti-neutrinos have exactly the properties that we have deduced in the previous chapters for the edge loops, under the condition that their Burgers vectors have a norm such that $|\vec{B}_{EL}| = 3a$, so that the curvature charge due to flexion has a norm equal to $|q_{\theta EL}| = 6\pi a$.

With respect to the electron and anti-electron, we already have emitted the hypothesis, that the twist disclination loop is a good candidate to represent them. In that case, to insure that the charges of rotation do correspond, it is necessary that the angle of rotation $\boldsymbol{\Omega}_{TL}$ between two consecutive planes be equal to $\pm 3\pi / 2$, so that the norm of its rotation charge satisfies the relation $|q_{\lambda TL}| = 3\pi^2 R_{TL}^2 / 2$. However, under this pure form, the twist loop was already identified as the particle W^- or W^+ (fig. 31.2, e and f). Furthermore, the electron and positron must present an asymmetry between matter and anti-matter, and they must satisfy the weak leptonic interactions (fig. C.3,a). To satisfy these requirements, we must again use a combination of a twist loop with angle $\pm 3\pi / 2$, which satisfies rule 3 and which does not possess color, with an edge loop which corresponds to the insertion of 3 consecutive planes of color ***R***, ***G***, ***B***, or to the subtraction of 3 consecutive planes $\overline{\boldsymbol{R}}$, $\overline{\boldsymbol{G}}$, $\overline{\boldsymbol{B}}$. In principle, there should be 4 different electrons, with charges $q_{\lambda TL} = \pm 3\pi^2 R_{TL}^2 / 2$ and $q_{\theta EL} = \pm 6\pi a$. However, the simplest way (and less energetic) to create an electron and a positron would be to compress the assembly of quarks ddd and $\overline{d}\,\overline{d}\,\overline{d}$, so as to collapse the 3 twist loops in one and collapse the 3 edge loops in one. We obtain then the electrons and positrons represented in figure 31.6, c et d.

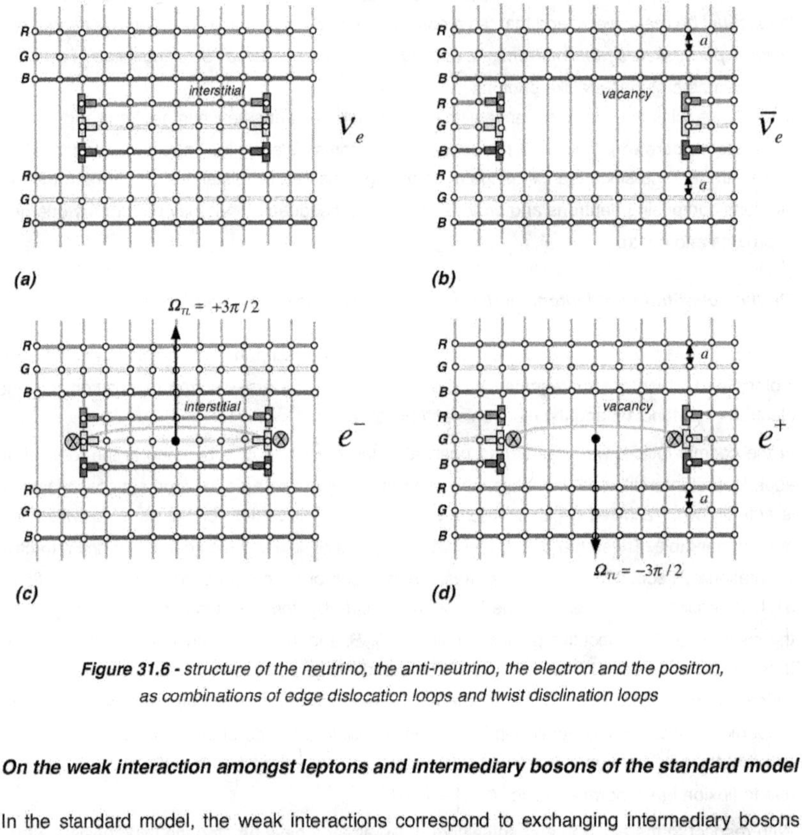

***Figure 31.6** - structure of the neutrino, the anti-neutrino, the electron and the positron, as combinations of edge dislocation loops and twist disclination loops*

On the weak interaction amongst leptons and intermediary bosons of the standard model

In the standard model, the weak interactions correspond to exchanging intermediary bosons $W^{\pm}$ or Z^0, which permits the exchange of electrical charge between the two particles (see fig. C.3). In order for the Feynman diagrams of figure C.3 to work with the dispirations of our model, it is necessary that the intermediary bosons be pure twist loops with angle of rotation Ω_{TL} respectively worth $\pm 3\pi/2$ ou 0, as represented in figure 31.2, e to g.
The intermediary bosons are then the only massive gauge bosons, which is readily understood if they are effectively pure twist loops. Experimentally we have found that their mass are much higher than that of the electron and positron, which could be understood by the fact that the rotation $\pm 3\pi/2$ must be done entirely on a distance a in the case of intermediary bosons, while in the case of electron and positron the rotation of $\pm 3\pi/2$ can be spread out over 3 successive planes, and thus over a distance $3a$, which must diminish considerably the local distortions of the lattice, and thus the energy of the particle. This could also be the reason why a gauge boson associates very fast with 3 PDLs, interstitial or vacancy, to diminish its energy, which could explain perfectly the weak interactions of figure C.3!
One should also note that, in the standard model of particles, the gauge bosons W^+ and Z^0

are of spin 1, and thus that they do not satisfy the principle of exclusion, which means that two gauge bosons can occupy the same state, and thus can be superposed, which in fact create a twist loop with angle Ω_{TL} which is equal to $\pm 3\pi$. On the other hand, the electron and positron are spin 1/2 particles, which satisfy the Pauli exclusion principle. They therefore cannot occupy the same state, which means that we cannot superimpose them, which becomes naively true if we consider the loop structure of electrons and of positrons as shown in figure 31.5.
We can then report the properties of leptons and gauge bosons in table 31.5.

symbol	Ω_{TL}	$q_{\lambda TL}$	**edge loop**	$q_{\theta EL}$
ν_e	$+3\pi/2$	0	*interstitial*	$-6\pi a$
e^-	0	$-3\pi^2 R_{TL}^2/2$	*interstitial*	$-6\pi a$
$\overline{\nu}_e$	$-3\pi/2$	0	*vacancy*	$6\pi a$
e^+	-3π	$+3\pi^2 R_{TL}^2/2$	*vacancy*	$6\pi a$
W^-	$+3\pi/2$	$-3\pi^2 R_{TL}^2/2$	-	0
W^+	$-3\pi/2$	$+3\pi^2 R_{TL}^2/2$	-	0
Z^0	$(+3\pi/2)+(-3\pi/2)$	0	-	0

Table31.5 *- The leptons of the first family and the intermediary gauge bosons*

31.3 – A tentative explanation of the 3 families of quarks and leptons of the standard model

In the standard model (Appendix C), there exist not only quarks and leptons as we just described them but there exists two additional families of quarks and leptons (fig. C.1), which are separated mostly by the increasing masses when we go from one family to the next. We show in figure 31.7 the progression of masses for the elementary particles of the standard model, by giving a multiplicative factor for the masses, in the vertical and horizontal directions. We notice that the multiplication factors shown are really high when going from one family to the other, while the multiplication factors to go from one particle to the next in each family are not that high, except in the case of the passage from neutrino to electron, which would indicate that the topological structure responsible for the large increase in mass changes from one family to the other but remains the same within the same family. It is also remarkable that the multiplying factors associated with neutrino V_e are a lot more high than all the other factors, which would indicate that it is the structures of the edge dislocation loops which change from one family to the other.

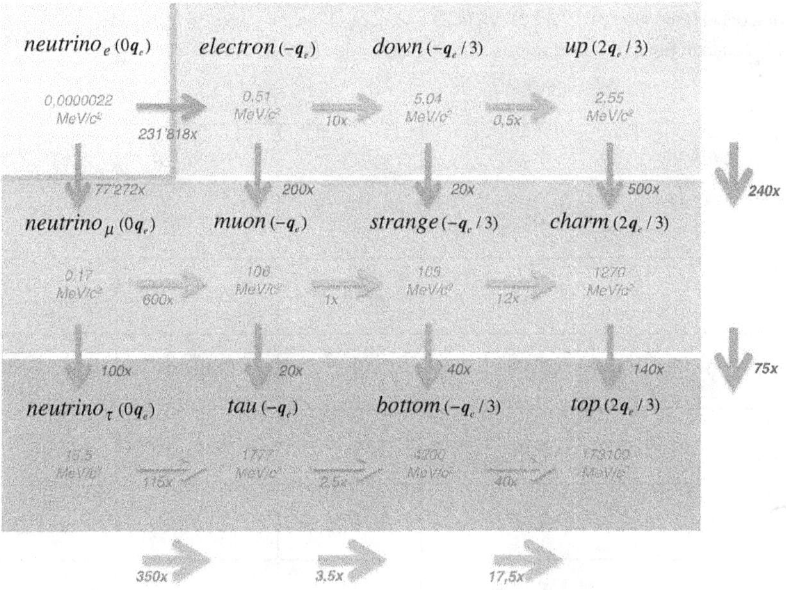

***Figure 31.7** - Masses of the particles of the standard model, as experimentally measured and expressed in MeV/c², with the multiplying factors*

With respect to the very large multiplicative factor to go from neutrino V_e to electron e^-, it can be explained by the large difference between an edge loop and a twist loop, as we explained in section 19.11 by relation *(19.118)*.

On the possibility to use edge disclination loops

If it is the structure of the edge parts of the topological singularity which must present changes from across families of leptons and quarks and justify such large energy variations from one family to the other, there exists within the theory of lattice singularities ideal candidates to fulfill these requirements: the edge disclination loops which were discussed in section 9.3, and notably the loops which could be realized with combinations of edge disclinations $C1$, $\overline{C1}$, $C2$ and $\overline{C2}$ (fig. 31.8).

Indeed, it is possible, by coupling two edge disclination loops $C1$ and $\overline{C1}$, or $C2$ and $\overline{C2}$, to form a topological structure which is complex and would correspond to an edge dislocation loop. The configuration of the two loops in relation to each other allows us to create pseudo loops of edge dislocations, either of vacancy or interstitial type, as shown in the two examples of figure

31.9 in the case of loops formed on the basis of disclinations $C1$ and $\bar{C}1$, represented here by additional planes.

$C1 : \Omega = +90° : \Theta = -\pi / 2$	$\bar{C}1 : \Omega = -90° : \Theta = +\pi / 2$
$C2 : \Omega = +180° : \Theta = -\pi$	$\bar{C}2 : \Omega = -180° : \Theta = +\pi$

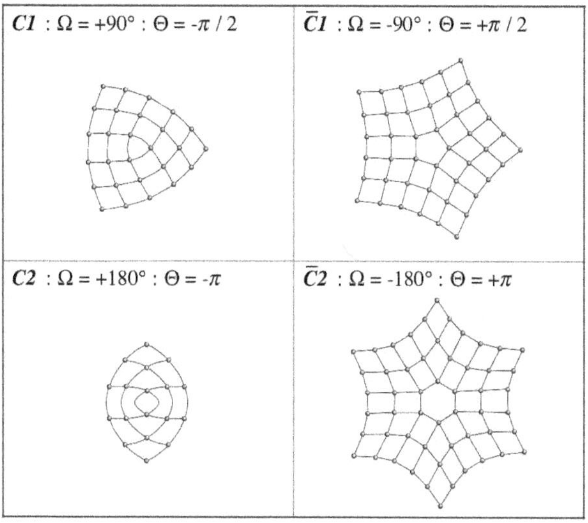

Tableau 31.8 - *Family of quantized edge disclinations in a cubic lattice*

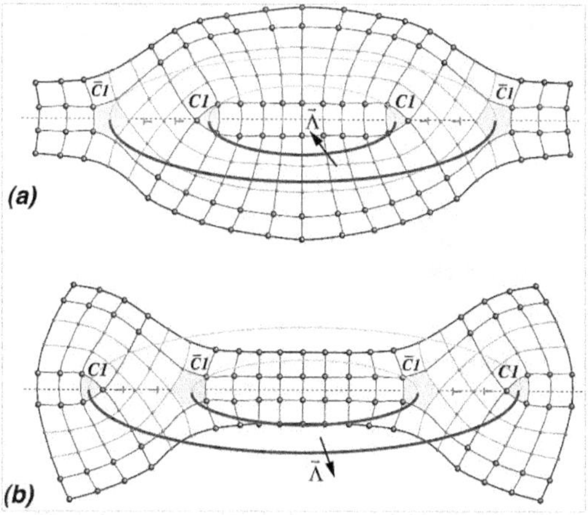

Figure 31.9 - *Formation of a loop and an anti-loop of type* $[C1,\bar{C}1]$

Even if the energy of the field of the edge dislocation is very small in the perfect lattice, the energy of a singularity made with disclination loops must be a lot higher, because resulting from very large distortions of rotation and shear in the lattice in the immediate vicinity of the loops!

On the construction of 3 families of quarks

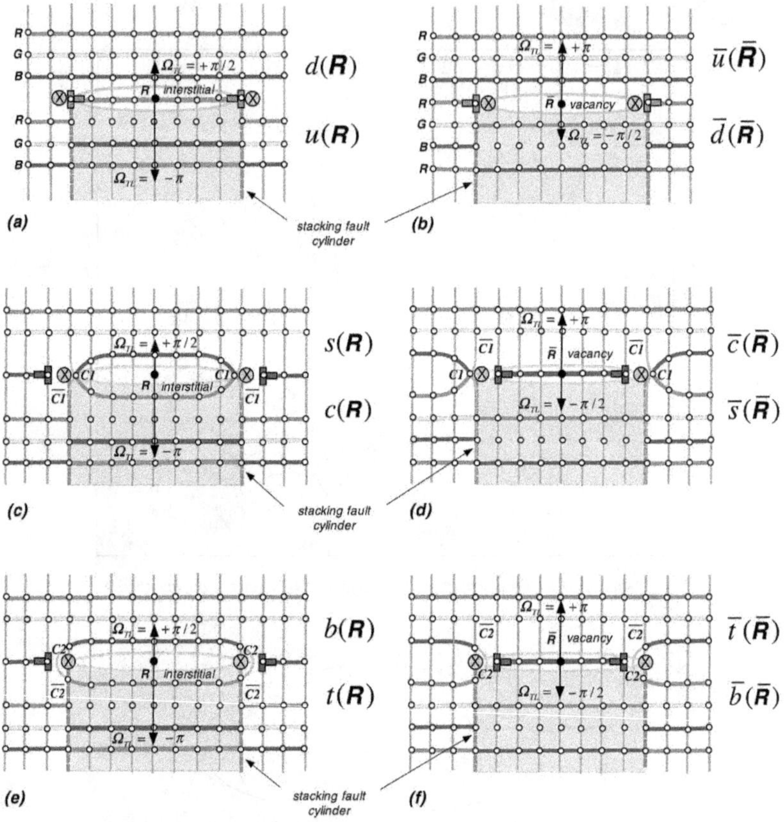

***Figure 31.10** - Structure of quarks as combinations of twist disclination loops, edge dislocation loops and pairs of edge disclination loops.*

In figure 31.10, we have shown the possible topological structures that could explain the quark families on the basis of the introduction of a pair of edge disclinations $C1$ and $\overline{C1}$ for the second family of quarks and a pair of edge disclinations $C2$ and $\overline{C2}$ for the third family of quarks. It should be noted that the structure of quarks is shown very schematically in this figure, since we have not reported the great distortions of the lattice implied by the presence of pairs of edge disclination loops. And it is precisely the much larger topological distortions of loops $C2$ and $\overline{C2}$ when compared to the loops $C1$ and $\overline{C1}$ which could explain the energy differences observed between particles of the second and third family. With respect to the first family of quarks, it does not contain any edge disclinations, which would explain the energy differences of particles in the first and second family of quarks.

With this explanation for the quark family, the gauge bosons $W^{\pm}$ and Z^0 described in figure 31.2 are not modified since they do not involve edge dislocations! Furthermore, all the mechanisms described for the first family remain valid with the 3 quark families describe in figure 31.10, be it the weak hadronic interactions (fig. C.3) which involve gauge bosons $W^{\pm}$ as that reported by figure 31.3 or the *strong interactions* involving the bicolor gluons of figure 31.5 !

On the possible realization of 3 families of leptons

The pair of edge disclanation loops $C1$ and $\overline{C1}$, or $C2$ and $\overline{C2}$, can also be used to explain the 3 families of leptons which are observed in the standard model.
We have reported in figure 31.11 the topological structure of the 3 families of leptons based on the introduction of the pair of loops $C1$ and $\overline{C1}$, or $C2$ and $\overline{C2}$.

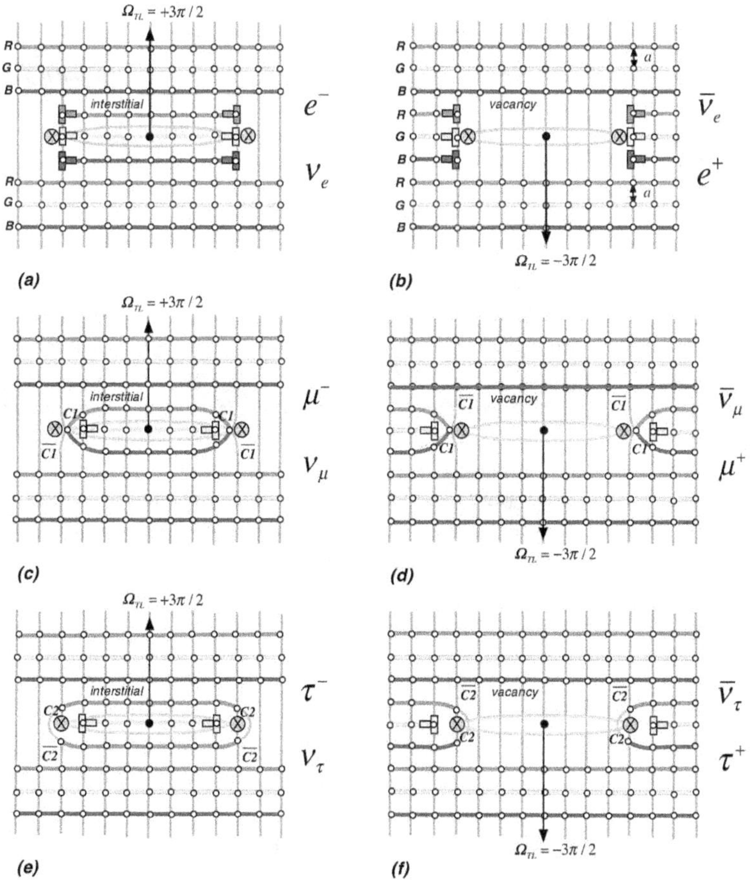

Figure 31.11 *- Structure of leptons, as assemblies of edge dislocation loops and pairs of edge disclination loops as well as twist disclination loops*

As in the case of the quarks of figure 31.10, the structures presented in figure 31.11, as well as all to topological structures reported in figures 31.2 to 31.9, strictly respect the rules of colors introduced in the beginning of the chapter, notably rule 1.
It is also easy to verify that all leptons of figure 31.11 will satisfy the weak leptonic and semi-leptonic interactions reported in figure C.3. Furthermore as there exist no cylinders of stacking faults in the structure of leptons, these leptons are not submitted to the strong force!

31.4 – On the interest of the analogy between colored cosmic lattice and the standard model of particles

The analogy of our model, the colored cosmic lattice with its topological singularities, with the standard model of particles is excellent, and it is rich in explanations to many mysterious aspects of the standard model that we will enumerate here.

On the structure of particles of the standard model in 3 families of quarks and leptons

The topological structures of the edge dislocation loops, of the twist disclination loops and of the pairs of edge disclination loops that we introduced in the colored cubic lattice and which possess strict stacking rules and rotation rules of the colored planes, allow us to build all the particles of the standard model of particle physics, namely the quarks and leptons, which present a structure in 3 families for which the masses are very different. These different masses can be explained by the elastic energy of distortion of the lattice in the immediate vicinity of the loops. These diverse quarks and leptons also satisfy all the properties of the weak interaction and the strong interaction using respectively the intermediary gauge bosons $W^{\pm}$ and Z^{0} and the gluons, which possess also their own topological structure in the colored cosmic lattice. With respect to the strong force, it possesses all the good asymptotic properties due to the fact that it is generated by a cylinder of stacking faults whose energy increases if it is elongated, and is responsible for the existence of baryons and mesons, which are the only local topological structures which is 'non-colored' that one can form on the basis of quarks! In this fashion, we can reconstruct all the particles of the standard model, as for example the baryons and mesons of figures C.4 et C.5, composed of quarks and anti-quarks u, d, s and or c.

On the fields of interaction by weak force and strong force

With respect to the field forces acting on topological loops, they have simple topological explanations:

- the *weak force* is essentially due to the diminution of energy of formation of a dispiration loop when we associate an edge loop with a twist loop as was seen in section 25.3. The short range of the interaction potential of this force (attractive) explains the radio-active disintegration of elementary particles, by overcoming the interaction potential via a QM tunnel effect.
There exists gauge bosons that are exchanged during the weak interaction: these are the intermediary bosons $W^{\pm}$ and Z^{0}, which have a well defined topological structure, reported in figure 31.2.

- the *strong force*, which links 2 or 3 quarks together, is due to a cylinder of stacking faults, generated by the fact that the twist disclination loops associated with the quarks have a charge $q_{\lambda TL}$ which are nothing else than the 1/3 or 2/3 of the charge of the twist loop associated with the electron. The dissociation distance between a doublet or a triplet of quarks depends essentially on the energy of stacking faults per unit of surface. If this energy is really strong, we can imagine that the loops are very close to each other as we have illustrated in the figures of this chapter. But if this energy if very weak, we could imagine fault tubes made of membranes whose diameter (equal to the diameter of the topological loops) is a lot smaller than the length, so that the topological singularities of doublet and triplets could look like *«long strings»* terminated at each extremity by topological loops!
There exists gauge bosons exchanged during the strong interaction: they are the bicolor gluons, which have a well defined topological structure, reported in figure 31.5.

On the possibility of calculating the energy of particles of the standard model

A first interesting consequence of this explanation of the particles of the standard model is linked to the fact that, in the case of dispiration loops and their interactions via the weak force and the strong force, the energies involved for the *formation* of multiplets of loops has a known origin, since it is the sum of the following energies:

(i) The formation energies associated with the strong local distortions of the lattice generated by the singularities and stored in the vicinity of the objects,

(ii) the connection faults which appear here because the lattice we imagined is a 'colored' one, linked to the fact that it should possess axial properties,

(iii) the energies involved in the *weak force* in the gravitational couplings between edge loops and twist loops as described in section 25.3,

(iv) the energies stored over longer distances, which are due to the long range distortions of the lattice linked to a global charge q_θ of curvature by flexion and the global charge q_λ of torsion by rotation of multiple loops, which are contained in the calculations of formation energies, of gravitation and of curvature charges as we have done in the previous chapters, as well as the self vibrational energies and self rotational energies as we have seen in chapters 28 and 29 respectively.

The total energy of formation of loop multiples can then be calculated in a rigorous fashion, on the condition of knowing the exact elastic properties (the modules K_0, K_1, K_2 and K_3) as well as the surface fault energy γ_0 of the cosmic lattice in which the objects appear. This energetic aspect is very important as, in the case of the standard model of particles, the origin and value of the energy of elementary particles (their mass) is still rather mysterious, and are introduced as parameters in the standard model, which are measured experimentally!
Furthermore, by observing the distortions of the lattice that are generated in the vicinity of the loops of different families, we can imagine that the progression of energy of multiple loops as a function of their 'family order number' must be largely non-linear, a phenomenon which is observed in the case of the energies of leptons and quarks of the different families of the standard model.

On the elementary nature of the particles of the standard model

Another interesting consequence of our conception of the standard model is that there is a different 'elementarity' to the topological loops of dispiration of our lattice as opposed to the leptons and quarks of the standard model. These are not elementary particles but the assembly of more elementary constructs, namely the edge dislocation loops, of interstitial or vacancy types, the twist disclination loops and the 4 loops formed out of the edge disclinations $C1$, $\overline{C1}$, $C2$ and $\overline{C2}$ of figure 31.8 linked to each other in the form of pairs of loops by a strong force purely of topological nature, the ribbon of virtual dislocations linking the two edge disclination loops.
To judge the other potentials coming out of this idea of how particles of the standard model come about, it should be verified if this approach would allow us to justify and explain the complicated set of selection rules which had to be introduced in particle physics to describe all the interactions which are observed experimentally. This is only a suggestion and its detailed development is not included in this treatise. It should be noted that similar approaches to the decomposition of the particles of the standard model have already been proposed in particle physics, but under different forms, such as the model based on pre-quarks called *«rishons»*. However these models have proven to not be fruitful.

On the role of the curvature charge in the standard model

The curvature charge q_θ plays an important role in the building of a model of "elementary topological singularities" to explain the standard model of particles. We notice relatively easily that this charge, for which there is no analogy in the standard model, satisfies a conservation principle during the loop interactions, be them weak or strong.
The question is then to know if the charge q_θ, which is conserved during the interactions between loops, has a correlation with one of the characteristic values or with one of the conservations relations of the standard model (such as the Gell-Mann-Nishijima relation for example). The answer to this question could assuredly give us strong hints for particle physics, as we have repeatedly shown that *it is the curvature charge q_θ which is responsible for the asymmetry between matter and anti-matter,* and consequently responsible for the evolution of matter and anti-matter, as well as the presence of dark matter, in the form of a repulsive sea of neutrinos around galaxies!

31.5 – On the open questions about the 'colored cosmic lattice' and its analogy with the standard model

The analogy developed in this chapter between topological dispiration loops in an imaginary cubic lattice and the standard model of particle physics is fruitful to understand many obscure points of particle physics, such as the topological nature of elementary particles, as well as the weak and strong forces, or the origin of mass in elementary particles.
However there exist many open questions, which merit to be studied in detail, amongst which the main ones are:

On the application of the concept of spin

As we have shown in chapter 29, the notion of spin seems to correspond to a real rotation of topological loops. But there are many questions which would warrant further study.

The first question is naturally to try and imagine how an edge dislocation loop and a twist disclination loop or an edge disclination loop can turn on themselves in a cubic cosmic lattice, knowing that additionally we need a tube of stacking faults to explain the strong force in the case of baryons and mesons. Is there a topological explanation for such a rotational movement, or should we imagine a lattice with even weirder properties?

The second question is linked to the value that one should assign to the spin of the topological loop. For example, why does the electron, which corresponds to a perfect coupling between a twist disclination loop and an edge dislocation loop, have a spin 1/2, while the gauge boson $W^{\pm}$, which would correspond to an isolated twist disclination loop, have a spin 1? Does there exist a topological reason for a given loop to possess a 1/2 or a spin 1, or should we again, invoke 'strange properties' of the colored cosmic lattice?

The third question would be to know how to apply more carefully the concept of spin developed in chapter 29 to the colored cosmic model that we have described in this chapter. The response to this question could allow us to find an explanation to the existence of particles composed of the same quarks, but with different spins, such as mesons π^{+} and ρ^{+} each composed of quarks $u\overline{d}$, but with respective spins 0 and 1, or the baryons p (proton) and Δ^{+} which are both composed of quarks uud, but with respective spins worth 1/2 and 3/2. Such a study could also allow to explain the exact origin of spin 1/2 of baryons and of spin 1 of mesons, which is still a rather obscure point of the standard model of particles, but which remains just as obscure in our theory of a 'colored' lattice!

On the QCD theory

It should be instructive and interesting to study more in depth the wave equations, the concepts of *bosons and fermions*, the *Pauli exclusion principle* developed in chapter 28, as well as the spin notion introduced in chapter 29, as well as the topological loop singularities which reconstruct the standard model, and see if such a study would not lead us eventually to a comprehensible physical explanation of QCD?

On the existence of supersymmetric models

A more in-depth study could not only explain why there exists *fermions* in the standard model (particles with spin 1/2 as quarks and leptons) and *bosons* (particles with spin 1 like the intermediary gauge bosons and the gluons), but it could also answer the question of knowing if it would be possible to create a zoology of particles identical to those that we have obtained in this chapter, but for which we would have inverted spins 1/2 and 1, which would give rise to a 'supersymmetric model'!

On the existence of a fourth family of quarks and leptons

When we introduced the edge disclinations $\boldsymbol{C1}$, $\overline{\boldsymbol{C1}}$, $\boldsymbol{C2}$ and $\overline{\boldsymbol{C2}}$ to explain the families of the standard model, we have consciously neglected the existence of loops of edge disclination $C3$ and $\overline{C3}$ reported in figure 9.20. If these disclinations could really exist in a colored cosmic lattice, there would be a *fourth family of quarks and leptons*, whose energy would then be humongous!

On the existence of «exotic leptons»

In our description of quarks of figure 31.10, the fact that quarks possess electric charges -1/3 and +2/3 of the charge of the electron, while the anti-quarks possess an electrical charge +1/3 and -2/3 of the electron charge, is easily explained by the hypothetic *rules of succession of the colors in the particulate planes.* However, for the leptons which we have introduce in figure 31.11, we have now chosen *arbitrarily* to associate to the neutrino an electrical charge -1 to obtain the *electron of matter* and to associate to the anti-neutrino an electrical charge +1 to obtain the anti-matter positron. However the rules about color that we have introduced would not prevent us from associating a charge +1 to the neutrino to obtain *an exotic positron of matter* and an electrical charge -1 to the anti-neutrino to obtain an exotic electron of antimatter. There is here another subject of reflexion. Indeed:

- either the *exotic leptons do not exist*, in which case one should surely find a more convincing explanation than the one we have proposed to explain that the electron be made of matter and the positron of anti-matter (we have proposed that the electron and the positron could be considered as the results of compression of assemblies of quarks ddd and $\overline{d}\,\overline{d}\,\overline{d}$, which would explain the existence of electrons of matter and positrons of anti-matter and the absence of electrons of anti-matter and the positrons of matter),

- either the *exotic leptons do exist*, in which case one should explain why these particles have never been observed experimentally!

Conclusion

In conclusion, we must admit that our colored cosmic model seems to have a lot of unresolved questions, which could lead to very interesting research!

PART II

E

Some other hypothetical consequences of the perfect cosmic lattice

Pure gravitational fluctuations of the cosmic lattice: quantum fluctuations of the vacuum, multiverses and gravitons

Chapter 32

Gravitational fluctuations: quantum vacuum fluctuations, multiverses and gravitons

In chapters 22 to 26, we have essentially focused on the static fields of expansion associated with topological singularities. We have also studied their interactions. In chapters 28 to 30, we have introduced the dynamical fluctuations of the field of expansion associated with topological singularities as well as transversal waves, which correspond to quantum mechanics. In this chapter we will focus on the dynamics of the field of expansion in the domain $\tau_0 < \tau_{0cr}$, by dealing with the problem of temporal fluctuations of the field of expansion which are not associated to topological singularities or transversal waves.

We will start by describing random fluctuations with null average energy, which presents strong analogies with the famous quantum vacuum fluctuations. We will then show that such fluctuations, which we will call "gravitational fluctuations" can be stable under the condition that they appear as a *quadruplet of fluctuations,* so that it presents a non null total energy which should not depend on time.

By considering these fluctuations stable, macroscopic and isotropic, it is possible to give another version of the cosmological expansion of the Universe, by introducing the notion of *multiverses* of expansion and contraction in a perfect infinite cosmic lattice.

We can also imagine stable longitudinal "gravitational fluctuations", microscopic and quantized, which would then correspond to hypothetical particles we will call *«gravitons»*.

32.1 – Local longitudinal "gravitational" fluctuations

At chapter 14, we have seen that, in a lattice in which the propagation of longitudinal waves is not possible, there can be local longitudinal vibrations, which we will call *gravitational fluctuations* $\tau^{(p)}(\vec{r},t)$ as they are fluctuations of the field of volume expansion. We will go deeper in this topic by describing the longitudinal fluctuations of a cosmic lattice containing no topological singularities nor transversal waves.

On gravitational fluctuations in the absence of singularities and transversal waves

In the absence of topological singularities and transversal waves, let's imagine the existence of fluctuations $\tau^{(p)}(\vec{r},t)$ of the field of expansion of a cosmic lattice in the domain $\tau_0 < \tau_{0cr}$

$$\tau(\vec{r},t) = \tau_0 + \tau^{(p)}(\vec{r},t) \qquad (32.1)$$

These fluctuations $\tau^{(p)}(\vec{r},t)$, if they exist, must satisfy the equation of Newton of the volume expansion. In the absence of topological singularities and transversal waves, and by neglecting the effects of vacancies and interstitials ($\vec{p} \cong m\vec{\phi}$), the equation of motion of Newton for the longitudinal perturbations is given by relation *(18.9)* in which we neglect all the fields besides $\tau^{(p)}$ and $\vec{\phi}^{(p)}$, and for which we directly perform the divergence

$$\operatorname{div}\left(nm\frac{d\vec{\phi}^{(p)}}{dt}\right)=\Delta\left[\left(\frac{4K_2}{3}+2K_1(1+\tau_0)-K_0\right)\tau^{(p)}+K_1\left(\tau^{(p)}\right)^2\right] \tag{32.2}$$

By considering the fluctuations $\tau^{(p)}(\vec{r},t)$ small, it is possible to linearize the equation, by neglecting the term $(\tau^{(p)})^2$ and taking the density n out of the divergence

$$nm\operatorname{div}\left(\frac{d\vec{\phi}^{(p)}}{dt}\right)=\left(\frac{4}{3}K_2+2K_1\left(1+\tau_0\right)-K_0\right)\Delta\tau^{(p)} \tag{32.3}$$

We can introduce a parameter α worth

$$\alpha=K_0-4K_2/3-2K_1\left(1+\tau_0\right) \tag{32.4}$$

which is positive if the cosmic lattice does not present longitudinal waves, meaning $\tau_0<\tau_{0cr}$, and we replace the total derivative by the partial derivative of time so that

$$\operatorname{div}\left(\frac{d\vec{\phi}^{(p)}}{dt}\right)\cong\operatorname{div}\left(\frac{\partial\vec{\phi}^{(p)}}{\partial t}\right)\cong\frac{\partial}{\partial t}\left(\operatorname{div}\vec{\phi}^{(p)}\right)\cong-\frac{\alpha}{nm}\Delta\tau^{(p)} \tag{32.5}$$

By using the geometro-kinetic equation for the volume expansion and no sources of lattice and by neglecting the total derivative, namely

$$\frac{S_n}{n}=0=-\frac{d\tau^{(p)}(\vec{r},t)}{dt}+\operatorname{div}\vec{\phi}^{(p)}\quad\Rightarrow\quad\frac{\partial\tau^{(p)}(\vec{r},t)}{\partial t}\cong\operatorname{div}\vec{\phi}^{(p)} \tag{32.6}$$

by combining relations *(32.5)* and *(32.6)*, we obtain the linearized Newton equation for weak gravitational fluctuations in the domain $\tau_0<\tau_{0cr}$

$$\frac{\partial^2\tau^{(p)}}{\partial t^2}\cong-\frac{\alpha}{nm}\Delta\tau^{(p)}\quad\text{with}\quad\alpha>0 \tag{32.7}$$

If we separate the spatial behavior from the temporal behavior of these $\tau^{(p)}(\vec{r},t)$, we can write them as the product of a spatial function $\psi(\vec{r})$ with an oscillatory term in $\mathrm{e}^{-i\omega t}$

$$\tau^{(p)}(\vec{r},t)\cong\psi(\vec{r})\mathrm{e}^{-i\omega t} \tag{32.8}$$

By introducing this writing of fluctuations in the Newton equation, we obtain the equation which describes the spatial component $\psi(\vec{r})$ when the fluctuations have a frequency ω

$$\psi(\vec{r})\cong\frac{|\alpha|}{mn\omega^2}\Delta\psi(\vec{r}) \tag{32.9}$$

For example, let's imagine a local fluctuation around the origin, along 3 axis in space. For such a fluctuation to satisfy the Newton equation, the spatial component $\psi(\vec{r})$ must be written under the following form

$$\psi(\vec{r})=\psi_0\,\mathrm{e}^{-\frac{|x_1|}{\delta_1}}\,\mathrm{e}^{-\frac{|x_2|}{\delta_2}}\,\mathrm{e}^{-\frac{|x_3|}{\delta_3}} \tag{32.10}$$

which, introduced in the Newton equation, allows us to link the frequency ω of the fluctuation to its spatial range δ_i along the three directions of space

$$\left[\frac{1}{\delta_1^2}+\frac{1}{\delta_2^2}+\frac{1}{\delta_3^2}\right]=\frac{mn\omega^2}{\alpha} \quad \Rightarrow \quad \omega=\sqrt{\frac{\alpha}{mn}}\sqrt{\frac{1}{\delta_1^2}+\frac{1}{\delta_2^2}+\frac{1}{\delta_3^2}} \tag{32.11}$$

We note then that the frequency of a gravitational fluctuation is inversely proportional to its spatial coordinates.

On the energy of an isolated gravitational fluctuation

Let's consider a unique gravitational fluctuation given by

$$\tau^{(p)}(\vec{r},t)\cong\psi_0 e^{-\frac{|x_1|}{\delta_1}} e^{-\frac{|x_2|}{\delta_2}} e^{-\frac{|x_3|}{\delta_3}} \cos\omega t \tag{32.12}$$

Let's calculate the elastic energy stored by this perturbation of the lattice. The density of elastic energy is given by the following expression if the background expansion τ_0 of the cosmic lattice is not null

$$e^{dist}(\vec{r},t)\cong K_1\left(\tau^{(p)}(\vec{r},t)\right)^2-\left(K_0-2K_1\tau_0\right)\tau^{(p)}(\vec{r},t) \tag{32.13}$$

The total elastic energy of the fluctuation is obtained by integrating over all the space

$$\begin{aligned} E^{dist}(t)&=\iiint_V\left[K_1\left(\tau^{(p)}(\vec{r},t)\right)^2-\left(K_0-2K_1\tau_0\right)\tau^{(p)}(\vec{r},t)\right]dV \\ &=K_1\psi_0^2\cos^2\omega t\iiint_V e^{-2\frac{|x_1|}{\delta_1}} e^{-2\frac{|x_2|}{\delta_2}} e^{-2\frac{|x_3|}{\delta_3}} dV-\left(K_0-2K_1\tau_0\right)\psi_0\cos\omega t\iiint_V e^{-\frac{|x_1|}{\delta_1}} e^{-\frac{|x_2|}{\delta_2}} e^{-\frac{|x_3|}{\delta_3}} dV \end{aligned} \tag{32.14}$$

hence

$$E^{dist}(t)=K_1\psi_0^2\delta_1\delta_2\delta_3\cos^2\omega t-8\left(K_0-2K_1\tau_0\right)\psi_0\delta_1\delta_2\delta_3\cos\omega t \tag{32.15}$$

let's also try to calculate the kinetic energy stored by this perturbation of the lattice. The velocity satisfies approximatively the following equation

$$\operatorname{div}\vec{\phi}^{(p)}=\frac{\partial\tau^{(p)}(\vec{r},t)}{\partial t}=\psi_0 e^{-\frac{|x_1|}{\delta_1}} e^{-\frac{|x_2|}{\delta_2}} e^{-\frac{|x_3|}{\delta_3}} \omega\sin\omega t \quad\Rightarrow\quad \frac{\partial\phi_i^{(p)}}{\partial x_i}=\frac{1}{3}\psi_0 e^{-\frac{|x_1|}{\delta_1}} e^{-\frac{|x_2|}{\delta_2}} e^{-\frac{|x_3|}{\delta_3}} \omega\sin\omega t \tag{32.16}$$

So that

$$\vec{\phi}^{(p)}=-\frac{1}{3}\psi_0\omega\sin\omega t\left(e^{-\frac{|x_1|}{\delta_1}} e^{-\frac{|x_2|}{\delta_2}} e^{-\frac{|x_3|}{\delta_3}}\right)\sum_i\left(\delta_i\vec{e}_i\right) \tag{32.17}$$

The density of kinetic energy of the fluctuation is thus worth

$$e^{kin}(\vec{r},t)=\frac{1}{2}mn\left(\vec{\phi}^{(p)}\right)^2=\frac{1}{18}mn\psi_0^2\left(\delta_1^2+\delta_2^2+\delta_3^2\right)e^{-2\frac{|x_1|}{\delta_1}} e^{-2\frac{|x_2|}{\delta_2}} e^{-2\frac{|x_3|}{\delta_3}} \omega^2\sin^2\omega t \tag{32.18}$$

The total kinetic energy of the fluctuation is obtained by integrating over space

$$E^{kin}(t)\cong\iiint_V e^{kin}(\vec{r},t)dV=\frac{1}{18}mn\psi_0^2\left(\delta_1^2+\delta_2^2+\delta_3^2\right)\delta_1\delta_2\delta_3\omega^2\sin^2\omega t \tag{32.19}$$

From which we deduce the total energy of the fluctuation is

$$\begin{aligned} E^{fluctuation}(t)&=E^{dist}(t)+E^{kin}(t)=-\left[8\left(K_0-2K_1\tau_0\right)\psi_0\delta_1\delta_2\delta_3\right]\cos\omega t \\ &+\left[K_1\psi_0^2\delta_1\delta_2\delta_3\right]\cos^2\omega t+\left[\frac{1}{18}mn\psi_0^2\delta_1\delta_2\delta_3\left(\delta_1^2+\delta_2^2+\delta_3^2\right)\omega^2\right]\sin^2\omega t \end{aligned} \tag{32.20}$$

32.2 – Random microscopic "gravitational fluctuations" and quantum fluctuations of the vacuum

Let's consider now microscopic longitudinal fluctuations, meaning gravitational fluctuations for which the amplitude ψ_0 is very weak.

Let's only consider here the very simple case of an isotropic gravitational fluctuation, meaning such that the range in the three directions of space is equal. In the case of the perfect cosmic lattice, for $\tau_0 < \tau_{0cr}$, we have that $K_0 >> K_1$ and $K_0 / mn = c_t^2$, so that

$$\begin{cases} \omega^2 \cong 3c_t^2 \dfrac{1}{\delta^2} \\ E^{fluctuation}(t) \cong K_0 \psi_0 \delta^3 \left[-8\cos\omega t + \dfrac{K_1}{K_0}\psi_0 \cos^2 \omega t + \dfrac{1}{2}\psi_0 \sin^2 \omega t \right] \end{cases} \tag{32.21}$$

We immediately notice that if this fluctuation is such that $\psi_0 << 1$, it is the energy of distorsion associated with K_0 which largely dominates the others so that

$$E^{fluctuation}(t) \cong E^{dist}(t) \cong -8 K_0 \psi_0 \delta^3 \cos\omega t \tag{32.22}$$

This results in an energy of fluctuation which can be positive or negative. It means that a lattice which does not have longitudinal waves, could be subject to a superposition of local fluctuations with various frequencies ω_k, various phases φ_k and various amplitudes ψ_{0k}, and for which the centers would be randomly located at positions $\vec{r}_k$, so that it would as

$$\tau^{(p)}(\vec{r},t) = \sum_k \psi_{0k}\, e^{-\frac{|x_1 - x_{1k}|}{\delta_{1k}}} e^{-\frac{|x_2 - x_{2k}|}{\delta_{2k}}} e^{-\frac{|x_3 - x_{3k}|}{\delta_{3k}}} e^{-i(\omega_k t + \varphi_k)} \tag{32.23}$$

with

$$\omega_k = \sqrt{\frac{\alpha}{mn}} \sqrt{\frac{1}{\delta_{1k}^2} + \frac{1}{\delta_{2k}^2} + \frac{1}{\delta_{3k}^2}} \tag{32.24}$$

As the energy of each of the fluctuations can be positive or negative, the global instantaneous energy of the field $\tau^{(p)}(\vec{r},t)$ would *always present a null average energy.* We can try to represent this schematically in the lattice, as was done in figure 32.1.

This *field of microscopic 'gravitational fluctuations'* is not formed from stable fluctuations in time, since their energy is not a constant. It is in fact constituted of 'vanishing' fluctuations, which appear and disappear spontaneously, while maintaining a null global energy of the lattice. As such the field of gravitational fluctuations is the perfect analog of the *quantum field of fluctuations of the vacuum,* which is also composed with quantum fluctuations at the microscopic level, with positive and negative energies, but where the average energy remains null.

In the presence of fluctuations *(32.23)*, let's calculate the product of $\tau^{(p)}(\vec{r},t)$ by it's complex conjugate. We have

$$\tau^{(p)}(\vec{r},t) \bullet \tau^{(p)*}(\vec{r},t) = \left(\sum_k \psi_{0k}\, e^{-\frac{|x_1 - x_{1k}|}{\delta_{1k}}} e^{-\frac{|x_2 - x_{2k}|}{\delta_{2k}}} e^{-\frac{|x_3 - x_{3k}|}{\delta_{3k}}} e^{-i(\omega_k t + \varphi_k)} \right) \left(\sum_k \psi_{0k}\, e^{-\frac{|x_1 - x_{1k}|}{\delta_{1k}}} e^{-\frac{|x_2 - x_{2k}|}{\delta_{2k}}} e^{-\frac{|x_3 - x_{3k}|}{\delta_{3k}}} e^{i(\omega_k t + \varphi_k)} \right) \tag{32.25}$$

Let's express this product by separating the group of terms for which $m = n$ from the group for which $m \neq n$

$$\tau^{(p)}(\vec{r},t)\bullet\tau^{(p)*}(\vec{r},t)=\sum_{m=n}\left(\psi_{0m}\,e^{-\frac{|x_1-x_{1m}|}{\delta_{1m}}}\,e^{-\frac{|x_2-x_{2m}|}{\delta_{2m}}}\,e^{-\frac{|x_3-x_{3m}|}{\delta_{3m}}}\right)^2$$
$$+\sum_{m\neq n}\psi_{0m}\psi_{0n}\,e^{-\frac{|x_1-x_{1m}|}{\delta_{1m}}}\,e^{-\frac{|x_1-x_{1n}|}{\delta_{1n}}}\,e^{-\frac{|x_2-x_{2m}|}{\delta_{2m}}}\,e^{-\frac{|x_1-x_{1n}|}{\delta_{1n}}}\,e^{-\frac{|x_3-x_{3m}|}{\delta_{3m}}}\,e^{-\frac{|x_3-x_{3n}|}{\delta_{3n}}}\,e^{i(\omega_n-\omega_m)t}\,e^{i(\varphi_n-\varphi_m)} \quad (32.26)$$

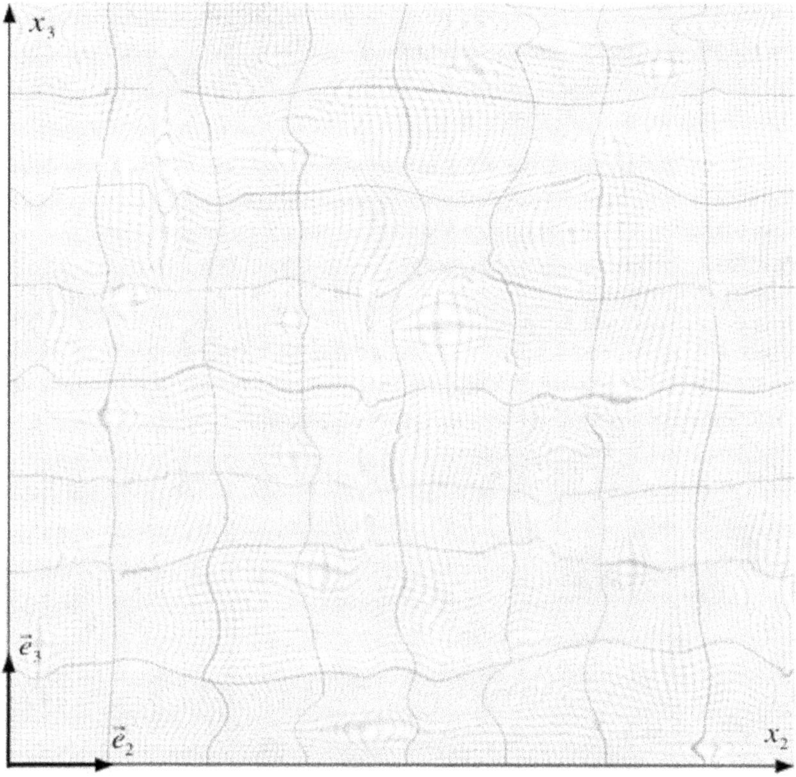

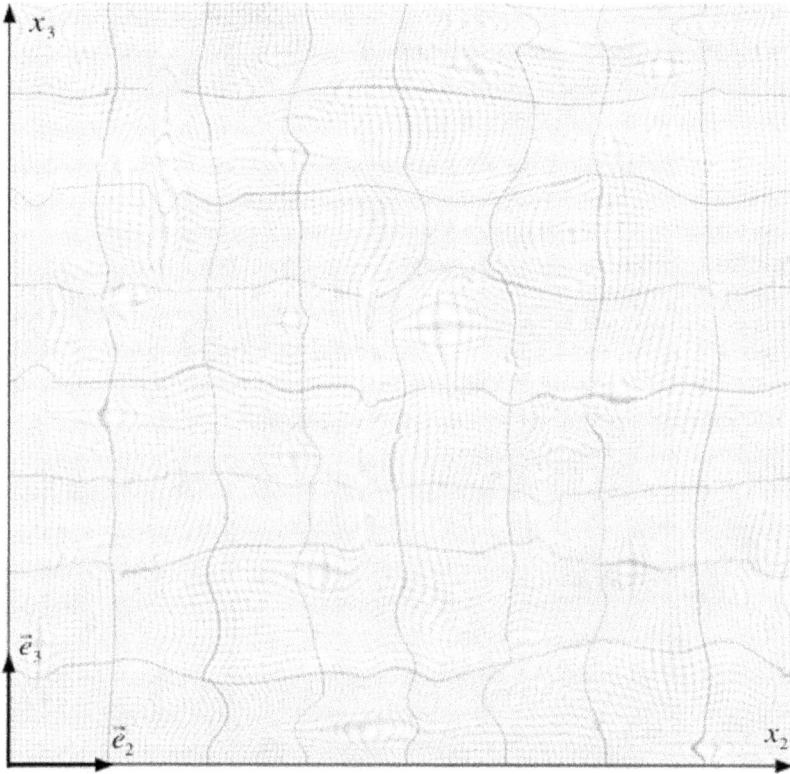

***Figure 32.1** - Schematic representation of the field $\tau^{(p)}(\vec{r},t)$ of elementary gravitational fluctuations*

The first term of the product clearly has a non-null value since it is the sum of squares of the amplitudes of each random fluctuations, while the second term can only have a null value due to the random positive and negative terms of the product $e^{i(\omega_n-\omega_m)t}e^{i(\varphi_n-\varphi_m)}$. We thus obtain a non null product which is nothing else than the instantaneous product of the wave function $\psi(\vec{r},t)$ by it's complex conjugate

$$\tau^{(p)}(\vec{r},t)\bullet\tau^{(p)*}(\vec{r},t)=\sum_{m=n}\left(\psi_{0m}\,e^{-\frac{|x_1-x_{1m}|}{\delta_{1m}}}\,e^{-\frac{|x_2-x_{2m}|}{\delta_{2m}}}\,e^{-\frac{|x_3-x_{3m}|}{\delta_{3m}}}\right)^2=\psi(\vec{r},t)\bullet\psi^*(\vec{r},t) \quad (32.27)$$

In chapter 28, we have interpreted the product $\underline{\psi}(\vec{r},t)\bullet\underline{\psi}^*(\vec{r},t)$ as the probability of presence of the topological singularity responsible for the wave function $\psi(\vec{r},t)$. We can thus apply here the probabilistic concept and imagine that the instantaneous value of $\psi(\vec{r},t)\bullet\psi^*(\vec{r},t)$ corresponds to a probability of presence of a virtual topological singularity, namely a topological singularity that does not really exist, which matches perfectly to the usual interpretation of quantum fluctuations of the vacuum in quantum mechanics!

32.3 – Stable «gravitational» oscillations

is it possible to form stable random gravitational oscillations?

To form stable gravitational fluctuations, as local longitudinal oscillations that have some permanence in the lattice, the total energy *(32.20)* of the unique fluctuation *(32.12)* is a complex problem. Indeed, if the fluctuation must be a localized vibration at frequency ω, it should in principle have an instantaneous energy of oscillation independent of time, which is manifestly not the case of the expression *(32.20)*. In this last one, we first have a term linked to the energy of distortion associated with the elastic module $K_0-2K_1\tau_0$, which presents a strong temporal dependency on $\cos\omega t$. This term is rather surprising as it has a null temporal value. For the energy of the fluctuation to not depend on this term, we must associate to fluctuation *(32.12)* a second fluctuation very close and very similar, so that it depends on $-\cos\omega t$, and thus that

$$\tau^{(p)}(\vec{r},t)\cong\psi_{0a}\,e^{-\frac{|x_1-x_{a1}(t)|}{\delta_{a1}}}\,e^{-\frac{|x_2-x_{a2}(t)|}{\delta_{a2}}}\,e^{-\frac{|x_3-x_{a3}(t)|}{\delta_{a3}}}\cos\omega t-\psi_{0b}\,e^{-\frac{|x_1-x_{b1}(t)|}{\delta_{b1}}}\,e^{-\frac{|x_2-x_{b2}(t)|}{\delta_{b2}}}\,e^{-\frac{|x_3-x_{b3}(t)|}{\delta_{b3}}}\cos\omega t \tag{32.28}$$

The total energy of this fluctuation will be written

$$E^{fluctuation}(t)=\begin{bmatrix}-\left[8\left(K_0-2K_1\tau_0\right)\psi_{a0}\delta_{a1}\delta_{a2}\delta_{a3}\right]\cos\omega t+\left[8\left(K_0-2K_1\tau_0\right)\psi_{b0}\delta_{b1}\delta_{b2}\delta_{b3}\right]\cos\omega t\\ +K_1\left(\psi_{a0}^2\delta_{a1}\delta_{a2}\delta_{a3}+\psi_{b0}^2\delta_{b1}\delta_{b2}\delta_{b3}\right)\cos^2\omega t\\ +(1/18)mn\left[\psi_{a0}^2\delta_{a1}\delta_{a2}\delta_{a3}\left(\delta_{a1}^2+\delta_{a2}^2+\delta_{a3}^2\right)\omega^2+\psi_{b0}^2\delta_{b1}\delta_{b2}\delta_{b3}\left(\delta_{b1}^2+\delta_{b2}^2+\delta_{b3}^2\right)\omega^2\right]\sin^2\omega t\end{bmatrix} \tag{32.29}$$

For the term in $\cos\omega t$ to disappear, we must then have that

$$\psi_{a0}\delta_{a1}\delta_{a2}\delta_{a3}=\psi_{b0}\delta_{b1}\delta_{b2}\delta_{b3} \tag{32.30}$$

But even if this condition is satisfied, we nevertheless have a term of energy of distortion associate with module K_1 and the kinetic energy term which are not independent of time, and which are of non null average, which signifies that they are two terms which provide the energy of fluctuation of the oscillation. But for the energy of oscillation to have meaning, it's instantaneous value must be independent of time. These two terms are the ones respectively in $\cos^2\omega t$ and $\sin^2\omega t$. We could for example imagine that the coefficients of these two temporal functions are equal, so that they could combine to erase the temporal dependence. It would be necessary that by using relation *(32.11)* giving the frequency of fluctuation, that

$$K_1\left[\psi_{a0}^2\delta_{a1}\delta_{a2}\delta_{a3}+\psi_{b0}^2\delta_{b1}\delta_{b2}\delta_{b3}\right]=\frac{1}{18}\alpha\begin{bmatrix}+\psi_{a0}^2\delta_{a1}\delta_{a2}\delta_{a3}\left(\delta_{a1}^2+\delta_{a2}^2+\delta_{a3}^2\right)\left(\dfrac{1}{\delta_{a1}^2}+\dfrac{1}{\delta_{a2}^2}+\dfrac{1}{\delta_{a3}^2}\right)\\ +\psi_{b0}^2\delta_{b1}\delta_{b2}\delta_{b3}\left(\delta_{b1}^2+\delta_{b2}^2+\delta_{b3}^2\right)\left(\dfrac{1}{\delta_{b1}^2}+\dfrac{1}{\delta_{b2}^2}+\dfrac{1}{\delta_{b3}^2}\right)\end{bmatrix} \tag{32.31}$$

By using condition (32.26), and after some calculation, this condition can be rewritten

$$\left[\begin{array}{l} +\dfrac{\psi_{a0}}{\psi_{a0}+\psi_{b0}}\left(3+\dfrac{\delta_{a1}^2+\delta_{a2}^2}{\delta_{a3}^2}+\dfrac{\delta_{a1}^2+\delta_{a3}^2}{\delta_{a2}^2}+\dfrac{\delta_{a2}^2+\delta_{a3}^2}{\delta_{a1}^2}\right) \\ +\dfrac{\psi_{b0}}{\psi_{a0}+\psi_{b0}}\left(3+\dfrac{\delta_{b1}^2+\delta_{b2}^2}{\delta_{b3}^2}+\dfrac{\delta_{b1}^2+\delta_{b3}^2}{\delta_{b2}^2}+\dfrac{\delta_{b2}^2+\delta_{b3}^2}{\delta_{b1}^2}\right) \end{array}\right]=\frac{18K_1}{K_0-4K_2/3-2K_1(1+\tau_0)}<1 \quad si \quad \tau_0<\tau_{0cr} \tag{32.32}$$

But since the term within brackets is larger than 3 and that is should at the same time be smaller than 1 if $\tau_0<\tau_{0cr}$, it is impossible to remove the dependencies in $\cos^2\omega t$ and $\sin^2\omega t$ by an addition of the potential energy term associated with module K_1 and the kinetic energy term.

We must subsequently imagine another ad hoc mechanism that would ensure the time independence of the energy associated with the terms in $\cos^2\omega t$ and $\sin^2\omega t$. As a matter of fact, it is possible to associate to the fluctuation represented by *(32.28)* another similar fluctuation, but which depends on $\sin\omega t$ instead of $\cos\omega t$, so that the total fluctuation is composed of four individual fluctuations *a, b, c and d* situated in different points of the lattice, respectively in $\vec{r}_a(t)$, $\vec{r}_b(t)$, $\vec{r}_c(t)$ *and* $\vec{r}_d(t)$, *so that*

$$\tau^{(p)}(\vec{r},t)\cong\left[\begin{array}{l} +\psi_{0a}\,e^{-\frac{|x_1-x_{a1}(t)|}{\delta_{a1}}}\,e^{-\frac{|x_2-x_{a2}(t)|}{\delta_{a2}}}\,e^{-\frac{|x_3-x_{a3}(t)|}{\delta_{a3}}}\cos\omega t-\psi_{0b}\,e^{-\frac{|x_1-x_{b1}(t)|}{\delta_{b1}}}\,e^{-\frac{|x_2-x_{b2}(t)|}{\delta_{b2}}}\,e^{-\frac{|x_3-x_{b3}(t)|}{\delta_{b3}}}\cos\omega t \\ \pm\psi_{0c}\,e^{-\frac{|x_1-x_{c1}(t)|}{\delta_{c1}}}\,e^{-\frac{|x_2-x_{c2}(t)|}{\delta_{c2}}}\,e^{-\frac{|x_3-x_{c3}(t)|}{\delta_{c3}}}\sin\omega t\mp\psi_{0d}\,e^{-\frac{|x_1-x_{d1}(t)|}{\delta_{d1}}}\,e^{-\frac{|x_2-x_{d2}(t)|}{\delta_{d2}}}\,e^{-\frac{|x_3-x_{d3}(t)|}{\delta_{d3}}}\sin\omega t \end{array}\right] \tag{32.33}$$

The energy of this fluctuation is then written

$$E^{fluctuation}(t)=\left[\begin{array}{l} -\left[8(K_0-2K_1\tau_0)\psi_{a0}\delta_{a1}\delta_{a2}\delta_{a3}\right]\cos\omega t+\left[8(K_0-2K_1\tau_0)\psi_{b0}\delta_{b1}\delta_{b2}\delta_{b3}\right]\cos\omega t \\ \mp\left[8(K_0-2K_1\tau_0)\psi_{c0}\delta_{c1}\delta_{c2}\delta_{c3}\right]\sin\omega t\ \pm\left[8(K_0-2K_1\tau_0)\psi_{d0}\delta_{d1}\delta_{d2}\delta_{d3}\right]\sin\omega t \\ +K_1\left[\psi_{a0}^2\delta_{a1}\delta_{a2}\delta_{a3}+\psi_{b0}^2\delta_{b1}\delta_{b2}\delta_{b3}\right]\cos^2\omega t+K_1\left[\psi_{c0}^2\delta_{c1}\delta_{c2}\delta_{c3}+\psi_{d0}^2\delta_{d1}\delta_{d2}\delta_{d3}\right]\sin^2\omega t \\ +(1/18)mn\left[\psi_{a0}^2\delta_{a1}\delta_{a2}\delta_{a3}\left(\delta_{a1}^2+\delta_{a2}^2+\delta_{a3}^2\right)\omega^2+\psi_{b0}^2\delta_{b1}\delta_{b2}\delta_{b3}\left(\delta_{b1}^2+\delta_{b2}^2+\delta_{b3}^2\right)\omega^2\right]\sin^2\omega t \\ +(1/18)mn\left[\psi_{c0}^2\delta_{c1}\delta_{c2}\delta_{c3}\left(\delta_{c1}^2+\delta_{c2}^2+\delta_{c3}^2\right)\omega^2+\psi_{d0}^2\delta_{d1}\delta_{d2}\delta_{d3}\left(\delta_{d1}^2+\delta_{d2}^2+\delta_{d3}^2\right)\omega^2\right]\cos^2\omega t \end{array}\right] \tag{32.34}$$

For all the terms that are dependent on time to disappear, we must have all the following equations satisfied

$$\left\{\begin{array}{l} \psi_{a0}\delta_{a1}\delta_{a2}\delta_{a3}=\psi_{b0}\delta_{b1}\delta_{b2}\delta_{b3} \\ \psi_{c0}\delta_{c1}\delta_{c2}\delta_{c3}=\psi_{d0}\delta_{d1}\delta_{d2}\delta_{d3} \\ \psi_{a0}^2\delta_{a1}\delta_{a2}\delta_{a3}+\psi_{b0}^2\delta_{b1}\delta_{b2}\delta_{b3}=\psi_{c0}^2\delta_{c1}\delta_{c2}\delta_{c3}+\psi_{d0}^2\delta_{d1}\delta_{d2}\delta_{d3} \\ \psi_{a0}^2\delta_{a1}\delta_{a2}\delta_{a3}\left(\delta_{a1}^2+\delta_{a2}^2+\delta_{a3}^2\right)+\psi_{b0}^2\delta_{b1}\delta_{b2}\delta_{b3}\left(\delta_{b1}^2+\delta_{b2}^2+\delta_{b3}^2\right) \\ \quad=\psi_{c0}^2\delta_{c1}\delta_{c2}\delta_{c3}\left(\delta_{c1}^2+\delta_{c2}^2+\delta_{c3}^2\right)+\psi_{d0}^2\delta_{d1}\delta_{d2}\delta_{d3}\left(\delta_{d1}^2+\delta_{d2}^2+\delta_{d3}^2\right) \end{array}\right. \tag{32.35}$$

$$\left\{\frac{mn}{\alpha}\omega^2=\left(\frac{1}{\delta_{a1}^2}+\frac{1}{\delta_{a2}^2}+\frac{1}{\delta_{a3}^2}\right)=\left(\frac{1}{\delta_{b1}^2}+\frac{1}{\delta_{b2}^2}+\frac{1}{\delta_{b3}^2}\right)=\left(\frac{1}{\delta_{c1}^2}+\frac{1}{\delta_{c2}^2}+\frac{1}{\delta_{c3}^2}\right)=\left(\frac{1}{\delta_{d1}^2}+\frac{1}{\delta_{d2}^2}+\frac{1}{\delta_{d3}^2}\right)\right.$$

By assuming that the amplitudes of the four fluctuations are equal

Hypothesis: $\psi_{a0}=\psi_{b0}=\psi_{c0}=\psi_{d0}=\psi_0$ (32.36)

it is necessary that the following conditions on the range of the fluctuations be satisfied

$$\begin{cases} \delta_{a1}\delta_{a2}\delta_{a3} = \delta_{b1}\delta_{b2}\delta_{b3} = \delta_{c1}\delta_{c2}\delta_{c3} = \delta_{d1}\delta_{d2}\delta_{d3} = ABC \\ \dfrac{1}{\delta_{a1}^2}+\dfrac{1}{\delta_{a2}^2}+\dfrac{1}{\delta_{a3}^2}=\dfrac{1}{\delta_{b1}^2}+\dfrac{1}{\delta_{b2}^2}+\dfrac{1}{\delta_{b3}^2}=\dfrac{1}{\delta_{c1}^2}+\dfrac{1}{\delta_{c2}^2}+\dfrac{1}{\delta_{c3}^2}=\dfrac{1}{\delta_{d1}^2}+\dfrac{1}{\delta_{d2}^2}+\dfrac{1}{\delta_{d3}^2}=\dfrac{1}{A^2}+\dfrac{1}{B^2}+\dfrac{1}{C^2} \\ \delta_{a1}^2+\delta_{a2}^2+\delta_{a3}^2=\delta_{b1}^2+\delta_{b2}^2+\delta_{b3}^2=\delta_{c1}^2+\delta_{c2}^2+\delta_{c3}^2=A^2+B^2+C^2 \end{cases} \tag{32.37}$$

where the product ABC is in fact proportional to the *«volume»* occupied by a fluctuation within the lattice.

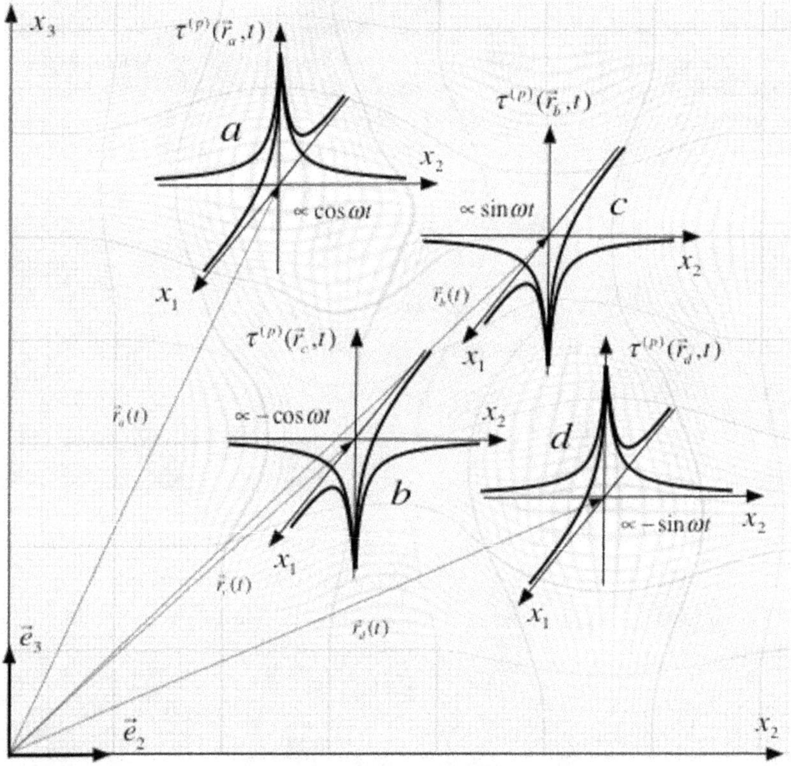

***Figure 32.2** - Schematic representation of the 4 fluctuations making the fluctuation $\tau^{(p)}(\vec{r},t)$ (32.40) with constant kinetic energy*

In this case, the global energy of the fluctuation becomes independent of time and is written

$$E^{fluctuation} = 2\psi_0^2 ABC\left[K_1 + \frac{\alpha}{18}\left(A^2+B^2+C^2\right)\left(\frac{1}{A^2}+\frac{1}{B^2}+\frac{1}{C^2}\right)\right] \tag{32.38}$$

But as $K_1 << \alpha \cong K_0$ for $\tau_0 << \tau_{0cr}$, we deduce that the energy of the fluctuation is almost exclusively <u>*kinetic energy*</u> *if* $\tau_0 << \tau_{0cr}$, so that the total energy and the frequency of the fluctuation is then worth

$$\begin{cases} E^{fluctuation} \cong E^{kin} \cong \dfrac{K_0}{9}\psi_0^2 ABC\left(A^2+B^2+C^2\right)\left(\dfrac{1}{A^2}+\dfrac{1}{B^2}+\dfrac{1}{C^2}\right) \\ \omega \cong c_t\sqrt{\dfrac{1}{A^2}+\dfrac{1}{B^2}+\dfrac{1}{C^2}} \end{cases} \quad \text{if} \quad \tau_0 << \tau_{0cr} \tag{32.39}$$

The fluctuation is then written

$$\tau^{(p)}(\vec{r},t) \cong \psi_0 \begin{bmatrix} +\mathrm{e}^{-\frac{|x_1-x_{a1}(t)|}{\delta_{a1}}}\mathrm{e}^{-\frac{|x_2-x_{a2}(t)|}{\delta_{a2}}}\mathrm{e}^{-\frac{|x_3-x_{a3}(t)|}{\delta_{a3}}}\cos\omega t - \mathrm{e}^{-\frac{|x_1-x_{b1}(t)|}{\delta_{b1}}}\mathrm{e}^{-\frac{|x_2-x_{b2}(t)|}{\delta_{b2}}}\mathrm{e}^{-\frac{|x_3-x_{b3}(t)|}{\delta_{b3}}}\cos\omega t \\ \pm\mathrm{e}^{-\frac{|x_1-x_{c1}(t)|}{\delta_{c1}}}\mathrm{e}^{-\frac{|x_2-x_{c2}(t)|}{\delta_{c2}}}\mathrm{e}^{-\frac{|x_3-x_{c3}(t)|}{\delta_{c3}}}\sin\omega t \mp \mathrm{e}^{-\frac{|x_1-x_{d1}(t)|}{\delta_{d1}}}\mathrm{e}^{-\frac{|x_2-x_{d2}(t)|}{\delta_{d2}}}\mathrm{e}^{-\frac{|x_3-x_{d3}(t)|}{\delta_{d3}}}\sin\omega t \end{bmatrix} \tag{32.40}$$

in which the range of the fluctuation must satisfy the following relations

$$\delta_{a1}\delta_{a2}\delta_{a3} = \delta_{b1}\delta_{b2}\delta_{b3} = \delta_{c1}\delta_{c2}\delta_{c3} = \delta_{d1}\delta_{d2}\delta_{d3} = ABC \tag{32.41}$$

Such a fluctuation is represented schematically in figure 32.2.

32.4 – Stable macroscopic gravitational oscillations in an infinite cosmic lattice and multiverse

In the domain $\tau_0 < \tau_{0cr}$ of the cosmic lattice in which there are no longitudinal waves, there is thus the possibility of seeing a stable macroscopic fluctuation formed of 4 elementary fluctuations *(32.33)* which represents local longitudinal vibrations with a given frequency ω , and such that the total energy of the fluctuation *(32.39)*, essentially of kinetic nature, does not depend on time.

Let's consider the case of a macroscopic fluctuation composed of 4 fluctuations which are isotropic, meaning $\delta_{x1} \cong \delta_{x2} \cong \delta_{x3} \cong \delta$. From relation *(32.39)* we deduce the frequency of oscillation ω of this macroscopic fluctuation, which is inversely proportional to the range δ , as well as the energy of the fluctuation which is essentially of kinetic nature

$$\begin{cases} E^{fluctuation} \cong E^{kin} \cong K_0\psi_0^2\delta^3 \\ \omega \cong \sqrt{3}\dfrac{c_t}{\delta} \end{cases} \quad \textit{if} \quad \tau_0 << \tau_{0cr} \tag{32.42}$$

We deduce that the macroscopic gravitational fluctuations with large amplitude ψ_0 in the domain $\tau_0 < \tau_{0cr}$ would have a frequency proportional to the inverse of their range δ and that their total energy is proportional to the product of the square ψ_0^2 of their amplitude and their volume δ^3 , and independent of their frequency ω .

On the possible existence of interleaved universes in contraction and expansion in a perfect infinite cosmic lattice

In a perfect cosmic lattice which would be infinite, we cannot develop the scenario of cosmic expansion as we have done in chapter 16 (fig. 16.8 et 16.11a) where the lattice was *finite*! However, we can imagine the appearance of macroscopic fluctuations just as the one we just described (fig. 32.2), which would have a very large volume δ^3 , so that its frequency of oscillations

(32.42) would be very low. And if its amplitude ψ_0 was also high enough, for the small ***HS*** observers, which would be placed within these elementary fluctuations, those would present all the characteristics of a universe which would oscillate between a maximum expansion and a maximum contraction with frequency ω . Thus, the set of 4 elementary fluctuations could represent a *multiverse*. And within each of these elementary fluctuations, meaning within each multi-verse, the observations of the ***HS*** would be very similar to those done by the ***HS*** which would exist within the universe as was described in chapter 16 (fig. 16.8 et 16.11a) in the case of the finite cosmic lattice!

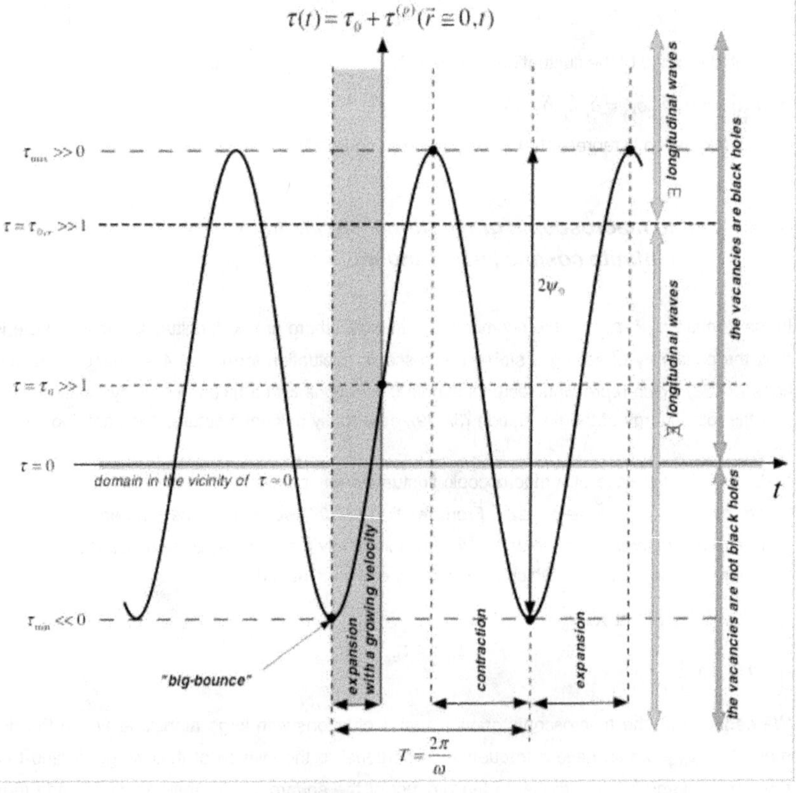

Figure 32.3 - *Expansions-contractions of a multi-verse. The behavior of gravitational force of interaction between topological singularities that corresponds is shown in figure 27.1*

But for that to be true, there are necessary conditions which can be deduced from figure 32.3 and which would be the following:

- *(i)* the infinite cosmic lattice has a background expansion such that $\tau_0 >> 1$ so that the formation of vacancy black holes appears in the domain (greyed in figure 32.3) where the expansion of the universe is done at increasing velocity,

- *(ii)* that the amplitude ψ_0 of the oscillation around $\tau = \tau_0$ be sufficient for that oscillation to go from the domain of values situated around $\tau \approx 0$ for the scenario of cosmology developed in section 27.2 to be applicable to these multi-verses!

Under these two conditions, each of these multi-verses presents an expansion starting in a «big-bounce», but it does not means that the volume expansion goes through a singularity of the type $\tau \rightarrow -\infty$, with, in its vicinity, a cosmic evolution similar to that described in section 27.2, which presents a phase with an expansion at increasing velocity during which there are vacancy black holes, and then a phase of expansion with a decreasing velocity taking us to the maximum expansion, which can or not go through a domain of expansion situated beyond the critical value of expansion τ_{0cr} where there appears longitudinal waves to the detriment of local longitudinal fluctuations. Then this Universe would go through a phase of contraction taking it to a new «big-bounce»!

However, it is clear that the full calculation of the behavior of such a multiverse it not as simple as what we just talked about, if anything because we have taken the hypothesis that $\tau_0 << \tau_{0cr}$ in our calculations, which would not be an appropriate hypothesis in the case of large gravitational fluctuations forming this multi-verse!

32.5 – Quantified microscopic gravitational oscillations: on hypothetical "gravitons" quasi-particles

Let's consider now *stable microscopic oscillations of fluctuations*, meaning gravitational fluctuations of the type described by relation *(32.40)*, of constant energy, but for which the amplitude ψ_0 would be very weak.

On a hypothetical quantification of stable gravitational fluctuations

We will here look only at the very simple case of a gravitational fluctuation that would isotropic, meaning that the ranges in the 3 directions of space are equal. Let's suppose very hypothetically the constant energy be *quantified* in the same way that the photons were quantified, by using conjecture 11 previously introduced. By taking again relations *(32.42)* and by introducing the quantified energy $2\hbar\omega$ corresponding to the four degrees of liberty of oscillation of the fluctuation, we obtain a stable isotropic fluctuation, local and quantized, with range δ, of frequency ω, of kinetic energy $E^{fluctuation}$ and of amplitude ψ_0 such that

$$\begin{cases} \omega \cong \sqrt{3}\dfrac{c_t}{\delta} \quad \Rightarrow \quad \delta \cong \sqrt{3}\dfrac{c_t}{\omega} \\ E^{fluctuation} \cong E^{kin} \cong K_0\psi_0^2\delta^3 = 2\hbar\omega \qquad \text{if} \qquad \tau_0 << \tau_{0cr} \\ \psi_0 \cong \sqrt{\dfrac{2\sqrt{3}\hbar c_t}{K_0}\dfrac{1}{\delta^2}} = \dfrac{1}{3}\sqrt{\dfrac{2\sqrt{3}\hbar c_t}{K_0}}\left(\dfrac{\omega}{c_t}\right)^2 \end{cases} \tag{32.43}$$

This stable quantized fluctuation would then present a spatial extension δ proportional to the inverse of the frequency, as well as an energy and amplitude proportional to frequency ω. This fluctuation would not need to move about the lattice, but could walk about the lattice. We here

have a quasi-particle, just like the photon in the case of transversal electromagnetic waves of rotation, but it is a quasi-particle which is associated to longitudinal 'gravitational' expansions of the lattice. We can then talk in this case of a quasi-particle of type *«gravitons»*.
As a matter of fact, these gravitons do not correspond to the gravitons of GR. The gravitons postulated in our theory are quasi-particles that are energetically stable and that can migrate within the lattice, but that do not have to move at the speed of transversal waves. On the contrary, the gravitons of GR are supposed to be moving about at the speed of light. Furthermore our gravitons do not cary the gravitational interaction between two singularities, but only the local quantized energetic fluctuations of the field of expansion, contrary to the gravitons of general relativity, which are considered as mediating particles of the gravitational interaction!
The configuration of the four elementary fluctuations composing the 'graviton' quasi-particle, can be very complex. The only condition is that the four elementary fluctuations be able to exchange energy between themselves in order to maintain the total kinetic energy constant. We could for example imagine axial gravitons, meaning quasi-particles for which the four elementary fluctuations are aligned along a preferential axis. In that case, we can give two extreme examples for those quasi-particles, depending on whether they are extended of contracted on the preferential axis.

Condensed axial "graviton" along axis Ox_2

Suppose a gravitational fluctuation condensed along axis Ox_2, with ranges such δ along the axis Ox_1 and Ox_3 equal and much larger that the range δ_2 along the axis Ox_2. We then deduce that

$$\delta >> \delta_2 \quad \Rightarrow \quad \begin{cases} \omega \cong \dfrac{c_t}{\delta_2} \quad \Rightarrow \quad \delta_2 \cong \dfrac{c_t}{\omega} \\ E^{fluctuation} \cong E^{kin} \cong \dfrac{2K_0}{9}\psi_0^2\delta^4 = 2\hbar\omega \\ \psi_0 = 3\sqrt{\dfrac{\hbar}{K_0}}\dfrac{\sqrt{\omega}}{\delta^2} = 3\sqrt{\dfrac{\hbar c_t}{K_0}}\dfrac{1}{\delta^2\sqrt{\delta_2}} \end{cases} \tag{32.44}$$

We notice that it is the spread δ_2 of the fluctuation along axis Ox_2 which is fixed and depends on the frequency ω of the quasi-particle and that the amplitude ψ_0 and the spread δ along the axis perpendicular to Ox_2 are linked but not fixed, so that a quasi-particle can deform by spreading of shrinking along the axis perpendicular to Ox_2 ! This effect is again an aspect of the famous *non-locality of quantum mechanics*, such as was described in the case of photons!

Axial "gravitons" spread along axis Ox_2

Suppose now a gravitational fluctuation spread along axis Ox_2, with range δ along axis Ox_1 and Ox_3 equal and much smaller than δ_2 along axis Ox_2. In this case, we notice that the range δ of the fluctuation along axis Ox_1 and Ox_3 are fixed and depend on the frequency ω of the quasi particle, and that the amplitude ψ_0 and the spread δ_2 along the axis Ox_2 are linked, but not fixed, so that the quasi-particle can deform by spreading or shrinking along axis

Ox_2 (this is also an effect of the non-locality of quantum mechanics)

$$\delta_2 >> \delta \quad \Rightarrow \quad \begin{cases} \omega \cong \sqrt{2} c_t \dfrac{1}{\delta} \quad \Rightarrow \quad \delta \cong \sqrt{2} c_t \dfrac{1}{\omega} \\ E^{fluctuation} \cong E^{kin} \cong \dfrac{2K_0}{9} \psi_0^2 \delta_2^4 = 2\hbar\omega \\ \psi_0 = 3\sqrt{\dfrac{\hbar}{K_0} \dfrac{\sqrt{\omega}}{\delta_2^2}} = 3\sqrt{\dfrac{\sqrt{2}\hbar c_t}{K_0} \dfrac{1}{\delta_2^2 \sqrt{\delta}}} \end{cases} \tag{32.45}$$

We deduce that an axial «graviton» with frequency ω , and thus of kinetic energy $\hbar\omega$, can oscillate between a condensed form along axis Ox_2 and spread out along axis Ox_1 and Ox_3, and an opposed form, very spread out along Ox_2 and shrunk along axis Ox_1 and Ox_3.

In figure 32.4, we have represented schematically the case of a hypothetical axial «graviton» in the case where it is spread out along axis Ox_2 . In this representation we have shown the instantaneous volume expansion of the «graviton», by specifying the oscillations of the four components in $\cos\omega t$ and $\sin\omega t$.

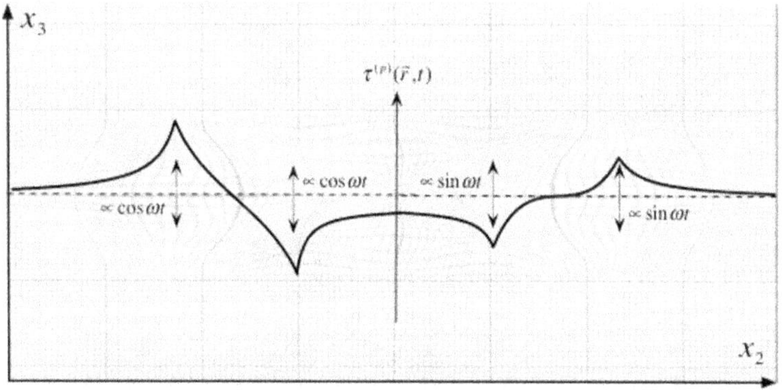

Figure 32.4 *- Schematic representation of an axial "graviton" spread out along* Ox_2

Let's remark, to finish, that the hypothetical quasi-particles we call "gravitons" are very different from hypothetical evanescent fluctuations of gravity and are analogous to the quantum fluctuations of the vacuum we described previously (fig. 32.1). Indeed, in the case of the evanescent gravitational fluctuations, the energy of the fluctuations is essentially *an energy of elastic distortions* associated to module K_0 of the cosmic lattice, oscillating between a positive and negative value, and with null average energy, while the energy of the hypothetical "graviton" quasi-particles is essentially of *kinetic nature* and possesses a constant value, which assures their long term stability. And to obtain this stability, the quasi-particle must be composed of four fluctuations strongly correlated and non-separable which assures the constancy of energy, in the same way that 'photon' quasi-particles are composed of two fluctuations of perpendicular rotation and out of phase, which gives them their helicity and assures the constant energy. There is a strong analogy between "photon" quasi-particles and "graviton" quasi-particles, not only by the constitution which assures a constant energy but also due to their non-local nature, namely their aptitude to spread in space while conserving their identity and energy, which is a typical quantum mechanical property!

Conclusion

As we already said in the introduction, this essay does not aim to present a *"Theory of Everything"* in its final form, but aims to show that a rigorous approach to elastic lattices via an Euler coordinates description gives us a *framework for investigation* that is a lot simpler than the differential geometries developed in General Relativity, but can be very rich and fertile as we were able to show relatively easily that it is possible to *(i)* obtain strong, or perfect, analogies with all the theories of physics, including Maxwell equations, special relativity, general relativity, newton gravity, modern cosmology, quantum mechanics, and the standard model of particles, and *(ii)* bring to light unifying bridges between these theories

On the central role played by the Newton equation in the cosmic lattice

Since the beginning of the second part of this treatise, we notice that the *Newton equation of motion (13.14)* which we have established in chapter 13 for an imaginary isotropic cosmic lattice, played a central role and that it describes most of the striking properties of the 'perfect cosmic lattice' amongst which:

- the *propagation of transversal waves coupled with longitudinal wavelets,* described by the Newton equation of motion *(13.14)*, which implies that the pure transversal waves cannot exist without a circular polarization (which is a fundamental property of photons),

- the existence of domains of expansion ($\tau_0 < \tau_{0cr}$) in which there are *no solutions of longitudinal waves to the Newton equation of motion (13.14)*, but only *quasi-static solutions* which give rise to the phenomena of gravitational interactions between topological singularities, or the *local longitudinal modes of vibration*, which are the basis for *quantum mechanics and the spin of local topological singularities,*

- the *curvature wave rays* in the gradients of volume expansion, which is also a direct consequence of the equation of Newton *(13.14)*, and which predicts the possibility of existence of black holes which capture the transversal perturbations,

- the *complete equations of Maxwell* (table 17.1) for the rotation vector $\vec{\omega}$ when the field of volume expansion is homogenous, which shows that the equation $\partial\vec{B}/\partial t = -\overrightarrow{\mathrm{rot}}\,\vec{E}$ of electromagnetism is nothing else than the expression of the equation of Newton *(13.14)* of the cosmic lattice applied to this particular case,

- *special relativity*, with the contraction of distances and the dilation of time for an observer in movement within the lattice, is a direct consequence of the first partial equation of Newton *(18.7)*,

- the *newtonian gravitation* and the *General Relativity* are direct consequences of the quasi-static solution of the second partial equation of Newton *(18.9)*,

- the *spatial curvature* for a ***GO*** observer exterior to the lattice and the *curvature of space time for the local observers **HS*** within the lattice, which implies a direct analogy between the divergence of the equation of Newton *(24.64)* and the famous equation of Einstein with *null diver-*

gence $\vec{\nabla} \bullet \boldsymbol{T} = 0$ *of the energy-momentum tensor (24.67),* which assures us that conservation of energy and momentum are respected,

- the *black holes*, static solutions of the second partial equation of Newton *(18.9)* in the presence of macroscopic vacancies of the lattice when $\tau_0 > 1$,

- the *neutron stars or pulsars*, static solutions of the second partial equation of Newton *(18.9)* in the presence of macroscopic interstitials in the lattice,

- the *weak interaction* between the edge dislocation and the twist disclinations, which is also a consequence of the second partial equation of Newton *(18.9)*,

- the *quantum mechanics,* the *wave function,* the *Schrödinger wave equation*, and the *notion of spin*, which are consequences of the second partial equation of Newton *(18.9)* in the dynamic case of gravitational perturbations associated with local topological singularities in the cosmic lattice

- the *photons*, quantas of the transversal solution with circular polarization of the Newton equation *(13.14)*

- and finally quantum concepts, such as *bosons, fermions* and *indistinguishability of topological singularities,* as well as the *Pauli exclusion principle*, which are all deduced from the second partial equation of Newton *(18.9)* applied to many local topological singularities.

On the importance of the microscopic structure of the cosmic lattice

The structure of the cosmic lattice plays a dominant role in the analogies we have developed, but it is at the level of microscopic loop singularities that they play a crucial role. It was necessary to imagine a cubic lattice with *planes (imaginarily «colored» in red, green and blue) which satisfy certain simple rules concerning their arrangement* to find topological loops which are remarkably similar to all the particles, including leptons, quarks, intermediary bosons and gluons of the first family of elementary particles of the Standard Model, as well as an asymptotic force between singularities analogous to quarks, which is similar to the strong force of the Standard Model, and which forces the formation of doublets or triplets of loop singularities similar to mesons and baryons of the Standard Model!

It was necessary to imagine a more complex structure of edge dislocation loops in that cosmic lattice, based on *doublets of edge disclination*, to try to explain the three families of the Standard model of particle physics.

From this enumeration of the important roles of the equation of Newton and the structure of the cosmic lattice, we can conclude that the *newtonian inertia of the lattice in the absolute space, its elasticity by shear, by rotation, and by volume expansion, and its microscopic structure are the principal ingredients* of the theory of the cosmic lattice and explain all the properties of this peculiar lattice, and thus, by analogy, of the real universe, even if there remains many obscure points in this analogy!

On the obscure points concerning the cosmic lattice

It is clear that the cosmic lattice that we developed in this book, in spite of it's obvious successes, is not perfect. There are many obscure points which would deserve to be studied, and if possible elucidated, amongst which we will name the more important ones, in the order in which they were seen in the book and in the form of questions:

- are there really 'corpuscles' in the cosmic lattice which would possess a purely newtonian inertial mass in the absolute space, and which are the relations existing between these bodies and the Higgs boson of the Standard Model?
- what is the physical nature of the elasticity of the cosmic lattice, which leads to modules $K_0 = K_3 > 0,\ 0 < K_1 << K_0,\ 0 \le K_2 << K_3$ which allow to express the elastic free energy of the lattice per unit volume?
- where does the kinetic energy which is necessary to the cosmological behavior of the cosmic lattice come from?
- what role would the 'lattice corpuscles' play in the cosmic lattice in hypothetical diffusion mechanisms within the cosmic lattice (which could explain certain properties of magnetism)?
- are there really vectorial electrical charges?
- what are the parameters $\zeta_{TL},\ \zeta_{EL},\ \zeta_{ML}$ introduced to calculate the energy of loops and are they truly independent of the expansion of the background?
- what could be the role of gliding mixed dislocation loops in our analogy with the elementary particles?
- what is the relevance of the analogy between the vacancy clusters and the black holes, as well as between the interstitial clusters and the neutron stars?
- what physical explanation and what numerical values should we give to parameters $\alpha_{TL},\ \beta_{TL},\ \alpha_{EL},\ \beta_{EL},\ \alpha_{ML},\ \beta_{ML}$ which were used for the derivation of the dependence of the energy of the loops on the background expansion.
- how do we physically explain the fact that parameters a and b, which were used to find the local clock and rods of the ***HS*** observer, must exactly have the value 1/4?
- what is the thermic dependence of the free energy of the cosmic lattice, and could it be justified by a kind of "liquid phase" of the lattice for values that are very small of the volume expansion, in the vicinity of the «big-bang»?
- how do we explain that, from a topological standpoint, a dislocation loop or a disclination loop can turn on itself?
- what is the reason a topological loop possesses a spin 1/2 or a spin 1?
- what are exactly the structures of the cosmic lattice and the nature of the colored planed of said lattice and the «corpuscles» composing the lattice?
- how do topological singularities analog to quarks possess a spin while they are linked to the others by a tube of fault energy?
- and many other questions, that are purely physical or philosophical!

On the unification power of our theory of the cosmic lattice

Indeed, if there are some points that are obscure, there appears a lot of strong analogies with many of the great theories of modern physics, and as such it has a very strong *unifying potential.* And this theory of the cosmic lattice is *very simple,* contrary to the superstring theories or M theory which are proposed by theoretical physicists to unify the different forces of the physical theories by *"quantifying gravity"* and by independently introducing the four elementary inter-

actions, which leads to mathematical theories which are *extremely complex*, in complicated spaces (n=11 for M theory!), and which lack any predictive power!
It is interesting to notice here that superstring theories need to invoke strings and branes in multidimensional spaces to quantify gravity, while our theory also uses strings, loops and branes, but which are simply topological singularities of the lattice, in 3D, with an additional dimension of absolute time, which is completely disconnected from the dimensions of space, since time can be measured by the universal clock of the ***GO*** in the absolute space.
And if the quasi-static volume expansion of this lattice at the macroscopic level is the expression of the phenomena attributed to gravitation, the dynamical fluctuations of the expansion of the lattice at the microscopic level are nothing else than expressions of phenomena attributed to quantum mechanics! It is then a false lead to try to quantify gravity, since quantum mechanics is precisely the expression of these dynamical fluctuations at the microscopic level!

On the epistemology and the consequences of this treatise

It is true that, in this treatise, nothing gives a definitive explanation to the existence of the universe, to why there is a big-bang, and why the universe could behave like a solid lattice. These points remain, at least for now, philosophical or of the domain of individual belief. But from a epistemological standpoint, this essay shows that it is possible to find an interesting framework to unify the various current physical theories, a framework in which the 'raison d'être' of the Universe remains the only mystery. This approach is based in fact on a very simple concept, which could be stated as, inspired by a citation of Feynman:

> *«It it possible to observe and measure from the outside, using an Eulerian referential which is absolute and has fixed rods and a universal clock, the spatial evolution of a solid lattice which possesses a certain microscopic structure, elastic properties and newtonian inertial properties. This sentence alone contains, as you will see, a large quantity of informations about the universe if you use a bit of imagination and reflexion.»*

APPENDICES

A) Elements of analytical mechanics

B) Elements of quantum mechanics

C) Standard Model of elementary particles

D) Conjectures of the cosmic lattice

E) Mathematical formulas

Appendix A

Elements of analytical mechanics

Analytical mechanics is a formalism which allows us to elegantly resolve problems of mechanics, but which intervenes as well in the formalism of quantum mechanics.

A.1 - Basics of analytical mechanics

To describe a mechanical system composed of N elements and submitted to links, meaning conditions that limit the possible movements of the system, it is always possible to introduce forces so that the equation of Newton of each element of the system is written

$$m_{(\alpha)}\vec{a}_{(\alpha)} = \vec{F}_{(\alpha)} + \vec{R}_{(\alpha)} \quad \text{where} \quad \alpha = 1, \dots, N \tag{A.1}$$

and in which $\vec{F}_{(\alpha)}$ represents the resulting applied forces and $\vec{R}_{(\alpha)}$ represent the linking forces.

Generalized coordinates

We can also introduce a set of k *generalized coordinates* q_i, so that the position $P_{(\alpha)}$ of each of the N elements of the system are uniquely determined by the relations

$$\vec{\xi}_{(\alpha)} = \vec{\xi}_{(\alpha)}\left(q_1, q_2, q_3, \dots, q_k, t\right) \tag{A.2}$$

We then call *virtual compatible displacement* $\delta\vec{\xi}_{(\alpha)}$ any displacement of the position $P_{(\alpha)}$ of the element α, such that it satisfies variations δq_i of the generalized coordinates q_i and we talk of a *holonomic system* with k degrees of liberty, if the choice of coordinates q_i and the linking forces are such that there are no further condition on variables $\{q_i, \dot{q}_i\}$, and

$$\delta\vec{\xi}_{(\alpha)} = \sum_{i=1}^{k} \frac{\partial\vec{\xi}_{(\alpha)}}{\partial q_i}\,\delta q_i \tag{A.3}$$

For all compatible virtual displacements $\delta\vec{\xi}_{(\alpha)}$, we talk about perfect link if they satisfy the condition of null virtual work

$$\sum_{\alpha=1}^{N} \vec{R}_{(\alpha)}\delta\vec{\xi}_{(\alpha)} = 0 \tag{A.4}$$

which implies that the sum of work done by the link forces is null, namely that the link forces implied are not dissipative.

For a mechanical system submitted to perfect links, relations *(A.1)* and *(A.4)* allow us to deduce the *D'Alembert equations* of the system for any compatible virtual displacement

$$\sum_{\alpha=1}^{N} \left(m_{(\alpha)}\vec{a}_{(\alpha)} - \vec{F}_{(\alpha)}\right)\delta\vec{\xi}_{(\alpha)} = 0 \tag{A.5}$$

We talk of a *holonomic system with links independent of time* if relations *(A.2)* do not explicitly involve time

$$\vec{\xi}_{(\alpha)} = \vec{\xi}_{(\alpha)}(q_1, q_2, q_3, \dots, q_k) \tag{A.6}$$

A.2 - Lagrangian formalism

The kinetic energy T of a mechanical system, represented by its generalized coordinates *(A.2)* is written

$$T = \sum_{\alpha=1}^{N} \frac{1}{2} m_{(\alpha)} \left(\frac{d\vec{\xi}_{(\alpha)}}{dt} \right)^2 = \sum_{\alpha=1}^{N} \frac{1}{2} m_{(\alpha)} \left(\sum_{i=1}^{k} \frac{\partial \vec{\xi}_{(\alpha)}}{\partial q_i} \dot{q}_i + \frac{\partial \vec{\xi}_{(\alpha)}}{\partial t} \right)^2 \tag{A.7}$$

which finally transforms as

$$T = \sum_{\alpha=1}^{N} m_{(\alpha)} \left[\frac{1}{2} \sum_{i,j} \frac{\partial \vec{\xi}_{(\alpha)}}{\partial q_i} \frac{\partial \vec{\xi}_{(\alpha)}}{\partial q_j} \dot{q}_i \dot{q}_j + \sum_i \frac{\partial \vec{\xi}_{(\alpha)}}{\partial q_i} \frac{\partial \vec{\xi}_{(\alpha)}}{\partial t} \dot{q}_i + \frac{1}{2} \left(\frac{\partial \vec{\xi}_{(\alpha)}}{\partial t} \right)^2 \right] \tag{A.8}$$

We note that the links are independent of time *(A.6)*, the kinetic energy is a positive quadratic form in the $\dot{q}_i$, which does not depend on time t. Starting from the kinetic energy, we easily calculate the following relation

$$\frac{d}{dt} \left(\frac{\partial T}{\partial \dot{q}_i} \right) = \frac{d}{dt} \left(\sum_{\alpha=1}^{N} m_{(\alpha)} \vec{v}_{(\alpha)} \frac{\partial \vec{\xi}_{(\alpha)}}{\partial q_i} \right) = \sum_{\alpha=1}^{N} m_{(\alpha)} \vec{a}_{(\alpha)} \frac{\partial \vec{\xi}_{(\alpha)}}{\partial q_i} + \frac{\partial T}{\partial q_i} \tag{A.9}$$

from which we deduce that for all compatible virtual displacements we have

$$\sum_{\alpha=1}^{N} m_{(\alpha)} \vec{a}_{(\alpha)} \delta \vec{\xi}_{(\alpha)} = \sum_{\alpha,i} m_{(\alpha)} \vec{a}_{(\alpha)} \frac{\partial \vec{\xi}_{(\alpha)}}{\partial q_i} \delta q_i = \sum_{i=1}^{k} \left[\frac{d}{dt} \left(\frac{\partial T}{\partial \dot{q}_i} \right) - \frac{\partial T}{\partial q_i} \right] \delta q_i \tag{A.10}$$

Furthermore, if the system is submitted to forces $\vec{F}_{(\alpha)}$ deriving from a potential V, we have

$$\sum_{\alpha=1}^{N} \vec{F}_{(\alpha)} \delta \vec{\xi}_{(\alpha)} = \sum_{\alpha=1}^{N} \vec{F}_{(\alpha)} \frac{\partial \vec{\xi}_{(\alpha)}}{\partial q_i} \delta q_i = -\sum_{\alpha,n,i} \frac{\partial V}{\partial \xi_{(\alpha)n}} \frac{\partial \xi_{(\alpha)n}}{\partial q_i} \delta q_i = -\sum_{i=1}^{k} \frac{\partial V}{\partial q_i} \delta q_i \tag{A.11}$$

where the $\xi_{(\alpha)n}$ are the components of the vector $\vec{\xi}_{(\alpha)}$. The last two relations allow us to re-write the equations of D'Alembert *(A.5)* under the following form

$$\sum_{i=1}^{k} \left[\frac{d}{dt} \left(\frac{\partial T}{\partial \dot{q}_i} \right) - \left(\frac{\partial T}{\partial q_i} - \frac{\partial V}{\partial q_i} \right) \right] \delta q_i = 0 \tag{A.12}$$

As we are considering a holonomic system, the δq_i are independent, and if we admit again that the potential V is independent of velocities $\dot{q}_i$, we can introduce the *lagrangian* L of the system and write the Lagrange equations of the system under the form

$$L(q_i, \dot{q}_i, t) = T(q_i, \dot{q}_i, t) - V(q_i, t) \tag{A.13}$$

$$\frac{d}{dt} \left(\frac{\partial L}{\partial \dot{q}_i} \right) - \left(\frac{\partial L}{\partial q_i} \right) = 0 \quad (i = 1, 2, 3, \dots, k\} \tag{A.14}$$

We can define a quantity p_i called *generalized momentum* or *conjugated momentum to* q_i, with the following relation

$$p_i(q_i, \dot{q}_i, t) = \frac{\partial L}{\partial \dot{q}_i} \tag{A.15}$$

If the lagrangian L does not depend explicitly on q_i, the conjugated momentum p_i is a constant of movement. Indeed, according to the equations of Lagrange *(A.14)*, we have

$$\frac{dp_i}{dt} = \frac{d}{dt}\left(\frac{\partial L}{\partial \dot{q}_i}\right) = \frac{\partial L}{\partial q_i} = 0 \qquad (A.16)$$

A.3 - Hamiltonian formalism

Thanks to the definition of the conjugated momentums p_i, we can define the following value H such that

$$H(q_i,\dot{q}_i,t) = \sum_{i=1}^{k} p_i(q_i,\dot{q}_i,t)\dot{q}_i - L(q_i,\dot{q}_i,t) \qquad (A.17)$$

By using the *Legendre transformation*, by inversion of relations *(A.15)*, meaning by extracting $\dot{q}_i = \dot{q}_i(q_j,p_j,t)$, we show the *hamiltonian* H of the system

$$H(q_i,p_i,t) = \sum_{i=1}^{k} p_i\dot{q}_i - L = H\left(q_i,\dot{q}_i\left(q_j,p_j,t\right),t\right) \qquad (A.18)$$

The hamiltonian of a system satisfies a set of relations that we can deduce from its definition, the Lagrange equation and the definition of generalized momenta

$$\begin{cases} \dfrac{\partial H}{\partial q_i} = \sum_j p_j \dfrac{\partial \dot{q}_j}{\partial q_i} - \dfrac{\partial L}{\partial q_i} - \sum_j \dfrac{\partial L}{\partial \dot{q}_j}\dfrac{\partial \dot{q}_j}{\partial q_i} = -\dfrac{\partial L}{\partial q_i} = -\dfrac{dp_i}{dt} = -\dot{p}_i \\ \dfrac{\partial H}{\partial p_i} = \dot{q}_i + \sum_j p_j \dfrac{\partial \dot{q}_j}{\partial p_i} - \sum_j \dfrac{\partial L}{\partial \dot{q}_j}\dfrac{\partial \dot{q}_j}{\partial p_i} = \dot{q}_i \end{cases} \qquad (A.19)$$

$$\begin{cases} \dfrac{dH}{dt} = \sum_j \left(\dfrac{\partial H}{\partial q_j}\dot{q}_j + \dfrac{\partial H}{\partial p_j}\dot{p}_j\right) + \dfrac{\partial H}{\partial t} = \dfrac{\partial H}{\partial t} \\ \dfrac{dH}{dt} = \sum_j \left(\dot{p}_j\dot{q}_j + p_j\ddot{q}_j\right) - \dfrac{\partial L}{\partial t} - \sum_j \left(\dfrac{\partial l}{\partial q_j}\dot{q}_j + \dfrac{\partial l}{\partial \dot{q}_j}\ddot{q}_j\right) = -\dfrac{\partial L}{\partial t} \end{cases} \qquad (A.20)$$

From which we deduce the *Hamilton equations of the system*

$$\begin{cases} \dot{q}_i = \dfrac{\partial H}{\partial p_i} \\ \dot{p}_i = -\dfrac{\partial H}{\partial q_i} \end{cases} \quad (i = 1,2,3,\ \dots\ ,k) \qquad (A.21)$$

as well as the following property

$$\frac{dH}{dt} = \frac{\partial H}{\partial t} = -\frac{\partial L}{\partial t} \qquad (A.22)$$

If the hamiltonian does not explicitly depend on q_i, then the impulse p_i is a constant of the movement. Indeed, following, *(A.21)*

$$\dot{p}_i = -\frac{\partial H}{\partial q_i} = 0 \qquad (A.23)$$

If the Lagrangian does not depend explicitly on time then the hamiltonian will not depend explicitly on time and it is a constant of movement. Indeed, following *(A.22)*

$$\frac{dH}{dt} = \frac{\partial H}{\partial t} = -\frac{\partial L}{\partial t} = 0 \tag{A.24}$$

If the kinetic energy T does not depend explicitly on time (links are independent of time), and as the potential energy V does not depend on velocities $\dot{q}_i$ by hypothesis, then the hamiltonian H represents the mechanical energy of the system. Indeed thanks to relation *(A.13)*, we verify that, if $\partial \vec{\xi}_{(\alpha)} / \partial t = 0$, we have

$$\sum_{i=1}^{k} p_i \dot{q}_i = \sum_{i=1}^{k} \dot{q}_i \frac{\partial L}{\partial \dot{q}_i} = \sum_{i=1}^{k} \dot{q}_i \frac{\partial T}{\partial \dot{q}_i} = 2T \tag{A.25}$$

so that

$$H = 2T - L = 2T - (T - V) = T + V \tag{A.26}$$

Appendix B

Elements of quantum mechanics

Quantum Mechanics is a formalism that allows us to calculate the behavior of microscopic physical systems.

B.1 - Wave function and Schrödinger equation

The wave function of a system

In quantum mechanics, it is the *complex wave function* $\Psi(\vec{\xi},t)$ which describes a system with one or more particles. One of the possible interpretations of the wave function is to assign to the square of the wave function a probability of presence of the particle in space. The probability P of finding the particles in a certain volume V is thus equal to the following expression

$$P = \iiint_V \Psi(\vec{\xi},t)\ \Psi^*(\vec{\xi},t)\ dv \qquad (B.1)$$

where $\Psi^*(\vec{\xi},t)$ represents the complex conjugate of the wave function $\Psi(\vec{\xi},t)$.
Any real observable A on this system of particles will be associated with an operator $\hat{A}$ acting on the wave function such that

$$\hat{A}\psi_n = a_n \psi_n \qquad (B.2)$$

where the a_n are the *eigenvalues* and the ψ_n are the eigenstates of the system. The eigenstates ψ_n form *a complete orthogonal set*, meaning that, on the total volume V_t of the system ,we have

$$\iiint_{V_t} \psi_n\ \psi_m^*\ dv = 1 \quad (n = m) \quad \text{and} \quad \iiint_{V_t} \psi_n\ \psi_m^*\ dv = 0 \quad (n \neq m) \qquad (B.3)$$

The complete wave function Ψ of the system can always be represented as a linear superposition of eigenstates ψ_n, weighted by the coefficients c_n

$$\Psi = \sum_n c_n \psi_n \qquad (B.4)$$

To the components of the momentum (quantity of movement) $\vec{p}$ and the position $\vec{\xi}$ of a particle corresponds the operators

$$\hat{p}_k \quad \rightarrow \quad -i\hbar\frac{\partial}{\partial \xi_k} \quad \text{and} \quad \hat{\xi}_k \quad \rightarrow \quad \xi_k \qquad (B.5)$$

where i is the complex number $\sqrt{-1}$ and $h = 2\pi\hbar$ is the *Planck's constant.*
The total energy E of a system at time t, is given by operators

$$\hat{E} \quad \rightarrow \quad i\hbar\frac{\partial}{\partial t} \quad \text{and} \quad \hat{t} \quad \rightarrow \quad t \qquad (B.6)$$

The operators represent the act of measuring, which do not necessarily commute, which corresponds to the fact that the order in which the operations are done plays a role in Quantum Mechanics. We define the *commutator* of observables A and B as

$$\left[\hat{A},\hat{B}\right]=\hat{A}\bullet\hat{B}-\hat{B}\bullet\hat{A} \qquad (B.7)$$

For example, the operators of momentum and position of a particle, as well as those of energy and time do not commute

$$\left[\hat{p}_k,\hat{\xi}_k\right]=\hat{p}_k\bullet\hat{\xi}_k-\hat{\xi}_k\bullet\hat{p}_k=-i\hbar \quad \text{and} \quad \left[\hat{E},\hat{t}\right]=\hat{E}\bullet\hat{t}-\hat{t}\bullet\hat{E}=i\hbar \qquad (B.8)$$

This is the expression of the *uncertainty principle of Heisenberg*, which states that the measurement of a pair of observables change one another, so that the variations in measure $\Delta\xi_k$, Δp_k, Δt, ΔE, ... are linked to each other by

$$\Delta\xi_k\bullet\Delta p_k\geq\hbar \quad \text{or} \quad \Delta t\bullet\Delta E\geq\hbar \qquad (B.9)$$

The Schrödinger equation of a system

The relations between quantum operators are the same as those intervening between classical observables. For a unique particle submitted to a potential V the classic expression of the hamiltonian is written

$$H=T+V=\frac{\vec{p}^{\,2}}{2m}+V(\vec{\xi}) \qquad (B.10)$$

Because the hamiltonian of the particle represents its total energy, by replacing the observables by the operators associated to it, we obtain the *Schrödinger equation* of the particle

$$i\hbar\frac{\partial\Psi}{\partial t}=\left(-\frac{\hbar^2\Delta}{2m}+V(\vec{\xi})\right)\Psi \qquad (B.11)$$

B.2 - The standing eigenstates of a particle

For a stationary movement, which would not depend on time, the hamiltonian H of a particle represents the constant energy E of the particle, and the Schrödinger equation which is independent of time is symbolically written

$$\hat{H}\Psi=E\Psi \qquad (B.12)$$

so that the eigenstates ψ_n of the wave function are deduced from the following equation

$$\left(-\frac{\hbar^2\Delta}{2m}+V(\vec{\xi})\right)\psi_n=\varepsilon_n\psi_n \qquad (B.13)$$

in which the ε_n are the *eigenvalues of the energy*, and thus the quantified energy levels of the particle. As an example, we will give some solutions to this stationary equation for various types of potentials.

Harmonic oscillator

Let's consider a particle under potential $V(\xi)=k\xi^2/2$ of a *harmonic oscillator*. This particle satisfies the classic hamiltonian

$$H = \frac{\vec{p}^2}{2m} + 2\pi^2 m\nu^2\xi^2 \quad \text{with} \quad \nu = \frac{1}{2\pi}\sqrt{\frac{k}{m}} \tag{B.14}$$

where k is a callback constant, m is the mass of the particle and ν the proper frequency of the oscillator.

The resolution of the equation of equation of Schrödinger *(B.13)* leads us to the following eigenvalues of the energy ε_n

$$\varepsilon_n = h\nu(n+1/2) \qquad (n = 0,1,2,3,\ ...\ ) \tag{B.15}$$

Anharmonic oscillator

Consider a particle under potential $V(\xi) = k\xi^2/2 - b\xi^3$ of an *anharmonic oscillator*, with coefficient b satisfying relation $b << k/2$. Such a potential has a maximum at $\xi = k/3b$, worth $k^3/(54b^2)$. If the particle gets a potential energy which is larger than this maximum value, it can escape from the potential.

The resolution of the Schrödinger equation *(B.13)* leads to the eigenvalues ε_n, given by a series expansion of the form

$$\varepsilon_n = h\nu(n+1/2) - h\nu x_e(n+1/2)^2 + h\nu y_e(n+1/2)^3 + ... \qquad (n = 0,1,2,3,\ ...\ ) \tag{B.16}$$

where the non-dimensional constants x_e, y_e and following are small compared to unity, and rapidly decreasing.

Particles in a box

Given a free particle ($V = 0$), submitted to the boundary conditions of a box with sides L_1, L_2, L_3. The resolution of the Schrödinger equation *(B.13)* leads to the following energies ε_{ijk}

$$\varepsilon_{ijk} = \frac{h^2}{8m}\left(\frac{i^2}{L_1^2} + \frac{j^2}{L_2^2} + \frac{k^2}{L_3^2}\right) \qquad (i,j,k = 0,1,2,\ ...) \tag{B.17}$$

The rotation of two linked particles

Consider a pair of point particles, separated by a distance r, submitted to *a rotation around their center of mass.* The classic hamiltonian is written

$$H = \frac{1}{2I}\left(p_\theta^2 + \frac{1}{\sin^2\theta}p_\varphi^2\right) \quad \text{with} \quad I = \frac{m_1 m_2}{m_1 + m_2} r^2 \tag{B.18}$$

where I is the inertial moment of the two particles of mass m_1 and m_2.

The resolution of the Schrödinger equation *(B.13)* leads to the eigenvalues ε_j

$$\varepsilon_j = \frac{h^2}{8\pi^2 I} j(j+1) \qquad (j = 0,1,2,\ ...) \tag{B.19}$$

For each value of energy ε_j corresponding classically to a different angular velocity, there are $2j+1$ eigenstates, which correspond classically to different orientations of the axis of rotation.

We say that the energy state ε_j is $2j+1$ times *degenerated.*

The particle in a central potential

Consider a particle with mass m under a central potential of type

$$V(\vec{\xi}) = -\frac{e^2}{|\vec{\xi}|} \tag{B.20}$$

where e is a constant.

The resolution of Schrödinger's equation *(B.13)* leads to following energy eigenvalues ε_n

$$\varepsilon_n = -\frac{e^4 m}{2\hbar^2}\frac{1}{n^2} \qquad (n = 1,2,3,\ \dots) \tag{B.21}$$

where n is a whole number called the *principal quantum number,* which characterizes the energetic level of the particle. The detailed resolution of this problem in polar coordinates r,θ,φ leads us to write the eigenstates as a product of a function of r and a function of θ,φ

$$\psi_{n,l,m_l} = R_{n,l}(r)Y_{l,m_l}(\theta,\varphi) \qquad (0 \le l \le n-1) \qquad \text{and} \qquad (-l \le m_l \le +l) \tag{B.22}$$

These two functions are parametrized by the quantum numbers n,l,m_l. The number l is an integer between 0 and $n-1$, called the *azimuthal quantum number,* which characterizes the quantification of the angular kinetic momentum $\vec{L}$ of the particle, the norm of which is worth $|\vec{L}| = \hbar\sqrt{l(l+1)}$, while m_l is the *magnetic quantum number,* worth $-l \le m_l \le +l$, and which characterizes the quantification of the projection of the angular kinetic momentum along an axis z, which is worth $L_z = \hbar m_l$. Indeed for each value of energy ε_n corresponds

$$\sum_{l=0}^{n-1}(2l+1) = n^2 \tag{B.23}$$

different eigenstates of the wave function. We say that the energy state ε_n is n^2 degenerated. As a convention, to the quantum azimuthal number we associate the letters $s,p,d,f,g,\dots$ depending on whether it is worth 0, 1, 2, 3, 4, ... and the different states of the particle are designed by the symbols $1s,2s,2p,3s,3p,3d,\dots$ etc., in which the number and the letter represent the principal quantum number and the azimuthal quantum number respectively.

The density of states in phase space

Let's consider the case of a particle trapped in a box with volume $V = L_1 \cdot L_2 \cdot L_3$. We have seen that the limit conditions imposed on Schrödinger's equation *(B.13)* imply a discrete distribution of energies ε_{ijk} *(B.17).* It is then interesting to calculate the phase volume Γ occupied by each energetic state of the particle. Let's start by counting the number N of states for which energy is inferior or equal to a given value E. For that, we calculate in 3D a number of points, separated by $1/L_1, 1/L_2, 1/L_3$ along axis $O\xi_1, O\xi_2, O\xi_3$, contained in the first octant of a sphere with radius R

$$N = \frac{1}{8}\frac{4\pi R^3}{3}\frac{1}{\frac{1}{L_1}\bullet\frac{1}{L_2}\bullet\frac{1}{L_3}} = \frac{\pi R^3 V}{6} \tag{B.24}$$

in which the radius R is calculated from expression *(B.17)* of the energy levels, for a maximum energy $\varepsilon_{ijk} = E$

$$R = \sqrt{\frac{i^2}{L_1^2} + \frac{j^2}{L_2^2} + \frac{k^2}{L_3^2}} = \sqrt{\frac{8mE}{h^2}} \tag{B.25}$$

With regards to the phase space volume V_Γ occupied by a particle whose energy is inferior or equal to the value E, and as a consequence for which the momentum is inferior or equal to $|\vec{p}| = \sqrt{2mE}$, it is given by

$$V_\Gamma = \iiint \iiint dp_1 dp_2 dp_3 d\xi_1 d\xi_2 d\xi_3 = \frac{4\pi|\vec{p}|^3}{3} V = \frac{4\pi(2mE)^{3/2}}{3} \underbrace{\frac{6Nh^3}{8\pi(2mE)^{3/2}}}_{V} = Nh^3 \tag{B.26}$$

We deduce that the phase space volume v_Γ occupied by each energetic state is given by V_Γ / N and is worth

$$v_\Gamma = V_\Gamma / N = h^3 \tag{B.27}$$

This result, namely that the phase space volume v_Γ occupied by each eigenstate of a particle possessing z degrees of liberty is worth h^z, is rather generic in quantum mechanics and is an important concept in statistical physics.

B.3 - On bosons and fermions

Let's consider a system with 2 independent particles contained in a box. The energy E of that system will be given by

$$E = \varepsilon_n + \varepsilon_m \tag{B.28}$$

where ε_n and ε_m represent the discrete energies of the two particles, given by relation *(B.17)*, and where n and m represent two particular combinations of three quantum numbers i, j, k.
The Schrödinger equation of that system is written

$$\hat{H}\Psi = E\Psi \quad \Rightarrow \quad -\frac{\hbar^2}{2m}(\Delta_1 + \Delta_2)\Psi = (\varepsilon_n + \varepsilon_m)\Psi \tag{B.29}$$

It is easy to verify that a solution to this equation is given by the product of wave functions of particles 1 and 2, by supposing that particle 1 has an energy $\varepsilon_n^{(1)}$ and that particle 2 has an energy $\varepsilon_m^{(2)}$

$$\Psi = \psi_n(\vec{\xi}_1)\psi_m(\vec{\xi}_2) \tag{B.30}$$

But for a same value of energy E of the system, there exists an other solution, which corresponds to the exchange of two identical particles, namely that particle 1 has energy $\varepsilon_m^{(1)}$ and that particle 2 has energy $\varepsilon_n^{(2)}$

$$\Psi = \psi_n(\vec{\xi}_2)\psi_m(\vec{\xi}_1) \tag{B.31}$$

However, one of the fundamental properties of linear homogeneous differential equations is that any linear combination of particular solutions is also a solution, so that the more general solution to the Schrödinger equation can be written as

$$\Psi = \alpha\psi_n(\vec{\xi}_1)\psi_m(\vec{\xi}_2) + \beta\psi_n(\vec{\xi}_2)\psi_m(\vec{\xi}_1) \tag{B.32}$$

This expression would indicate that there exists a vast number of stationary states for a system with 2 particles. However, we must take into account the fact that due to the uncertainty principle in quantum mechanics, identical particles lose their individuality. We say that the quantum particles are non discernible, which means that it is not possible to follow the trajectory of a given particle over time. If we consider the wave function Ψ of the system, we know that Ψ^2 determines the probability to find two particles in a certain portion of space. If we exchange the two particles, it is clear that Ψ^2 must remain unchanged. However the phase of Ψ can be modified by this exchange so that $\Psi \to \Psi e^{i\eta}$. If we proceed to a second exchange of particles, we have that $\Psi \to \Psi e^{2i\eta}$ and we find ourselves in the initial state Ψ, if $e^{2i\eta} = 1$. To satisfy this last condition, it is sufficient that $e^{2i\eta} = +1$ or $e^{2i\eta} = -1$.
In the case where the wave function Ψ transforms as $\Psi \to +\Psi$ during the exchange of two particles, the wave function is said to be *symmetric*, and the particles are called *bosons*. The wave function Ψ is written with a normalization factor α

$$\Psi = \alpha\left(\psi_n(\vec{\xi}_1)\psi_m(\vec{\xi}_2) + \psi_n(\vec{\xi}_2)\psi_m(\vec{\xi}_1)\right) \tag{B.33}$$

If the wave function Ψ transforms as $\Psi \to -\Psi$, the wave function is said to be *antisymmetric*, and the particles are called *fermions*. The wave function Ψ is written with a normalization factor β as

$$\Psi = \beta\left(\psi_n(\vec{\xi}_1)\psi_m(\vec{\xi}_2) - \psi_n(\vec{\xi}_2)\psi_m(\vec{\xi}_1)\right) \tag{B.34}$$

The non-discernibility of two particles is then very clearly seen in the two previous expressions for the antisymmetric function, it is not possible for two particles to be in the same state since the quantum function Ψ would be null: that is the mathematical expression of the Pauli exclusion principle, which states that two fermions cannot occupy the same state simultaneously.
In the case of a system with N identical particles, the previous concepts are easily generalized. In the case of bosons, the symmetric wave function Ψ_{sym} of the system is written under the form

$$\Psi_{sym} = \alpha\sum_P \psi_n(\vec{\xi}_1)\psi_m(\vec{\xi}_2)\,\ldots\,\psi_k(\vec{\xi}_N) \tag{B.35}$$

where the sum is on all the possible permutations of the system. If the system possesses n_1 particles in energy n, n_2 particles in energy m, n_3 particles in energy k, etc., the number of terms that correspond to the wave function Ψ_{sym} is given by

$$P = \frac{N!}{n_1!n_2!n_3!\,\ldots} \tag{B.36}$$

In the case of fermions, the antisymmetric wave function Ψ_{anti} of the system can be written under the form of a determinant

$$\Psi_{anti} = \beta\begin{vmatrix} \psi_n(\vec{\xi}_1) & \psi_n(\vec{\xi}_2) & \ldots \\ \psi_m(\vec{\xi}_1) & \psi_m(\vec{\xi}_2) & \ldots \\ \ldots & \ldots & \ldots \end{vmatrix} \tag{B.37}$$

Indeed the permutation of two columns of a determinant changes the sign of the determinant, which assures the anti-symmetry of the wave function Ψ_{anti} under the exchange of two particles. Furthermore, we also know that the determinant is null if two lines are identical, which corresponds here to the expression of the Pauli exclusion principle, namely that a quantum state cannot be occupied by more than one fermion.

B.4 - Particle spin

In general, the wave function Ψ of a system is composed of two parts: *a spatial component and a spin component.* The notion of *spin* is a purely quantum characteristic of particles, associated with the classical notion of angular kinetic momentum of the particle. If the angular momentum of the spin of a particle is $\vec{S}$ and its angular orbital momentum is $\vec{L}$, the totally angular momentum $\vec{J}$ of the particle will correspond to the vectorial sum of the two: $\vec{J} = \vec{S} + \vec{L}$.
The norm of $\vec{S}$ is quantized as $\hbar\sqrt{s(s+1)}$ and the projection of $\vec{S}$ on axis z is worth $S_z = \hbar m_z$, where s is the quantum number of spin corresponding to an integer in the case of boson or a half integer in the case of fermions and m_z is a quantum number that can span $-s$ and $+s$ in integer steps. We can both recognize that the norm of $\vec{S}$ is one of these projections, for example S_z. The azimuthal angle is always undetermined as the operators S_x and S_y do not commute with the operator associated to S_z.
The spin component of the wave function Ψ of the two particles is symmetric when the spin of the two particles are anti-parallel. We have the following possibilities for the wave function Ψ of the two particles:
• *Fermions:* the wave function is antisymmetric $\Rightarrow$ parallel spins and anti-symmetric spatial component, or anti-parallel spins and spatial symmetric components.
• *Bosons:* symmetric wave function $\Rightarrow$ parallel spins and symmetric spatial components, or anti-parallel spins and antisymmetric spatial components.

B.5 - The Dirac equation

The Dirac equation, formulated in 1928, is born out of an attempt to incorporate special relativity to a quantum formulation in the framework of relativistic quantum mechanics of elementary particles with half integer spins, such as electrons. For that, Dirac sought to transform the Schrödinger equation in order to render it invariant under the Lorentz transform, namely to render it compatible with the principles of special relativity.
To insure the compatibility of a quantum wave equation with the Lorentz transform, it is necessary that the derivations of time and space contribute in a symmetric fashion in the wave equation, which is not the case of the Schrödinger equation. One way to go about it is to use the relativistic relation

$$E^2 / c^2 - \vec{p}^2 = m^2 c^2 \quad (B.38)$$

and use the operators defined in *(B.5)* and *(B.6)* to obtain a wave equation

$$\Delta\psi - \frac{1}{c^2}\frac{\partial^2\psi}{\partial t^2} = \frac{m^2c^2}{\hbar^2}\psi \quad (B.39)$$

This equation is symmetric in the derivations of time and space, but it is a second order equation, which makes it resolution hard as we must then specify the initial conditions of the function and it's derivatives to find a solution!

Dirac then has the idea of writing relation *(B.39)* under the following form

$$\Delta-\frac{1}{c^2}\frac{\partial^2}{\partial t^2}=\left(A\frac{\partial}{\partial x_1}+B\frac{\partial}{\partial x_2}+C\frac{\partial}{\partial x_3}+\frac{i}{c}D\frac{\partial}{\partial t}\right)\left(A\frac{\partial}{\partial x_1}+B\frac{\partial}{\partial x_2}+C\frac{\partial}{\partial x_3}+\frac{i}{c}D\frac{\partial}{\partial t}\right) \qquad (B.40)$$

and to look for the values of A,B,C,D to have finishing cross terms in $\partial x_i \partial x_j$. For that to be the case, one must introduce 4x4 matrices which satisfy a series of relations such as $AB+BA=0,...$ and $A^2=B^2=...=1$.

This factorization allows us to write

$$\left(A\frac{\partial}{\partial x_1}+B\frac{\partial}{\partial x_2}+C\frac{\partial}{\partial x_3}+\frac{i}{c}D\frac{\partial}{\partial t}\right)\psi=\frac{mc}{\hbar}\psi \qquad (B.41)$$

where $\psi(\vec{r},t)$ is now a wave function with ***four components***.

The formulation given by Dirac to this equation is the following

$$i\hbar\frac{\partial\psi}{\partial t}(\vec{r},t)=\left(\alpha_0 mc^2-i\hbar c\sum_{j=1}^{3}\alpha_j\frac{\partial}{\partial x_j}\right)\psi(\vec{r},t) \qquad (B.42)$$

where m is the mass of the particle, c is the speed of light, $\hbar$ the reduced Planck constant, x_i and t the coordinates of space and time, and $\psi(\vec{r},t)$ the wave function with four components. The fact that the wave function is now formulated in terms of a spinor with four components rather than a simple scalar is due to the demands placed by special relativity. In this formulation, the matrices A,B,C,D become Dirac matrices α_i, with $i=1,...,4$, of dimension 4x4, which act on the spinor $\psi(\vec{r},t)$. We can write these matrices in the Dirac representation, in terms of the Pauli matrices, under the form

$$\alpha_0=\begin{pmatrix}1&0\\0&-1\end{pmatrix}\ ;\ \vec{\alpha}_i=\begin{pmatrix}0&\vec{\sigma}_i\\\vec{\sigma}_i&0\end{pmatrix} \qquad (B.43)$$

in which the complex 2x2 Pauli matrices are written

$$\vec{\sigma}_1=\vec{\sigma}_x=\begin{pmatrix}0&1\\1&0\end{pmatrix}\ ;\ \vec{\sigma}_2=\vec{\sigma}_y=\begin{pmatrix}0&-i\\i&0\end{pmatrix}\ ;\ \vec{\sigma}_3=\vec{\sigma}_z=\begin{pmatrix}1&0\\0&-1\end{pmatrix} \qquad (B.44)$$

It is common in quantum mechanics to consider the operator of the momentum $\hat{\vec{p}}$ and in this case the equation of Dirac is written in condensed form as

$$\hat{\vec{p}}=-\sum_j\vec{e}_j i\hbar\frac{\partial}{\partial x_j}\ \Rightarrow\ i\hbar\frac{\partial\psi}{\partial t}(\vec{r},t)=\left(mc^2\alpha_0+c\vec{\alpha}\bullet\hat{\vec{p}}\right)\psi(\vec{r},t) \qquad (B.45)$$

The Dirac equation *(B.42)* takes into account quite naturally the notion of spin and allows us to predict the existence of antiparticles. Indeed, besides the solution corresponding to the electron, there exist another solution corresponding to a particle with negative energy and opposed charge, the positron, which was identified later in 1932, by Carl Anderson by using a fog chamber.

Appendix C

Standard Model of elementary particles

As we speak, particle physics explains the intimate structure of matter thanks to a model called the Standard Model. This model uses *fermions*, particles of matter that present two rather different families, the family of the *leptons* and the family of the *quarks*. It also has 3 types of interaction between these fermions: the electromagnetic interaction, the weak interaction and the strong interaction.

The interactions between *fermions* of matter happen in the exchange of particles called *gauge bosons*, corresponding to the quantas of the quantum fields of interaction. The electromagnetic interaction uses the *photon* γ, the weak interaction uses 3 *gauge bosons* Z^0, W^+ and W^-, and the strong interaction uses 8 gauge bosons which are called *gluons*.

With respect to the mass of particle, it is introduced in the standard model by a new interaction associated with the Higgs quantum field, whose interaction particle is called the *Higgs boson*.

The Standard Model of particles, in spite of its success, leaves many unresolved questions among which the problem of the gravitational interaction which is not present. We will also list a series of unresolved problems in the Standard Model.

C.1 – On leptons and quarks

The family of *leptons* (fig. C.1) is composed of three generations and two types of particles: three particles that are electrically neutral called *electronic neutrino* (ν_e), *muonic neutrino* (ν_μ) and *tau neutrino* (ν_τ), and three particles that are electrically charged, called *electron* (e^-), *muon* (μ^-) and *tau* (τ^-). Each of these 6 particles possesses in principle an anti-particle ($\overline{\nu}_e$, $\overline{\nu}_\mu$, $\overline{\nu}_\tau$, e^+, μ^+ et τ^+) which is essentially characterized by an opposed electrical charge, which already begs the question about the existence of anti-particles for neutrinos! The leptons are quasi-punctual particles which are sensible to the electromagnetic interaction and the weak interaction, but not to the strong interaction. It was thought for a long time that neutrinos do not posses mass but recent measurements show differently.

The family of *quarks* (fig. C.1) is also composed of 3 generations with two types of particles electrically charged: a first generation is formed of quarks *down (d)* and *up (u)*, respectively with electrical charges -1/3 and +2/3 of the electrical charge of an electron, a second generation composed of quarks *strange (s)* and *charm (c)*, respectively with electrical charges -1/3 et +2/3 of the electrical charge of an electron, and a third generation composed of *bottom (b)* and *up (t)*, respectively with electrical charges -1/3 and +2/3 of the electrical charge of the electron. Each quark possesses its anti-particle with an opposed electrical charge ($\overline{d}$, $\overline{u}$, $\overline{s}$, $\overline{c}$, $\overline{b}$ et $\overline{t}$). The quarks are sensible to the electromagnetic interaction, to the weak interaction and to the strong interaction. The quarks are not free particles, rather they exist in the form of assemblies of

quarks called *hadrons.* The quarks are linked within the hadrons by the strong force.
The hadrons are under two forms: the *mesons* composed of a quark and an anti-quark, such as particles $\Pi^+(u\bar{d})$, $\Pi^-(d\bar{u})$ or $\Pi^0(u\bar{u})$, and the *baryons* composed of 3 quarks, such as the *proton* (uud) and the *neutron* (ddu), or 3 anti-quarks.
Each particle of matter, whether it be a lepton or a quark, possesses a non-null mass and a spin spin 1/2, which gives it the status of a *fermion.*

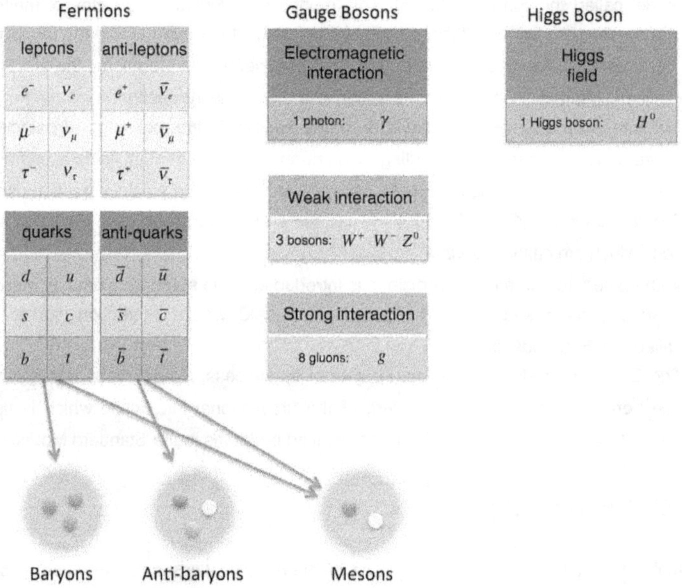

***Figure C.1** - The particles of the Standard Model*

C.2 - On the fundamental interactions and the gauge bosons

In the Standard Model, we consider the following 3 possible interactions between particles: *electromagnetic interaction, weak interaction and strong interaction.* These interactions are described by the theories of the quantum field (except the gravitational interaction which can never be introduced in the Standard Model, despite intense research for the *graviton* gauge boson which would be associated to it). Each interaction calls upon a field which is dedicated to it, and is done via the exchange of a particle called a *gauge boson* (fig. C.1), which corresponds to the quantum of the field considered. The electromagnetic interaction calls upon the *photon* (γ *),* a gauge boson with null mass. The weak interaction calls upon the 3 gauge bosons Z^0, W^+ and W^-, particles with non-null mass, electrically neutral, positive and negative respectively. With regards to the strong interaction, it involves 8 gauge bosons called *gluons,* which are actually null mass particles. All the gauge bosons associated with these interactions are spin 1 particles, which explains their *bosons* name.

C3 - On the electromagnetic interaction and the quantum electrodynamics

The quantum theory which describe the electromagnetic interaction is called *quantum electrodynamics*. It is a quantization of the electro-magnetic field: charged particles interact with the exchange of field quanta, the *photons*. It is a relativistic theory, as it takes into account the propagation time of the interaction, namely the velocity of the vector boson, the photon. In that theory, we can represent an interaction in a very simple and expedient way, thanks to *Feynman diagrams*. In figure C.2, we have represented the example of interactions between two electrons by the exchange of a *virtual photon*, which we qualify here as "virtual", as it cannot be detected experimentally.

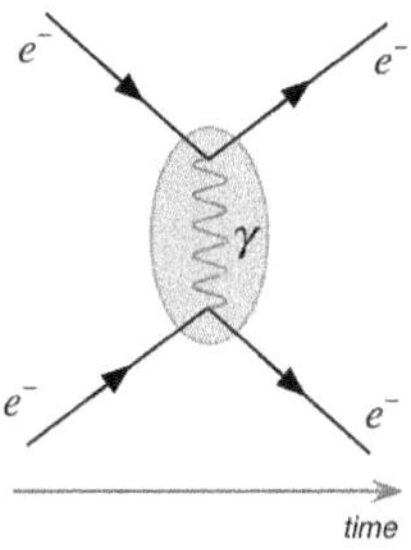

Figure C.2 *- Feynman diagram of interaction between two electrons and a photon*

C4 - On the weak interaction and the electro-weak theory

The weak interaction acts on all the elementary fermions, be they leptons or quarks. It is actually the only interaction acting on neutrinos. It is responsible for nuclear decay. This interaction has two aspects: the weak interaction via *charged currents*, whose vectors are the gauge bosons W^+ and W^-, and the weak interaction by *neutral current*, whose vector is the gauge boson Z^0. The gauge bosons of the weak interaction are the only ones which present non null masses, and in fact very elevated masses, which imposes, when combined with the uncertainty relation of Heisenberg $\Delta E \bullet \Delta t \approx \hbar$ and the relation of Einstein $\Delta E = mc^2$, a lifetime Δt which is very short and, as a consequence, since the velocity of light is a limit, a reach of the interaction which is small (on the order of $10^{-15}[m]$), which explains that this interaction only manifest at the scale of the atomic core.

As the gauge bosons W^+ and W^- have a non-null electrical charge. The fermions can change their electrical charge during an interaction via the exchange of W^+ or W^-, which changes their *flavor* (we call *fermionic flavor* the nature of the fermion: electron, neutrino, quark *u*, quark *d*, etc.). For example, beta radio-activity is explained by the emission of a W^- by a quark *d* of the neutron, which changes then its flavor and becomes a quark u. Then, the W^- materializes in the form of an electron and an electronic anti-neutrino (figure C.1,b).

The gauge boson Z^0 does not possess an electrical charge, and cannot induce a change in flavor during a weak interaction. The weak interaction by neutral current is similar to the exchange of a photon. Two fermions that can exchange a photon can also exchange a Z^0, with the exception of a neutrino which can exchange a Z^0 but cannot exchange a photon as it is a neutral particle.

There exists many types of weak interaction depending on the fermions that interact: the leptonic interactions, the semi-leptonic interactions and the hadronic interactions, for which examples of Feynman diagrams are reported in figure C.3.

The electromagnetic interaction and the weak l'interaction were unified in a quantum theory called the *electro-weak theory.*

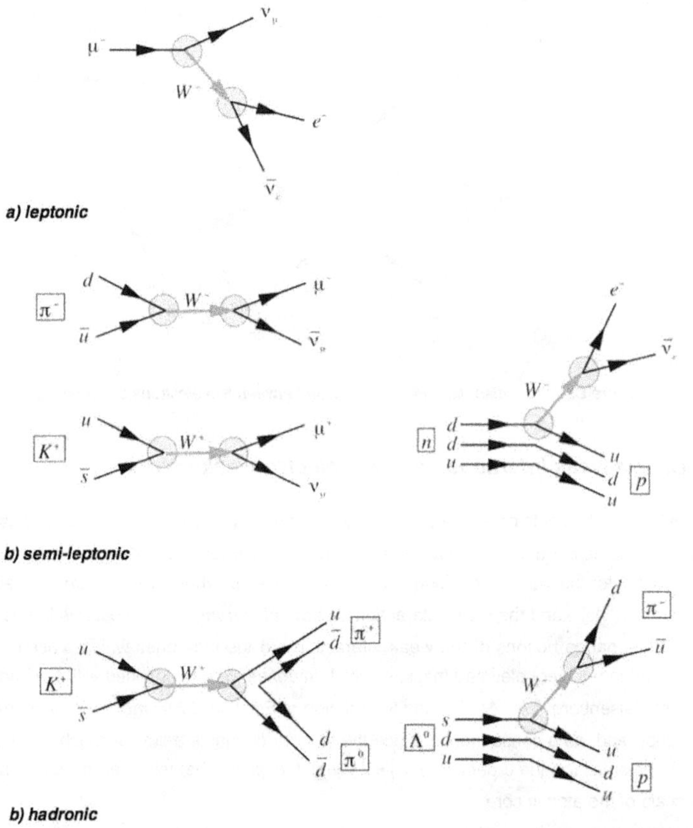

Figure C.3 - *examples of Feynman diagrams*
for the weak leptonic, semi-leptonic and hadronic interactions

C5 - On the strong interaction and the quantum chromodynamics

The strong interaction is a short interaction between quarks by the intermediary of gluons, the vector gauge bosons of this interaction. This interaction allows us to explain not only the *mesons* composed of a quark and an anti-quark and the *baryons* composed of 3 quarks, but also how neutrons and protons link up to form atomic nuclei.

To elaborate a quantum theory of the strong interaction, it was necessary to use a new type of charge, called the *color charge*, hence the name *Quantum Chromodynamic Theory.* Each quark possesses a color charge, red (R), green (G) or blue (B), and the anti-quarks possess a charge with the complementary colors ($\overline{R}$), ($\overline{G}$) or ($\overline{B}$).

The strong interaction is then explained by the exchange of colored gluons between elementary fermions with a color charge, which allows us to exchange the color charges between fermions. There exist 8 gluons with different colors, corresponding to the 8 different combinations of a color and an anti-color. Thus, during the exchange of a gluon between 2 quarks, they can interact which is not the case for other gauge bosons! The leptons which do not possess a color charge, are not subject to the strong force.

While the mass of gluons is null, the strong interaction is of very short range, on the order of $10^{-15}[m]$, and it possesses a characteristic that is strange: it gets stronger the further away the quarks are! And if they are infinitely close they do not interact anymore. This property bears the name of *asymptotic freedom*, and it is responsible for *quark confinement* inside hadrons: this implies that *quarks cannot exist by themselves*!

The particles formed of quarks are hadrons, meaning linked states with many quarks via gluons. The hadrons are 'white' meaning have a color combination which is null. We can then consider

- *the baryons*, combinations of three quarks respectively red, green and blue, or *anti-baryons*, combinations of three anti-quarks respectively anti-red, anti-green and anti-blue. The triplets formed of three quarks (picked in the set *u, d, s* or *c*) are represented by the diagrams in figure C.4, in the cases where the global spin is $1/2$ and $3/2$, with the name given to the particle corresponding to this triplet,
- *the mesons*, which contain a color quark (red, green and blue) and an anti-quark with the corresponding anti-color (anti-red, anti-green or anti-blue). The doublets formed of a quark and an anti-quark (from quarks *u, d, s* or *c*) are represented in the diagram of figure C.5, in the case of a global spin 1 and 0, with the name given to the particle corresponding to the doublet.

C6 - On the mass of the particles and the Higgs boson

In a first version of the Standard Model, all the particles (leptons and quarks) we described had no mass, which could not be right. To patch that fact, the theoreticians imagined a fifth interaction, different from all four others (electromagnetic, weak, strong and gravity), by invoking a field whose quantum is a particle of spin 0: The *Higgs boson* H^0 (fig. 29.1). It is then the interaction between elementary fermions with null mass and the Higgs field via Higgs bosons which confer a mass to the fermions of the Standard Model! The existence of the Higgs was recently experimentally verified at CERN.

C7 - On the issues and open questions of the Standard Model

Despite the success of the Standard Model, there are many unresolved questions in this model,

amongst which we can cite quickly the following:
- the absence of gravitational interaction in this theory,
- the existence of 12 fermions (leptons and quarks) and 4 fundamental forces in the theory,
- the existence of three generations of fermions (leptons and quarks), which allows us to account for the violation of the ***CP*** invariance (charge/parity), which we estimate is the cause of the asymmetry between matter and anti-matter in the actual universe,
- the prediction of the masses of the various fermions and bosons,

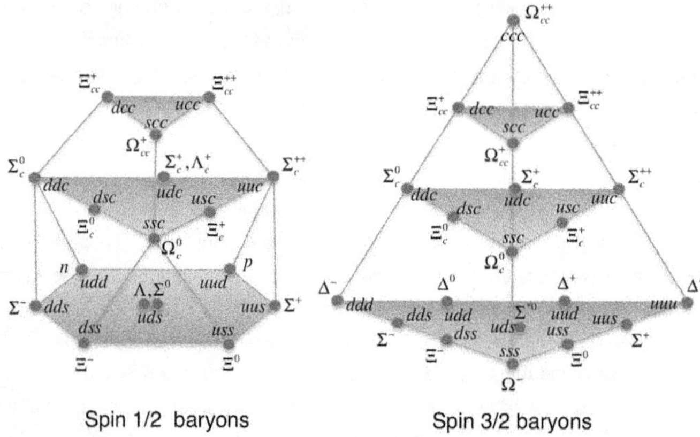

Figure C.4 *- the baryons composed of triplets of quarks (u, d, s or c)*

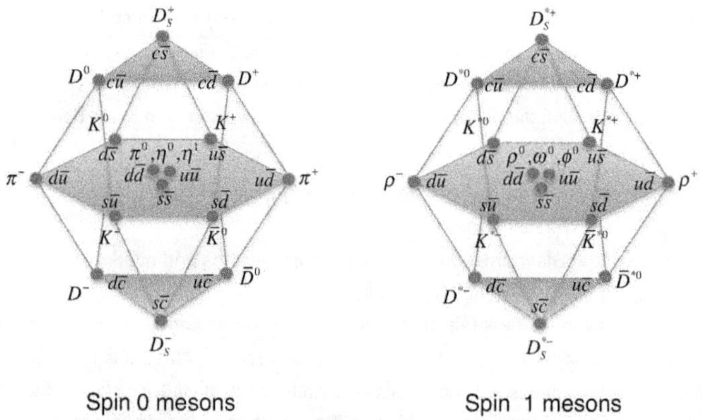

Figure C.5 *- the mesons composed of doublets of quark-antiquark (u, d, s or c)*

- the necessity to adjust experimentally a large number of parameters (26 parameters are needed in the case where the neutrinos have a mass!) to obtain a functional theory, such as the masses of particle (table C.1) and the intensities of the various forces (tableau C.2), which must necessarily be set by the experimental results,

- the existence of 3 massive gauge bosons in the case of the weak interaction,
- the confinement due to the strong force,
- the quantized electrical values of 1, 1/3 and 2/3,
- the absence of an explanation for dark energy and dark matter in that model.

Fermions ***(spin 1/2)***		**Electrical charge (q_e)**	**Mass (MeV/c^2)**
lepton electron	e^-	-1	0,51
lepton muon	μ^-	-1	106
lepton tau	τ^-	-1	1777
lepton electron neutrino	ν_e	0	< 0,0000022
lepton muon neutrino	ν_μ	0	< 0,17
lepton tau neutrino	ν_τ	0	< 15,5
quark Up	u	+2/3	2,55
quark Down	d	-1/3	5,04
quark Charm	c	+2/3	1270
quark Strange	s	-1/3	105
quark Top	t	+2/3	173100
quark Bottom	b	-1/3	4200

Table C.1 *- the charges and masses of the leptons and the quarks.*

Gauge bosons ***(spin 1)***		**Electrical charge (q_e)**	**Mass (MeV/c^2)**
photon	γ	0	0
boson intermediary	W^+	-1	80398
boson intermediary	W^-	+1	80398
neutral boson intermediary	Z^0	0	90187
8 gluons	g	0	0

Bosons de jauge ***(spin 0)***		**Charge électrique (q_e)**	**Masse (MeV/c^2)**
boson de Higgs	H^0	0	125000

Table C.2 *- the charges and masses of the gauge bosons*

Appendix D

Conjectures of the cosmic lattice

We will list here the different axioms that were postulated *à priori* in our theory of the cosmic lattice and that are chosen to find *à posteriori* strong analogies with all the other theories of physics.

Conjecture 0: *the free energy of the "cosmic lattice" per unit volume of lattice is*

$$F^{déf} = F^{déf}\left[\tau, \tau^2, (\vec{\alpha}_i^{él})^2, (\vec{\omega}^{él})^2, (\vec{\alpha}_i^{an})^2, (\vec{\omega}^{an})^2\right] \quad (13.5)$$

Conjecture 1: *for the cosmic lattice to present analogies with Einstein Gravitation, with electromagnetism and with the photons of quantum mechanics, the following must hold:*

- $\exists$ *pure transverse waves circularly polarized* $\Leftrightarrow K_2 + K_3 > 0$
- $\nexists$ *longitudinal waves* $\Leftrightarrow \begin{cases} \tau_0 < \tau_{0cr} = \dfrac{K_0}{2K_1} - \dfrac{2K_2}{3K_1} - 1 \quad (K_1 > 0) \\ \tau_0 > \tau_{0cr} = \dfrac{K_0}{2K_1} - \dfrac{2K_2}{3K_1} - 1 \quad (K_1 < 0) \end{cases}$ (14.30)

Conjecture 2: *the usual singularities of the field of expansion must be 'negative' for them to correspond to the usual gravitational field* (15.5)

Conjecture 3: *It seems more 'reasonable' to imagine cosmic lattices with* $K_1 > 0$, *so as to have a finite expansion* (16.19)

Conjecture 4: *the modules* K_2 *and* K_3 *must satisfy the relation* $|K_2| << |K_3|$, *<u>or else</u> the module* K_2 *could be null* $(K_2 = 0)$ (17.41)

Conjecture 5: *modulus* K_1 *satisfies relation* $|K_1| << |K_3|$ (19.52)

Conjecture 6: $K_0 > 0$ (19.56)

Conjectures 0 à 6: *the «perfect cosmic lattice:* $\begin{cases} K_0 = K_3 > 0, \\ 0 < K_1 << K_0 = K_3 \\ 0 \le K_2 << K_3 = K_0 \end{cases}$ (19.58)

Conjecture 7: *the radius of a twist disclination loop is much greater than the cosmic lattice step:*

$$\ln\left(A_{TL} R_{TL} / a\right) >> 1 \quad (19.92)$$

Conjecture 8: *the singularities of vacancy type correspond by analogy to anti-matter and the singularities of interstitial type to matter* (22.115)

Conjecture 9: *the metric of our theory in a weak gravitational field must be the same than the Schwarzschild metric in General Relativity* (24.49)

Conjecture 10: *There exists two operators which, when applied to the wave function $\underline{\psi}(\vec{r},t)$ of a mobile singularity in the lattice, allows us to measure its relativistic energy and its relativistic momentum, namely*

$$\begin{cases} i\hbar\dfrac{\partial}{\partial t}\underline{\psi} \to E_v\underline{\psi} \\ -i\hbar\dfrac{\partial}{\partial x_i}\underline{\psi} \to \mathbf{P}_{vi}\,\underline{\psi} \end{cases} \qquad (28.15)$$

Conjecture 11: K_1 *satisfies the following equation in the cosmic lattice*

$$K_1 \geq K_{1cr} = K_0\,2\pi^4\,R_{TL}^2\,/\,q_{\lambda TL}^2 \qquad (29.9)$$

Conjecture 12: The angle Ω_{TL} takes discrete values linked to the symmetry of the lattice and independent in volume expansion $(\beta_{TL} = cste)$ (31.1)

Conjecture 13: The stacking of planes ***R, G, B*** follows three elementary rules: (31.2)

Rule 1: the *alternation of planes **R**, **G**, **B*** cannot be broken (either by impossibility or by a very large energy associated with a surface stacking fault energy γ_1),

Rule 2: in a given direction of space, there may appear a stacking fault corresponding to a shift in the alternation of planes ***R, G, B***, which possesses a surface stacking fault energy γ_0 *which is not null* (fig. 31.1, image 4).

Rule 3: if a plane with a given color undergoes a rotation by an angle $\pm\pi/2$, $\pm\pi$ or $\pm 3\pi/2$, *it changes color* according to table 31.1, which corresponds to the existence of a given axial property of the lattice.

Appendix E

Mathematical formulas

The mathematical concepts needed for the comprehension of this book are summed up in this appendix, written up as a review of differential calculus. We establish in this annex the formulas for integral derivatives on mobile domains.

E.1 - Vectorial calculus

The scalar product

The *scalar product of* two vectors $\vec{u}$ and $\vec{v}$ gives us a scalar a . In a orthonormal framework, the scalar product is calculated as the sum of the products of the components

$$a = \vec{u}\,\vec{v} = u_1 v_1 + u_2 v_2 + u_3 v_3 = \sum_i u_i v_i \tag{E.1}$$

The *intrinsic properties* of the scalar product are the following

$$\begin{cases} a = \vec{u}\,\vec{v} = |\vec{u}||\vec{v}|\cos\alpha \\ a = \vec{u}\,\vec{v} = \vec{v}\,\vec{u} \\ \vec{u} \perp \vec{v} \;\Rightarrow\; \vec{u}\,\vec{v} = 0 \\ |\vec{u}| = \sqrt{\vec{u}\,\vec{u}} = \sqrt{\vec{u}^2} = L_u \quad (norm) \end{cases} \tag{E.2}$$

The vectorial product

The *vectorial product* of two vectors $\vec{u}$ and $\vec{v}$ gives us a vector $\vec{w}$. In a orthonormal framework, the *vectorial product* of two vectors can be calculated by using the formalism of determinant calculation in the following way

$$\vec{w} = \vec{u} \wedge \vec{v} = \begin{vmatrix} \vec{e}_1 & \vec{e}_2 & \vec{e}_3 \\ u_1 & u_2 & u_3 \\ v_1 & v_2 & v_3 \end{vmatrix} = \sum_i \vec{e}_i \left(u_j v_k - u_k v_j \right) \tag{E.3}$$

The *intrinsic properties* of the vectorial product are the following

$$\begin{cases} |\vec{w}| = |\vec{u} \wedge \vec{v}| = |\vec{u}||\vec{v}|\sin\alpha = S \quad (area) \\ \vec{w} = \vec{u} \wedge \vec{v} = -\vec{v} \wedge \vec{u} \\ \vec{w} \perp \vec{u} \quad et \quad \vec{w} \perp \vec{v} \\ \vec{u} \,||\, \vec{v} \;\Rightarrow\; \vec{u} \wedge \vec{v} = 0 \end{cases} \tag{E.4}$$

The mixed product

The *mixed product* of three vectors $\vec{u}$, $\vec{v}$ and $\vec{w}$ is composed of the scalar product and the vectorial product which gives us a scalar b .

In a orthonormal framework, the *mixed product* of three vectors is equal to the determinant of the components of the three vectors

$$b=\left[\vec{u},\vec{v},\vec{w}\right]=\vec{u}\left(\vec{v}\wedge\vec{w}\right)=\begin{vmatrix}u_1 & u_2 & u_3\\ v_1 & v_2 & v_3\\ w_1 & w_2 & w_3\end{vmatrix}=\sum_i u_i\left(v_j w_k-v_k w_j\right) \tag{E.5}$$

The *intrinsic properties* of the mixed product are the following:

$$\begin{cases} b=\left[\vec{u},\vec{v},\vec{w}\right]=\vec{u}\left(\vec{v}\wedge\vec{w}\right)=V \quad (volume)\\ b=\vec{u}\left(\vec{v}\wedge\vec{w}\right)=\vec{v}\left(\vec{w}\wedge\vec{u}\right)=\vec{w}\left(\vec{u}\wedge\vec{v}\right)\end{cases} \tag{E.6}$$

Compositions of products

The three products of vectors, the scalar product, the vectorial product and the mixed product can be composed together. These compositions verify the following equations

$$\begin{cases} \vec{u}\wedge\left(\vec{v}\wedge\vec{w}\right)=\left(\vec{u}\,\vec{w}\right)\vec{v}-\left(\vec{u}\,\vec{v}\right)\vec{w}\\ \left(\vec{u}\wedge\vec{t}\right)\left(\vec{v}\wedge\vec{w}\right)=\left(\vec{u}\,\vec{v}\right)\left(\vec{t}\,\vec{w}\right)-\left(\vec{u}\,\vec{w}\right)\left(\vec{t}\,\vec{v}\right)\\ \left(\vec{u}\wedge\vec{t}\right)\wedge\left(\vec{v}\wedge\vec{w}\right)=\left[\vec{u},\vec{t},\vec{w}\right]\vec{v}-\left[\vec{u},\vec{t},\vec{v}\right]\vec{w}\\ \left[\vec{u},\vec{v},\vec{w}\right]\vec{t}=\left[\vec{t},\vec{v},\vec{w}\right]\vec{u}+\left[\vec{u},\vec{t},\vec{w}\right]\vec{v}+\left[\vec{u},\vec{v},\vec{t}\right]\vec{w}\end{cases} \tag{E.7}$$

E.2 - Vectorial Analysis

The gradient operator

The *gradient of a scalar field* f is a vectorial field $\vec{u}$ defined in each point A of space by the following limit taken on the surface S limiting a volume V taken around point A (figure E.1)

$$\vec{u}\big|_A=\overline{\mathrm{grad}}\,f\big|_A=\lim_{V\to 0}\frac{1}{V}\oint\!\!\!\oint_S f\,d\vec{S} \tag{E.8}$$

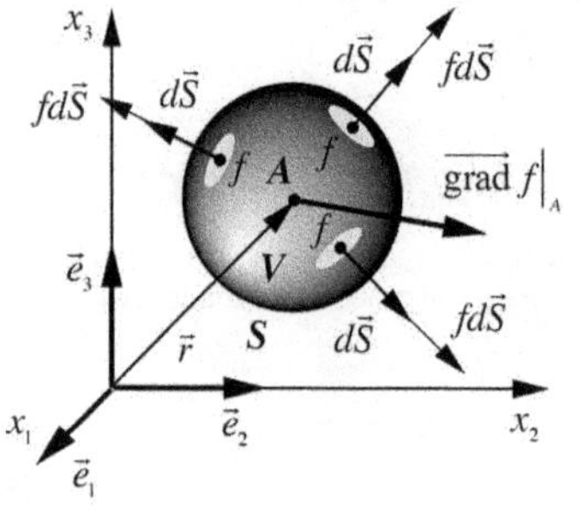

Figure E.1 *- gradient operator*

The direction of vector $\vec{u}$ is consequently perpendicular to the level surfaces of the function f in space and its norm is proportional to the velocity of variation of the function f in that direction. This intrinsic *definition* shows that the gradient is an *invariant* of field f, meaning that it does not depend on the system of coordinates that one chooses. In a orthonormal framework, the gradient can be calculated thanks to the following relation

$$\vec{u}(\vec{r}) = \overrightarrow{\text{grad}}\, f = \frac{\partial f}{\partial x_1}\vec{e}_1 + \frac{\partial f}{\partial x_2}\vec{e}_2 + \frac{\partial f}{\partial x_3}\vec{e}_3 = \sum_i \frac{\partial f}{\partial x_i}\vec{e}_i \tag{E.9}$$

The *intrinsic properties* of the gradient operator are the following:

- the derivative of f in a given direction $\vec{s}$ in space is given by the scalar product

$$\frac{df}{ds} = \vec{s}\,\overrightarrow{\text{grad}} f \tag{E.10}$$

- if $\overrightarrow{\text{grad}}\, f = \overrightarrow{\text{grad}}\, g$, then $f = g + constant$ (E.11)

- if $\vec{u} = \overrightarrow{\text{grad}}\, f$, $\vec{u}$ is called a *conservative vectorial field* and it can be derived from the *scalar potential* f. In this case, if the field $\vec{u}$ is known, the field f can be easily found, by resolution of $df = u_1 dx_1 + u_2 dx_2 + u_3 dx_3$, modulo an additive constant. If $\vec{u} = \overrightarrow{\text{grad}}\, f$, we also have

$$\int_{P_1}^{P_2} \vec{u}\, d\vec{r} = f(\boldsymbol{P}_2) - f(\boldsymbol{P}_1) \tag{E.12}$$

$$\oint_C \vec{u}\, d\vec{r} = 0 \quad ; \quad \forall \; closed\ contour\ \boldsymbol{C} \tag{E.13}$$

- the *gradient theorems*

$$\oiint_S f\, d\vec{S} = \iiint_V \overrightarrow{\text{grad}}\, f\, dV \tag{E.14}$$

$$\oint_C f\, d\vec{r} = -\iint_S \overrightarrow{\text{grad}}\, f \wedge d\vec{S} \tag{E.15}$$

The curl (rotational) operator

The *curl (rotational) of a vectorial field* $\vec{u}$ is another vectorial field $\vec{v}$ which satisfies in all point $\boldsymbol{A}$ of space the following limit taken on a contour $\boldsymbol{C}$ around point $\boldsymbol{A}$ and perpendicular to any direction $\vec{\boldsymbol{n}}$ in space (figure E.2)

$$\vec{\boldsymbol{n}}\,\vec{v}\big|_A = \vec{\boldsymbol{n}}\,\overrightarrow{\text{rot}}\,\vec{u}\big|_A = \lim_{S \to 0} \frac{1}{\boldsymbol{S}} \oint_C \vec{u}\, d\vec{r} \tag{E.16}$$

The direction of vector $\vec{v}$ is consequently perpendicular to the surface of maximum circulation of a vector $\vec{u}$ around $\boldsymbol{A}$ and its norm is proportional to the velocity of circulation of $\vec{u}$ around $\boldsymbol{A}$ in that direction.

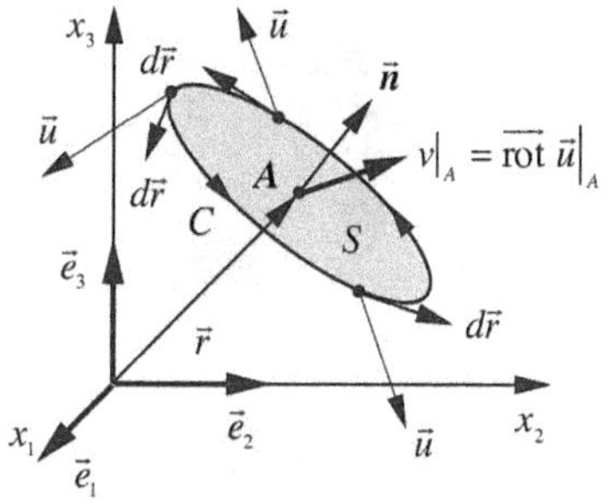

Figure E.2 *- rotational operator*

This *intrinsic definition* shows that rotational is an *invariant* of field $\vec{u}$, meaning that it does not depend on the system of coordinates we have chosen. In a orthonormal framework, the rotational can be calculated thanks to the formalism of the determinant, in the following fashion

$$\vec{v}(\vec{r}) = \overrightarrow{\mathrm{rot}}\,\vec{u} = \begin{vmatrix} \vec{e}_1 & \vec{e}_2 & \vec{e}_3 \\ \partial/\partial x_1 & \partial/\partial x_2 & \partial/\partial x_3 \\ u_1 & u_2 & u_3 \end{vmatrix} \tag{E.17}$$

The *intrinsic properties* of the rotational operator are the following:

- $\overrightarrow{\mathrm{rot}}\,\overrightarrow{\mathrm{grad}}\,f \equiv 0$ (E.18)
- if $\overrightarrow{\mathrm{rot}}\,\vec{u} \equiv 0$, $\vec{u}$ is said to be *irrotational vectorial field.* In this case, the previous property implies that $\vec{u} = \overrightarrow{\mathrm{grad}}\,f$
- if $\overrightarrow{\mathrm{rot}}\,\vec{u} = \overrightarrow{\mathrm{rot}}\,\vec{v}$, then $\vec{u} = \vec{v} + \overrightarrow{\mathrm{grad}}\,f$ (E.19)
- if $\vec{v} = \overrightarrow{\mathrm{rot}}\,\vec{u}$, $\vec{v}$ is said to be a *rotational vectorial field* and it derives from the *vector potential* $\vec{u}$. In this case, if the vectorial field $\vec{v}$ is known, the vectorial field $\vec{u}$ can be found up to the gradient of a scalar field
- rotational theorems

$$\oiint_S d\vec{S} \wedge \vec{v} = \iiint_V \overrightarrow{\mathrm{rot}}\,\vec{v}\,dV \tag{E.20}$$

$$\oint_C \vec{v}\,d\vec{r} = \iint_S \overrightarrow{\mathrm{rot}}\,\vec{v}\,d\vec{S} \tag{E.21}$$

The divergence operator

The *divergence of a vectorial field* $\vec{u}$ is a scalar field g satisfying in every point A of space the following limit taken on surface S limiting volume V taken around point A (figure E.3)

$$g|_A = \mathrm{div}\,\vec{u}|_A = \lim_{V \to 0} \frac{1}{V} \oiint_S \vec{u}\,d\vec{S} \tag{E.22}$$

The scalar g represents as a consequence the limit of the field flux $\vec{u}$ through the surface S and is only different from zero if the field $\vec{u}$ diverges locally around a point A . This *intrinsic definition* shows that the divergence is an *invariant* of the field $\vec{u}$, meaning that it does not depend on the system of coordinates that was chosen.

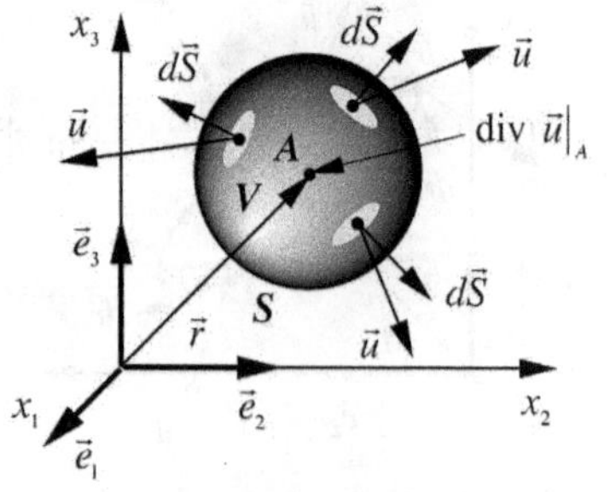

Figure E.3 - *divergence operator*

In an orthonormal framework, the divergence can be calculated thanks to the following relation

$$g(\vec{r}) = \operatorname{div} \vec{u} = \frac{\partial u_1}{\partial x_1} + \frac{\partial u_2}{\partial x_2} + \frac{\partial u_3}{\partial x_3} = \sum_i \frac{\partial u_i}{\partial x_i} \quad (E.23)$$

The *intrinsic properties* of the divergence operator are the following:

- $\operatorname{div} \overrightarrow{\operatorname{rot}} \vec{u} \equiv 0$ (E.24)
- if $\operatorname{div} \vec{v} \equiv 0$, $\vec{v}$ is said to be a *conservative vectorial field or solenoidal field* and the previous property implies that $\vec{v} = \overrightarrow{\operatorname{rot}} \vec{u}$
- if $\operatorname{div} \vec{u} = \operatorname{div} \vec{v}$, then $\vec{u} = \vec{v} + \overrightarrow{\operatorname{rot}} \vec{w}$ (E.25)
- if $g = \operatorname{div} \vec{u}$, $\vec{u}$ is said to be a *divergent vectorial field.* In this case, if the scalar g is known, the vectorial field $\vec{u}$ can be calculated up to a rotational vectorial field
- the *divergence theorem*

$$\oiint_S \vec{v}\, d\vec{S} = \iiint_V \operatorname{div} \vec{v}\, dV \quad (E.26)$$

The Laplacian operator

The *laplacian of a scalar field* f, or the *laplacian of a vectorial field* $\vec{u}$, is a scalar field g, or a vectorial field $\vec{v}$, defined intrinsically by applying twice the vectorial operators div, $\overrightarrow{\operatorname{grad}}$ or $\overrightarrow{\operatorname{rot}}$ by the following relations

$$\begin{cases} g(\vec{\xi}) = \Delta f = \operatorname{div} \overrightarrow{\operatorname{grad}} f \\ \vec{v}(\vec{\xi}) = \Delta \vec{u} = \overrightarrow{\operatorname{grad}} \operatorname{div} \vec{u} - \overrightarrow{\operatorname{rot}}\, \overrightarrow{\operatorname{rot}} \vec{u} \end{cases} \quad (E.27)$$

These *intrinsic definitions* show that the Laplacian is an *invariant* of field f or $\vec{u}$, meaning that it does not depend on the system of coordinates that were chosen, since it is defined in terms of the operators div, $\overrightarrow{\operatorname{grad}}$ or $\overrightarrow{\operatorname{rot}}$. In an orthonormal framework, the Laplacian can be calculated thanks to the following relations

$$\begin{cases} g(\vec{r}) = \Delta f = \operatorname{div} \overrightarrow{\operatorname{grad}} f = \dfrac{\partial^2 f}{\partial x_1^2} + \dfrac{\partial^2 f}{\partial x_2^2} + \dfrac{\partial^2 f}{\partial x_3^2} = \sum_i \dfrac{\partial^2 f}{\partial x_i^2} \\ \vec{v}(\vec{r}) = \Delta \vec{u} = \overrightarrow{\operatorname{grad}} \operatorname{div} \vec{u} - \overrightarrow{\operatorname{rot}}\, \overrightarrow{\operatorname{rot}} \vec{u} = \dfrac{\partial^2 \vec{u}}{\partial x_1^2} + \dfrac{\partial^2 \vec{u}}{\partial x_2^2} + \dfrac{\partial^2 \vec{u}}{\partial x_3^2} = \sum_i \dfrac{\partial^2 \vec{u}}{\partial x_i^2} \end{cases} \quad (E.28)$$

The *intrinsic properties* of the laplacian operator are the following:

- $\begin{cases} \Delta \vec{u} = \overrightarrow{\operatorname{grad}} \operatorname{div} \vec{u} & \textit{if} \quad \overrightarrow{\operatorname{rot}} \vec{u} = 0 \\ \Delta \vec{u} = -\overrightarrow{\operatorname{rot}}\, \overrightarrow{\operatorname{rot}} \vec{u} & \textit{if} \quad \operatorname{div} \vec{u} = 0 \end{cases}$ (E.29)
- $\Delta \vec{u} = \vec{e}_1 \Delta u_1 + \vec{e}_2 \Delta u_2 + \vec{e}_3 \Delta u_3$ (E.30)
- The *Green formulas*

$$\iiint_V f \Delta g\, dV = \oiint_S f\, \overrightarrow{\operatorname{grad}} g\, d\vec{S} - \iiint_V \overrightarrow{\operatorname{grad}} f\, \overrightarrow{\operatorname{grad}} g\, dV \quad (E.31)$$

$$\iiint_V (f \Delta g - g \Delta f)\, dV = \oiint_S \left(f\, \overrightarrow{\operatorname{grad}} g - g\, \overrightarrow{\operatorname{grad}} f \right) d\vec{S} \quad (E.32)$$

The « del » operator

The formal operator *"del"* $\vec{\nabla}$ is an operator defined by the relation

$$\vec{\nabla} = \vec{e}_1 \frac{\partial}{\partial x_1} + \vec{e}_2 \frac{\partial}{\partial x_2} + \vec{e}_3 \frac{\partial}{\partial x_3} = \sum_i \vec{e}_i \frac{\partial}{\partial x_i} \tag{E.33}$$

To use this operator, some simple rules are observed including the following:

- *"del"* behaves as a vector,
- if *"del"* works on a product, the result is the sum of expressions obtained by considering successively the derivation of each of the terms (derivation rule of a product),
- *"del"* must come before the terms on which it acts. If, after transformation, *"del"* acts on a term on which it did not before, we consider this term as a constant as far as *"del"* is concerned,
- The classic operators of vectorial analysis are then expressed in the following form:

$$\begin{cases} \overrightarrow{\text{grad}}\, f = \vec{\nabla} f \\ \text{div}\, \vec{u} = \vec{\nabla} \vec{u} \\ \overrightarrow{\text{rot}}\, \vec{u} = \vec{\nabla} \wedge \vec{u} \end{cases} \quad \text{and} \quad \begin{cases} \Delta f = (\vec{\nabla}\vec{\nabla}) f = \vec{\nabla}^2 f \\ \Delta \vec{u} = (\vec{\nabla}\vec{\nabla}) \vec{u} = \vec{\nabla}^2 \vec{u} \end{cases} \tag{E.34}$$

On the linearity, distributivity and iteration of the vectorial operators

The vectorial operators satisfy a set of rules that are very convenient, linked to *linearity*, *distributivity* and *iteration* of these diverse operators as shown by the following formulas, where α et β are constants

$$\begin{cases} \overrightarrow{\text{grad}}(\alpha f + \beta g) = \alpha\, \overrightarrow{\text{grad}}\, f + \beta\, \overrightarrow{\text{grad}}\, g \\ \text{div}(\alpha \vec{u} + \beta \vec{v}) = \alpha\, \text{div}\, \vec{u} + \beta\, \text{div}\, \vec{v} \\ \overrightarrow{\text{rot}}(\alpha \vec{u} + \beta \vec{v}) = \alpha\, \overrightarrow{\text{rot}}\, \vec{u} + \beta\, \overrightarrow{\text{rot}}\, \vec{v} \\ \Delta(\alpha f + \beta g) = \alpha \Delta f + \beta \Delta g \\ \Delta(\alpha \vec{u} + \beta \vec{v}) = \alpha \Delta \vec{u} + \beta \Delta \vec{v} \end{cases} \tag{E.35}$$

$$\begin{cases} \overrightarrow{\text{grad}}(fg) = f\, \overrightarrow{\text{grad}}\, g + g\, \overrightarrow{\text{grad}}\, f \\ \overrightarrow{\text{grad}}\, (\vec{u}\vec{v}) = \vec{u} \wedge \overrightarrow{\text{rot}}\, \vec{v} + \vec{v} \wedge \overrightarrow{\text{rot}}\, \vec{u} + (\vec{u}\vec{\nabla})\vec{v} + (\vec{v}\vec{\nabla})\vec{u} \\ \text{div}(f\vec{v}) = f\, \text{div}\, \vec{v} + \vec{v}\, \overrightarrow{\text{grad}}\, f \\ \text{div}(\vec{u} \wedge \vec{v}) = \vec{v}\, \overrightarrow{\text{rot}}\, \vec{u} - \vec{u}\, \overrightarrow{\text{rot}}\, \vec{v} \\ \overrightarrow{\text{rot}}(f\vec{v}) = f\, \overrightarrow{\text{rot}}\, \vec{v} + \overrightarrow{\text{grad}}\, f \wedge \vec{v} \\ \overrightarrow{\text{rot}}(\vec{u} \wedge \vec{v}) = \vec{u}\, \text{div}\, \vec{v} - \vec{v}\, \text{div}\, \vec{u} + (\vec{v}\vec{\nabla})\vec{u} - (\vec{u}\vec{\nabla})\vec{v} \\ \Delta(fg) = f \Delta g + g \Delta f + 2\, \overrightarrow{\text{grad}}\, f\, \overrightarrow{\text{grad}}\, g \end{cases} \tag{E.36}$$

$$\begin{cases} \overrightarrow{\text{rot}}\ \overrightarrow{\text{grad}}\, f \equiv 0 \\ \text{div}\ \overrightarrow{\text{rot}}\, \vec{v} \equiv 0 \\ \text{div}\ \overrightarrow{\text{grad}}\, f = \Delta f \\ \overrightarrow{\text{rot}}\, \overrightarrow{\text{rot}}\ \vec{v} = \overrightarrow{\text{grad}}\, \text{div}\, \vec{v} - \Delta \vec{v} \end{cases} \tag{E.37}$$

E.3 - Derivatives of integrals on volumes, surfaces and mobile contours

The derivative of an integral on a mobile volume

Given the integral of a function $f(\vec{r},t)$ on a volume V_m mobile with velocity $\vec{\phi}(\vec{r},t)$ in space and the problem of calculation of temporal variation of this integral. To carry out this calculation, we must consider the volume V_m at instant t , which becomes volume V'_m at instant $t+\Delta t$ (figure E.4). By definition the temporal derivative of the integral is equal to the following limit, calculated for Δt going to zero

$$\begin{aligned}\frac{d}{dt}\iiint_{V_m} f(\vec{r},t)dV &= \lim_{\Delta t\to 0}\frac{1}{\Delta t}\left(\iiint_{V'_m} f(\vec{r},t+\Delta t)dV - \iiint_{V_m} f(\vec{r},t)dV\right)\\ &= \lim_{\Delta t\to 0}\frac{1}{\Delta t}\left(\iiint_{V_m}\left[f(\vec{r},t+\Delta t)-f(\vec{r},t)\right]dV + \iiint_{V'_m-V_m} f(\vec{r},t+\Delta t)dV\right)\end{aligned} \quad (E.38)$$

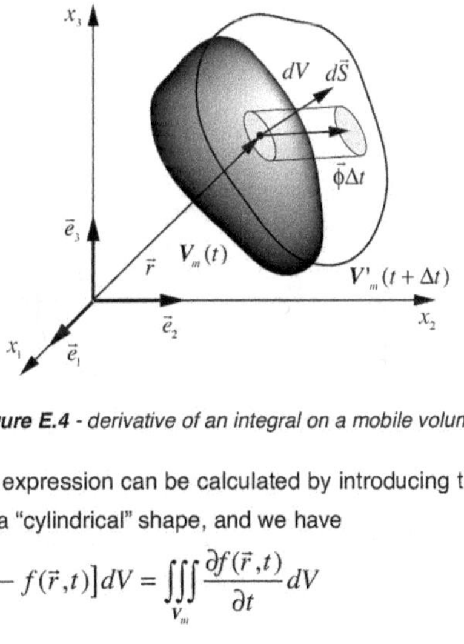

Figure E.4 *- derivative of an integral on a mobile volume*

The two limits of this last expression can be calculated by introducing the volume $dV = \vec{\phi}\Delta t d\vec{s}$ defined in the figure with a "cylindrical" shape, and we have

$$\lim_{\Delta t\to 0}\frac{1}{\Delta t}\iiint_{V_m}\left[f(\vec{r},t+\Delta t)-f(\vec{r},t)\right]dV = \iiint_{V_m}\frac{\partial f(\vec{r},t)}{\partial t}dV \quad (E.39)$$

$$\lim_{\Delta t\to 0}\frac{1}{\Delta t}\iiint_{V'_m-V_m} f(\vec{r},t+\Delta t)dV = \lim_{\Delta t\to 0}\frac{1}{\Delta t}\oiint_{S_m} f(\vec{r},t+\Delta t)\vec{\phi}\Delta t d\vec{S} = \oiint_{S_m}\vec{\phi}f(\vec{r},t)d\vec{S} = \iiint_{V_m}\mathrm{div}\left(f(\vec{r},t)\vec{\phi}\right)dV \quad (E.40)$$

so that the formula we seek is written under the form

$$\frac{d}{dt}\iiint_{V_m} f\,dV = \iiint_{V_m}\left[\frac{\partial f}{\partial t}+\mathrm{div}\left(f\vec{\phi}\right)\right]dV \quad (E.41)$$

Just as the argument of the integral can be written

$$\frac{\partial f}{\partial t}+\mathrm{div}\left(f\vec{\phi}\right) = \frac{\partial f}{\partial t}+(\vec{\phi}\vec{\nabla})f + f\,\mathrm{div}\vec{\phi} \quad (E.42)$$

the formula for the calculation of the derivative of the integral on a field $\vec{g}$ is deduced from *(E.*

68)

$$\frac{d}{dt}\iiint_{V_m} \vec{g}\,dV = \iiint_{V_m}\left[\frac{\partial \vec{g}}{\partial t} + (\vec{\phi}\vec{\nabla})\vec{g} + \vec{g}\,\mathrm{div}\vec{\phi}\right]dV \qquad (E.43)$$

The derivative of an integral on a mobile surface

Given the field integral $\vec{g}(\vec{r},t)$ on a surface S_m mobile with velocity $\vec{\phi}(\vec{r},t)$ and the problem of the calculation of the temporal variations of this integral. To carry out this calculation, we must consider the surface S_m at instant t, which becomes the surface S'_m at instant $t+\Delta t$ (figure E.5). By definition, the temporal derivative of the integral is equal to the following limit, calculated for Δt tending to zero

$$\frac{d}{dt}\iint_{S_m}\vec{g}(\vec{r},t)d\vec{S} = \lim_{\Delta t\to 0}\frac{1}{\Delta t}\left(\iint_{S'_m}\vec{g}(\vec{r},t+\Delta t)d\vec{S}' - \iint_{S_m}\vec{g}(\vec{r},t)d\vec{S}\right) \qquad (E.44)$$

$$\frac{d}{dt}\iint_{S_m}\vec{g}(\vec{r},t)d\vec{S} = \lim_{\Delta t\to 0}\frac{1}{\Delta t}\left(\iint_{S_m}[\vec{g}(\vec{r},t+\Delta t)-\vec{g}(\vec{r},t)]d\vec{S}\right) + \lim_{\Delta t\to 0}\frac{1}{\Delta t}\left(\iint_{S'_m}\vec{g}(\vec{r},t+\Delta t)d\vec{S}' - \iint_{S_m}\vec{g}(\vec{r},t+\Delta t)d\vec{S}\right) \qquad (E.45)$$

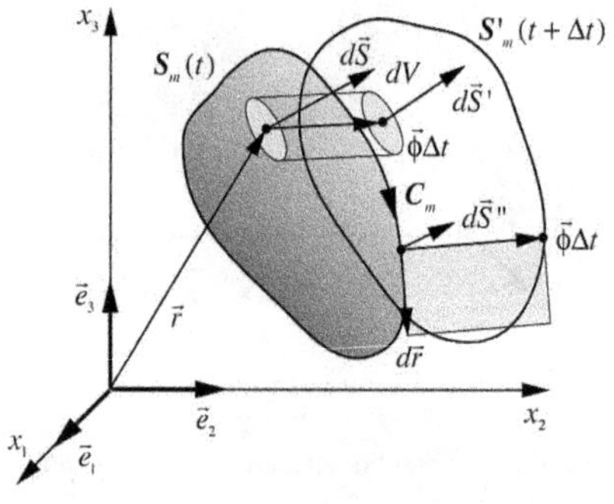

Figure E.5 *- derivative of an integral on a mobile surface*

The first limit is deduced immediately

$$\lim_{\Delta t\to 0}\frac{1}{\Delta t}\left(\iint_{S_m}[\vec{g}(\vec{r},t+\Delta t)-\vec{g}(\vec{r},t)]d\vec{S}\right) = \iint_{S_m}\frac{\partial \vec{g}(\vec{r},t)}{\partial t}d\vec{S} \qquad (E.46)$$

To calculate the second limit, we must define S''_m as a lateral surface linking the contours of surfaces S_m and S'_m, S_t as the closed surface made up of the surfaces S_m, S'_m and S''_m, V_m as the volume comprised inside the closed surfaces S_t and C_m as the contour around the surface S_m, then we effect the following transformations, by using the divergence and rotational theorems

$$\begin{aligned}\iint_{S'_m}\vec{g}\,d\vec{S}' - \iint_{S_m}\vec{g}\,d\vec{S} &= \iint_{S_t}\vec{g}\,d\vec{S} - \iint_{S''_m}\vec{g}\,d\vec{S}'' = \iiint_{V_m}\mathrm{div}\,\vec{g}\,dV - \oint_{C_m}\vec{g}\left(d\vec{r}\wedge\vec{\phi}\Delta t\right)\\ &= \iint_{S_m}\mathrm{div}\,\vec{g}\left(\vec{\phi}\Delta t\,d\vec{S}\right) - \Delta t\oint_{C_m}d\vec{r}\left(\vec{\phi}\wedge\vec{g}\right) = \Delta t\iint_{S_m}\vec{\phi}\,\mathrm{div}\,\vec{g}\,d\vec{S} - \Delta t\iint_{S_m}\overrightarrow{\mathrm{rot}}\left(\vec{\phi}\wedge\vec{g}\right)d\vec{S}\end{aligned} \qquad (E.47)$$

We have for the second limit

$$\lim_{\Delta t\to 0}\frac{1}{\Delta t}\left(\iint_{S'_m}\vec{g}(\vec{r},t+\Delta t)d\vec{S}\,'-\iint_{S_m}\vec{g}(\vec{r},t+\Delta t)d\vec{S}\right)=\iint_{S_m}\left[\vec{\phi}\,\mathrm{div}\,\vec{g}-\overrightarrow{\mathrm{rot}}\left(\vec{\phi}\wedge\vec{g}\right)\right]d\vec{S} \qquad (E.48)$$

so that the formula we seek is written

$$\frac{d}{dt}\iint_{S_m}\vec{g}\,d\vec{S}=\iint_{S_m}\left[\frac{\partial\vec{g}}{\partial t}+\vec{\phi}\,\mathrm{div}\,\vec{g}-\overrightarrow{\mathrm{rot}}\left(\vec{\phi}\wedge\vec{g}\right)\right]d\vec{S} \qquad (E.49)$$

If the integral is calculated on a closed surface, the previous relationship is written

$$\oiint_{S_m}\overrightarrow{\mathrm{rot}}\left(\vec{\phi}\wedge\vec{g}\right)d\vec{S}=\iiint_{V_m}\mathrm{div}\left[\overrightarrow{\mathrm{rot}}\left(\vec{\phi}\wedge\vec{g}\right)\right]dV\equiv 0 \qquad (E.50)$$

so that the derivation formula becomes

$$\frac{d}{dt}\oiint_{S_m}\vec{g}\,d\vec{S}=\oiint_{S_m}\left(\frac{\partial\vec{g}}{\partial t}+\vec{\phi}\,\mathrm{div}\,\vec{g}\right)d\vec{S} \qquad (E.51)$$

This last relation could have been derived directly from formula *(E.43)* by using the divergence theorem

$$\begin{aligned}\frac{d}{dt}\oiint_{S_m}\vec{g}\,d\vec{S}&=\frac{d}{dt}\iiint_{V_m}\mathrm{div}\,\vec{g}\,dV=\iiint_{V_m}\left[\frac{\partial\,\mathrm{div}\,\vec{g}}{\partial t}+\mathrm{div}\left(\vec{\phi}\,\mathrm{div}\,\vec{g}\right)\right]dV\\&=\iiint_{V_m}\mathrm{div}\left(\frac{\partial\vec{g}}{\partial t}+\vec{\phi}\,\mathrm{div}\,\vec{g}\right)dV=\oiint_{S_m}\left(\frac{\partial\vec{g}}{\partial t}+\vec{\phi}\,\mathrm{div}\,\vec{g}\right)d\vec{S}\end{aligned} \qquad (E.52)$$

The derivative of an integral on a mobile contour

Given the integral of a field $\vec{g}(\vec{r},t)$ on a contour C_m mobile with velocity $\vec{\phi}(\vec{r},t)$ and the problem of calculating the temporal variation of that integral. To carry that calculation, we must consider a contour C_m at instant t, which becomes contour C'_m at instant $t+\Delta t$ (figure E.6). By definition, the temporal derivative of the integral is equal to the following limit, calculated for Δt tending to zero

$$\begin{aligned}\frac{d}{dt}\int_{C_m}\vec{g}(\vec{r},t)d\vec{r}&=\lim_{\Delta t\to 0}\frac{1}{\Delta t}\left(\int_{C'_m}\vec{g}(\vec{r},t+\Delta t)d\vec{r}\,'-\int_{C_m}\vec{g}(\vec{r},t)d\vec{r}\right)\\&=\lim_{\Delta t\to 0}\frac{1}{\Delta t}\left\{\int_{C_m}\left[\vec{g}(\vec{r},t+\Delta t)-\vec{g}(\vec{r},t)\right]d\vec{r}\right\}\\&+\lim_{\Delta t\to 0}\frac{1}{\Delta t}\left[\int_{C'_m}\vec{g}(\vec{r},t+\Delta t)d\vec{r}\,'-\int_{C_m}\vec{g}(\vec{r},t+\Delta t)\,d\vec{r}\right]\end{aligned} \qquad (E.53)$$

The first term of the limit transforms by giving a partial derivative

$$\lim_{\Delta t\to 0}\frac{1}{\Delta t}\left\{\int_{C_m}\left[\vec{g}(\vec{r},t+\Delta t)-\vec{g}(\vec{r},t)\right]d\vec{r}\right\}=\int_{C_m}\frac{\partial\vec{g}(\vec{r},t)}{\partial t}\,d\vec{r} \qquad (E.54)$$

To calculate the second term of the limit, we have to define C_l like a closed contour

$A_1A_2B_2B_1A_1$ and S as a surface surrounded by the contour C_t, and then to effect the following transformations, by applying the Stokes theorem of the rotational

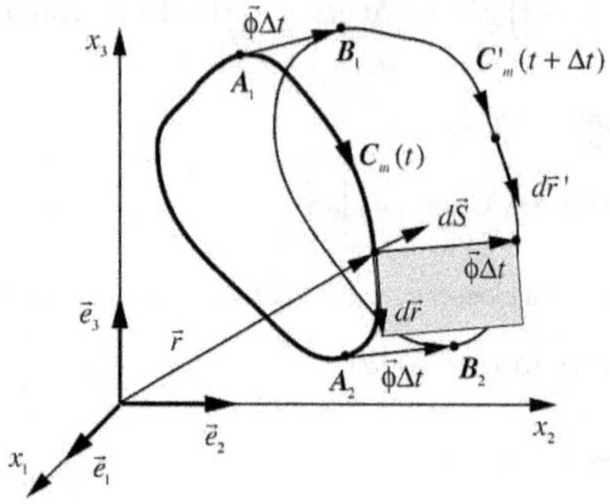

Figure E.6 *- derivative of an integral on a mobile contour*

$$\begin{aligned}\int_{C'_m} \vec{g}\,d\vec{r}\,' - \int_{C_m} \vec{g}\,d\vec{r} &= -\oint_{C_t} \vec{g}\,d\vec{r} - \int_{A_1}^{B_1} \vec{g}\,d\vec{r} + \int_{A_2}^{B_2} \vec{g}\,d\vec{r} \\ &= -\iint_S \overrightarrow{\text{rot}}\,\vec{g}\,d\vec{s} - \left[\vec{g}\vec{\phi}\right]_{A_1}\Delta t + \left[\vec{g}\vec{\phi}\right]_{A_2}\Delta t\end{aligned} \tag{E.55}$$

$$\begin{aligned}\int_{C'_m} \vec{g}\,d\vec{r}\,' - \int_{C_m} \vec{g}\,d\vec{r} &= -\int_{C_m} \overrightarrow{\text{rot}}\,\vec{g}\left(d\vec{r} \wedge \vec{\phi}\Delta t\right) - \left[\vec{g}\vec{\phi}\right]_{A_1}\Delta t + \left[\vec{g}\vec{\phi}\right]_{A_2}\Delta t \\ &= -\Delta t \int_{C_m} \left(\vec{\phi} \wedge \overrightarrow{\text{rot}}\,\vec{g}\right) d\vec{r} - \left[\vec{g}\vec{\phi}\right]_{A_1}\Delta t + \left[\vec{g}\vec{\phi}\right]_{A_2}\Delta t\end{aligned} \tag{E.56}$$

The second term of the limit is then

$$\lim_{\Delta t \to 0} \frac{1}{\Delta t}\left[\int_{C'_m} \vec{g}\,d\vec{r}\,' - \int_{C_m} \vec{g}\,d\vec{r}\right] = \left[\vec{g}\vec{\phi}\right]_{A_2} - \left[\vec{g}\vec{\phi}\right]_{A_1} - \int_{C_m} \left(\vec{\phi} \wedge \overrightarrow{\text{rot}}\,\vec{g}\right) d\vec{r} \tag{E.57}$$

so that the derivation formula sought after is written

$$\frac{d}{dt}\int_{C_m} \vec{g}\,d\vec{r} = \left[\vec{g}\vec{\phi}\right]_{A_2} - \left[\vec{g}\vec{\phi}\right]_{A_1} + \int_{C_m} \left(\frac{\partial \vec{g}}{\partial t} - \vec{\phi} \wedge \overrightarrow{\text{rot}}\,\vec{g}\right) d\vec{r} \tag{E.58}$$

On a closed contour, the points A_1 and A_2 are equal, so that it becomes in this case

$$\frac{d}{dt}\oint_{C_m} \vec{g}\,d\vec{r} = \oint_{C_m} \left(\frac{\partial \vec{g}}{\partial t} - \vec{\phi} \wedge \overrightarrow{\text{rot}}\,\vec{g}\right) d\vec{r} \tag{E.59}$$

Furthermore, this last formula could have been directly established by relation *(E.52)*. Indeed

$$\begin{aligned}\frac{d}{dt}\oint_{C_m} \vec{g}\,d\vec{r} &= \frac{d}{dt}\iint_S \overrightarrow{\text{rot}}\,\vec{g}\,d\vec{S} = \iint_{S_m}\left[\frac{\partial \overrightarrow{\text{rot}}\,\vec{g}}{\partial t} + \vec{\phi}\,\text{div}\,\overrightarrow{\text{rot}}\,\vec{g} + \overrightarrow{\text{rot}}\left(\overrightarrow{\text{rot}}\,\vec{g} \wedge \vec{\phi}\right)\right] d\vec{S} \\ &= \iint_{S_m} \overrightarrow{\text{rot}}\left[\frac{\partial \vec{g}}{\partial t} + \overrightarrow{\text{rot}}\,\vec{g} \wedge \vec{\phi}\right] d\vec{S} = \oint_{C_m}\left(\frac{\partial \vec{g}}{\partial t} - \vec{\phi} \wedge \overrightarrow{\text{rot}}\,\vec{g}\right) d\vec{r}\end{aligned} \tag{E.60}$$

www.ingramcontent.com/pod-product-compliance
Lightning Source LLC
Chambersburg PA
CBHW070346200326
41518CB00012B/2155